Structure and Reactions of Light Exotic Nuclei

Structure and Reactions of Light Exotic Nuclei

Yasuyuki Suzuki
Department of Physics, Faculty of Science, Niigata University, Japan

Rezső G. Lovas
Institute of Nuclear Research of the Hungarian Academy of Sciences, Debrecen, Hungary

Kazuhiro Yabana
Institute of Physics, University of Tsukuba, Japan

and

Kálmán Varga
Institute of Nuclear Research of the Hungarian Academy of Sciences, Debrecen, Hungary

CRC Press
Taylor & Francis Group
Boca Raton London New York

CRC Press is an imprint of the
Taylor & Francis Group, an **informa** business
A TAYLOR & FRANCIS BOOK

First published 2003 by Taylor & Francis

Published 2019 by CRC Press
Taylor & Francis Group
6000 Broken Sound Parkway NW, Suite 300
Boca Raton, FL 33487-2742

CRC Press is an imprint of Taylor & Francis Group, an Informa business

First issued in paperback 2019

ISBN-13: 978-0-367-45458-6 (pbk)
ISBN 13: 978-0-415-30872-4 (hbk)

Visit the Taylor & Francis Web site at
http://www.taylorandfrancis.com

and the CRC Press Web site at
http://www.crcpress.com

British Library Cataloguing in Publication Data
A catalogue record for this book is available from the British Library

Library of Congress Cataloging in Publication Data
A catalog record for this title has been requested

Contents

Preface

This book is intended to give an introduction to the rapidly developing research field of light exotic nuclei. *Exotic* are the nuclei of unusual composition. Since the middle of the 1980s more and more effort has been spent on studying these exotic objects. Our knowledge on the nuclei of *normal* composition is by no means exhaustive. The interest of researchers has turned to exotic nuclei partly because studying them sheds light on normal nuclei as well. The research of exotic nuclei has been started with the advent of accelerated beams of such nuclei. This technical novelty has revitalized nuclear physics, and the facilities producing radioactive ion beams and their theoretical hinterland now offer students the experience of pioneering research.

First light nuclei have been studied, and a number of their common traits have been discovered. To pinpoint their peculiarity summarily, one can say that their basic mode of motion is not single-particle or collective motion but few-body motion between clusters and individual nucleons. Few-cluster dynamics is also shown, to a lesser extent, by normal nuclei. The description of this more complicated behaviour requires more general tools. Light exotic nuclei now promise the clue to really understanding all light nuclei, normal or exotic. The heavier a nucleus, the more it tends to conform to the usual pattern and the more challenging it will be to describe any anomalous behaviour.

An introductory text is expected to provide a certain amount of knowledge and to lead to a certain level of understanding. The primary objective of the book is to help the reader understand the subject. While containing a certain stock of illustrative facts, this book is not intended to be complete in factual matters. It is not meant to be a review or an objective report of the state of the art, nor is it hoped to give a full set of current references.

This book is on theoretical physics for both experimental and theoretical physicists. The intention of the authors is to provide material for different manners of reading. This is a *textbook* since it starts from a level that is mastered by a first-year graduate student and proceeds in steps that are hopefully small enough to take by a student of experimental physics as well. But it is meant to reach an advanced level in two senses of the word. First, it is to confront the reader with problems that are counted as hot issues at the time of writing the book and to prepare him or her for treating problems that are expected to emerge in this field in the foreseeable future. And, second, the dedicated reader will be guided to climb some challenging heights.

What is the minimum that a student of this subject should know of the field in general? First of all, he should know from what and how our knowledge on light exotic nuclei derives. Unlike in traditional nuclear physics, virtually all information that we have of these nuclei comes from reaction experiments. Nuclear collisions are complicated phenomena, and our knowledge of exotic nuclei must be preceded by understanding the mechanisms of their collisions. Thus theory is an inseparable part of gaining information, and even an experimentalist working on exotic nuclei must be familiar with the rudiments of this theory. Therefore, Part I of this book is devoted to the theory of collisions of light exotic nuclei.

The basic phenomena linked with exotic nuclei are well-known commonplaces. Neutron halos, α-clustering, the enhanced low-energy electric dipole strength, anomalous level sequences, β-delayed particle emission etc. are easily comprehensible at a phenomenological level. But it is not so trivial to tell the difference between a neutron halo and neutron skin or to tell whether the low-energy strength concentration is caused by a resonance or a threshold effect. Without such a distinction, however, one cannot conjecture in which particular systems these phenomena should be looked for. The answer to these questions is in the realm of structure theory, and Part II of this book is to give an introduction to the structure aspects of light exotic nuclei. Our way of doing this is not repeating conventional nuclear structure theory by changing some affirmative statements to negative or *vice versa*, nor do we bombard the reader with many *ad hoc* novel approaches. What we attempt instead is to elaborate one approach, which seems general enough to treat any phenomena and to provide a scheme which, in principle, should be suited for explaining everything relevant. In this respect the book *resembles a monograph*. Even though only a single line is pursued, the reader will encounter most of the relevant ideas and concepts of current use in the literature.

The approach adopted in Part II is based on a multicluster model, in which the intercluster motion is treated accurately. The main ingredients of this model are an ingenious basis construction and an efficient basis optimization. The basis is called the correlated Gaussian basis, and the optimization is called the stochastic variational method.

Although the approach is specified in such technical terms, its basic concepts are simple, and in the book they are introduced slowly, step by step. The subject is presented with technical details but in such a way that the reader need not go deep into these details if he does not want to. The most gruelling details are relegated to Appendices.

The prior knowledge required depends to some extent on the depth of reading. Familiarity with basic quantum mechanics at an undergraduate level and with nuclei at the level of an introductory nuclear physics course is certainly assumed, but perhaps nothing else. The main line of concepts is elaborated by explaining all ingredients in elementary terms, but side-references to some alternative theoretical methods and techniques are, in some places, unavoidable. In an attempt to make the text self-contained, plausible brief explanations are given for these concepts as well.

Angular-momentum algebra is administered sparingly. The most special technique to be applied plentifully is a generating function technique for integration. This is explained carefully, but a superficial reading is possible without checking the minute details. The derivations of formulae are decomposed into elementary steps, and each step is written down. So the formidable-looking multiline mathematical derivations to be found are just to serve the reader's convenience.

Parts I and II can be read independently. The theoretical frameworks of reaction theory and structure theory are not very closely connected. The main link between them is that they rely on the same data. This fact, though not very satisfactory in principle, facilitates the use of this book. For example, it can be used as a text for two consecutive graduate courses.

The logical skeleton of Part I is shown in Fig. P.1. The arrows show which chapter relies on which other chapter. Thus it is recommended to read the first three chapters before reading any of the rest. Chapters 4 and 5 can be read more or less independently, but the last two chapters, which are quite independent of each other, rely on both of them.

The logical scheme of Part II is plotted in Fig. P.2. The structure theory of light exotic nuclei rests on three pillars: on the correlated Gaussian basis (Chap. 9), on the stochastic variational method (Chap. 10) and on the cluster model (Chap. 11). The full theory of light exotic nuclei is contained in Chap. 12, and it requires the results of all of Chaps. 9–11. The concrete applications of the theory to light exotic nuclei (Chap. 13) hinges on Chap. 12.

For Part II, alternative reading strategies are also conceivable. Chapters 10 and 11 are readable by themselves for the benefit of those who want to learn the stochastic variational method or the cluster models, respectively. For those who are familiar with these subjects, it may be

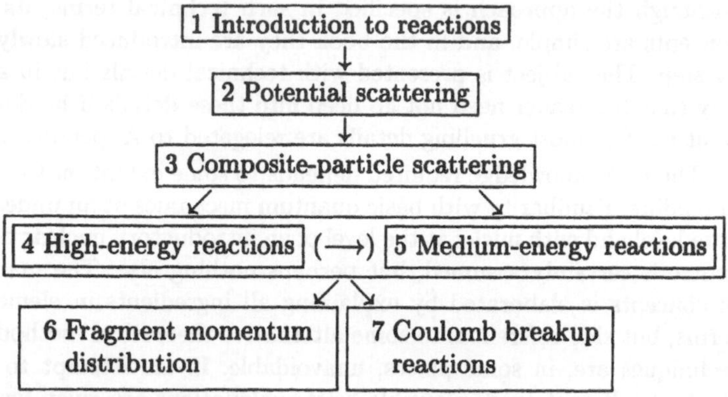

Figure P.1: Logical chart of Part I

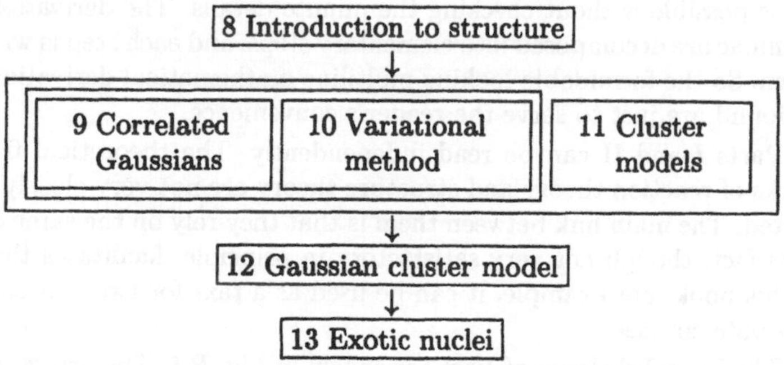

Figure P.2: Logical chart of Part II.

advisable, conversely, to skip these chapters, at least for first reading. For those interested mainly in the structure of individual nuclei keeping aloof from the foundation of the models, it may suffice to read the relevant section in Chap. 13, after perusing Sect. 12.2 about the specific ingredients and implementation of the approach.

The appendices are not just collections of formulae; they are more or less self-contained and didactically arranged articles on the details of the formalism and of the physical background. Reading them is useful to attain deeper understanding or skill for microscopic nuclear structure calculations. Appendix D on antisymmetrization, App. C on the peculiarities of loosely bound states of multiparticle systems and App. H on the actual implementation of the multicluster formalism are recommended even for non-theoreticians. Their reading may elucidate what sometimes appears arcane juggling in the literature.

At the end of the volume a glossary of symbols and a glossary of the abbreviations used in the book can be found.

This work has emerged from a long-standing co-operation between the nuclear theory groups at the Faculty of Science and Graduate School of Science and Technology, Niigata University and at the Institute of Nuclear Research of the Hungarian Academy of Sciences, Debrecen. This co-operation has been supported by Grants-in-aid for Scientific Research, Monbusho (Japan) and OTKA and OMFB grants (Hungary) and cooperative exchanges between the Japan Society for the Promotion of Science and the Hungarian Academy of Sciences. All these supports are gratefully acknowledged.

Figure 1.3 has been taken over from Physics Letters B, Fig. 6.1 from Physical Review Letters, Figs. 5.1, 5.2 and 5.3 from Progress of Theoretical Physics Supplement and Fig. 13.23 from Progress of Theoretical Physics, with the kind permission of Elsevier Science, the American Physical Society and the Progress of Theoretical Physics, respectively.

The authors are thankful to all those with whom they discussed the issues contained in the book and who have helped them by providing them with some material or have read part of the text: Dr. B. Abu-Ibrahim, Dr. K. Arai, Dr. P. Descouvemont, Prof. K. Katō, Dr. D. J. Millener, Dr. H. Nemura, Dr. Y. Ogawa, Dr. T. Ohtsubo, Dr. A. Ozawa, Dr. K. F. Pál, Dr. P. Roussel-Chomaz, Prof. T. Suzuki, Mr. M. Takahashi, Prof. I. Tanihata and Dr. J. Usukura. Special thanks are due to Prof. D. Baye, Prof. B. Gyarmati and Prof. K. T. Hecht, each of whom has undertaken a critical reading of most of the text at some stage and to Prof. P. E. Hodgson for his encouragement and stimulating attention.

Niigata
Debrecen
July 2001

Y. Suzuki
R. G. Lovas
K. Yabana
K. Varga

Introduction

Light exotic nuclei are light nuclei whose nucleonic composition is unusual. An exotic isotope contains many more or many fewer neutrons than a stable isotope of the same element. In the chart of the nuclei they lie far from the stability line, so they are mostly very unstable. From a theoretical point of view, the hallmark of light exotic nuclei is that their shell structure is loosened up or, indeed, overwhelmed by few-body or few-cluster effects.

Exotic nuclei can be produced by nuclear reactions. Because of their rapid decay, macroscopic amounts cannot be collected, and the exotic nuclei can be best studied by forming secondary beams from the primary reaction product. Exotic nuclei came to the focus of interest with the advent of secondary radioactive beams.

In the primary collision they are mostly formed in excited states, but they get de-excited to their ground states well before they collide with the targets. The ground states are usually unstable with respect to β-decay. Their lifetime is, typically, of the order of a millisecond to a second, which is much longer than the time scale ('the period') of nucleonic motion ($\sim 10^{-23}$ s) inside the nucleus. Therefore, these exotic nuclei have long enough lifetimes to possess well-defined many-body structures as bound (or quasibound) systems of nucleons.

In the first experiments with the secondary beams the interest was focussed on what is the probability that the projectile enters into interaction with the target so that it does not emerge as the same nucleus. For some neutron-rich projectiles, like ^{6}He, ^{11}Li, ^{11}Be and ^{14}Be, the cross section of such processes, the 'interaction cross section', shows a steep increase, signalling that these nuclei have larger radii than their neighbours. Much of the increased cross section comes from neutron-removal channels. Moreover, for these processes, the momentum distribution of the charged fragment has been found to be anomalously narrow, from which it can be inferred that the neutrons are removed from orbits whose momentum distribution is narrow. According to the uncertainty principle, this implies large spatial distribution of the neutron(s) removed, with respect to the residue. These observations can be accounted for by the hypothesis that these nuclei have 'neutron halos'. The whereabouts of the halo nuclei on the nuclear chart, as they are known now, is shown by Fig. 1.

These 'halo nuclei' can be imagined as composed of a spatially extensive dilute 'halo' of a few (typically one or two) nucleons and a normal-density compact 'core'. The binding of the halo nucleons to the core is extremely weak, so that they are likely to be torn off in collisions. The halo structure has been confirmed by other experimental observations as well, and it is the most prominent phenomenon observed in light exotic nuclei.

Nuclear structure studies are primarily concerned with the properties of discrete states, i.e., bound states and resonances. In reactions, however, nuclear systems are in scattering states, which are much more complicated than discrete states. That is why the traditional arrange-

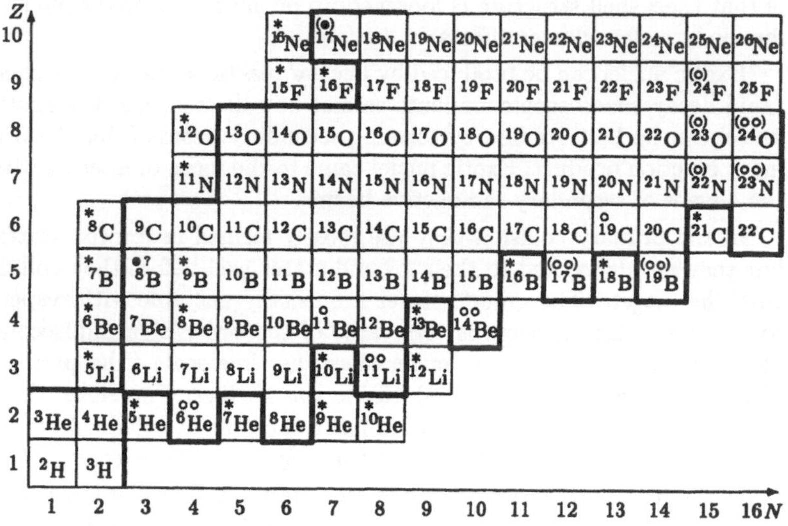

Figure 1: The lowest section of the nuclear chart with the drip lines drawn in thick lines. The nuclei marked by * decay by emitting nucleon(s). Those with open circles and full circles are supposed to be neutron-halo and proton-halo nuclei, respectively (one circle for one-nucleon halo and two for two-nucleon halo). The mark '?' for ^{8}B expresses that it is a limiting case. Parentheses sandwiching symbols mean that there is some evidence or at least some conjecture, but further confirmation is yet required. The nucleus ^{19}B may rather have a thick neutron skin, i.e., a neutron envelope formed collectively by a number of neutrons (see footnote on p. 181), than a halo.

ment of books on nuclear physics is that nuclear structure comes first and is followed by the presentation of reactions. Our thinking follows the same pattern whether we are concerned with ordinary or exotic nuclei, and that is expressed by the title of this book. Nevertheless, the presentation of the structure and reaction parts is reversed. The main reason for this is that virtually all information that we have on light exotic nuclei comes from reactions. In fact, the reliability of this information hinges very much on the reliability of the reaction models and analyses. Much of the information comes from 'high-energy' reactions (i.e., reactions with bombarding energies of several hundred MeV per nucleon), whose mechanism is fairly simple, which enables one to make the analyses reliable. Correspondingly, the ideas used in the reaction theory are not oversophisticated. The picture of the light exotic nuclei that emerges from the reaction analyses is, on the contrary, pretty intricate, and requires sophisticated theoretical concepts and techniques. This contrast also shows that it is desirable to change the traditional sequence.

The reactions with halo nuclei are reactions with weakly bound projectiles, and that requires specific treatment. In conventional reaction theory, a weakly bound projectile poses a formidable problem: the coupling to continuum states should be taken into account because the breakup effects can become important in each channel. In spite of considerable progress in this field in recent decades, this problem is still challenging both conceptually and practically. When, however, the projectile–target relative motion is much faster than that of the halo nucleons within the projectile, the situation becomes much simpler. In describing such reactions one can essentially neglect the inner motion, which is a kind of adiabatic approximation. Moreover, the high incident energy makes it possible to treat the kinetic energy operator in a simplified manner, which leads to the eikonal approximation. The eikonal approximation applied to composite-particle scattering is called the Glauber approximation. The Glauber approximation treats the nuclear collisions as a succession of many elementary collisions. A combination of the eikonal and adiabatic approximations provides an especially simple, practical description of the reaction.

Part I of this book lays down the foundations for understanding the mechanisms of the reactions of exotic nuclei by using these approximations. Since the structure information on exotic nuclei is primarily obtained through reaction experiments using the secondary beams, the theory of reactions with exotic nuclei is of interest not only for its own sake, i.e., for studies of the reaction mechanisms, but also as a framework to extract structure information reliably.

All nuclei that show prominent interaction cross sections lie close to the line, on the neutron-rich side of the chart of nuclei, beyond which no bound system can be formed. This line is called the neutron drip line (see Fig. 1) since any unbound system formed beyond this line decays with neutron emission. Most nuclei with prominent interaction cross sections have the property that the nucleus obtained by removing one of their neutrons is unstable against ejecting one more neutron. For example, ^{11}Li exhibits the enhanced interaction cross section, and ^{10}Li is unbound with respect to neutron emission. The lowest threshold of ^{11}Li is thus not ^{10}Li+n but ^{9}Li+2n, and this threshold is just 250–400 keV above the ground state of ^{11}Li. It is tempting to consider ^{11}Li to be a ^{9}Li+n+n three-body system. It is interesting that none of the pairs of the constituents, ^{9}Li+n or n+n, forms a bound state. A three-body system with this property is called 'Borromean'.[1] Viewed as an $\alpha(^{4}\mathrm{He})$+n+n bound state, ^{6}He is another example of a Borromean nucleus.

The number of three-body bound states depends sensitively on the properties of the two-body subsystems. When at least two of the two-body subsystems have just one bound state whose energy is exactly zero, then, under certain conditions, the three-particle system will have an infinite number of bound states in the vicinity of zero energy. These states, called 'Efimov states', are very loosely bound and their wave functions extend far beyond the range of the potential. By making the two-body interactions more or less attractive, the number of bound states becomes finite, and then the Efimov states disappear. Although in a Borromean system there is no two-body bound state, there is still some similarity to the Efimov case. The ground-state density of the Borromean nucleus may also reach to distances much beyond the interaction range, and the formation of such a state may be helped by near-threshold two-body resonances.

To describe the halo structure, one may attempt to divide the nucleus into a core and halo nucleons. In this way one may concentrate on the asymptotics of the weakly-bound system. A realistic description of the halo structure, however, requires furthermore a fair consideration of all correlations between the core and the halo nucleons (including that implied by the Pauli principle) as well as the distortion of the core effected by the halo nucleons. A state of a composite system can always be expanded in terms of the states of the free constituents, and in

[1] A system is called Borromean if it is bound but can be decomposed into three subsystems, any two of which cannot form a bound state. This term comes from the coat of arms of the famous Milanese family Borromeo. This contains three rings each of which is interlaced with the other two such that by breaking any of them, the other two also release each other. This symbol is used in several forms in Japanese coats of arms as well. The cover illustration shows one of these, the three 'bamboo rings'.

such an expansion the core distortion makes excited core states appear. Thus core distortion and core excitation are one and the same effect.

The behaviour of the nucleus as a composition of subsystems is a clustering phenomenon. It is known that the clustering plays an important role in light nuclei. In particular, α-clustering is prominent owing to the large binding energy of the α-particle. For example, the ground state of ^{8}Be is like two α-clusters, and the first excited state of ^{16}O at 6.05 MeV is like $\alpha+^{12}$C. Clustering leads to a spatial localization of the nucleons, and thereby breaks the dominance of the mean field.[2] A realistic picture of exotic nuclei is a multicluster picture. A model based on such a picture involves a few-body problem with some composite bodies, α-cluster, triton (^{3}H or t) and helion (^{3}He or h) clusters, and, in addition, some unclustered nucleons, to be treated as 'single-nucleon clusters'.

The simplest model for the halo nuclei is a phenomenological macroscopic model. A cluster model is said to be macroscopic if the clusters are treated as structureless and phenomenological if the interaction potentials between them are determined phenomenologically by fitting to some (mostly scattering) data. The macroscopic models can reproduce the basic features of the loosely bound nucleus reasonably well, but mostly fail to produce sufficient binding. This problem shows that the basic assumptions of this model are too restrictive: there is no way to take account of the excitation (or distortion) of the clusters. Moreover, the Pauli principle between the clusters cannot be taken into account properly. Some variants of semimicroscopic models (such as the cluster-orbital shell model) suffer from similar drawbacks. The halo neutrons must overlap with the surface region of the core, so the description of the tail of the core may influence the dynamics of the halo neutrons considerably. In this respect it is important to mention that all cores that come into consideration, e.g., the ^{9}Li core for ^{11}Li, the ^{10}Be core for ^{11}Be and the ^{12}Be core for ^{14}Be, have open shells, except the α-cluster as a core. The configurations of these cores certainly deviate from the simplest shell-model states and are distortable. All these problems can

[2]The term 'mean field' means a single-particle substitute for a field that depends on multiparticle variables. In this book it is used in a restricted sense. It is meant to be a single-particle potential, acting on each nucleon, generated by averaging over the motions of all others. An actual realization of this idea is the Hartree–Fock self-consistent field. In the mean-field picture the nucleons orbit in single-particle states generated by the mean field. There are, strictly speaking, as many mean fields in a nucleus as the number of nucleons since the mean field is bound to depend on the label of the particle that does not act on the particular particle considered, i.e., on the state of the very particle considered. But the mean-field picture makes sense only if the state dependence of the mean field is not strong. In a clustered system, however, the single-particle states may correspond to different cluster centres, thus the mean field is likely to be strongly state-dependent.

be eliminated in a straightforward way if all nucleonic degrees of freedom are taken into account, i.e., in fully microscopic models.

The standard microscopic nuclear models also have their own shortcomings. The Hartree–Fock model is a routine approximation based on the mean-field assumption but may be too restricted to describe the clustered distribution of the nucleons and the core-plus-halo structure in exotic nuclei. The application of the shell model is hampered by the loosening up of the shell structure as established by the study of stable nuclei. Empirical indications of this are, e.g., that the $N=8$ and $N=20$ shell closures seem to break down in ^{11}Li and ^{12}Be, and ^{32}Mg, respectively; e.g., in ^{11}Li the ground state is likely to contain a mixture of $(1s_{1/2})^2$ and $(0p_{1/2})^2$ components. Furthermore, in Borromean nuclei even the lowest-lying valence single-particle orbits are unbound, and the conventional bound-state bases would require huge dimensions.

The model of fermionic or antisymmetrized molecular dynamics has been extensively applied to light exotic nuclei. In this model the wave function is approximated by one Slater determinant of Gaussian single-particle wave packets, whose centres are variational parameters, or by a combination of a few such determinants. When a single Slater determinant is used, the Gaussian centres are just the mean positions of the nucleons, hence showing up cluster structure appealingly. Such a model is restricted, however, in describing exotic nuclei in two respects: it does not treat the correlations adequately, and it cannot adapt itself to the long-tail asymptotics because the width of the Gaussians has to be fixed. The remedy of these shortcomings by combining many Gaussian wave packets is straightforward but costly.

For the time being *ab initio* calculations with realistic nucleon–nucleon interactions are possible for nuclei with mass number $A \leq 8$, but they are very difficult for $A > 4$. Thus, to study light exotic nuclei, it is a best compromise to use a microscopic cluster model. In such applications typically three or more clusters should be taken and, unlike in the macroscopic models, the clusters do have structures.

Part II of this book presents a fully microscopic multicluster model as a theory that should be suitable for the description of light exotic nuclei. The main features of this model are the following: (i) there is no compromise on the Pauli principle and on the treatment of the centre of mass, (ii) the dynamics is governed by effective nucleon–nucleon interactions, thus no phenomenological cluster–cluster or cluster–nucleon potential is introduced, (iii) there is no need to use a passive-core[3] plus valence-nucleons approximation, and (iv) the distortion of the clusters can be allowed for.

[3]The core appearing in shell models may be identified with one cluster or with a system of clusters.

The dynamics of the clusters is rather complicated even for few-cluster systems, and its realistic description requires a variational approach with flexible trial functions. The trial function has to represent the correlated relative motion of the clusters at short and long distances as well as their inner motion. These requirements are satisfied by a linear combination of multivariable Gaussian functions called the correlated Gaussians. In the theory to be presented in Part II of this book correlated Gaussians are used. The terms of the trial function, referred to as basis states, can be chosen in an infinite number of ways, and the selection procedure is part of the variational solution. A most practical procedure is stochastic, and the solution involving this selection procedure is called the stochastic variational method. The trial function thus constructed is found to provide virtually exact solutions for a number of few-body bound-state problems. In most of the examples to be presented the stochastic variational method with correlated Gaussians has been used.

The study of light exotic nuclei is an introduction to the broader research field of unstable nuclei, which is yet to be explored. It is expected that it will reveal formerly hidden aspects of nuclear structure.

In the sd-shell and beyond, for nuclei that are imbalanced in protons and neutrons, the proton and neutron Fermi levels are expected to split. For neutron-rich nuclei this will lead to the formation of thick neutron skins or cluster structures with the binding mediated by neutrons. The strength of the spin-orbit force of the single-particle motion is expected to be anomalous, and that changes the arrangements of the orbits into shells. This should cause the well-known magic numbers to be superseded by new ones in these regions. For proton-rich nuclei proton–neutron pairing is expected to appear.

The other field to which exotic nuclei give clues is nuclear astrophysics, especially stellar nucleosynthesis. The physics of radioactive nuclear beams will also be invaluable in illuminating the nuclear aspects of supernova collapse and explosion and providing information on the equation of state of the asymmetric nuclear matter.

The layout of Parts I and II will be outlined in their own introductory chapters.

Part I

Reactions with light exotic nuclei

Chapter 1

Introduction to Part I

The advances of experimental techniques of radioactive secondary ion beams have given rise to rapid progress in the research of unstable nuclei in the last one and a half decade of the twentieth century [1–6]. The light exotic nuclei of our concern decay by β-decay, and the typical lifetime is of the order of a millisecond to one second. They are mostly stable against nucleon emission, so they can be described as bound states. Their lifetimes are, however, rather short when considered from the viewpoint of measuring the nuclear properties.

The experimental study of unstable nuclei with radioactive secondary beams was started at the BEVALAC accelerator in Berkeley in the middle of the 1980s [7]. The radioactive beams were produced as projectile-like fragments in violent nucleus–nucleus collisions. At high incident energies, i.e., when the nucleus–nucleus relative motion is faster than the internal motion of the nucleons in the colliding nuclei, the nucleus–nucleus collisions can be visualized in a participant–spectator picture. Some nucleons of the projectile collide violently with some of the nucleons of the target. As a result, these nucleons get detached from both of the colliding partners and form a 'hot spot', which is called the participant. The rest of the projectile, which is called the projectile-like fragment, proceeds forward with a velocity that is close to the incident velocity. This part of the projectile does not carry too much excitation, and the target also leaves behind a similarly passive fragment, the so-called target-like fragment. Both of these passive fragments may be called spectators.

The projectile-like fragments produced by any particular colliding system show a great variety since the number and type of the nucleons lost by the projectile during the collision may vary randomly. Even extremely neutron-rich or proton-rich nuclei may be produced occasionally. After the collision, the projectile-like fragments immediately cool down to their ground states (g.s.) through emitting particles and electromagnetic radiation. The secondary beams may be formed out of these projectile-like fragments. They are collimated on some stable target nuclei, and the properties of the projectile nuclei are explored by observing these collisions.

In the first experiments with such secondary beams various interaction cross sections have been measured systematically at the incident energy of 800 A MeV [8–10]. The interaction cross section is the sum of the cross sections for all channels in which the nucleonic composition of the projectile is changed.

The measured $A+^{12}\mathrm{C}$ interaction cross sections for some light nuclei A with charge numbers Z=2–5 are shown in Fig. 1.1. As the neutron number increases along an isotopic chain, the cross section first increases regularly, following the geometrical estimate

$$\sigma_{\mathrm{reac}} = \pi(R_{\mathrm{P}} + R_{\mathrm{T}})^2, \tag{1.1}$$

where the radii are $R_{\mathrm{x}} = r_0 A_{\mathrm{x}}^{1/3}$ (x = P, T) fm, with $r_0 = 1.1 \sim 1.2$ fm and A_{P}, A_{T} are the projectile and target mass numbers, respectively. This is actually the area around the target centre within which the centre of the projectile has to impinge in order to hit the target. The interaction cross section in high-energy collisions is believed to be very close to the total reaction cross section (which includes all processes except for elastic scattering). Figure 1.1 shows that this formula reproduces the interaction cross section except for some neutron-rich nuclei. Furthermore, the validity of this geometrical cross section suggests that, if the two densities overlap during the collision, then at least one of them is excited with a probability close to unity. According to this interpretation, an abrupt increase of the interaction cross section indicates a drastic change of the nuclear radius, a departure from the simple rule $R \propto A^{1/3}$.

For nuclei close to the neutron drip line, such as ^{11}Li, the interaction cross section does show an abrupt increase. This has been the first hint of halo structure [8, 11]. Such a sudden increase is also seen in other interaction cross sections, like those for ^{6}He, ^{11}Be, ^{14}Be. They are also supposed to have halo structure. The measurements are now extended to much heavier nuclei, e.g., to Na isotopes [12]. There are indications for halo structure, e.g., in ^{19}C [13–16] and ^{23}O [16].

Measurements have been performed for the fragment production cross sections [21] and charge-exchange cross sections [22] as well. It was observed that the large interaction cross section is accompanied by large cross sections of the removal of one or two neutrons.

All nuclei with prominent interaction cross sections lie close to the neutron drip line. Most of them have the property that, when one of

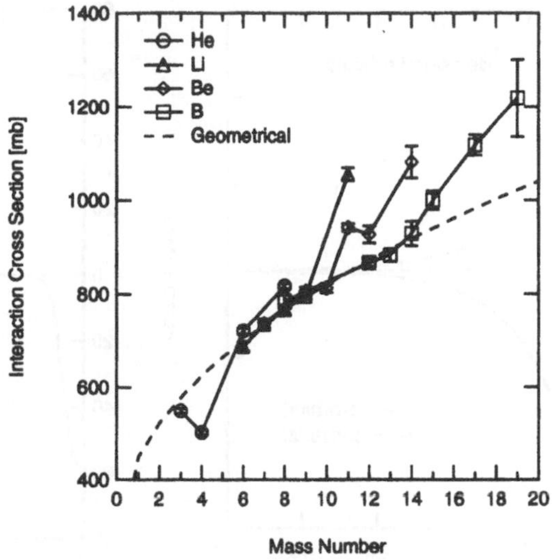

Figure 1.1: Interaction cross sections measured at 800 A MeV [8–10, 17–20] for the collisions of light projectiles with a carbon target plotted against the projectile mass number. The cross sections belonging to isotopes of the same element are connected by lines. The geometrical cross section, Eq. (1.1) with $r_0 = 1.15$ fm, is shown for comparison (1 barn $\equiv$ 1 b = 100 fm^2).

their neutrons is removed, they become unstable and eject one more neutron. (They are Borromean, cf. footnote on p. 4.) The binding of the last two neutrons is very weak, typically around 1 MeV or even less. This should be compared with the average single-nucleon separation energy of stable nuclei, which is about 8 MeV. The small one- or two-neutron separation energies and the large nuclear radii lead to the idea of halo structure: one or two neutrons of these nuclei orbit around a compact core in extremely extensive orbits, like in a halo, tied to the core very weakly. The large spatial distribution of halo neutrons is a quantum effect. Due to the small separation energy, the exponential wave-function tail extends far into the classically forbidden region, which enhances the nuclear radius with respect to the $A^{1/3}$ systematics.

Let us consider a single neutron bound in an s-orbit in a spherically symmetric potential well as a simple model for the halo structure. Closest to this model is the nucleus ^{11}Be, whose g.s. is supposed to have a predominantly s-wave one-neutron halo. In Fig. 1.2 we show 1s- and 0p-wave single-neutron radial wave functions in a Woods–Saxon potential, which is also plotted. The shape of the potential is tailored for ^{11}Be, and its depth is adjusted to reproduce the neutron separation

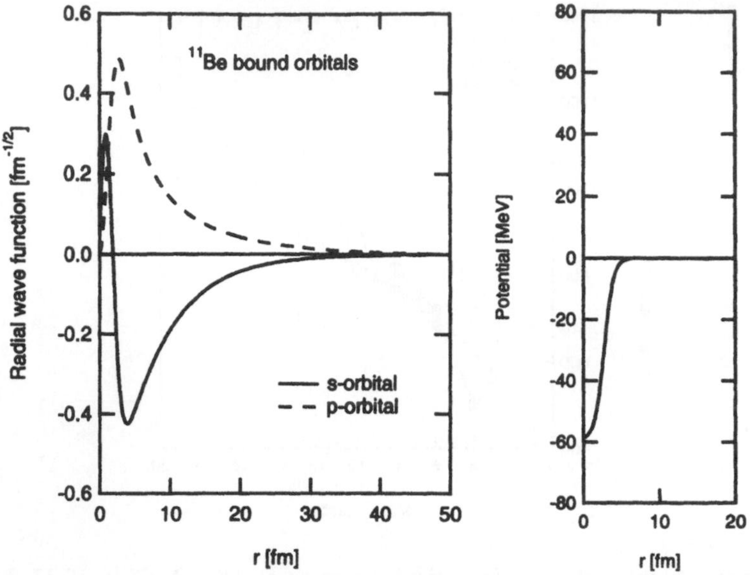

Figure 1.2: Radial wave functions of weakly bound single-particle (s.p.) s- (solid curve) and p-wave (dashed curve) orbits in a Woods–Saxon potential, imitating the g.s. and first excited state of ^{11}Be in a ^{10}Be+n model, with energies −0.503 MeV and −0.187 MeV, respectively. The potential shape is also shown in a separate panel. The classical turning points are around 6 fm for both partial waves.

energy of the g.s., 0.503 MeV, and then, separately, of the first excited state, 0.187 MeV.

The figure shows that it has a large probability that the neutron stays outside the region that is prohibited in classical mechanics. A one-particle halo is nothing but an orbit with a classically forbidden long tail. The neutron wave function in the classically forbidden region is proportional to $\frac{1}{r}e^{-\kappa r}$ where κ is related to the neutron separation energy $|\varepsilon|$ by $|\varepsilon| = \hbar^2\kappa^2/2\mu$, with μ being the reduced mass of the neutron and the core. Taking $|\varepsilon| = 0.503$ MeV, we obtain $\kappa^{-1} \simeq 7$ fm, which is much larger than the radius of light nuclei, which accounts for the long tail. Since all nuclei with large interaction cross sections are characterized by small neutron separation energies, it is natural to attribute the large interaction cross section to the large spatial extension of their halo.

Additional clear evidence for the halo structure has been observed in the momentum distribution of fragments [21]. It is well-known that the momentum of a fragment obtained by removing a nucleon from the projectile in a high-energy collision reflects their relative momentum in the g.s. of the projectile [23]. The dynamics of the collision only distorts the momentum distribution to a little extent, especially for the momentum component parallel to the incident beam (longitudinal direction). In systematic measurements of the fragment momenta of unstable nuclei it was found that, for projectiles that are known to have halo structure, the core momentum distribution shows a very narrow peak [21]. In Fig. 1.3 we show the measured transverse momentum distribution (which is the distribution of the momentum component perpendicular to the beam line) of fragment ^{9}Li of projectile ^{11}Li. The distribution of the ^{9}Li momentum is found to be fitted by a superposition of two Gaussians, $\exp[-\frac{1}{2}(p/\sigma)^2]$. The widths of the two Gaussians are 21 ± 3 MeV/c and 80 ± 4 MeV/c, respectively. This should be compared with the momentum distribution in a stable nucleus, whose width is typically 50–200 MeV/c [24].

The narrow momentum distribution may be intuitively understood by considering the uncertainty principle between the position and momentum. A large spatial spread implies large uncertainty in the position, as we see in Fig. 1.2, which, in turn, corresponds to a narrow momentum distribution. To be more quantitative, let us try to estimate the contribution of the wave-function tail $\frac{1}{r}e^{-\kappa r}$ to the

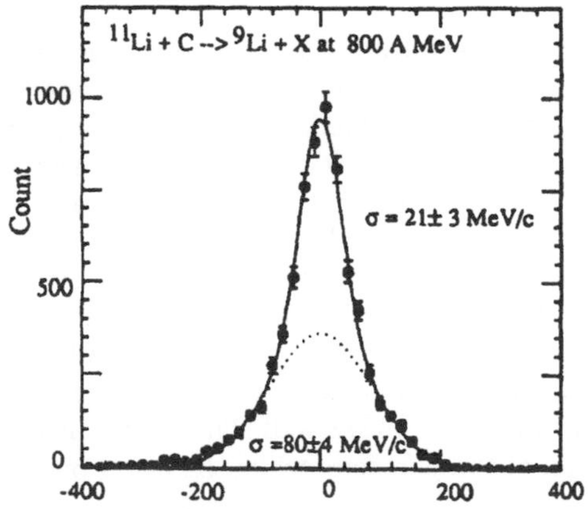

Figure 1.3: Transverse momentum distribution of the ^{9}Li fragment removed from ^{11}Li in a collision with ^{12}C at $E = 800\,A$ MeV. The measured momentum distribution can be fitted by a superposition of two Gaussians. The narrower one has a width of 21 ± 3 MeV/c, which should be compared with the momentum width of normal nuclei, 50–200 MeV/c. Taken from Ref. [17].

momentum distribution. The Fourier transform of $\frac{1}{r}e^{-\kappa r}$ is proportional to a Lorentzian, $[p^2 + (\hbar\kappa)^2]^{-1}$. The half width of this function is just $\hbar\kappa$, while the width of the spatial extension is characterized by κ^{-1}. So the two widths are inversely proportional to each other, in accord with the uncertainty principle. The Gaussian width σ may be approximately related to the Lorentzian width $\hbar\kappa$ by $\sigma = \hbar\kappa/\sqrt{2\ln 2}$. If we assume a dineutron (as a single particle, with zero binding) coupled to the ^{9}Li core in ^{11}Li, the spatial width is given by $\kappa^{-1} = 6.5$ fm and yields the momentum width $\sigma = 25.8$ MeV/c, which is in reasonable agreement with the experimental value, 21 MeV/c.

Another observation characteristic of halo nuclei has been found in their reactions with heavy target nuclei. The interaction cross section has been found to be much larger than originally expected based on the *nuclear* reaction mechanism [25]. The enhancement comes from the increase of the neutron-removal cross section, and originates from a breakaway of the halo induced by the strong Coulomb force.[1] Though the halo neutrons do not feel the Coulomb force directly, the Coulomb repulsion between the charged core and target distorts the whole projectile nucleus and induces this breakup. In Fig. 1.4, we show the interaction cross section and the two-neutron-removal cross section of ^{11}Li impinging on various target nuclei. The dashed curves show estimates for the cross section induced by the nuclear force alone. The measured cross sections are much larger than those expected from the nuclear interaction.

In high-energy experiments the measured cross sections are inclusive, which means that not all final fragments are detected, thus leaving the final state partially undefined.[2] Since the energy transferred from the projectile–target relative motion to the internal excitations is large, it is not possible to distinguish the individual states of the outgoing fragments. As the energy gets lower, it becomes possible to measure exclusive cross sections, in which the final internal states of both partners are identified. The simplest channel is elastic scattering, in which the colliding partners emerge as they entered into the process, in their g.s. [27–29]. There are measurements for inelastic scattering [30,31] and charge-exchange reactions [32] as well. The former provides information on the excitation of the partners, while the latter on their isospin multiplets.

[1] In the energy region of interest the dominant mechanism of neutron removal is breakup, which can be regarded as projectile excitation into the continuum. In this book neutron removal and breakup will be used as synonyms, to emphasize either the outcome or the mechanism of the process.

[2] In quantum-mechanical terms this is an instance of incomplete measurement [26]. The measured cross section should be identified with an incoherent sum over a number of exit channels.

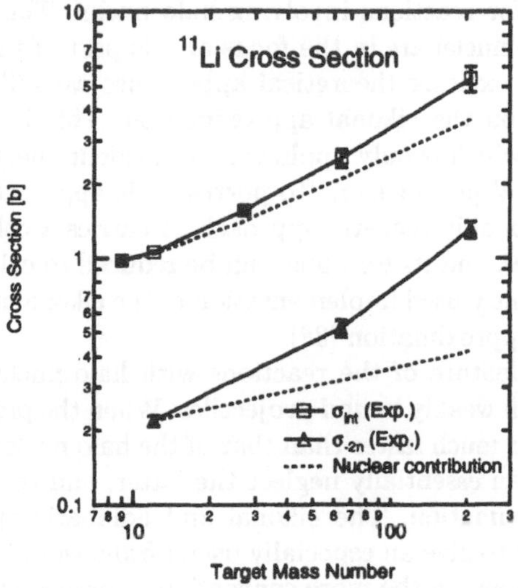

Figure 1.4: Experimental interaction cross section and two-neutron-removal cross section of the collision of ^{11}Li with various target nuclei at $E = 800\,A$ MeV [25]. The estimated contributions induced by the nuclear force are indicated by dashed curves.

Finally, at low incident energies, the cross section is dominated by fusion reactions. Since the fusion cross section also grows with the size of the projectile, it may be expected that the fusion cross section is largest for neutron-rich unstable nuclei in each isotopic chain. Furthermore, since for heavier nuclei the valley of stability turns towards the neutron-rich side due to the Coulomb force, a neutron-rich beam is expected to favour the fusion reaction. These considerations lead us to the intriguing idea of employing neutron-rich radioactive secondary beams to synthesize superheavy elements. Extremely neutron-rich projectiles have been used to investigate fusion reactions of halo nuclei experimentally [33,34], and these cases have been studied theoretically as well [35–37].

In this part of the book, we will expound a theoretical framework for the description of various reaction mechanisms of unstable nuclei. We especially concentrate on the descriptions of the reactions of halo nuclei. We have two reasons for that. First, halo structure is an interesting new nuclear phenomenon, and its significance is central in the field of light exotic nuclei. In fact, information on halo structure comes mainly from reaction experiments. Second, it turns out that the theoretical framework that has proved successful for reactions of stable nuclei needs to be modified for reactions involving halo nuclei. Thus the reaction theories for halo nuclei are in the focus of this part of the book.

From among existing theoretical approaches we will mainly focus on those based on the eikonal approximation. This is a high-energy approximation, which is only applicable to incident energies of at least a few tens of MeV per nucleon. To portray this approach, one can say that it is a reasonably realistic approach, it carries a coherent picture of these processes, and its formulae can be reduced to calculable forms. The most frequently used implementation of the eikonal approximation is the Glauber approximation [38].

The specific feature of the reactions with halo nuclei is that they are reactions with weakly bound projectiles. When the projectile–target relative motion is much faster than that of the halo nucleons within the projectile, one can essentially neglect the latter, and that is called the adiabatic approximation. The eikonal and adiabatic approximations can be combined to give an especially useful framework for the description of the reactions. In the more conventional coupled-channel formalism, on the contrary, a weakly bound projectile poses the formidable problem of treating the coupling to continuum states. In spite of considerable progress in this field in recent decades, the problem is still challenging both conceptually and practically.

Thus the high bombarding energy makes it possible to use the eikonal approximation, and the strong continuum effects make it virtually necessary. That accounts for the central role of the eikonal and Glauber approximations here, and that is why the subject of this part of the book differs from those of the standard textbooks on direct reaction theories [39,40]. For completeness, the relationship between the present approach and conventional direct reaction theory is clarified in Appendix A.

Chapter 2

Potential scattering

2.1 Elements of scattering theory

2.1.1 Scattering wave function

In this chapter some basic concepts and results of quantum-mechanical scattering theory for a potential problem will be summarized. More detailed explanation can be found in the standard textbooks of quantum mechanics [26, 41]. In the sections to come we present an approximate treatment of the problem, the so-called eikonal approximation, which will be extensively used in the subsequent chapters.

Let us consider a particle with mass m scattered by a central potential $V(r)$. The stationary scattering wave function $\psi(\boldsymbol{r})$ satisfies the static Schrödinger equation

$$\left[-\frac{\hbar^2}{2m}\boldsymbol{\nabla}^2 + V(r)\right]\psi(\boldsymbol{r}) = E\psi(\boldsymbol{r}). \tag{2.1}$$

The potential $V(r)$ is assumed to be non-zero only in a finite region. The inclusion of the Coulomb potential requires appropriate modifications. In the asymptotic region, far from the scattering centre, the wave function $\psi(\boldsymbol{r})$ satisfies a free-particle Schrödinger equation and a boundary condition corresponding to the scattering situation: it has to be a sum of a plane wave and of an outgoing spherical wave. When the incident beam impinges upon the scattering centre from the negative z-axis, the solution satisfies the boundary condition

$$\psi(\boldsymbol{r}) = \mathrm{e}^{ikz} + f(\theta)\frac{\mathrm{e}^{ikr}}{r} \qquad (r \to \infty). \tag{2.2}$$

The wave number k is related to the incident energy E as $E = \hbar^2 k^2/2m$. When the potential is central, the wave function is axially symmetric, and the scattering amplitude f only depends on the polar angle θ. Because of energy conservation, the incident and outgoing wave numbers are equal.

The differential scattering cross section is the modulus square of the scattering amplitude:

$$\frac{d\sigma}{d\Omega} = |f(\theta)|^2, \tag{2.3}$$

where Ω is a solid angle around direction θ. The total (or integrated) elastic scattering cross section is then given by

$$\sigma_{\text{el}} = \int d\Omega \frac{d\sigma}{d\Omega} = \int d\Omega\, |f(\theta)|^2. \tag{2.4}$$

2.1.2 Integral equation for scattering

The equation of motion for the wave function $\psi(\boldsymbol{r})$ can be cast into the form of an integral equation. For this purpose, we introduce the free-particle Green's functions $G(\boldsymbol{r}-\boldsymbol{r}')$, which, by definition, are the solutions of

$$\left[E - \left(-\frac{\hbar^2}{2m}\boldsymbol{\nabla}^2\right)\right] G(\boldsymbol{r}-\boldsymbol{r}') = \delta(\boldsymbol{r}-\boldsymbol{r}'). \tag{2.5}$$

This equation has the following two independent solutions,

$$G^{(\pm)}(\boldsymbol{r}-\boldsymbol{r}') = -\frac{2m}{\hbar^2}\frac{e^{\pm ik|\boldsymbol{r}-\boldsymbol{r}'|}}{4\pi|\boldsymbol{r}-\boldsymbol{r}'|}. \tag{2.6}$$

To prove Eq. (2.6), one should use the well-known mathematical identity

$$\boldsymbol{\nabla}^2 \frac{1}{|\boldsymbol{r}-\boldsymbol{r}'|} = -4\pi\delta(\boldsymbol{r}-\boldsymbol{r}'). \tag{2.7}$$

We can rewrite the Schrödinger equation (2.1) into an integral equation in terms of any of the Green's functions $G = \alpha G^{(+)} + (1-\alpha)G^{(-)}$, where each α implies a well-defined boundary condition. The solution that satisfies the scattering boundary condition is that containing just $G = G^{(+)}(\boldsymbol{r}-\boldsymbol{r}')$:

$$\psi(\boldsymbol{r}) = e^{ikz} + \int d\boldsymbol{r}'\, G^{(+)}(\boldsymbol{r}-\boldsymbol{r}')V(\boldsymbol{r}')\psi(\boldsymbol{r}'). \tag{2.8}$$

This equation is known as the Lippmann–Schwinger equation (with outgoing boundary condition). We may easily check that this $\psi(\boldsymbol{r})$ satisfies the Schödinger equation by applying the operator $\left[E - (-\hbar^2\boldsymbol{\nabla}^2/2m)\right]$ to both sides. That the (unique) solution of Eq. (2.8) satisfies the boundary condition (2.2) can be verified by analysing the asymptotic behaviour of the second term.

To this end, let us take the limit $\boldsymbol{r} \to \infty$. For $\boldsymbol{r} \gg \boldsymbol{r}'$, the Green's function $G^{(+)}(\boldsymbol{r}-\boldsymbol{r}')$ behaves like

$$G^{(+)}(\boldsymbol{r}-\boldsymbol{r}') \sim -\frac{2m}{\hbar^2}\frac{1}{4\pi r}\mathrm{e}^{ikr-ik\hat{\boldsymbol{r}}\cdot\boldsymbol{r}'}, \tag{2.9}$$

where $\hat{\boldsymbol{r}}$ denotes the unit vector pointing to the $\boldsymbol{r}$ direction. Substituting Eq. (2.9) into Eq. (2.8), we obtain the following asymptotic form

$$\psi(\boldsymbol{r}) = \mathrm{e}^{ikz} - \frac{2m}{\hbar^2}\frac{\mathrm{e}^{ikr}}{4\pi r}\int d\boldsymbol{r}'\,\mathrm{e}^{-ik\hat{\boldsymbol{r}}\cdot\boldsymbol{r}'}V(\boldsymbol{r}')\psi(\boldsymbol{r}') \quad (r \to \infty), \tag{2.10}$$

which can be identified with Eq. (2.2). In this way we obtain, for the scattering amplitude,

$$f(\theta) = -\frac{2m}{\hbar^2}\frac{1}{4\pi}\int d\boldsymbol{r}'\,\mathrm{e}^{-i\boldsymbol{k}'\cdot\boldsymbol{r}'}V(\boldsymbol{r}')\psi(\boldsymbol{r}'), \tag{2.11}$$

where $\boldsymbol{k}'$ stands for the vector $k\hat{\boldsymbol{r}}$. This vector may be called the vector of propagation of the outgoing wave; $|\boldsymbol{k}'| = k$ and its angle with the z-axis is θ.

To calculate the scattering amplitude with Eq. (2.11), we need to know the exact scattering wave function $\psi(\boldsymbol{r})$. Since $\psi(\boldsymbol{r})$ contains $f(\theta)$ in the asymptotic region, one might think that Eq. (2.11) is not useful. One should note, however, that the exact wave function is actually required only in the region in which the potential $V(r)$ does not vanish. As will be seen, the eikonal approximation provides an approximate wave function, which is valid where the potential does not vanish, but does not have the correct asymptotic behaviour. Expression (2.11) will then be used to obtain a scattering amplitude in the eikonal approximation.

2.1.3 Flux conservation and optical theorem

A collision event between two nuclei may lead to outcomes other than a simple deflection. For example, it may give rise to excitations or fragmentations of the colliding nuclei. An outcome with a well-defined set of outgoing nuclei of well-defined intrinsic states may be called a channel.

The quantities characterizing a nuclear collision in such a general case are related by flux conservation, and that is the subject of the present subsection. When only the elastic channel is open (i.e., when there is not enough energy to reach the other channels), the total cross section, σ_{tot}, is equal to the total elastic scattering cross section, σ_{el}, and the elastic scattering may be described as a potential scattering. Because of flux conservation, the scattered flux is equal to that taken away from the incoming plane wave, which is observable as an interference between the plane wave and the spherical wave at zero degree. This relationship is called the optical theorem, and gives another formula for the total cross section, an alternative of Eq. (2.4). When other channels are also open, part of the flux goes into other channels, and the total cross section calculated from the forward scattering amplitude will be larger than that calculated from Eq. (2.4). The cross section related to the forward scattering amplitude must be interpreted as the total cross section of all possible processes, whereas Eq. (2.4) is just the total elastic cross section. Their difference, which is related to the flux 'absorbed' by other channels, is called the total absorption cross section. When non-elastic channels are open, the scattering process can be described as scattering from a complex potential. We now formulate flux conservation, derive the optical theorem, and then extend its interpretation to the absorptive case.

The probability current density, defined by

$$\boldsymbol{j} = \frac{\hbar}{2mi}(\psi^* \boldsymbol{\nabla}\psi - \psi\boldsymbol{\nabla}\psi^*), \tag{2.12}$$

plays an important role in the derivation of a cross-section formula from the wave function. This current density is conserved for a stationary state when the Hamiltonian is Hermitean. To see this, we can use the equation of continuity,

$$\frac{\partial\rho}{\partial t} + \boldsymbol{\nabla}\cdot\boldsymbol{j} = \frac{2}{\hbar}\rho\,\mathrm{Im}\,V(r), \tag{2.13}$$

which can be derived from the time-dependent Schrödinger equation. Here ρ is the density: $\rho = |\psi(\boldsymbol{r})|^2$. For a static solution, the density is time-independent, so that $\partial\rho/\partial t = 0$. If the potential is real and hence the Hamiltonian is Hermitean, the divergence of the current density, $\boldsymbol{\nabla}\cdot\boldsymbol{j}$, is seen to vanish,

$$\boldsymbol{\nabla}\cdot\boldsymbol{j} = 0, \tag{2.14}$$

and that is the flux conservation. This meaning becomes obvious, if we enclose the system into a volume V of surface S and integrate Eq. (2.14) over this volume. From the Gauss theorem it follows that

$$\int_V d\boldsymbol{r}\,\boldsymbol{\nabla}\cdot\boldsymbol{j} = \int_S dS\,\hat{\boldsymbol{r}}\cdot\boldsymbol{j} = 0, \tag{2.15}$$

and that indeed expresses that the sum of all incoming and outgoing fluxes are in exact balance.

We will, however, often use the nuclear optical model, in which the potential generally includes a negative imaginary part. In that case the flux is not conserved but is absorbed since its divergence is negative. Referring to this effect, we say that the imaginary part of the potential is absorptive.

The scattering differential cross section is defined as the ratio of the flux of outgoing particles per unit solid angle to the flux of incoming particles per unit area. To obtain the cross section, we will calculate the asymptotic fluxes belonging to the terms of Eq. (2.2). A precise analysis of the flux in the forward direction will give us the optical theorem, which relates the forward scattering amplitude with the total cross section.

In defining the cross section, the interference between the asymptotic plane wave and scattering wave is disregarded. Underlying this treatment is an idea that the time-independent framework we are using is a limiting case of a time-dependent theory, in which the scattering is described by a wave packet, whose spreading is slower than its propagation. Then the incoming and scattered wave packets can be approximated in the space–time regions before and after the collisions by a plane wave and by a spherical wave, respectively. The space–time regions of the (approximate) asymptotic plane wave and of the asymptotic spherical wave are separated, and there is no interference, except sharp at zero scattering angle, where the scattered waves cannot be separated from the incident wave. The time-independent framework is equivalent to the time-dependent framework in all predictions [42].

The incident plane wave, $\psi = \mathrm{e}^{ikz}$, gives the probability current density $\boldsymbol{j}_{\text{inc}} = (\hbar k/m)\hat{\boldsymbol{z}} = v\hat{\boldsymbol{z}}$, where $\hat{\boldsymbol{z}}$ denotes the unit vector pointing to the z-direction. This shows that the flux per unit area and time in the z-direction implied by the incident plane wave e^{ikz} is v. The current density belonging to the outgoing scattering wave $\psi = f(\theta)\frac{1}{r}\mathrm{e}^{ikr}$ in the $\hat{\boldsymbol{r}}$ direction for large r is

$$\boldsymbol{j}_{\text{out}} = v\hat{\boldsymbol{r}}\frac{|f(\theta)|^2}{r^2}. \tag{2.16}$$

This expression implies that the outgoing flux per unit solid angle and time is $v|f(\theta)|^2$. The differential scattering cross section is then $|f(\theta)|^2$ as was given in Eq. (2.3). Let us integrate the outgoing flux over the whole solid angle on a large spherical surface S. Then we see that the outgoing flux is proportional to the total elastic cross section defined in Eq. (2.4):

$$\int_S dS\,\hat{\boldsymbol{r}}\cdot\boldsymbol{j}_{\text{out}} = v\sigma_{\text{el}}. \tag{2.17}$$

It follows from the conservation of the total flux, Eq. (2.14), that the flux of the incident plane wave, which is directed to the z-axis, should decrease by the same amount as goes into the scattered wave. Since, according to Eq. (2.17), the outgoing flux is proportional to the integrated cross section, the flux conservation provides an expression for the integrated cross section. That is the optical theorem mentioned above. To derive it, let us choose the normal vector $\hat{\boldsymbol{r}}$ to point outwards and apply the Gauss theorem (2.15) to the asymptotic form of the wave function, Eq. (2.2). The radial component of the total flux, $\hat{\boldsymbol{r}}\cdot\boldsymbol{j}$, is calculated with the use of $\hat{\boldsymbol{r}}\cdot\boldsymbol{\nabla}=\partial/\partial r$:

$$\hat{\boldsymbol{r}}\cdot\boldsymbol{j} = v\cos\theta + v\frac{|f(\theta)|^2}{r^2} + v\,\mathrm{Re}\left[\frac{1+\cos\theta}{r}f(\theta)\mathrm{e}^{ikr(1-\cos\theta)}\right] \\ -v\,\mathrm{Im}\left[\frac{1}{kr^2}f(\theta)\mathrm{e}^{ikr(1-\cos\theta)}\right]. \tag{2.18}$$

The first and second terms are the fluxes of the incident plane wave and of the outgoing scattering wave, respectively. The third and fourth terms are the interference terms between them. The decrease of the flux in the forward direction should appear in the interference terms. Each term of this expression is integrated over the surface S of large radius R. The integral of the first term is zero since in a plane wave the outgoing flux is equal to the incoming flux. The integral of the second term gives, as before, the total elastic cross section, Eq. (2.17). The integral of the third and fourth terms needs careful treatment since the exponential function $\mathrm{e}^{ikr(1-\cos\theta)}$ oscillates rapidly as a function of the angle θ when the radial coordinate r is large. As an illustration, the integral of $R^{-1}(1+\cos\theta)f(\theta)\mathrm{e}^{ikR(1-\cos\theta)}$ is considered explicitly:

$$\int_S dS\,\frac{1+\cos\theta}{R}f(\theta)\mathrm{e}^{ikR(1-\cos\theta)} \\ = 2\pi R\int_{-1}^{+1}dt\,(1+t)\tilde{f}(t)\mathrm{e}^{ikR(1-t)} \\ = \frac{4\pi i}{k}f(0) + \frac{2\pi}{ik}\int_{-1}^{+1}dt\left[\frac{d}{dt}(1+t)\tilde{f}(t)\right]\mathrm{e}^{ikR(1-t)}. \tag{2.19}$$

In the first equality the surface integral is expressed as $dS = R^2 d\Omega = 2\pi R^2\sin\theta d\theta$ and $t=\cos\theta$ and $\tilde{f}(t)=f(\theta)$ are introduced. To get an integrand whose behaviour at large R is more apparent, an integration by parts has been performed in the last line. The second term will vanish in the limit $R\to\infty$, as becomes even more apparent if another integration by parts of the same pattern is performed. That will provide another factor R^{-1}, which makes the whole expression vanish, provided $\tilde{f}(t)$ is a smooth function as is normally the case. For similar reasons, the contribution of the fourth term of Eq. (2.18) to the surface integral will vanish when $R \to \infty$. The result is

$$\int_S dS\,\hat{\boldsymbol{r}} \cdot \boldsymbol{j} = v \left[\int d\Omega\, |f(\theta)|^2 - \frac{4\pi}{k} \operatorname{Im} f(0) \right], \tag{2.20}$$

which is zero, when the potential is real. That leads to the optical theorem:

$$\sigma_{\text{tot}} = \sigma_{\text{el}} = \frac{4\pi}{k} \operatorname{Im} f(0). \tag{2.21}$$

If the potential $V(\boldsymbol{r})$ is not real, $\boldsymbol{\nabla} \cdot \boldsymbol{j} = \frac{2}{\hbar} \rho \operatorname{Im} V(\boldsymbol{r})$, owing to absorption [cf. Eq. (2.13)]. The absorption cross section may be defined by the relation

$$v\sigma_{\text{abs}} = -\int_S dS\,\hat{\boldsymbol{r}} \cdot \boldsymbol{j} = -\int_V d\boldsymbol{r}\, \boldsymbol{\nabla} \cdot \boldsymbol{j} = \frac{2}{\hbar} \int_V d\boldsymbol{r}\, \rho(\boldsymbol{r}) \left[-\operatorname{Im} V(\boldsymbol{r}) \right]. \tag{2.22}$$

Then the optical theorem can be generalized as

$$\sigma_{\text{tot}} = \sigma_{\text{el}} + \sigma_{\text{abs}} = \frac{4\pi}{k} \operatorname{Im} f(0). \tag{2.23}$$

2.2 The eikonal approximation

2.2.1 Derivation

For problems of scattering by spherically symmetric potentials, the practical calculation of the scattering amplitude is not so difficult. Usually a partial-wave decomposition of the Schrödinger equation is first introduced. The radial Schrödinger equation obtained for each partial wave is a second-order ordinary differential equation. These equations should be solved for the required energy with the boundary condition that the solution vanish at the origin. The phase shifts are then extracted from the wave functions in the outer spatial region where the potential vanishes. The scattering amplitude can also be expanded into partial waves, and can thus be expressed in terms of phase shifts.

For a scattering problem of composite particles, however, the exact scattering solution of the Schrödinger equation is extremely difficult even for a three-body problem. Therefore, appropriate approximation methods are indispensable. In almost all methods antisymmetrization effects are neglected or treated implicitly. One of the simplest approximations to the scattering problem is the Born approximation [see Eq. (A.2)], which is a first-order term in a perturbative expansion with respect to the potential. However, since the nuclear force is strong, it is well-known that the naive Born approximation is useless in nuclear reactions.

Unlike the Born approximation, the eikonal approximation does not require an expansion with respect to the strength of the interaction. Therefore, it can be used widely in nuclear reactions. The condition for its validity is that the incident energy should be high enough. As a preparation for the composite-particle scattering case, the eikonal approximation is introduced here for potential scattering.

When the incident energy is high, the wave function $\psi(\boldsymbol{r})$ will oscillate rapidly, and its deviation from a plane wave can be expected to be small. This will be substantiated later by semiclassical arguments. To formulate this assumption, let us write the wave function as

$$\psi(\boldsymbol{r}) = \mathrm{e}^{ikz}\hat{\psi}(\boldsymbol{r}). \tag{2.24}$$

Then the function $\hat{\psi}(\boldsymbol{r})$ is assumed not to oscillate rapidly. To derive an equation for $\hat{\psi}(\boldsymbol{r})$, let us substitute this expression for $\psi(\boldsymbol{r})$ into the Schrödinger equation:

$$-\frac{\hbar^2}{2m}\boldsymbol{\nabla}^2\mathrm{e}^{ikz}\hat{\psi}(\boldsymbol{r}) + V(\boldsymbol{r})\mathrm{e}^{ikz}\hat{\psi}(\boldsymbol{r}) = E\mathrm{e}^{ikz}\hat{\psi}(\boldsymbol{r}). \tag{2.25}$$

Multiplying both sides by e^{-ikz} from the left leads to

$$-\frac{\hbar^2}{2m}\left(\mathrm{e}^{-ikz}\boldsymbol{\nabla}^2\mathrm{e}^{ikz}\right)\hat{\psi}(\boldsymbol{r}) + V(\boldsymbol{r})\hat{\psi}(\boldsymbol{r}) = E\hat{\psi}(\boldsymbol{r}). \tag{2.26}$$

To move $\boldsymbol{\nabla}^2$ through e^{ikz}, one has to use the identity $\boldsymbol{\nabla}_\mu\mathrm{e}^{ikz}f(\boldsymbol{r}) = \mathrm{e}^{ikz}(ik\delta_{\mu z} + \boldsymbol{\nabla}_\mu)f(\boldsymbol{r})$ repeatedly ($\mu = x, y, z$). One obtains

$$\left(vp_z + \frac{\boldsymbol{p}^2}{2m}\right)\hat{\psi}(\boldsymbol{r}) + V(\boldsymbol{r})\hat{\psi}(\boldsymbol{r}) = 0. \tag{2.27}$$

It should be noted that $\boldsymbol{p}$ and p_z are still operators, while $v = \hbar k/m$ is just a parameter.

The function $\hat{\psi}(\boldsymbol{r})$ is expected to be smooth, and $(\boldsymbol{p}^2/2m)\hat{\psi}(\boldsymbol{r})$ is bound to be much smaller than $vp_z\hat{\psi}(\boldsymbol{r})$ because the latter contains the incident momentum mv. It is then justifiable to neglect the kinetic-energy operator. This is the basic assumption in the eikonal approximation.

Let us consider more precisely on what conditions this assumption is valid. To put it simply, the term $(\boldsymbol{p}^2/2m)\hat{\psi}(\boldsymbol{r})$ may not be small enough, unless both the momentum change and the momentum spread of the wave function are relatively small throughout the collision. Starting with the momentum spread, we can put it as follows:

1. The momentum uncertainty δp caused by the scattering should be much smaller than the incident momentum $\hbar k$. The function $\hat{\psi}$ may

vary with a length scale that is equal to the range of the potential, to be denoted by a. This produces a momentum uncertainty $\delta p \sim \hbar/a$, and that is what should be smaller than the incident momentum $\hbar k$. Therefore, the condition is

$$\frac{\delta p}{p} \approx \frac{1}{ka} \ll 1. \tag{2.28}$$

This amounts to requiring that the wave length of the incident wave be much smaller than the spatial range of the potential. This condition is equivalent to that of the validity of the semiclassical picture for particle scattering. Moreover, it is analogous to the condition for the validity of geometrical optics on a microscopic scale.

2. The momentum transfer Δp by the particle in the collision must be smaller than the incident momentum, $p=\hbar k=mv$. Let us denote the average depth of the potential by V. The force (in the classical sense) acting on the particle may be estimated to be V/a. This force acts on the particle for a period of time, $\Delta t \simeq 2a/v$, and the momentum transfer is given by $\Delta p = F\Delta t \simeq 2V/v$. The condition can be formulated as

$$\frac{\Delta p}{p} \approx \frac{2V}{mv^2} = \frac{V}{E} \ll 1. \tag{2.29}$$

This condition is equivalent to assuming that the incident energy is much higher than the depth of the potential. It should be noted that, although $ka \gg 1$ and $V/E \ll 1$ are assumed, there is no condition for the product, $ka \cdot V/E = 2Va/\hbar v$, which will have some significance later when a comparison is made with the Born approximation.

Under these assumptions, the function $\hat{\psi}(\boldsymbol{r})$ satisfies

$$\frac{\partial}{\partial z}\hat{\psi}(x,y,z) = \frac{1}{i\hbar v}V(x,y,z)\hat{\psi}(x,y,z). \tag{2.30}$$

At $z \to -\infty$, the wave function $\psi(\boldsymbol{r})$ is a plane wave, which implies, for $\hat{\psi}(\boldsymbol{r})$, the boundary condition $\hat{\psi}(x,y,z\to-\infty)=1$. In Eq. (2.30) the coordinates x and y appear just as parameters. The equation thus reduces to a first-order ordinary differential equation, with two parameters, which can be solved for a set of the x, y parameters independently.

The solution of the above differential equation can obviously be given as

$$\hat{\psi}(x,y,z) = \exp\left[\frac{1}{i\hbar v}\int_{-\infty}^{z} dz'\, V(x,y,z')\right]. \tag{2.31}$$

Substitution of this into Eq. (2.24) gives the wave function in the eikonal approximation:

$$\psi(\boldsymbol{r}) = \exp\left[ikz + \frac{1}{i\hbar v}\int_{-\infty}^{z} dz'\, V(x,y,z')\right]. \tag{2.32}$$

It should be noted that the approximate solution (2.32) does not have the correct asymptotic form as described in Eq. (2.2). Indeed, in the outer region $x^2+y^2>a^2$ the integrand vanishes, so that the scattering wave function is nothing but the incident plane wave, without any scattered wave $\frac{1}{r}e^{ikr}$. But this is not fatal; it merely means that the wave function is unrealistic only out of the range of the potential, and so the scattering amplitude cannot be extracted from the asymptotic wave function. The scattering amplitude can still be calculated from Eq. (2.11) since, as was mentioned before, this expression only requires the wave function in the spatial region where the potential does not vanish. The approximate scattering amplitude is thus given by

$$f(\theta) = -\frac{2m}{\hbar^2}\frac{1}{4\pi}\int d\boldsymbol{r}\, \exp\left[-i\boldsymbol{k}'\cdot\boldsymbol{r} + ikz + \frac{1}{i\hbar v}\int_{-\infty}^{z} dz'\, V(x,y,z')\right] V(\boldsymbol{r}). \tag{2.33}$$

The momentum transfer vector $\boldsymbol{q}$ is conventionally defined by

$$\boldsymbol{q} = \boldsymbol{k}' - \boldsymbol{k}. \tag{2.34}$$

(This is actually a wave number vector.) For elastic scattering the length of the momentum transfer is related to the scattering angle by $q = 2k\sin\frac{\theta}{2}$. At high incident energies we expect that the scattering mainly occurs in the forward direction. Then, in the directions of appreciable cross section, q is small compared with the incident wave number k, and $\boldsymbol{q}$ is approximately orthogonal to the incident direction $\boldsymbol{k}$. The latter can be seen by writing

$$\boldsymbol{k}'^2 - \boldsymbol{k}^2 = (\boldsymbol{k}'+\boldsymbol{k})\cdot(\boldsymbol{k}'-\boldsymbol{k}) = (\boldsymbol{k}'+\boldsymbol{k})\cdot\boldsymbol{q} \tag{2.35}$$

and noting that $\boldsymbol{k}'^2 - \boldsymbol{k}^2 = 0$ because of the energy conservation. The approximate orthogonality $\boldsymbol{k}\cdot\boldsymbol{q}\approx 0$ is valid even for inelastic scattering to low-lying states, when $|\boldsymbol{k}'^2 - \boldsymbol{k}^2|$ is still small.

Since $\boldsymbol{k}$ is parallel to the z-axis, $\boldsymbol{q}$ lies approximately in the xy-plane: $\boldsymbol{q} = (q_x, q_y, 0)$. In this approximation $-i\boldsymbol{k}'\cdot\boldsymbol{r} + ikz = -i\boldsymbol{q}\cdot\boldsymbol{r} = -iq_x x - iq_y y$ holds. With this inserted, the z-integration in Eq. (2.33) can be simplified as

$$\begin{aligned} f(\theta) &= -\frac{2m}{\hbar^2}\frac{1}{4\pi}\int dx \int dy\, e^{-iq_x x - iq_y y}\int_{-\infty}^{+\infty} dz V(x,y,z) e^{\frac{1}{i\hbar v}\int_{-\infty}^{z} dz'\, V(x,y,z')} \\ &= -\frac{2m}{\hbar^2}\frac{1}{4\pi} i\hbar v \int dx \int dy\, e^{-iq_x x - iq_y y}\left[e^{\frac{1}{i\hbar v}\int_{-\infty}^{+\infty} dz\, V(x,y,z)} - 1\right], \end{aligned} \tag{2.36}$$

where use is made of the identity

$$\begin{aligned} &\int_{-\infty}^{+\infty} dz\, V(x,y,z) e^{\frac{1}{i\hbar v}\int_{-\infty}^{z} dz'\, V(x,y,z')} \\ &\qquad = i\hbar v \left[e^{\frac{1}{i\hbar v}\int_{-\infty}^{z} dz'\, V(x,y,z')}\right]_{z=-\infty}^{z=+\infty}. \end{aligned} \tag{2.37}$$

The xy-projection of $\boldsymbol{r}$ is called the impact parameter vector and is denoted by $\boldsymbol{b} = (x, y, 0)$. As was noted below Eq. (2.30), the impact parameter $\boldsymbol{b}$ appears as a parameter in the eikonal approximation. The vector $\boldsymbol{r}$ is given by $\boldsymbol{r} = \boldsymbol{b} + z\hat{\boldsymbol{z}}$, and the scattering amplitude is an integral, over the impact parameter plane $\boldsymbol{b}$, of a function containing an independent z-integration. This fact may be interpreted by saying that in the eikonal approximation a straight-line trajectory is assumed.[1] This interpretation will be elaborated further in the next subsection.

Let us introduce the so-called phase-shift function

$$\chi(\boldsymbol{b}) = -\frac{1}{\hbar v}\int_{-\infty}^{+\infty} dz\, V(\boldsymbol{b} + z\hat{\boldsymbol{z}}). \tag{2.38}$$

With this, the scattering amplitude is written as

$$f(\theta) = \frac{ik}{2\pi}\int d\boldsymbol{b}\, e^{-i\boldsymbol{q}\cdot\boldsymbol{b}}\left[1 - e^{i\chi(\boldsymbol{b})}\right]. \tag{2.39}$$

This is the usual formula for the scattering amplitude in the eikonal approximation.

For a spherically symmetric potential $V(r)$, the phase-shift function depends only on the length of the impact parameter vector $b = |\boldsymbol{b}|$. In that case, expressing $\boldsymbol{b}$ in polar coordinates (b, ϕ) ($b_x = b\cos\phi$, $b_y = b\sin\phi$), we can carry out the integration over the angle ϕ:

$$f(\theta) = ik\int_0^{\infty} db\, b\, J_0(qb)\left[1 - e^{i\chi(b)}\right], \tag{2.40}$$

with $q = 2k\sin\frac{\theta}{2}$. Here $J_0(x)$ is the zeroth-order Bessel function [43], which has the following integral form:

$$J_0(x) = \frac{1}{2\pi}\int_0^{2\pi} d\phi\, e^{ix\cos\phi}. \tag{2.41}$$

An interesting correspondence can be found between this expression and the standard quantum-mechanical partial-wave expansion of the scattering amplitude, which sheds light on the semiclassical nature of the eikonal approximation. The partial-wave formula for the amplitude of scattering by a spherically symmetric potential reads

$$f(\theta) = \frac{1}{2ik}\sum_{l=0}^{\infty}(2l+1)(e^{2i\delta_l} - 1)P_l(\cos\theta), \tag{2.42}$$

[1]Nevertheless, there is scattering into directions other than that of the incident wave since the contributions of the individual straight trajectories add up coherently in Eq. (2.36). This trajectory concept does not coincide with that of the geometrical optics, in which the gradient lines of the wave front are called trajectories.

where δ_l is the phase shift.

The correspondence between the quantum-mechanical and classical angular momenta, $l+\frac{1}{2} = kb$, is well-known, and it is therefore reasonable to replace the sum over angular momentum l by the integral $k\int db$. Furthermore, there exists an asymptotic formula [43]

$$P_l\left(\cos\frac{x}{l}\right) \sim J_0(x) \qquad (l \to \infty). \tag{2.43}$$

It is then clear that the argument x of the Bessel function corresponds to that of the Legendre polynomial by $x = l\theta \approx 2(l+\frac{1}{2})\sin\frac{\theta}{2} = 2kb\sin\frac{\theta}{2}$. What remains is to relate the phase shift to the phase-shift function $\chi(b)$, with $b = k^{-1}(l+\frac{1}{2})$, such that $\chi(b) = 2\delta_l$. This last correspondence explains the term 'phase-shift function' and completes the correspondence between the two formulae.

It is very instructive to relate the eikonal approximation to two of the standard approximation methods, the Wentzel–Kramers–Brillouin (WKB) approximation and the Born approximation.

The WKB method [26] is one of the systematic means to relate quantum mechanics to classical mechanics. Via the WKB method the phase-shift function $\chi(b)$ can be related to the phase shift δ_l. For a spherically symmetric potential $V(r)$, the WKB approximation to the phase shift δ_l is given by [26]

$$\delta_l = \int_{r_0}^{\infty} dr\,\sqrt{k^2 - \frac{2m}{\hbar^2}V(r) - \frac{l(l+1)}{r^2}} - \int_{r_1}^{\infty} dr\,\sqrt{k^2 - \frac{l(l+1)}{r^2}}, \tag{2.44}$$

where r_0 and r_1 are zeros of the respective integrands. Introducing the impact parameter b by the substitution $l(l+1) \to (kb)^2$, one obtains an approximate phase shift as a function of the impact parameter b:

$$\delta_{\mathrm{WKB}}(b) \approx k\int_{r_0}^{\infty} dr\,\sqrt{1 - \frac{b^2}{r^2} - \frac{2m}{\hbar^2 k^2}V(r)} - k\int_{b}^{\infty} dr\,\sqrt{1 - \frac{b^2}{r^2}}. \tag{2.45}$$

If one assumes that $|V(r)|$ is much smaller than the bombarding energy $\hbar^2k^2/2m$, an expansion can be performed, which reduces the phase shift, in the first order in the potential, to one half of the phase-shift function:

$$\delta_{\mathrm{WKB}}(b) \approx -\frac{1}{\hbar v}\int_{b}^{\infty} dr\,\frac{V(r)}{\sqrt{1 - \frac{b^2}{r^2}}} = -\frac{1}{2\hbar v}\int_{-\infty}^{+\infty} dz\, V(\sqrt{b^2 + z^2}). \tag{2.46}$$

This is one of the ways to provide systematic corrections to the eikonal approximation [44].

To derive the Born approximation from the eikonal approximation, we should consider a case in which $|\chi(\boldsymbol{b})| \ll 1$. Then we may expand $e^{i\chi(\boldsymbol{b})}$ into a Taylor series. With only the first-order term retained, the scattering amplitude reduces to

$$f(\theta) = \frac{ik}{2\pi}\int d\boldsymbol{b}\, e^{-i\boldsymbol{q}\cdot\boldsymbol{b}}[-i\chi(\boldsymbol{b})] = -\frac{m}{2\pi\hbar^2}\int d\boldsymbol{r}\, e^{-i\boldsymbol{q}\cdot\boldsymbol{r}}V(\boldsymbol{r}). \tag{2.47}$$

The last expression is nothing but the first-order Born approximation to the scattering amplitude (2.11).

An order-of-magnitude estimate for the phase-shift function $\chi(\boldsymbol{b})$ is given by $Va/\hbar v$. Since Eq. (2.47) is valid if $|\chi(\boldsymbol{b})| \ll 1$, it is obvious that the condition for the validity of the Born approximation will be ensured when the condition $|Va/\hbar v| \ll 1$ is satisfied.[2] As was mentioned after Eq. (2.29), the quality of the eikonal approximation does not depend on $Va/\hbar v$. As will be seen later in Sect. 2.3.1, the condition $|Va/\hbar v| \ll 1$ is not satisfied by the usual nuclear potentials for elastic scattering. This fact makes the eikonal approximation superior to the simple Born approximation in describing nuclear scattering problems.

Let us now calculate the integrated cross sections in the eikonal approximation. It is convenient to transform the integration over the solid angle $\Omega = (\theta, \phi)$ to an integration over the momentum transfer $\boldsymbol{q}$. With the components $q_x = q\cos\phi$ and $q_y = q\sin\phi$ of $\boldsymbol{q}$, whose length is $q = 2k\sin\frac{\theta}{2}$, we can write

$$dq_x dq_y = \begin{vmatrix} \frac{\partial q_x}{\partial\theta} & \frac{\partial q_y}{\partial\theta} \\ \frac{\partial q_x}{\partial\phi} & \frac{\partial q_y}{\partial\phi} \end{vmatrix} d\theta d\phi = k^2\sin\theta d\theta d\phi. \tag{2.48}$$

With this, the integration over the solid angle can be expressed in terms of an integration over q_x and q_y as

$$d\Omega = \sin\theta d\theta d\phi = \frac{1}{k^2}dq_x dq_y = \frac{d\boldsymbol{q}}{k^2}. \tag{2.49}$$

[2]That is so only in the high-energy limit $ka \gg 1$. For the Born approximation to be valid in general, the first-order Born correction to the plane wave should be much smaller than the plane wave in magnitude, i.e., $\left|\int d\boldsymbol{r}'\, G^{(+)}(\boldsymbol{r}-\boldsymbol{r}')V(\boldsymbol{r}')e^{i\boldsymbol{k}\cdot\boldsymbol{r}'}\right| \ll 1$ [cf. Eq. (2.8)]. This expression may be estimated at the potential centre, $\boldsymbol{r}=0$, by using Eq. (2.6):

$$\left|\int d\boldsymbol{r}\, G^{(+)}(\boldsymbol{r})V(\boldsymbol{r})e^{i\boldsymbol{k}\cdot\boldsymbol{r}}\right| \approx \frac{mV}{\hbar^2 k}\left|\int_0^a dr\left(e^{2ikr}-1\right)\right| = \frac{mV}{2\hbar^2 k^2}\left|e^{2ika} - 2ika - 1\right|.$$

The behaviour of the right-hand side depends on the value of ka. If $ka \gg 1$, the second term dominates, and the expression goes to $Va/\hbar v$. If, however, $ka \ll 1$, the second-order term coming from the MacLaurin series of e^{2ika} dominates, and the expression behaves like $mVa^2/\hbar^2$.

The total elastic scattering cross section $\sigma_{\rm el}$ can then be calculated from Eq. (2.39) as

$$\begin{aligned}\sigma_{\rm el} &= \int d\Omega\, |f(\theta)|^2 \\ &= \int \frac{d\boldsymbol{q}}{k^2} \left| \frac{ik}{2\pi} \int d\boldsymbol{b}\, \mathrm{e}^{-i\boldsymbol{q}\cdot\boldsymbol{b}} \left[1 - \mathrm{e}^{i\chi(\boldsymbol{b})}\right] \right|^2 = \int d\boldsymbol{b} \left|1 - \mathrm{e}^{i\chi(\boldsymbol{b})}\right|^2, \end{aligned} \tag{2.50}$$

where use has been made of the identity

$$\int d\boldsymbol{q}\, \mathrm{e}^{-i\boldsymbol{q}\cdot(\boldsymbol{b}-\boldsymbol{b}')} = (2\pi)^2 \delta(\boldsymbol{b} - \boldsymbol{b}'). \tag{2.51}$$

Formula (2.50) indicates that the integrand $P_{\rm el}(\boldsymbol{b}) = |1 - \mathrm{e}^{i\chi(\boldsymbol{b})}|^2$ may be interpreted as the probability that the particle incident with the impact parameter vector $\boldsymbol{b}$ is elastically scattered by the potential $V(\boldsymbol{r})$. When b is large, the phase-shift function $\chi(\boldsymbol{b})$ vanishes and the probability $P_{\rm el}(\boldsymbol{b})$ also vanishes.

When the potential $V(\boldsymbol{r})$ is complex, there appears the absorption cross section defined by Eq. (2.22). It can be evaluated in the eikonal approximation by substituting the wave function (2.32) into Eq. (2.22):

$$\begin{aligned}\sigma_{\rm abs} &= \frac{2}{\hbar v} \int_V d\boldsymbol{r}\, [-\mathrm{Im}\, V(\boldsymbol{r})] \left| \mathrm{e}^{\frac{1}{i\hbar v}\int_{-\infty}^{z} dz'\, V(x,y,z')} \right|^2 \\ &= \int dx \int dy \left[-\mathrm{e}^{\frac{2}{\hbar v}\int_{-\infty}^{z} dz'\, \mathrm{Im}\, V(x,y,z')} \right]_{z=-\infty}^{z=+\infty} \\ &= \int d\boldsymbol{b} \left[1 - |\mathrm{e}^{i\chi(\boldsymbol{b})}|^2\right]. \end{aligned} \tag{2.52}$$

To verify this, one has to recognize that only the imaginary part of V contributes to the modulus square in the first line, and then Eq. (2.37) can be used. The integrand, $P_{\rm abs}(\boldsymbol{b}) = 1 - |\mathrm{e}^{i\chi(\boldsymbol{b})}|^2$, may be interpreted as an absorption probability, namely, that of the particle incident with the impact parameter $\boldsymbol{b}$.

As was seen in Sect. 2.1.3, the exact total cross section can be calculated in two alternative ways: either by the optical theorem, $\sigma_{\rm tot} = (4\pi/k)\, \mathrm{Im}\, f(0)$ [Eq. (2.21)], or by $\sigma_{\rm tot} = \sigma_{\rm el} + \sigma_{\rm abs}$ [Eq. (2.23)]. In the eikonal approximation the first formula gives, with the use of Eq. (2.39),

$$\sigma_{\rm tot} = \int d\boldsymbol{b}\, 2 \left[1 - \mathrm{Re}\, \mathrm{e}^{i\chi(\boldsymbol{b})}\right]. \tag{2.53}$$

If the potential $V(\boldsymbol{r})$ is real and hence the phase-shift function $\chi(\boldsymbol{b})$ is also real, this formula coincides with that of the total elastic cross section, Eq. (2.50). When the potential $V(\boldsymbol{r})$ and the phase-shift function

$\chi(\boldsymbol{b})$ are complex, the sum of the total elastic cross section, Eq. (2.50), and the absorption cross section, Eq. (2.52), coincides with Eq. (2.53) derived from the optical theorem. Thus the scattering amplitude (2.39) of the eikonal approximation satisfies the optical theorem. This is a non-trivial result, which shows the coherence of the eikonal approximation and the high degree of its correspondence with exact quantum mechanics.

2.2.2 Treatment of the spin-orbit potential

The preceding derivation assumed that the potential only depends on the distance from the scattering centre. It is straightforward but non-trivial to generalize it to a potential containing a spin-orbit term. That is the aim of this subsection. Let us consider the scattering of a spin-$\frac{1}{2}$ particle by a potential which includes a spin-orbit force. The Schrödinger equation looks like

$$\left[-\frac{\hbar^2}{2m}\nabla^2 + V_c(\boldsymbol{r}) + V_{\text{s.o.}}(\boldsymbol{r})\boldsymbol{\sigma}\cdot\boldsymbol{L}\right]\psi(\boldsymbol{r}) = E\psi(\boldsymbol{r}), \tag{2.54}$$

where $\boldsymbol{\sigma} = 2\hbar^{-1}\boldsymbol{s}$ is the vector of the Pauli spin matrices and $\boldsymbol{L}$ is the (dimensionless) orbital angular-momentum operator defined as $\boldsymbol{L}=\hbar^{-1}\boldsymbol{r}\times\boldsymbol{p}$. The scattering amplitude for a particle with spin is an operator acting on the spinor. A general form of the scattering amplitude for a spin-$\frac{1}{2}$ particle is expressed as [40]

$$\hat{f}(\theta,\phi) = f(\theta) + \boldsymbol{\sigma}\cdot\hat{\boldsymbol{n}}\, g(\theta), \tag{2.55}$$

where $\hat{\boldsymbol{n}}$ is a unit vector perpendicular to the scattering plane, which contains the incident and outgoing wave vectors $\boldsymbol{k}$ and $\boldsymbol{k}'$. It is expressed as

$$\hat{\boldsymbol{n}} = \frac{\boldsymbol{k}\times\boldsymbol{k}'}{|\boldsymbol{k}\times\boldsymbol{k}'|}. \tag{2.56}$$

To derive the eikonal approximation for this case, the orbital angular momentum operator $\boldsymbol{L}$ has to be treated on the same footing as the kinetic energy. Since the wave function $\psi(\boldsymbol{r})$ is assumed not to change much with respect to the incident plane wave, the angular momentum operator may be approximated by

$$\boldsymbol{L} = \frac{1}{\hbar}\boldsymbol{r}\times\boldsymbol{p} \to k\boldsymbol{b}\times\hat{\boldsymbol{z}}, \tag{2.57}$$

where $\boldsymbol{b}$ is the xy-component of the coordinate $\boldsymbol{r}$. With this, the eikonal approximation can be carried through by analogy with the derivation of Eq. (2.32). Equation (2.24) should again be used to separate the plane wave, and the result for the remaining factor $\hat{\psi}$ is

$$\hat{\psi}(\boldsymbol{r}) = \exp\left\{\frac{1}{i\hbar v}\int_{-\infty}^{z} dz' \; [V_c(\boldsymbol{b}+z'\hat{\boldsymbol{z}}) + V_{s.o.}(\boldsymbol{b}+z'\hat{\boldsymbol{z}})k\boldsymbol{\sigma}\cdot(\boldsymbol{b}\times\hat{\boldsymbol{z}})]\right\}\chi_{\frac{1}{2}m_s}, \tag{2.58}$$

where $\chi_{\frac{1}{2}m_s}$ is the spin state of the incident wave [not to be mixed up with the phase-shift function $\chi(\boldsymbol{b})$]. The scattering amplitude can also be derived in the same way as Eq. (2.39):

$$\hat{f}(\theta,\phi) = \frac{ik}{2\pi}\int d\boldsymbol{b}\, \mathrm{e}^{-i\boldsymbol{q}\cdot\boldsymbol{b}}\left[1 - \mathrm{e}^{i\chi_c(b)+i\chi_{s.o.}(b)k\boldsymbol{\sigma}\cdot(\boldsymbol{b}\times\hat{\boldsymbol{z}})}\right], \tag{2.59}$$

where $\chi_c(b)$ and $\chi_{s.o.}(b)$ are phase-shift functions calculated from V_c and $V_{s.o.}$, respectively. Since this expression includes the Pauli matrix $\boldsymbol{\sigma}$, it is an operator acting on the spin wave function.

With the identity

$$\mathrm{e}^{i\gamma\boldsymbol{\sigma}\cdot\hat{\boldsymbol{u}}} = \cos\gamma + i\boldsymbol{\sigma}\cdot\hat{\boldsymbol{u}}\sin\gamma, \tag{2.60}$$

where $\hat{\boldsymbol{u}}$ is an arbitrary unit vector, the scattering amplitude (2.59) can be expressed as

$$\begin{aligned}\hat{f}(\theta,\phi) &= \frac{ik}{2\pi}\int d\boldsymbol{b}\, \mathrm{e}^{-i\boldsymbol{q}\cdot\boldsymbol{b}}\left\{1 - \mathrm{e}^{i\chi_c(b)}\cos\left[kb\chi_{s.o.}(b)\right]\right\}\\ &\quad - \frac{ik}{2\pi}\int d\boldsymbol{b}\, \mathrm{e}^{-i\boldsymbol{q}\cdot\boldsymbol{b}+i\chi_c(b)}\sin\left[kb\chi_{s.o.}(b)\right] i\boldsymbol{\sigma}\cdot\left(\hat{\boldsymbol{b}}\times\hat{\boldsymbol{z}}\right),\end{aligned} \tag{2.61}$$

where $\hat{\boldsymbol{b}} = b^{-1}\boldsymbol{b}$. This expression can be reduced to the form (2.55) after integration over the azimuthal angle. The first term corresponds to $f(\theta)$ of Eq. (2.55) and is expressed as [cf. Eq. (2.40)]

$$f(\theta) = ik\int_0^{\infty} db\, bJ_0(qb)\left\{1 - \mathrm{e}^{i\chi_c(b)}\cos\left[kb\chi_{s.o.}(b)\right]\right\}. \tag{2.62}$$

To perform the integration of the second term over the angle ϕ, one should first note that $\boldsymbol{q}$ and $\boldsymbol{b}$ span a plane perpendicular to the incident direction $\hat{\boldsymbol{z}}$. We choose the direction of the x-axis to be that of the vector $\boldsymbol{q}$, with ϕ the azimuthal angle of $\hat{\boldsymbol{b}}$. As a consequence, $\boldsymbol{\sigma}\cdot(\hat{\boldsymbol{b}}\times\hat{\boldsymbol{z}}) = \sigma_x\sin\phi - \sigma_y\cos\phi$, and with this the second term of Eq. (2.61) can be rewritten as

$$-\frac{ik}{2\pi}\int_0^{\infty} db\, b\int_0^{2\pi} d\phi\, \mathrm{e}^{-iqb\cos\phi+i\chi_c(b)}\sin[kb\chi_{s.o.}(b)]i(\sigma_x\sin\phi - \sigma_y\cos\phi). \tag{2.63}$$

The integral containing the term $\sigma_x \sin\phi$ vanishes identically. The integral containing the term $\sigma_y \cos\phi$ may be expressed in terms of the first-order Bessel function $J_1(x)$, which is defined by

$$J_1(x) = -\frac{d}{dx}J_0(x) = -\frac{i}{2\pi}\int_0^{2\pi} d\phi\, \cos\phi\, e^{ix\cos\phi}. \tag{2.64}$$

By using $J_1(-x) = -J_1(x)$ and $\sigma_y = \boldsymbol{\sigma}\cdot\hat{\boldsymbol{n}}$, which follows from $\hat{\boldsymbol{n}} \parallel \boldsymbol{k}\times(\boldsymbol{k}'-\boldsymbol{k}) \parallel \boldsymbol{k}\times\boldsymbol{q}$, the function $g(\theta)$ of Eq. (2.55) can be written as

$$g(\theta) = ik\int_0^{\infty} db\, bJ_1(qb)\, e^{i\chi_c(b)} \sin[kb\chi_{s.o.}(b)]. \tag{2.65}$$

The 'partial' scattering amplitude for elastic scattering from a state with spin projection m_s to m_s' is $\langle\chi_{\frac{1}{2}m_s'}|\hat{f}|\chi_{\frac{1}{2}m_s}\rangle$. The differential scattering cross section can be calculated by averaging $|\langle\chi_{\frac{1}{2}m_s'}|\hat{f}|\chi_{\frac{1}{2}m_s}\rangle|^2$ over the initial spin projections m_s and summing over the final projections m_s'. The result is

$$\frac{d\sigma}{d\Omega} = |f(\theta)|^2 + |g(\theta)|^2. \tag{2.66}$$

The polarization caused by the scattering process is characterized by other combinations of $f(\theta)$ and $g(\theta)$ [45].

2.2.3 Projectile-rest frame

In deriving the eikonal approximation, we wrote the wave function in the form $\psi(\boldsymbol{r}) = e^{ikz}\hat{\psi}(\boldsymbol{r})$. We get an insight into the physical meaning of this factorization by transforming the problem from the laboratory frame (or 'target system') to the inertial frame in which the projectile is at rest initially (projectile-rest frame or 'beam system'). To this end, let us start from the time-dependent Schrödinger equation and subject it and its scattering solution to a Galilean transformation.

The Galilean transformation connects two inertial systems, K and K'. Let system K' move with a constant velocity $\boldsymbol{v}$ with respect to system K. The coordinate $\boldsymbol{r}$, wave number vector $\boldsymbol{k}$, and time t in the system K are related to $\boldsymbol{r}'$, $\boldsymbol{k}'$, and t' in the system K' as

$$\begin{aligned} \boldsymbol{r} &= \boldsymbol{r}' + \boldsymbol{v}t, \\ \boldsymbol{k} &= \boldsymbol{k}' + \frac{1}{\hbar}m\boldsymbol{v}, \\ t &= t'. \end{aligned} \tag{2.67}$$

It is simplest to imagine the effect of the Galilean transformation between plane wave states in systems K and K',

$$\psi(\boldsymbol{r},t) = \mathrm{e}^{i\boldsymbol{k}\cdot\boldsymbol{r} - \frac{i}{\hbar}\frac{\hbar^2 k^2}{2m}t}, \tag{2.68}$$

$$\psi'(\boldsymbol{r}',t) = \mathrm{e}^{i\boldsymbol{k}'\cdot\boldsymbol{r}' - \frac{i}{\hbar}\frac{\hbar^2 k'^2}{2m}t}. \tag{2.69}$$

By substitution of Eqs. (2.67) into Eq. (2.68), one obtains the transformation rule for these wave functions:

$$\begin{aligned}\psi(\boldsymbol{r},t) &= \mathrm{e}^{\frac{i}{\hbar}m\boldsymbol{v}\cdot\boldsymbol{r}' + \frac{i}{\hbar}\frac{1}{2}mv^2 t}\psi'(\boldsymbol{r}',t) \\ &= \mathrm{e}^{\frac{i}{\hbar}m\boldsymbol{v}\cdot\boldsymbol{r} - \frac{i}{\hbar}\frac{1}{2}mv^2 t}\psi'(\boldsymbol{r} - \boldsymbol{v}t, t).\end{aligned} \tag{2.70}$$

By representing a general state in terms of a plane-wave basis, it is easy to show that the transformation rule (2.70) holds for any general state ψ.

For a particle scattered by a potential $V(\boldsymbol{r})$, it is natural to choose the system K to be that in which the potential is at rest and the incident particle moves with velocity $\boldsymbol{v}$ initially and the system K' to be that which moves with velocity $\boldsymbol{v}$ with respect to K. In this system K' the particle is at rest initially, and the potential moves with velocity $-\boldsymbol{v}$. With the notation $\boldsymbol{k} = m\boldsymbol{v}/\hbar$, the relation between the wave functions in the two systems looks like

$$\psi(\boldsymbol{r},t) = \mathrm{e}^{i\boldsymbol{k}\cdot\boldsymbol{r} - \frac{i}{\hbar}\frac{\hbar^2 k^2}{2m}t}\psi'(\boldsymbol{r} - \boldsymbol{v}t, t). \tag{2.71}$$

From this formula it is obvious that the function $\hat{\psi}(\boldsymbol{r})$ introduced in Eq. (2.24) is nothing but the wave function of the particle in the projectile-rest frame. The incident plane-wave state in system K, i.e., $\psi_{\text{inc}}(\boldsymbol{r},t) = \mathrm{e}^{i\boldsymbol{k}\cdot\boldsymbol{r} - \frac{i}{\hbar}\frac{\hbar^2 k^2}{2m}t}$, is apparently transformed to $\psi'(\boldsymbol{r},t) = 1$.

The Schrödinger equation must have the same form in any inertial system. Indeed, by substituting the transformation (2.67) into the Schrödinger equation in the laboratory frame, we get a similar equation, with the time-dependent potential, moving with velocity $-\boldsymbol{v}$:

$$i\hbar\frac{\partial}{\partial t}\psi'(\boldsymbol{r},t) = \left[-\frac{\hbar^2}{2m}\boldsymbol{\nabla}^2 + V(\boldsymbol{r} + \boldsymbol{v}t)\right]\psi'(\boldsymbol{r},t). \tag{2.72}$$

The eikonal approximation in the projectile-rest frame amounts to supposing that the passage of the potential is so fast that the particle does not move much during the collision. We may then ignore the kinetic-energy operator in Eq. (2.72). This is precisely the same assumption as introduced in deriving Eq. (2.30), and the resultant equation does indeed have the same form, with $z = vt$. The solution to this approximate Schrödinger equation will look like

$$\psi'(\boldsymbol{r},t) = \exp\left[\frac{1}{i\hbar}\int_{-\infty}^{t} dt'\, V(\boldsymbol{r}+\boldsymbol{v}t')\right]. \tag{2.73}$$

This function reduces to $\psi'(\boldsymbol{r},t)=1$ at $t\to-\infty$ as it should take the form of the initial plane wave. Taking the time limit $t\to\infty$ gives the final-state wave function in terms of the phase-shift function (2.38):

$$\psi'(\boldsymbol{r},t\to\infty) = \mathrm{e}^{i\chi(\boldsymbol{b})}, \tag{2.74}$$

where $\boldsymbol{b}$ is the projection of the vector $\boldsymbol{r}$ to the xy-plane, i.e., the impact parameter. The function $\mathrm{e}^{i\chi(\boldsymbol{b})}$ is thus the wave function after collision in the projectile-rest frame.

Though the wave function $\psi'(\boldsymbol{r}) = \mathrm{e}^{i\chi(\boldsymbol{b})}$ is obtained in the time limit $t\to\infty$, it can only be a good approximation for a certain period of time after the collision. The kinetic-energy operator, which has been neglected, will make the 'ripple' caused by the collision more diffuse. We may describe the time development of the wave function after the collision by the free-particle Schrödinger equation with the initial condition $\psi(\boldsymbol{r})=\mathrm{e}^{i\chi(\boldsymbol{b})}$. Since the wave function is independent of the z-coordinate and has a cylindrical symmetry, the momentum transfer obviously includes xy-components only. This finding tallies with our previous assumption that the momentum transfer vector $\boldsymbol{q}$ is approximately transverse.

2.3 Illustrative examples

A few standard examples will now be shown to illustrate the performance of the eikonal approximation.

2.3.1 Square-well potential

Strongly absorptive square-well potential

This problem is reminiscent of the elastic scattering of two composite nuclei. Since the nuclei are easily excited when penetrating each other, the probability that at least one of them is excited is very close to unity if they overlap in the course of the collision. The loss of the incoming flux into excitation channels is represented by the absorptive (negative imaginary) potential term.

Let us assume a strong absorption in a sphere of a certain radius R. The phase-shift function then satisfies the following relation

$$\mathrm{e}^{i\chi(b)} = \Theta(b-R) \equiv \begin{cases} 0 & b<R, \\ 1 & b\geq R. \end{cases} \tag{2.75}$$

The scattering amplitude is given by Eq. (2.40):

$$f(\theta) = ik \int_0^R db\, bJ_0(qb) = ikR^2 \frac{J_1(qR)}{qR}, \tag{2.76}$$

where the scattering angle θ enters through $q = 2k \sin \frac{\theta}{2}$. The elastic differential cross section, $|f(\theta)|^2$, vanishes at the zeros of the Bessel function $J_1(qR)$. The first zero of $J_1(x)$ is at $x_1 \simeq 3.83$, and the corresponding angle is given by $kR\theta \simeq 3.83$.

For the integrated cross sections Eqs. (2.50), (2.52) and (2.53) predict

$$\begin{aligned}
\sigma_{\text{el}} &= \int d\boldsymbol{b}\, |1 - e^{i\chi(b)}|^2 = \pi R^2, \\
\sigma_{\text{abs}} &= \int d\boldsymbol{b} \left[1 - |e^{i\chi(b)}|^2\right] = \pi R^2, \\
\sigma_{\text{tot}} &= \sigma_{\text{el}} + \sigma_{\text{abs}} = 2\pi R^2.
\end{aligned} \tag{2.77}$$

Both the elastic cross section and the absorption cross section are equal to the geometric cross section, πR^2. Although the potential is so absorptive that the real part plays no role, the elastic cross section is as large as the absorption cross section itself. This is a diffraction effect of wave mechanics caused by the edge of the absorptive potential. Diffraction, this important effect of wave mechanics, does appear, whether the incident energy is high or low.

Complex-valued square-well potential

This potential is a simplified version of the optical potential for the scattering of neutrons from nuclei. The potential is parametrized as

$$V(r) = (V_0 + iW_0)\Theta(R - r), \tag{2.78}$$

where V_0 and W_0 are (usually negative) constants. For this potential Eq. (2.38) gives, for the phase-shift function,

$$\chi(b) = -\frac{1}{\hbar v}(V_0 + iW_0) 2\sqrt{R^2 - b^2}\,\Theta(R - b). \tag{2.79}$$

The scattering amplitude cannot be given in a closed form, but the integrated cross sections can be. The absorption cross section looks like

$$\begin{aligned}
\sigma_{\text{abs}} &= \int d\boldsymbol{b} \left[1 - |e^{i\chi(b)}|^2\right] = 2\pi \int_0^R db\, b \left[1 - \exp\left(x\sqrt{1 - \frac{b^2}{R^2}}\right)\right] \\
&= \pi R^2 C(x),
\end{aligned} \tag{2.80}$$

where $x = 4W_0R/\hbar v$ and $C(x)$ is defined by

$$C(x) = 1 - \frac{2}{x^2}\left[(x-1)e^x + 1\right]. \tag{2.81}$$

We note that the absorption cross section is determined solely by the imaginary part of the potential. This is because in the eikonal approximation straight-line trajectories are assumed, i.e., the bending of the trajectories by the real part of the potential is neglected.

To elucidate the role of parameter x, it is convenient to view the problem as neutron scattering by a nucleus. Then the absorptive potential depth W_0 is related to the mean free path λ of the neutron in the nucleus. The wave function of a neutron of energy E propagating in a uniform medium represented by the potential $V_0 + iW_0$ is e^{ikz}, with the complex wave number k given by $\hbar^2k^2/2m = E - (V_0 + iW_0)$. It is easy to show that, when $E \gg |V_0|$, the probability density decreases as $|e^{ikz}|^2 = e^{-z/\lambda}$, which shows that $\lambda = \hbar v/2(-W_0)$ is the mean free path. The parameter x is expressed with λ as $|x| = 2R/\lambda$, thus it is the ratio of the nuclear diameter to the mean free path. For moderate neutron energies in neutron–nucleus scattering, this parameter is of the order of unity. The limiting cases are

$$\sigma_{\text{abs}} \simeq \begin{cases} \frac{2}{3}|x|\pi R^2 & |x| \to 0, \\ \pi R^2 & |x| \to \infty. \end{cases} \tag{2.82}$$

The strong absorption limit of Eq. (2.77) belongs to large $|x|$. The total cross section is given by

$$\sigma_{\text{tot}} = \int d\boldsymbol{b}\, 2\left[1 - \text{Re}\, e^{i\chi(\boldsymbol{b})}\right] = 2\pi R^2\, \text{Re}\, C(z), \tag{2.83}$$

where $z = 2(W_0 - iV_0)R/\hbar v$.

To assess the validity of the eikonal approximation for nuclear reactions, we compare some scattering cross sections calculated with full quantum mechanics to cross sections calculated with the eikonal approximation. The radius of the potential is $R = 5$ fm, which is about the radius of a mass number 40 nucleus. The potential depths are taken $V_0 = -40$ MeV and $W_0 = -10$ MeV. The parameter $Va/\hbar v$, which appeared in the discussion of the Born approximation in Eq. (2.47), is about 3 at the incident energy of 50 MeV. Since this is larger than unity, the Born approximation cannot be used in this case. The eikonal approximation is, however, expected to be valid whenever $E > |V_0|$ [cf. Eq. (2.29)]. Figure 2.1 shows the integrated cross sections of the elastic scattering, σ_{el}, and of the absorption, σ_{abs}. We see that, though around 100 MeV both cross sections are somewhat underestimated by

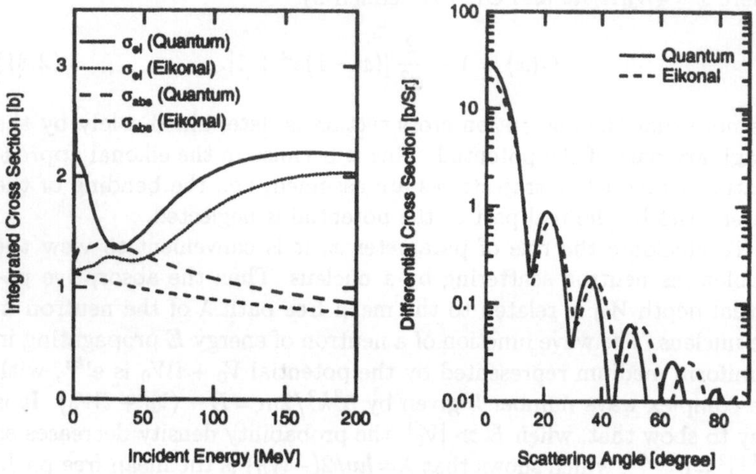

Figure 2.1: Left: Energy dependence of the integrated neutron cross sections for a model nucleus, calculated by exact quantum mechanics and by the eikonal approximation. Right: Differential cross section of the elastic neutron scattering at $E = 100$ MeV for the same case.

about 20%, the energy dependence and the ratio between the elastic and absorption cross sections are well reproduced by the eikonal approximation even at $E \sim |V_0|$. As expected, the approximation becomes better for higher incident energies. In the right panel of Fig. 2.1, differential elastic cross sections are compared. Again, the cross section calculated by the eikonal approximation shows a good correspondence with the exact one. A realistic neutron optical potential has a smooth surface, and in that case the exact and eikonal calculations agree even better.

2.3.2 Coulomb scattering

As a last illustration, the scattering of a particle by a Coulomb potential, $V(r) = Ze^2/r$, is considered. The phase-shift function of the Coulomb potential,

$$\chi_C(b) = -\frac{Ze^2}{\hbar v}\int_{-\infty}^{+\infty} dz\, \frac{1}{\sqrt{b^2+z^2}}, \tag{2.84}$$

is divergent because of the $\frac{1}{r}$ behaviour of the potential. To avoid this, the tail of the Coulomb potential is cut off beyond a radius a:

$$V(r) = \frac{Ze^2}{r}\Theta(a-r). \tag{2.85}$$

In this way the phase-shift function will be

$$\chi_C(b) = -\frac{2Ze^2}{\hbar v}\Theta(a-b)\ln\frac{a+\sqrt{a^2-b^2}}{b}. \tag{2.86}$$

This formula indicates that b must not exceed a, so a similar cutoff should be introduced for b. The cutoff only influences scattering to extreme forward angles, and may be chosen to simulate the natural cutoff coming from electron screening[3] in the target. Moreover, it is expedient to choose a so large as to make the ratio $\frac{b}{a}$ small within the cutoff radius for b. The phase-shift function can then be expanded as

$$\chi_C(b) = \frac{2Ze^2}{\hbar v}\left[\ln\frac{b}{2a} + \left(\frac{b}{2a}\right)^2 + \cdots\right] \approx 2\eta\ln\frac{kb}{2ka}, \tag{2.87}$$

where $\eta = Ze^2/\hbar v$ is the Coulomb or Sommerfeld parameter and the MacLaurin series of $\ln\left(1+\sqrt{1-x^2}\right)$ (with $x = \frac{b}{a}$) has been truncated after the second-order term. Using Eq. (2.87) and excluding $\theta = 0$, one can obtain a closed form for the scattering amplitude:

$$f_C(\theta) = -\frac{2k\eta}{q^2}\mathrm{e}^{-2i\eta\ln(qa)+2i\sigma_0}, \tag{2.88}$$

with $\sigma_0 = \arg\Gamma(1+i\eta)$. This differs from the exact Coulomb (or Rutherford) scattering amplitude,

$$F_C(\theta) = -\frac{2k\eta}{q^2}\mathrm{e}^{-2i\eta\ln(\sin\theta/2)+2i\sigma_0}, \tag{2.89}$$

just in the phase factor $\mathrm{e}^{-2i\eta\ln(2ka)}$, thus the differential cross section, $|f_C(\theta)|^2$, reproduces the Rutherford formula irrespective of the angle.

Let us next study the scattering of a particle by a potential composed of a Coulomb term and a short-range term $V_N(r)$. The truncated Coulomb potential is used as above:

$$V(r) = \frac{Ze^2}{r}\Theta(a-r) + V_N(r). \tag{2.90}$$

The phase-shift function (2.38) is additive with respect to the potentials, so in the scattering amplitude of the eikonal approximation,

[3]The Coulomb field of the target atom is of finite range since the atom is neutral.

Eq. (2.39), the Coulomb phase-shift function $\chi_C(b)$ and the nuclear phase-shift function $\chi_N(b)$ are added up:

$$f(\theta) = \frac{ik}{2\pi} \int d\boldsymbol{b}\, e^{-i\boldsymbol{q}\cdot\boldsymbol{b}} \left[1 - e^{i\chi_C(b)+i\chi_N(b)}\right]. \tag{2.91}$$

It is convenient to rewrite the scattering amplitude into a sum of two terms, the Coulomb scattering amplitude and the remainder:

$$\begin{aligned} f(\theta) &= \frac{ik}{2\pi} \int d\boldsymbol{b}\, e^{-i\boldsymbol{q}\cdot\boldsymbol{b}} \left[1 - e^{i\chi_C(b)}\right] + \frac{ik}{2\pi} \int d\boldsymbol{b}\, e^{-i\boldsymbol{q}\cdot\boldsymbol{b}} e^{i\chi_C(b)} \left[1 - e^{i\chi_N(b)}\right] \\ &= f_C(\theta) + \frac{ik}{2\pi} \int d\boldsymbol{b}\, e^{-i\boldsymbol{q}\cdot\boldsymbol{b}} e^{2i\eta[\ln(kb)-\ln(2ka)]} \left[1 - e^{i\chi_N(b)}\right] \\ &= e^{-2i\eta\ln(2ka)} \left\{ F_C(\theta) + \frac{ik}{2\pi} \int d\boldsymbol{b}\, e^{-i\boldsymbol{q}\cdot\boldsymbol{b}+2i\eta\ln(kb)} \left[1 - e^{i\chi_N(b)}\right] \right\}, \end{aligned} \tag{2.92}$$

where, in the second step, Eqs. (2.88) and (2.87), and then, in the third step, Eq. (2.89) have been used. It is rewarding to see in Eq. (2.92) that the same phase factor, $e^{-2i\eta\ln(2ka)}$, is factored out from the second term as from the pure Columb term. Thus the differential cross section $d\sigma/d\Omega = |f(\theta)|^2$ is again independent of the cutoff parameter a.

The Coulomb potential between two heavy nuclei bends the classical trajectory substantially. This is the case when $\eta \gg 1$. In such a case the approximation with straight-line trajectories breaks down, but some effects of the trajectory bending may be taken into account in the eikonal approximation [46, 47] by a simple prescription.

When the trajectory is a straight line, the impact parameter b is equal to the distance of the closest approach. For a classical Coulomb trajectory, however, the distance of closest approach, b_{ca}, is related to the impact parameter b by

$$kb_{ca} = \eta + \sqrt{\eta^2 + k^2b^2}. \tag{2.93}$$

The simple prescription to allow for trajectory bending is to replace b with b_{ca} in evaluating the phase-shift function $\chi_N(b)$ induced by the nuclear potential:

$$\chi_N(b) = -\frac{1}{\hbar v} \int_{-\infty}^{+\infty} dz\, V_N\left(\sqrt{b_{ca}^2 + z^2}\right). \tag{2.94}$$

It has been shown [46] that the angular distributions of elastic collisions between heavy nuclei can be described accurately in the eikonal approximation combined with this prescription.

Chapter 3

Glauber theory for composite-particle scattering

3.1 Some kinematics

In this chapter we extend the eikonal approximation to scattering problems including composite particles. The eikonal approximation is to be introduced as a many-body description or, in the language of Part II, as a microscopic description. In this framework the reaction will look like a succession of multiple collisions of nucleons. This description is known as the Glauber theory [38]. The Glauber theory is a microscopic reaction theory of high-energy collisions based on the eikonal approximation and on the bare nucleon–nucleon interaction.

Before expounding the theory, we briefly review some kinematical aspects of nuclear collisions. The centre-of-mass (c.m.) motion of a whole colliding system is always irrelevant to us. Experiments are performed in the laboratory frame, where the target nucleus is usually at rest. A theoretical analysis, on the other hand, is convenient in the c.m. frame, in which the total momentum is zero. The collision of two point particles can always be described by separating the c.m. and relative motions, with a one-body-like equation for the relative motion. (The c.m. and relative motions can be separated in non-relativistic kinematics but not in a relativistic framework.) Similar statement applies to the collisions of composite particles: the c.m. motion can be separated from the internal motion provided non-relativistic kinematics is used. The relative motion of projectile and target before the collision can be described by a one-body equation.

We shall denote the mass numbers of the projectile and target as A_P and A_T, respectively. In the (one-body) equation of relative motion the mass is always the reduced mass, $M_{PT}=m_N A_P A_T/(A_P+A_T)$, where m_N is an average mass of the nucleon in the nucleus, i.e., the total mass of the system divided by the sum of the mass numbers. (Except for the nucleon itself and for very light nuclei, m_N is found to be $m_N \simeq 931.5$ MeV/c^2.) The energy of the relative motion is

$$E_{rel} = \frac{A_T}{A_P + A_T} E_P, \tag{3.1}$$

where E_P is the energy of the incident projectile in the laboratory frame. The velocity v and the wave number[1] K of the relative motion are related to E_{rel} by

$$E_{rel} = \frac{1}{2} M_{PT} v^2 = \frac{\hbar^2 K^2}{2M_{PT}}. \tag{3.2}$$

In non-relativistic mechanics the velocity v is equal to the velocity of the incident projectile v_P. This is expressible as

$$\frac{v_P}{c} = \left(\frac{2}{m_N c^2}\frac{E_P}{A_P}\right)^{1/2} = 0.0463\left(\frac{E_P}{A_P}\right)^{1/2}, \tag{3.3}$$

where E_P is measured in units of MeV. This expression is valid in the non-relativistic regime, i.e., when $v_P/c \ll 1$. When the velocity is comparable to the velocity of light, relativistic kinematics[2] must be

[1] The wave number of the relative motion between nuclei is to be denoted by K, while k is used for potential scattering and nucleon scattering.

[2] The relativistic kinematics of the collision of two particles can be formulated as follows. Let us assume that particle 1 of mass m_1 is incident, with momentum p_1 along the z-axis, on particle 2 of mass m_2, which is at rest in the laboratory frame. The total energy of particle 1 is $\epsilon_1 = (m_1^2c^4 + c^2p_1^2)^{1/2}$, thus its velocity is $v_1 = c^2p_1/\epsilon_1$ and its kinetic energy is $E_1 = (\gamma-1)m_1c^2 = p_1^2/[(\gamma+1)m_1]$, with the Lorentz factor $\gamma=[1-(v_1/c)^2]^{-1/2}=[1+(p_1/m_1c)^2]^{1/2}$.

The c.m. frame is defined as the frame in which the total momentum is zero. Viewed in the laboratory frame, it moves with the velocity $v_{c.m.} = c^2p_1/(\epsilon_1+m_2c^2) = v_1\epsilon_1/(\epsilon_1+m_2c^2)$. The four-momenta of particles 1 and 2 in the c.m. frame are thus given by

$$\begin{aligned} \epsilon_1^* &= \gamma_{c.m.}\left(\epsilon_1 - \tfrac{v_{c.m.}}{c}cp_1\right), & cp_1^* &= \gamma_{c.m.}\left(-\tfrac{v_{c.m.}}{c}\epsilon_1 + cp_1\right), \\ \epsilon_2^* &= \gamma_{c.m.}m_2c^2, & cp_2^* &= -\gamma_{c.m.}v_{c.m.}m_2c, \end{aligned}$$

where $\gamma_{c.m.} = [1-(v_{c.m.}/c)^2]^{-1/2}$ and the x- and y-components are suppressed. (The quantities with respect to the c.m. frame are distinguished by asterisks.) The velocities in the c.m. frame are given by $v_1^* = c^2p_1^*/\epsilon_1^* = (v_1 - v_{c.m.})/[1-(v_1v_{c.m.}/c^2)]$, $v_2^* = c^2p_2^*/\epsilon_2^* = -v_{c.m.}$. The relative velocity may be defined by $v^* = v_1^* - v_2^* = v_1[1-(v_{c.m.}/c)^2]/[1-(v_1v_{c.m.}/c^2)]$. The energy of the relative motion, E^*, may be defined by the sum of the kinetic energies of particles: $E^* = \epsilon_1^* - m_1c^2 + \epsilon_2^* - m_2c^2$. It is easy to show that, in the non-relativistic limit, $v^* \simeq v_1$ and $E^* \simeq E_1m_2/(m_1+m_2)$.

used. Then the total energy of the projectile, $\epsilon_{\rm P}$, is

$$\epsilon_{\rm P} = \frac{A_{\rm P} m_{\rm N} c^2}{\sqrt{1 - \left(\frac{v_{\rm P}}{c}\right)^2}} = E_{\rm P} + A_{\rm P} m_{\rm N} c^2, \tag{3.4}$$

from which $v_{\rm P}/c$ can be expressed as

$$\frac{v_{\rm P}}{c} = \left[1 - \left(\frac{m_{\rm N} c^2}{\frac{E_{\rm P}}{A_{\rm P}} + m_{\rm N} c^2}\right)^2\right]^{1/2}. \tag{3.5}$$

The relativistic treatment is important when the incident energy of the projectile per nucleon, $E_{\rm P}/A_{\rm P}$, is comparable with the nucleon mass $m_{\rm N}c^2$. For example, for $E_{\rm P}/A_{\rm P} = 60$ MeV, $v_{\rm P}/c = 0.343$, and for $E_{\rm P}/A_{\rm P} = 800$ MeV, $v_{\rm P}/c = 0.843$. Roughly speaking, below 100 MeV per nucleon, the non-relativistic kinematics is adequate, but above 100 MeV, the relativistic expression is required.

We shall use, nevertheless, the non-relativistic framework even for high-energy reactions for the sake of the eikonal approximation, which is formulated in non-relativistic quantum mechanics. The main reason for this is that the many-body scattering effects may be conveniently taken into account by the eikonal theory. The scattering amplitude of the collision of composite partners will be expressed in terms of nucleon–nucleon scattering amplitudes. The non-relativistic expressions are employed both for the nucleon–nucleon and for the nucleus–nucleus collisions. We may expect that some important part of the relativistic effects is included in the nucleon–nucleon scattering amplitudes, which have been extracted by ***non-relativistic analyses*** from nucleon–nucleon cross sections ***measured at the same range of energy per nucleon.*** If so, then the amplitudes of scattering of composite nuclei will also carry relativistic effects, even though the non-relativistic Schrödinger equation is used.

The condition (2.28) for the validity of the eikonal approximation involves the wave number. Since now there are two radii involved, it has to be rewritten into $KR \gg 1$, where R is a length scale of the collision, the radius of the volume in which the nucleon–nucleon collisions take place. It may be as large as a few tens of fm, but it is at least the sum of the radii of the two nuclei. The wave number of the relative motion is given by

$$K = \frac{M_{\rm PT} v_{\rm P}}{\hbar} = 4.72 \frac{A_{\rm P} A_{\rm T}}{A_{\rm P} + A_{\rm T}} \frac{v_{\rm P}}{c} \quad [\text{fm}^{-1}]. \tag{3.6}$$

Except when the incident energy is very low, $KR \gg 1$ is satisfied. When both nuclei are composite, not only the radii but also the wave numbers become large due to the large reduced mass, and the condition is especially easily fulfilled.

3.2 Glauber theory

3.2.1 Formal treatment

Consider first the reaction of a nucleon with a target nucleus. The coordinates of the target nucleons are to be denoted by $\boldsymbol{r}_i$ $(i=1,\ldots,A)$, where A is the number of nucleons. The spin coordinates will be suppressed, but if any spin-dependent interactions make it necessary, it is easy to generalize the formalism to include spins (cf. Sect. 2.2.2). For simplicity, protons and neutrons will not be distinguished. The formulae to be derived will be readily applicable if the differences between the proton–proton, proton–neutron and neutron–neutron interactions are ignored. This difference is not so significant in high-energy collisions.

Let the intrinsic Hamiltonian of the target nucleus be H_{T} and a complete set of its eigenfunctions be Φ_α:

$$H_{\mathrm{T}}\Phi_\alpha(\boldsymbol{r}_1,\boldsymbol{r}_2,\ldots,\boldsymbol{r}_A)=E_\alpha\Phi_\alpha(\boldsymbol{r}_1,\boldsymbol{r}_2,\ldots,\boldsymbol{r}_A). \tag{3.7}$$

Their explicit forms will not be required. The position vector of the incident nucleon will be denoted by $\boldsymbol{r}$. The total Hamiltonian, which includes the kinetic energy of the incident nucleon and the interaction V between the incident nucleon and the target nucleons, is

$$H=\frac{\boldsymbol{p}^2}{2m}+H_{\mathrm{T}}+\sum_{i=1}^{A}V(\boldsymbol{r}-\boldsymbol{r}_i). \tag{3.8}$$

For simplicity, we assume that the target nucleus is so heavy that the total c.m. of the system is at rest and measure the coordinates of the target nucleons $\boldsymbol{r}_i$ from the spatially fixed c.m. position, thus treating all the coordinates independent. The proper treatment would be to replace the projectile mass m with the reduced mass, as was indicated in Sect. 3.1, and to introduce $A-1$ independent intrinsic coordinates for the target nucleus as will be done in Part II.[3] (The inverse setup of

[3]The c.m. problem is inessential for the reactions discussed in Part I but is very serious for the structure problems to be tackled in Part II. Therefore, it is treated as simply as possible here and as precisely as possible there. In this way the notations applied in Part I and Part II can only be mixed with extreme caution. The main difference is this: In Part I the coordinate vectors $\{\boldsymbol{r}_1,\ldots,\boldsymbol{r}_A\}$ denote A independent Cartesian vectors in the fixed-c.m. approximation but the same symbol is used for the exact treatment of the c.m., in which case they are coordinates with respect to the (moving) c.m. A set of coordinates with respect to the c.m. has to

a composite projectile and a nucleon target, can, of course, be treated analogously in the projectile-rest frame.)

The scattering wave function $\Psi(\boldsymbol{r}, \boldsymbol{r}_1, \boldsymbol{r}_2, \ldots, \boldsymbol{r}_A)$ is assumed to satisfy the Schrödinger equation

$$H\Psi(\boldsymbol{r}, \boldsymbol{r}_1, \boldsymbol{r}_2, \ldots, \boldsymbol{r}_A) = E\Psi(\boldsymbol{r}, \boldsymbol{r}_1, \boldsymbol{r}_2, \ldots, \boldsymbol{r}_A), \tag{3.9}$$

with the $r \to \infty$ boundary condition

$$\begin{aligned} &\Psi(\boldsymbol{r}, \boldsymbol{r}_1, \boldsymbol{r}_2, \ldots, \boldsymbol{r}_A) \\ &= e^{ikz}\Phi_0(\boldsymbol{r}_1, \ldots, \boldsymbol{r}_A) + \sum_\alpha f_\alpha(\theta, \phi)\frac{e^{ik_\alpha r}}{r}\Phi_\alpha(\boldsymbol{r}_1, \ldots, \boldsymbol{r}_A). \end{aligned} \tag{3.10}$$

Here Φ_0 is the g.s. wave function of the target with energy E_0 and Φ_α (for $\alpha \neq 0$) are excited states with energy E_α. The incident wave number vector $\boldsymbol{k}$ and the outgoing one, $\boldsymbol{k}_\alpha = k_\alpha \hat{\boldsymbol{r}}$, satisfy the energy conservation relation

$$E = \frac{\hbar^2 k^2}{2m} + E_0 = \frac{\hbar^2 k_\alpha^2}{2m} + E_\alpha. \tag{3.11}$$

Only energetically allowed channels ($E_\alpha < E$) are included. A comment is in order on the restrictions implied by the asymptotic form (3.10).

First we neglect the identity of the incident nucleon with the target nucleons. A correct treatment of the fermionic statistics requires antisymmetrization, which induces particle-exchange processes. These are rather important at low incident energies. When, however, the incident energy is much higher than the kinetic energy of the Fermi motion (cf. p. 130) in the nucleus, which is approximately 40 MeV, the exchange effects become insignificant.

---

satisfy $\sum_{i=1}^{A} \boldsymbol{r}_i = 0$, which shows that, when the c.m. is not fixed, $\{\boldsymbol{r}_1, \ldots, \boldsymbol{r}_A\}$ is a linearly dependent set. In Part II, on the other hand, $\{\boldsymbol{r}_1, \ldots, \boldsymbol{r}_A\}$ always denote Cartesian vectors with respect to a fixed external frame, and the vectors with respect to the c.m. are denoted by $\{\bar{\boldsymbol{r}}_1, \ldots, \bar{\boldsymbol{r}}_A\}$ [see Eq. (9.124)], so that $\bar{\boldsymbol{r}}_i \equiv \boldsymbol{r}_i - \boldsymbol{R}_{\text{c.m.}}$, where $\boldsymbol{R}_{\text{c.m.}}$ is the vector pointing to the c.m. from the fixed origin. The coordinate sets used in Part II are summed up in Sect. 9.1.3 and in App. B.

The advantage of the notation used in Part I is that almost all formulae written down in the approximate scheme can be reinterpreted in the exact scheme and are found to be valid. One notable modification is required, however: when $\{\boldsymbol{r}_1, \ldots, \boldsymbol{r}_A\}$ are interpreted as coordinates with respect to the c.m., then $\int d\boldsymbol{r}_1 \ldots \int d\boldsymbol{r}_A$ should actually be understood as an integration over $A-1$ independent variables! The formulae written in the external frame of reference and those in the c.m. frame are so similar since they are fairly schematic. For a detailed explicit formulation, it is indispensable to distinguish $\{\bar{\boldsymbol{r}}_1, \ldots, \bar{\boldsymbol{r}}_A\}$ from $\{\boldsymbol{r}_1, \ldots, \boldsymbol{r}_A\}$ clearly as is done in Part II. Although structure problems are always formulated in intrinsic frames of reference like the c.m. frame, it is still often convenient to introduce coordinates with respect to an extrinsic frame since the evaluation of matrix elements is much simpler in terms of them. For a detailed precise treatment of the c.m. problem in a case relevant to reactions, see App. E.4.

We neglect the degrees of freedom other than the nucleonic degrees of freedom. At high incident energies hadrons (mostly mesons) can be produced. These processes, however, cannot be allowed for explicitly in the non-relativistic framework. In practice, hadron production is a major component of the cross section at high incident energies. Since we are interested in the nucleonic channels, the particle-production channels matter only as causing loss of the incident flux, so they will be taken into account by an absorptive term in the nucleon–nucleon interaction.

Finally, we neglect all reaction processes which cannot be described adequately by the asymptotic form (3.10). For example, pickup reactions, in which the incident nucleon carries away a few nucleons from the target to form a composite bound state, are excluded. For a pickup process to be likely to occur, the momentum of the target nucleon to be picked up must be comparable with those of the projectile nucleons. Thus the pickup process may only be important at low incident energies, not exceeding the kinetic energy of the target nucleons. Processes leading to the fragmentation of the target nucleus are, however, taken into account as continuum states of the target.

For processes covered by Eq. (3.10), the cross section of the scattering to the final state α is given by

$$\frac{d\sigma_\alpha}{d\Omega} = \frac{v_\alpha}{v}|f_\alpha|^2. \tag{3.12}$$

This formula follows from the definition that the cross section is the ratio of the outgoing fluxes with respect to the incoming fluxes. The ratio of the velocities, v_α/v, appears since the flux is proportional to the velocity.

Corresponding to the Schrödinger equation, a Lippmann–Schwinger equation can be introduced as was done in Eq. (2.8) for potential scattering. For this purpose, an appropriate Green's function is needed. Take the Hamiltonian H_0 with no interaction between the colliding partners:

$$H_0 = \frac{\boldsymbol{p}^2}{2m} + H_{\mathrm{T}}. \tag{3.13}$$

Upon projecting the Schödinger equation (3.9) to Φ_α, it reduces to a free-particle relative-motion Schrödinger equation for channel α, which implies a free-particle Green's function of the type of Eq. (2.6). The Green's function belonging to the Hamiltonian (3.13) satisfies

$$(E - H_0)G^{(+)}(\boldsymbol{r} - \boldsymbol{r}') = \delta(\boldsymbol{r} - \boldsymbol{r}')\mathbf{1}. \tag{3.14}$$

It is an operator with respect to the target intrinsic degrees of freedom, and the operator $\mathbf{1} = \sum_\alpha |\Phi_\alpha\rangle\langle\Phi_\alpha|$ is the unit operator in that state space. This multichannel Green's function can be constructed from the free-particle Green's functions belonging to each channel. Indeed, Eq. (3.14) is satisfied by the operator

$$G^{(+)}(\boldsymbol{r}-\boldsymbol{r}') = \sum_{\alpha(E_\alpha<E)} -\frac{2m}{\hbar^2}\frac{\mathrm{e}^{ik_\alpha|\boldsymbol{r}-\boldsymbol{r}'|}}{4\pi|\boldsymbol{r}-\boldsymbol{r}'|}|\Phi_\alpha\rangle\langle\Phi_\alpha| + \sum_{\alpha(E_\alpha>E)} -\frac{2m}{\hbar^2}\frac{\mathrm{e}^{-\kappa_\alpha|\boldsymbol{r}-\boldsymbol{r}'|}}{4\pi|\boldsymbol{r}-\boldsymbol{r}'|}|\Phi_\alpha\rangle\langle\Phi_\alpha|, \tag{3.15}$$

with

$$E = \begin{cases} E_\alpha + \frac{\hbar^2 k_\alpha^2}{2m} & (E_\alpha < E), \\ E_\alpha - \frac{\hbar^2 \kappa_\alpha^2}{2m} & (E_\alpha > E). \end{cases} \tag{3.16}$$

As we see, the multichannel Green's function is a combination of products of relative-motion Green's functions (2.6) and intrinsic-state projection operators.

The Green's function can be used to set up the Lippmann–Schwinger equation for the scattering wave function Ψ:

$$\Psi(\boldsymbol{r},\boldsymbol{r}_1,\ldots,\boldsymbol{r}_A) = \mathrm{e}^{ikz}\Phi_0(\boldsymbol{r}_1,\ldots,\boldsymbol{r}_A) + \int d\boldsymbol{r}'\, G^{(+)}(\boldsymbol{r}-\boldsymbol{r}')\sum_{i=1}^{A} V(\boldsymbol{r}'-\boldsymbol{r}_i)\Psi(\boldsymbol{r}',\boldsymbol{r}_1,\ldots,\boldsymbol{r}_A). \tag{3.17}$$

In the limit $r \to \infty$, the terms of the Green's function belonging to closed channels go to zero, so its asymptotic form is

$$G^{(+)}(\boldsymbol{r}-\boldsymbol{r}') \sim \sum_{\alpha(E_\alpha<E)} -\frac{2m}{\hbar^2}\frac{1}{4\pi r}\mathrm{e}^{ik_\alpha r - ik_\alpha\hat{\boldsymbol{r}}\cdot\boldsymbol{r}'}|\Phi_\alpha\rangle\langle\Phi_\alpha|. \tag{3.18}$$

Substitution of this in the Lippmann–Schwinger equation yields the formal expression for the scattering amplitude:

$$f_\alpha(\theta,\phi) = -\frac{2m}{\hbar^2}\frac{1}{4\pi}\int d\boldsymbol{r}'\int d\boldsymbol{r}_1\cdots\int d\boldsymbol{r}_A\,\mathrm{e}^{-i\boldsymbol{k}'_\alpha\cdot\boldsymbol{r}'}\Phi^*_\alpha(\boldsymbol{r}_1,\ldots,\boldsymbol{r}_A) \times \sum_{i=1}^{A} V(\boldsymbol{r}'-\boldsymbol{r}_i)\Psi(\boldsymbol{r}',\boldsymbol{r}_1,\ldots,\boldsymbol{r}_A), \tag{3.19}$$

where $\boldsymbol{k}'_\alpha$ stands for the propagation vector $k_\alpha\hat{\boldsymbol{r}}$ of the outgoing wave. Since this formula requires Ψ only in the interaction region, it may be useful for calculating $f_\alpha(\theta,\phi)$. This formula will actually be used to obtain the scattering amplitude in the eikonal approximation.

3.2.2 Eikonal approximation

The eikonal approximation introduced in Sect. 2.2 for potential scattering will now be extended to composite-particle scattering. The logic of the presentation will follow the pattern of Sect. 2.2. The fully microscopic description of composite-particle scattering in the eikonal approximation is known as the Glauber theory. (A description is said to be 'microscopic' if all nucleons of the system described are treated explicitly.)

As in Sect. 2.2 for potential scattering, we assume now that, if the rapid oscillation of the incident plane wave is removed, the remainder of the wave function does not oscillate much. Let us therefore put the scattering wave function as

$$\Psi(\boldsymbol{r},\boldsymbol{r}_1,\ldots,\boldsymbol{r}_A) = \mathrm{e}^{ikz}\hat{\Psi}(\boldsymbol{r},\boldsymbol{r}_1,\ldots,\boldsymbol{r}_A). \tag{3.20}$$

At $z \to -\infty$ the relative wave function should be a pure incoming wave, which implies that $\hat{\Psi}$ must be just the target g.s., i.e., $\hat{\Psi} \to \Phi_0$. The equation for $\hat{\Psi}$ reads as [cf. Eq. (2.27)]

$$\left[vp_z + \frac{\boldsymbol{p}^2}{2m} + (H_\mathrm{T} - E_0) + \sum_{i=1}^{A} V(\boldsymbol{r}-\boldsymbol{r}_i)\right]\hat{\Psi}(\boldsymbol{r},\boldsymbol{r}_1,\ldots,\boldsymbol{r}_A) = 0. \tag{3.21}$$

When the incident energy is high, it is reasonable to assume that the function $\hat{\Psi}$ does not oscillate much as a function of $\boldsymbol{r}$, thus the effect of the kinetic-energy operator $\boldsymbol{p}^2/2m$ will be negligible compared with that of vp_z. The resulting equation still carries the complexity of the Schrödinger equation for the target internal degrees of freedom.

The eigenvalues of the operator $H_\mathrm{T}-E_0$ are the excitation energies of the target. When the incident energy is high, one can argue, however, that $H_\mathrm{T}-E_0$ is also negligible compared with vp_z. The argument is the following.

Imagine that the reaction proceeds as a single collision between the incident nucleon and a nucleon in the target nucleus and neglect the motion of the target nucleons before the collision and the interaction between the target nucleons. Thus the process is regarded as a free collision between the incident nucleon and a target nucleon. The two nucleons get the same amount of momentum transfer in opposite directions, which implies that the excitation energy of the target is similar to the change of the kinetic energy of the projectile, which is the contribution of $\boldsymbol{p}^2/2m$ to Eq. (3.21). Thus it is consistent to neglect them simultaneously.

Since the neglect of the excitation energy amounts to assuming that the internal motion of the target is slow compared with the fast motion of the incident particle, it is sometimes called an adiabatic approximation (cf. App. A). The theory to be outlined is based on this approximation. It should be noted, however, that the adiabatic picture fails in the Coulomb breakup reaction, which will be discussed in Chap. 7.

Once both $p^2/2m$ and $H_T - E_0$ are omitted, Eq. (3.21) reduces to [cf. Eq. (2.30)]

$$\frac{\partial}{\partial z}\hat{\Psi}(\boldsymbol{r},\boldsymbol{r}_1,\ldots,\boldsymbol{r}_A) = \frac{1}{i\hbar v}\sum_{i=1}^{A} V(\boldsymbol{r}-\boldsymbol{r}_i)\hat{\Psi}(\boldsymbol{r},\boldsymbol{r}_1,\ldots,\boldsymbol{r}_A), \qquad (3.22)$$

and the initial condition is that $\hat{\Psi}(\boldsymbol{r},\boldsymbol{r}_1,\ldots,\boldsymbol{r}_A) = \Phi_0(\boldsymbol{r}_1,\ldots,\boldsymbol{r}_A)$ for $z \to -\infty$. Introducing the impact parameter vector $\boldsymbol{b}$ through $\boldsymbol{r} = (\boldsymbol{b}, z)$, one obtains [cf. Eq. (2.31)]

$$\begin{aligned}\hat{\Psi}&(\boldsymbol{r},\boldsymbol{r}_1,\ldots,\boldsymbol{r}_A)\\ &= \exp\left[\frac{1}{i\hbar v}\int_{-\infty}^{z} dz' \sum_{i=1}^{A} V(\boldsymbol{b}+z'\hat{\boldsymbol{z}}-\boldsymbol{r}_i)\right]\Phi_0(\boldsymbol{r}_1,\ldots,\boldsymbol{r}_A). \qquad (3.23)\end{aligned}$$

Substitution of this result into Eq. (3.20) leads to an approximate scattering wave function. In conformity with the potential scattering case, this function does not have the correct outgoing asymptotic form. It should be regarded valid in the region where the interaction potential is non-zero, and, therefore, can be used in Eq. (3.19) to derive an approximate expression for the scattering amplitude.

To be consistent with the neglect of the internal excitation energy, we neglect the difference of the incident and outgoing velocities. Then the momentum transfer vector $\boldsymbol{q} = \boldsymbol{k}' - \boldsymbol{k}$ is approximately orthogonal to the incident direction at forward angles and is a two-dimensional vector in the plane perpendicular to the incident direction. Substitution of Eqs. (3.20) and (3.23) into the scattering amplitude (3.19) yields [cf. Eq. (2.33)]

$$\begin{aligned}f_\alpha(\theta,\phi) = &-\frac{2m}{\hbar^2}\frac{1}{4\pi}\int d\boldsymbol{b}\int dz\, \mathrm{e}^{-i\boldsymbol{q}\cdot\boldsymbol{b}}\langle\Phi_\alpha|\sum_{i=1}^{A} V(\boldsymbol{b}+z\hat{\boldsymbol{z}}-\boldsymbol{r}_i)\\ &\times \exp\left[\frac{1}{i\hbar v}\int_{-\infty}^{z} dz' \sum_{i=1}^{A} V(\boldsymbol{b}+z'\hat{\boldsymbol{z}}-\boldsymbol{r}_i)\right]|\Phi_0\rangle. \qquad (3.24)\end{aligned}$$

Here $\langle\Phi_\alpha|\cdots|\Phi_0\rangle$ indicates integration over the coordinates of the target nucleons. The integration over the z-coordinate of the incident nucleon can be carried out:

$$\int_{-\infty}^{+\infty} dz \sum_{i=1}^{A} V(\boldsymbol{b}+z\hat{\boldsymbol{z}}-\boldsymbol{r}_i)\exp\left[\frac{1}{i\hbar v}\int_{-\infty}^{z} dz' \sum_{i=1}^{A} V(\boldsymbol{b}+z'\hat{\boldsymbol{z}}-\boldsymbol{r}_i)\right]$$

$$= i\hbar v \left\{ \exp\left[\sum_{i=1}^{A} \frac{1}{i\hbar v} \int_{-\infty}^{+\infty} dz\, V(\boldsymbol{b} + z\hat{\boldsymbol{z}} - \boldsymbol{r}_i) \right] - 1 \right\}. \tag{3.25}$$

It is interesting to observe that the right-hand side of this equation does not actually depend on the z-component, z_i, of the target nucleon coordinate $\boldsymbol{r}_i = (\boldsymbol{s}_i, z_i)$, where $\boldsymbol{s}_i$ is the projection of $\boldsymbol{r}_i$ onto the xy-plane. The coordinate system is shown in Fig. 3.1.

By defining the phase-shift function describing the nucleon–nucleon scattering [cf. Eq. (2.38)] as

$$\chi(\boldsymbol{b}) = -\frac{1}{\hbar v} \int_{-\infty}^{+\infty} dz\, V(\boldsymbol{b} + z\hat{\boldsymbol{z}}), \tag{3.26}$$

we finally obtain

$$f_\alpha(\theta, \phi) = \frac{ik}{2\pi} \int d\boldsymbol{b}\, e^{-i\boldsymbol{q}\cdot\boldsymbol{b}} \langle \Phi_\alpha | 1 - e^{i\sum_{i=1}^{A} \chi(\boldsymbol{b} - \boldsymbol{s}_i)} | \Phi_0 \rangle. \tag{3.27}$$

This is the scattering amplitude for the transition to the target excited state α.

The approximation scheme elaborated is essentially the Glauber theory. It is a microscopic many-body scattering framework, whose main input is the nucleon–nucleon interaction. The basic assumption underlying the Glauber theory is that the incident nucleon energy is so high that the momentum transfer becomes negligible in comparison with the incident momentum. This results in the emergence of the concept of the straight-line trajectory discussed in Sect. 2.2.1. The neglect of the momentum transfer in comparison with the incident momentum entails that the excitation energy of the target nucleus should also be neglected. Note, however, that no perturbative concepts have been employed. Consequently, the scattering amplitude of the Glauber theory includes the effects of multiple nucleon scattering to all orders.

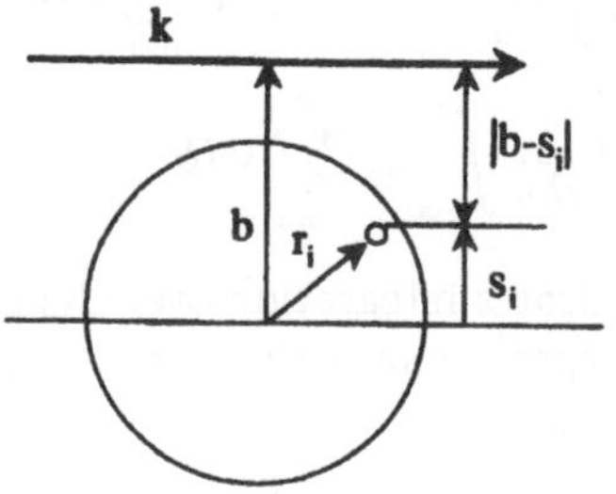

Figure 3.1: Coordinate system for nucleon–nucleus scattering. The symbols are explained in the text.

In some practical calculations, we may need to distinguish the phase-shift functions (3.26) for the proton–proton, proton–neutron and neutron–neutron systems. The generalization to such a case is straightforward.

3.2.3 Cross sections and reaction probabilities

In high-energy collisions one is often not interested in the cross sections of transitions to specific final target states. Although the energy transferred to the target is small in comparison with the incident energy, it is large in the scale of the excitation energies, so that many final target states are reached, and the cross section for each final state is rather small. Rather than the cross section for any particular final state, the sum of cross sections over a certain class of final target states is mostly measured. Such summed-up cross sections are called inclusive. The inclusive cross sections are evaluated conveniently by employing closure relations, which give an insight into their nature. This section is devoted to the calculation of inclusive cross sections in the Glauber theory. We also discuss a generalization of the optical theorem to collisions with composite target nuclei.

Let us consider any particular cross section integrated over the scattering solid angle. This integration is performed, similarly to the case of potential scattering, by replacing it with the one over the two-dimensional momentum transfer, $d\Omega = d\boldsymbol{q}/k^2$ [cf. Eq. (2.49)]. To be consistent with the neglect of the excitation energy, we neglect the difference between the incident and outgoing velocities, $v_\alpha \simeq v$. The integration can then be performed and gives

$$\begin{aligned}\sigma_\alpha &= \int d\Omega \, \frac{v_\alpha}{v} |f_\alpha(\theta,\phi)|^2 \\ &\simeq \int \frac{d\boldsymbol{q}}{k^2} \left| \frac{ik}{2\pi} \int d\boldsymbol{b} \, \mathrm{e}^{-i\boldsymbol{q}\cdot\boldsymbol{b}} \langle \Phi_\alpha | 1 - \mathrm{e}^{i\sum_{i=1}^{A} \chi(\boldsymbol{b}-\boldsymbol{s}_i)} | \Phi_0 \rangle \right|^2 \\ &= \int d\boldsymbol{b} \left| \langle \Phi_\alpha | 1 - \mathrm{e}^{i\sum_{i=1}^{A} \chi(\boldsymbol{b}-\boldsymbol{s}_i)} | \Phi_0 \rangle \right|^2 . \end{aligned} \tag{3.28}$$

The integrand of this expression,

$$P_\alpha(\boldsymbol{b}) = \left| \langle \Phi_\alpha | 1 - \mathrm{e}^{i\sum_{i=1}^{A} \chi(\boldsymbol{b}-\boldsymbol{s}_i)} | \Phi_0 \rangle \right|^2 , \tag{3.29}$$

is the probability that the target, which is initially in the g.s., Φ_0, is excited to the state Φ_α by the incident nucleon moving at a trajectory of impact parameter $\boldsymbol{b}$. In particular, $\alpha = 0$ means the elastic scattering, and its cross section is given by

$$\sigma_{\mathrm{el}} = \int d\boldsymbol{b} \left| 1 - \langle \Phi_0 | \mathrm{e}^{i\sum_{i=1}^{A} \chi(\boldsymbol{b}-\boldsymbol{s}_i)} | \Phi_0 \rangle \right|^2 . \tag{3.30}$$

The cross section summed over all possible final target states is called the total cross section. It is evaluated with the help of the closure relation for the target intrinsic states:

$$1 = \sum_{\alpha} |\Phi_\alpha\rangle\langle\Phi_\alpha|. \tag{3.31}$$

With this relation, the total cross section is

$$\sigma_{\text{tot}} = \sum_{\alpha} \sigma_\alpha = \int d\boldsymbol{b}\, \langle\Phi_0| \left|1 - e^{i\sum_{i=1}^{A}\chi(\boldsymbol{b}-\boldsymbol{s}_i)}\right|^2 |\Phi_0\rangle. \tag{3.32}$$

It is remarkable that the formula for the total cross section contains just one wave function, Φ_0.

The total reaction cross section σ_{reac} represents all processes in which the target is left behind excited. It is obtained by

$$\sigma_{\text{reac}} = \sigma_{\text{tot}} - \sigma_{\text{el}} = \int d\boldsymbol{b}\, P_{\text{reac}}(\boldsymbol{b}), \tag{3.33}$$

where

$$\begin{aligned} P_{\text{reac}}(\boldsymbol{b}) &= \langle\Phi_0| \left|e^{i\sum_{i=1}^{A}\chi(\boldsymbol{b}-\boldsymbol{s}_i)}\right|^2 |\Phi_0\rangle - \left|\langle\Phi_0|e^{i\sum_{i=1}^{A}\chi(\boldsymbol{b}-\boldsymbol{s}_i)}|\Phi_0\rangle\right|^2 \\ &= 1 - \left|\langle\Phi_0|e^{i\sum_{i=1}^{A}\chi(\boldsymbol{b}-\boldsymbol{s}_i)}|\Phi_0\rangle\right|^2 \quad \text{for real } \chi, \end{aligned} \tag{3.34}$$

with $P_{\text{reac}}(\boldsymbol{b})$ being the probability that the nucleon incident with the impact parameter $\boldsymbol{b}$ excites the target nucleus irrespective of its final state. Again this cross section requires only the g.s. wave function of the target nucleus. The second line of Eq. (3.34) holds if and only if the nucleon–nucleon potential V and, consequently, the phase-shift function $\chi(\boldsymbol{b})$, is real. The second term of $P_{\text{reac}}(\boldsymbol{b})$ is, of course, the probability that the target hit by a projectile with impact parameter $\boldsymbol{b}$ remains in its g.s.

We next derive these cross sections again, but now through the optical theorem (2.21). We learned from Sect. 2.2.1 that the eikonal theory of potential scattering bears out the optical theorem. We shall see immediately that the Glauber theory of composite-particle scattering follows suit.

In the collision of a nucleon with a target nucleus the outgoing flux in all channels is gained at the expense of the incoming flux, which enters at $\theta = 0$ in the elastic channel and suffers loss via interference with the outgoing wave (cf. Sect. 2.1.3). The balance of fluxes is expressed by a generalized optical theorem [cf. Eq. (2.21)]:

$$\sigma_{\text{tot}} = \frac{4\pi}{k}\text{Im} f_{\alpha=0}(0). \tag{3.35}$$

The right-hand side of the above equation is given by Eq. (3.27) as

$$\sigma_{\text{tot}} = 2 \int d\boldsymbol{b}\, \langle \Phi_0 | 1 - \text{Re}\, e^{i \sum_{i=1}^{A} \chi(\boldsymbol{b}-\boldsymbol{s}_i)} | \Phi_0 \rangle. \tag{3.36}$$

If the potential V and the phase-shift function χ of nucleon–nucleon collisions are real, this expression does indeed coincide precisely with the total cross section, Eq. (3.32), which was obtained by summing over all final states. The optical theorem is thus satisfied in the Glauber theory of the *multichannel* nucleon–nucleus collision.

When the incident energy is high enough, the nucleon–nucleon collisions give rise to hadron production. It is not possible to fully describe such processes based on the Schrödinger equation. However, we may employ a complex nucleon–nucleon potential V and, consequently, a complex phase-shift function χ. The loss of the flux induced by the imaginary part of the potential is then interpreted as an allowance for the hadron production channels. When V is not real, the total cross section derived from the optical theorem does not coincide with that obtained by summation over the target states, Eq. (3.32). Since the total cross section in the optical theorem is derived from the loss of the flux, it includes processes that involve hadron production. On the other hand, the sum over the final states does not include those. Therefore, the difference between these two cross sections has to represent the processes in which at least one hadron is produced. Since the hadron production processes are described as absorption of the flux, this difference may be called the absorption cross section:

$$\sigma_{\text{abs}} = \sigma_{\text{tot}} - \sum_{\alpha} \sigma_\alpha = \int d\boldsymbol{b}\, P_{\text{abs}}(\boldsymbol{b}), \tag{3.37}$$

where

$$P_{\text{abs}}(\boldsymbol{b}) = 1 - \langle \Phi_0 | \left| e^{i \sum_{i=1}^{A} \chi(\boldsymbol{b}-\boldsymbol{s}_i)} \right|^2 | \Phi_0 \rangle. \tag{3.38}$$

What happens to the reaction probability when hadron production occurs? The reaction probability $P_{\text{reac}}(\boldsymbol{b})$ in the first line of Eq. (3.34) now represents the probability of finding the target nucleus left behind by a collision of impact parameter $\boldsymbol{b}$ in any of its excited states as a result of any process *except* those involving hadron production. The sum $P_{\text{reac}}(\boldsymbol{b})+P_{\text{abs}}(\boldsymbol{b})$ is the same but *includes* hadron production processes. It is therefore appropriate to call this quantity also reaction probability:

$$P_{\text{reac}}(\boldsymbol{b}) = 1 - \left| \langle \Phi_0 | e^{i \sum_{i=1}^{A} \chi(\boldsymbol{b}-\boldsymbol{s}_i)} | \Phi_0 \rangle \right|^2. \tag{3.39}$$

This expression is nothing but the second line of Eq. (3.34), which was found to be valid there provided there is no hadron production.

Comparing Eqs. (3.34) and (3.39), we see that the formula of $P_{reac}(\boldsymbol{b})$ does not depend on whether or not there is absorption; what depend on it are the values of the ingredients.

The various cross sections and reaction probabilities introduced above include the state $e^{i\sum_i \chi(\boldsymbol{b}-\boldsymbol{s}_i)}|\Phi_0\rangle$. This state is analogous to $\psi'(\boldsymbol{r}, t\to\infty) = e^{i\chi(\boldsymbol{b})}$ given in Eq. (2.74), and may also be regarded as the wave function after collision in the projectile-rest frame since it is the $z\to\infty$ limit of Eq. (3.23):

$$\hat{\Psi}(\boldsymbol{r}, \boldsymbol{r}_1, \ldots, \boldsymbol{r}_A) = e^{i\sum_{i=1}^{A} \chi(\boldsymbol{b}-\boldsymbol{s}_i)}\Phi_0(\boldsymbol{r}_1, \ldots, \boldsymbol{r}_A) \quad (z\to\infty), \tag{3.40}$$

where $\boldsymbol{r} = \boldsymbol{b} + z\hat{\boldsymbol{z}}$. Subtracting the initial state Φ_0 from $\hat{\Psi}$, one obtains a function which can be regarded as the 'scattered wave function'. This may be decomposed into various target states:

$$\begin{aligned}
\Psi_{scat} &= \hat{\Psi} - \Phi_0 = \left[e^{i\sum_{i=1}^{A} \chi(\boldsymbol{b}-\boldsymbol{s}_i)} - 1\right]\Phi_0 \\
&= \sum_{\alpha} \Phi_\alpha \langle \Phi_\alpha | e^{i\sum_{i=1}^{A} \chi(\boldsymbol{b}-\boldsymbol{s}_i)} - 1 | \Phi_0 \rangle \\
&= \Phi_0 \langle \Phi_0 | e^{i\sum_{i=1}^{A} \chi(\boldsymbol{b}-\boldsymbol{s}_i)} - 1 | \Phi_0 \rangle + \sum_{\alpha\neq 0} \Phi_\alpha \langle \Phi_\alpha | e^{i\sum_{i=1}^{A} \chi(\boldsymbol{b}-\boldsymbol{s}_i)} | \Phi_0 \rangle.
\end{aligned} \tag{3.41}$$

It is apparent from this decomposition that the reaction probability for a state α, given by Eq. (3.29), is the probability of finding the state Φ_α in the scattered wave function Ψ_{scat} as a result of a collision with impact parameter $\boldsymbol{b}$. The norm square of Ψ_{scat} is given by [cf. Eq. (3.31)]

$$\begin{aligned}
\langle \Psi_{scat} | \Psi_{scat} \rangle &= \sum_{\alpha} \left| \langle \Phi_\alpha | e^{i\sum_{i=1}^{A} \chi(\boldsymbol{b}-\boldsymbol{s}_i)} - 1 | \Phi_0 \rangle \right|^2 \\
&= \langle \Phi_0 | \left| e^{i\sum_{i=1}^{A} \chi(\boldsymbol{b}-\boldsymbol{s}_i)} - 1 \right|^2 | \Phi_0 \rangle,
\end{aligned} \tag{3.42}$$

which, according to Eq. (3.32), enters into the formula for the total cross section as $\sigma_{tot} = \int d\boldsymbol{b}\, \langle \Psi_{scat} | \Psi_{scat} \rangle$.

When there is hadron production, then $\langle \hat{\Psi} | \hat{\Psi} \rangle < 1$, and the deficit of the norm square is the probability of hadron production. It shows the consistency of these considerations that the absorption probability, Eq. (3.38), which is the same as the hadron production probability, is equal to $1 - \langle \hat{\Psi} | \hat{\Psi} \rangle$.

3.2.4 Nucleus+nucleus collision

The extension of the Glauber formalism presented in Sects. 3.2.1–3.2.3 to the case of two colliding nuclei is straightforward.

We label the projectile by P and the target by T. The Hamiltonian of the nucleus P is $H_{\rm P}$ and the Schrödinger equation for P is $H_{\rm P}\Phi_\alpha^{\rm P} = E_\alpha^{\rm P}\Phi_\alpha^{\rm P}$, and similar notations apply to T. The relative coordinate and momentum of the two nuclei are denoted by $\boldsymbol{R}$ and $\boldsymbol{P}$. The intrinsic coordinates of nucleus P measured from the c.m. of P are $\boldsymbol{r}_i^{\rm P}$ ($i = 1, \ldots, A_{\rm P}$) and, similarly, those of T are $\boldsymbol{r}_i^{\rm T}$ ($i = 1, \ldots, A_{\rm T}$). Not all of the coordinates $\boldsymbol{r}_i^{\rm P}$ are independent as they satisfy $\sum_i \boldsymbol{r}_i^{\rm P} = 0$. The total Hamiltonian of the colliding system is then given by

$$H = \frac{\boldsymbol{P}^2}{2M_{\rm PT}} + H_{\rm P} + H_{\rm T} + \sum_{i\in{\rm P}}\sum_{j\in{\rm T}} V(\boldsymbol{R} + \boldsymbol{r}_i^{\rm P} - \boldsymbol{r}_j^{\rm T}), \tag{3.43}$$

where $M_{\rm PT}$ is the reduced mass of the two nuclei.

The scattering wave function $\Psi(\boldsymbol{R}, \boldsymbol{r}_1^{\rm P}, \ldots, \boldsymbol{r}_{A_{\rm P}}^{\rm P}, \boldsymbol{r}_1^{\rm T}, \ldots, \boldsymbol{r}_{A_{\rm T}}^{\rm T})$ satisfies the Schrödinger equation, $H\Psi = E\Psi$. The asymptotic wave function should look like

$$\begin{aligned}
&\Psi(\boldsymbol{R}, \boldsymbol{r}_1^{\rm P}, \ldots, \boldsymbol{r}_{A_{\rm P}}^{\rm P}, \boldsymbol{r}_1^{\rm T}, \ldots, \boldsymbol{r}_{A_{\rm T}}^{\rm T}) \\
&= {\rm e}^{iKZ}\Phi_0^{\rm P}(\boldsymbol{r}_1^{\rm P}, \ldots, \boldsymbol{r}_{A_{\rm P}}^{\rm P})\Phi_0^{\rm T}(\boldsymbol{r}_1^{\rm T}, \ldots, \boldsymbol{r}_{A_{\rm T}}^{\rm T}) \\
&+ \sum_{\alpha\beta} f_{\alpha\beta}(\theta, \phi)\frac{{\rm e}^{iK_{\alpha\beta}R}}{R}\Phi_\alpha^{\rm P}(\boldsymbol{r}_1^{\rm P}, \ldots, \boldsymbol{r}_{A_{\rm P}}^{\rm P})\Phi_\beta^{\rm T}(\boldsymbol{r}_1^{\rm T}, \ldots, \boldsymbol{r}_{A_{\rm T}}^{\rm T}),
\end{aligned} \tag{3.44}$$

where K is the wave number of the relative motion in the incident channel, $K_{\alpha\beta}$ is the wave number in the channel belonging to projectile state α and target state β, and Z is the z-component of the relative coordinate $\boldsymbol{R}$. The magnitude of $K_{\alpha\beta}$ is determined by the energy conservation:

$$E = \frac{\hbar^2 K^2}{2M_{\rm PT}} + E_0^{\rm P} + E_0^{\rm T} = \frac{\hbar^2 K_{\alpha\beta}^2}{2M_{\rm PT}} + E_\alpha^{\rm P} + E_\beta^{\rm T}. \tag{3.45}$$

Outgoing waves are only included for open channels, which satisfy $E_\alpha^{\rm P} + E_\beta^{\rm T} < E$.

The assumptions implied by this asymptotic form are similar to those involved in Eq. (3.10). For example, transfer reactions are excluded since they are unimportant at high incident energies. The theory disregards all processes that cannot be described by this asymptotic form as none of them has any significance.

By analogy with Sect. 3.2.2, one can introduce $\hat{\Psi}$ by

$$\begin{aligned}
&\Psi(\boldsymbol{R}, \boldsymbol{r}_1^{\rm P}, \ldots, \boldsymbol{r}_{A_{\rm P}}^{\rm P}, \boldsymbol{r}_1^{\rm T}, \ldots, \boldsymbol{r}_{A_{\rm T}}^{\rm T}) \\
&\qquad = {\rm e}^{iKZ}\hat{\Psi}(\boldsymbol{R}, \boldsymbol{r}_1^{\rm P}, \ldots, \boldsymbol{r}_{A_{\rm P}}^{\rm P}, \boldsymbol{r}_1^{\rm T}, \ldots, \boldsymbol{r}_{A_{\rm T}}^{\rm T}),
\end{aligned} \tag{3.46}$$

which satisfies the equation of motion

$$\left[vP_z + \frac{\boldsymbol{P}^2}{2M_{\rm PT}} + (H_{\rm P} - E_0^{\rm P}) + (H_{\rm T} - E_0^{\rm T}) + \sum_{i\in{\rm P}}\sum_{j\in{\rm T}} V(\boldsymbol{R} + \boldsymbol{r}_i^{\rm P} - \boldsymbol{r}_j^{\rm T})\right] \times \hat{\Psi}(\boldsymbol{R}, \boldsymbol{r}_1^{\rm P}, \ldots, \boldsymbol{r}_{A_{\rm P}}^{\rm P}, \boldsymbol{r}_1^{\rm T}, \ldots, \boldsymbol{r}_{A_{\rm T}}^{\rm T}) = 0. \tag{3.47}$$

Just as in the nucleon–nucleus collision, the kinetic-energy operator $\boldsymbol{P}^2/2M_{\rm PT}$ as well as the internal Hamiltonians $H_{\rm P} - E_0^{\rm P}$ and $H_{\rm T} - E_0^{\rm T}$ can be omitted in front of the wave function. With integration of the approximate equation, and putting the resulting $\hat{\Psi}$ into Eq. (3.46), one arrives at the approximate wave function valid in the interaction region:

$$\Psi(\boldsymbol{R}, \boldsymbol{r}_1^{\rm P}, \ldots, \boldsymbol{r}_{A_{\rm P}}^{\rm P}, \boldsymbol{r}_1^{\rm T}, \ldots, \boldsymbol{r}_{A_{\rm T}}^{\rm T}) = \exp\left[iKZ + \frac{1}{i\hbar v}\int_{-\infty}^{Z} dZ' \sum_{i\in P}\sum_{j\in T} V(\boldsymbol{b} + Z'\hat{\boldsymbol{z}} + \boldsymbol{r}_i^{\rm P} - \boldsymbol{r}_j^{\rm T})\right] \times \Phi_0^{\rm P}(\boldsymbol{r}_1^{\rm P}, \ldots, \boldsymbol{r}_{A_{\rm P}}^{\rm P})\Phi_0^{\rm T}(\boldsymbol{r}_1^{\rm T}, \ldots, \boldsymbol{r}_{A_{\rm T}}^{\rm T}), \tag{3.48}$$

where $\boldsymbol{R}$ is written as $\boldsymbol{R} = (\boldsymbol{b}, Z)$.

To get the nucleus–nucleus scattering amplitude in the Glauber theory, one should generalize Eq. (3.19) to the nucleus–nucleus collision and substitute Eq. (3.48) for the wave function. The result is [cf. Eq. (3.27)]

$$f_{\alpha\beta}(\theta, \phi) = \frac{iK}{2\pi}\int d\boldsymbol{b}\, {\rm e}^{-i\boldsymbol{q}\cdot\boldsymbol{b}} \langle \Phi_\alpha^{\rm P}\Phi_\beta^{\rm T}|1 - {\rm e}^{i\sum_{i\in{\rm P}}\sum_{j\in{\rm T}}\chi(\boldsymbol{b}+\boldsymbol{s}_i^{\rm P}-\boldsymbol{s}_j^{\rm T})}|\Phi_0^{\rm P}\Phi_0^{\rm T}\rangle, \tag{3.49}$$

where $\langle\cdots|\cdots|\cdots\rangle$ denotes integration over all (independent) intrinsic coordinates and $\boldsymbol{s}_i^{\rm P}$ ($\boldsymbol{s}_j^{\rm T}$) is the projection of $\boldsymbol{r}_i^{\rm P}$ ($\boldsymbol{r}_j^{\rm T}$) onto the xy-plane perpendicular to the incident direction. The coordinate system adopted is shown in Fig. 3.2.

The integrated cross sections and the optical theorem can be derived in exactly the same way as in the nucleon–nucleus case. The integrated cross section of the transition leading to the state α of the projectile and state β of the target is obtained as

$$\sigma_{\alpha\beta} = \int d\boldsymbol{b}\, P_{\alpha\beta}(\boldsymbol{b}), \tag{3.50}$$

where the reaction probability $P_{\alpha\beta}(\boldsymbol{b})$ is defined by

$$P_{\alpha\beta}(\boldsymbol{b}) = \left|\langle \Phi_\alpha^{\rm P}\Phi_\beta^{\rm T}|1 - {\rm e}^{i\sum_{i\in{\rm P}}\sum_{j\in{\rm T}}\chi(\boldsymbol{b}+\boldsymbol{s}_i^{\rm P}-\boldsymbol{s}_j^{\rm T})}|\Phi_0^{\rm P}\Phi_0^{\rm T}\rangle\right|^2. \tag{3.51}$$

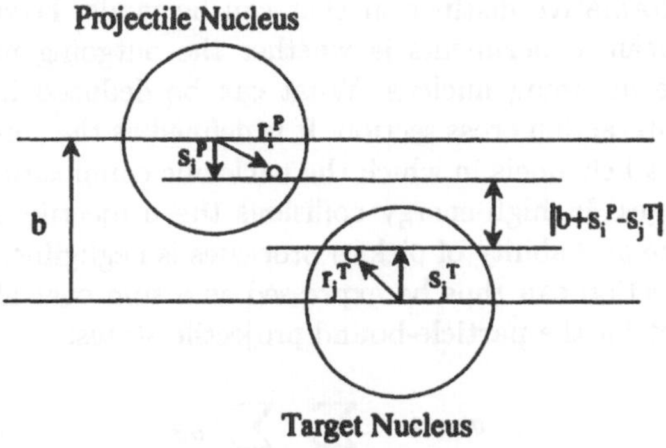

Figure 3.2: Coordinate system for nucleus–nucleus scattering. The symbols are explained in the text.

The total reaction cross section is then calculated by summing up over all final states $(\alpha\beta)$ except for the pair of the g.s., $(\alpha\beta) = (00)$, and using the appropriate closure relation:

$$\sigma_{\text{reac}} = \sum_{\alpha\beta} \sigma_{\alpha\beta} - \sigma_{00} = \int d\boldsymbol{b}\, P_{\text{reac}}(\boldsymbol{b}), \tag{3.52}$$

where

$$\begin{aligned} P_{\text{reac}}(\boldsymbol{b}) &= \langle \Phi_0^P \Phi_0^T | \left| e^{i \sum_{i\in P} \sum_{j\in T} \chi(\boldsymbol{b}+\boldsymbol{s}_i^P - \boldsymbol{s}_j^T)} \right|^2 | \Phi_0^P \Phi_0^T \rangle \\ &\quad - \left| \langle \Phi_0^P \Phi_0^T | e^{i \sum_{i\in P} \sum_{j\in T} \chi(\boldsymbol{b}+\boldsymbol{s}_i^P - \boldsymbol{s}_j^T)} | \Phi_0^P \Phi_0^T \rangle \right|^2 \\ &= 1 - \left| \langle \Phi_0^P \Phi_0^T | e^{i \sum_{i\in P} \sum_{j\in T} \chi(\boldsymbol{b}+\boldsymbol{s}_i^P - \boldsymbol{s}_j^T)} | \Phi_0^P \Phi_0^T \rangle \right|^2, \end{aligned} \tag{3.53}$$

where the second line is seen to hold if no absorption (hadron production) takes place, thus χ is real. When there is hadron production, σ_{reac} comprises the contributions from all processes that excite the projectile or the target or both *except* from hadron production processes. The total reaction cross section *including* hadron production may be derived from the optical theorem. The result for the reaction probability that includes hadron production again has the same form as in the no-absorption case, the last line of Eq. (3.53). The difference between the no-absorption and with-absorption cases is merely in the nucleon–nucleon phase-shift function.

In high-energy collisions there are a great number of open channels, and it is difficult to identify them experimentally. A simplest and yet very informative distinction that can be made, however, even in secondary-beam experiments is whether the outgoing nucleus is the same as the incoming nucleus. What can be deduced in this way is called the interaction cross section. It is defined as the sum of the cross sections for all channels in which the nucleonic composition of the projectile changes. In high-energy collisions the projectile can only lose nucleons; the probability of pickup processes is negligible. The interaction cross section can thus be expressed as a sum over all final states (α, β) except for the particle-bound projectile states:

$$\sigma_{\text{int}} = \sum_{\alpha \notin \text{bound}} \sum_{\beta} \sigma_{\alpha\beta}. \tag{3.54}$$

Since σ_{reac} includes all reactions, while σ_{int} does not, the inequality $\sigma_{\text{int}} \leq \sigma_{\text{reac}}$ holds. In high-energy collisions, however, their difference is estimated to be very small: $\sigma_{\text{int}} \simeq \sigma_{\text{reac}}$. As will be discussed later in Sect. 3.4, the high-energy nuclear collisions are dominated by hard nucleon–nucleon collisions, which excite both the projectile and the target highly, and that induces nucleon emission from the projectile. This scenario does not apply at low incident energies; there direct transitions to low-lying excited projectile states are important, which contribute to σ_{reac} but not to σ_{int}.

These reaction probabilities and integrated cross sections can again be interpreted in terms of the wave function after collision, which can be defined by taking the $Z \to \infty$ limit in Eq. (3.48) and dropping e^{iKZ}:

$$\begin{aligned} &\hat{\Psi}(\boldsymbol{R}, \boldsymbol{r}_1^{\text{P}}, \ldots, \boldsymbol{r}_{A_{\text{P}}}^{\text{P}}, \boldsymbol{r}_1^{\text{T}}, \ldots, \boldsymbol{r}_{A_{\text{T}}}^{\text{T}}) \\ &\to e^{i \sum_{i \in \text{P}} \sum_{j \in \text{T}} \chi(\boldsymbol{b} + \boldsymbol{s}_i^{\text{P}} - \boldsymbol{s}_j^{\text{T}})} \Phi_0^{\text{P}}(\boldsymbol{r}_1^{\text{P}}, \ldots, \boldsymbol{r}_{A_{\text{P}}}^{\text{P}}) \Phi_0^{\text{T}}(\boldsymbol{r}_1^{\text{T}}, \ldots, \boldsymbol{r}_{A_{\text{T}}}^{\text{T}}), \end{aligned} \tag{3.55}$$

where $\boldsymbol{R} = \boldsymbol{b} + Z\hat{\boldsymbol{z}}$.

3.2.5 Profile function

The scattering amplitude and cross sections in the preceding sections were written in terms of the nucleon–nucleon phase-shift function and the projectile and target intrinsic wave functions. These results form Glauber's multiple scattering theory, and have been applied to the analyses of high-energy nuclear reactions extensively.

The phase-shift function $\chi(\boldsymbol{b})$ for the nucleon–nucleon scattering calculated from the bare nuclear force is too complicated to employ in a practical calculation because of its strong quantum-number dependence and complicated form. It is usual instead to employ a simple parametrization.[4] The function that lends itself to such a parametrization is the profile function defined by

$$\Gamma(\boldsymbol{b}) = 1 - e^{i\chi(\boldsymbol{b})}. \tag{3.56}$$

The function $\Gamma(\boldsymbol{b})$ is a two-dimensional Fourier transform of the scattering amplitude of the nucleon–nucleon collision [cf. Eq. (2.39)]:

$$f(\theta) = \frac{ik}{2\pi}\int d\boldsymbol{b}\, e^{-i\boldsymbol{q}\cdot\boldsymbol{b}}\Gamma(\boldsymbol{b}). \tag{3.57}$$

If the process does not involve the production of hadrons, the profile function should satisfy $|1-\Gamma(\boldsymbol{b})| = 1$, which we call the unitarity condition. With the hadron production described as an absorption, the phase-shift function $\chi(\boldsymbol{b})$ becomes complex, with $\mathrm{Im}\,\chi(\boldsymbol{b}) \geq 0$ [cf. Eq. (3.26)], which entails the inequality $|1-\Gamma(\boldsymbol{b})| < 1$ for the profile function. Since the nuclear force is short-ranged, the phase-shift function $\chi(\boldsymbol{b})$ as well as the profile function $\Gamma(\boldsymbol{b})$ corresponding to the nucleon–nucleon force will also be short-ranged, and the Coulomb force is usually neglected.

In practical analyses of high-energy collisions, the profile function is most often parametrized as

$$\Gamma(\boldsymbol{b}) = \frac{1-i\alpha}{4\pi\beta^2}\sigma_{\mathrm{tot}}^{\mathrm{NN}}\exp\left(-\frac{b^2}{2\beta^2}\right). \tag{3.58}$$

Here $\sigma_{\mathrm{tot}}^{\mathrm{NN}}$ is the total cross section of the nucleon–nucleon collisions at the relevant incident energy. The parameters α and β are to be determined from other observables in the nucleon–nucleon collision. The parameter β is related to the range of the nuclear force and may be determined so as to reproduce the angular distribution of the nucleon–nucleon elastic scattering. The elastic scattering cross section of the nucleon–nucleon collision reads

$$\sigma_{\mathrm{el}}^{\mathrm{NN}} = \frac{1+\alpha^2}{16\pi\beta^2}\left(\sigma_{\mathrm{tot}}^{\mathrm{NN}}\right)^2. \tag{3.59}$$

At high incident energies the elastic cross section is appreciably smaller than the total cross section because of the hadron production processes.

[4]In this way the bare nucleon–nucleon interaction itself is indirectly replaced by a simple function that has a similar effect. This parametrization is yet not to be mixed up with using an effective interaction as is done in the description of lower-energy collisions (cf. Chap. 5). The effect of the effective force to be used there is quite different from the effect of the bare nucleon–nucleon force. The effective interaction mocks up some effects of the nuclear medium that reaction theory cannot treat explicitly, while the parametrization here is meant to produce the same effect as the bare nucleon–nucleon force. (At high energies the medium effects are negligible.)

For example, at the incident energy of 800 MeV per nucleon, the parameters $\sigma_{\text{tot}}^{\text{NN}} = 43$ mb, $\beta^2 = 0.20$ fm^2, and $\alpha = -0.10$ are often employed. This parametrization gives the elastic cross section of 19 mb, while the measured value is 25 mb. The difference between the total cross section and the elastic cross section is the hadron production cross section. A systematic trend of the parameters determined from the nucleon-nucleon scattering data is to be found in Ref. [47, 48].

Though the above parametrization does not satisfy the condition $|1-\Gamma(\boldsymbol{b})| < 1$ in a certain impact-parameter region, it is successful in describing various reaction processes in the Glauber theory.

A zero-range limit of the profile function is also often employed. That is obtained by taking the $\beta \to 0$ limit in Eq. (3.58):

$$\Gamma(\boldsymbol{b}) = \frac{1}{2}(1 - i\alpha)\sigma_{\text{tot}}^{\text{NN}}\delta(\boldsymbol{b}). \tag{3.60}$$

3.3 Optical-limit approximation to the phase-shift function

3.3.1 Nucleon–nucleus case

As seen in Eqs. (3.27) and (3.49), the calculation of the scattering amplitude of the Glauber theory requires multiple integrations with respect to the intrinsic coordinates. Let us take the elastic scattering amplitude of a nucleon–nucleus reaction as an example. It is useful to define a nucleon–nucleus phase-shift function $\chi_{\text{el}}(\boldsymbol{b})$ as

$$\begin{aligned} e^{i\chi_{\text{el}}(\boldsymbol{b})} &= \langle\Phi_0|e^{i\sum_{i=1}^{A}\chi(\boldsymbol{b}-\boldsymbol{s}_i)}|\Phi_0\rangle \\ &= \langle\Phi_0|\Pi_{i=1}^{A}\left[1 - \Gamma(\boldsymbol{b}-\boldsymbol{s}_i)\right]|\Phi_0\rangle. \end{aligned} \tag{3.61}$$

This is a basic ingredient of the reaction probability (3.34). For small targets composed of a few nucleons, it may be possible to calculate the multidimensional integration explicitly, but not without lengthy calculations even in that case. For the sake of systematic studies and intuitive understanding, it is highly desirable to work out an approximate scheme for the evaluation of such integrals.

A frequently used approximation is an expansion with respect to correlation effects in the target wave function. Its lowest-order approximation is known as the optical-limit approximation (OLA) [38].

Let us first evaluate the matrix element (3.61) neglecting the correlation effects in the target wave function. A pure no-correlation density function is a product of independent s.p. probability densities:

$$|\Phi_0(\boldsymbol{r}_1,\ldots,\boldsymbol{r}_A)|^2 = \prod_{i=1}^{A} n_i(\boldsymbol{r}_i), \tag{3.62}$$

where $n_i(\boldsymbol{r})$ stands for the normalized density distribution of an ith nucleon.

This expression is consistent with a many-particle wave function, which is a product of s.p. states, and thus obviously violates the symmetry under the permutation of identical particles. If Φ_0 were a Slater determinant, the square of the wave function could not be expressed in the form of Eq. (3.62). Though the Slater determinant wave function is said to represent independent-particle motion, it does include the correlations coming from the Pauli principle, which are sometimes called Pauli correlations (cf. Sect. 12.3.1 in Part II). Moreover, such a product wave function violates translational symmetry as well, and so do even the Slater determinants. A proper treatment of the c.m. motion results in another departure from the product form, sometimes called the c.m. correlations. We will never use such crude approximations in Part II, but we use them here to obtain more transparent formulae.

For a general many-body wave function, the g.s. density $\rho(\boldsymbol{r})$ is usually defined as

$$\rho(\boldsymbol{r}) = \langle\Phi_0|\sum_{i=1}^{A}\delta(\boldsymbol{r}-\boldsymbol{r}_i)|\Phi_0\rangle = \sum_{i=1}^{A}\int d\boldsymbol{r}_1\cdots\int d\boldsymbol{r}_{i-1}\int d\boldsymbol{r}_{i+1}\cdots\int d\boldsymbol{r}_A$$
$$\times|\Phi_0(\boldsymbol{r}_1,\ldots,\boldsymbol{r}_{i-1},\boldsymbol{r},\boldsymbol{r}_{i+1},\ldots,\boldsymbol{r}_A)|^2. \tag{3.63}$$

This would be entirely correct only if $\boldsymbol{r}_i$ were measured from the c.m. [cf. Eq. (9.127) of Part II]. Applied to the product wave function implied by Eq. (3.62), this gives

$$\rho(\boldsymbol{r}) = \sum_{i=1}^{A} n_i(\boldsymbol{r}). \tag{3.64}$$

For an uncorrelated wave function satisfying Eq. (3.62), the nucleon–nucleus phase-shift function $\chi_{\rm el}(\boldsymbol{b})$ defined by Eq. (3.61) has the form

$$e^{i\chi_{\rm el}(\boldsymbol{b})} = \prod_{i=1}^{A}\left[1-\int d\boldsymbol{r}\, n_i(\boldsymbol{r})\Gamma(\boldsymbol{b}-\boldsymbol{s})\right]. \tag{3.65}$$

Let us make an estimate for the integral $\int d\boldsymbol{r}\, n_i(\boldsymbol{r})\Gamma(\boldsymbol{b}-\boldsymbol{s})$ in a simplified situation. Let $n_i(\boldsymbol{r})$ be uniform in a sphere of radius R, i.e., $n_i(\boldsymbol{r}) = (3/4\pi R^3)\Theta(R-r)$. In a strong-absorption limit $e^{i\chi(\boldsymbol{b})} \approx \Theta(b-a)$, where

a is the nucleon–nucleon interaction range [cf. Eq. (2.75)]. In this case the profile function $\Gamma(\boldsymbol{b}) = 1 - e^{i\chi(\boldsymbol{b})}$ is simply given by $\Gamma(\boldsymbol{b}) = \Theta(a - b)$. According to Eq. (2.77), in the strong-absorption limit the total cross section is related to the interaction range a by $\sigma_{\text{tot}}^{\text{NN}} = 2\pi a^2$. At high energies the total cross section is about 40 mb, which gives $a = 0.8$ fm. Since the nuclear radius R is much larger than a except for very light nuclei, the integral may be evaluated approximately as a ratio between the volume of a cylinder of radius a and length $2\sqrt{R^2 - b^2}$ and the volume of a sphere of radius R (see Fig. 3.3):

$$\int d\boldsymbol{r}\, n(\boldsymbol{r})\Gamma(\boldsymbol{b} - \boldsymbol{s}) \approx \frac{\pi a^2 \cdot 2\sqrt{R^2 - b^2}}{\frac{4}{3}\pi R^3} = \frac{3}{2}\left(\frac{a}{R}\right)^2 \sqrt{1 - \frac{b^2}{R^2}}. \tag{3.66}$$

Because of the factor $(a/R)^2$, the integral is much smaller than unity even in the central collision, i.e., for $b = 0$. Then the integral $1 - \int d\boldsymbol{r}\, n_i(\boldsymbol{r})\Gamma(\boldsymbol{b} - \boldsymbol{s})$ can be approximated by $e^{-\int d\boldsymbol{r}\, n_i(\boldsymbol{r})\Gamma(\boldsymbol{b}-\boldsymbol{s})}$. This is the OLA:

$$\begin{aligned} e^{i\chi_{\text{el}}^{\text{OLA}}(\boldsymbol{b})} &= \prod_{i=1}^{A} e^{-\int d\boldsymbol{r}\, n_i(\boldsymbol{r})\Gamma(\boldsymbol{b}-\boldsymbol{s})} = e^{-\sum_{i=1}^{A} \int d\boldsymbol{r}\, n_i(\boldsymbol{r})\Gamma(\boldsymbol{b}-\boldsymbol{s})} \\ &= e^{-\int d\boldsymbol{r}\, \rho(\boldsymbol{r})\Gamma(\boldsymbol{b}-\boldsymbol{s})}, \end{aligned} \tag{3.67}$$

with the phase-shift function

$$\chi_{\text{el}}^{\text{OLA}}(\boldsymbol{b}) = i\int d\boldsymbol{r}\, \rho(\boldsymbol{r})\Gamma(\boldsymbol{b} - \boldsymbol{s}). \tag{3.68}$$

The multidimensional integration has now been reduced to a single integration, a convolution of the profile function with the density of the target nucleus.

The meaning of the approximate phase-shift function $\chi_{\text{el}}^{\text{OLA}}(\boldsymbol{b})$ can be elucidated, again, by a simple model. Let the nuclear density $\rho(\boldsymbol{r})$

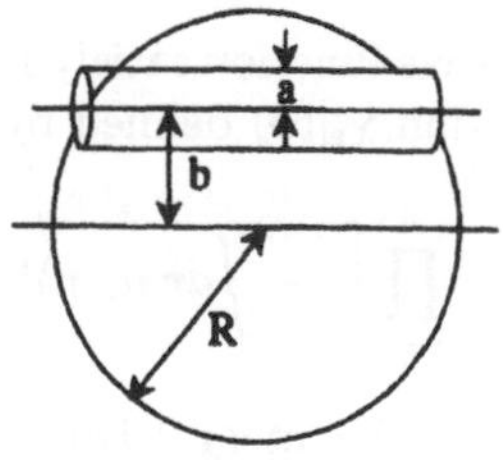

Figure 3.3: Interaction region (cylinder of radius a) for a nucleon impinging on a target nucleus (sphere of radius R) with impact parameter b.

be constant inside a sphere of radius R, i.e., $\rho_0 = A/(\frac{4}{3}\pi R^3)$ and let $a \ll R$. Then, for a central collision ($\boldsymbol{b} \simeq 0$), the integral reduces to

$$\chi_{\rm el}^{\rm OLA}(0) \simeq 2iR\rho_0 \int d\boldsymbol{s}\,\Gamma(\boldsymbol{s}) = 2R\rho_0 \frac{2\pi}{k} f(0), \tag{3.69}$$

where $f(0)$ is the forward scattering amplitude of the nucleon–nucleon scattering and Eq. (3.57) is used in the second step. With the optical theorem used for the nucleon–nucleon scattering, one obtains

$$\left|{\rm e}^{i\chi_{\rm el}^{\rm OLA}(0)}\right|^2 = {\rm e}^{-2{\rm Im}\,\chi_{\rm el}^{\rm OLA}(0)} \approx {\rm e}^{-2R\rho_0\sigma_{\rm tot}^{\rm NN}}. \tag{3.70}$$

This approximation to Eq. (3.61), implied by the OLA, can be used to calculate the total reaction cross section for the nucleon–nucleus scattering as seen from Eqs. (3.34) and (3.39). The quantity $\left|{\rm e}^{i\chi_{\rm el}(0)}\right|^2$ gives the probability that the nucleon does not excite the target in passing through its centre. If the nucleons of the target are considered classical particles, it is meaningful to introduce the mean free path of the incident nucleon in the target. It is given by $\lambda = 1/(\rho_0\sigma_{\rm tot}^{\rm NN})$. The right-hand side of Eq. (3.70) is then the probability that the incident nucleon passes through the centre of the target nucleus of diameter $2R$ without any collision with target nucleons. The OLA phase-shift function is thus found to give the target excitation probability evaluated in a classical statistical picture of nucleon–nucleon collisions.

It is, however, possible to set up a systematic approximate scheme to calculate the phase-shift function, which reduces to the OLA in the lowest order. In this scheme the higher-order terms incorporate the correlation effects. Let us define a function $G(\boldsymbol{b}, \lambda)$, which contains a parameter λ:

$$G(\boldsymbol{b}, \lambda) = \langle\Phi_0| \prod_{i=1}^{A} [1 - \lambda\Gamma(\boldsymbol{b} - \boldsymbol{s}_i)]\, |\Phi_0\rangle. \tag{3.71}$$

The phase-shift function is given by $i\chi_{\rm el}(\boldsymbol{b}) = \ln G(\boldsymbol{b}, 1)$. We expand $\ln G(\boldsymbol{b}, \lambda)$ around $\lambda=0$, and then put $\lambda=1$:

$$\ln G(\boldsymbol{b}, \lambda) = \lambda \left[\frac{\partial}{\partial\lambda} \ln G(\boldsymbol{b}, \lambda)\right]_{\lambda=0} + \frac{1}{2}\lambda^2 \left[\frac{\partial^2}{\partial\lambda^2} \ln G(\boldsymbol{b}, \lambda)\right]_{\lambda=0} + \ldots, \tag{3.72}$$

where $G(\boldsymbol{b}, 0) = 1$ has been used. The first term is

$$\left[\frac{\partial}{\partial\lambda} \ln G\right]_{\lambda=0} = -\langle\Phi_0| \sum_{i=1}^{A} \Gamma(\boldsymbol{b} - \boldsymbol{s}_i)|\Phi_0\rangle = -\int d\boldsymbol{r}\, \rho(\boldsymbol{r})\Gamma(\boldsymbol{b} - \boldsymbol{s}), \tag{3.73}$$

which coincides with the OLA for the phase-shift function given in Eqs. (3.67) and (3.68). It should be observed that the product ansatz for the square of the wave function, Eq. (3.62), has not been assumed in the present derivation.[5]

The next term in the expansion is

$$\left[\frac{\partial^2}{\partial\lambda^2}\ln G\right]_{\lambda=0} = \left[\frac{\partial^2}{\partial\lambda^2}G\right]_{\lambda=0} - \left[\frac{\partial}{\partial\lambda}G\right]^2_{\lambda=0}$$
$$= 2\langle\Phi_0|\sum_{i<j}^{A}\Gamma(\boldsymbol{b}-\boldsymbol{s}_i)\Gamma(\boldsymbol{b}-\boldsymbol{s}_j)|\Phi_0\rangle - \left[\langle\Phi_0|\sum_{i=1}^{A}\Gamma(\boldsymbol{b}-\boldsymbol{s}_i)|\Phi_0\rangle\right]^2. \quad (3.74)$$

By introducing the two-body (or two-particle) density [cf. Eq. (9.160) in Part II] as

$$\rho(\boldsymbol{r},\boldsymbol{r}') = \langle\Phi_0|\sum_{i\neq j}^{A}\delta(\boldsymbol{r}-\boldsymbol{r}_i)\delta(\boldsymbol{r}'-\boldsymbol{r}_j)|\Phi_0\rangle$$
$$= A(A-1)\int d\boldsymbol{r}_3\cdots\int d\boldsymbol{r}_A\,|\Phi_0(\boldsymbol{r},\boldsymbol{r}',\boldsymbol{r}_3,\ldots,\boldsymbol{r}_A)|^2, \quad (3.75)$$

the second-order contribution is now expressed as

$$2\langle\Phi_0|\sum_{i<j}^{A}\Gamma(\boldsymbol{b}-\boldsymbol{s}_i)\Gamma(\boldsymbol{b}-\boldsymbol{s}_j)|\Phi_0\rangle - \left[\langle\Phi_0|\sum_{i}\Gamma(\boldsymbol{b}-\boldsymbol{s}_i)|\Phi_0\rangle\right]^2$$
$$= \int d\boldsymbol{r}_1\int d\boldsymbol{r}_2\,\rho(\boldsymbol{r}_1,\boldsymbol{r}_2)\Gamma(\boldsymbol{b}-\boldsymbol{s}_1)\Gamma(\boldsymbol{b}-\boldsymbol{s}_2) - \left[\int d\boldsymbol{r}\,\rho(\boldsymbol{r})\Gamma(\boldsymbol{b}-\boldsymbol{s})\right]^2. \quad (3.76)$$

When the nucleons move independently in the target nucleus, the two-particle density factors into s.p. densities: $\rho(\boldsymbol{r}_1,\boldsymbol{r}_2) = \rho(\boldsymbol{r}_1)\rho(\boldsymbol{r}_2)$. Then the two terms in Eq. (3.76) cancel exactly. The fact that the second-order contribution vanishes when there is no correlation indicates that the λ-expansion can be regarded as an 'expansion with respect to the correlations' among the nucleons.

It should be remarked that the assumption $\rho(\boldsymbol{r}_1,\boldsymbol{r}_2)=\rho(\boldsymbol{r}_1)\rho(\boldsymbol{r}_2)$ is never satisfied strictly in any many-body wave function. This becomes apparent by noting that $\int d\boldsymbol{r}_1\int d\boldsymbol{r}_2\,\rho(\boldsymbol{r}_1,\boldsymbol{r}_2) = A(A-1)$ by definition, whereas $\int d\boldsymbol{r}_1\int d\boldsymbol{r}_2\,\rho(\boldsymbol{r}_1)\rho(\boldsymbol{r}_2) = A^2$. The larger the system, the less important this normalization problem becomes.

[5] In App. E.4 this derivation is repeated with treating the c.m. motion as well as antisymmetrization correctly. The result has the form of Eq. (3.73), but the definition of the coordinates and integrations implicit in the density (3.63) have to be reinterpreted as coordinates with respect to the c.m.

As was stated above, an important correlation effect arises purely from the fermionic statistics of the particles. A simplest wave function satisfying the antisymmetrization requirement is a Slater determinant. For the Slater determinant $\Phi_0 = (1/\sqrt{A!})\det\{\phi_i(\boldsymbol{r}_j)\}$ constructed from a set of orthonormal functions $\{\phi_i\}$, the two-body density is given by

$$\rho(\boldsymbol{r}_1, \boldsymbol{r}_2) = \sum_{ij} |\phi_i(\boldsymbol{r}_1)|^2 |\phi_j(\boldsymbol{r}_2)|^2 - \sum_{ij} \phi_i^*(\boldsymbol{r}_1)\phi_j^*(\boldsymbol{r}_2)\phi_j(\boldsymbol{r}_1)\phi_i(\boldsymbol{r}_2). \tag{3.77}$$

The first term coincides with the product of the one-body densities, $\rho(\boldsymbol{r}_1)\rho(\boldsymbol{r}_2)$. The second term is an exchange term, and originates from the antisymmetrization of the wave function. We thus see that the correlation effects contained in a Slater determinant come merely from the antisymmetrization.

There are other types of correlations among the nucleons. One is the short-range correlations, which arise from the strong repulsion in the two-nucleon force at short distances. A broader class are the dynamic correlations, which are not allowed for by the Hartree–Fock approximation. Another special class of the dynamic correlations are cluster correlations, which will be widely discussed in Part II. These correlations may come into play through the higher order terms as we shall see in Sect. 13.10 (Part II).

3.3.2 Nucleus–nucleus case

Our concern in this subsection is again the OLA for elastic scattering. The phase-shift function $\chi_{\rm el}(\boldsymbol{b})$ relevant to the nucleus–nucleus case is defined [cf. Eq. (3.61)] by

$$e^{i\chi_{\rm el}(\boldsymbol{b})} = \langle \Phi_0^{\rm P}\Phi_0^{\rm T}| \prod_{i\in{\rm P}} \prod_{j\in{\rm T}} [1 - \Gamma(\boldsymbol{b} + \boldsymbol{s}_i^{\rm P} - \boldsymbol{s}_j^{\rm T})] |\Phi_0^{\rm P}\Phi_0^{\rm T}\rangle. \tag{3.78}$$

For $\chi_{\rm el}(\boldsymbol{b})$ the OLA can be derived by a straightforward generalization of Eqs. (3.71)–(3.73). The function G, which contains the expansion parameter λ, is to be defined by

$$G(\boldsymbol{b}, \lambda) = \langle \Phi_0^{\rm P}\Phi_0^{\rm T}| \prod_{i\in{\rm P}} \prod_{j\in{\rm T}} [1 - \lambda\Gamma(\boldsymbol{b} + \boldsymbol{s}_i^{\rm P} - \boldsymbol{s}_j^{\rm T})] |\Phi_0^{\rm P}\Phi_0^{\rm T}\rangle. \tag{3.79}$$

Expanding $\ln G(\boldsymbol{b}, \lambda)$ around $\lambda = 0$ and taking the lowest-order term in Eq. (3.72) gives

$$\chi_{\rm el}^{\rm OLA}(\boldsymbol{b}) = i \int d\boldsymbol{r}^{\rm P}\, \rho_{\rm P}(\boldsymbol{r}^{\rm P}) \int d\boldsymbol{r}^{\rm T}\, \rho_{\rm T}(\boldsymbol{r}^{\rm T}) \Gamma(\boldsymbol{b} + \boldsymbol{s}^{\rm P} - \boldsymbol{s}^{\rm T}), \tag{3.80}$$

where $\rho_{\rm P(T)}(\boldsymbol{r})$ is the g.s. density distribution of the projectile (target) nucleus.

In contrast with the nucleon–nucleus scattering case, however, an analogous approximation goes beyond an expansion in terms of the correlations. As an illustration, let us evaluate the phase-shift function for a pair of uncorrelated wave functions like Eq. (3.62):

$$|\Phi_0^{\mathrm{P}}(\boldsymbol{r}_1^{\mathrm{P}},\ldots,\boldsymbol{r}_{A_{\mathrm{P}}}^{\mathrm{P}})|^2 = \prod_{i\in\mathrm{P}} n_i^{\mathrm{P}}(\boldsymbol{r}_i^{\mathrm{P}}), \quad |\Phi_0^{\mathrm{T}}(\boldsymbol{r}_1^{\mathrm{T}},\ldots,\boldsymbol{r}_{A_{\mathrm{T}}}^{\mathrm{T}})|^2 = \prod_{i\in\mathrm{T}} n_i^{\mathrm{T}}(\boldsymbol{r}_i^{\mathrm{T}}). \tag{3.81}$$

The phase-shift function (3.78) reduces to

$$\begin{aligned} e^{i\chi_{\mathrm{el}}(\boldsymbol{b})} &= \prod_{i\in\mathrm{P}}\int d\boldsymbol{r}_i^{\mathrm{P}}\, n_i^{\mathrm{P}}(\boldsymbol{r}_i^{\mathrm{P}}) \prod_{j\in\mathrm{T}}\int d\boldsymbol{r}_j^{\mathrm{T}}\, n_j^{\mathrm{T}}(\boldsymbol{r}_j^{\mathrm{T}})\,[1-\Gamma(\boldsymbol{b}+\boldsymbol{s}_i^{\mathrm{P}}-\boldsymbol{s}_j^{\mathrm{T}})] \\ &= \left\langle \prod_{i\in\mathrm{P}}\prod_{j\in\mathrm{T}}[1-\Gamma_{ij}]\right\rangle, \end{aligned} \tag{3.82}$$

where $\langle\cdots\rangle$ means integration over all projectile and target nucleon coordinates and Γ_{ij} is a shorthand for $\Gamma(\boldsymbol{b}+\boldsymbol{s}_i^{\mathrm{P}}-\boldsymbol{s}_j^{\mathrm{T}})$. To get a more explicit expression for the function $\chi_{\mathrm{el}}^{\mathrm{OLA}}(\boldsymbol{b})$ of Eq. (3.80), one has to resort first to the following approximation:

$$\begin{aligned} &\left\langle \prod_{i\in\mathrm{P}}\prod_{j\in\mathrm{T}}[1-\Gamma_{ij}]\right\rangle \\ &\quad= \left\langle 1-\sum_{i\in\mathrm{P},j\in\mathrm{T}}\Gamma_{ij}+\sum_{\substack{i\in\mathrm{P},j\in\mathrm{T},k\in\mathrm{P},l\in\mathrm{T}\\(ij)\neq(kl)}}\Gamma_{ij}\Gamma_{kl}-\cdots\right\rangle \\ &\quad\approx 1-\sum_{i\in\mathrm{P},j\in\mathrm{T}}\langle\Gamma_{ij}\rangle+\sum_{\substack{i\in\mathrm{P},j\in\mathrm{T},k\in\mathrm{P},l\in\mathrm{T}\\(ij)\neq(kl)}}\langle\Gamma_{ij}\rangle\langle\Gamma_{kl}\rangle-\cdots \\ &\quad= \prod_{i\in\mathrm{P}}\prod_{j\in\mathrm{T}}[1-\langle\Gamma_{ij}\rangle]. \end{aligned} \tag{3.83}$$

To reach the third line from the second line, one must make a drastic approximation. In the term $\Gamma_{ij}\Gamma_{kl}$ with $i=k$ and $j\neq l$, e.g., the following has to be done:

$$\begin{aligned} &\int d\boldsymbol{r}_i^{\mathrm{P}}\int d\boldsymbol{r}_j^{\mathrm{T}}\int d\boldsymbol{r}_l^{\mathrm{T}}\, n_i^{\mathrm{P}}(\boldsymbol{r}_i^{\mathrm{P}})n_j^{\mathrm{T}}(\boldsymbol{r}_j^{\mathrm{T}})n_l^{\mathrm{T}}(\boldsymbol{r}_l^{\mathrm{T}})\Gamma(\boldsymbol{b}+\boldsymbol{s}_i^{\mathrm{P}}-\boldsymbol{s}_j^{\mathrm{T}})\Gamma(\boldsymbol{b}+\boldsymbol{s}_i^{\mathrm{P}}-\boldsymbol{s}_l^{\mathrm{T}}) \\ &\quad\longrightarrow \int d\boldsymbol{r}_i^{\mathrm{P}}\int d\boldsymbol{r}_j^{\mathrm{T}}\, n_i^{\mathrm{P}}(\boldsymbol{r}_i^{\mathrm{P}})n_j^{\mathrm{T}}(\boldsymbol{r}_j^{\mathrm{T}})\Gamma(\boldsymbol{b}+\boldsymbol{s}_i^{\mathrm{P}}-\boldsymbol{s}_j^{\mathrm{T}}) \\ &\quad\times \int d\boldsymbol{r}_i^{\mathrm{P}}\int d\boldsymbol{r}_l^{\mathrm{T}}\, n_i^{\mathrm{T}}(\boldsymbol{r}_i^{\mathrm{P}})n_l^{\mathrm{T}}(\boldsymbol{r}_l^{\mathrm{T}})\Gamma(\boldsymbol{b}+\boldsymbol{s}_i^{\mathrm{P}}-\boldsymbol{s}_l^{\mathrm{T}}). \end{aligned} \tag{3.84}$$

It is not obvious under what conditions this approximation is justifiable.

If the above replacement can be justified, we can proceed in the same way as we did in Eq. (3.67) for the elastic scattering amplitude of nucleon–nucleus scattering:

$$\begin{aligned} e^{i\chi_{el}(b)} &\approx \prod_{i\in P}\prod_{j\in T}\left[1-\int d\boldsymbol{r}^{P}\, n_i^{P}(\boldsymbol{r}^{P})\int d\boldsymbol{r}^{T}\, n_j^{T}(\boldsymbol{r}^{T})\Gamma(\boldsymbol{b}+\boldsymbol{s}^{P}-\boldsymbol{s}^{T})\right] \\ &\approx e^{-\sum_{i\in P}\sum_{j\in T}\int d\boldsymbol{r}^{P}\, n_i^{P}(\boldsymbol{r}^{P})\int d\boldsymbol{r}^{T}\, n_j^{T}(\boldsymbol{r}^{T})\Gamma(\boldsymbol{b}+\boldsymbol{s}^{P}-\boldsymbol{s}^{T})} \\ &= e^{-\int d\boldsymbol{r}^{P}\,\rho_P(\boldsymbol{r}^{P})\int d\boldsymbol{r}^{T}\,\rho_T(\boldsymbol{r}^{T})\Gamma(\boldsymbol{b}+\boldsymbol{s}^{P}-\boldsymbol{s}^{T})} \equiv e^{i\chi_{el}^{OLA}}. \end{aligned} \tag{3.85}$$

The final expression indeed agrees with Eq. (3.80) since $\rho_P(\boldsymbol{r}^P) = \sum_{i\in P} n_i^P(\boldsymbol{r}^P)$ and similarly for the target. The condition is that each term of the integral $\int d\boldsymbol{r}_i^P \int d\boldsymbol{r}_j^T\, n_i^P(\boldsymbol{r}_i^P) n_j^T(\boldsymbol{r}_j^T)\Gamma(\boldsymbol{b}+\boldsymbol{s}_i^P-\boldsymbol{s}_j^T)$ has to be small compared with unity. This condition is well satisfied as was discussed in the previous section. Thus, the OLA for the nucleus–nucleus collision implies an approximation, Eq. (3.83), in addition to the smallness of the correlation effects in the two colliding nuclei.

The OLA is nevertheless known to work quite well for the collisions between stable nuclei. However, it has been found that the approximation is rather poor when applied to the reactions of halo nuclei. Since the approximation in Eq. (3.85) is as reasonable for halo nuclei as for stable nuclei, Eq. (3.83) is likely to be responsible for the failure. This point will be discussed in detail later in Sect. 4.2.3.

Calculations of the phase-shift function for the nucleus–nucleus collision will again be taken up in Sect. 3.5, and an attempt to go beyond the naive OLA will be made there.

3.4 Total reaction cross section

In this section the total reaction cross section in the OLA will be discussed, starting with the formalism and then turning to empirical facts, which illustrate the performance of the approach. A review on the theoretical treatment of the total reaction cross section is given in Ref. [49].

The total reaction cross section can be expressed as $\sigma_{reac} = \int d\boldsymbol{b}\,[1 - |e^{i\chi_{el}(\boldsymbol{b})}|^2]$ [cf. Eqs. (3.52), (3.53), (3.61) and (3.78)]. In the OLA the function $\chi_{el}^{OLA}(\boldsymbol{b})$ has to be substituted for $\chi_{el}(\boldsymbol{b})$, which yields

$$\sigma_{reac} = \int d\boldsymbol{b}\left[1 - e^{-2\mathrm{Im}\chi_{el}^{OLA}(\boldsymbol{b})}\right]. \tag{3.86}$$

The function $\chi_{el}^{OLA}(\boldsymbol{b})$ is conveniently expressed in terms of the thickness functions defined by

$$T(\boldsymbol{s}) = \int dz\,\rho(\boldsymbol{s} + z\hat{\boldsymbol{z}}). \tag{3.87}$$

For the nucleon–nucleus case, Eq. (3.68) takes the form

$$\chi_{\rm el}^{\rm OLA}(\boldsymbol{b}) = i\int d\boldsymbol{s}\, T(\boldsymbol{s})\Gamma(\boldsymbol{b}-\boldsymbol{s}), \tag{3.88}$$

while, for the nucleus–nucleus case, Eq. (3.80) reduces to

$$\chi_{\rm el}^{\rm OLA}(\boldsymbol{b}) = i\int d\boldsymbol{s}^{\rm P}\int d\boldsymbol{s}^{\rm T}\, T_{\rm P}(\boldsymbol{s}^{\rm P})T_{\rm T}(\boldsymbol{s}^{\rm T})\Gamma(\boldsymbol{b}+\boldsymbol{s}^{\rm P}-\boldsymbol{s}^{\rm T}). \tag{3.89}$$

If we furthermore use a zero-range profile function (3.60), we obtain, for the nucleon–nucleus cross section,

$$\sigma_{\rm reac} = \int d\boldsymbol{b}\left[1-{\rm e}^{-\sigma_{\rm tot}^{\rm NN}T(\boldsymbol{b})}\right], \tag{3.90}$$

and, for the nucleus–nucleus case,

$$\sigma_{\rm reac} = \int d\boldsymbol{b}\left[1-{\rm e}^{-\sigma_{\rm tot}^{\rm NN}\int d\boldsymbol{s}\, T_{\rm P}(\boldsymbol{s})T_{\rm T}(\boldsymbol{b}+\boldsymbol{s})}\right]. \tag{3.91}$$

These expressions can be interpreted in an intuitive picture borrowed from classical mechanics as was discussed below Eq. (3.70). For a nucleon–nucleus collision, the nucleus may be viewed as a gas composed of independent nucleons with density $\rho(\boldsymbol{r})$. The incident nucleon then collides with the target nucleons randomly, with cross section $\sigma_{\rm tot}^{\rm NN}$. The probability that the incident nucleon collides at least once with the target nucleons is given by the integrand of Eq. (3.90). The Glauber theory provides the foundation for this formula, and clarifies its derivation from quantum theory.

As seen in Fig. 3.4, the total nucleon–nucleon cross section $\sigma_{\rm tot}^{\rm NN}$ does not depend so much on the incident energy beyond a few hundred MeV: it is about 40 mb (4 fm^2).[6] Since the density inside the nucleus is about ρ=0.17 fm^{-3}, the mean free path of a high-energy nucleon in the nucleus is $\lambda=1/(\rho\sigma_{\rm tot}^{\rm NN})=1.5$ fm. The mean free path is thus much shorter than the diameter of the nucleus, except for very light nuclei. This means that, for the nucleon–nucleus collisions, the probability that the nucleon penetrating the nucleus makes collisions at least once with

[6]The nucleon–nucleon cross section is large at low incident energies because it is determined mainly by the long-range attraction of the nuclear force. The difference between the proton–neutron and the proton–proton cross sections comes from the Pauli principle and the isospin dependence of the nuclear force. While two like nucleons can only take $T=1$, the proton–neutron system can take both $T=0$ and $T=1$, and the nuclear force is more attractive in the $T=0$ state. The cross sections show a minimum around $E_{\rm lab}=200$ MeV, where the long-range attractive and short-range repulsive parts of the nuclear force compensate for each other. At high energies the cross sections are dominated by meson production processes.

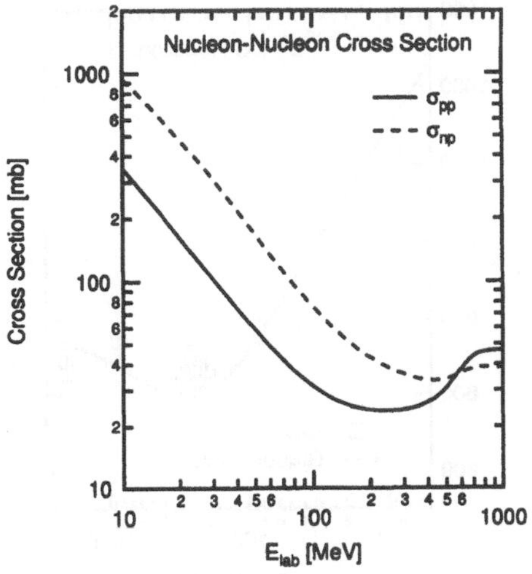

Figure 3.4: Total nucleon–nucleon cross section as a function of the laboratory incident energy [50].

the target nucleons is close to unity, except for peripheral collisions, in which the impact parameter b is close to the target radius R. The total reaction cross section is then close to the geometrical cross section, $\sigma_{\text{reac}} \simeq \pi R^2$.

In the nucleus–nucleus collisions we need to consider collisions of all pairs of the nucleons of the projectile and target. Even if only one pair makes a collision, reaction has occurred. So, when two nuclei overlap, the probability of reaction is much closer to unity than in the nucleon–nucleus case. So the reaction probability $P_{\text{reac}}(b)$ is very close to unity when the impact parameter is smaller than the sum of the radii, $b < R_{\text{P}} + R_{\text{T}}$. Thus our first guess for the total reaction cross section is the geometrical cross section [cf. Eq. (1.1)]:

$$\sigma_{\text{reac}} = \pi(R_{\text{P}} + R_{\text{T}})^2. \tag{3.92}$$

It emerges from a theoretical analysis of the nucleus–nucleus reaction cross sections [51] that the total reaction cross sections of diverse systems can be systematically described by Eq. (3.86) or (3.91). The analysis was extended to as low incident energies as a few tens of MeV per nucleon [52, 53]. Although the applicability of the Glauber theory is questionable at so low energies, Eq. (3.91) was still found to give a reasonable description for the energy dependence of the total reaction cross section. In the zero-range limit of OLA the energy dependence of the reaction cross section comes entirely from the total nucleon–nucleon

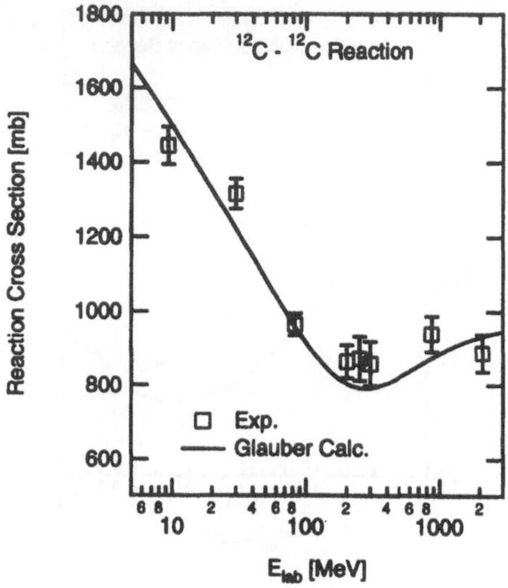

Figure 3.5: Energy dependence of the total reaction cross section of the $^{12}C+^{12}C$ collision. The experimental data are from Ref. [53] and references therein. The theoretical curve is calculated by Eq. (3.91) with a single Gaussian shape for the density distribution [51]. The energy $E_{\rm lab}$ on the abscissa is the energy per nucleon.

cross section, $\sigma_{\rm NN}^{\rm tot}$.

Figure 3.5 shows the energy dependence of the total reaction cross section of the $^{12}C+^{12}C$ collision. The measured cross section increases at low incident energies, and this trend as well as the absolute value of the cross section are well reproduced by Eq. (3.91), which is shown as solid curve. A Gaussian shape is assumed for the density distribution of ^{12}C.

Although Eq. (3.91) works well for a wide energy region, the uncertainty of the foundations of the Glauber theory at low incident energies should not be forgotten. Effects that go beyond the Glauber theory are the bending of the relative-motion orbit by the Coulomb [52, 53] and nuclear forces [54], the Fermi motion and the Pauli blocking effects [55].

There exists an empirical formula which describes the total reaction cross section for various pairs of nuclei and for a wide energy range with a few parameters [53]. This is somewhat similar to the geometrical cross section formula, Eq. (3.92):

$$\sigma_{\rm reac} = \pi(R_{\rm vol} + R_{\rm surf})^2 \left(1 - \frac{B_{\rm C}}{E_{\rm cm}}\right), \tag{3.93}$$

where B_C is the height of the Coulomb barrier:

$$B_C = \frac{Z_P Z_T e^2}{r_C \left(A_P^{1/3} + A_T^{1/3}\right)}, \tag{3.94}$$

with Z_P, Z_T the atomic numbers and $r_C = 1.3$ fm. The factor $(1 - B_C/E_{cm})$ is to allow for the reduction of the cross section by the Coulomb repulsion at low incident energies. The nuclear radii are decomposed into 'volume' and 'surface' terms:

$$R_{vol} = r_0 \left(A_P^{1/3} + A_T^{1/3}\right), \quad R_{surf} = r_0 \left(a \frac{A_P^{1/3} A_T^{1/3}}{A_P^{1/3} + A_T^{1/3}} - c\right). \tag{3.95}$$

It was found that the measured cross sections of various pairs of nuclei can be reproduced at various incident energies with energy-independent parameters $r_0 = 1.1$ fm and $a = 1.85$ and with parameter c, which increases from $c = 0.65$ at $E = 30\,A$ MeV to $c \approx 2.0$ at high incident energies.

The geometrical formula (3.92) was used to determine the radii of unstable nuclei from measurements of the total interaction cross section [8]. It should be clear from the above discussions that the nuclear radii thus deduced must depend on the incident energy. A comparison of the

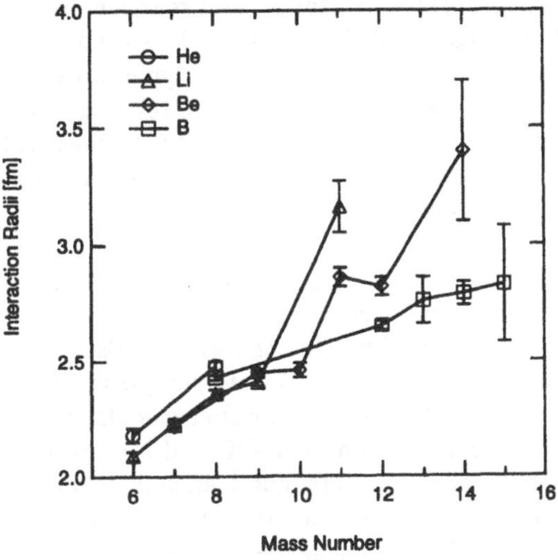

Figure 3.6: Nuclear radii determined from measured interaction cross sections with a geometrical cross-section formula. The dots belonging to isotopes of the same element are connected by lines.

radii determined in this way at a fixed incident energy is still useful to get information on the relative sizes. In Fig. 3.6 the radii of some light stable and unstable nuclei, determined from interaction cross section measurements at $800\,A$ MeV [8, 10] are shown. Following the trend of the anomalously large interaction cross sections at the neutron drip line as shown in Fig. 1.1, the nuclear radius also becomes anomalously large there.

Although the abrupt increase of the radius does demonstrate the anomalous behaviour of some drip-line nuclei, these radii may not be accurate since the geometrical formula for the total reaction cross section, Eq. (3.92), is meaningful only when the nucleus has a well-defined surface. When a nucleus has a diffuse density distribution, such as a halo structure, it is difficult to define the nuclear surface and the radius. Therefore, one should be cautious in making any quantitative comparison of such radii with theory.

3.5 Phase-shift function revisited

3.5.1 Complete calculation

As was seen in the previous sections, the phase-shift function for elastic scattering defined by Eqs. (3.61) and (3.78) plays a key role in calculating both the elastic scattering and reaction cross sections. The calculation of the phase-shift function involves a multidimensional integration, and its explicit evaluation is in general extremely tedious. The commonly used approximation called OLA takes into account the leading term in terms of consecutive nucleon–nucleon collisions, thus neglecting the correlation effects. Its validity is not self-evident and requires careful examination.

It is possible to compare the OLA with full Glauber calculations for very small systems, such as p+α and α+α [58]. Figure 3.7 shows the p+α elastic differential cross section at $E_{\rm p} = 1.05$ GeV [59]. The wave function of ^{4}He was taken to be the simplest 0s harmonic-oscillator shell-model (HOSM) state specified by the size parameter $\beta = 0.52\,{\rm fm}^{-2}$ [cf. Eq. (9.52) in Part II]. The full Glauber calculation reproduces the elastic differential cross section. The OLA also reproduces it at small angles reasonably well but is not good enough to describe the cross section for larger values of the momentum transfer. Figure 3.8 compares two full Glauber calculations for the α+α elastic scattering cross section to test the effect of the nuclear wave function. In one calculation the α-particle is described by the $(0{\rm s})^4$ HOSM, while in the other by an exact calculation with a realistic nuclear force (cf. Sect. 13.9.3). Though for small momentum transfers both wave functions reproduce the data

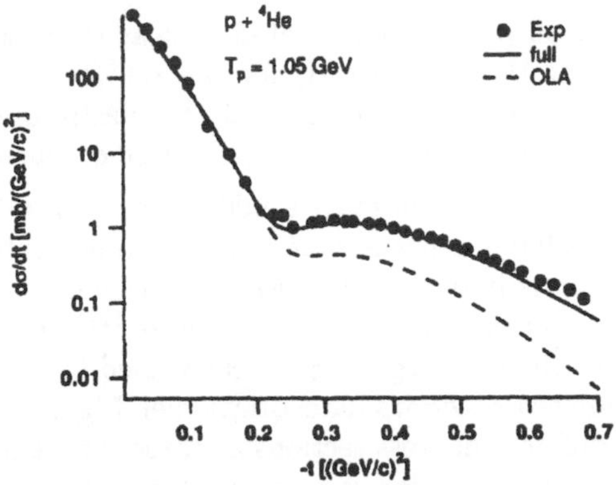

Figure 3.7: The p+^{4}He elastic differential cross section at $E_p = 1.05$ GeV as a function of the square of the four-momentum transfer. [For elastic scattering this is just $-t = \hbar^2 q^2$ in the c.m. frame.] The solid curve is the complete Glauber calculation, and the dashed curve is the OLA. The data are taken from Ref. [56].

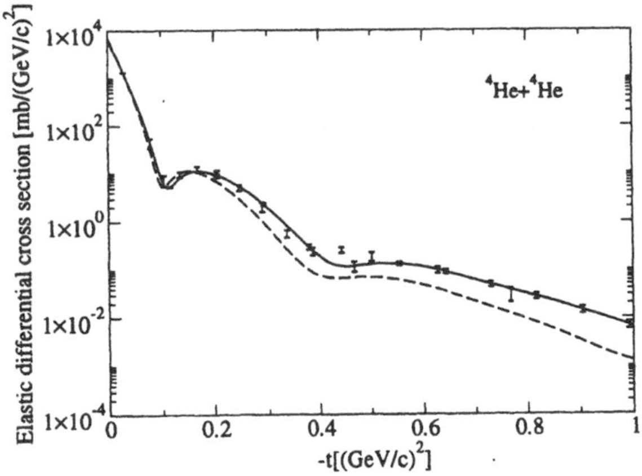

Figure 3.8: ^{4}He+^{4}He elastic differential cross section at $E_\alpha = 2.57$ GeV as a function of the square of the four-momentum transfer. The solid curve results from a complete Glauber calculation with a realistic wave function, while the dashed curve is with a 0s HOSM for the α-particle. The two wave functions give the same rms radius, 1.46 fm. The data are taken from Ref. [57].

well, it is interesting to see that they predict an appreciable difference beyond the second peak. An example for a complete calculation of the phase-shift function with a sophisticated nuclear wave function will be presented in Sect. 13.10.2 (Part II) for the p+^{6}He elastic scattering at $E_p = 717$ MeV [58]. We shall see that the elastic differential cross sections are reproduced very well by a full Glauber calculation.

A complete Glauber calculation would be hard even for a nucleon–nucleus case for a larger nucleus. One thus has to be content with OLA calculations, so it is important to assess its soundness or limitations before studying a nucleus–nucleus case. The p+^{12}C collision is a good example because the ^{12}C target is frequently used in the measurement of the interaction cross sections for exotic nuclei. Figure 3.9 displays the p+^{12}C elastic differential cross sections at three different energies [63]. The OLA phase-shift function is calculated by Eq. (3.67), where the parameters of $\Gamma(\boldsymbol{b})$ are taken from Ref. [48] and the ^{12}C density is parametrized by a combination of two Gaussians of different ranges. The OLA predictions are the dashed curves. It is seen that the OLA calculation produces a very reasonable fit to experiment at $E_p = 800$ MeV. The fit is particularly satisfactory up to about 15 degrees. The OLA

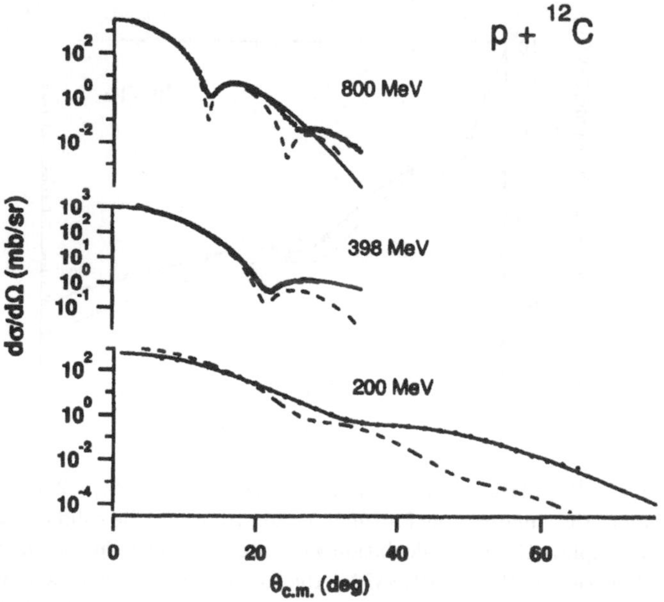

Figure 3.9: Elastic differential cross sections for the p+^{12}C collision at three different energies. The solid curve is a phenomenological fit by Eq. (3.96), while the dashed curve is the OLA prediction, Eq. (3.100). The data [60–62] are plotted in dots.

is still fairly good at $E_p = 400$ MeV. As the incident energy decreases to $E_p = 200$–300 MeV, $\sigma_{\text{tot}}^{\text{NN}}$ reaches its minimum (see Fig. 3.4), indicating the increase of the mean free path or the nuclear transparency. Therefore, larger medium effects and Pauli correlations [55, 64] are expected here. In fact, the OLA phase-shift function fails to reproduce the experimental data even at forward angles.

This test of the OLA indicates that it is important to seek the possibilities of going beyond the OLA, and Sect. 3.5.2 as well as Sect. 13.10.2 in Part II will be concerned with that.

3.5.2 Effective profile function

It is certainly desirable to describe the nucleon–(target-)nucleus (NT) scattering well. To this end let us explore the possibility of using the nucleon–nucleus scattering as an elementary vehicle in the Glauber theory, assuming the target nucleus as a scatterer and introduce an effective profile function Γ_{NT} for the nucleon–nucleus scattering [65]. In this way various complicated effects, neglected otherwise, would be automatically included in an effective manner through the nucleon–nucleus amplitude. The function Γ_{NT} may be parametrized in accord with the nucleon–nucleon profile function (3.58) as

$$\Gamma_{\text{NT}}(\boldsymbol{b}) = \sum_j \frac{1 - i\alpha_j}{4\pi\beta_j^2}\sigma_j \exp\left(-\frac{b^2}{2\beta_j^2}\right), \tag{3.96}$$

and the parameters σ_j, β_j and α_j can be determined by fitting the experimental elastic angular distribution and the total cross section of nucleon–nucleus scattering. The solid curve in Fig. 3.9 is a two-term fit by Eq. (3.96). The parameters, e.g., at $E_p = 800$ MeV (for $j = 1, 2$) are $\sigma_j = 52.89$, -18.78 fm^2, $\beta_j^2 = 1.970$, 1.074 fm^2 and $\alpha_j = -0.1117$, 0.01495 [63, 65]. The reaction cross section of the p+^{12}C collision is predicted to be 249 mb, which is in good agreement with 255 mb, extrapolated from the data [66] at 870 MeV.

The phase-shift function for a nucleus–nucleus scattering, defined by Eq. (3.78), may now be approximated by [cf. Eq. (3.61)]

$$e^{i\tilde{\chi}_{\text{el}}(\boldsymbol{b})} = \langle \Phi_0^{\text{P}} | \prod_{i \in \text{P}} [1 - \Gamma_{\text{NT}}(\boldsymbol{b} + \boldsymbol{s}_i^{\text{P}})] | \Phi_0^{\text{P}} \rangle. \tag{3.97}$$

The validity of this approximation will be tested for the ^{4}He+^{12}C and ^{6}He+^{12}C systems in Sect. 13.10.3. If the OLA[7] is used on the right-

[7] As is explained after Eq. (3.66), the OLA is justifiable when the range of the profile function is much shorter than the nuclear radius. In Eq. (3.97), however, the range of the profile function is comparable with the nuclear radius. Nevertheless, the formula works well in practice as will be shown in Table 3.1.

hand side of Eq. (3.97), the following simple expression is obtained for the phase-shift function:

$$\tilde{\chi}_{\rm el}^{\rm OLA}(\boldsymbol{b}) = i \int d\boldsymbol{r}\, \rho_{\rm P}(\boldsymbol{r}^{\rm P})\Gamma_{\rm NT}(\boldsymbol{b}+\boldsymbol{s}^{\rm P}), \tag{3.98}$$

where we need only the density of the projectile and the effective profile function $\Gamma_{\rm NT}$ of Eq. (3.96). This expression looks very appealing because of its simplicity. Its utility will be shown below.

Once $\Gamma_{\rm NT}$ is determined to fit the nucleon–nucleus scattering data at a given energy, one may approximately obtain the phase-shift function by Eq. (3.98) and then can systematically calculate the reaction cross sections of various projectiles incident on that target at the same incident energy per nucleon. When no appropriate data are available, however, one cannot determine $\Gamma_{\rm NT}$ in this way, and one may think that the utility of $\Gamma_{\rm NT}$ is lost. To get an effective profile function without data, it is desirable to find a way of calculating it from the elementary nucleon–nucleon profile function $\Gamma(\boldsymbol{b})$. The nucleon–nucleus profile function should be related to the nucleon–nucleus phase-shift function (3.61) as the nucleon–nucleon profile function is to the nucleon–nucleon phase-shift function [cf. Eqs. (3.56)]. Thus

$$\Gamma_{\rm NT}(\boldsymbol{b}) = 1 - \langle \Phi_0^{\rm T} | \prod_{j\in{\rm T}} [1-\Gamma(\boldsymbol{b}-\boldsymbol{s}_j^{\rm T})]\, |\Phi_0^{\rm T}\rangle \tag{3.99}$$

should hold and Eq. (3.97) or (3.98) gives us an alternative to calculate the phase-shift function through $\Gamma(\boldsymbol{b})$.

The use of the OLA simplifies the calculation of the phase-shift function further. By approximating the right-hand side of Eq. (3.99) as

$$\Gamma_{\rm NT}(\boldsymbol{b}) \approx 1 - \exp\left[-\int d\boldsymbol{r}^{\rm T}\, \rho_{\rm T}(\boldsymbol{r}^{\rm T})\Gamma(\boldsymbol{b}-\boldsymbol{s}^{\rm T})\right] \tag{3.100}$$

and substituting it into Eq. (3.98), one obtains [63, 65]

$$\chi_{\rm eff}(\boldsymbol{b}) = i\int d\boldsymbol{r}^{\rm P}\, \rho_{\rm P}(\boldsymbol{r}^{\rm P})\left\{1-\exp\left[-\int d\boldsymbol{r}^{\rm T}\, \rho_{\rm T}(\boldsymbol{r}^{\rm T})\Gamma(\boldsymbol{b}+\boldsymbol{s}^{\rm P}-\boldsymbol{s}^{\rm T})\right]\right\}. \tag{3.101}$$

The input into this formula is just the densities and Γ, and, in this respect, it is similar to the original OLA formula (3.80).

If the convolution of $\rho_{\rm T}$ and Γ is small enough compared to unity, Eq. (3.101) reduces to the usual OLA formula; otherwise some multiple-scattering effects are included. To elucidate this point, we note that Eq. (3.97) may be derived by starting with the original definition (3.78) and replacing $\prod_{j\in\mathrm{T}}\left[1-\Gamma(\boldsymbol{b}+\boldsymbol{s}_i^{\mathrm{P}}-\boldsymbol{s}_j^{\mathrm{T}})\right]|\Phi_0^{\mathrm{T}}\rangle$ by

$$|\Phi_0^{\mathrm{T}}\rangle\langle\Phi_0^{\mathrm{T}}|\prod_{j\in\mathrm{T}}\left[1-\Gamma(\boldsymbol{b}+\boldsymbol{s}_i^{\mathrm{P}}-\boldsymbol{s}_j^{\mathrm{T}})\right]|\Phi_0^{\mathrm{T}}\rangle=\left[1-\Gamma_{\mathrm{NT}}(\boldsymbol{b}+\boldsymbol{s}_i^{\mathrm{P}})\right]|\Phi_0^{\mathrm{T}}\rangle \tag{3.102}$$

for each nucleon i in the projectile. This amounts to discarding the possibility that the target could get back to its initial state Φ_0^{T} by a succession of nucleon–nucleon collisions. This looks to be a reasonable assumption, indeed.

Now there are three different approximations available to the phase-shift function, Eqs. (3.80), (3.98) and (3.101), which all require the projectile density. Table 3.1 compares the reaction cross sections calculated by these different phase-shift functions for different projectiles, incident on the ^{12}C target, with experiment. Most of the projectile densities were generated by the variational microscopic multicluster model to be presented in Part II. For very weakly bound projectiles, whose density must have a long tail, this tail was corrected to follow an exponential fall-off corresponding to the separation energy of the lowest-lying threshold. The densities were constrained to reproduce the empirical matter root

Table 3.1: Comparison of theoretical reaction cross sections with interaction cross sections for collisions with ^{12}C measured at 800 A MeV.

Projectile	r_{m} (fm)	σ_{reac} (mb)			Exp. [8–10]
		$\chi_{\mathrm{el}}^{\mathrm{OLA}}$	$\tilde{\chi}_{\mathrm{el}}^{\mathrm{OLA}}$	χ_{eff}	
^{6}He	2.49	782	717	707	722±6
^{7}Li	2.35	789	742	734	736±6
^{7}Be	2.31	780	735	726	738±9
^{8}He	2.44	848	791	781	817±6
^{8}Li	2.37	824	780	771	768±9
^{8}B	2.38	829	783	772	798±6 [19]
^{9}Li	2.32	841	801	791	796±6
^{9}Be	2.41	854	814	804	806±9
^{9}C	2.49	887	837	827	(834±18) [18]
^{10}Be	2.28	851	817	806	813±13
^{12}C	2.33	896	869	856	856±9

Note: The phase-shift functions are calculated in three different approximations: Eq. (3.80) ($\chi_{\mathrm{el}}^{\mathrm{OLA}}$), Eq. (3.98) ($\tilde{\chi}_{\mathrm{el}}^{\mathrm{OLA}}$) and Eq. (3.101) ($\chi_{\mathrm{eff}}$). The effects of the Coulomb interaction are neglected. The symbol r_{m} is the matter rms radius calculated from the projectile density. The cross section in parenthesis has been taken at 730 A MeV.

mean square (rms) radii.[8] As is clearly seen, the OLA always overestimates the reaction cross sections, whereas the simple formula (3.98) gives results much closer to the measured values. This strongly supports the utility of the effective profile function Γ_{NT}. It is worthwhile to note that the predictions of the simple formula (3.101) are also close to those of Eq. (3.98). We thus have a tool at hand, which enables one to predict the reaction cross sections fairly accurately, and, based on that, the nuclear sizes can be extracted from experiment more reliably than earlier.

The roles of the projectile and the target are interchangeable in the calculation of the elastic scattering amplitudes and reaction cross sections, so that it may be possible to symmetrize Eq. (3.101) as follows:

$$\chi_{\text{effs}}(\mathbf{b}) = \frac{i}{2}\int d\boldsymbol{r}^{\mathrm{P}}\, \rho_{\mathrm{P}}(\boldsymbol{r}^{\mathrm{P}}) \left\{1 - \exp\left[-\int d\boldsymbol{r}^{\mathrm{T}}\, \rho_{\mathrm{T}}(\boldsymbol{r}^{\mathrm{T}})\Gamma(\boldsymbol{b}+\boldsymbol{s}^{\mathrm{P}}-\boldsymbol{s}^{\mathrm{T}})\right]\right\} + \frac{i}{2}\int d\boldsymbol{r}^{\mathrm{T}}\, \rho_{\mathrm{T}}(\boldsymbol{r}^{\mathrm{T}}) \left\{1 - \exp\left[-\int d\boldsymbol{r}^{\mathrm{P}}\, \rho_{\mathrm{P}}(\boldsymbol{r}^{\mathrm{P}})\Gamma(\boldsymbol{b}+\boldsymbol{s}^{\mathrm{T}}-\boldsymbol{s}^{\mathrm{P}})\right]\right\}. \tag{3.103}$$

The symmetrical treatment is especially desirable when the nucleon–projectile scattering and the nucleon–target scattering are equally well described in terms of the densities. It is found that the difference between the cross sections obtained with Eq. (3.101) and with Eq. (3.103) is indeed very small.

[8] The rms proton, neutron and matter radii are defined as $\langle n^{-1}\sum_{i=1}^{n} \boldsymbol{r}_i^2\rangle^{1/2}$, where $\boldsymbol{r}_i$ are coordinates with respect to the c.m., $\langle\cdots\rangle$ is an expectation value and $n = Z$, N and $A = Z + N$, respectively.

Chapter 4

High-energy reactions of halo nuclei

4.1 Simple model for halo nuclei

In this chapter we apply the Glauber theory to the reactions of halo nuclei at high incident energy. The reactions of halo nuclei are particular in several respects, which distinguish them from the reactions of other nuclei. Because of the spatially extended, weakly bound nature of the halo neutrons, the reaction mechanisms may be quite different, which any theoretical framework should allow for. An assumption or approximation found to be valid for the reactions of stable nuclei may not be valid for halo nuclei. The OLA, which has proved quite acceptable and broadly applicable in the reactions of stable nuclei, fails for halo nuclei. We will present a more realistic framework, which has now been widely used to analyse the reactions of halo nuclei, and discuss the origin of the failure of the OLA.

Some halo nuclei, like ^{11}Li, ^{6}He and ^{14}Be, have two-neutron halos. These nuclei can emit another neutron if one of their neutrons is removed: they are Borromean. This fact suggests the important role of the neutron–neutron correlation in the halo structure. There exists also single-neutron halo structure, e.g., in ^{11}Be.

To describe halo structure, a schematic model will be used. Let us call the residual nucleus of halo-neutron-removal processes the core (nucleus). For example, ^{11}Li is composed of two halo neutrons and the ^{9}Li core. Since the halo neutrons are bound very weakly to the core, it seems a reasonable approximation to identify the core state involved in the halo structure with the g.s. of the core as an isolated nucleus. It is common to assume a product form for the g.s. of the halo nuclei, such as $\Phi_0^P = \phi_0 \Phi_0^C$, where Φ_0^C is the g.s. wave function of the core nucleus and ϕ_0 is the wave function of the halo neutrons.

To simplify the formalism, we will elaborate the theory for the case of a single-neutron halo nucleus. The extension to two-neutron halo nuclei is straightforward. The intrinsic coordinates of the projectile are to be measured from the c.m. of the projectile. The vectors pointing to the nucleons and to the core will be denoted by $\boldsymbol{r}_1^P, \ldots, \boldsymbol{r}_{A_P}^P$ and $\boldsymbol{R}_C = A_C^{-1} \sum_{i \in C} \boldsymbol{r}_i^P$, respectively, where $A_C = A_P - 1$. The coordinate system is shown in Fig. 4.1. Then the total wave function of the projectile in the g.s. is assumed to be

$$\Phi_0^P(\boldsymbol{r}_1^P, \ldots, \boldsymbol{r}_{A_P}^P) = \phi_0(\boldsymbol{r}_1^P - \boldsymbol{R}_C)\Phi_0^C(\boldsymbol{r}_2^P, \ldots, \boldsymbol{r}_{A_P}^P). \tag{4.1}$$

If the wave function Φ_0^C is constructed by a mean-field method referring to the c.m. of the core as the origin, the coordinates $\boldsymbol{r}_i^P$ $(i = 2, \ldots, A_P)$ in Eq. (4.1) should read as $\boldsymbol{r}_i^P - \boldsymbol{R}_C$. If the core nucleus is so heavy that the difference between the c.m. of the projectile and the c.m. of the core is negligible, the vector pointing to the core c.m. can be set to zero: $\boldsymbol{R}_C = 0$. For simplicity, we will formulate the theory by neglecting this difference. Then the halo-neutron wave function ϕ_0 includes only the coordinate of the halo neutron measured from the projectile c.m., $\phi_0(\boldsymbol{r}_1^P)$. We will later indicate how the formulae should be modified to treat the coordinates correctly. The correct treatment of the c.m. coordinate is indispensable for describing the reactions induced by the Coulomb field. This issue will be discussed in Chap. 7.

The wave function (4.1) also ignores the indistinguishability between the halo neutron and the neutrons in the core nucleus, and with that, the antisymmetry requirement under the exchange of nucleons. In this respect, it is similar to a macroscopic core+n model. In Sect. 11.8.2 of

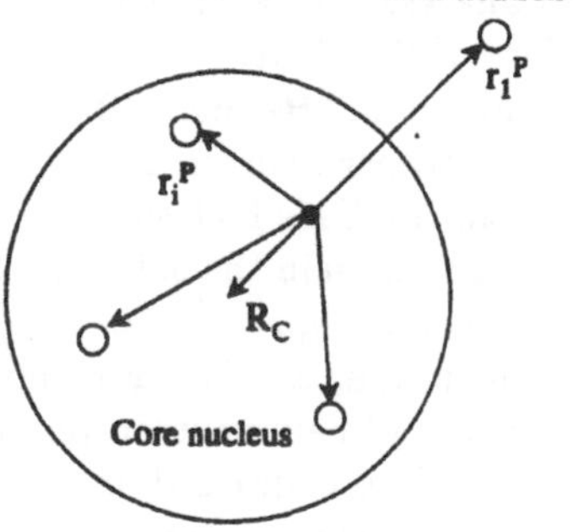

Figure 4.1: Coordinate system for a projectile composed of a halo neutron and a core, with the c.m. at the origin.

Part II it will be pointed out that a macroscopic cluster model can be derived from a microscopic model under certain conditions and limitations. The function $\phi_0(\boldsymbol{r}_1^{\rm P}-\boldsymbol{R}_{\rm C})$ should obey the constraints imposed in general on the intercluster relative wave function in a macroscopic cluster model.

In Part II all nuclei will be treated fully microscopically, and, from that point of view, this model looks naive, but embedded in a reaction theory it must be as simple as this.

The wave functions of the excited states of the halo nucleus will also be assumed to have the product form (4.1). The Hamiltonian of the halo nucleus is a sum of the core intrinsic Hamiltonian and the Hamiltonian of the relative motion between the core and the halo neutron. Thus the halo nucleus is characterized by two quantum numbers: those specifying the relative motion and the core intrinsic state. The halo nuclei usually have few particle-stable excited states, since the particle-emission threshold is very low. In the following, it will be assumed that ϕ_0 is the only bound state of the neutron–core relative motion. All excited states $\phi_{\boldsymbol{k}}$ are then unbound and are specified by the asymptotic momentum $\boldsymbol{k}$. The relative-motion states satisfy the closure relation

$$|\phi_0\rangle\langle\phi_0| + \int d\boldsymbol{k}|\phi_{\boldsymbol{k}}\rangle\langle\phi_{\boldsymbol{k}}| = 1. \tag{4.2}$$

The states of the core, both the particle-bound and continuum states, are to be denoted by $\Phi_\gamma^{\rm C}$. The state space of the projectile is then a direct product of those of the neutron–core relative motion and the core intrinsic motion:

$$\{\Phi_\alpha^{\rm P}\} = \{|\phi_0\rangle, |\phi_{\boldsymbol{k}}\rangle\} \times \{\Phi_\gamma^{\rm C}\}, \tag{4.3}$$

where $\alpha = (\boldsymbol{k}, \gamma)$ specifies the quantum state of the projectile nucleus, and comprises two quantum numbers, 0 or $\boldsymbol{k}$, for the neutron–core relative motion and the quantum number γ for the intrinsic states of the core. The state $\Phi_0^{\rm P} = \phi_0\Phi_0^{\rm C}$ is the only bound state of the nucleus.

4.2 Glauber theory for halo nuclei

4.2.1 Cross section formulae

The measurements of inclusive cross sections at high incident energies provide important information on the structure of unstable nuclei. The interaction cross section is the sum of the cross sections of all processes that change the composition of the projectile nucleus, i.e., the number of protons and/or neutrons. The halo structure was first discovered in the systematic measurements of the interaction cross sections of reactions induced by He and Li isotopes. Since most reaction events change the composition of the projectile, the interaction cross section is hardly smaller than the total reaction cross section. For a range of nuclei it is given, to a good accuracy, by a simple geometrical formula (3.92), $\sigma_{\rm int} \simeq \sigma_{\rm reac} = \pi(R_{\rm P}+R_{\rm T})^2$. The dependence of the radius on the mass number gives a systematic dependence on the nuclear partners. Large enhancement has, however, been observed with respect to this systematics for nuclei near the drip line. This has led to the idea of halo structure.

The neutron-removal cross sections have also been found to be enhanced systematically for the same nuclei. It is obvious that these are the channels that are responsible for the increase of the interaction cross section.

The subject of this section is to derive formulae for some inclusive cross sections of halo nuclei in the Glauber theory [67]. There is nothing special in the scattering amplitude. The final states are specified by the quantum numbers α of the projectile and β of the target. The projectile state α comprises the quantum state of the halo and of the core as $\alpha = (\{0\,\text{or}\,\boldsymbol{k}\}, \gamma)$ [cf. Eq. (4.3)]. The integrated cross section $\sigma_{\alpha\beta}$ for a specific final state (α, β) is given by Eqs. (3.50) and (3.51).

The total reaction cross section, which is the sum of the cross sections $\sigma_{\alpha\beta}$ over all final states α and β except for the elastic scattering $(\alpha\beta) = (00)$, is given in Eqs. (3.52) and (3.53) (which have been shown to hold whether or not there is absorption):

$$\sigma_{\rm reac} = \int d\boldsymbol{b} \left[1 - |\langle \Phi_0^{\rm P} \Phi_0^{\rm T} | \prod_{i \in {\rm P}} \prod_{j \in {\rm T}} (1 - \Gamma_{ij}) | \Phi_0^{\rm P} \Phi_0^{\rm T} \rangle|^2 \right], \tag{4.4}$$

where the profile function is

$$1 - \Gamma_{ij} = e^{i\chi(\boldsymbol{b} + \boldsymbol{s}_i^{\rm P} - \boldsymbol{s}_j^{\rm T})}. \tag{4.5}$$

The wave functions $\Phi_0^{\rm P}$ and $\Phi_0^{\rm T}$ are the respective g.s. Next we consider the interaction cross section, which is conveniently measured in transmission type experiments, i.e., in experiments in which the sole aim is to identify the impinging projectile and the transmitted nucleus. Since the projectile has just one bound state, any reactions leading to the excitation of the projectile contribute to the interaction cross section irrespective of the target states. Therefore, the interaction cross section is the sum of the cross sections $\sigma_{\alpha\beta}$ over final states in which the projectile state α is different from the g.s., that is, $\sigma_{\rm int} = \sum_{\alpha \neq 0, \beta} \sigma_{\alpha\beta}$.

The sum over the final target states β can be performed by using the closure relation $\sum_\beta |\Phi_\beta^{\rm T}\rangle\langle\Phi_\beta^{\rm T}|=1$. The summation results in

$$\sigma_{\rm int} = \int d\boldsymbol{b} \left[1-\langle\Phi_0^{\rm P}\Phi_0^{\rm T}|\prod_{i\in {\rm P}}\prod_{j\in {\rm T}}(1-\Gamma_{ij}^*)|\Phi_0^{\rm P}\rangle\langle\Phi_0^{\rm P}|\prod_{i\in {\rm P}}\prod_{j\in {\rm T}}(1-\Gamma_{ij})|\Phi_0^{\rm P}\Phi_0^{\rm T}\rangle\right]. \tag{4.6}$$

The first bra and the last ket in Eq. (4.6) contain both the projectile and target wave functions, while the intermediate ket-bra includes only the projectile wave function. With the integrations written explicitly, the matrix element in Eq. (4.6) looks like

$$\begin{aligned}
&\langle\Phi_0^{\rm P}\Phi_0^{\rm T}|\prod_{i\in {\rm P}}\prod_{j\in {\rm T}}(1-\Gamma_{ij}^*)|\Phi_0^{\rm P}\rangle\langle\Phi_0^{\rm P}|\prod_{i\in {\rm P}}\prod_{j\in {\rm T}}(1-\Gamma_{ij})|\Phi_0^{\rm P}\Phi_0^{\rm T}\rangle\\
&= \int d\boldsymbol{r}_1^{\rm T}\cdots\int d\boldsymbol{r}_{A_{\rm T}}^{\rm T}\,|\Phi_0^{\rm T}(\boldsymbol{r}_1^{\rm T},\ldots,\boldsymbol{r}_{A_{\rm T}}^{\rm T})|^2\\
&\times \int d\boldsymbol{r}_1^{\rm P}\cdots\int d\boldsymbol{r}_{A_{\rm P}}^{\rm P}\,|\Phi_0^{\rm P}(\boldsymbol{r}_1^{\rm P},\ldots,\boldsymbol{r}_{A_{\rm P}}^{\rm P})|^2\prod_{i\in {\rm P}}\prod_{j\in {\rm T}}[1-\Gamma^*(\boldsymbol{b}+\boldsymbol{s}_i^{\rm P}-\boldsymbol{s}_j^{\rm T})]\\
&\times \int d\boldsymbol{r}_1^{\rm P\prime}\cdots\int d\boldsymbol{r}_{A_{\rm P}}^{\rm P\,\prime}\,|\Phi_0^{\rm P}(\boldsymbol{r}_1^{\rm P\prime},\ldots,\boldsymbol{r}_{A_{\rm P}}^{\rm P\,\prime})|^2\prod_{i\in {\rm P}}\prod_{j\in {\rm T}}[1-\Gamma(\boldsymbol{b}+\boldsymbol{s}_i^{\rm P\prime}-\boldsymbol{s}_j^{\rm T})].
\end{aligned} \tag{4.7}$$

As was discussed below Eq. (3.54), the above two cross sections satisfy $\sigma_{\rm int} \le \sigma_{\rm reac}$ because, by definition, $\sigma_{\rm reac}$ includes more processes. The difference is assessed to be small in high-energy collisions. This is true especially for a projectile with halo structure. The difference between them comes from processes in which the halo nucleus remains in the g.s. while the target nucleus is excited, i.e., $(\alpha=0,\ \beta\neq 0)$. Such transitions must be very weak since the halo nucleus is much more easily torn apart than the target. At any rate, the Glauber theory depicts a high-energy reaction as a sequence of nucleon–nucleon collision processes. Since any nucleon–nucleon collision must affect both the projectile and target simultaneously, it favours simultaneous excitation of both. On these grounds, it is justifiable to assume that, for a halo nucleus, the difference between the interaction cross section and the reaction cross section is very small, indeed [67].

Next let us consider the neutron-removal cross section. What is observed experimentally is a fragment of the projectile, deprived of some nucleons, but the target nucleus is not observed. It is plausible that, if a neutron is removed from a single-neutron-halo nucleus, then it is the halo neutron, and, indeed, the core is detected mostly as a fragment. It is also reasonable to assume that the core nucleus remains in its g.s., thus the projectile final state is $\alpha=(\boldsymbol{k},0)$. Since the target final state β is not identified, summation must be performed over all of them. Under these assumptions the neutron-removal cross section is given by

$$\sigma_{-\mathrm{n}} = \sum_{\beta} \int d\boldsymbol{k}\, \sigma_{\alpha=(\boldsymbol{k}0),\beta}. \tag{4.8}$$

Substituting Eqs. (3.50) and (3.51) into Eq. (4.8) yields

$$\sigma_{-\mathrm{n}} = \sum_{\beta} \int d\boldsymbol{k} \left| \langle \phi_{\boldsymbol{k}} \Phi_0^{\mathrm{C}} \Phi_\beta^{\mathrm{T}} | \prod_{i\in \mathrm{P}} \prod_{j \in \mathrm{T}} (1-\Gamma_{ij}) | \phi_0 \Phi_0^{\mathrm{C}} \Phi_0^{\mathrm{T}} \rangle \right|^2. \tag{4.9}$$

The integration over $\boldsymbol{k}$ and the summation over β can be performed with the help of the closure relations (4.2) and $\sum_\beta |\Phi_\beta^{\mathrm{T}}\rangle\langle\Phi_\beta^{\mathrm{T}}| = 1$:

$$\begin{aligned}\sum_{\beta} \int d\boldsymbol{k} |\phi_{\boldsymbol{k}} \Phi_0^{\mathrm{C}} \Phi_\beta^{\mathrm{T}}\rangle\langle \phi_{\boldsymbol{k}} \Phi_0^{\mathrm{C}} \Phi_\beta^{\mathrm{T}}| &= \int d\boldsymbol{k} |\phi_{\boldsymbol{k}} \Phi_0^{\mathrm{C}}\rangle\langle\phi_{\boldsymbol{k}} \Phi_0^{\mathrm{C}}| \\ &= (1 - |\phi_0\rangle\langle\phi_0|)\, |\Phi_0^{\mathrm{C}}\rangle\langle\Phi_0^{\mathrm{C}}| = |\Phi_0^{\mathrm{C}}\rangle\langle\Phi_0^{\mathrm{C}}| - |\Phi_0^{\mathrm{P}}\rangle\langle\Phi_0^{\mathrm{P}}|. \end{aligned} \tag{4.10}$$

With this, Eq. (4.9) reduces to

$$\begin{aligned}\sigma_{-\mathrm{n}} = \int d\boldsymbol{b} \Bigg[&\langle \Phi_0^{\mathrm{P}} \Phi_0^{\mathrm{T}} | \prod_{i\in \mathrm{P}} \prod_{j \in \mathrm{T}} (1-\Gamma_{ij}^*) | \Phi_0^{\mathrm{C}} \rangle \langle \Phi_0^{\mathrm{C}} | \prod_{i\in \mathrm{P}} \prod_{j \in \mathrm{T}} (1-\Gamma_{ij}) | \Phi_0^{\mathrm{P}} \Phi_0^{\mathrm{T}} \rangle \\ &- \langle \Phi_0^{\mathrm{P}} \Phi_0^{\mathrm{T}} | \prod_{i\in \mathrm{P}} \prod_{j \in \mathrm{T}} (1-\Gamma_{ij}^*) | \Phi_0^{\mathrm{P}} \rangle \langle \Phi_0^{\mathrm{P}} | \prod_{i\in \mathrm{P}} \prod_{j \in \mathrm{T}} (1-\Gamma_{ij}) | \Phi_0^{\mathrm{P}} \Phi_0^{\mathrm{T}} \rangle \Bigg]. \end{aligned} \tag{4.11}$$

The second term in the square brackets is exactly the same as appeared in the interaction cross section (4.7). In the first term the integration over the halo-neutron coordinate $\boldsymbol{r}_1^{\mathrm{P}}$ gives unity if the unitarity condition, $|1-\Gamma|^2 = 1$, holds:

$$\int d\boldsymbol{r}_1 |\phi_0(\boldsymbol{r}_1)|^2 \prod_{j\in T} |1-\Gamma_{1j}|^2 = 1. \tag{4.12}$$

This simplifies the first term as

$$\begin{aligned} &\langle \Phi_0^{\mathrm{P}} \Phi_0^{\mathrm{T}} | \prod_{i\in \mathrm{P}} \prod_{j \in \mathrm{T}} (1-\Gamma_{ij}^*) | \Phi_0^{\mathrm{C}} \rangle \langle \Phi_0^{\mathrm{C}} | \prod_{i\in \mathrm{P}} \prod_{j \in \mathrm{T}} (1-\Gamma_{ij}) | \Phi_0^{\mathrm{P}} \Phi_0^{\mathrm{T}} \rangle \\ &= \langle \Phi_0^{\mathrm{C}} \Phi_0^{\mathrm{T}} | \prod_{i\in \mathrm{C}} \prod_{j \in \mathrm{T}} (1-\Gamma_{ij}^*) | \Phi_0^{\mathrm{C}} \rangle \langle \Phi_0^{\mathrm{C}} | \prod_{i\in \mathrm{C}} \prod_{j \in \mathrm{T}} (1-\Gamma_{ij}) | \Phi_0^{\mathrm{C}} \Phi_0^{\mathrm{T}} \rangle \\ &= \int d\boldsymbol{r}_1^{\mathrm{T}} \cdots \int d\boldsymbol{r}_{A_{\mathrm{T}}}^{\mathrm{T}} |\Phi_0^{\mathrm{T}}(\boldsymbol{r}_1^{\mathrm{T}}, \ldots, \boldsymbol{r}_{A_{\mathrm{T}}}^{\mathrm{T}})|^2 \end{aligned}$$

$$\times \int d\boldsymbol{r}_2^{\mathrm{P}} \cdots \int d\boldsymbol{r}_{A_{\mathrm{P}}}^{\mathrm{P}} |\Phi_0^{\mathrm{C}}(\boldsymbol{r}_2^{\mathrm{P}}, \ldots, \boldsymbol{r}_{A_{\mathrm{P}}}^{\mathrm{P}})|^2 \prod_{i \in \mathrm{C}} \prod_{j \in \mathrm{T}} [1 - \Gamma^*(\boldsymbol{b} + \boldsymbol{s}_i^{\mathrm{P}} - \boldsymbol{s}_j^{\mathrm{T}})]$$
$$\times \int d\boldsymbol{r}_2^{\mathrm{P}'} \cdots \int d\boldsymbol{r}_{A_{\mathrm{P}}}^{\mathrm{P}}{}' |\Phi_0^{\mathrm{C}}(\boldsymbol{r}_2^{\mathrm{P}'}, \ldots, \boldsymbol{r}_{A_{\mathrm{P}}}^{\mathrm{P}}{}')|^2 \prod_{i \in \mathrm{C}} \prod_{j \in \mathrm{T}} [1 - \Gamma(\boldsymbol{b} + \boldsymbol{s}_i^{\mathrm{P}'} - \boldsymbol{s}_j^{\mathrm{T}})]. \tag{4.13}$$

This should be substituted into Eq. (4.11). For simplicity, we just insert the less explicit second line of Eq. (4.13), to obtain

$$\sigma_{-\mathrm{n}} = \int d\boldsymbol{b} \Bigg[\langle \Phi_0^{\mathrm{C}} \Phi_0^{\mathrm{T}} | \prod_{i \in \mathrm{C}} \prod_{j \in \mathrm{T}} (1 - \Gamma_{ij}^*) | \Phi_0^{\mathrm{C}} \rangle \langle \Phi_0^{\mathrm{C}} | \prod_{i \in \mathrm{C}} \prod_{j \in \mathrm{T}} (1 - \Gamma_{ij}) | \Phi_0^{\mathrm{C}} \Phi_0^{\mathrm{T}} \rangle$$
$$- \langle \Phi_0^{\mathrm{P}} \Phi_0^{\mathrm{T}} | \prod_{i \in \mathrm{P}} \prod_{j \in \mathrm{T}} (1 - \Gamma_{ij}^*) | \Phi_0^{\mathrm{P}} \rangle \langle \Phi_0^{\mathrm{P}} | \prod_{i \in \mathrm{P}} \prod_{j \in \mathrm{T}} (1 - \Gamma_{ij}) | \Phi_0^{\mathrm{P}} \Phi_0^{\mathrm{T}} \rangle \Bigg]. \tag{4.14}$$

To sum up, in deriving (4.14) from Eq. (4.11) the unitarity condition has been assumed. Since the unitarity condition holds below the hadron (mostly meson) production threshold, the two formulae coincide there. When hadron production takes place, Eq. (4.11) gives the cross section of neutron removal that is not accompanied by hadron production. On the other hand, since over the hadron production threshold the unitarity holds for the ensemble of all channels including the hadron-production channels, Eq. (4.14) includes processes in which hadrons are produced.

4.2.2 Relations between cross sections

In this section general relations will be derived from the Glauber theory between some interaction cross sections and neutron-removal cross sections and they will be compared with experiment.

Consider the collisions of a halo nucleus and its core nucleus with some target. It will be shown immediately that the difference between the two interaction cross sections is equal to the neutron-removal cross section from the halo nucleus [67].

Applying Eq. (4.6) to the collision of the core nucleus C with the target T gives

$$\sigma_{\mathrm{int}}(\mathrm{C}) = \int d\boldsymbol{b} \Bigg[1 - \langle \Phi_0^{\mathrm{C}} \Phi_0^{\mathrm{T}} | \prod_{i \in \mathrm{C}} \prod_{j \in \mathrm{T}} (1 - \Gamma_{ij}^*) | \Phi_0^{\mathrm{C}} \rangle$$
$$\times \langle \Phi_0^{\mathrm{C}} | \prod_{i \in \mathrm{C}} \prod_{j \in \mathrm{T}} (1 - \Gamma_{ij}) | \Phi_0^{\mathrm{C}} \Phi_0^{\mathrm{T}} \rangle \Bigg]. \tag{4.15}$$

By subtracting Eq. (4.15) from Eq. (4.6) applied to P = C+n, the right-hand side becomes equal to that of Eq. (4.14). Equating the left-hand sides gives the simple relationship

$$\sigma_{\rm int}(\mathrm{C}+\mathrm{n}) - \sigma_{\rm int}(\mathrm{C}) = \sigma_{-\rm n}(\mathrm{C}+\mathrm{n}). \tag{4.16}$$

It is worthwhile to recall on what conditions the above relation holds. The Glauber theory was used as a reaction theory. The main ingredients of this are the eikonal approximation and the neglect of the excitation energies, which should be a reasonable approximation at high incident energies. Furthermore, as a structure model, the wave function of the halo nucleus is assumed to be the product of the core–neutron relative wave function and of the internal core wave function, and the latter is taken to be the same as that of the isolated core nucleus.

This relation can be generalized to the reactions of Borromean two-neutron-halo nuclei (i.e., P = C+2n nuclei for which C+n is neutron-unstable):

$$\sigma_{\rm int}(\mathrm{C}+2\mathrm{n}) - \sigma_{\rm int}(\mathrm{C}) = \sigma_{-2\rm n}(\mathrm{C}+2\mathrm{n}). \tag{4.17}$$

In deriving this, one has to assume that the g.s. of the halo nucleus is given by the product of the halo-neutron wave function and the core wave function, which is the same as that in the g.s. of nucleus C. The two-neutron-halo wave function, however, need not factorize into s.p. wave functions.

Independent measurements are available for some cross sections related by Eq. (4.17), so the accuracy of the relation can be checked. Since the most fragile assumption is the product ansatz, the comparison will be relevant to that.

The measured cross sections are summarized in Table 4.1 for some reactions on ^{12}C target at the incident energy of 800 A MeV. The

Table 4.1: Interaction cross sections, one-neutron-removal, two-neutron-removal and four-neutron-removal cross sections (in mb) of the collision of some projectiles with the ^{12}C target at 800 A MeV [2,8,10,25,68].

P	$\sigma_{\rm int}$(P)	$\sigma_{\rm int}$(P−2n)	$\sigma_{-2\rm n}$(P)	$\sigma_{-4\rm n}$(P)
^{11}Li	1060 ± 10	796 ± 6	220 ± 10	
^{6}He	722 ± 5	503 ± 5	189 ± 14	
^{8}He	817 ± 6	722 ± 5	202 ± 17	95 ± 9
P	$\sigma_{\rm int}$(P)	$\sigma_{\rm int}$(P−n)	$\sigma_{-\rm n}$(P)	
^{11}Be	942 ± 8	813 ± 10	169 ± 4	

difference between the interaction cross sections of ^{11}Li and ^{9}Li is 260±16 mb, and that is close to the two-neutron-removal cross section of 220 ± 10 mb. A similar statement can be made for the ^{6}He case. The relationship, however, does not hold for the two-neutron-removal cross section of ^{8}He as considered to be composed of a ^{6}He core and a two-neutron halo. This failure indicates that it is not correct to assume that there is a ^{6}He inert core in ^{8}He. Indeed, the two-neutron separation energy of ^{8}He is 2.13 MeV, which is even larger than the two-neutron separation energy of ^{6}He which is 0.97 MeV. The ^{6}He nucleus has a two-neutron-halo structure in the g.s.. The addition of two more neutrons may have a strong influence on the halo neutrons in ^{6}He. Therefore, it is plausible that the assumption of the inert ^{6}He core does not hold. This problem will be further discussed in Sect. 13.2.

It is, however, possible to derive another relation involving the reactions of ^{8}He, without forcing ^{6}He to be a core by extending the Glauber theory to up to four-neutron removal [69]:

$$\sigma_{\rm int}(^{8}{\rm He}+{\rm T}) = \sigma_{-2{\rm n}}(^{8}{\rm He}+{\rm T})+\sigma_{-4{\rm n}}(^{8}{\rm He}+{\rm T})+\sigma_{\rm int}(^{4}{\rm He}+{\rm T}). \quad (4.18)$$

The notations are now self-explanatory. This relation is derived based on the fact that ^{7}He and ^{5}He are not stable against neutron emission, which implies that neutron removal from either ^{6}He or ^{8}He is actually a two-neutron-removal process. The only assumption about the structure of ^{8}He is that its g.s. is a product of the four-neutron wave function and of the ^{4}He core wave function. No relationship is assumed between the two-neutron wave function in ^{6}He and the four-neutron wave function in ^{8}He.

The ^{4}He+^{12}C interaction cross section is $\sigma_{\rm int}(^{4}{\rm He}+^{12}{\rm C}) = 503 \pm 5$ mb [9]. With this and with the data in Table 4.1, the right-hand side of Eq. (4.18) adds up to 800 ± 31 mb, in good agreement with the interaction cross section measured for ^{8}He+^{12}C, 817±6 mb.

The difference between the ^{11}Be+^{12}C and ^{10}Be+^{12}C interaction cross sections is 129±18 mb (see Table 4.1). This is smaller than the measured cross section of one-neutron removal from ^{11}Be, 169±4 mb. The discrepancy may indicate that the structure of the g.s. of ^{11}Be is not simply a ^{10}Be g.s. core coupled to a neutron. Some theoretical investigations also indicate that it is actually a mixture of configurations. A 2^+ excited core coupled to a $0d_{5/2}$ neutron seems to be admixed appreciably to the 0^+-core–$1s_{1/2}$-neutron configuration. This seems to be corroborated by analyses of medium-energy reactions [70–72]. The structure of the ^{11}Be g.s. will be discussed in more detail in Sect. 13.6.

These comparisons have confirmed some reasonable structure assumptions in an extremely direct way and, indirectly, indicate the soundness of the underlying reaction theory.

4.2.3 Optical-limit approximation for halo nuclei

The evaluation of the matrix elements involved in the Glauber theory requires the computation of complex multidimensional integrals. Though their exact evaluation may not be prohibitively complicated for light projectiles and targets, it is by no means useless to look for reasonable approximations. A possible approximation could employ the effective profile function $\Gamma_{\mathrm{NT}}(\boldsymbol{b})$ as defined in Eq. (3.96). The subject of this section is the adaptation of the OLA, which was introduced in Chap. 3, to halo nuclei. The approximation consists in an expansion with respect to the many-body correlation effects. The lowest-order expansion works fairly well for high-energy reactions of stable nuclei, but, as we shall see, that is not so for the halo nuclei [67,73,74].

As an example, let us consider a matrix element which appears in the amplitude of the elastic scattering of a halo nucleus:

$$\begin{aligned} e^{i\chi_{\mathrm{el}}(\boldsymbol{b})} &= \langle \phi_0 \Phi_0^{\mathrm{C}} \Phi_0^{\mathrm{T}} | \prod_{i\in \mathrm{P}} \prod_{j \in \mathrm{T}} (1-\Gamma_{ij}) | \phi_0 \Phi_0^{\mathrm{C}} \Phi_0^{\mathrm{T}} \rangle \\ &= \int d\boldsymbol{r}_1^{\mathrm{P}} \left| \phi_0(\boldsymbol{r}_1^{\mathrm{P}}) \right|^2 \langle \Phi_0^{\mathrm{C}} \Phi_0^{\mathrm{T}} | \prod_{i\in \mathrm{P}} \prod_{j \in \mathrm{T}} (1-\Gamma_{ij}) | \Phi_0^{\mathrm{C}} \Phi_0^{\mathrm{T}} \rangle, \end{aligned} \tag{4.19}$$

where the matrix element in the last expression implies integration over the core and target coordinates.

Equation (4.19) contains a folding with the density of the halo orbit. Since this is very different from that of other nucleons, the validity of any approximate method here should be thoroughly examined. A full-fledged OLA will be tested against a calculation in which it is introduced just for the core and target coordinates, for which it is likely to be valid.

Suppressing Φ_0^{C} and Φ_0^{T}, one can write

$$\begin{aligned} &\langle \Phi_0^{\mathrm{C}} \Phi_0^{\mathrm{T}} | \prod_{i\in \mathrm{P}} \prod_{j \in \mathrm{T}} (1-\Gamma_{ij}) | \Phi_0^{\mathrm{C}} \Phi_0^{\mathrm{T}} \rangle \equiv \left\langle \prod_{i\in \mathrm{P}} \prod_{j \in \mathrm{T}} (1-\Gamma_{ij}) \right\rangle \\ &= \left\langle \prod_{j\in \mathrm{T}} (1-\Gamma_{1j}) \prod_{i\in \mathrm{C}} \prod_{j \in \mathrm{T}} (1-\Gamma_{ij}) \right\rangle \approx \prod_{j\in \mathrm{T}} (1-\langle\Gamma_{1j}\rangle) \prod_{i\in \mathrm{C}} \prod_{j \in T} (1-\langle\Gamma_{ij}\rangle) \\ &= \prod_{j\in \mathrm{T}} \left[1 - \int d\boldsymbol{r}^{\mathrm{T}} n_j^{\mathrm{T}}(\boldsymbol{r}^{\mathrm{T}}) \Gamma(\boldsymbol{b}+\boldsymbol{s}_1^{\mathrm{P}}-\boldsymbol{s}^{\mathrm{T}}) \right] \\ &\quad \times \prod_{i\in \mathrm{C}} \prod_{j \in \mathrm{T}} \left[1 - \int d\boldsymbol{r}^{\mathrm{P}} \int d\boldsymbol{r}^{\mathrm{T}} n_i^{\mathrm{P}}(\boldsymbol{r}^{\mathrm{P}}) n_j^{\mathrm{T}}(\boldsymbol{r}^{\mathrm{T}}) \Gamma(\boldsymbol{b}+\boldsymbol{s}^{\mathrm{P}}-\boldsymbol{s}^{\mathrm{T}}) \right] \\ &\approx \exp\left[- \int d\boldsymbol{r}^{\mathrm{T}} \rho_{\mathrm{T}}(\boldsymbol{r}^{\mathrm{T}}) \Gamma(\boldsymbol{b}+\boldsymbol{s}_1^{\mathrm{P}}-\boldsymbol{s}^{\mathrm{T}}) \right] \end{aligned}$$

$$\times \exp\left[-\int d\boldsymbol{r}^{\mathrm{P}} \int d\boldsymbol{r}^{\mathrm{T}} \rho_{\mathrm{C}}(\boldsymbol{r}^{\mathrm{P}})\rho_{\mathrm{T}}(\boldsymbol{r}^{\mathrm{T}})\Gamma(\boldsymbol{b}+\boldsymbol{s}^{\mathrm{P}}-\boldsymbol{s}^{\mathrm{T}})\right]$$

$$= \mathrm{e}^{i\chi_{n\mathrm{T}}(\boldsymbol{b}+\boldsymbol{s}_1^{\mathrm{P}})+i\chi_{\mathrm{CT}}(\boldsymbol{b})}. \tag{4.20}$$

The approximation in the second line is the same as in Eq. (3.83). The phase-shift functions

$$\chi_{n\mathrm{T}}(\boldsymbol{b}) = i\langle\Phi_0^{\mathrm{T}}|\sum_{j\in\mathrm{T}}\Gamma(\boldsymbol{b}-\boldsymbol{s}_j^{\mathrm{T}})|\Phi_0^{\mathrm{T}}\rangle, \tag{4.21}$$

$$\chi_{\mathrm{CT}}(\boldsymbol{b}) = i\langle\Phi_0^{\mathrm{C}}\Phi_0^{\mathrm{T}}|\sum_{i\in\mathrm{C}}\sum_{j\in\mathrm{T}}\Gamma(\boldsymbol{b}+\boldsymbol{s}_i^{\mathrm{C}}-\boldsymbol{s}_j^{\mathrm{T}})|\Phi_0^{\mathrm{C}}\Phi_0^{\mathrm{T}}\rangle, \tag{4.22}$$

represent, respectively, the neutron–target and core–target elastic scattering in the OLA. These definitions are the same as those introduced in Eqs. (3.88) and (3.89). Note that in Eq. (4.22) $\boldsymbol{s}_i^{\mathrm{C}}$ is used instead of $\boldsymbol{s}_i^{\mathrm{P}}$ because, in the core–target phase-shift function $\chi_{\mathrm{CT}}(\boldsymbol{b})$, the position vector of a core nucleon should be measured from the c.m. of the core.

Substitution of Eq. (4.20) into Eq. (4.19) gives the phase-shift function in which the OLA is only applied to the integrations over the C and T coordinates:

$$\mathrm{e}^{i\chi_{\mathrm{el}}(\boldsymbol{b})} = \mathrm{e}^{i\chi_{\mathrm{CT}}(\boldsymbol{b})}\langle\phi_0|\mathrm{e}^{i\chi_{n\mathrm{T}}(\boldsymbol{b}+\boldsymbol{s}_1^{\mathrm{P}})}|\phi_0\rangle. \tag{4.23}$$

If the OLA is used for all nucleon coordinates, including the halo-neutron coordinates, the phase-shift function reduces to

$$\mathrm{e}^{i\chi_{\mathrm{el}}(\boldsymbol{b})} \approx \mathrm{e}^{i\chi_{\mathrm{CT}}(\boldsymbol{b})+i\langle\phi_0|\chi_{n\mathrm{T}}(\boldsymbol{b}+\boldsymbol{s}_1^{\mathrm{P}})|\phi_0\rangle}. \tag{4.24}$$

Thus the application of the OLA with respect to the halo-neutron coordinate consists in the following replacement:

$$\langle\phi_0|\mathrm{e}^{i\chi_{n\mathrm{T}}(\boldsymbol{b}+\boldsymbol{s}_1^{\mathrm{P}})}|\phi_0\rangle \longrightarrow \mathrm{e}^{i\langle\phi_0|\chi_{n\mathrm{T}}(\boldsymbol{b}+\boldsymbol{s}_1^{\mathrm{P}})|\phi_0\rangle}. \tag{4.25}$$

This replacement could be justifiable for a sharp density distribution $|\phi_0(\boldsymbol{r}_1^{\mathrm{P}})|^2$ as it is exact for a Dirac-δ. In the opposite limit of a smooth extensive distribution that is not the case. The halo-neutron density is spread out more than the range of $\mathrm{e}^{i\chi_{n\mathrm{T}}(\boldsymbol{b})}$, which is similar to that of the target nucleus. This situation corresponds to a case of Eq. (3.66) and Fig. 3.3 in which $R\sim a$. Therefore, the OLA does not seem to be justifiable for the halo neutron.

To expose this problem in a different way, let us make the so-called cumulant expansion for the phase-shift function of Eq. (4.23). With the notation $\langle\cdots\rangle$ for an expectation value in ϕ_0, the cumulant expansion looks like $[\chi = \chi_{nT}(\boldsymbol{b}+\boldsymbol{s}_1^{\mathrm{P}})]$

$$\begin{aligned}\langle \mathrm{e}^{i\chi}\rangle &= \mathrm{e}^{i\langle\chi\rangle}\langle \mathrm{e}^{i(\chi-\langle\chi\rangle)}\rangle \\ &= \mathrm{e}^{i\langle\chi\rangle}\left\langle 1 - i(\chi-\langle\chi\rangle) - \frac{1}{2}(\chi-\langle\chi\rangle)^2 - \frac{i}{6}(\chi-\langle\chi\rangle)^3 + \cdots \right\rangle \\ &= \mathrm{e}^{i\langle\chi\rangle}\left[1 - \frac{1}{2}\langle(\chi-\langle\chi\rangle)^2\rangle - \frac{i}{6}\langle(\chi-\langle\chi\rangle)^3\rangle + \cdots\right] \\ &= \exp\left[\langle i\chi\rangle - \frac{1}{2}\langle(\chi-\langle\chi\rangle)^2\rangle - \frac{i}{6}\langle(\chi-\langle\chi\rangle)^3\rangle + \cdots\right]. \end{aligned} \tag{4.26}$$

where, in the last equality, $\mathrm{e}^x \approx 1 - x$ for small $|x|$ has been used. The lowest-order term gives Eq. (4.25). The higher-order terms, which are neglected in the OLA, are related to the variance of the neutron–target phase-shift function with respect to the halo-neutron density. The first neglected term contains the second-order variance

$$\left\langle(\chi-\langle\chi\rangle)^2\right\rangle = \langle\phi_0|\left(\chi_{nT}\right)^2|\phi_0\rangle - \langle\phi_0|\chi_{nT}|\phi_0\rangle^2. \tag{4.27}$$

Since larger extension implies larger variance, the error of the OLA in the treatment of the extra neutron is enhanced by the halo nature of the neutron orbit. This explains why the OLA cannot cope with the spatially extended halo structure.

The phase-shift function (4.23) contains another approximation, the neglect of the difference between the projectile and core c.m., which can be overcome. As in Eq. (4.1), the correct argument of the halo wave function is $\phi_0(\boldsymbol{r}_1^{\mathrm{P}} - \boldsymbol{R}_{\mathrm{C}})$. Since the coordinates are measured from the projectile c.m., $\boldsymbol{r}_1^{\mathrm{P}} + A_{\mathrm{C}}\boldsymbol{R}_{\mathrm{C}} = 0$ must be satisfied. Therefore, the correct variable which describes the relative motion between the halo neutron and the core is

$$\boldsymbol{r}_1^{\mathrm{P}\prime} \equiv \boldsymbol{r}_1^{\mathrm{P}} - \boldsymbol{R}_{\mathrm{C}} = \frac{A_{\mathrm{P}}}{A_{\mathrm{C}}}\boldsymbol{r}_1^{\mathrm{P}}. \tag{4.28}$$

Some of the arguments of the functions Γ involved in the phase-shift function (4.19) contain the halo-neutron coordinate $\boldsymbol{b} + \boldsymbol{s}_1^{\mathrm{P}} - \boldsymbol{s}_j^{\mathrm{T}}$, but others do not ($\boldsymbol{b} + \boldsymbol{s}_i^{\mathrm{P}} - \boldsymbol{s}_j^{\mathrm{T}}$, $i = 2, \ldots, A_{\mathrm{P}}$). The former can be rewritten as

$$\boldsymbol{b} + \boldsymbol{s}_1^{\mathrm{P}} - \boldsymbol{s}_j^{\mathrm{T}} = \boldsymbol{b}_{\mathrm{C}} + (\boldsymbol{s}_1^{\mathrm{P}} - \boldsymbol{S}_{\mathrm{C}}) - \boldsymbol{s}_j^{\mathrm{T}} = \boldsymbol{b}_{\mathrm{C}} + \boldsymbol{s}_1^{\mathrm{P}\prime} - \boldsymbol{s}_j^{\mathrm{T}}, \tag{4.29}$$

where $\boldsymbol{b}_{\mathrm{C}} = \boldsymbol{b} + \boldsymbol{S}_{\mathrm{C}} = \boldsymbol{b} - A_{\mathrm{C}}^{-1}\boldsymbol{s}_1^{\mathrm{P}} = \boldsymbol{b} - A_{\mathrm{P}}^{-1}\boldsymbol{s}_1^{\mathrm{P}\prime}$ is the impact parameter of the core–target relative motion. The latter can be similarly cast into the form

$$\boldsymbol{b} + \boldsymbol{s}_i^{\mathrm{P}} - \boldsymbol{s}_j^{\mathrm{T}} = \boldsymbol{b}_{\mathrm{C}} + \boldsymbol{s}_i^{\mathrm{P}\prime} - \boldsymbol{s}_j^{\mathrm{T}} \quad (i = 2, \ldots, A_{\mathrm{P}}), \tag{4.30}$$

where $s_i^{P\prime} = s_i^P - S_C$ for $i = 2, \ldots, A_P$. Equations (4.29) and (4.30) indicate that the correct treatment requires the following replacement:

$$\begin{aligned} e^{i\chi_{el}(b)} &= e^{i\chi_{CT}(b)}\langle\phi_0|e^{i\chi_{nT}(b+s_1^P)}|\phi_0\rangle \\ &\longrightarrow \langle\phi_0|e^{i\chi_{CT}(b_C)+i\chi_{nT}(b_C+s_1^P)}|\phi_0\rangle, \end{aligned} \tag{4.31}$$

where $b_C = b - A_P^{-1}s_1^P$. Since the argument of χ_{CT} now includes the coordinate of the halo neutron, it cannot be factored out, as before, in front of the integral.

The total reaction cross section for the collision with a halo nucleus can now be calculated following Eqs. (3.52) and (3.53):

$$\sigma_{reac} = \int db \left[1 - \left|\langle\phi_0|e^{i\chi_{CT}(b-\frac{1}{A_P}s_1^P)+i\chi_{nT}(b+\frac{A_P-1}{A_P}s_1^P)}|\phi_0\rangle\right|^2\right]. \tag{4.32}$$

The second term in the integrand is the probability that the halo nucleus as well as the target nucleus remain in their g.s. after collision (with impact parameter b). If we ignore the c.m. correction discussed above, this term reduces to a product of two factors: the first is $e^{-2\mathrm{Im}\chi_{CT}(b)}$, which is the probability $P_{CT\to CT}(b)$ that both the core and the target remain in their g.s. in the C+T collision, while the second factor is $|\langle\phi_0|e^{i\chi_{nT}(b+s_1^P)}|\phi_0\rangle|^2$, which is the probability $P_{nT\to nT}(b)$ that both the neutron–core relative motion and the target nucleus remain in their g.s. as a result of the n+T collision. Thus, with the approximate treatment of the c.m., the collision is considered as a coincidence of two independent events, so that the product gives the probability that the neutron–core relative motion, the core nucleus as well as the target nucleus remain in their g.s. after the P+T collision. It is then obvious that $1 - P_{CT\to CT}(b)P_{nT\to nT}(b)$ gives the probability that at least one of them is excited.

In Sect. 4.2.2 it was shown that the neutron-removal cross section is equal to the difference between the interaction cross sections in the collisions of the halo nucleus and of the core nucleus with the same target. Since, for a halo nucleus, the interaction cross section is very close to the total reaction cross section, one may calculate the neutron-removal cross section as the difference between the two total reaction cross sections.

The total reaction cross section formula $\sigma_{reac} = \int db\,[1 - e^{-2\mathrm{Im}\chi_{el}(b)}]$ [cf. Eq. (3.61) or (3.78)] applied to a collision with the core nucleus gives $\sigma_{reac}(C+T) = \int db\,[1 - e^{-2\mathrm{Im}\chi_{CT}(b)}]$. The neutron-removal cross section is then $\sigma_{-n} = \sigma_{reac}(P+T) - \sigma_{reac}(C+T)$, i.e.,

$$\sigma_{-n} = \int db \left[e^{-2\mathrm{Im}\chi_{CT}(b)} - \left|\langle\phi_0|e^{i\chi_{CT}(b-\frac{1}{A_P}s_1^P)+i\chi_{nT}(b+\frac{A_P-1}{A_P}s_1^P)}|\phi_0\rangle\right|^2\right]$$

$$\approx \int d\boldsymbol{b}\, \mathrm{e}^{-2\mathrm{Im}\chi_{\mathrm{CT}}(\boldsymbol{b})} \left[1 - \left| \langle \phi_0 | \mathrm{e}^{i\chi_{\mathrm{nT}}(\boldsymbol{b}+\boldsymbol{s}_1^{\mathrm{P}})} | \phi_0 \rangle \right|^2 \right]. \tag{4.33}$$

The second expression is obtained if the c.m. effects are ignored. The integrand of this cross section is the probability of the neutron removal for a given impact parameter. In the latter expression it is given as a product of two probabilities: $\mathrm{e}^{-2\mathrm{Im}\chi_{\mathrm{CT}}(\boldsymbol{b})}$, the probability that both the core and target remain in their g.s., and $1-|\langle \phi_0 | \mathrm{e}^{i\chi_{\mathrm{nT}}(\boldsymbol{b}+\boldsymbol{s}_1^{\mathrm{P}})} | \phi_0 \rangle|^2$, which is the probability that the halo neutron is removed from the g.s. orbit, irrespective of whether the target is excited or not. The simultaneous occurrence of these events is indeed neutron removal without change in the core.

The phase-shift function $\chi_{\mathrm{CT}}(\boldsymbol{b} - A_{\mathrm{P}}^{-1} \boldsymbol{s}_1^{\mathrm{P}})$ represents the core–target elastic scattering, so the first line of Eq. (4.33) shows that the neutron removal is influenced by the core–target elastic scattering. This effect may be called a shakeoff mechanism due to the core–target elastic scattering. This influence disappears, as is seen in the second line, if the difference between the projectile and core c.m. is neglected.

It is possible to decompose the neutron removal into two further processes, elastic and inelastic neutron removal, depending on whether the target is excited. Such a classification will be useful in describing the momentum distribution of the core [cf. Chap. 6]. These cross sections are defined by

$$\sigma_{-\mathrm{n}}^{\mathrm{el}} = \int d\boldsymbol{k}\, \sigma_{\alpha=(\boldsymbol{k},0),\beta=0},$$
$$\sigma_{-\mathrm{n}}^{\mathrm{inel}} = \sum_{\beta \neq 0} \int d\boldsymbol{k}\, \sigma_{\alpha=(\boldsymbol{k},0),\beta}. \tag{4.34}$$

An explicit form of the elastic neutron-removal cross section is obtained similarly to the derivation of Eq. (4.14):

$$\sigma_{-\mathrm{n}}^{\mathrm{el}} = \int d\boldsymbol{b} \left[\langle \Phi_0^{\mathrm{P}} \Phi_0^{\mathrm{T}} | \prod_{i \in \mathrm{C}} \prod_{j \in \mathrm{T}} (1 - \Gamma_{ij}^*) | \Phi_0^{\mathrm{C}} \Phi_0^{\mathrm{T}} \rangle \right.$$
$$\times \langle \Phi_0^{\mathrm{C}} \Phi_0^{\mathrm{T}} | \prod_{i \in \mathrm{C}} \prod_{j \in \mathrm{T}} (1 - \Gamma_{ij}) | \Phi_0^{\mathrm{P}} \Phi_0^{\mathrm{T}} \rangle$$
$$\left. - \left| \langle \Phi_0^{\mathrm{P}} \Phi_0^{\mathrm{T}} | \prod_{i \in \mathrm{P}} \prod_{j \in \mathrm{T}} (1 - \Gamma_{ij}) | \Phi_0^{\mathrm{P}} \Phi_0^{\mathrm{T}} \rangle \right|^2 \right]. \tag{4.35}$$

Applying the OLA to the integrations, except to that over the halo-neutron coordinate, we obtain

$$\sigma_{-\mathrm{n}}^{\mathrm{el}} = \int d\boldsymbol{b} \Bigg[\mathrm{e}^{-2\mathrm{Im}\chi_{\mathrm{CT}}(\boldsymbol{b})} \langle \phi_0 | \mathrm{e}^{-2\mathrm{Im}\chi_{\mathrm{nT}}(\boldsymbol{b}+\boldsymbol{s}_1^{\mathrm{P}})} | \phi_0 \rangle$$

$$- \left| \langle \phi_0 | \mathrm{e}^{i\chi_{\mathrm{CT}}(\boldsymbol{b}-\frac{1}{A_{\mathrm{P}}}\boldsymbol{s}_1^{\mathrm{P}}) + i\chi_{\mathrm{nT}}(\boldsymbol{b}+\frac{A_{\mathrm{P}}-1}{A_{\mathrm{P}}}\boldsymbol{s}_1^{\mathrm{P}})} | \phi_0 \rangle \right|^2 \Bigg] \qquad (4.36)$$

$$\approx \int d\boldsymbol{b}\, \mathrm{e}^{-2\mathrm{Im}\chi_{\mathrm{CT}}(\boldsymbol{b})} \Bigg[\langle \phi_0 | \mathrm{e}^{-2\mathrm{Im}\chi_{\mathrm{nT}}(\boldsymbol{b}+\boldsymbol{s}_1^{\mathrm{P}})} | \phi_0 \rangle$$

$$- \left| \langle \phi_0 | \mathrm{e}^{i\chi_{\mathrm{nT}}(\boldsymbol{b}+\boldsymbol{s}_1^{\mathrm{P}})} | \phi_0 \rangle \right|^2 \Bigg], \qquad (4.37)$$

where, in the last expression, the difference between the c.m. of the projectile and of the core is ignored. The inelastic neutron-removal cross section is obtained by subtracting $\sigma_{-\mathrm{n}}^{\mathrm{el}}$ from $\sigma_{-\mathrm{n}}$ given by Eq. (4.33):

$$\sigma_{-\mathrm{n}}^{\mathrm{inel}} = \int d\boldsymbol{b}\, \mathrm{e}^{-2\mathrm{Im}\chi_{\mathrm{CT}}(\boldsymbol{b})} \left[1 - \langle \phi_0 | \mathrm{e}^{-2\mathrm{Im}\chi_{\mathrm{nT}}(\boldsymbol{b}+\boldsymbol{s}_1^{\mathrm{P}})} | \phi_0 \rangle \right]. \qquad (4.38)$$

These results can be understood intuitively in the approximation in which the difference between the two c.m. is ignored. Then the probability of the neutron removal in the (C+n)+T three-body scattering can be factorized into those of two independent two-body processes, and one of the processes can be decomposed into two non-overlapping events. The two independent two-body processes are the elastic scattering of subsystem C+T and the excitation of the neutron orbit in the n+T collision. The probability of the simultaneous occurrence of these processes is $P_{\mathrm{CT}\to\mathrm{CT}}(\boldsymbol{b})P_{-\mathrm{n}}(\boldsymbol{b})$, where $P_{\mathrm{CT}\to\mathrm{CT}}(\boldsymbol{b}) = \mathrm{e}^{-2\mathrm{Im}\chi_{\mathrm{CT}}(\boldsymbol{b})}$ [see the paragraph below Eq. (4.32)] and it is $P_{-\mathrm{n}}(\boldsymbol{b})$ that can be decomposed into two terms representing two different outcomes of the n+T collision. As discussed in Eq. (3.40), the state $\mathrm{e}^{i\chi_{\mathrm{nT}}(\boldsymbol{b}+\boldsymbol{s}_1^{\mathrm{P}})}|\phi_0\rangle$ is the wave function of the halo neutron after collision with the target nucleus in the projectile-rest frame. The norm of this wave function is the probability that the target remains unexcited. The probability that the neutron is removed by inelastic neutron–target collision is thus

$$P_{-\mathrm{n}}^{\mathrm{inel}}(\boldsymbol{b}) = 1 - \mathrm{Norm}\left[\mathrm{e}^{i\chi_{\mathrm{nT}}(\boldsymbol{b}+\boldsymbol{s}_1^{\mathrm{P}})} | \phi_0 \rangle \right]$$

$$= 1 - \langle \phi_0 | \mathrm{e}^{-2\mathrm{Im}\chi_{\mathrm{nT}}(\boldsymbol{b}+\boldsymbol{s}_1^{\mathrm{P}})} | \phi_0 \rangle. \qquad (4.39)$$

In the state $\mathrm{e}^{i\chi_{\mathrm{nT}}(\boldsymbol{b}+\boldsymbol{s}_1^{\mathrm{P}})}|\phi_0\rangle$ not only the halo-neutron g.s. ϕ_0 but also the continuum states are contained [cf. Eq. (4.2)]. The continuum components correspond to the process in which the neutron is removed by being scattered elastically by the target nucleus. The norm of the continuum components is obtained by eliminating the component of ϕ_0 from $e^{i\chi_{nT}(b+s_1^P)}|\phi_0\rangle$. The probability of neutron removal by elastic scattering from the target is thus

$$P_{-n}^{el}(b) = \text{Norm}\left[(1-|\phi_0\rangle\langle\phi_0|)\, e^{i\chi_{nT}(b+s_1^P)}|\phi_0\rangle\right]$$
$$= \langle\phi_0|e^{-2\text{Im}\chi_{nT}(b+s_1^P)}|\phi_0\rangle - \left|\langle\phi_0|e^{i\chi_{nT}(b+s_1^P)}|\phi_0\rangle\right|^2. \quad (4.40)$$

The total neutron-removal probability is given by the sum of the above two probabilities:

$$P_{-n}(b) = P_{-n}^{el}(b) + P_{-n}^{inel}(b)$$
$$= 1 - \left|\langle\phi_0|e^{i\chi_{nT}(b+s_1^P)}|\phi_0\rangle\right|^2. \quad (4.41)$$

The quantity $|\langle\phi_0|e^{i\chi_{nT}(b+s_1^P)}|\phi_0\rangle|^2$ is the probability that the halo neutron remains in the g.s. orbit after collision with the target. The neutron-removal probability is given by subtracting this from unity, in accord with the discussion below Eq. (4.33).

The formulae for the total reaction cross section, Eq. (4.32), and for the neutron-removal cross section, Eq. (4.33), can be simply extended to two-neutron-halo nuclei. Equation (4.32) is changed as

$$\sigma_{reac} = \int db \left[1 - \left|\langle\phi_0|e^{i\chi_{CT}(b_C)+i\chi_{nT}(b_C+s_1^P)+i\chi_{nT}(b_C+s_2^P)}|\phi_0\rangle\right|^2\right], \quad (4.42)$$

where $b_C = b - A_P^{-1}(s_1^P + s_2^P)$ and ϕ_0 is the g.s. wave function of the two-neutron-halo nucleus. If the halo nucleus is Borromean, the two-neutron-removal cross section is obtained from Eq. (4.33) by changing it in the same way as Eq. (4.32) has been changed into Eq. (4.42).

4.3 Applications

As illustrative applications of the Glauber theory to the reactions of halo nuclei, the reactions of the nucleus ^{11}Be at the incident energy of 800 A MeV will be discussed.

In the g.s. of ^{11}Be the neutron separation energy is as small as $|\varepsilon_n| = 0.503$ MeV. In the standard shell model (SM) the odd neutron of ^{11}Be should sit on the $0p_{1/2}$ level. However, the spin–parity of the g.s. is $J^\pi = \frac{1}{2}^+$. This is usually expressed by saying that the order of the 1s and 0p orbits is inverted in this nucleus, and the last neutron occupies the $1s_{1/2}$ orbit (cf. Sect. 13.6). The core is ^{10}Be. It is assumed to be in its $J^\pi = 0^+$ state, which is indeed the g.s. of the isolated core nucleus. There is a bound excited state with negative parity at the very low excitation energy of 0.316 MeV. According to the usual SM, in this state the last neutron occupies the $0p_{1/2}$ orbit.

In the reaction theory of halo nuclei the halo nucleus was assumed to have just one bound state. Although it is possible to allow for the second bound state, it substantially complicates the formulation. We assume, instead, that the processes in which the excited state comes into play are negligible, and use the framework elaborated above without modification. This approximation may be very reasonable.

The ingredients of the reaction probabilities and cross sections are the core–target and the neutron–target phase-shift functions and the halo-neutron wave function $\phi_0(\boldsymbol{r})$. The phase-shift functions are calculated according to Eqs. (3.88) and (3.89), with the zero-range form, Eq. (3.60), used for the profile function, and a Gaussian shape for the nuclear density. The wave function of the halo neutron is constructed by solving a s.p. Schrödinger equation for the relative motion between the neutron and the core. A Woods–Saxon shape, $V(r) = -V_0\{1+\exp[(r-R)/a]\}^{-1}$, is assumed for the neutron–core potential, with $R = 2.67$ fm, $a = 0.6$ fm, and V_0 chosen so as to put the 1s state at -0.503 MeV. The wave function that is shown in Fig. 1.2 is this.

The total reaction cross section for the halo nucleus is calculated according to Eq. (4.32). For comparison, alternative calculations, in which the OLA treatment includes the halo orbit, will also be shown. In these full OLA calculations the total cross section has been calculated according to Eq. (3.91). The results are compared with measured data in Fig. 4.2. Cross sections for three target nuclei, ^{12}C, ^{64}Cu and ^{208}Pb, are shown. Their density distributions are described by Gaussians [51]. For ^{208}Pb the cross section calculated in the most accurate way, by Eq. (4.32), underestimates the measured cross section. This is because the neutron-removal cross section induced by the Coulomb field is not included. The calculations with the full OLA [Eq. (3.91)], on the contrary, overestimate the cross section for the ^{64}Cu as well as for the ^{208}Pb target.

The one-neutron-removal cross section can be calculated as a difference between the total reaction cross sections for ^{11}Be and for ^{10}Be [cf. Eq. (4.33)]. The measured total reaction cross section for ^{10}Be is well reproduced with Eq. (3.91). Figure 4.2 also shows the one-neutron-removal cross sections. The calculations with Eq. (4.33) underestimate the neutron-removal cross sections. The discrepancy becomes considerable for the target ^{208}Pb, and that, again, comes from the Coulomb breakup mechanism, which will be explained in Chap. 7.

The one-neutron-removal cross section in the full OLA is also shown in the figure. The cross section is much larger than the cross section without applying OLA to the halo neutron. The difference amounts to more than a factor of three for ^{208}Pb. This large difference clearly demonstrates that the OLA for the halo neutron causes a large error. This error contributes both to the neutron-removal cross section and the total reaction cross section.

There have been several attempts to extract the density distribution of the halo nuclei employing Eq. (3.91). However, such analyses are hampered by the inadequacy of the OLA for halo nuclei. As an accurate relation between the cross section and the density distribution, Eqs. (4.32) and (4.33) should be used. That the OLA overestimates the cross section was pointed out in Refs. [67,73] and was stressed in Ref. [74].

The neutron-removal cross section is underestimated by the theoretical calculations even for the target ^{12}C. For such a light target, the Coulomb breakup mechanism is believed to be unimportant. In fact, as is discussed in Sect. 4.2.2, the measured cross sections do not satisfy the relationship (4.16) very well, while the theoretical cross sections do satisfy it exactly. Thus the Glauber theory apparently fails to reproduce

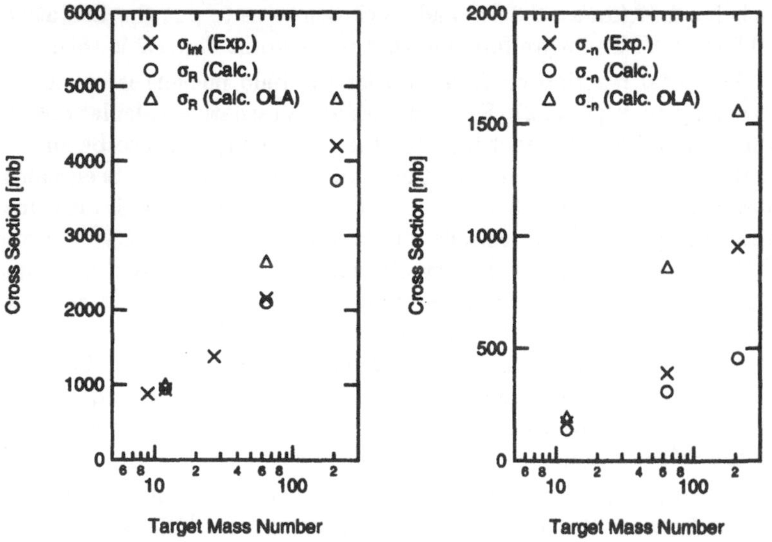

Figure 4.2: Interaction cross section (left) and one-neutron-removal cross section (right) in the collision of ^{11}Be with ^{12}C, ^{64}Cu and ^{208}Pb at the incident energy of 800 A MeV. Glauber calculations are compared with measured cross sections [2,8,10,75]. Cross sections in which the OLA is applied to the halo neutron as well are also shown. The Coulomb breakup is not taken into account in the calculations.

the interaction and neutron-removal cross sections simultaneously. The reason for this discrepancy should be traced back to the assumption used for the ^{11}Be wave function in the calculation.

Finally, we give an anatomy of the reaction mechanism by presenting some reaction probabilities $P(\boldsymbol{b})$ for the ^{11}Be+^{12}C collision and for the corresponding C+T collision, ^{10}Be+^{12}C. The formulae used are summarized in Table 4.2, and the results are plotted in Fig. 4.3. The total reaction probability for ^{11}Be plotted as a thick solid curve in the left panel is close to unity for the small impact parameter region. It has a long tail for large impact parameters owing to the neutron-removal reaction. The reaction probability for ^{10}Be, shown by a dashed curve, is also close to unity at small impact parameter values, but goes close to zero around $b \sim 7$ fm, wherefrom the nuclear densities of the core and target cease to overlap. The difference between the solid and dashed curves is the neutron-removal probability, which is shown as a thin solid curve. In the right panel the neutron-removal probability is decomposed into two terms, the elastic and inelastic breakup probabilities. The inelastic breakup probability is much larger than the elastic one. This indicates that the halo neutron is dominantly removed by a hard collision of the halo neutron with the target nucleus. Around impact parameters equalling the sum of the core and target radii and beyond, the neutron-removal probability is overwhelming. It should be noted that the cross section is given by $2\pi \int db\, bP(b)$, which amplifies

Table 4.2: Formulae for reaction probabilities used in the results to be shown in Fig. 4.3. The last two formulae slightly differ from the integrands of Eqs. (4.36) and (4.38) because the integration variables $\boldsymbol{b}$ used there are not the same as the impact parameter $\boldsymbol{b}$ used here.

$P_{\rm reac}(\boldsymbol{b})$	$1 - \left\| \langle\phi_0\|{\rm e}^{i\chi_{\rm CT}(\boldsymbol{b}-\frac{1}{A_{\rm P}}\boldsymbol{s}_1^{\rm P})+i\chi_{\rm nT}(\boldsymbol{b}+\frac{A_{\rm P}-1}{A_{\rm P}}\boldsymbol{s}_1^{\rm P})}\|\phi_0\rangle \right\|^2$
$P_{\rm reac}^{\rm core}(\boldsymbol{b})$	$1 - {\rm e}^{-2{\rm Im}\chi_{\rm CT}(\boldsymbol{b})}$
$P_{\rm reac}^{\rm OLA}(\boldsymbol{b})$	$1 - \left\| {\rm e}^{\langle\phi_0\|i\chi_{\rm CT}(\boldsymbol{b}-\frac{1}{A_{\rm P}}\boldsymbol{s}_1^{\rm P})+i\chi_{\rm nT}(\boldsymbol{b}+\frac{A_{\rm P}-1}{A_{\rm P}}\boldsymbol{s}_1^{\rm P})\|\phi_0\rangle} \right\|^2$
$P_{\rm -n}^{\rm el}(\boldsymbol{b})$	$\langle\phi_0\|{\rm e}^{-2{\rm Im}\chi_{\rm CT}(\boldsymbol{b}-\frac{1}{A_{\rm P}}\boldsymbol{s}_1^{\rm P})-2{\rm Im}\chi_{\rm nT}(\boldsymbol{b}+\frac{A_{\rm P}-1}{A_{\rm P}}\boldsymbol{s}_1^{\rm P})}\|\phi_0\rangle$
	$- \left\| \langle\phi_0\|{\rm e}^{i\chi_{\rm CT}(\boldsymbol{b}-\frac{1}{A_{\rm P}}\boldsymbol{s}_1^{\rm P})+i\chi_{\rm nT}(\boldsymbol{b}+\frac{A_{\rm P}-1}{A_{\rm P}}\boldsymbol{s}_1^{\rm P})}\|\phi_0\rangle \right\|^2$
$P_{\rm -n}^{\rm inel}(\boldsymbol{b})$	$\langle\phi_0\|{\rm e}^{-2{\rm Im}\chi_{\rm CT}(\boldsymbol{b}-\frac{1}{A_{\rm P}}\boldsymbol{s}_1^{\rm P})}\|\phi_0\rangle$
	$-\langle\phi_0\|{\rm e}^{-2{\rm Im}\chi_{\rm CT}(\boldsymbol{b}-\frac{1}{A_{\rm P}}\boldsymbol{s}_1^{\rm P})-2{\rm Im}\chi_{\rm nT}(\boldsymbol{b}+\frac{A_{\rm P}-1}{A_{\rm P}}\boldsymbol{s}_1^{\rm P})}\|\phi_0\rangle$

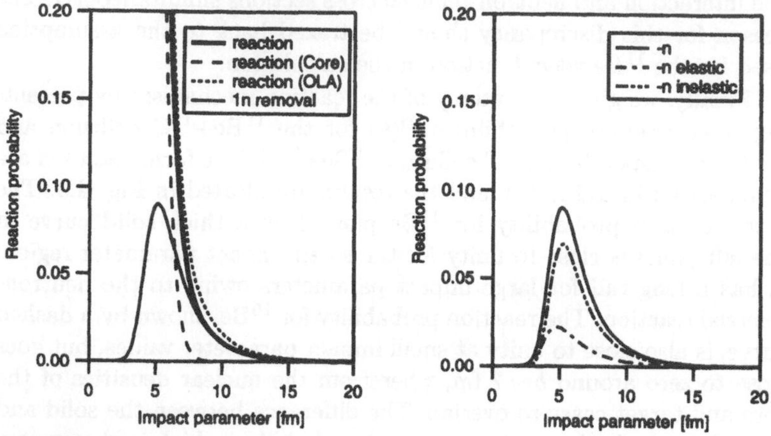

Figure 4.3: Reaction probabilities of the ^{11}Be+^{12}C collision as functions of the impact parameter at the incident energy of 800 A MeV. In the left panel the total reaction probabilities for the ^{11}Be and ^{10}Be projectiles, the total reaction probability for ^{11}Be in the OLA and the single-neutron-removal probability for ^{11}Be are shown. In the right panel the single-neutron-removal probability is decomposed into elastic and inelastic components. The formulae for the probabilities are summarized in Table 4.2.

the contribution coming from large impact parameters. The OLA reaction probability is shown as well. It is too large in the large impact parameter region.

Although the description of high-energy reactions with two-neutron-halo nuclei contains inessential complications from the point of view of reaction theory, it is appropriate to mention them here because of their importance. Since the first clear evidence was found for the anomalously large radius of ^{11}Li, much effort has been devoted to the theoretical description of the reactions with ^{11}Li. The Glauber theory presented in this chapter was developed and applied first to this reaction. In the early stage simplified models were adopted for ^{11}Li, such as the independent-particle model [73, 76, 77] and a dineutron cluster model [78]. At the next stage the cross sections were calculated with more realistic three-body wave functions [79–81].

Analyses were then performed with the aim of extracting information on the halo structure from the reaction cross sections. A discussion of the relation between the nuclear radii and the cross sections is given in Ref. [74]. The influence of correlations in the two-neutron wave function on the reaction observables is discussed in Ref. [82].

Chapter 5

Medium-energy reactions of halo nuclei

5.1 Elastic scattering of stable nuclei

5.1.1 Optical model

This chapter is devoted to the reactions of halo nuclei at medium incident energy, a few tens of MeV per nucleon. In this energy region it is possible to make so-called 'exclusive measurements' of the reactions, i.e., measurements in which the final states of the projectile and target can be identified. The simplest single channel is the elastic scattering. There are other kinds of exclusive measurements currently under progress, including inelastic and transfer reactions to explore the properties of the g.s. and excited states of exotic nuclei.

As the incident energy gets lower, the Glauber theory loses its foundation. Various many-body effects which are neglected in the Glauber theory become significant. They include the Pauli exclusion effects between the projectile and target nucleons and other many-body effects. Below the incident energy of a few hundred MeV per nucleon, one needs to develop another theory as a substitute for the Glauber theory.

In this energy region the many-body effects are more or less uniform for all systems, which makes it possible to describe them as effects of the nuclear matter or nuclear medium. These medium effects can then be formulated in a form, in which the nuclear collisions again look like a number of collisions between individual nucleons, but with a modified 'effective' nucleon–nucleon interaction. The theory of effective nuclear interactions in the nuclear medium is known as the G-matrix theory [83].

Owing to its reliance on 'bare' nucleon–nucleon collisions, the Glauber theory fails in describing medium-energy reactions, but the eikonal approximation is still applicable. In Sect. 2.3 it was pointed out that the fully quantum-mechanical description of the neutron–nucleus scattering described by an optical potential can be reasonably well approximated by the eikonal approximation for incident energies higher than a few tens of MeV, comparable with the depth of the optical potential.

For the elastic scattering of stable nuclei, the description in terms of the phenomenological complex optical potential is successful in a wide region of incident energy and for various projectile nuclei, including nucleons. The optical model also serves as a starting point for the theories of other direct reactions.

The optical model reproduces the measured differential cross sections well. The optical potential is usually constructed phenomenologically and has a smooth dependence on the incident energy. The imaginary part of the potential describes the loss of flux from the elastic channel. The physical origin of this is the excitation of either the projectile or the target or transfer of nucleons between them. For nucleon scattering, the real part of the potential has the shape and the magnitude similar to the SM potential for bound nucleons. The imaginary part of the potential is negative to represent the absorption of flux. At low incident energies, the absolute value of the imaginary part is smaller than that of the real part.

In Sect. 2.3 it was shown that the strength of the imaginary part of the nucleon–nucleus optical potential is related to the mean free path λ of the nucleon in the nucleus. In a uniform complex potential $V+iW$ the wave function of a nucleon of energy E may be a plane wave e^{ikz} with a complex wave number satisfying $\hbar^2k^2/2m = E-(V+iW)$. The probability density decreases along z due to the absorptive potential W as $|e^{ikz}|^2 = e^{-2z\mathrm{Im}k}$. The mean free path is proportional to the inverse of the imaginary part of the wave number: $\lambda = 1/(2\mathrm{Im}k)$. For a typical potential depth of $V = -50$ MeV and $W = -10$ MeV, the mean free path for the incident nucleon of $E = 50$ MeV is estimated to be about 5 fm, which is approximately the same as the diameter of a light nucleus. Therefore, the bombarding nucleon can penetrate fairly deep in the target without exciting it.

According to a statistical estimate for collisions between the incident and target nucleons, the mean free path is expressed as $\lambda = 1/(\rho\sigma)$, where ρ is the density of the target nucleus and σ is the cross section of their collision. The normal nuclear density, $\rho = 0.17$ fm^{-3}, and the mean free path of 5 fm yield a cross section of $\sigma \simeq 1.2$ fm^2 = 12 mb, which is to be compared with the nucleon–nucleon cross section at low energies, which amounts to a few hundred mb as seen in Fig. 3.4. Thus the in-medium cross section can only be an 'effective' cross section. The long mean free path, or, equivalently, the small effective cross section of the nucleon–nucleon collision, is understood as a consequence of the prohibition of collisions due to the Pauli exclusion principle. The nucleon–nucleon collision in the nuclear medium is allowed only when the final states of the nucleons are not blocked by other nucleons. The G-matrix effective interaction has been devised just to describe this situation [83], and the G-matrix theory succeeds in explaining the behaviour of the nucleon optical potential discussed above [84, 85].

5.1.2 Folding model

The real part of the optical potential for nucleon–nucleus scattering can be derived very well from the effective interaction by the so-called folding model. It is obtained as the folding of the effective nuclear force V with the density distribution of the target $\rho(\boldsymbol{r})$:

$$V^{\mathrm{F}}(\boldsymbol{r}) = \int d\boldsymbol{r}'\rho(\boldsymbol{r}')V(\boldsymbol{r}-\boldsymbol{r}'). \tag{5.1}$$

The imaginary part can also be deduced from the G-matrix effective force similarly, but less successfully.

The optical potential is also useful to describe the nucleus–nucleus elastic scattering [86]. The double-folding model is known to work quite well for a systematic description of the real part of the optical potential for various pairs of the colliding nuclei [87]:

$$V^{\mathrm{DF}}(\boldsymbol{R}) = \int d\boldsymbol{r}^{\mathrm{P}} \int d\boldsymbol{r}^{\mathrm{T}} \rho_{\mathrm{P}}(\boldsymbol{r}^{\mathrm{P}})\rho_{\mathrm{T}}(\boldsymbol{r}^{\mathrm{T}})V(\boldsymbol{R}+\boldsymbol{r}^{\mathrm{P}}-\boldsymbol{r}^{\mathrm{T}}), \tag{5.2}$$

where V is the effective nuclear force, similar to that used to describe nucleon scattering. [A derivation of this formula in a microscopic framework is given in Eq. (11.67).] The nucleus–nucleus optical potential is, however, much more absorptive in the surface region than the nucleon potential. Where the surfaces of the colliding nuclei start to overlap, the probability that at least one of the nuclei gets excited becomes very close to unity. Therefore, only the peripheral collisions in which the impact parameter b is close to the sum of the radii contribute to the elastic scattering and direct reactions. Accordingly, the physically relevant part of the optical potential is the region around and outside the sum of the radii.

We show that the folding potential can be derived from the eikonal approximation by employing a cumulant expansion [38] for the phase-shift function of the elastic scattering. The scattering amplitude of the nucleon–nucleus elastic scattering was given in Eq. (3.61) as

$$\mathrm{e}^{i\chi_{\mathrm{el}}(\boldsymbol{b})} = \langle \Phi_0 | \mathrm{e}^{i\sum_{i=1}^{A}\chi(\boldsymbol{b}-\boldsymbol{s}_i)} | \Phi_0 \rangle. \tag{5.3}$$

The phase-shift function of the nucleon–nucleon scattering, $\chi(\boldsymbol{b})$, is now constructed from the effective nuclear potential V instead of the bare force [cf. Eq. 3.26]:

$$\chi(\boldsymbol{b}) = -\frac{1}{\hbar v}\int_{-\infty}^{\infty} dz\, V(\boldsymbol{b}+z\hat{\boldsymbol{z}}). \tag{5.4}$$

The (near) singularity of the bare nuclear force is eliminated in the effective nuclear force, and we may evaluate the phase-shift function $\chi_{\rm el}(\boldsymbol{b})$ in a perturbative expansion. The notations $\chi = \sum_{i=1}^{A}\chi(\boldsymbol{b}-\boldsymbol{s}_i)$ and $\langle\Phi_0|\sum_{i=1}^{A}\chi(\boldsymbol{b}-\boldsymbol{s}_i)|\Phi_0\rangle = \langle\chi\rangle$ will help to make the formulae concise. With these the Taylor expansion of the phase-shift function in Eq. (5.3) is written as

$$\begin{aligned}
\chi_{\rm el}(\boldsymbol{b}) &= -i\ln\langle e^{i\chi}\rangle \\
&= -i\ln\left[e^{i\langle\chi\rangle}\left\langle e^{i(\chi-\langle\chi\rangle)}\right\rangle\right] \\
&= \langle\chi\rangle - i\ln\left\langle 1+i(\chi-\langle\chi\rangle)-\frac{1}{2}(\chi-\langle\chi\rangle)^2-\frac{i}{6}(\chi-\langle\chi\rangle)^3+\cdots\right\rangle \\
&= \langle\chi\rangle + \frac{i}{2}\langle(\chi-\langle\chi\rangle)^2\rangle - \frac{1}{6}\langle(\chi-\langle\chi\rangle)^3\rangle \\
&\quad -\frac{i}{24}\langle(\chi-\langle\chi\rangle)^4\rangle + \frac{i}{8}\langle(\chi-\langle\chi\rangle)^2\rangle^2+\cdots
\end{aligned} \tag{5.5}$$

This is the same cumulant expansion as in the evaluation of the phase-shift function for halo nuclei in Eq. (4.26), where the expansion did not work well as an approximation to the integration over the coordinates of the halo neutron.

Taking the lowest order of the expansion gives

$$\chi_{\rm el}^{(0)}(\boldsymbol{b}) = \langle\Phi_0|\sum_{i=1}^{A}\chi(\boldsymbol{b}-\boldsymbol{s}_i)|\Phi_0\rangle = \int d\boldsymbol{r}\,\rho(\boldsymbol{r})\chi(\boldsymbol{b}-\boldsymbol{s}). \tag{5.6}$$

This is not the same as that derived previously in Eq. (3.67); that was obtained by neglecting the correlation effects in the target wave function. To obtain formula (5.6) from Eq. (3.67), we need to make a Taylor expansion for the nucleon–nucleon profile function $\Gamma(\boldsymbol{b}) = 1-e^{i\chi(\boldsymbol{b})} \simeq -i\chi(\boldsymbol{b})$.

For a real effective nuclear force V, the function $\chi(\boldsymbol{b})$ is real, and so the above phase-shift function $\chi_{\rm el}(\boldsymbol{b})$ is also real. In this scheme the excitations of the target nucleus arise from successive nucleon–nucleon collisions, and thus the imaginary part comes from higher-order terms of Eq. (5.5).

We still have to show the relation between the folding-model potential (5.1) and the cumulant expansion for the phase-shift function,

Eq. (5.6). The phase-shift function belonging to the folding potential of Eq. (5.1) is

$$\chi^{\mathrm{F}}(\boldsymbol{b}) = -\frac{1}{\hbar v}\int_{-\infty}^{\infty} dz\, V^{\mathrm{F}}(\boldsymbol{b}+z\hat{\boldsymbol{z}}) = \int d\boldsymbol{r}\,\rho(\boldsymbol{r})\chi(\boldsymbol{b}-\boldsymbol{s}), \tag{5.7}$$

where Eqs. (5.1) and (5.4) were used. The right-hand sides of Eqs. (5.7) and (5.6) coincide, which indicates that the description of the elastic scattering in the folding model is equivalent to the leading term in the cumulant expansion of the phase-shift function.

A similar relation holds for the double-folding potential (5.2) and the OLA phase-shift function for the nucleus–nucleus case, Eq. (3.80). The leading term of the perturbative expansion for the nucleon–nucleon profile function is

$$\begin{aligned}
\chi_{\mathrm{el}}(\boldsymbol{b}) &= -i\ln\langle\Phi_0^{\mathrm{P}}\Phi_0^{\mathrm{T}}|e^{i\sum_{i\in\mathrm{P}}\sum_{j\in\mathrm{T}}\chi(\boldsymbol{b}+\boldsymbol{s}_i^{\mathrm{P}}-\boldsymbol{s}_j^{\mathrm{T}})}|\Phi_0^{\mathrm{P}}\Phi_0^{\mathrm{T}}\rangle \\
&\approx \langle\Phi_0^{\mathrm{P}}\Phi_0^{\mathrm{T}}|\sum_{i\in\mathrm{P}}\sum_{j\in\mathrm{T}}\chi(\boldsymbol{b}+\boldsymbol{s}_i^{\mathrm{P}}-\boldsymbol{s}_j^{\mathrm{T}})|\Phi_0^{\mathrm{P}}\Phi_0^{\mathrm{T}}\rangle \\
&= \int d\boldsymbol{r}^{\mathrm{P}}\int d\boldsymbol{r}^{\mathrm{T}}\rho_{\mathrm{P}}(\boldsymbol{r}^{\mathrm{P}})\rho_{\mathrm{T}}(\boldsymbol{r}^{\mathrm{T}})\chi(\boldsymbol{b}+\boldsymbol{s}^{\mathrm{P}}-\boldsymbol{s}^{\mathrm{T}}).
\end{aligned} \tag{5.8}$$

For a real nucleon–nucleon phase-shift function, this phase-shift function is also real, and no excitations are included up to this order. The phase-shift function belonging to the double-folding potential (5.2) is the same as the right-hand side of Eq. (5.8).

The folding potential derived from an appropriate real effective nuclear force, combined with a phenomenological imaginary term has been successful in describing the elastic scattering of various pairs of nuclei [87]. A substantial reduction of the real strength has, however, been found necessary with respect to the folding potential for weakly bound projectiles like ^{6,7}Li and ^{9}Be [87, 88]. This effect was explained by coupled-channel calculations in which projectile excitations were explicitly taken into account. It was revealed that the excitation of a weakly bound projectile into continuum states has an important influence on the elastic scattering, and in the optical model this has to be mocked up by a substantial reduction of the depth of the real potential term. Since a halo nucleus is an extreme case of a weakly bound projectile, a similar effect is expected for halo-nucleus projectiles. Therefore, the applicability of the folding model to halo nuclei is questionable. This is a confirmation of the finding that the first order of the cumulant expansion, which is equivalent to the folding model, is not a good approximation for halo nuclei.

As an example for the performance of the folding model, let us compare some optical-model and coupled-channel calculations which

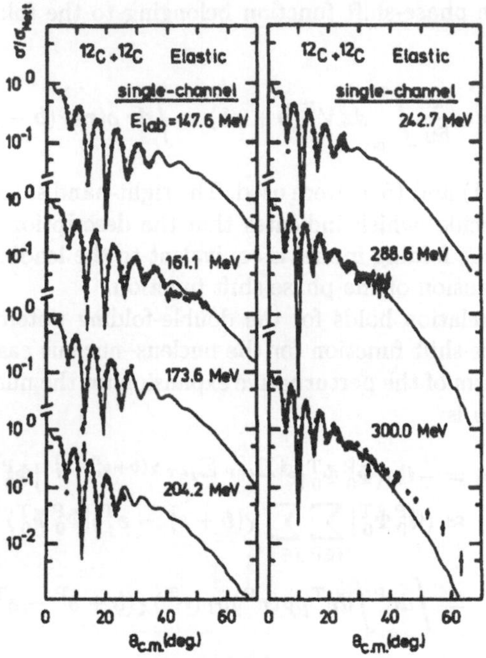

Figure 5.1: $^{12}C+^{12}C$ elastic scattering cross sections (divided by $\sigma_{\rm Mott}$) at $E_{\rm lab}$=148–300 MeV. The curves are obtained with a double-folding real plus a phenomenological imaginary potential. The real part of the potential is multiplied by a scaling factor. Taken from Ref. [88].

use the double-folding potential for the scattering of two projectiles, ^{12}C and ^{6}Li.

The $^{12}C+^{12}C$ elastic scattering, divided by the corresponding Mott cross section[1] $\sigma_{\rm Mott}$, is shown in Fig. 5.1 for several incident energies [88]. The double-folding real term of the potential was multipled by an energy-dependent constant to achieve good fit to the data, while the imaginary part was chosen to have the Woods–Saxon shape and its parameters were also adjusted to the data. The scaling factor of the real term was found to be 0.75–0.9, except for 300 MeV case, where it was 0.66.

[1]The Mott scattering is the Coulomb scattering of two identical particles. The differential cross section of the Mott scattering in the c.m. frame is symmetrical with respect to $\theta=\frac{\pi}{2}$ and oscillatory due to the interference of Rutherford amplitudes [Eq. (2.89)] $F_C(\theta)$ and $F_C(\pi-\theta)$ due to the indistinguishability of the particles flying apart with angles θ and $\pi-\theta$: $d\sigma_{\rm Mott}/d\Omega=|F_C(\theta)+\epsilon F_C(\pi-\theta)|^2$, where $\epsilon=1$ for bosons and -1 for fermions [26].

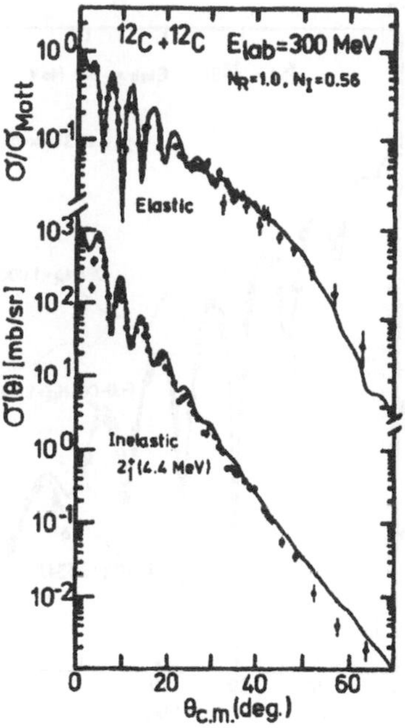

Figure 5.2: $^{12}C+^{12}C$ elastic and inelastic scattering cross section at $E_{\rm lab}=300$ MeV. (The elastic cross section is divided by $\sigma_{\rm Mott}$.) The curves are results of a coupled-channel model, with coupling both to discrete excited states and to the three-α continuum. Taken from Ref. [88].

The results of the coupled-channel analysis [88] are shown in Fig. 5.2 for the elastic and inelastic differential cross sections. Both the discrete excitations of the carbon nucleus and its breakup into the three-α continuum were included. The diagonal potentials as well as the nondiagonal (coupling) potentials[2] were constructed with the folding model, without any renormalization of their strengths. From the success of the coupled-channel calculation it can be inferred that the role of the reduction of the real part of the potential in the single-channel calculation is to simulate the effects of the coupling to excited states.

Figure 5.3 shows the analysis of the ^{6}Li+^{28}Si elastic scattering at 99 MeV incident energy. The threshold for the breakup of ^{6}Li into an

[2] A coupling potential can also be written as a double folding of the nucleon–nucleon force with some s.p. densities. When the transition leads, e.g., from $\Phi_0^{\rm P}$ to $\Phi_i^{\rm P}$, the ordinary projectile density $\rho(\boldsymbol{r})=\langle\Phi_0^{\rm P}|\sum_i\delta(\boldsymbol{r}_i-\boldsymbol{r})|\Phi_0^{\rm P}\rangle$ [cf. Eq. (3.63) or (9.127) later] is to be replaced by the transition density $\rho^{\rm Tr}(\boldsymbol{r})=\langle\Phi_i^{\rm P}|\sum_i\delta(\boldsymbol{r}_i-\boldsymbol{r})|\Phi_0^{\rm P}\rangle$.

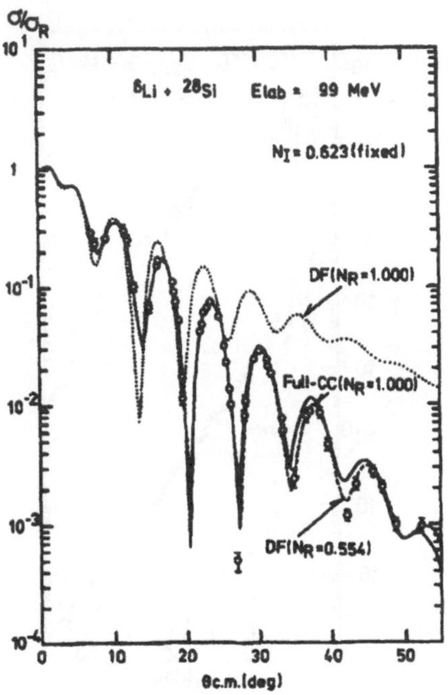

Figure 5.3: ^{6}Li+^{28}Si elastic scattering cross section at $E_{\rm lab} = 99$ MeV. The solid curve is obtained with a coupled-channel calculation, with coupling to the deuteron–α continuum of ^{6}Li. The dashed and dotted curves are double-folding calculations with and without a normalization factor N_R. A double-folding potential is used for the imaginary part, scaled by the factor N_I shown. Taken from Ref. [88].

α-particle and a deuteron is low, $|\varepsilon_{\alpha d}| = 1.475$ MeV. The calculation with the double-folding potential without renormalization gives poor fit at large scattering angles. The potential with a substantial reduction factor of 0.55, however, gives reasonable description. The coupled-channel calculation for this system includes the projectile breakup into the α–deuteron continuum channels. The double-folding potential for the coupling as well as for the diagonal potentials reproduces the elastic scattering cross section very well, without renormalization. Thus, one can say that the need for the strong reduction of the potential in the single-channel calculation is indeed an effect of the coupling to the continuum.

In the framework of the eikonal approximation with phase-shift functions (5.3) and (5.8) it is possible to construct optical potentials that incorporate effects that go beyond the folding model. To come to that point, one has to see how to invert the relation between the potential and the phase-shift function; i.e., how to construct an optical potential from the phase-shift function. The resulting potential will be a unique spherically symmetric potential [38].

Consider an axially symmetric case, in which the phase-shift function depends only on the length of the impact parameter b. Let us look for a spherically symmetric potential $V_{\rm opt}(r)$ which gives the phase-shift function $\chi_{\rm el}(b)$,

$$\chi_{\rm el}(b) = -\frac{1}{\hbar v}\int_{-\infty}^{+\infty} dz V_{\rm opt}(\sqrt{b^2+z^2}). \tag{5.9}$$

To come to the inversion, let us first denote the impact parameter vector by $\boldsymbol{b}=(x,y)$ and integrate both sides of Eq. (5.9) with respect to y over the whole axis:

$$\int_{-\infty}^{+\infty} dy\, \chi_{\rm el}(\sqrt{x^2+y^2}) = -\frac{1}{\hbar v}\int_{-\infty}^{+\infty} dy \int_{-\infty}^{+\infty} dz\, V_{\rm opt}(\sqrt{x^2+y^2+z^2}). \tag{5.10}$$

The two-dimensional integration on the right-hand side can be transformed into the polar coordinate system $\{y,z\}=\{\rho\cos\theta,\ \rho\sin\theta\}$ with the angular integral giving a trivial factor 2π. By introducing a variable r through $r^2=x^2+\rho^2$, one has

$$\begin{aligned}\int_{-\infty}^{+\infty} dy\, \chi_{\rm el}(\sqrt{x^2+y^2}) &= -\frac{2\pi}{\hbar v}\int_0^{\infty} d\rho\, \rho V_{\rm opt}(\sqrt{x^2+\rho^2}) \\ &= -\frac{2\pi}{\hbar v}\int_x^{\infty} dr\, r V_{\rm opt}(r).\end{aligned} \tag{5.11}$$

Differentiating both sides with respect to x and introducing r instead of x results in

$$\begin{aligned}V_{\rm opt}(r) &= \frac{\hbar v}{2\pi}\frac{1}{r}\frac{d}{dr}\int_{-\infty}^{+\infty} dy\, \chi_{\rm el}(\sqrt{r^2+y^2}) \\ &= \frac{\hbar v}{\pi}\frac{1}{r}\frac{d}{dr}\int_r^{\infty} db \frac{b}{\sqrt{b^2-r^2}}\chi_{\rm el}(b).\end{aligned} \tag{5.12}$$

This procedure will be used later as a test of the influence of the coupling to the continuum on the optical potential.

5.2 Few-body direct reaction model

5.2.1 The model

To proceed, we focus our attention again on the reactions involving a single-neutron-halo nucleus [79, 80]. In the medium-energy range we must treat the three-body dynamics of the halo neutron, the core nucleus, and the target nucleus. Our prime concern is to show up the influence of the halo neutron on the dynamics of elastic scattering.

The three constituents are treated as point particles. The neutron–target as well as the core–target scattering can be well described in the phenomenological optical model, so these potentials will be used as effective potentials among the constituents.

The coordinate and momentum operators for the relative motion between the projectile and the target are denoted by $\boldsymbol{R}$ and $\boldsymbol{P}$, respectively, and the relative coordinate between the halo neutron and the core nucleus is denoted by $\boldsymbol{r}$. Then the three-body Schrödinger equation looks like

$$\left[\frac{\boldsymbol{P}^2}{2M_{\mathrm{PT}}} + H_{\mathrm{nC}} + V_{\mathrm{CT}}\left(\boldsymbol{R} - \frac{\boldsymbol{r}}{A_{\mathrm{P}}}\right) + V_{\mathrm{nT}}\left(\boldsymbol{R} + \frac{A_{\mathrm{P}}-1}{A_{\mathrm{P}}}\boldsymbol{r}\right)\right]\Psi(\boldsymbol{R},\boldsymbol{r}) = E\Psi(\boldsymbol{R},\boldsymbol{r}), \quad (5.13)$$

where M_{PT} is the projectile–target reduced mass and H_{nC} is the intrinsic Hamiltonian for the neutron–core relative motion. As before, the neutron is bound with a very small binding energy $|\varepsilon_0|$: $H_{\mathrm{nC}}\phi_0(\boldsymbol{r}) = \varepsilon_0\phi_0(\boldsymbol{r})$ and that is the only bound state of H_{nC}. All excited states of H_{nC} are continuum states, which are specified by their asymptotic momenta $\boldsymbol{k}$: $H_{\mathrm{nC}}\phi_{\boldsymbol{k}}(\boldsymbol{r}) = (\hbar^2k^2/2\mu)\phi_{\boldsymbol{k}}(\boldsymbol{r})$ where μ is the nucleon–core reduced mass. The potentials V_{CT} and V_{nT} are the core–target and the neutron–target optical potentials, respectively. The potential V_{CT} is assumed to describe the elastic scattering between the core and target at the same incident energy per nucleon. The vector $\boldsymbol{R}-A_{\mathrm{P}}^{-1}\boldsymbol{r}$ is the relative coordinate between the core and the target and $\boldsymbol{R}+A_{\mathrm{P}}^{-1}(A_{\mathrm{P}}-1)\boldsymbol{r}$ is the coordinate between the halo neutron and the target nucleus. For simplicity, we suppress the spin degree of freedom and disregard the spin-dependence of the interactions.

The wave function must satisfy the following scattering boundary condition at $R\to\infty$:

$$\Psi(\boldsymbol{R},\boldsymbol{r}) = \mathrm{e}^{iKZ}\phi_0(\boldsymbol{r}) + f_{\mathrm{el}}(\theta)\frac{\mathrm{e}^{iKR}}{R}\phi_0(\boldsymbol{r}) + \int d\boldsymbol{k}\, f_{\boldsymbol{k}}(\theta)\frac{\mathrm{e}^{iK'R}}{R}\phi_{\boldsymbol{k}}(\boldsymbol{r}), \quad (5.14)$$

where the wave numbers are related by the energy conservation:

$$E = \frac{\hbar^2K^2}{2M_{\mathrm{PT}}} + \varepsilon_0 = \frac{\hbar^2K'^2}{2M_{\mathrm{PT}}} + \frac{\hbar^2k^2}{2\mu}. \quad (5.15)$$

The internal structures of the core and target are not involved explicitly. The optical potential is supposed to describe the situation in which the participants are in their g.s. All processes involving the excitations of the constituents are taken into account, as loss of flux, through absorptive potentials. Therefore, the present model can only describe the dynamics of processes in which the core and target nuclei remain in their g.s. The only inelastic process considered in this model explicitly is the tearing off of the halo neutron.

5.2.2 Eikonal approximation

To obtain an exact solution of a quantum-mechanical three-body equation is a difficult task. The present three-body problem is complicated because the halo neutron easily breaks away. The aim is therefore just to find a reliable approximation.

In Sect. 2.3 it was shown that the eikonal approximation performs fairly well for the neutron–nucleus elastic scattering at medium incident energies. Although the Glauber theory, which is based on the collisions of bare nucleons, is not applicable here, the eikonal approximation is still a good approximation.

Considering the projectile as a composite two-body system, the eikonal approximation can be worked out as in Sect. 3.2.2. Expressing the three-body wave function $\Psi(\boldsymbol{R},\boldsymbol{r})$ as an incident plane wave times a function $\hat{\Psi}(\boldsymbol{R},\boldsymbol{r})$ slowly varying in $\boldsymbol{R}$,

$$\Psi(\boldsymbol{R},\boldsymbol{r}) = \mathrm{e}^{iKZ}\hat{\Psi}(\boldsymbol{R},\boldsymbol{r}), \tag{5.16}$$

yields, for $\hat{\Psi}(\boldsymbol{R},\boldsymbol{r})$,

$$\begin{aligned}\Bigg[vP_z + \frac{\boldsymbol{P}^2}{2M_{PT}} + (H_{\mathrm{nC}} - \varepsilon_0) + V_{\mathrm{CT}}\left(\boldsymbol{R} - \frac{\boldsymbol{r}}{A_{\mathrm{P}}}\right) \\ + V_{\mathrm{nT}}\left(\boldsymbol{R} + \frac{A_{\mathrm{P}}-1}{A_{\mathrm{P}}}\boldsymbol{r}\right)\Bigg]\hat{\Psi}(\boldsymbol{R},\boldsymbol{r}) = 0.\end{aligned} \tag{5.17}$$

The kinetic-energy operator $\boldsymbol{P}^2/2M_{\mathrm{PT}}$ is again neglected on the grounds that the momentum transfer is assumed to be small. Then the equation will contain the transverse component $\boldsymbol{b} = \boldsymbol{R} - Z\hat{\boldsymbol{z}}$ as a parameter, which facilitates its treatment as an impact *parameter* and a visualization of the process as propagation along straight-line trajectories. The equation is, however, still fairly complicated since the operator of the internal excitation energy, $H_{\mathrm{nC}} - \varepsilon_0$, is yet to be neglected.

The neglect of the internal excitation energy is, again, tantamount to assuming that the halo-neutron state is frozen during the reaction. This is only valid if the motion of the halo neutron in the projectile is much slower than the projectile–target relative motion. To see whether this condition is fulfilled, let us compare the time scales of the two motions.

If v is the relative velocity and R_{H} is the halo radius, the target nucleus passes through the diameter of the halo in a time interval $\Delta t \simeq 2R_{\mathrm{H}}/v$. For a $50\,A$ MeV incident energy and $R_{\mathrm{H}} = 10$ fm, we have $\Delta t = 2 \times 10^{-22}$ s. This is to be compared with the time in which the projectile breakup takes place. The smaller the excitation energy, the larger the probability of excitation; it is therefore reasonable to assess the time scale by taking an excitation just over the threshold as typical. Then the period of the halo-neutron breakup can be estimated to be $\Delta t_{\mathrm{H}} = \hbar/|\varepsilon_0|$. With $\varepsilon_0 = -0.5$ MeV, the time interval is $\Delta t_H = 1.3 \times 10^{-21}$ s. We see that the excitation of the halo neutron is several times slower than the projectile–target relative motion. It follows then that the halo-neutron excitation can be neglected, i.e., the operator of the internal excitation energy, $H_{\mathrm{nC}} - \varepsilon_0$, can be dropped. This approximation is known as an adiabatic approximation in the direct reaction theory (cf. App. A), and has been successfully applied to deuteron scattering [89]. The adiabatic approximation is introduced in the Glauber theory as well (cf. Sect. 3.2). The 'eikonal and adiabatic approximation' to be used for medium-energy reactions shows close resemblance to the Glauber theory, the only difference being in the nature of the nucleon–nucleon force used.

Neglecting both the kinetic-energy operator and the internal excitation energy in Eq. (5.17) leads to the solution

$$\hat{\Psi}(\boldsymbol{R}, \boldsymbol{r}) = \exp\left\{\frac{1}{i\hbar v}\int_{-\infty}^{Z} dZ' \left[V_{\mathrm{CT}}\left(\boldsymbol{b} + Z\hat{\boldsymbol{z}} - \frac{\boldsymbol{r}}{A_{\mathrm{P}}}\right) + V_{\mathrm{nT}}\left(\boldsymbol{b} + Z\hat{\boldsymbol{z}} + \frac{A_{\mathrm{P}}-1}{A_{\mathrm{P}}}\boldsymbol{r}\right)\right]\right\}\phi_0(\boldsymbol{r}). \tag{5.18}$$

In the limit of $Z \to \infty$, this solution is nothing but the wave function after the collision in the projectile-rest frame [cf. Eq. (3.40)], $\hat{\Psi}(\boldsymbol{b}, \boldsymbol{r})$. It can be written in terms of the phase-shift functions as

$$\hat{\Psi}(\boldsymbol{b}, \boldsymbol{r}) = \mathrm{e}^{i\chi_{\mathrm{CT}}\left(\boldsymbol{b} - \frac{\boldsymbol{s}}{A_{\mathrm{P}}}\right) + i\chi_{\mathrm{nT}}\left(\boldsymbol{b} + \frac{A_{\mathrm{P}}-1}{A_{\mathrm{P}}}\boldsymbol{s}\right)}\phi_0(\boldsymbol{r}). \tag{5.19}$$

The phase-shift functions $\chi_{\mathrm{CT}}(\boldsymbol{b})$ and $\chi_{\mathrm{nT}}(\boldsymbol{b})$ can be calculated from the optical potentials $V_{\mathrm{CT}}(\boldsymbol{R})$ and $V_{\mathrm{nT}}(\boldsymbol{r})$, respectively.

The scattering amplitudes are obtained as shown in Chap. 3 [cf. Eq. (3.49)]. The scattering amplitude for elastic scattering is given by

$$f_{\mathrm{el}}(\theta) = \frac{iK}{2\pi}\int d\boldsymbol{b}\,\mathrm{e}^{-i\boldsymbol{q}\cdot\boldsymbol{b}}\left[1 - \langle\phi_0|\mathrm{e}^{i\chi_{\mathrm{CT}}\left(\boldsymbol{b} - \frac{\boldsymbol{s}}{A_{\mathrm{P}}}\right) + i\chi_{\mathrm{nT}}\left(\boldsymbol{b} + \frac{A_{\mathrm{P}}-1}{A_{\mathrm{P}}}\boldsymbol{s}\right)}|\phi_0\rangle\right]. \tag{5.20}$$

The amplitude of transition to a continuum state in which the core and the target are not excited but the halo neutron is ejected to a state with core–neutron asymptotic momentum $\boldsymbol{k}$ is

$$f_{\boldsymbol{k}}(\theta) = -\frac{iK}{2\pi}\int d\boldsymbol{b}\, \mathrm{e}^{-i\boldsymbol{q}\cdot\boldsymbol{b}} \langle \phi_{\boldsymbol{k}} | \mathrm{e}^{i\chi_{\mathrm{CT}}\left(\boldsymbol{b}-\frac{\boldsymbol{s}}{A_{\mathrm{P}}}\right)+i\chi_{\mathrm{nT}}\left(\boldsymbol{b}+\frac{A_{\mathrm{P}}-1}{A_{\mathrm{P}}}\boldsymbol{s}\right)} |\phi_0\rangle. \quad (5.21)$$

This is nothing but Eq. (3.27) applied to a one-neutron-halo nucleus.

These formulae show close resemblance to those derived in the Glauber theory with the OLA in Sect. 4.2.3. The only difference is that the phase-shift functions are calculated from the optical potentials between the constituent particles instead of the nucleon–nucleon profile function.

5.2.3 Cross sections and reaction probabilities

The integrated cross sections can be calculated in exactly the same way as derived previously, with some modification of the interpretation. In particular, the core- and target-excitation channels are not treated explicitly now; they are included, through the absorption, in the optical potentials. Thus these cross sections can be accessed via the total cross section provided by the optical theorem.

The integrated cross section of the elastic scattering can be obtained from the scattering amplitude (5.20):

$$\begin{aligned}\sigma_{\mathrm{el}} &= \int \frac{d\boldsymbol{q}}{K^2}\, |f_{\mathrm{el}}(\theta)|^2 \\ &= \int d\boldsymbol{b} \left| 1 - \langle \phi_0 | \mathrm{e}^{i\chi_{\mathrm{CT}}\left(\boldsymbol{b}-\frac{\boldsymbol{s}}{A_{\mathrm{P}}}\right)+i\chi_{\mathrm{nT}}\left(\boldsymbol{b}+\frac{A_{\mathrm{P}}-1}{A_{\mathrm{P}}}\boldsymbol{s}\right)} |\phi_0\rangle \right|^2. \end{aligned} \quad (5.22)$$

The optical theorem (3.35) enables one to obtain the total cross section:

$$\begin{aligned}\sigma_{\mathrm{tot}} &= \frac{4\pi}{K}\mathrm{Im} f_{\mathrm{el}}(0) \\ &= 2\int d\boldsymbol{b}\, \mathrm{Re}\left[1 - \langle \phi_0 | \mathrm{e}^{i\chi_{\mathrm{CT}}\left(\boldsymbol{b}-\frac{\boldsymbol{s}}{A_{\mathrm{P}}}\right)+i\chi_{\mathrm{nT}}\left(\boldsymbol{b}+\frac{A_{\mathrm{P}}-1}{A_{\mathrm{P}}}\boldsymbol{s}\right)} |\phi_0\rangle \right] \\ &= 2\int d\boldsymbol{b}_{\mathrm{C}}\, \mathrm{Re}\left[1 - \mathrm{e}^{i\chi_{\mathrm{CT}}(\boldsymbol{b}_{\mathrm{C}})} \langle \phi_0 | \mathrm{e}^{i\chi_{\mathrm{nT}}(\boldsymbol{b}_{\mathrm{C}}+\boldsymbol{s})} |\phi_0\rangle \right], \end{aligned} \quad (5.23)$$

where, in the second line, the elastic scattering amplitude (5.20) is used. In the last expression the cross section is expressed as an integral over the impact parameter of core nucleus, $\boldsymbol{b}_{\mathrm{C}}$, which is related to the impact parameter $\boldsymbol{b}$ of the projectile by $\boldsymbol{b}_{\mathrm{C}} = \boldsymbol{b} - A_{\mathrm{P}}^{-1}\boldsymbol{s}$.

The total reaction cross section[3] is given by subtracting the elastic scattering cross section, Eq. (5.22), from the total cross section,

[3]When the Coulomb potential is included in the phase-shift function, σ_{el} and σ_{tot} diverge, as they should, owing to the contributions to the elastic channel from infinity in the integration over the impact parameter. But σ_{reac} remains finite since the infinite contributions cancel out in $\sigma_{\mathrm{reac}} = \sigma_{\mathrm{tot}} - \sigma_{\mathrm{el}}$.

Eq. (5.23):

$$\sigma_{\text{reac}} = \int d\boldsymbol{b} \left[1 - \left| \langle \phi_0 | e^{i\chi_{\text{CT}}\left(\boldsymbol{b} - \frac{\boldsymbol{s}}{A_{\text{P}}}\right) + i\chi_{\text{nT}}\left(\boldsymbol{b} + \frac{A_{\text{P}}-1}{A_{\text{P}}}\boldsymbol{s}\right)} | \phi_0 \rangle \right|^2 \right]. \tag{5.24}$$

There is no way to simplify this by changing the variable of integration.

The neutron-removal process described by the scattering amplitude $f_{\boldsymbol{k}}(\theta)$ is elastic neutron removal since the model assumes that both core and target are in their g.s. throughout. The elastic neutron-removal cross section is calculated by integrating over the solid angle and the final momentum of the neutron as follows:

$$\begin{aligned}
\sigma^{\text{el}}_{-n} &= \int \frac{d\boldsymbol{q}}{K^2} \int d\boldsymbol{k}\, |f_{\boldsymbol{k}}(\theta)|^2 \\
&= \int d\boldsymbol{k} \int d\boldsymbol{b}\, \langle \phi_0 | e^{-i\chi^*_{\text{CT}} - i\chi^*_{\text{nT}}} | \phi_{\boldsymbol{k}} \rangle \langle \phi_{\boldsymbol{k}} | e^{i\chi_{\text{CT}} + i\chi_{\text{nT}}} | \phi_0 \rangle \\
&= \int d\boldsymbol{b} \left[\langle \phi_0 | e^{-2\text{Im}\chi_{\text{CT}}\left(\boldsymbol{b} - \frac{\boldsymbol{s}}{A_{\text{P}}}\right) - 2\text{Im}\chi_{\text{nT}}\left(\boldsymbol{b} + \frac{A_{\text{P}}-1}{A_{\text{P}}}\boldsymbol{s}\right)} | \phi_0 \rangle \right. \\
&\qquad \left. - \left| \langle \phi_0 | e^{i\chi_{\text{CT}}\left(\boldsymbol{b} - \frac{\boldsymbol{s}}{A_{\text{P}}}\right) + i\chi_{\text{nT}}\left(\boldsymbol{b} + \frac{A_{\text{P}}-1}{A_{\text{P}}}\boldsymbol{s}\right)} | \phi_0 \rangle \right|^2 \right],
\end{aligned} \tag{5.25}$$

where use is made of the closure relation

$$\int d\boldsymbol{k}\, |\phi_{\boldsymbol{k}}\rangle \langle \phi_{\boldsymbol{k}}| = 1 - |\phi_0\rangle\langle\phi_0|. \tag{5.26}$$

This cross section is, of course, included in the total reaction cross section (5.24). Since the elastic neutron-removal is the only reaction process treated explicitly in the present three-body model, the remainder of the reaction cross section, $\sigma_{\text{reac}} - \sigma^{\text{el}}_{-n}$, represents the absorption cross section due either to the excitation of the core or of the target or both:

$$\begin{aligned}
\sigma_{\text{abs}} &= \sigma_{\text{reac}} - \sigma^{\text{el}}_{-n} \\
&= \int d\boldsymbol{b} \left[1 - \langle \phi_0 | e^{-2\text{Im}\chi_{\text{CT}}\left(\boldsymbol{b} - \frac{\boldsymbol{s}}{A_{\text{P}}}\right) - 2\text{Im}\chi_{\text{nT}}\left(\boldsymbol{b} + \frac{A_{\text{P}}-1}{A_{\text{P}}}\boldsymbol{s}\right)} | \phi_0 \rangle \right] \\
&= \int d\boldsymbol{b}_{\text{C}} \left[1 - e^{-2\text{Im}\chi_{\text{CT}}(\boldsymbol{b}_{\text{C}})} \langle \phi_0 | e^{-2\text{Im}\chi_{\text{nT}}(\boldsymbol{b}_{\text{C}} + \boldsymbol{s})} | \phi_0 \rangle \right].
\end{aligned} \tag{5.27}$$

The cross section $\sigma^{\text{inel}}_{-n}$ of inelastic neutron removal in which the target is excited is included in the absorption cross section. Just as in the high-energy case, the neutron-removal cross section is assumed to be equal to the difference between the total reaction cross sections of collisions of the halo nucleus and of the core with the same target. Then $\sigma_{-n}^{\rm inel}$ is the difference between $\sigma_{\rm abs}$ of Eq. (5.27) and the total reaction cross section belonging to the core nucleus, $\int d\boldsymbol{b}_{\rm C}\,[1-{\rm e}^{-2{\rm Im}\chi_{\rm CT}(\boldsymbol{b}_{\rm C})}]$:

$$\sigma_{-n}^{\rm inel} = \int d\boldsymbol{b}_{\rm C}\, {\rm e}^{-2{\rm Im}\chi_{\rm CT}(\boldsymbol{b}_{\rm C})} \left[1 - \langle\phi_0|{\rm e}^{-2{\rm Im}\chi_{n\rm T}(\boldsymbol{b}_{\rm C}+\boldsymbol{s})}|\phi_0\rangle\right]. \quad (5.28)$$

5.3 Applications

5.3.1 Deuteron reactions

The binding energy of the deuteron is about 2.2 MeV, which is much smaller than the usual nucleon separation energies, and the reactions of the deuteron do show the effects of weak binding. Before discussing the reactions of halo nuclei, we consider the application of the eikonal and adiabatic approximation to deuteron reactions. Since detailed and accurate three-body calculations are available for them, they lend themselves to tests of the applicability of the eikonal approximation to medium-energy reactions of loosely bound nuclei [79,80].

A deuteron reaction should be described as a three-body reaction of the proton, neutron and target nucleus. The Schrödinger equation reads as

$$\left[\frac{\boldsymbol{P}^2}{2M_{\rm PT}} + H_{\rm pn} + U_{\rm p}\left(\boldsymbol{R}+\frac{\boldsymbol{r}}{2}\right) + U_{\rm n}\left(\boldsymbol{R}-\frac{\boldsymbol{r}}{2}\right)\right]\Psi(\boldsymbol{R},\boldsymbol{r}) = E\Psi(\boldsymbol{R},\boldsymbol{r}), \quad (5.29)$$

where $\boldsymbol{R}$ and $\boldsymbol{r}$ stand for the relative coordinate between the deuteron and the target and between the proton and the neutron, respectively, $H_{\rm pn}$ is the Hamiltonian of the deuteron, and the two potentials U are phenomenological optical potentials, which describe the elastic scattering of the two nucleons from the target at the same incident energy per nucleon.

The elastic scattering amplitude is given by [cf. Eq. (5.20)]

$$f_{\rm el}(\theta) = \frac{iK}{2\pi}\int d\boldsymbol{b}\,{\rm e}^{-i\boldsymbol{q}\cdot\boldsymbol{b}}\left[1 - e^{i\chi_{\rm el}(\boldsymbol{b})}\right], \quad (5.30)$$

where the phase-shift function of the elastic scattering is to be calculated as

$$e^{i\chi_{\rm el}(\boldsymbol{b})} = \langle\phi_0|e^{i\chi_{\rm p}(\boldsymbol{b}+\frac{\boldsymbol{s}}{2})+i\chi_{\rm n}(\boldsymbol{b}-\frac{\boldsymbol{s}}{2})}|\phi_0\rangle. \quad (5.31)$$

Here $\boldsymbol{s}$ represents the projection of the relative vector $\boldsymbol{r}$ onto the plane perpendicular to the incident direction, and $\chi_{\rm p}(\boldsymbol{b})$, $\chi_{\rm n}(\boldsymbol{b})$ are the phase-shift functions calculated with the optical potentials $U_{\rm p}$ and $U_{\rm n}$, respectively.

Figure 5.4 shows the differential cross section of the d+^{58}Ni elastic scattering at deuteron energy $E_d = 80$ MeV. Three calculations are compared with the measured data. In the quantum-mechanical three-body calculation [90] (dotted) the three-body Schrödinger equation is solved by the coupled-channel method, with the continuum states approximately treated as a set of discrete states [continuum-discretized coupled-channel (CDCC) method].

The differential cross section calculated by the eikonal and adiabatic approximation, Eq. (5.30), is shown by the dashed curve. The solid curve is a deuteron optical-model result calculated with the optical potential constructed from the phase-shift function $\chi_{el}(\boldsymbol{b})$ of Eq. (5.31), through Eq. (5.12). With this procedure, the breakup effect is taken into account by the construction of the optical potential.

We see that the quantum-mechanical calculation with the three-body model describes the elastic scattering of the deuteron accurately, apart from a slight overshooting at large scattering angles. The scattering amplitude (5.30) gives too large cross section at large scattering angles, but the eikonal-based optical model comes closer to the three-body results. It is thus plausible that the deuteron optical poten-

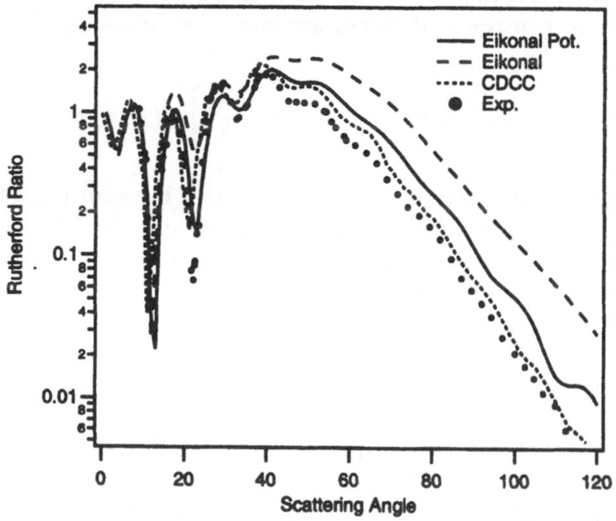

Figure 5.4: The d+^{58}Ni elastic scattering cross section at incident deuteron energy $E_d = 80$ MeV. The measured cross section is shown by dots, the eikonal and adiabatic approximation (5.30) by the dashed curve and the CDCC calculation [90] by the dotted curve; eikonal-based optical model [Eqs. (5.31), (5.12)] by the solid curve.

tial constructed from the nucleonic phase-shift functions includes the breakup effects fairly well. The discrepancy at large scattering angles can be attributed to the inadequate treatment of deflection of larger angles in a description in terms of straight-line trajectories.

For the total reaction and elastic breakup cross sections the eikonal and adiabatic approximation yields $\sigma_{\text{reac}} = 1458$ mb and $\sigma^{\text{el}}_{-n} = 150$ mb. The elastic breakup cross section occupies about 10% of the total reaction cross section, so 90% of the cross section comes from breakup accompanied by target excitation. The cross sections calculated in the CDCC method are $\sigma_{\text{reac}} = 1571$ mb and $\sigma^{\text{el}}_{-n} = 126$ mb. For these quantities the eikonal and adiabatic approximation implies an error not exceeding 20%. The deuteron optical-model gives the total reaction cross section $\sigma_{\text{reac}} = 1519$ mb. The difference from 1458 mb originates from the eikonal approximation for the relative motion.

The d+^{58}Ni elastic scattering is suited for displaying the breakup effects on the optical potential. The folding potential for the deuteron,

$$V^{\text{F}}(R) = \int d\boldsymbol{r} |\phi_0(\boldsymbol{r})|^2 \left[U_{\text{p}} \left(\boldsymbol{R} + \frac{\boldsymbol{r}}{2} \right) + U_{\text{n}} \left(\boldsymbol{R} - \frac{\boldsymbol{r}}{2} \right) \right], \tag{5.32}$$

includes the deuteron g.s. only, without any effect of the deuteron breakup. The potential (5.12) constructed from the phase-shift function (5.31), on the contrary, does comprise dynamic effects from the deuteron breakup. The difference of an optical potential that allows for excitation effects from the folding potential is called the dynamic polarization potential.

The optical potential $V_{\text{opt}}(r)$ calculated with Eq. (5.12) and the folding potential $V^{\text{F}}(r)$ are shown in Fig. 5.5. Though their gross features are similar, their difference is characteristic of a weakly bound projectile. The dynamic polarization gives rise to a smaller real part and a stronger imaginary part *at the potential surface*. In physical terms, these breakup effects are repulsive and absorptive in the surface region, in keeping with the behaviour of any weakly bound projectiles.

To get insight into the origin of this behaviour, let us expand the phase-shift function (5.31) into Taylor series like in Eq. (5.5):

$$\chi_{\text{el}}(\boldsymbol{b}) = \langle \chi_{\text{pn}} \rangle + \frac{i}{2} \left[\langle \chi^2_{\text{pn}} \rangle - \langle \chi_{\text{pn}} \rangle^2 \right] + \cdots = \chi^{(1)} + \chi^{(2)} + \cdots, \tag{5.33}$$

where $\chi_{\text{pn}} = \chi_{\text{p}}(\boldsymbol{b} + \frac{1}{2}\boldsymbol{s}) + \chi_{\text{n}}(\boldsymbol{b} - \frac{1}{2}\boldsymbol{s})$. The symbol $\langle \cdots \rangle$ means an expectation value in the deuteron wave function ϕ_0. The optical potential constructed with the first-order term $\chi^{(1)}$ via the procedure (5.12) is the folding potential, Eq. (5.32). The next term, $\chi^{(2)}$, gives the leading correction to the folding potential. To get a qualitative idea of the breakup effects, one should examine this term.

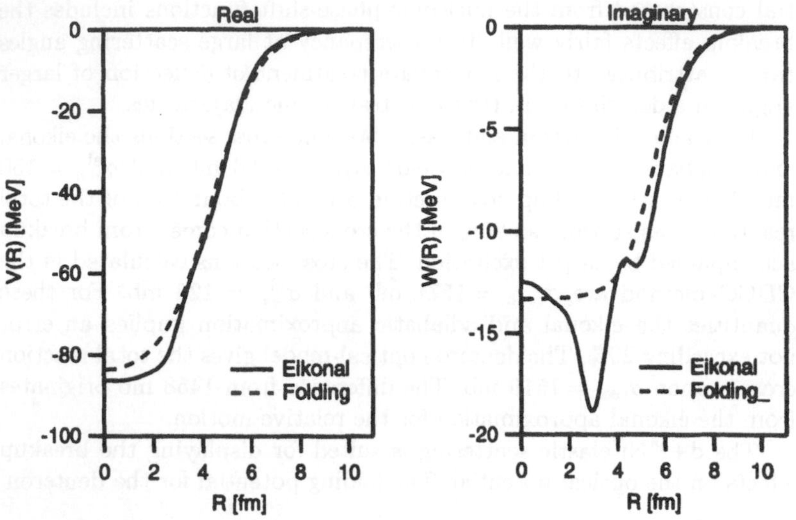

Figure 5.5: Real part (left panel) and imaginary part (right panel) of the deuteron potential. Dashed curve: folding model [Eq. (5.32)], solid curve: eikonal [Eq. (5.12) with Eq. (5.31)].

Let us assume that the proton and neutron optical potentials are the same and that their real and imaginary parts have the same shape: $U_p = U_n = (V_0 + iW_0)U(r)$ Then the phase-shift function χ_{pn} becomes $\chi_{pn}(b, s) = (V_0 + iW_0) f(b, s)$, where

$$f(b,s) \equiv -\frac{1}{\hbar v}\int_{-\infty}^{\infty} dZ \left[U\left(b + Z\hat{z} + \frac{r}{2}\right) + U\left(b + Z\hat{z} - \frac{r}{2}\right)\right]. \quad (5.34)$$

The second-order contribution to the phase-shift function, $\chi^{(2)}$, is

$$\begin{aligned} \chi^{(2)}(b) &= \frac{i}{2}(V_0 + iW_0)^2 \langle f - \langle f\rangle\rangle^2 \\ &= \left[-V_0W_0 + \frac{i}{2}(V_0^2 - W_0^2)\right] \langle f - \langle f\rangle\rangle^2. \end{aligned} \quad (5.35)$$

Since the variance of any function f is positive, the sign of the real part of $\chi^{(2)}$ is the same as that of $-V_0W_0$ and the sign of the imaginary part agrees with that of $V_0^2 - W_0^2$. For not very high incident energies, both the real and imaginary parts of the optical potential are negative, and the real part is stronger than the imaginary part. The sign of the potential is opposite to the sign of the phase-shift function at the potential surface. These facts account for the sign of the dynamic polarization potential; it is repulsive since $V_0W_0 > 0$ and absorptive since $-(V_0^2 - W_0^2) < 0$.

Note that the second-order contribution to the dynamic polarization potential is proportional to the variance of a function f with respect to the g.s. density of the relative motion of the constituents of the projectile, $|\phi_0|^2$. The more this density is smeared out, the larger this variance is, and that is the mechanism through which the weakly bound nature of the projectile manifests itself in the dynamic polarization term.

5.3.2 Reactions with ^{11}Be: integrated cross sections

In this section the medium-energy ^{11}Be+^{12}C collision will be discussed. The eikonal and adiabatic approximation is likely to be more reliable here than for the deuteron-induced reactions for several reasons. Because of the larger reduced mass of the ^{11}Be+target system, the wave number is larger. Then the semiclassical picture which is assumed in the eikonal approximation becomes adequate down to lower incident energies. Furthermore, because of a stronger absorption of the ^{10}Be core, for small impact parameters, virtually all flux is simply absorbed, and, for larger impact parameters, the straight-line trajectory assumption implicit in the eikonal approximation is more applicable. Finally, the neutron separation energy in ^{11}Be is smaller than the binding energy of the deuteron. Since the adiabatic approximation is found to be valid even for the deuteron reaction, the freezing of the neutron motion should be less strong an approximation for a reaction with a halo nucleus.

Let us first consider the reaction probabilities for various channels. They are given by the same formulae as in the high-incident-energy case, as is summarized in Table 4.2. The difference is that now the phase-shift functions that enter into the reaction probabilities have to be calculated from optical potentials. In the calculations to be shown below optical potentials of Woods–Saxon shape have been employed.

The reaction probabilities for the ^{11}Be+^{12}C collision at an incident energy of 50 A MeV are shown in Fig. 5.6. The solid curve shows the neutron-removal probability and the dashed line represents the reaction probability of the collision of the ^{10}Be core with the same target. The neutron-removal probability is largest at around $b \sim 7$ fm, where the reaction probability of reaction between the core and the target decreases rapidly from unity to zero. The neutron-removal probability has a long tail as a result of the neutron halo. The neutron-removal probability is decomposed into elastic and inelastic components.

It is instructive to compare the reaction probabilities at a high (800 A MeV) incident energy, Fig. 4.3, with those at medium (50 A MeV) energies, Fig. 5.6. The neutron-removal probability at 50 A MeV is larger than that at 800 A MeV. As for its composition, it is conspic-

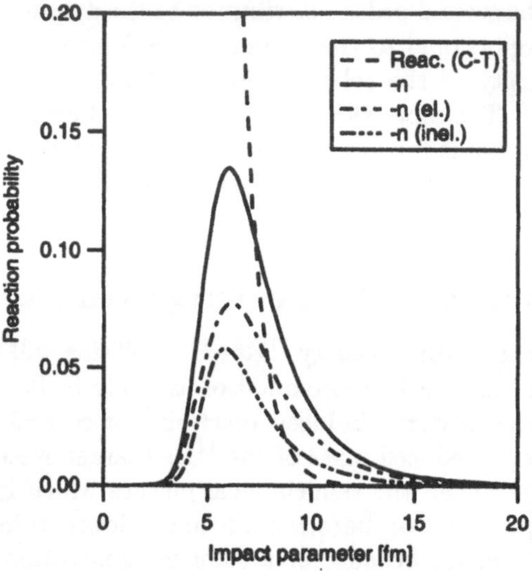

Figure 5.6: Reaction probabilities for the ^{11}Be+^{12}C collision at 50 A MeV calculated by the eikonal and adiabatic approximation. Solid curve: probability of single-neutron removal (breakup); dot-dashed curve: elastic breakup probability; dot-dot-dashed curve: inelastic breakup probability; dashed curve: reaction probability for the ^{10}Be+^{12}C collision.

uous that the inelastic breakup probability does not change much between high and medium energies, while the elastic breakup probability is much larger at medium energy. At 50 A MeV the elastic breakup probability exceeds even the inelastic breakup probability. This change of the neutron-removal mechanism is closely related to the change of the mechanism of the nucleon–nucleus scattering. In low- and medium-energy reactions, the nucleon–nucleus optical potential is dominated by its real part. The refraction of the halo neutron by the target potential contributes to the elastic breakup process. As the incident nucleon energy gets higher, the real part of the nucleon optical potential decreases, and the imaginary part becomes dominant. The inelastic scattering of the halo neutron from the target nucleus is an absorption process, allowed for by the imaginary part of the optical potential, and leads to inelastic breakup.

Direct measurements of elastic and inelastic breakup cross sections at incident energies of a few MeV per nucleon have been reported for ^{11}Be [91] and for ^{8}B [92] projectiles. It was observed that the elastic breakup cross section is indeed larger than the inelastic one in this energy region.

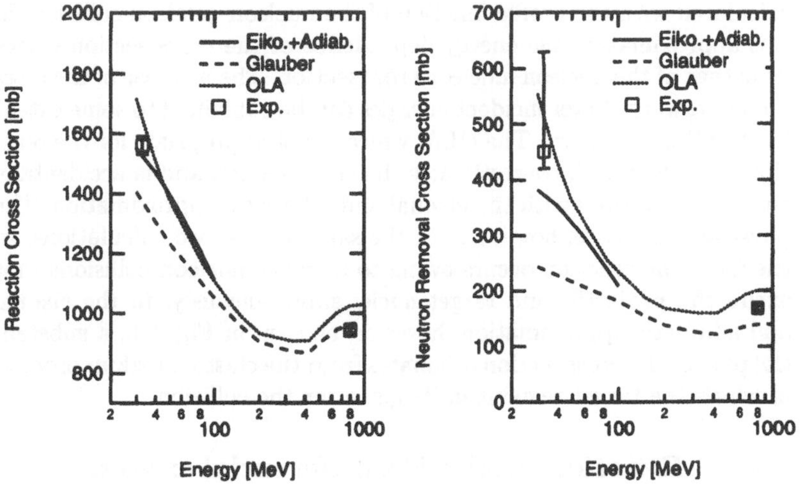

Figure 5.7: Energy dependence of ^{11}Be+^{12}C cross sections. The left and right panels show the reaction cross section and the neutron-removal cross section, respectively. Experimental cross sections are shown at 800 A MeV [2, 10] and at 33 A MeV [93]. The solid, dashed and dotted curves, respectively, result from the eikonal and adiabatic approximation, from the Glauber theory and from the OLA to the Glauber theory.

Figure 5.7 shows the energy dependence of the ^{11}Be+^{12}C cross sections. The measured total reaction cross section at 33 A MeV is much larger than the interaction cross section at 800 A MeV. A similar energy dependence of the total reaction cross section can also be seen for normal nuclei, as was shown in Fig. 3.5 for the ^{12}C+^{12}C reaction. This energy dependence is especially significant for the neutron-removal cross section.

The cross sections calculated in the eikonal and adiabatic approximation are shown by solid curves. The total reaction cross section is calculated with Eq. (5.24). The neutron-removal cross section is given by the sum of the elastic and inelastic breakup cross sections, Eqs. (5.25) and (5.28). The energy dependence of the cross section comes from two sources. First, the definition of the phase-shift function $\chi(\boldsymbol{b})$ contains the incident energy through the velocity. Second, the optical potential itself depends on the energy. Both effects tend to increase the cross sections at low incident energies.

The cross sections calculated in the Glauber theory are also shown, although their use cannot be validated at low incident energies, so they should be taken as qualitative estimates. The phase-shift functions are calculated with the profile function of the nucleon–nucleon collision. In this approximation the energy dependence of the cross section comes from that of the nucleon–nucleon cross section. The increase of the cross section towards lower incident energies can be explained to some extent by the Glauber theory. The OLA, which is not appropriate for the reactions of halo nuclei, overestimates the cross section and is accidentally close to the results with the eikonal and adiabatic approximation. The physical causes are, however, not the same in the two calculations. In the OLA the reaction occurs owing to nucleon–nucleon collisions that excite the projectile and target nuclei simultaneously. In the eikonal and adiabatic approximation, however, as seen in Fig. 5.6, a substantial part of the cross section originates from the elastic breakup process, in which the target remains in its g.s. after the collision.

5.3.3 Reactions with ^{11}Be: differential cross sections

The behaviour of the differential cross sections may reveal interesting aspects of the process studied and of its descriptions. Our example for this purpose is, again, the $^{11}\mathrm{Be}+^{12}\mathrm{C}$ elastic scattering at $E = 49.3\,A$ MeV. The measured elastic scattering cross section [94] is compared with theoretical curves in Fig. 5.8. Three theoretical curves are shown. The solid curve is the eikonal–adiabatic cross section calculated from the scattering amplitude (5.20). The dashed curve is the cross section by the folding model. The folding-model amplitude for a collision involving a halo nucleus is given by [cf. Eqs. (5.7) and (5.30)]

$$f_{\mathrm{el}}(\theta) = \frac{iK}{2\pi}\int d\boldsymbol{b}\, \mathrm{e}^{-i\boldsymbol{q}\cdot\boldsymbol{b}}\left[1 - \mathrm{e}^{\langle\phi_0|i\chi_{\mathrm{CT}}\left(\boldsymbol{b}-\frac{\boldsymbol{s}}{A_{\mathrm{P}}}\right)+i\chi_{\mathrm{nT}}\left(\boldsymbol{b}+\frac{A_{\mathrm{P}}-1}{A_{\mathrm{P}}}\boldsymbol{s}\right)|\phi_0\rangle}\right], \tag{5.36}$$

where the phase-shift functions are deduced from the folding potential

$$V^{\mathrm{F}}(R) = \int d\boldsymbol{r}\,\rho_{\mathrm{n}}(\boldsymbol{r})\left[V_{\mathrm{CT}}\left(\boldsymbol{R}-\frac{\boldsymbol{r}}{A_{\mathrm{P}}}\right) + V_{\mathrm{nT}}\left(\boldsymbol{R}+\frac{A_{\mathrm{P}}-1}{A_{\mathrm{P}}}\boldsymbol{r}\right)\right] \tag{5.37}$$

via Eq. (5.7). (Here $A_{\mathrm{P}} = 11$.) In the above expression $\rho_{\mathrm{n}}(\boldsymbol{r})$ is the density of the halo neutron. The folding model overestimates the cross section at large scattering angles. Finally, the dotted curve is the cross section in which the neutron–target interaction is neglected in the scattering amplitude (5.36):

$$f_{\mathrm{el}}(\theta) = \frac{iK}{2\pi}\int d\boldsymbol{b}\, \mathrm{e}^{-i\boldsymbol{q}\cdot\boldsymbol{b}}\left[1 - \langle\phi_0|\mathrm{e}^{i\chi_{\mathrm{CT}}\left(\boldsymbol{b}-\frac{\boldsymbol{s}}{A_{\mathrm{P}}}\right)}|\phi_0\rangle\right]. \tag{5.38}$$

As seen in the figure, the cross section is rather indifferent to whether or not the neutron–target potential is taken into account, except at large

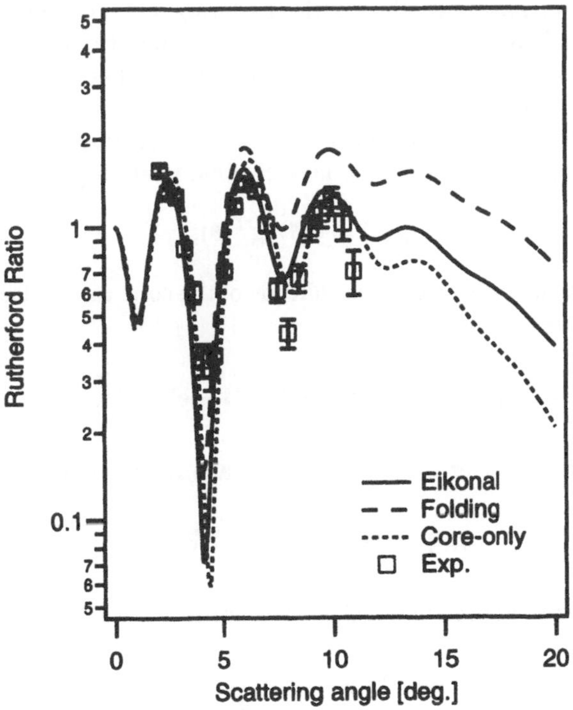

Figure 5.8: Differential cross section of the ^{11}Be+^{12}C elastic scattering at 49.3 A MeV incident energy. The measured data are from Ref. [94]. Solid curve: calculation with eikonal and adiabatic approximation; dashed curve: eikonal approximation with the folding potential; dotted curve: the same as the model producing the solid curve but with the neutron–target potential neglected.

scattering angles. For the ^{11}Be projectile, the neutron–target interaction does not contribute too much to the elastic scattering. This is just the opposite of the behaviour of the reaction cross sections. One of the main criteria of a halo nucleus is the large neutron-removal cross section in its collisions, and that is induced by the neutron–target potential [cf. Eq. (4.33)].

To be scattered elastically, the halo nucleus must get a substantial jerk from the target nucleus via the core–target and neutron–target potentials. In a perturbative picture it is meaningful to imagine such jerks one by one. Because of the mass ratio between the core and the neutron, it is much more likely that the core deflected by such a shock can carry away the neutron than that the neutron deflected could take away the core. Therefore, the elastic scattering will receive much more contribution from the core–target interaction than from the neutron–target interaction, while for the neutron-removal channels this is just the other way round.

The elastic scattering amplitude, with the neutron–target interaction neglected, satisfies a simple factorization relation [95]:[4]

$$f_{\rm el}(\theta) = g(\theta) f_{\rm el}^{\rm CT}(\theta), \tag{5.39}$$

where $f_{\rm el}^{\rm CT}(\theta)$ is the scattering amplitude of the core–target elastic scattering given by

$$f_{\rm el}^{\rm CT}(\theta) = \frac{iK}{2\pi}\int d\boldsymbol{b}\, \mathrm{e}^{-i\boldsymbol{q}\cdot\boldsymbol{b}}\left[1 - \mathrm{e}^{i\chi_{\rm CT}(\boldsymbol{b})}\right] \tag{5.40}$$

and $g(\theta)$ is the Fourier transform of the halo-neutron density:

$$g(\theta) = \int d\boldsymbol{r}\, |\phi_0(\boldsymbol{r})|^2 \mathrm{e}^{-i\frac{1}{A_{\rm P}}\boldsymbol{q}\cdot\boldsymbol{s}}. \tag{5.41}$$

This provides an approximate relation between the two differential cross sections:

$$\frac{d\sigma_{\rm el}}{d\Omega}[^{11}\mathrm{Be+T}] \simeq |g(\theta)|^2 \frac{d\sigma_{\rm el}}{d\Omega}[^{10}\mathrm{Be+T}]. \tag{5.42}$$

The factor $|g(\theta)|^2$ allows for the fact that the c.m. position of the core nucleus ^{10}Be does not coincide with that of the projectile ^{11}Be. Since the halo neutron is smeared out in a broad region, this difference can be significant, which then reduces the factor $|g(\theta)|^2$ substantially below unity.

---

[4] This can be proven as follows:

$$\begin{aligned}
&\frac{iK}{2\pi}\int d\boldsymbol{b}\,\mathrm{e}^{-i\boldsymbol{q}\cdot\boldsymbol{b}}\left[1 - \langle\phi_0|\mathrm{e}^{i\chi_{\rm CT}\left(\boldsymbol{b}-\frac{\boldsymbol{s}}{A_{\rm P}}\right)}|\phi_0\rangle\right]\\
&= \frac{iK}{2\pi}\left[\int d\boldsymbol{b}\,\mathrm{e}^{-i\boldsymbol{q}\cdot\boldsymbol{b}} - \int d\boldsymbol{b}\int d\boldsymbol{r}\,\mathrm{e}^{-i\boldsymbol{q}\cdot\boldsymbol{b}}|\phi_0(\boldsymbol{r})|^2\mathrm{e}^{i\chi_{\rm CT}\left(\boldsymbol{b}-\frac{\boldsymbol{s}}{A_{\rm P}}\right)}\right]\\
&= \frac{iK}{2\pi}\left[\int d\boldsymbol{b}\,\mathrm{e}^{-i\boldsymbol{q}\cdot\boldsymbol{b}} - \int d\boldsymbol{b}'\int d\boldsymbol{r}\,\mathrm{e}^{-i\boldsymbol{q}\cdot\left(\boldsymbol{b}'+\frac{\boldsymbol{s}}{A_{\rm P}}\right)}|\phi_0(\boldsymbol{r})|^2\mathrm{e}^{i\chi_{\rm CT}(\boldsymbol{b}')}\right]\\
&= \frac{iK}{2\pi}\int d\boldsymbol{b}\,\mathrm{e}^{-i\boldsymbol{q}\cdot\boldsymbol{b}}\left[1 - \mathrm{e}^{i\chi_{\rm CT}(\boldsymbol{b})}g(\theta)\right]\\
&= \frac{iK}{2\pi}g(\theta)\int d\boldsymbol{b}\,\mathrm{e}^{-i\boldsymbol{q}\cdot\boldsymbol{b}}\left[1 - \mathrm{e}^{i\chi_{\rm CT}(\boldsymbol{b})}\right] + \frac{iK}{2\pi}[1-g(\theta)]\int d\boldsymbol{b}\,\mathrm{e}^{-i\boldsymbol{q}\cdot\boldsymbol{b}}\\
&= g(\theta)f_{\rm el}^{\rm CT}(\theta) + \frac{iK}{2\pi}[1-g(\theta)](2\pi)^2\delta(\boldsymbol{q})\\
&= g(\theta)f_{\rm el}^{\rm CT}(\theta).
\end{aligned}$$

In the last step $g(\theta)\delta(\boldsymbol{q}) = \delta(\boldsymbol{q})$ is used, which follows from the identity $x\delta(x) \equiv 0$.

Thus, in some energy region the contribution of the neutron–target potential to the elastic scattering may be negligible. There the simple relationship (5.42) holds, and, at least in principle, (the Fourier transform of) the density distribution of the halo neutron can be extracted from the elastic scattering cross sections.

5.3.4 Reactions with other nuclei

Since exclusive reactions are very useful to explore the properties of nuclei in their g.s. and excited states, measurements of diverse reactions have been performed in the medium-energy region. In this section some salient experiments and the progress in the relevant theory are briefly surveyed.

The early experiments were elastic or quasielastic[5] scattering experiments with halo-nucleus projectiles both on composite targets [27, 29, 96] and on proton target [28, 97]. They were analysed with phenomenological methods [98–101] to find some signatures of halo structure. Analyses with microscopic approaches, in which the effects of the projectile breakup are taken into account, have also been developed. In the early studies simple wave functions were used for the two-neutron-halo nuclei [79,80,102]. Later appeared the calculations employing more realistic wave functions obtained from three-body models [103–105].

Progress in theory includes various ways of getting rid of, or improving on, the eikonal and adiabatic approximation in the framework of a few-body direct-reaction approach. Higher-order corrections to the eikonal approximation were introduced [106]. A calculation in the adiabatic approximation, without the eikonal approximation, was then reported [107]. Non-adiabatic corrections were discussed in Ref. [108]. Furthermore, a time-dependent coupled-channel approach was developed to go beyond the adiabatic approximation [109]. Finally, a CDCC calculation for the exact solution of the few-body coupled equations was presented for ^{8}B breakup reactions [110].

The elastic scattering of a halo nucleus on proton is expected to provide information on the density distribution of the halo nucleus in a direct way since the nucleon optical potential is supposed to have a shape similar to the nuclear density. For a microscopic description, standard multiple-scattering theory can be used [101]. In fact, unambiguous information on the density distribution is expected from high-energy proton elastic scattering, and such measurements have indeed been made [111]. A microscopic analysis of this experiment in the Glauber theory will be discussed in Sect. 13.10.2.

[5]The elastic channel measured together with some of the inelastic scattering channels, belonging to target excitations that cannot be separated from the elastic channel.

Inelastic scattering experiments with proton target have already provided appreciable information on the excited states of unstable nuclei. The excited states of ^{8}He [30] and of some Be isotopes [31] have been observed in this reaction. The observation of a dipole excited state in ^{11}Li has been reported in Ref. [112]. There is, however, an alternative explanation of these findings in terms of breakup into the continuum, without resort to an excited state [113]. Inelastic scattering experiments with composite targets have also been performed. An observation to a transition to an excited states of ^{6}He has been reported in Ref. [114].

Chapter 6

Fragment momentum distribution in reactions with halo nuclei

6.1 Momentum distribution of projectile fragments

In the medium- and high-energy breakup reactions of some halo nuclei measurements have been performed for the momentum distributions of the core fragments as well as of the halo neutrons. As the process of the detachment of the halo nucleons from the projectile is not a violent perturbation at high energies, the core fragment is found to have almost the same velocity as the incident projectile in the laboratory frame.

The observation of nuclear fragments in the forward direction with velocity close to the incident velocity was first discussed in conjunction with deuteron stripping reactions. In this reaction one of the nucleons of the deuteron is captured by the target nucleus, and the other is observed in the forward direction with approximately the same speed as that of the projectile deuteron. In the projectile-rest frame each nucleon has a finite momentum spread corresponding to their relative motion in the g.s. of the deuteron. This aspect of the stripping reaction was first discussed by Serber [115]. A theoretical framework to be outlined below is an extension of that approach [115,116], and is sometimes called the Serber model.

The measurement of the fragment momentum distribution, $d\sigma/dk$, has been a prime objective of high-energy collision experiments with various composite nuclei [24]. The observations made in systematic investigations can be summarized in broad terms as follows: (1) The momentum distribution does not depend on the incident energy and target nucleus. (2) The momentum distribution depends on the projectile and the fragment. (3) The shape of the momentum distribution is a Gaussian. Because of properties (1) and (2), the momentum distribution must reflect the structure of the projectile nucleus prior to the collision.

The components of the momentum $\boldsymbol{k}$ of the fragment in the projectile-rest frame are the component parallel to the incident (longitudinal) direction, $k_{\parallel}$, and the transverse component, $\boldsymbol{k}_{\perp}$. The Gaussian distribution of the fragment momentum can be characterized by its mean value and width. The mean value of the longitudinal momentum is denoted by $\langle k_{\parallel} \rangle$. From systematic measurements this has the values of -10 to -130 MeV/c where the negative sign indicates that the fragment flies more slowly than the projectile. Let $\sigma_{\parallel}$ and $\sigma_{\perp}$ denote the rms half widths (standard deviation) of the longitudinal and transverse momentum distributions, respectively. In high-energy reactions the width is nearly the same for the longitudinal and transverse directions and has a value of 50 to 200 MeV/c for normal nuclei. For lower incident energies, the transverse width $\sigma_{\perp}$ is substantially larger than the longitudinal width $\sigma_{\parallel}$. This behaviour can be explained by considering the bending of the trajectory in the projectile–target relative motion [117]. The momentum distribution in the frame of the average fragment momentum can thus be parametrized as

$$P(\boldsymbol{k}) \propto \exp\left[-\frac{k_{\parallel}^2}{2\sigma_{\parallel}^2} - \frac{\boldsymbol{k}_{\perp}^2}{2\sigma_{\perp}^2}\right]. \tag{6.1}$$

As an illustration, Fig. 6.1 shows the longitudinal momentum distribution of the ^{10}Be fragment emerging from the collision of ^{12}C with a Be target at 2.1 A GeV incident energy [118].

The momentum distribution may also be characterized by its full width at half maximum (FWHM). If the Gaussian shape (6.1) is assumed for the longitudinal momentum distribution, the FWHM is given by $\sqrt{8\ln 2}\,\hbar\sigma_{\parallel}$. If the Lorentzian shape $[k_{\parallel}^2 + \sigma_{\parallel}^2]^{-1}$ is used, the FWHM is given by $2\,\hbar\sigma_{\parallel}$.

The momentum distribution has been interpreted in terms of the nucleonic Fermi motion in the projectile [24, 118, 119]. Later, however, it was reconsidered in a different cluster fragmentation model, in which the basic ingredients are the fragment separation energy and the radius [120]. We present the former interpretation in the following.

The half width $\sigma_{\parallel}$ is found to have a simple dependence on the

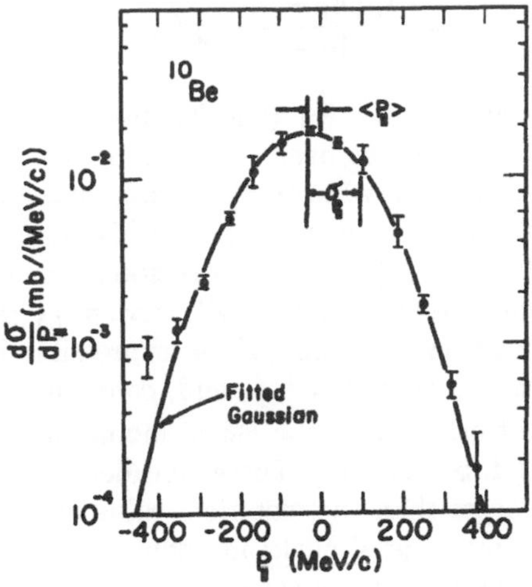

Figure 6.1: Longitudinal momentum distribution of ^{10}Be fragments in the collision of ^{12}C on a Be target at 2.1 A GeV. Taken from Ref. [118].

projectile mass number A_{P} and the fragment mass number A_{F}:

$$\sigma_{\|} = \sigma_0 \sqrt{\frac{A_{\mathrm{F}}(A_{\mathrm{P}} - A_{\mathrm{F}})}{A_{\mathrm{P}} - 1}}. \tag{6.2}$$

The symbol σ_0 is the width of the fragment momentum distribution in a single-nucleon removal. This has been empirically found to be $\hbar\sigma_0 \simeq 70-90$ MeV/c. This expression shows that the square of the width has a parabolic dependence on the fragment mass number and takes a maximum where the fragment mass number is one half of the projectile mass number. This parabolic dependence can be understood by a simple argument based on a statistical assumption [119]. Suppose that the momentum of the whole projectile nucleus, $\boldsymbol{P}_{\mathrm{P}} = \sum_{i=1}^{A_{\mathrm{P}}} \boldsymbol{p}_i$, is equal to zero, which is the case in the projectile-rest frame. Let the mean values of $\boldsymbol{p}_i^2$ and $\boldsymbol{p}_i \cdot \boldsymbol{p}_j$ be denoted by $\langle \boldsymbol{p}^2 \rangle$ and $\langle \boldsymbol{p} \cdot \boldsymbol{p}' \rangle$. Then we have

$$\begin{aligned} 0 &= \sum_i \boldsymbol{p}_i^2 + \sum_{i \neq j} \boldsymbol{p}_i \cdot \boldsymbol{p}_j \\ &= A_{\mathrm{P}} \langle \boldsymbol{p}^2 \rangle + A_{\mathrm{P}}(A_{\mathrm{P}} - 1)\langle \boldsymbol{p} \cdot \boldsymbol{p}' \rangle. \end{aligned} \tag{6.3}$$

The mean square of the fragment momentum $\boldsymbol{P}_{\mathrm{F}} = \sum_{i \in \mathrm{fragment}} \boldsymbol{p}_i$ is, similarly,

$$\langle \boldsymbol{P}_{\mathrm{F}}^2 \rangle = A_{\mathrm{F}} \langle \boldsymbol{p}^2 \rangle + A_{\mathrm{F}}(A_{\mathrm{F}} - 1)\langle \boldsymbol{p} \cdot \boldsymbol{p}' \rangle$$

$$= \frac{A_\mathrm{F}(A_\mathrm{P} - A_\mathrm{F})}{A_\mathrm{P} - 1} \langle \boldsymbol{p}^2 \rangle, \tag{6.4}$$

where, in the second equality, Eq. (6.3) has been employed. With this, the parabolic dependence has been shown to be a simple consequence of a statistical assumption, which derives from quantum mechanics.[1]

The elementary width σ_0 is related to the Fermi momentum of nucleon in the projectile [119]. Since σ_0 corresponds to the width of one Cartesian component, it is related to $\langle \boldsymbol{p}^2 \rangle$ as $\hbar^2 \sigma_0^2 = \frac{1}{3} \langle \boldsymbol{p}^2 \rangle$. In the Fermi gas model the mean momentum $\langle \boldsymbol{p}^2 \rangle$ is expressible in terms of the Fermi momentum p_F, as $\langle \boldsymbol{p}^2 \rangle = \frac{3}{5} p_\mathrm{F}^2$, and, consequently, $\hbar \sigma_0 = \frac{1}{\sqrt{5}} p_\mathrm{F}$. From electron scattering data, the Fermi momentum is known to be $p_\mathrm{F} \sim 260$ MeV/c (cf. p. 47) [121]. These considerations yield $\hbar \sigma_0 = 110$ MeV/c, slightly larger than the empirical value of 90 MeV/c. Taking into account the finite size of nuclei in defining the Fermi momentum would reduce the discrepancy to some extent [23].

For the fragment momentum distributions, systematic theoretical analyses have been performed in the Glauber theory [23, 122]. In this chapter this theory will be adapted and extended to the special case of halo nuclei. The fragment momentum distribution is related to the distribution of the momentum of the fragment with respect to the rest of the nucleons within the projectile. The information on this intranuclear momentum distribution is, however, entangled with the distorting effects of reaction dynamics, and the information coming from the nuclear surface can be disentangled best. The Glauber theory is an efficient tool to understand the reaction dynamics and thereby to unfold the reaction effects from the nuclear-structure information. The most important finding in this respect is that the longitudinal momentum distribution is influenced much less by the reaction effects than the transverse momentum distribution.

The momentum distribution of the core fragment in the reaction of halo nuclei has first been reported in Ref. [21] for the projectile ^{11}Li. The measured transverse momentum distribution of the core fragment ^{9}Li has been found much narrower than that measured in the reactions of normal nuclei. This observation played an important role in ascertaining the existence of the halo structure. The narrow momentum distribution was interpreted as arising from the large extension of the projectile nucleus, which, in turn, is a consequence of the position–momentum uncertainty relation.

For normal nuclei, electron quasielastic knockout reactions also offer an opportunity to investigate the momentum distribution [121]. For

[1] In the simple statistical picture, however, the average of the fragment momentum $\langle \boldsymbol{P}_\mathrm{F} \rangle$ is assumed to be identically zero. This is at variance with the observation of a small finite mean value of the longitudinal momentum $k_\parallel$.

unstable nuclei, however, no such electron scattering data are yet available since it would require difficult cross-beam collision experiments of electrons with secondary-beam unstable nuclei. Therefore, the measurement of the fragment momentum distribution in nuclear collisions is the only method to examine the momentum distribution of the loosely bound nucleons in the g.s. of unstable nuclei.

Detailed measurements have been reported on the halo momentum distributions in various experimental conditions. They include measurements of the momentum distributions of halo nucleons and of core fragments; the latter both in longitudinal and transverse directions; both with light targets, which induce nuclear breakup mostly, and with heavy targets, which also favour the Coulomb breakup mechanism.

These investigations have demonstrated that the longitudinal momentum distribution of the core fragment is almost independent of the reaction conditions [123]. This fact corroborates that the effects of the reaction mechanism are small and that the longitudinal momentum distribution reflects most faithfully the intrinsic structure of the halo nucleus. The theoretical treatment of momentum distributions has also been developed for lower incident energies; that approach is based on direct reaction theory. In this book the approach based on the Glauber theory will be tackled.

6.2 Formalism

In this section a formalism is derived for the description of the fragment momentum distributions based on the Glauber theory. To begin with, in Sect. 6.2.1 some basic formulae will be given for the neutron-removal cross sections as functions of the fragment momenta. To proceed, the formalism will be derived for elastic and inelastic breakup processes separately. The elastic breakup process is considered in Sect. 6.2.2. The formula relevant to inelastic breakup processes is first derived in a heuristic way in Sect. 6.2.3 and then in a rigorous way in Sect. 6.2.4.

6.2.1 Fundamentals

Let us again consider a single-neutron-halo nucleus and assume that the halo neutron occupies an s-orbit in the g.s. The Glauber scattering amplitude for a transition in which the neutron takes momentum (i.e., wave number vector) $\boldsymbol{k}$ relative to the core, the core remains in its g.s. and the target gets excited into state β is [cf. Eq. (3.49)]

$$f_{\boldsymbol{k}\beta}(\boldsymbol{q}) = -\frac{iK}{2\pi}\int d\boldsymbol{b}\, \mathrm{e}^{-i\boldsymbol{q}\cdot\boldsymbol{b}}\langle \phi_{\boldsymbol{k}}\Phi_0^{\mathrm{C}}\Phi_\beta^{\mathrm{T}}|\prod_{i\in P}\prod_{j\in T}(1-\Gamma_{ij})|\phi_0\Phi_0^{\mathrm{C}}\Phi_0^{\mathrm{T}}\rangle, \quad (6.5)$$

where $\boldsymbol{q}$ is the two-dimensional transverse vector representing the momentum transferred from the target nucleus to the whole projectile. The differential cross section per unit solid angle and per unit volume of the momentum space around $\boldsymbol{k}$ is given by

$$\frac{d\sigma_\beta}{d\Omega d\boldsymbol{k}} = \frac{v_{k\beta}}{v}|f_{\boldsymbol{k}\beta}(\boldsymbol{q})|^2 \approx |f_{\boldsymbol{k}\beta}(\boldsymbol{q})|^2, \tag{6.6}$$

where, in the approximate equality, the difference in the velocities of the incident and outgoing channels is ignored as before. The solid angle is related to $\boldsymbol{q}$ by $d\Omega = d\boldsymbol{q}/K^2$. The neutron-removal cross section per unit volume of the momentum space around $\boldsymbol{q}$ and around $\boldsymbol{k}$, $d\sigma_{-\mathrm{n}}/d\boldsymbol{q}d\boldsymbol{k}$, is a sum of $d\sigma_\beta/d\Omega d\boldsymbol{k}$ over the final target states β:

$$\frac{d\sigma_{-\mathrm{n}}}{d\boldsymbol{q}d\boldsymbol{k}} = \frac{1}{K^2}\sum_\beta \frac{d\sigma_\beta}{d\Omega d\boldsymbol{k}} = \frac{1}{K^2}\sum_\beta |f_{\boldsymbol{k}\beta}(\boldsymbol{q})|^2. \tag{6.7}$$

The momenta of the core fragment and of the halo neutron are denoted by $\boldsymbol{P}$ and $\boldsymbol{p}_\mathrm{n}$, respectively. (In the rest of this chapter the momentum vector is measured in units of $\hbar$, so it actually stands for the wave number vector.) They are related to $\boldsymbol{k}$ and $\boldsymbol{q}$ by $A_\mathrm{P}^{-1}(A_\mathrm{C}\boldsymbol{p}_\mathrm{n} - \boldsymbol{P}) = \boldsymbol{k}$ and $\boldsymbol{p}_\mathrm{n} + \boldsymbol{P} = \boldsymbol{q}$, which yields

$$\boldsymbol{P} = \frac{A_\mathrm{P}-1}{A_\mathrm{P}}\boldsymbol{q} - \boldsymbol{k}. \tag{6.8}$$

The neutron-removal cross section per unit core momentum $\boldsymbol{P}$ is by definition

$$\frac{d\sigma_{-\mathrm{n}}}{d\boldsymbol{P}} = \int d\boldsymbol{q}\int d\boldsymbol{k}\,\delta\left(\boldsymbol{P} - \frac{A_\mathrm{P}-1}{A_\mathrm{P}}\boldsymbol{q} + \boldsymbol{k}\right)\frac{d\sigma_{-\mathrm{n}}}{d\boldsymbol{q}d\boldsymbol{k}}. \tag{6.9}$$

Since the momentum transfer $\boldsymbol{q}$ is a transverse vector, the momentum in the longitudinal direction remains unaffected by the momentum transfer. Indeed, the above expression can be made simpler if it is integrated over the transverse momentum $\boldsymbol{P}_\perp$. This yields the longitudinal momentum distribution:

$$\begin{aligned}\frac{d\sigma_{-\mathrm{n}}}{dP_\parallel} &= \int d\boldsymbol{P}_\perp \int d\boldsymbol{q}\int d\boldsymbol{k}\,\delta\left(\boldsymbol{P} - \frac{A_\mathrm{P}-1}{A_\mathrm{P}}\boldsymbol{q} + \boldsymbol{k}\right)\frac{d\sigma_{-\mathrm{n}}}{d\boldsymbol{q}d\boldsymbol{k}}\\ &= \int d\boldsymbol{k}\,\delta(P_\parallel + k_z)\frac{d\sigma_{-\mathrm{n}}}{d\boldsymbol{k}},\end{aligned} \tag{6.10}$$

where the neutron-removal cross section per unit relative momentum $\boldsymbol{k}$, $d\sigma_{-\mathrm{n}}/d\boldsymbol{k}$, is obtained by integrating Eq. (6.7) over the momentum transfer $\boldsymbol{q}$ as follows:

$$\begin{aligned}
\frac{d\sigma_{-\mathrm{n}}}{d\boldsymbol{k}} &= \int d\boldsymbol{P}_{\perp} \int d\boldsymbol{q}\, \delta\left(\boldsymbol{P}_{\perp} - \frac{A_{\mathrm{P}}-1}{A_{\mathrm{P}}}\boldsymbol{q} + \boldsymbol{k}_{\perp}\right) \frac{d\sigma_{-\mathrm{n}}}{d\boldsymbol{q}d\boldsymbol{k}} \\
&= \int \frac{d\boldsymbol{q}}{K^2} \sum_{\beta} |f_{\boldsymbol{k}\beta}(\boldsymbol{q})|^2 \\
&= \int d\boldsymbol{b}\, \langle \phi_0 \Phi_0^{\mathrm{C}} \Phi_0^{\mathrm{T}} | \prod_{i\in P} \prod_{j\in T} (1 - \Gamma_{ij}^*) | \phi_{\boldsymbol{k}} \Phi_0^{\mathrm{C}} \rangle \\
&\quad \times \langle \phi_{\boldsymbol{k}} \Phi_0^{\mathrm{C}} | \prod_{i\in P} \prod_{j\in T} (1 - \Gamma_{ij}) | \phi_0 \Phi_0^{\mathrm{C}} \Phi_0^{\mathrm{T}} \rangle,
\end{aligned} \tag{6.11}$$

where the closure property, $\sum_{\beta} |\Phi_{\beta}^{\mathrm{T}}\rangle\langle\Phi_{\beta}^{\mathrm{T}}| = 1$, is used. Equations (6.10) and (6.11) show that the longitudinal momentum distribution of the core fragment obviously depends on the distribution of the relative momentum between the neutron and core in the projectile g.s. The relation between the halo wave function ϕ_0 and the fragment momentum distribution is, however, not so simple.

As is discussed in Sects. 4.2.3 and 5.2.3, the halo nuclei can be broken up by two distinct processes: by elastic and inelastic breakup, which differ in whether or not the target nucleus gets excited as a result of the collision. It was shown in Sect. 5.3.2 that, in the incident energy region of a few tens of MeV per nucleon, the cross sections of elastic and inelastic breakup processes are comparable. At higher incident energies, however, the inelastic breakup process becomes predominant. The breakup induced by the target Coulomb field belongs to the elastic breakup processes. The Coulomb breakup becomes important for heavy target nuclei and will be discussed in Chap. 7.

The elastic and inelastic neutron-removal processes show qualitative differences as far as the final neutron–core relative momentum $\boldsymbol{k}$ is concerned. In the elastic breakup process the halo neutron and the core are scattered elastically by the target, and the breakup occurs because they are diverted differently. Since the momentum transfer in such an elastic collision is small, the processes with small relative momentum $\boldsymbol{k}$ are preponderant. When the relative momentum is small, the relative wave function $\phi_{\boldsymbol{k}}$ substantially deviates from a plane wave owing to the halo-neutron–core interaction. Therefore, $\phi_{\boldsymbol{k}}$ must be constructed by solving the Schrödinger equation for the neutron–core relative motion with appropriate outgoing boundary conditions.

On the other hand, the inelastic breakup process is dominated by violent nucleon–nucleon collisions between the halo neutron and the nucleons in the target. The halo neutron thus suffers large momentum transfers, which makes the neutron–core relative momentum $\boldsymbol{k}$ large.

Then, the neutron–core interaction may be neglected and the final wave function $\phi_{\boldsymbol{k}}$ can be taken a plane wave. The elastic and inelastic breakup processes thus require different treatments for the final wave function, $\phi_{\boldsymbol{k}}$. We work out appropriate theoretical treatments for the two cases separately.

6.2.2 Elastic breakup

The scattering amplitude of the elastic breakup process has been derived in the few-body direct-reaction model [cf. Eq. (5.21)]:

$$f_{\boldsymbol{k}}(\boldsymbol{q}) = -\frac{iK}{2\pi}\int d\boldsymbol{b}\ \mathrm{e}^{-i\boldsymbol{q}\cdot\boldsymbol{b}}\langle\phi_{\boldsymbol{k}}|\mathrm{e}^{i\chi_{\mathrm{CT}}\left(\boldsymbol{b}-\frac{\boldsymbol{s}}{A_{\mathrm{P}}}\right)+i\chi_{\mathrm{nT}}\left(\boldsymbol{b}+\frac{A_{\mathrm{P}}-1}{A_{\mathrm{P}}}\boldsymbol{s}\right)}|\phi_0\rangle. \tag{6.12}$$

The same expression can be derived from the scattering amplitude of Eq. (6.5) in the Glauber theory. To see this, we apply the OLA to the coordinates of the core and the target. The matrix element in Eq. (6.5) can be written more explicitly as

$$\begin{aligned}\langle\phi_{\boldsymbol{k}}\Phi_0^{\mathrm{C}}\Phi_0^{\mathrm{T}}|\prod_{i\in P}\prod_{j\in T}(1-\Gamma_{ij})|\phi_0\Phi_0^{\mathrm{C}}\Phi_0^{\mathrm{T}}\rangle \\ \approx \int d\boldsymbol{r}\ \phi_{\boldsymbol{k}}^*(\boldsymbol{r})\phi_0(\boldsymbol{r})\mathrm{e}^{i\chi_{\mathrm{nT}}\left(\boldsymbol{b}+\frac{A_{\mathrm{P}}-1}{A_{\mathrm{P}}}\boldsymbol{s}\right)+i\chi_{\mathrm{CT}}\left(\boldsymbol{b}-\frac{\boldsymbol{s}}{A_{\mathrm{P}}}\right)},\end{aligned} \tag{6.13}$$

where the correct treatment of the c.m. is important and is done as explained in Eq. (4.31). Substituting this result into Eq. (6.5) produces Eq. (6.12), but, of course, unlike the original Eq. (5.21), it involves the interaction between the bare nucleons. The neutron-removal cross section in elastic breakup per unit core momentum $\boldsymbol{q}$ and $\boldsymbol{k}$ is given [cf. Eq. (6.7)] by

$$\frac{d\sigma_{-\mathrm{n}}^{\mathrm{el}}}{d\boldsymbol{q}d\boldsymbol{k}} = \frac{1}{K^2}|f_{\boldsymbol{k}}(\boldsymbol{q})|^2, \tag{6.14}$$

and the neutron-removal cross section per unit core momentum $\boldsymbol{P}$ is given by Eq. (6.9).

To obtain a more explicit expression for the elastic breakup cross section in terms of the longitudinal momentum distribution, one should follow Eqs. (6.10) and (6.11). Substituting Eq. (6.13) into Eq. (6.11) results in

$$\frac{d\sigma_{-\mathrm{n}}^{\mathrm{el}}}{d\boldsymbol{k}} = \int d\boldsymbol{b}\left|\langle\phi_{\boldsymbol{k}}|\mathrm{e}^{i\chi_{\mathrm{CT}}\left(\boldsymbol{b}-\frac{\boldsymbol{s}}{A_{\mathrm{P}}}\right)+i\chi_{\mathrm{nT}}\left(\boldsymbol{b}+\frac{A_{\mathrm{P}}-1}{A_{\mathrm{P}}}\boldsymbol{s}\right)}|\phi_0\rangle\right|^2. \tag{6.15}$$

As was mentioned in Sect. 6.2.1, the neutron–core interaction after the collision can be taken into account by using the eigenfunction of the relative Hamiltonian H_{nC} as $\phi_{\boldsymbol{k}}$.

Although the relation between the fragment momentum distribution and the halo wave function ϕ_0 is still not transparent, we cannot go further without the explicit construction of the final wave function $\phi_{\boldsymbol{k}}$. This function can be obtained by solving the Schrödinger equation for the neutron–core relative motion. For the reactions of two-neutron-halo nuclei, the final state is to be constructed by solving a continuum three-body problem [124, 125].

6.2.3 Inelastic breakup

Before giving a precise but involved derivation for the expression of the inelastic breakup cross section in terms of a momentum distribution, we give a heuristic derivation now, which will be more illuminating in respect of the physical content. The heuristic derivation is based on a spectator approximation of the breakup process [126–129]. The rigorous derivation, which does not resort to assumptions going beyond the Glauber theory will be given in Sect. 6.2.4. A similar result can also be obtained in the framework of the direct reaction theory [130].

Let us start with the wave function after collision in the projectile-rest frame. For the few-body direct reaction theory based on the eikonal approximation, this is given in Eq. (5.19). In the spectator approximation to be introduced the neutron is the spectator of the core–target collision, and one can allot it this role by neglecting the neutron–target interaction. With this approximation, Eq. (5.19) reduces to

$$\hat{\Psi}(\boldsymbol{b},\boldsymbol{r}) = \mathrm{e}^{i\chi_{\mathrm{CT}}\left(\boldsymbol{b}-\frac{\boldsymbol{s}}{A_{\mathrm{P}}}\right)}\phi_0(\boldsymbol{r}). \tag{6.16}$$

This wave function should now be expressed in terms of the coordinates $\boldsymbol{r}_{\mathrm{n}}$ of the neutron and $\boldsymbol{R}_{\mathrm{C}}$ of the core nucleus. These are related to the projectile–target and neutron–core relative coordinates $\boldsymbol{R}$ and $\boldsymbol{r}$ by

$$\boldsymbol{R}_{\mathrm{C}} = \boldsymbol{R} - \frac{\boldsymbol{r}}{A_{\mathrm{P}}}, \qquad \boldsymbol{r}_{\mathrm{n}} = \boldsymbol{R} + \frac{A_{\mathrm{P}}-1}{A_{\mathrm{P}}}\boldsymbol{r}. \tag{6.17}$$

The transverse component of $\boldsymbol{R}$ is the impact parameter vector $\boldsymbol{b}$. The wave function (6.16) is now written as $\hat{\Psi} = \mathrm{e}^{i\chi_{\mathrm{CT}}(\boldsymbol{S}_{\mathrm{C}})}\phi_0(\boldsymbol{r}_{\mathrm{n}}-\boldsymbol{R}_{\mathrm{C}})$, where $\boldsymbol{S}_{\mathrm{C}}$ is the transverse component of $\boldsymbol{R}_{\mathrm{C}}$. The Fourier transform of this wave function with respect to the coordinate of the core, $\boldsymbol{R}_{\mathrm{C}}$, gives the amplitude of the momentum distribution of the core for any particular position $\boldsymbol{r}_{\mathrm{n}}$ of the spectator neutron.

The momentum distribution of the core fragment, $g_{\mathrm{C}}(\boldsymbol{b}_{\mathrm{n}},\boldsymbol{P})$, may be defined as $g_{\mathrm{C}}(\boldsymbol{b}_{\mathrm{n}},\boldsymbol{P}) = (2\pi)^{-3}|\int d\boldsymbol{R}_{\mathrm{C}}\,\mathrm{e}^{-i\boldsymbol{P}\cdot\boldsymbol{R}_{\mathrm{C}}}\hat{\Psi}|^2$. With the transverse component of $\boldsymbol{r}_{\mathrm{n}}$ denoted by $\boldsymbol{b}_{\mathrm{n}}$, this takes the form

$$g_{\mathrm{C}}(\boldsymbol{b}_{\mathrm{n}},\boldsymbol{P}) = \frac{1}{(2\pi)^3}\left|\int d\boldsymbol{R}_{\mathrm{C}}\,\mathrm{e}^{-i\boldsymbol{P}\cdot\boldsymbol{R}_{\mathrm{C}}+i\chi_{\mathrm{CT}}(\boldsymbol{S}_{\mathrm{C}})}\phi_0(\boldsymbol{r}_{\mathrm{n}}-\boldsymbol{R}_{\mathrm{C}})\right|^2$$

$$= \frac{1}{(2\pi)^3} \left| \int d\boldsymbol{r}\, e^{i\boldsymbol{P}\cdot\boldsymbol{r}+i\chi_{CT}(\boldsymbol{b}_n-\boldsymbol{s})} \phi_0(\boldsymbol{r}) \right|^2 . \tag{6.18}$$

Note that the halo-neutron coordinate $\boldsymbol{r}_n$ appears only through its transverse component $\boldsymbol{b}_n$, which may be regarded as an impact parameter of the halo neutron.

We now assign a role to the halo neutron, which has been treated as a spectator and has been ignored in deriving Eq. (6.18) so far. For the core fragment to be observed in an inelastic breakup process, the halo neutron must excite the target nucleus. The probability of this process is given by $1-e^{-2\mathrm{Im}\chi_{nT}(\boldsymbol{b}_n)}$, as is stated after Eq. (4.38). To yield the inelastic breakup cross section, this probability must be multiplied by the momentum distribution (6.18) and integrated over the impact parameter $\boldsymbol{b}_n$. The inelastic breakup cross section per unit momentum $\boldsymbol{P}$ is then

$$\frac{d\sigma_{-n}^{\mathrm{inel}}}{d\boldsymbol{P}} = \int d\boldsymbol{b}_n \left[1 - e^{-2\mathrm{Im}\chi_{nT}(\boldsymbol{b}_n)}\right] \frac{1}{(2\pi)^3} \left| \int d\boldsymbol{r} e^{i\boldsymbol{P}\cdot\boldsymbol{r}+i\chi_{CT}(\boldsymbol{b}_n-\boldsymbol{s})} \phi_0(\boldsymbol{r}) \right|^2 . \tag{6.19}$$

Integrating this over the momentum $\boldsymbol{P}$ yields

$$\begin{aligned} \sigma_{-n}^{\mathrm{inel}} &= \int d\boldsymbol{P} \frac{d\sigma_{-n}^{\mathrm{inel}}}{d\boldsymbol{P}} \\ &= \int d\boldsymbol{b}_n \int d\boldsymbol{r} |\phi_0(\boldsymbol{r})|^2 e^{-2\mathrm{Im}\chi_{CT}(\boldsymbol{b}_n-\boldsymbol{s})} \left[1 - e^{-2\mathrm{Im}\chi_{nT}(\boldsymbol{b}_n)}\right] \\ &= \int d\boldsymbol{b}_C\, e^{-2\mathrm{Im}\chi_{CT}(\boldsymbol{b}_C)} \langle \phi_0 | 1 - e^{-2\mathrm{Im}\chi_{nT}(\boldsymbol{b}_C+\boldsymbol{s})} | \phi_0 \rangle. \end{aligned} \tag{6.20}$$

In the second line $(2\pi)^{-3}\int d\boldsymbol{P}\, |\int d\boldsymbol{r} e^{i\boldsymbol{P}\cdot\boldsymbol{r}} f(\boldsymbol{r})|^2 = \int d\boldsymbol{r} |f(\boldsymbol{r})|^2$ has been used, and in the third line the integration variable has been changed from $\boldsymbol{b}_n$ to $\boldsymbol{b}_C = \boldsymbol{b}_n - \boldsymbol{s}$, which is the impact parameter of the core. The result coincides with Eqs. (4.38) and (5.28).

The formula (6.19) indicates that the momentum distribution of the core fragment in the inelastic breakup is the squared Fourier transform of the initial orbit ϕ_0 provided the phase-shift function of the core–target collision χ_{CT} can be neglected. If this condition is fulfilled, the momentum distribution of the neutron–core relative motion in the projectile may be directly compared with the measured fragment momentum distribution. This assumption is called the transparent limit in the Serber model for deuteron stripping reaction. Simple and intuitive as it is, the core–target phase-shift function $\chi_{CT}(\boldsymbol{b}_n - \boldsymbol{s})$ may cause a significant difference between the Fourier transform and the full expression, Eq. (6.19) as will be seen in Sect. 6.3.2. Moreover, the integrated cross section becomes too large in the transparent limit.

Since χ_{CT} depends only on the transverse coordinates, it does not affect the longitudinal momentum distribution. This can easily be understood intuitively as follows: Since the core is assumed to be in its g.s. throughout, it can only be scattered elastically by the target. In the eikonal approximation, the momentum transfer in the elastic scattering is in the transverse direction. The longitudinal momentum is therefore not affected by the collision.

The longitudinal momentum distribution is obtained by integrating Eq. (6.19) over the transverse momentum $\boldsymbol{P}_{\perp}$:

$$\begin{aligned}\frac{d\sigma_{-n}^{\mathrm{inel}}}{dP_{\parallel}} &= \int d\boldsymbol{P}_{\perp} \frac{d\sigma_{-n}^{\mathrm{inel}}}{d\boldsymbol{P}} \\ &= \int d\boldsymbol{b}_{\mathrm{C}} \int d\boldsymbol{s}\, \mathrm{e}^{-2\mathrm{Im}\chi_{\mathrm{CT}}(\boldsymbol{b}_{\mathrm{C}})} \left[1 - \mathrm{e}^{-2\mathrm{Im}\chi_{n\mathrm{T}}(\boldsymbol{b}_{\mathrm{C}}+\boldsymbol{s})}\right] \\ &\quad \times \frac{1}{2\pi} \left| \int dz\, \mathrm{e}^{iP_{\parallel} z} \phi_0(\boldsymbol{r}) \right|^2, \qquad (6.21)\end{aligned}$$

where the relative coordinate $\boldsymbol{r}$ is $\boldsymbol{r} = (\boldsymbol{s}, z)$. The longitudinal momentum distribution of the core fragment is nearly proportional to the momentum distribution of the halo neutron in the projectile, and that is the anticipated direct relationship. The contributions coming from different values of the transverse coordinate $\boldsymbol{s}$ are piled up, and, as a fingerprint of the reaction mechanism, the phase-shift function assigns different weights to different values of $\boldsymbol{s}$. This causes some departure from the proportionality. It is clear that the longitudinal direction is distinguished, and no similar formula can be derived for the transverse momentum distribution.

Let us now examine what is the effect of the phase-shift functions

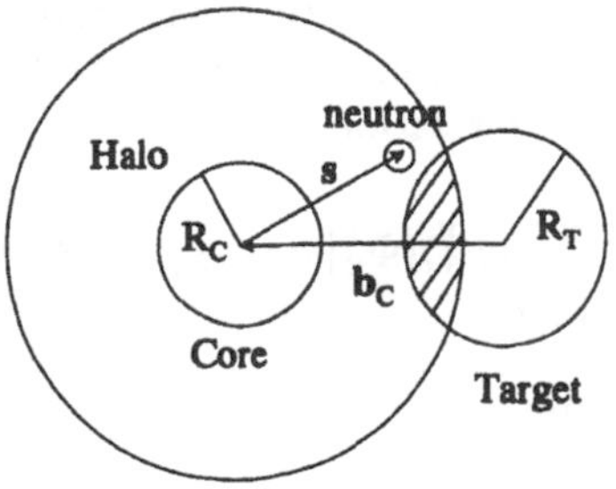

Figure 6.2: Coordinate system for the collision of a halo nucleus with a target. The halo neutron in the shaded area contributes to the momentum distribution of the core.

on the momentum distribution. Figure 6.2 shows a configuration of the core, the halo neutron and the target nucleus in the xy-plane, which is perpendicular to the incident beam. The core and target represented by circles of radius $R_{\rm C}$ and $R_{\rm T}$, respectively, are separated by the impact parameter $b_{\rm C}$. The distribution of the halo neutron is indicated by a large circle around the core. The neutron–core distance in transverse direction is denoted by s. The phase-shift functions act in the following way: The core–target phase-shift function $e^{-2{\rm Im}\chi_{\rm CT}(b_{\rm C})}$ behaves like a step function $\Theta(b_{\rm C}-R_{\rm C}-R_{\rm T})$; it is close to unity when the core and target do not overlap, i.e., $b_{\rm C}>R_{\rm C}+R_{\rm T}$. It soon approaches zero where the two nuclei overlap. This factor guarantees that the core nucleus does not interact strongly with the target so that it remains in the g.s. after collision. For the factor $1-e^{-2{\rm Im}\chi_{\rm nT}(\boldsymbol{b}_{\rm C}+\boldsymbol{s})}$ to be non-zero, the halo neutron must hit the target so that it can excite it; i.e., the condition $|\boldsymbol{b}_{\rm C}+\boldsymbol{s}|<R_{\rm T}$ must be satisfied. Both conditions $b_{\rm C}>R_{\rm C}+R_{\rm T}$ and $|\boldsymbol{b}_{\rm C}+\boldsymbol{s}|<R_{\rm T}$ are satisfied in the shaded region. The phase-shift functions have a significant effect only when the neutron is inside the shaded area. These conditions indicate that the longitudinal momentum distribution of the core picks up contributions from the wave function of the halo neutron only from regions that are outside the core.

6.2.4 Derivation in the Glauber theory

The subject of this section is the derivation of the cross section of the inelastic breakup, Eq. (6.19), from Eq. (6.11) within the usual confines of the Glauber theory for halo nuclei. The integration with respect to the core and target coordinates will be performed with the OLA, but no extra approximation will be introduced to treat the motion of the halo neutron. The cross section per unit momentum transfer $\boldsymbol{q}$ and unit relative momentum $\boldsymbol{k}$ of the inelastic breakup, $d\sigma_{-\rm n}^{\rm inel}/d\boldsymbol{q}d\boldsymbol{k}$, is obtained by omitting the g.s., $\beta=0$, in the sum (6.7). Substituting $f_{\boldsymbol{k}\beta}(\boldsymbol{q})$ of Eq. (6.5) and the closure relation

$$\sum_{\beta\neq 0}|\Phi_\beta^{\rm T}\rangle\langle\Phi_\beta^{\rm T}| = 1-|\Phi_0^{\rm T}\rangle\langle\Phi_0^{\rm T}| \tag{6.22}$$

into $d\sigma_{-\rm n}^{\rm inel}/d\boldsymbol{q}d\boldsymbol{k}$ results in

$$\begin{aligned}\frac{d\sigma_{-\rm n}^{\rm inel}}{d\boldsymbol{q}d\boldsymbol{k}} &= \frac{1}{K^2}\sum_{\beta\neq 0}|f_{\boldsymbol{k}\beta}(\boldsymbol{q})|^2\\ &= \frac{1}{(2\pi)^2}\int d\boldsymbol{b}\int d\boldsymbol{b}'\, e^{-i\boldsymbol{q}\cdot(\boldsymbol{b}-\boldsymbol{b}')}\end{aligned}$$

$$\times\left[\langle\phi_0\Phi_0^{\rm C}\Phi_0^{\rm T}|\prod_{i\in P}\prod_{j\in T}(1-\Gamma'^{*}_{ij})|\phi_k\Phi_0^{\rm C}\rangle\right.$$
$$\times\langle\phi_k\Phi_0^{\rm C}|\prod_{i\in P}\prod_{j\in T}(1-\Gamma_{ij})|\phi_0\Phi_0^{\rm C}\Phi_0^{\rm T}\rangle$$
$$-\langle\phi_0\Phi_0^{\rm C}\Phi_0^{\rm T}|\prod_{i\in P}\prod_{j\in T}(1-\Gamma'^{*}_{ij})|\phi_k\Phi_0^{\rm C}\Phi_0^{\rm T}\rangle$$
$$\left.\times\ \langle\phi_k\Phi_0^{\rm C}\Phi_0^{\rm T}|\prod_{i\in P}\prod_{j\in T}(1-\Gamma_{ij})|\phi_0\Phi_0^{\rm C}\Phi_0^{\rm T}\rangle\right]. \quad (6.23)$$

The profile function Γ'_{ij} is primed to indicate that its argument contains $\boldsymbol{b}'$.

The second term of Eq. (6.23) is a product of two matrix elements, whose OLA approximants are given in Eq. (6.13). The first term, however, has a structure which we have not encountered before. To apply the OLA to this, the following function has to be defined:

$$G(\boldsymbol{b},\boldsymbol{b}';\boldsymbol{s}_1^{\rm P},\boldsymbol{s}_1^{\rm P'};\lambda)=\int d\boldsymbol{r}_1^{\rm T}\cdots\int d\boldsymbol{r}_{A_{\rm T}}^{\rm T}\,|\Phi_0^{\rm T}(\boldsymbol{r}_1^{\rm T},\ldots,\boldsymbol{r}_{A_{\rm T}}^{\rm T})|^2$$
$$\times\int d\boldsymbol{r}_2^{\rm P}\cdots\int d\boldsymbol{r}_{A_{\rm P}}^{\rm P}\,|\Phi_0^{\rm C}(\boldsymbol{r}_2^{\rm P},\ldots,\boldsymbol{r}_{A_{\rm P}}^{\rm P})|^2$$
$$\times\int d\boldsymbol{r}_2^{\rm P'}\cdots\int d\boldsymbol{r}_{A_{\rm P}}^{\rm P}{}'\left|\Phi_0^{\rm C}(\boldsymbol{r}_2^{\rm P'},\ldots,\boldsymbol{r}_{A_{\rm P}}^{\rm P}{}')\right|^2$$
$$\times\prod_{i\in P}\prod_{j\in T}\left\{1-\lambda\left[\Gamma^*(\boldsymbol{b}'+\boldsymbol{s}_i^{\rm P'}-\boldsymbol{s}_j^{\rm T})+\Gamma(\boldsymbol{b}+\boldsymbol{s}_i^{\rm P}-\boldsymbol{s}_j^{\rm T})\right.\right.$$
$$\left.\left.-\Gamma^*(\boldsymbol{b}'+\boldsymbol{s}_i^{\rm P'}-\boldsymbol{s}_j^{\rm T})\Gamma(\boldsymbol{b}+\boldsymbol{s}_i^{\rm P}-\boldsymbol{s}_j^{\rm T})\right]\right\}. \quad (6.24)$$

The function $\ln G$ should be expanded around $\lambda = 0$, and only the zeroth- and first-order terms should be retained. Putting $\lambda = 1$, one obtains

$$G(\boldsymbol{b},\boldsymbol{b}';\boldsymbol{s}_1^{\rm P},\boldsymbol{s}_1^{\rm P'};\lambda=1)$$
$$\approx\exp\left\{-\int d\boldsymbol{r}_1^{\rm T}\cdots\int d\boldsymbol{r}_{A_{\rm T}}^{\rm T}\,|\Phi_0^{\rm T}(\boldsymbol{r}_1^{\rm T},\ldots,\boldsymbol{r}_{A_{\rm T}}^{\rm T})|^2\right.$$
$$\times\int d\boldsymbol{r}_2^{\rm P}\cdots\int d\boldsymbol{r}_{A_{\rm P}}^{\rm P}\,|\Phi_0^{\rm C}(\boldsymbol{r}_2^{\rm P},\ldots,\boldsymbol{r}_{A_{\rm P}}^{\rm P})|^2$$
$$\times\int d\boldsymbol{r}_2^{\rm P'}\cdots\int d\boldsymbol{r}_{A_{\rm P}}^{\rm P}{}'\left|\Phi_0^{\rm C}(\boldsymbol{r}_2^{\rm P'},\ldots,\boldsymbol{r}_{A_{\rm P}}^{\rm P}{}')\right|^2$$
$$\times\sum_{i\in P}\sum_{j\in T}\left[\Gamma^*(\boldsymbol{b}'+\boldsymbol{s}_i^{\rm P'}-\boldsymbol{s}_j^{\rm T})+\Gamma(\boldsymbol{b}+\boldsymbol{s}_i^{\rm P}-\boldsymbol{s}_j^{\rm T})\right.$$

$$-\Gamma^*(\boldsymbol{b}'+\boldsymbol{s}_i^{\mathrm{P}'}-\boldsymbol{s}_j^{\mathrm{T}})\Gamma(\boldsymbol{b}+\boldsymbol{s}_i^{\mathrm{P}}-\boldsymbol{s}_j^{\mathrm{T}})\Big]\Big\}. \tag{6.25}$$

Distinguishing the halo neutron from the core nucleons and introducing χ_{nT}, χ_{CT} [cf. Eqs. (4.21) and (4.22)] and two analogous quantities, χ_{nnT} and χ_{CCT}, one obtains

$$\begin{aligned}G(\boldsymbol{b},\boldsymbol{b}';\boldsymbol{s}_1^{\mathrm{P}},\boldsymbol{s}_1^{\mathrm{P}'};\lambda=1) &\approx \exp[-i\chi_{\mathrm{nT}}^*(\boldsymbol{b}'+\boldsymbol{s}_1^{\mathrm{P}'})\\ &+i\chi_{\mathrm{nT}}(\boldsymbol{b}+\boldsymbol{s}_1^{\mathrm{P}})+\chi_{\mathrm{nnT}}(\boldsymbol{b}'+\boldsymbol{s}_1^{\mathrm{P}'},\boldsymbol{b}+\boldsymbol{s}_1^{\mathrm{P}})\\ &-i\chi_{\mathrm{CT}}^*(\boldsymbol{b}')+i\chi_{\mathrm{CT}}(\boldsymbol{b})+\chi_{\mathrm{CCT}}(\boldsymbol{b}',\boldsymbol{b})].\end{aligned} \tag{6.26}$$

The function χ_{nnT} is defined as

$$\begin{aligned}\chi_{\mathrm{nnT}}(\boldsymbol{b}',\boldsymbol{b}) &= \langle\Phi_0^{\mathrm{T}}|\sum_{j\in T}\Gamma^*(\boldsymbol{b}'-\boldsymbol{s}_j^{\mathrm{T}})\Gamma(\boldsymbol{b}-\boldsymbol{s}_j^{\mathrm{T}})|\Phi_0^{\mathrm{T}}\rangle\\ &= \int d\boldsymbol{r}^{\mathrm{T}}\rho_{\mathrm{T}}(\boldsymbol{r}^{\mathrm{T}})\Gamma^*(\boldsymbol{b}'-\boldsymbol{s}^{\mathrm{T}})\Gamma(\boldsymbol{b}-\boldsymbol{s}^{\mathrm{T}}).\end{aligned} \tag{6.27}$$

Since $\chi_{\mathrm{nnT}}(\boldsymbol{b},\boldsymbol{b}')$ includes a product of the profile functions, its value must differ appreciably from zero for $|\boldsymbol{b}-\boldsymbol{b}'|\sim a$, where a is the range of the nuclear force. The function $\chi_{\mathrm{CCT}}(\boldsymbol{b},\boldsymbol{b}')$ is defined as

$$\begin{aligned}\chi_{\mathrm{CCT}}(\boldsymbol{b}',\boldsymbol{b}) &= \sum_{i\in C}\sum_{j\in T}\langle\Phi_0^{\mathrm{C}}\Phi_0^{\mathrm{T}}|\Gamma^*(\boldsymbol{b}'+\boldsymbol{s}_i^{\mathrm{C}}-\boldsymbol{s}_j^{\mathrm{T}})|\Phi_0^{\mathrm{C}}\rangle\\ &\quad\times\langle\Phi_0^{\mathrm{C}}|\Gamma(\boldsymbol{b}+\boldsymbol{s}_i^{\mathrm{C}}-\boldsymbol{s}_j^{\mathrm{T}})|\Phi_0^{\mathrm{C}}\Phi_0^{\mathrm{T}}\rangle\\ &= \frac{1}{A_{\mathrm{C}}}\int d\boldsymbol{r}^{\mathrm{T}}\rho_{\mathrm{T}}(\boldsymbol{r}^{\mathrm{T}})\int d\boldsymbol{r}^{\mathrm{C}}\rho_{\mathrm{C}}(\boldsymbol{r}^{\mathrm{C}})\int d\boldsymbol{r}^{\mathrm{C}'}\rho_{\mathrm{C}}(\boldsymbol{r}^{\mathrm{C}'})\\ &\quad\times\Gamma^*(\boldsymbol{b}'+\boldsymbol{s}^{\mathrm{C}'}-\boldsymbol{s}^{\mathrm{T}})\Gamma(\boldsymbol{b}+\boldsymbol{s}^{\mathrm{C}}-\boldsymbol{s}^{\mathrm{T}}).\end{aligned} \tag{6.28}$$

Note that the integration variables $\boldsymbol{s}_i^{\mathrm{P}}$ and $\boldsymbol{s}_i^{\mathrm{P}'}$ have been expressed in terms of $\boldsymbol{s}_i^{\mathrm{C}}$ and $\boldsymbol{s}_i^{\mathrm{C}'}$ since the coordinates of the core nucleons should be measured from the core c.m. rather than from the projectile c.m.

The result (6.26) has been obtained by neglecting the difference between the c.m. of the projectile and the core. However, for the momentum distribution of the core fragment, a proper treatment of the difference in the c.m. is important. The c.m. can be correctly treated along the lines of Sect. 4.2.3. For example, in Eq. (6.26) the impact parameter $\boldsymbol{b}$ is to be replaced by the core–target impact parameter $\boldsymbol{b}-A_{\mathrm{P}}^{-1}\boldsymbol{s}_1^{\mathrm{P}}$, where $\boldsymbol{s}_1^{\mathrm{P}}$ denotes the transverse component of the halo-neutron–core relative vector $\boldsymbol{r}_1^{\mathrm{P}}$. These changes amount to replacing Eq. (6.26) by

$$G(\boldsymbol{b},\boldsymbol{b}';\boldsymbol{s}_1^{\mathrm{P}},\boldsymbol{s}_1^{\mathrm{P}'};\lambda=1)$$

$$\approx \exp\left[-i\chi^*_{\rm nT}\left(b' + \frac{A_P - 1}{A_P}s_1^{P'}\right) + i\chi_{\rm nT}\left(b + \frac{A_P - 1}{A_P}s_1^{P}\right)\right.$$
$$+ \chi_{\rm nnT}\left(b' + \frac{A_P - 1}{A_P}s_1^{P'}, b + \frac{A_P - 1}{A_P}s_1^{P}\right)$$
$$\left. - i\chi^*_{\rm CT}\left(b' - \frac{s_1^{P'}}{A_P}\right) + i\chi_{\rm CT}\left(b - \frac{s_1^{P}}{A_P}\right) + \chi_{\rm CCT}\left(b' - \frac{s_1^{P'}}{A_P}, b - \frac{s_1^{P}}{A_P}\right)\right]. \tag{6.29}$$

The term $\chi_{\rm CCT}$ in the exponent of G is known to influence the results insignificantly, and will be dropped. Substituting Eq. (6.29) in Eq. (6.23) gives

$$\frac{d\sigma_{-n}^{\rm inel}}{dqdk} = \frac{1}{(2\pi)^2}\int db \int db'\, e^{-iq\cdot(b-b')}\int dr \int dr' \phi_0(r)\phi_k^*(r)\phi_k(r')\phi_0^*(r')$$
$$\times \exp\left[-i\chi^*_{\rm nT}\left(b' + \frac{A_P - 1}{A_P}s'\right) + i\chi_{\rm nT}\left(b + \frac{A_P - 1}{A_P}s\right)\right.$$
$$\left. - i\chi^*_{\rm CT}\left(b' - \frac{s'}{A_P}\right) + i\chi_{\rm CT}\left(b - \frac{s}{A_P}\right)\right]$$
$$\times\left\{\exp\left[\chi_{\rm nnT}\left(b' + \frac{A_P - 1}{A_P}s', b + \frac{A_P - 1}{A_P}s\right)\right] - 1\right\}. \tag{6.30}$$

As was noted earlier, the final wave function ϕ_k may be approximated by a plane wave for inelastic breakup process: $\phi_k \approx (2\pi)^{-3/2}e^{ik\cdot r}$. To obtain the cross section per unit momentum of the core fragment $\boldsymbol{P}$, integration must be performed over $\boldsymbol{q}$ and $\boldsymbol{k}$ while taking into account the constraint (6.8) coming from the conservation of the total momentum. In performing the integrations over $\boldsymbol{q}$ and $\boldsymbol{k}$ the following relation will be used:

$$\int dq \int dk\, \delta\left(P - \frac{A_P - 1}{A_P}q + k\right) e^{-iq\cdot(b-b')}\phi_k^*(r)\phi_k(r')$$
$$= \frac{1}{2\pi}\delta\left[b - b' + \frac{A_P - 1}{A_P}(s - s')\right] e^{iP\cdot(r-r')}. \tag{6.31}$$

With this, the cross section can be expressed as

$$\frac{d\sigma_{-n}^{\rm inel}}{dP} = \int dq \int dk\, \delta\left(P - \frac{A_P - 1}{A_P}q + k\right)\frac{d\sigma_{-n}^{\rm inel}}{dqdk}$$
$$= \int db_n\, e^{-2{\rm Im}\chi_{\rm nT}(b_n)}\left[e^{\chi_{\rm nnT}(b_n, b_n)} - 1\right]$$
$$\times \frac{1}{(2\pi)^3}\int dr \int dr' \phi_0(r)\phi_0^*(r')e^{iP\cdot(r-r') - i\chi^*_{\rm CT}(b_n - s') + i\chi_{\rm CT}(b_n - s)}$$

$$= \int d\boldsymbol{b}_{\rm n} e^{-2{\rm Im}\chi_{\rm nT}(\boldsymbol{b}_{\rm n})} \left[e^{\chi_{\rm nnT}(\boldsymbol{b}_{\rm n},\boldsymbol{b}_{\rm n})} - 1 \right]$$

$$\times \frac{1}{(2\pi)^3} \left| \int d\boldsymbol{r} \phi_0(\boldsymbol{r}) e^{i\boldsymbol{P}\cdot\boldsymbol{r} + i\chi_{\rm CT}(\boldsymbol{b}_{\rm n}-\boldsymbol{s})} \right|^2 . \qquad (6.32)$$

The function $\chi_{\rm nnT}(\boldsymbol{b},\boldsymbol{b}')$ with $\boldsymbol{b}=\boldsymbol{b}'$ appearing in Eq. (6.32) can be expressed in terms of $\chi_{\rm nT}(\boldsymbol{b})$. Employing the unitarity relation of the profile function, $|1-\Gamma|^2=1$, one obtains

$$\chi_{\rm nnT}(\boldsymbol{b},\boldsymbol{b}) = 2{\rm Im}\chi_{\rm nT}(\boldsymbol{b}). \qquad (6.33)$$

Putting this relation into Eq. (6.32) leads to Eq. (6.19) as envisaged.

This meticulous derivation verifies the heuristic derivation presented in Sect. 6.2.3. All in all, the Glauber theory readily provides the convenient relationship between the fragment momentum distribution of the inelastic breakup and the neutron–core momentum distribution within the projectile, without resort to any obscure spectator assumption.

6.3 Applications

6.3.1 A reaction with ^{11}Be

The single-neutron halo in ^{11}Be occupies mainly the $1s_{1/2}$ orbit. The formula derived in Sect. 6.2 can be readily applied to this case. Figure 6.3 shows the calculated longitudinal and transverse momentum distribution of ^{10}Be in the ^{11}Be+^{12}C collision at $E=800\,A$ MeV. The calculation only includes the inelastic breakup process since that is dominant at high incident energies as has been demonstrated in Fig. 4.3. The longitudinal momentum distribution has been calculated according to Eq. (6.21). The transverse momentum distribution has been calculated from Eq. (6.19) by integrating over P_y and P_z. The halo-neutron wave function has again been constructed from a Woods–Saxon potential. The potential parameters are set to reproduce the binding energy (0.503 MeV) of the halo neutron. The phase-shift functions have been constructed in the Glauber theory with the nucleon–nucleon profile function and the relevant nuclear densities. The squared Fourier transform of the halo wave function is shown by the solid curve. The longitudinal and transverse momentum distributions including the reaction effects are shown by dotted and dashed curves, respectively. It is apparent that the reaction dynamics makes the momentum distribution of the core fragment in the reaction slightly narrower than it is, relative to the halo neutron, within the nucleus. The transverse momentum distribution has a high-momentum component, reflecting the

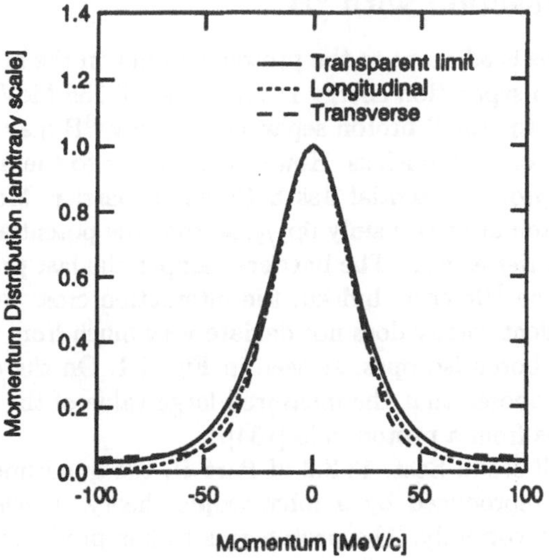

Figure 6.3: Calculated momentum distribution of the core nucleus ^{10}Be in the single-neutron removal from ^{11}Be in the ^{11}Be+^{12}C collision at the incident energy of 800 A MeV. The longitudinal (dotted curve) and the transverse (dashed curve) momentum distributions are compared with the distribution produced by the squared Fourier transform of the halo wave function (solid curve).

distortion by the core–target scattering. The rescaling of the contributions of the different domains of the transverse coordinate caused by the reaction dynamics has little effect, indeed.

At lower incident energies, say around and below 100 A MeV, the elastic breakup cross section becomes comparable in magnitude with the inelastic breakup cross section (cf. Sect. 5.3.3). The measured longitudinal momentum distribution of ^{10}Be in the ^{11}Be+^{9}Be collision at $E = 63\,A$ MeV has been analysed, with the elastic breakup included according to Eq. (6.10). It has been found that the shape of the momentum distribution obtained by including elastic as well as the inelastic breakup is similar to that extracted from the inelastic breakup alone [129].

The neutron momentum distribution coming from the ^{11}Be projectile has also been measured [131, 132]. Since the removed halo neutron suffers distortion by the target, a theory incorporating neutron–target final-state interactions is required to describe the neutron momentum distribution [133].

6.3.2 A reaction with ^{8}B

The ^{8}B nucleus is adjacent to the proton drip line in the nuclear chart, and the proton separation energy is very small, 0.138 MeV.

Because of the small proton separation energy, ^{8}B may be a candidate for a proton-halo nucleus. However, contrary to the halo neutron in ^{11}Be, the proton potential has a Coulomb barrier. Moreover, the dominant proton orbit is mainly $0p_{3/2}$, so that the potential contains a centrifugal barrier as well. The barriers hamper the last proton's staying far from the ^{7}Be core. Indeed, the interaction cross section of ^{8}B at a high incident energy does not deviate very much from the systematics of other boron isotopes, as seen in Fig. 1.1. On the other hand, it has been proposed that the measured large value of the quadrupole moment comes from a proton halo [134].

As we shall see in Sect. 13.3.3 of Part II, the quadrupole moment is fairly well reproduced by a microscopic theory, which treats the Coulomb force correctly. The proton rms radius produced is 0.49 fm larger than the neutron radius, and the proton spectroscopic amplitude[2] has a moderately long tail. As for the longitudinal momentum distribution of the ^{7}Be fragment, a very narrow distribution has been observed [135], but, as we shall see soon, that is partly an artefact produced by the reaction mechanism.

For a halo nucleon in an l-orbit, the formula (6.21) for the longitudinal momentum distribution has to be modified so as to include an averaging over the magnetic substates:

$$\frac{d\sigma_{-n}^{\mathrm{inel}}}{dP_{\parallel}} = \int d\boldsymbol{b}_{\mathrm{C}} \int d\boldsymbol{s}\, \mathrm{e}^{-2\mathrm{Im}\chi_{\mathrm{CT}}(\boldsymbol{b}_{\mathrm{C}})} \left[1 - \mathrm{e}^{-2\mathrm{Im}\chi_{\mathrm{nT}}(\boldsymbol{b}_{\mathrm{C}}+\boldsymbol{s})}\right] \times \frac{1}{2\pi}\frac{1}{2l+1}\sum_{m=-l}^{l}\left|\int dz\, \mathrm{e}^{iP_{\parallel}z}\phi_0^{(m)}(\boldsymbol{r})\right|^2, \tag{6.34}$$

where $\phi_0^{(m)}$ represents the halo-nucleon orbit of magnetic substate m. In this formula the reaction dynamics gives a specific weight to each magnetic substate through the integration over $\boldsymbol{s}$. This makes the magnetic substates behave differently [128, 136, 137]. For the ^{8}B=p+^{7}Be system, the width of the fragment momentum distribution turns out to be about half of that of the halo orbit.

The density distribution of the halo proton given by $|\phi_0^{(m)}|^2$ depends on the orientation of the p-orbit. The z-direction is that of the incident beam, in other words, the longitudinal direction. The density as well as the momentum distribution of the $m = 0$ orbit is elongated along the z-axis due to the nodal structure of the wave function. Since the

[2] The s.p. spectroscopic amplitude can be viewed roughly as a s.p. orbit.

density distribution decreases rapidly along the xy-direction, this orbit gives a small contribution to the proton-removal cross section. The orbits of $m=\pm 1$ show opposite behaviour. Their densities spread far in the transverse direction and show a narrow momentum distribution in the longitudinal direction. These orbits contribute dominantly to the longitudinal momentum distribution of the ^{7}Be fragment. Since the longitudinal momentum distribution of these orbits is narrower than that of the $m=0$ orbit, the reaction dynamics sharpens the longitudinal momentum distribution.

Squared Fourier transforms of the wave functions with different m-values are compared in the left-hand panel of Fig. 6.4. As was mentioned above, the momentum distributions in the longitudinal direction of the $m=\pm 1$ orbits are narrow, while the $m=0$ orbit shows a nodal behaviour and a wider distribution. Here the wave function has been constructed from a Woods–Saxon potential with Coulomb and centrifugal terms. The potential depth is set to reproduce the proton separation energy.

The right-hand panel of Fig. 6.4 displays the longitudinal momentum distribution of the ^{7}Be fragment in the ^{8}B+^{12}C collision at the incident energy of 1471 A MeV. Only the inelastic breakup contribution is included in the calculation. The momentum distribution of the

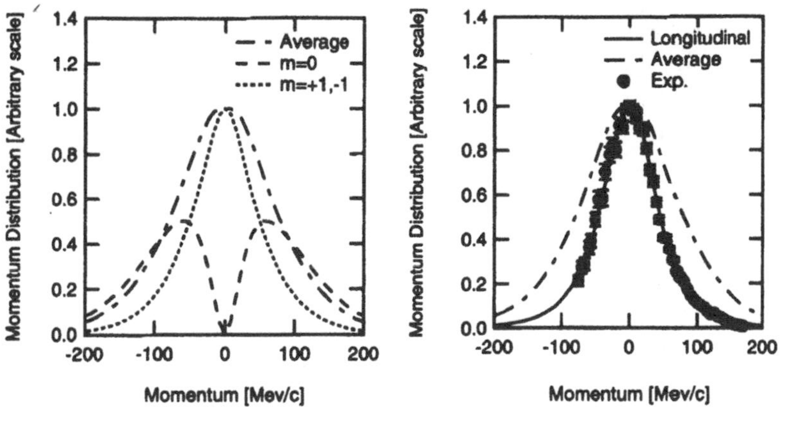

Figure 6.4: Left: p–^{7}Be momentum distributions as given by the p–^{7}Be relative-motion functions, with different m-values, plotted as functions of the momentum. Their average is also shown. Right: calculated longitudinal momentum distribution of the ^{7}Be core (solid curve) in the ^{8}B+^{12}C collision at $E=1471\,A$ MeV. Only the contribution of the inelastic breakup process is included. The average of the p–^{7}Be momentum distributions of the orbits with different m, the same as in the left panel, is also shown. Experiment: Ref. [138].

proton–core relative motion in ^{8}B averaged over m (dash-dotted curve) is about twice as broad as the fragment momentum distribution produced by the reaction dynamics (solid curve). This substantial difference is due mainly to the large contribution of the $m = \pm 1$ orbits to the breakup process. Another effect may come from the phase-shift functions, which pick up contributions from the transverse coordinate regions selectively. The measured fragment momentum distribution has an FWHM of 81±6 MeV/c [139] and 91±5 MeV/c [138], in fair agreement with theory, 92 MeV/c.

This example indicates that there may be appreciable distorting effects from the reaction mechanism, but they can be properly taken into account in theoretical analyses. So the distribution of the relative momentum coming from a structure calculation for the projectile can be compared with experiment by imitating the distorting effects of the reaction by theoretical calculations. In particular, the analysis for the p–^{7}Be momentum distribution in ^{8}B seems to be reliable. The FWHM of the intrinsic p–^{7}Be momentum distribution in ^{8}B is about 160 MeV/c (dash-dotted curve), which is three times as large as that of the n–^{10}Be distribution in ^{11}Be. This width is not spectacularly small, and it may be concluded that ^{8}B is not as good a halo nucleus as ^{11}Be if it can be called such at all.[3]

[3]During the preparation of the manuscript the authors have learned [140] of a ^{8}B+^{208}Pb Coulomb breakup experiment at Darmstadt for the ^{7}Be–p longitudinal momentum distribution, which gives $\hbar\sigma_{\parallel} = 38$ MeV/c. This corresponds to an FWHM of about 90 MeV/c, so this new result is quite consistent with the previous data [138, 139].

Chapter 7

Coulomb breakup reactions of halo nuclei

7.1 Soft dipole mode

7.1.1 Dipole strength function

The interaction cross section of the collision of halo nuclei with heavy target nuclei is much larger than expected from reactions induced by the nuclear force. The main cause of the large cross section is halo-neutron removal induced by the strong Coulomb field of the target. Although the Coulomb field does not act on the neutron, the core is pushed by the Coulomb field, and this shock shakes off the halo neutron easily, owing to its weak binding. This mechanism cannot be effective for deeply bound systems, and the probability of projectile breakup in the target Coulomb field is, indeed, substantial only for halo nuclei.

In a perturbative description the electromagnetic transition is characterized by a definite term in the multipole expansion of the electromagnetic field [141]. Thus it can be either electric or magnetic and has multipolarity 2^l. Like any electromagnetic transition, the Coulomb excitation from state i to f can be viewed as an absorption (or emission) of a photon and can be described in terms of the reduced transition probability $B(\pi l, i \to f)$ [cf. Eq. (9.86) and footnote on p. 423 later], where π stands for either electric or magnetic. The photon seen by the projectile as a result of the target Coulomb field passing by is virtual.[1]

[1] A travelling charged particle interacts with its environment by photon exchange; yet it does not emit light and does not change its state of motion unless such interaction occurs. Without interaction, the photons are absorbed by the emitter itself. It is in this sense that the photons emitted by the target in the projectile-rest frame are said to be virtual.

It will be discussed in Sect. 7.2 that the target Coulomb field can be considered a superposition of electromagnetic plane waves of various frequencies. For sufficiently large impact-parameter values, the field is approximately uniform, and the nuclear response is predominantly of electric dipole (E1) nature. Since the field is dominated by the low-frequency component, the Coulomb excitation is mostly a low-energy excitation, and the breakup of a halo nucleus also goes to the low-energy continuum.

To describe the dipole excitation of a halo nucleus, one should construct the electric dipole operator itself. Applying the general definition to the projectile, one has[2]

$$\boldsymbol{D} = e \sum_{i \in \text{proton}} (\boldsymbol{r}_i - \boldsymbol{R}_\text{P}), \tag{7.1}$$

where $\boldsymbol{r}_i$ is the coordinate of a proton (relative to a fixed origin) and $\boldsymbol{R}_\text{P}$ is the total c.m. coordinate of the projectile. Only charged particles respond to the electric field, but the neutron coordinates are also contained in the c.m. coordinate. This lends itself to reformulation, in which, as it were, the neutrons are also assigned 'effective' charges. Let us consider a one-neutron-halo nucleus. The coordinates of the halo neutron and of the core c.m. are denoted, as before, by $\boldsymbol{r}_\text{n}$ and $\boldsymbol{R}_\text{C}$, respectively, and the neutron–core relative coordinate by $\boldsymbol{r}=\boldsymbol{r}_\text{n}-\boldsymbol{R}_\text{C}$. Thus the c.m. coordinate $\boldsymbol{R}_\text{P}$ of the projectile is given by $\boldsymbol{R}_\text{P}=A_\text{P}^{-1}[(A_\text{P}-1)\boldsymbol{R}_\text{C}+\boldsymbol{r}_\text{n}]=\boldsymbol{R}_\text{C}+A_\text{P}^{-1}\boldsymbol{r}$, where A_P is the mass number of the projectile. Then the dipole operator can be expressed as

$$\boldsymbol{D} = e \sum_{i \in \text{proton}} (\boldsymbol{r}_i - \boldsymbol{R}_\text{C}) - \frac{Z_\text{C}}{A_\text{P}} e\boldsymbol{r}, \tag{7.2}$$

where Z_C is the charge number of the core nucleus. The first term of this expression is nothing but the dipole operator of the core nucleus itself, while the second term is the dipole operator related to the neutron–core relative motion. This looks like a dipole operator for a system with two opposite charges, and that is why the halo neutron is said to have an effective charge of $-(Z_\text{C}/A_\text{P})e$.

The dipole electromagnetic field excites the states of a nucleus with varying probabilities. The distribution of the excitation probabilities is called the (electric-dipole) strength function. The shape of the strength function is remarkably uniform and changes from one nucleus to the other smoothly. Namely, it is usually concentrated around a peak, which is called the giant dipole resonance (cf. 184). Although this is not a

[2]In Part II the summations over certain types of nucleons will be written more elegantly in terms of projection operators.

single resonance state, but the envelope of a great number of states, one can construct a discrete model state which comprises all dipole strength, and may be considered as an idealized giant dipole or 'coherent' state. This state is usually interpreted as a state in which all protons oscillate against all neutrons. The number of actual states that can be reached by dipole excitation from the g.s. is infinite since the giant dipole resonance lies in the continuum. These states can be considered as having a share of the model state, thus having some of its dipole strength. The excitation energy of the giant dipole state is usually well over 10 MeV (between 20 and 30 MeV for $A \leq 16$ nuclei) and its width is well below 10 MeV, thus there is very little dipole strength at low excitation energies.

In neutron-rich nuclei, however, a considerable dipole strength has been observed at low excitation energies. As an explanation for this, a new mode of dipole oscillation, the so-called soft dipole mode has been proposed [11, 25, 142–144]. This mode is visualized in a picture, in which the excess neutrons form a neutron halo or a neutron skin covering the nuclear surface and oscillate against the core. The dipole oscillation between the weakly bound neutrons and the core nucleus may be expected to lie at low excitation energies, because of a weak restoring force, and to have a considerable dipole strength [143–146]. It was hoped that a well-defined resonance state can be found, and that is responsible for this phenomenon.

It was, however, also observed that the dipole strength appears at low excitation energies for states involving weakly bound nucleons even if no resonance states can be found. The strong dipole strength appears just above the threshold energy. A lesser amount of strength decoupling occurs around the threshold of neutron removal also in heavy nuclei that support low-l s.p. neutron states near the threshold, and that is called the pygmy resonance. It can be explained as a threshold effect [147]. For the single-neutron-halo nucleus ^{11}Be, the observed dipole strength can also be well described by a threshold mechanism [148]. For other halo nuclei with two-neutron-halo structure, the dipole strength appears just above the two-neutron threshold [149–151] and may also originate from a similar mechanism. Thus the strength concentration at low energy may not be a genuine resonance [152–154] but just an effect caused by the weak binding.

To formulate the problem more precisely, let us consider the dipole strength distribution of a neutron bound in a spherically symmetric potential. We shall see the appearance of the strength just above the threshold when the neutron is bound weakly.

Let the neutron be bound in an s-orbit $\phi_0(\boldsymbol{r})$ and let it be the only bound state of Hamiltonian H_{nC}; all other states are continuum states:

$$H_{\text{nC}}\phi_0(\boldsymbol{r}) = \varepsilon_0\phi_0(\boldsymbol{r}), \quad H_{\text{nC}}\phi_{Elm}(\boldsymbol{r}) = E\phi_{Elm}(\boldsymbol{r}). \tag{7.3}$$

The continuum states $\phi_{Elm}(\boldsymbol{r})$ can be specified by the energy $E = \hbar^2k^2/2\mu$ (μ is the reduced mass) and the angular momentum lm, and may be normalized asymptotically as

$$\phi_{Elm}(\boldsymbol{r}) \to \sqrt{\frac{2\mu}{\pi\hbar^2 k}}\frac{\sin(kr - \frac{1}{2}l\pi + \delta_l)}{r}Y_{lm}(\hat{\boldsymbol{r}}), \tag{7.4}$$

where δ_l is the phase shift. The continuum wave functions constructed in this way satisfy the orthonormality relation

$$\int d\boldsymbol{r}\phi^*_{Elm}(\boldsymbol{r})\phi_{E'l'm'}(\boldsymbol{r}) = \delta(E - E')\delta_{ll'}\delta_{mm'} \tag{7.5}$$

and, consequently, the closure relation

$$|\phi_0\rangle\langle\phi_0| + \sum_{lm}\int_0^\infty dE\ |\phi_{Elm}\rangle\langle\phi_{Elm}| = 1. \tag{7.6}$$

With the complete set of states $\{|\phi_0\rangle, |\phi_{Elm}\rangle\}$, the spectral decomposition of the Hamiltonian looks like

$$H_{\text{nC}} = \varepsilon_0|\phi_0\rangle\langle\phi_0| + \sum_{lm}\int_0^\infty dE\ E|\phi_{Elm}\rangle\langle\phi_{Elm}|. \tag{7.7}$$

With the dipole operator for the neutron [cf. Eq. (7.2)],

$$M_{1m} = -\frac{Z_{\text{C}}}{A_{\text{P}}}erY_{1m}(\hat{\boldsymbol{r}}), \tag{7.8}$$

the dipole strength function per unit energy takes the form

$$\begin{aligned} \frac{dB(\text{E1})}{dE}\bigg|_E &= \int_0^\infty dE_f\,\delta(E_f - E)\sum_{l_f m_f m}|\langle\phi_{E_f l_f m_f}|M_{1m}|\phi_0\rangle|^2 \\ &= \sum_m|\langle\phi_{E1m}|M_{1m}|\phi_0\rangle|^2. \end{aligned} \tag{7.9}$$

The sum runs over all values, but for an $l = 0$ state ϕ_0, only $l_f = 1$ and $m_f = m$ contribute. The symbol $|_E$, to be omitted hereafter, indicates that the dipole strength function is a function of E. This expression may be interpreted as follows. The dipole operator M_{1m} causes a transition from the initial state ϕ_0 to the state[3] $M_{1m}\phi_0$. The

[3]The state $M_{1m}\phi_0$ is a strict analogue of the giant dipole coherent state; that is also defined as $M_{1m}\phi_0$, but in that case M_{1m} and ϕ_0 are defined for the whole nucleus.

state $M_{1m}\phi_0$ has angular momentum $l=1$ and z-component m, hence is orthogonal to ϕ_0: $\langle\phi_0|M_{1m}|\phi_0\rangle = 0$. Since there is no bound orbit with $l = 1$ by assumption, the function $M_{1m}\phi_0$ is a superposition of continuum wave functions. The dipole strength function $dB(\mathrm{E1})/dE$ is thus its probability distribution over the basis states.

The reduced transition probability $B(\mathrm{E1})$ is defined as the integral of Eq. (7.9) over the excitation energy, and is expressed in terms of the mean square radius of the g.s. as

$$\begin{aligned} B(\mathrm{E1}) &= \int_0^\infty dE\,\frac{dB(\mathrm{E1})}{dE} \\ &= \langle\phi_0|\sum_m M_{1m}^\dagger M_{1m}|\phi_0\rangle = \frac{3}{4\pi}\left(\frac{Z_\mathrm{C}}{A_\mathrm{P}}\right)^2 e^2\langle\phi_0|r^2|\phi_0\rangle. \end{aligned} \quad (7.10)$$

Here the symbol † means Hermitean conjugation. The $B(\mathrm{E1})$ value may be very large for a weakly bound orbit because the mean square radius can be pretty large.

7.1.2 Sum rule

The total multipole strength is limited as any quantity that is proportional to a mathematical probability. The sum rules are relations that set these limits. The major ingredient of such a sum rule, the so-called energy-weighted dipole sum S_1, is defined and evaluated as follows:

$$\begin{aligned} S_1 &= \int_0^\infty dE\,(E-\varepsilon_0)\frac{dB(\mathrm{E1})}{dE} \\ &= \sum_m \langle\phi_0|M_{1m}^\dagger \sum_{l_f m_f}\int_0^\infty dE\,(E-\varepsilon_0)|\phi_{El_f m_f}\rangle\langle\phi_{El_f m_f}|M_{1m}|\phi_0\rangle \\ &= \sum_m \langle\phi_0|M_{1m}^\dagger(H_\mathrm{nC}-\varepsilon_0)M_{1m}|\phi_0\rangle \\ &= \frac{1}{2}\sum_m \langle\phi_0|\left[M_{1m}^\dagger,[H_\mathrm{nC},M_{1m}]\right]|\phi_0\rangle. \end{aligned} \quad (7.11)$$

In the third equality the spectral representation of the Hamiltonian, Eq. (7.7) and the closure relation, Eq. (7.6), have been used, and, in the last step, the relation

$$\sum_m \langle\phi_0|M_{1m}(H_\mathrm{nC}-\varepsilon_0)M_{1m}^\dagger|\phi_0\rangle = \sum_m \langle\phi_0|M_{1m}^\dagger(H_\mathrm{nC}-\varepsilon_0)M_{1m}|\phi_0\rangle \quad (7.12)$$

has been employed, which holds because of $M_{1m}^\dagger = (-1)^m M_{1-m}$.

The double commutator is non-zero because the kinetic-energy operator in $H_{\rm nC}$ does not commute with M_{1m}. This, put into Eq. (7.11), yields a sum rule for the dipole strength belonging to the neutron–core relative motion:

$$S_1 \equiv \int_0^\infty dE\,(E-\varepsilon_0)\frac{dB(\mathrm{E1})}{dE} = \frac{9}{4\pi}\left(\frac{Z_C}{A_P}\right)^2 \frac{\hbar^2 e^2}{2\mu}. \tag{7.13}$$

Using Eq. (7.13), one can make an estimate for the properties of the dipole excitation. If the dipole strength were concentrated in a narrow excitation energy region, this excitation energy should be

$$E_{\rm av} \sim \frac{S_1}{B(\mathrm{E1})} = \frac{3\hbar^2}{2\mu}\frac{1}{\langle\phi_0|r^2|\phi_0\rangle}. \tag{7.14}$$

The mean square radius $\langle\phi_0|r^2|\phi_0\rangle$ of a weakly bound neutron orbit may be estimated to be about[4] 30 fm^2. The dipole strength due to the neutron–core relative motion thus appears at a low excitation energy of 2–2.5 MeV.

The energy-weighted sum rule for the dipole operator (7.1) of the whole nucleus is known as the Thomas–Reiche–Kuhn sum rule. It reads as

$$S_1(\mathrm{TRK}) \equiv \sum_f (E_f - E_{\rm g.s.})B(\mathrm{E1}, \mathrm{g.s.} \to f) = \frac{9}{4\pi}\frac{NZ}{A}\frac{\hbar^2 e^2}{2m}, \tag{7.15}$$

where N, Z and A are the neutron, proton and mass numbers, respectively, and m is the nucleon mass. The energy-weighted sum S_1 is contained in this total sum S_1(TRK), and the sum rule (7.13) is called a molecular sum rule [155]. The ratio of S_1 to the total sum is given by

$$\frac{S_1}{S_1(\mathrm{TRK})} = \frac{Z_{\rm C}}{A_{\rm P}N_{\rm P}}\frac{m}{\mu}. \tag{7.16}$$

This value is about 0.06 for ^{11}Be. The sum rule was originally presented in this form for the two-neutron-halo nucleus ^{11}Li in a schematic dineutron model [78] and in a three-body model [156–158].

7.1.3 Zero-range potential model

There is a simple model in which an analytic expression is available for the dipole strength distribution. The model is a neutron bound in a square-well potential whose radius is vanishingly small. In this limit

[4] This can be confirmed, e.g., by using the estimate of the spread of the tale of the neutron orbit on p. 14.

there are only $l=0$ bound states since, for $l>0$, the strong attraction is overwhelmed by the infinite repulsion coming from the centrifugal potential around the origin. Let the radius and depth of the potential be denoted by a and $-V$ ($V>0$). For an s-wave bound state with eigenenergy ε_0 ($\varepsilon_0<0$), the wave function inside the well is given by $\phi(r)\propto\frac{1}{r}\sin k_0 r$ with $\hbar^2k_0^2/2\mu=V-|\varepsilon_0|$, and the wave function outside the well is $\phi(r)\propto\frac{1}{r}e^{-\kappa r}$ with $\hbar^2\kappa^2/2\mu=|\varepsilon_0|$. The matching condition (the continuity of the logarithmic derivative) at $r=a$ determines the energy ε_0 through

$$\tan k_0 a = -\frac{k_0}{\kappa}. \tag{7.17}$$

Let us take the limit $a\to 0$, $V\to\infty$ such that Va^2 and ε_0 be finite. We shall see that these can be finite simultaneously. In this limit the right-hand side of Eq. (7.17) becomes $-\infty$ and $k_0a\to\frac{1}{2}(2n-1)\pi+\delta$ where n is a positive integer and δ is an infinitesimal positive number. The quantum number n is a kind of generalized node number. Let the potential bind just the lowest state, that with $n=1$. The normalized s-wave bound-state wave function $\phi_0(r)$ is given by

$$\phi_0(r)=\sqrt{2\kappa}\frac{e^{-\kappa r}}{r}Y_{00}(\hat{r}). \tag{7.18}$$

As to the scattering states, only the s-wave is influenced by the potential. Thus the continuum state with energy E and $l\neq 0$ is just the l-component of the plane wave, whose radial part is the lth order spherical Bessel function:

$$\phi_{Elm}(\boldsymbol{r})=\sqrt{\frac{2\mu k}{\pi\hbar^2}}j_l(kr)Y_{lm}(\hat{\boldsymbol{r}}), \tag{7.19}$$

where $\hbar^2k^2/2\mu=E$.

For completeness, let us determine the s-wave continuum wave function as well, although it is not necessary for the evaluation of the dipole strength. The interior solution $\phi(r)\propto\frac{1}{r}\sin k_1 r$ with $\hbar^2k_1^2/2\mu=E+V$ should be matched smoothly to the exterior solution with phase shift δ_0, $\phi(r)\propto\frac{1}{r}\sin(kr+\delta_0)$ with $\hbar^2k^2/2\mu=E$. The matching condition at $r=a$ determines the phase shift δ_0. From $V\gg E$ it follows that $k_1\simeq k_0$ and $ka\simeq 0$. The phase shift δ_0 satisfies $\tan\delta_0=-k/\kappa=-\sqrt{E/|\varepsilon_0|}$. The s-wave scattering solution at energy E is thus

$$\phi_{E00}(\boldsymbol{r})=\sqrt{\frac{2\mu k}{\pi\hbar^2}}\frac{\sin(kr+\delta_0)}{kr}Y_{00}(\hat{\boldsymbol{r}}). \tag{7.20}$$

The wave functions constructed in this way must satisfy the orthonormality and closure relations (7.5) and (7.6).

It is easy to show that the dipole matrix element is given by

$$\langle \phi_{E1m}|M_{1m}|\phi_0\rangle = -\frac{1}{\pi}\frac{Z_C}{A_P}e\sqrt{\frac{\hbar^2}{\mu}\frac{E^{3/4}|\varepsilon_0|^{1/4}}{(E+|\varepsilon_0|)^2}}, \quad (7.21)$$

where the integration formula

$$\int_0^\infty dr\, r^2 e^{-\kappa r} j_1(kr) = \frac{2k}{(k^2+\kappa^2)^2} \quad (7.22)$$

has been employed. Substitution of this into Eq. (7.9) gives the dipole strength function:

$$\frac{dB(\mathrm{E1})}{dE} = B(\mathrm{E1})\frac{1}{|\varepsilon_0|}f\left(\frac{E}{|\varepsilon_0|}\right), \quad (7.23)$$

where $B(\mathrm{E1})$ of Eq. (7.10) is given by

$$B(\mathrm{E1}) = \frac{3}{4\pi}\left(\frac{Z_C}{A_P}\right)^2\frac{e^2}{2\kappa^2}, \quad (7.24)$$

with the mean square radius,

$$\langle\phi_0|r^2|\phi_0\rangle = \frac{1}{2\kappa^2}. \quad (7.25)$$

The function $f(x)$ in Eq. (7.23), which gives the energy distribution of the strength, reads

$$f(x) = \frac{16}{\pi}\frac{x^{3/2}}{(x+1)^4}. \quad (7.26)$$

This function is normalized as $\int_0^\infty dx f(x) = 1$.

The dipole strength function of this model is displayed in Fig. 7.1 for the cases of three different binding energies: $|\varepsilon_0| = 0.5$, 1.0, and 4.0 MeV. At the threshold energy $E = 0$ it starts to increase and shows a peak at $E/|\varepsilon_0| = \frac{3}{5}$. It then decreases and tends to its asymptotic form, $\sim (E/|\varepsilon_0|)^{-5/2}$. As the binding is made deeper, the strength distribution shifts towards the higher excitation energies. The absolute value of the dipole strength in the low excitation-energy region decreases rapidly as the binding energy is increased. This is caused by two circumstances. First, the integrated strength, $B(\mathrm{E1})$, decreases with the binding energy increased [cf. Eq. (7.24)]. Second, the strength is distributed over a wider energy region, so that the strength per unit energy region has to decrease.

It is remarkable that, even though the p-wave has no resonance at all and is actually described by the plane wave, the dipole strength

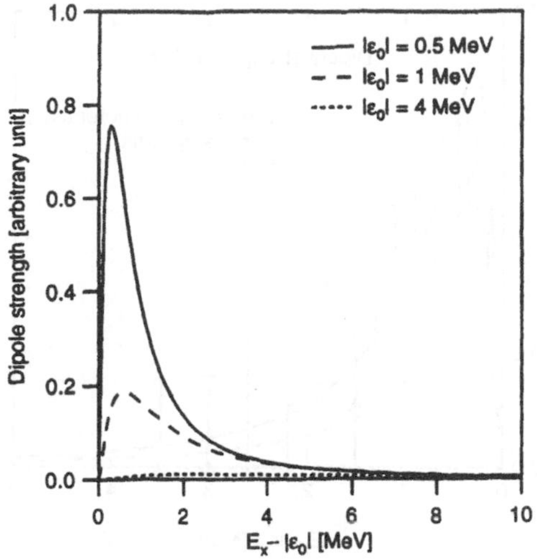

Figure 7.1: Dipole strength given by Eq. (7.23) for three different binding energies, $|\varepsilon_0|$ = 0.5, 1.0, and 4.0 MeV. The energy along the abscissa is measured from the threshold.

function does show a peak. This strength function describes rather well the dipole strength of ^{11}Be extracted from the measurement of a neutron-removal process. The nucleus ^{11}Be is known to possess a bound excited state with a low excitation energy, which can be excited by the dipole field from the g.s. In the excitation to the continuum there is a substantial dipole strength in the low-energy region [148].

The dipole strength function of the nucleus ^{11}Be calculated with the zero-range model (7.23) is compared in Fig. 7.2 with that calculated according to Eq. (7.9) in a Woods–Saxon potential model. Both models are set to give the s-orbit energy at $\varepsilon_0 = -0.503$ MeV, and the same potentials were used for the p-wave. The dipole strength function in the Woods–Saxon potential has a shape similar to that in the zero-range potential, but its magnitude is substantially larger. The measured dipole strength [148] is also included. The Woods–Saxon potential model describes the measured dipole strength fairly well. The slight overshooting of the strength in the Woods–Saxon model may be explained by the fact that there is a fairly strong dipole transition in ^{11}Be between the $\frac{1}{2}^+$ g.s. and the $\frac{1}{2}^-$ first excited state (see Sect. 13.6), which is ignored in the present treatment.

Though the wave function (7.18) gives the same correct asymptotic form as in the Woods–Saxon potential, its magnitude outside the

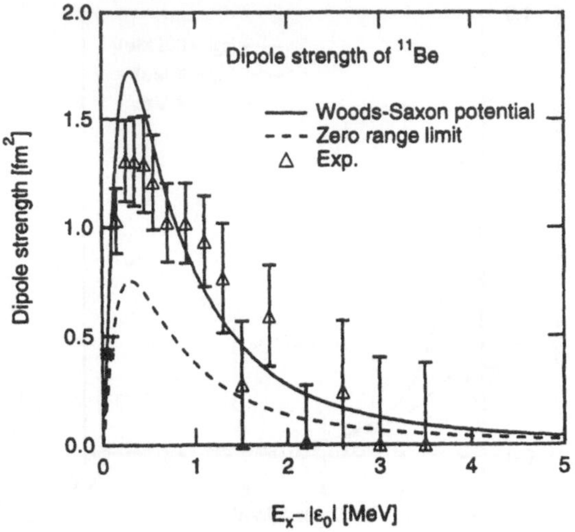

Figure 7.2: Dipole strength (7.9) of ^{11}Be calculated in a Woods–Saxon potential model (solid curve) and in the zero-range potential model (7.23) (dashed curve). The measured dipole strength [148] is also included.

potential is quite different from that in the Woods–Saxon model, and that is mainly responsible for the difference. To give an estimate for the difference, let us take a square-well potential of radius R instead of the Woods–Saxon potential. The wave function outside the potential is then [11][5]

$$\phi_0(\boldsymbol{r}) = \sqrt{2\kappa}\frac{e^{\kappa R}}{\sqrt{1+\kappa R}}\frac{e^{-\kappa r}}{r}Y_{00}(\hat{\boldsymbol{r}}), \tag{7.27}$$

which is to be compared with Eq. (7.18) for the δ-potential. The extra factor $e^{2\kappa R}/(1+\kappa R)$ accounts for the difference between the two models.

[5]The s-wave state $\phi_0(\boldsymbol{r}) = \frac{u(r)}{r}Y_{00}(\hat{\boldsymbol{r}})$ in the square-well potential is given by

$$u(r) = \begin{cases} A\sin kr & r < R, \\ Be^{\kappa(R-r)} & r \geq R, \end{cases}$$

where κ is related to the binding energy. For an orbit of vanishingly small binding energy, kR satisfies $kR = \frac{1}{2}(2n-1)\pi$ with n a positive integer. Then the continuity requirement and the normalization condition for the wave function give $A = (-1)^{n-1}B$ and $B = \sqrt{2\kappa/(1+\kappa R)}$.

7.2 Equivalent-photon method

There are two equivalent theoretical approaches to the description of the excitation of the projectile nucleus by the target Coulomb field. One is just an adaptation of the framework developed for the reactions induced by the nuclear force, and the other is called the equivalent-photon method [159]. In this section the latter approach will be briefly introduced, and the former approach will be the subject of Sect. 7.3.

In this approach the reaction is described in the projectile-rest frame, in which the Coulomb field of the target can be treated as a time-dependent electromagnetic field. The target nucleus is assumed to follow a straight-line trajectory with a constant velocity. This assumption should be good when the incident velocity is high enough. The target Coulomb field can then be regarded as a superposition of electromagnetic plane waves with various frequencies, and the excitation of the projectile can be attributed to an absorption of virtual photons created by the time-dependent field.

Suppose that the target nucleus moves along the $-z$ direction with velocity $\boldsymbol{v}=(0,0,-v)$, $v>0$ and impact parameter $\boldsymbol{b}$. It is regarded as a point particle with charge $Z_{\mathrm{T}}e$. The Lienard–Wiechert potential produced by the target nucleus is given by the scalar and vector potentials

$$\phi(\boldsymbol{r},t)=\frac{Z_{\mathrm{T}}e}{u}, \qquad \boldsymbol{A}(\boldsymbol{r},t)=\frac{Z_{\mathrm{T}}e}{u}\frac{\boldsymbol{v}}{c}, \tag{7.28}$$

with

$$u=\sqrt{\frac{1}{\gamma^2}(\boldsymbol{s}-\boldsymbol{b})^2+(z+vt)^2}, \qquad \gamma=\frac{1}{\sqrt{1-\left(\frac{v}{c}\right)^2}}, \tag{7.29}$$

where $\boldsymbol{r}$, the spatial variable of the functions $\phi(\boldsymbol{r},t)$ and $\boldsymbol{A}(\boldsymbol{r},t)$, is decomposed as $\boldsymbol{r}=(\boldsymbol{s},z)$. The electric and magnetic fields belonging to these potentials are

$$\boldsymbol{E}(\boldsymbol{r},t)=-\boldsymbol{\nabla}\phi-\frac{1}{c}\frac{\partial\boldsymbol{A}}{\partial t}=Z_{\mathrm{T}}e\frac{\boldsymbol{r}-\boldsymbol{R}(t)}{\gamma^2u^3}, \tag{7.30}$$

$$\boldsymbol{B}(\boldsymbol{r},t)=\boldsymbol{\nabla}\times\boldsymbol{A}=-\frac{\boldsymbol{v}}{c}\times\boldsymbol{E}(\boldsymbol{r},t), \tag{7.31}$$

where $\boldsymbol{R}(t)=\boldsymbol{b}+\boldsymbol{v}t$ is the coordinate of the target nucleus.

In the non-relativistic limit, $v\ll c$ ($\gamma\to 1$), these electric and magnetic fields reduce to those given by the Coulomb law and the Biot–Savart law, respectively. On the other hand, in the relativistic limit, $v\to c$ ($\gamma\gg 1$), the electric field becomes perpendicular to the incident direction. At the time of the closest approach, i.e., when $z + vt \sim 0$ and $u \sim |\boldsymbol{s} - \boldsymbol{b}|/\gamma$, the electric field is

$$\boldsymbol{E} \simeq Z_{\mathrm{T}} e\gamma \frac{\boldsymbol{s} - \boldsymbol{b}}{|\boldsymbol{s} - \boldsymbol{b}|^3} = \frac{Z_{\mathrm{T}} e\gamma}{|\boldsymbol{s} - \boldsymbol{b}|^2} (\widehat{\boldsymbol{s} - \boldsymbol{b}}). \tag{7.32}$$

One thus sees that, in a relativistic collision, the electric field is approximately transverse. It follows from Eq. (7.31) that the magnetic field is perpendicular both to the z-direction and to the electric field. Therefore, the Coulomb field of the rapidly moving target does look like an electromagnetic plane wave. The field strength is proportional to γ, and at $\boldsymbol{r} = 0$ (and at any other fixed point) the field persists only for a limited time interval, $\Delta t = b/\gamma v$ [cf. (7.29)].

This field of a short pulse can be considered a superposition of fields with various frequencies. A Fourier analysis of the electric field,

$$\boldsymbol{E}(\boldsymbol{r}, \omega) = \frac{1}{2\pi} \int_{-\infty}^{+\infty} dt\, e^{i\omega t} \boldsymbol{E}(\boldsymbol{r}, t), \tag{7.33}$$

gives a frequency spectrum of virtual photons. The Fourier amplitude can be expressed in terms of the modified Bessel functions K_0 and K_1, where

$$K_n(x) = \frac{(2n-1)!!}{x^n} \int_0^{\infty} dt \frac{\cos xt}{(t^2+1)^{n+\frac{1}{2}}} \quad (x > 0), \tag{7.34}$$

and by using $K_1(x) = -K_0'(x)$ as follows:

$$\begin{aligned} \boldsymbol{E}(\boldsymbol{r}, \omega) &= \frac{Z_{\mathrm{T}} e}{\pi} \frac{\xi}{bv} e^{-i\frac{\omega}{v} z} \\ &\times \left[(\widehat{\boldsymbol{s} - \boldsymbol{b}}) K_1 \left(\frac{\xi}{b} |\boldsymbol{s} - \boldsymbol{b}| \right) + i\hat{\boldsymbol{z}} \frac{1}{\gamma} K_0 \left(\frac{\xi}{b} |\boldsymbol{s} - \boldsymbol{b}| \right) \right], \end{aligned} \tag{7.35}$$

where $\xi = b\omega/\gamma v$. The magnitude of the electric field at the origin is

$$|\boldsymbol{E}(\boldsymbol{r}, \omega)|^2_{\boldsymbol{r}=0} = \left(\frac{Z_T e}{\pi} \right)^2 \frac{\xi^2}{b^2 v^2} \left[K_1^2(\xi) + \frac{1}{\gamma^2} K_0^2(\xi) \right]. \tag{7.36}$$

The first term in the square brackets on the right-hand side of Eq. (7.36) comes from the transverse component of the electric field, while the second term comes from the longitudinal component. The dominance of the transverse field at high velocity (i.e., large γ, small ξ) is shown by the factor γ^{-2} in the second term and by the asymptotic behaviour of the modified Bessel functions, $K_1(\xi) \simeq 1/\xi$ and $K_0(\xi) \simeq -\ln \xi$, for small ξ.

Both $K_1(\xi)$ and $K_0(\xi)$ behave as $e^{-\xi}$ for $\xi \gg 1$, which indicates that the electromagnetic field does not include components of frequency with $\omega \gg \gamma v/b$. This fact is consistent with the estimate for the time interval of the existence of the electromagnetic field, $\Delta t = b/\gamma v$.

In this picture the reaction probability for the Coulomb excitation is a product of the number of virtual photons created by the target motion and the cross section of photon absorption by the nucleus.

The number of photons induced by the electromagnetic field with a given frequency can be calculated as follows. The energy density of the electromagnetic field is given by

$$\mathcal{E} = \frac{1}{8\pi}\left[\boldsymbol{E}^2(\boldsymbol{r},t) + \boldsymbol{B}^2(\boldsymbol{r},t)\right]. \tag{7.37}$$

An integration of the energy density $\mathcal{E}$ over the coordinate z gives the energy of the electromagnetic wave per unit area, to be denoted by I. Since the electromagnetic wave propagates in the z-direction with velocity c, the integration over z is equivalent to an integration over time, multiplied by c. The energy density per unit area, I, should be expressed in terms of the Fourier component of the electric field $\boldsymbol{E}(\boldsymbol{r},\omega)$ defined in Eq. (7.33). Using the fact that the electric and magnetic terms in the energy density give equal contributions, one obtains

$$I = \frac{c}{4\pi}\int_{-\infty}^{+\infty} dt\, \boldsymbol{E}^2(\boldsymbol{r},t) = \int_0^{\infty} d\omega\, c\,|\boldsymbol{E}(\boldsymbol{r},\omega)|^2. \tag{7.38}$$

The integrand $c|\boldsymbol{E}(\boldsymbol{r},\omega)|^2$ is the energy of the electromagnetic field per unit area per unit frequency. Since the energy of a photon of frequency ω is $\hbar\omega$, the number of photons per unit area per unit frequency, $N(\boldsymbol{b},\omega)$, is given by

$$\hbar\omega N(\boldsymbol{b},\omega)d\omega = c|\boldsymbol{E}(\boldsymbol{r},\omega)|^2 d\omega. \tag{7.39}$$

The energy density for the whole projectile volume can be approximated by that at $\boldsymbol{r}=0$, which is

$$\begin{aligned} N(\boldsymbol{b},\omega) &= \frac{c}{\hbar\omega}|\boldsymbol{E}(\boldsymbol{r},\omega)|^2_{\boldsymbol{r}=0} \\ &= \frac{Z_T^2\alpha}{\pi^2}\left(\frac{c}{v}\right)^2 \frac{\xi^2}{\omega b^2}\left[K_1^2(\xi) + \frac{1}{\gamma^2}K_0^2(\xi)\right], \end{aligned} \tag{7.40}$$

where $\alpha = e^2/\hbar c$ is the fine structure constant.

The reaction probability for the Coulomb excitation can then be expressed with the photoabsorption cross section[6] $\sigma_\gamma(\omega)$ as

$$P(b) = \int_0^\infty d\omega N(b,\omega)\sigma_\gamma(\omega). \tag{7.41}$$

If there is only one bound state as is often the case for halo nuclei, the photoabsorption leads to the disintegration of the nucleus. In that case, the probability (7.41) is the breakup reaction probability due to the Coulomb interaction.

The extension of the equivalent photon method to transitions of higher electric and magnetic multipoles is discussed in Refs. [159, 160].

The inverse process of photodisintegration is radiative capture. An intriguing application of the equivalent photon method is the calculation of radiative capture cross sections. The radiative capture by light nuclei at low incident energies (below the Coulomb barrier) is very important in astrophysical nucleosynthesis. A direct measurement of these cross sections is, however, extremely difficult because their magnitude is very small. For example, the ^{7}Be(p,γ)^{8}B reaction cross section at low energies below the Coulomb barrier is essential for estimating the high-energy solar-neutrino flux. The reports on the direct measurements are controversial. The inverse process of this reaction is the photodisintegration of ^{8}B. Measuring the cross section of the Coulomb breakup of ^{8}B on a heavy target offers an interesting indirect method to determine the photodisintegration cross section through the equivalent-photon picture. Employing the relationship of detailed balance, one may thus extract the radiative capture cross section from the Coulomb breakup experiment [161, 162]. However, not only the (electric) dipole but also the quadrupole virtual photons contribute to the cross section, and the relative strength of the dipole and quadrupole transitions is not the same for capture as for Coulomb breakup processes. Therefore, the analysis must include both electric dipole and quadrupole transitions [163, 164].

7.3 Theory of the Coulomb breakup

7.3.1 Eikonal approximation

In this section the approach developed in Sect. 5.2 for reactions induced by the nuclear force is generalized to include the reactions induced by

[6]The photoabsorption cross section is related to the dipole strength function by

$$\sigma_\gamma(\omega) = \frac{16\pi^3}{9}\frac{\omega}{c}\frac{dB(\mathrm{E1})}{dE}.$$

the Coulomb field. In keeping with Sect. 5.2, a neutron+core+target three-body model will be discussed as a simplest representative of a reaction with a halo-nucleus projectile. The validity of the theory to be outlined now is, however, not restricted to the medium-energy region since the Coulomb force can be treated on the same footing in a broad energy region. Its limitations can also be discussed together for a broad energy range. The core and the target are treated as point particles and their excitations are not taken into consideration.

The starting point is, again, a three-body Schrödinger equation,

$$\left[\frac{\boldsymbol{P}^2}{2M_{\mathrm{PT}}} + H_{\mathrm{nC}} + V_{\mathrm{CT}}^{N}\left(\boldsymbol{R} - \frac{\boldsymbol{r}}{A_{\mathrm{P}}}\right) + V_{\mathrm{nT}}\left(\boldsymbol{R} + \frac{A_{\mathrm{P}}-1}{A_{\mathrm{P}}}\boldsymbol{r}\right) + \frac{Z_{\mathrm{C}}Z_{\mathrm{T}}e^2}{|\boldsymbol{R} - \frac{\boldsymbol{r}}{A_{\mathrm{P}}}|}\right]\Psi(\boldsymbol{R},\boldsymbol{r}) = E\Psi(\boldsymbol{R},\boldsymbol{r}), \tag{7.42}$$

where $\boldsymbol{R}$ and $\boldsymbol{r}$ are the projectile–target and the halo-neutron–core relative coordinate, respectively, $\boldsymbol{P}$ is the momentum operator of the projectile–target relative motion and Z_{C} and Z_{T} are charge numbers of the core and the target, respectively [cf. Eq. (5.13)]. A new piece of interaction is the Coulomb interaction between the core and the target. When the core and target overlap, the Coulomb interaction deviates from the point Coulomb potential, and, if necessary, this deviation can be included as part of the nuclear interaction. The neutron–target potential V_{nT} and the core–target potential V_{CT}^{N} are taken to be optical potentials.

As before, let us introduce the function $\hat{\Psi}(\boldsymbol{R},\boldsymbol{r})$ by $\Psi(\boldsymbol{R},\boldsymbol{r}) = e^{iKZ}\hat{\Psi}(\boldsymbol{R},\boldsymbol{r})$. The equation for $\hat{\Psi}(\boldsymbol{R},\boldsymbol{r})$ is similar to Eq. (5.17). By assuming that the momentum transfer between the projectile and the target is small compared with the incident relative momentum, the kinetic-energy operator of the relative motion, $\boldsymbol{P}^2/2M_{\mathrm{PT}}$, can be omitted, and the equation reduces to

$$\left[-i\hbar v\frac{\partial}{\partial Z} + (H_{\mathrm{nC}} - \varepsilon_0) + V_{\mathrm{CT}}^{N}\left(\boldsymbol{R} - \frac{\boldsymbol{r}}{A_{\mathrm{P}}}\right) + V_{\mathrm{nT}}\left(\boldsymbol{R} + \frac{A_{\mathrm{P}}-1}{A_{\mathrm{P}}}\boldsymbol{r}\right) + \frac{Z_{\mathrm{C}}Z_{\mathrm{T}}e^2}{|\boldsymbol{R} - \frac{\boldsymbol{r}}{A_{\mathrm{P}}}|}\right]\hat{\Psi}(\boldsymbol{R},\boldsymbol{r}) = 0, \tag{7.43}$$

where the relative coordinate $\boldsymbol{R}$ is related to the impact parameter $\boldsymbol{b}$ and Z by $\boldsymbol{R} = \boldsymbol{b} + Z\hat{\boldsymbol{z}}$. It is now useful to adopt the projectile-rest frame explained in Sect. 2.2.3 and introduce a time variable instead of Z via $Z = vt$. With the notation $\tilde{\Psi}(\boldsymbol{b},\boldsymbol{r},t) \equiv \hat{\Psi}(\boldsymbol{R},\boldsymbol{r})$, the equation for $\tilde{\Psi}(\boldsymbol{b},\boldsymbol{r},t)$ takes the form

$$i\hbar\frac{\partial}{\partial t}\tilde{\Psi}(\boldsymbol{b},\boldsymbol{r},t) = \left[(H_{\mathrm{nC}} - \varepsilon_0) + V_{\mathrm{CT}}^{N}\left(\boldsymbol{b} + vt\hat{\boldsymbol{z}} - \frac{\boldsymbol{r}}{A_{\mathrm{P}}}\right)\right.$$

$$+V_{\mathrm{nT}}\left(\boldsymbol{b}+vt\hat{\boldsymbol{z}}+\frac{A_{\mathrm{P}}-1}{A_{\mathrm{P}}}\boldsymbol{r}\right)+\frac{Z_{\mathrm{C}}Z_{\mathrm{T}}e^2}{|\boldsymbol{b}+vt\hat{\boldsymbol{z}}-\frac{\boldsymbol{r}}{A_{\mathrm{P}}}|}\Bigg]\tilde{\Psi}(\boldsymbol{b},\boldsymbol{r},t). \quad (7.44)$$

This is an approximate Schrödinger equation for the neutron–core relative motion in the projectile-rest frame with the nuclear and Coulomb fields of the target treated as time-dependent external fields. This is the same viewpoint as is adopted in the equivalent-photon method of Sect. 7.2. The initial condition is

$$\tilde{\Psi}(\boldsymbol{b},\boldsymbol{r},t\to-\infty)=\phi_0(\boldsymbol{r}). \quad (7.45)$$

To simplify the problem, the bound state of the halo neutron, ϕ_0, is, again, assumed to be an s-state.

Let us further proceed by applying the adiabatic approximation in exactly the same way as in Sect. 5.2.2. The excitation energy of the neutron–core relative motion, namely, the operator $(H_{\mathrm{nC}}-\varepsilon_0)$ is neglected in Eq. (7.44). The internal wave function of the halo neutron ϕ_0 is then frozen during the collision in this approximation. It will turn out, however, that this approximation is not justifiable for collision events with large impact parameters.

Nevertheless, let us temporarily adopt all these approximations. Then, after the collision, the wave function evolves into

$$\tilde{\Psi}(\boldsymbol{b},\boldsymbol{r},t\to\infty)=\mathrm{e}^{i\chi_{\mathrm{CT}}^{N}\left(\boldsymbol{b}-\frac{\boldsymbol{s}}{A_{\mathrm{P}}}\right)+i\chi_{\mathrm{nT}}\left(\boldsymbol{b}+\frac{A_{\mathrm{P}}-1}{A_{\mathrm{P}}}\boldsymbol{s}\right)+i\chi^{C}\left(\boldsymbol{b}-\frac{\boldsymbol{s}}{A_{\mathrm{P}}}\right)}\phi_0(\boldsymbol{r}). \quad (7.46)$$

The core–target phase-shift function χ_{CT}^{N} is to be constructed from the potential V_{CT}^{N}. The phase-shift function for the Coulomb potential[7] $\chi^{C}(\boldsymbol{b})$ was given in Sect. 2.3:

$$\chi^{C}(\boldsymbol{b})=2\eta\ln\frac{b}{2a}, \quad (7.47)$$

where $\eta=Z_{\mathrm{C}}Z_{\mathrm{T}}e^2/\hbar v$ is the Sommerfeld parameter and a is a cut-off radius. For later convenience, this Coulomb phase-shift function is decomposed as

$$\chi^{C}\left(\boldsymbol{b}-\frac{\boldsymbol{s}}{A_{\mathrm{P}}}\right)=\chi^{C}(\boldsymbol{b})+\Delta\chi^{C}(\boldsymbol{b},\boldsymbol{s}). \quad (7.48)$$

The first term on the right-hand side is the projectile–target Coulomb phase-shift function, with the Coulomb interaction acting between the c.m. of the projectile and the target. The second term arises because

[7] To avoid confusion with subscript C for 'core', here we use superscripts for 'Coulomb' and 'nuclear', rather than subscripts, and italicize them.

of the difference between the c.m. of the projectile and the core, and is given by

$$e^{i\Delta\chi^C(\boldsymbol{b},\boldsymbol{s})} = e^{i\chi^C\left(\boldsymbol{b}-\frac{\boldsymbol{s}}{A_P}\right)-i\chi^C(\boldsymbol{b})} = \left(\frac{|\boldsymbol{b}-\frac{\boldsymbol{s}}{A_P}|}{b}\right)^{2i\eta}. \tag{7.49}$$

Note that $\Delta\chi^C(\boldsymbol{b},\boldsymbol{s})$ does not depend on the cutoff parameter a.

The elastic scattering amplitude, e.g., is given by

$$f_{\text{el}}(\theta) = \frac{iK}{2\pi}\int d\boldsymbol{b}\, e^{-i\boldsymbol{q}\cdot\boldsymbol{b}}$$
$$\times\left[1 - e^{i\chi^C(\boldsymbol{b})}\langle\phi_0|e^{i\chi^N_{\text{CT}}\left(\boldsymbol{b}-\frac{\boldsymbol{s}}{A_P}\right)+i\chi_{n\text{T}}\left(\boldsymbol{b}+\frac{A_P-1}{A_P}\boldsymbol{s}\right)+i\Delta\chi^C(\boldsymbol{b},\boldsymbol{s})}|\phi_0\rangle\right]. \tag{7.50}$$

This expression is the same as Eq. (5.20) if the core–target phase-shift function χ_{CT} there consists of nuclear and Coulomb parts.

The phase-shift function $\Delta\chi^C(\boldsymbol{b},\boldsymbol{s})$ distorts the halo wave function ϕ_0 and may cause neutron breakup. Because of the long-range nature of the Coulomb force, the distortion by the phase-shift function $\Delta\chi^C$ is also long-ranged. For sufficiently large impact parameters, where the nuclear phase-shift function vanishes, the neutron breakup can take place only through $\Delta\chi^C$. This is contrary to the region of small impact parameters, where the nuclear phase-shift function dominates and the Coulomb effects play a minor role.

The overlap of the function $\tilde{\Psi}$ in Eq. (7.46) with ϕ_0 gives the probability amplitude of the nucleus' remaining in the g.s. The reaction probability is thus

$$P_{\text{reac}}(\boldsymbol{b}) = 1 - \left|\langle\phi_0|e^{i\chi^N_{\text{CT}}\left(\boldsymbol{b}-\frac{\boldsymbol{s}}{A_P}\right)+i\chi_{n\text{T}}\left(\boldsymbol{b}+\frac{A_P-1}{A_P}\boldsymbol{s}\right)+i\Delta\chi^C(\boldsymbol{b},\boldsymbol{s})}|\phi_0\rangle\right|^2. \tag{7.51}$$

For small impact parameters, the reaction probability is very close to unity since the core–target nuclear phase-shift function χ^N_{CT} contains a large negative imaginary part. This means that the core and/or target nuclei are excited by the nuclear force with very high probability there. In the impact-parameter region $b > R_{\text{C}} + R_{\text{T}}$ the function χ^N_{CT} is vanishingly small. Nevertheless, reactions of two types may occur: nuclear breakup of the halo neutron described by $\chi_{n\text{T}}$ and Coulomb breakup of the halo neutron via $\Delta\chi^C$. The nuclear breakup through $\chi_{n\text{T}}$ is limited to the impact parameter region within the sum of the radii of the halo-neutron orbit and of the target, $b < R_{\text{H}} + R_{\text{T}}$. Beyond this region, only the Coulomb breakup induced by $\Delta\chi^C$ persists.

The neutron-removal probability can be obtained by subtracting the reaction probability due to the core nucleus, $1 - e^{-2\text{Im}\chi^N_{CT}(\boldsymbol{b})}$, from

Eq. (7.51):

$$P_{-n}(\boldsymbol{b}) = e^{-2\mathrm{Im}\chi^{N}_{\mathrm{CT}}(\boldsymbol{b})} - \left|\langle\phi_0|e^{i\chi^{N}_{\mathrm{CT}}\left(\boldsymbol{b}-\frac{\boldsymbol{s}}{A_{\mathrm{P}}}\right)+i\chi_{\mathrm{nT}}\left(\boldsymbol{b}+\frac{A_{\mathrm{P}}-1}{A_{\mathrm{P}}}\boldsymbol{s}\right)+i\Delta\chi^{C}(\boldsymbol{b},\boldsymbol{s})}|\phi_0\rangle\right|^2. \tag{7.52}$$

The neutron-removal probability can again be decomposed into elastic and inelastic parts as is shown in Table 4.2 according to whether or not the target is excited. The Coulomb breakup mechanism is included in the elastic breakup probability.

To get a physical insight into the breakup process induced by the Coulomb force through $\Delta\chi^C$, let us examine how the reaction probability behaves as a function of b using a Taylor series expansion of $e^{i\Delta\chi^C}$. Let the impact parameter b be much larger than the radius of the halo-neutron orbit so as to exclude all mechanisms but the Coulomb breakup. Then it follows that

$$\begin{aligned}\langle\phi_0|e^{i\Delta\chi(\boldsymbol{b},\boldsymbol{s})}|\phi_0\rangle &= \langle\phi_0|\left(\frac{|\boldsymbol{b}-\frac{\boldsymbol{s}}{A_{\mathrm{P}}}|}{b}\right)^{2i\eta}|\phi_0\rangle \\ &= \langle\phi_0|\left[1-\frac{2i\eta}{A_{\mathrm{P}}}\frac{\hat{\boldsymbol{b}}\cdot\boldsymbol{s}}{b}+\frac{i\eta}{A_{\mathrm{P}}^2}\left(\frac{s}{b}\right)^2+\frac{2i\eta(i\eta-1)}{A_{\mathrm{P}}^2}\left(\frac{\hat{\boldsymbol{b}}\cdot\boldsymbol{s}}{b}\right)^2+\cdots\right]|\phi_0\rangle \\ &= 1-\frac{2\eta^2}{3A_{\mathrm{P}}^2b^2}\langle\phi_0|r^2|\phi_0\rangle+\cdots,\end{aligned} \tag{7.53}$$

where $\hat{\boldsymbol{b}}=b^{-1}\boldsymbol{b}$ is a unit vector and, in evaluating the matrix elements, the spherical symmetry of the s-wave state ϕ_0 is utilized; for example, $\langle\phi_0|s^2|\phi_0\rangle=\frac{2}{3}\langle\phi_0|r^2|\phi_0\rangle$.

At large impact parameters the reaction probability is equal to the neutron-removal probability. Substitution of Eq. (7.53) into Eq. (7.51) leads to

$$P_{-n}(\boldsymbol{b}) \approx \frac{4\eta^2}{3A_{\mathrm{P}}^2b^2}\langle\phi_0|r^2|\phi_0\rangle. \tag{7.54}$$

The expectation value of r^2 is proportional to the $B(\mathrm{E1})$ value given in Eq. (7.10). Therefore, this formula for the neutron-removal probability can be written alternatively as

$$P_{-n}(\boldsymbol{b}) \approx \frac{16\pi}{9}\left(\frac{Z_{\mathrm{T}}e}{\hbar v}\right)^2\frac{1}{b^2}B(\mathrm{E1}). \tag{7.55}$$

This formula shows that, for large impact parameter values, the breakup probability is proportional to the inverse square of the impact parameter, and the integrated cross section $\sigma_{-n} = 2\pi\int db\, bP_{-n}(\boldsymbol{b})$ diverges logarithmically. It will be shown in Sect. 7.3.2 that the neutron-removal probability decreases exponentially for large impact parameter values if the operator $H_{nC} - \varepsilon_0$ in Eq. (7.44) is not omitted. Therefore, the logarithmic divergence of the cross section is a spurious result, arising from the adiabatic approximation. Although the adiabatic approximation allows one to discuss the neutron-removal process without recourse to a perturbative expansion and to treat the nuclear and Coulomb mechanisms on the same footing, its applicability is limited to a certain region of the impact parameter.

The adiabatic approximation was advocated on the grounds that the reaction time is much shorter than the time scale of the resulting nuclear transition. In the nuclear breakup reaction the reaction time is estimated by the interval in which the projectile and the target pass through each other, $\Delta t \simeq 2R/v$, where R is the radius of the nucleus. In the Coulomb breakup reaction the reaction proceeds with much longer time scale because of the long-range nature of the Coulomb force. For a collision event in which the Coulomb breakup can occur with appreciable probability, b may be much larger than the typical nuclear radius, and the time scale may be estimated by $\Delta t \simeq b/v$ instead of $2R/v$. The time interval of the nuclear transition, on the other hand, has been estimated through the uncertainty principle by $\hbar/|\varepsilon_0|$. The validity of the adiabatic approximation hinges on the ratio of the two time scales, $\xi_0 = |\varepsilon_0|b/\hbar v$. The breakup probablity (7.51) may be reliable for the impact parameter region in which $\xi_0 \ll 1$.

Let us estimate the critical impact parameter for a reaction of ^{11}Be. The binding energy is about 0.5 MeV. At 800 A MeV incident energy, $\hbar v$ is about 170 MeV·fm, so the critical impact parameter $b_c \sim 340$ fm. The formula (7.51) may be reliable for impact parameters much smaller than 340 fm. For small impact parameters, Eq. (7.51) is thus suitable for describing the neutron-removal process induced by both the nuclear and the Coulomb mechanism and their interference as well. The validity of the adiabatic approximation will be discussed in conjunction with practical examples in Sect. 7.4.

7.3.2 Perturbative theory

The divergence problem of the adiabatic cross section at large impact parameters can be avoided by the use of the time-dependent perturbation theory. In the perturbative approach the intrinsic Hamiltonian H_{nC} is taken into account, which promises to cure this problem. The perturbative approach will turn out to be closely related to the equivalent-photon method of Sect. 7.2. While meeting the expectations in respect of the divergence problem, it fails to describe the strong excitation in collisions with small impact parameters accurately. In particular, it fails to describe the nuclear and Coulomb mechanism of the neutron removal simultaneously.

Since there is no reason to give up the eikonal approximation, the time-dependent Schrödinger equation (7.44) of that approximation may serve as a starting point. The difference is that the operator $H_{nC}-\varepsilon_0$ is now retained, thus the adiabatic approximation is not used any more. As a simplification, only collisions with large impact parameters are considered and only the Coulomb interaction is taken into account.

The wave function $\tilde{\Psi}(\boldsymbol{b},\boldsymbol{r},t)$ is expanded in terms of the eigenfunctions ϕ_0 and ϕ_{Elm} of H_{nC} [cf. Eqs. (7.4)–(7.6)] as

$$\tilde{\Psi}(\boldsymbol{b},\boldsymbol{r},t) = c_0(\boldsymbol{b},t)\phi_0(\boldsymbol{r})e^{-\frac{i}{\hbar}\varepsilon_0 t} + \sum_{lm}\int_0^\infty dE\, c_{Elm}(\boldsymbol{b},t)\phi_{Elm}(\boldsymbol{r})e^{-\frac{i}{\hbar}Et}, \qquad (7.56)$$

with the initial conditions $c_i(\boldsymbol{b},t\to-\infty)=\delta_{i0}$. In the limit of $t\to\infty$, all coefficients should converge. In the first-order perturbation theory, the coefficients $c_{Elm}(\boldsymbol{b})=c_{Elm}(\boldsymbol{b},t\to\infty)$ are given by

$$c_{Elm}(\boldsymbol{b}) = \frac{1}{i\hbar}\int_{-\infty}^{+\infty} dt\,\langle\phi_{Elm}|\frac{Z_C Z_T e^2}{|\boldsymbol{b}+vt\hat{\boldsymbol{z}}-\frac{\boldsymbol{r}}{A_P}|}|\phi_0\rangle e^{\frac{i}{\hbar}(E-\varepsilon_0)t}. \qquad (7.57)$$

The matrix element here is evaluated by using the integral formula

$$\int_{-\infty}^{+\infty} dt\,\frac{e^{i\omega t}}{|\boldsymbol{b}+vt\hat{\boldsymbol{z}}-\frac{\boldsymbol{r}}{A_P}|} = \frac{2}{v}e^{i\frac{\omega}{v}\frac{z}{A_P}}K_0\left(\frac{\omega}{v}\left|\boldsymbol{b}-\frac{\boldsymbol{s}}{A_P}\right|\right), \qquad (7.58)$$

where $K_0(x)$ is the zeroth-order modified Bessel function defined in Eq. (7.34). Substituting Eq. (7.58) into Eq. (7.57) leads to

$$c_{Elm}(\boldsymbol{b}) = \frac{2Z_C Z_T e^2}{i\hbar v}\langle\phi_{Elm}|e^{i\frac{E-\varepsilon_0}{\hbar v}\frac{z}{A_P}}K_0\left(\frac{E-\varepsilon_0}{\hbar v}\left|\boldsymbol{b}-\frac{\boldsymbol{s}}{A_P}\right|\right)|\phi_0\rangle. \qquad (7.59)$$

Let us examine the asymptotic behaviour of this expression when both $b \gg A_P^{-1}|\boldsymbol{s}|$ and $b \gg A_P^{-1}|z|$ are satisfied simultaneously. The convenient notations $E^* = E-\varepsilon_0$ and $\xi = E^* b/\hbar v$ will now be used. The operator in the matrix element can be expanded as

$$\begin{aligned} e^{i\frac{E^*}{\hbar v}\frac{z}{A_P}}K_0\left(\frac{E^*}{\hbar v}\left|\boldsymbol{b}-\frac{\boldsymbol{s}}{A_P}\right|\right) &= e^{i\xi\frac{z}{bA_P}}K_0\left(\xi\left|\hat{\boldsymbol{b}}-\frac{\boldsymbol{s}}{bA_P}\right|\right) \\ &= \left(1+i\frac{\xi}{b}\frac{z}{A_P}+\cdots\right)\left[K_0(\xi)-K_0'(\xi)\xi\frac{\hat{\boldsymbol{b}}\cdot\boldsymbol{s}}{bA_P}+\cdots\right] \\ &= K_0(\xi)+i\xi K_0(\xi)\frac{1}{b}\frac{z}{A_P}+\xi K_1(\xi)\frac{\hat{\boldsymbol{b}}\cdot\boldsymbol{s}}{bA_P}+\cdots \end{aligned} \qquad (7.60)$$

Here the relation $K_0'(x)=-K_1(x)$ has been used.

Let us define the spherical components (A_{+1}, A_0, A_{-1}) of a vector $\boldsymbol{A}$ as

$$A_{\pm 1} = -K_1(\xi)\frac{b_{\pm 1}}{b}, \quad A_0 = iK_0(\xi), \tag{7.61}$$

where $b_{\pm 1} = \mp(b_x \pm ib_y)/\sqrt{2}$. With these definitions, Eq. (7.60) is reduced to

$$K_0(\xi) + \frac{\xi}{A_P b}\sqrt{\frac{4\pi}{3}}r\sum_m A_{-m}Y_{1m}(\hat{\boldsymbol{r}}). \tag{7.62}$$

The first term is a constant and does not contribute to the transition matrix element in Eq. (7.59). The next term defines a dipole approximation, in which only the $l=1$ components of ϕ_{Elm} give non-vanishing contribution, viz.

$$c_{E1m}(b) = -\frac{2}{i\hbar v}Z_T e\sqrt{\frac{4\pi}{3}}\frac{\xi}{b}A_{-m}\langle\phi_{E1m}|M_{1m}|\phi_0\rangle, \tag{7.63}$$

where M_{1m} is the electric dipole transition operator of the halo neutron defined in Eq. (7.8).

The matrix element $\langle\phi_{E1m}|M_{1m}|\phi_0\rangle$ is independent of m. The halo-neutron-removal probability for a given relative energy E between the halo neutron and the core nucleus is now given by

$$\begin{aligned}\frac{dP_{-\mathrm{n}}(b)}{dE} &= \sum_m |c_{E1m}(b)|^2 \\ &= \frac{16\pi}{9}\left(\frac{Z_T e}{\hbar v}\right)^2\left(\frac{\xi}{b}\right)^2[K_1^2(\xi) + K_0^2(\xi)]\frac{dB(\mathrm{E1})}{dE},\end{aligned} \tag{7.64}$$

where $dB(\mathrm{E1})/dE$, defined in Eq. (7.9), denotes the dipole transition strength per unit excitation energy. The structure of this expression is similar to that obtained in the equivalent-photon method, Eqs. (7.40) and (7.41). Since non-relativistic kinematics is used in the present derivation, $\gamma=1$ is implied. For later reference, it is useful to express Eq. (7.64) in terms of the electric field strength (7.36) or the number of equivalent photons (7.40) taken in this non-relativistic limit:

$$\begin{aligned}\frac{dP_{-\mathrm{n}}(b)}{dE} &= \frac{16\pi^3}{9\hbar^2}|\boldsymbol{E}(\boldsymbol{r},\omega)|^2_{\boldsymbol{r}=0}\frac{dB(\mathrm{E1})}{dE} \\ &= \frac{16\pi^3}{9\hbar c}\omega N(b,\omega)\frac{dB(\mathrm{E1})}{dE}.\end{aligned} \tag{7.65}$$

We now examine the impact parameter dependence of the probability (7.64) and compare it with the probability (7.51) derived in the adiabatic approximation. The dipole strength $dB(\mathrm{E1})/dE$ of the halo nucleus is assumed to be concentrated in the low excitation energy region. For not very large impact parameters b, the parameter $\xi = (E - \varepsilon_0)b/\hbar v$ can be smaller than unity if the incident velocity v is sufficiently large. When the excitation energy E is small and can be neglected, this parameter reduces to $\xi_0 = |\varepsilon_0|b/\hbar v$, which was discussed at the end of Sect. 7.3.1 in the comparison between the two time scales. The assumption $\xi \ll 1$ is equivalent to assuming the validity of the adiabatic approximation.

When $\xi \ll 1$, the modified Bessel functions can be approximated as $K_0(\xi) \simeq -\ln \xi$ and $K_1(\xi) \simeq 1/\xi$. In this case, K_0 is negligible compared with K_1, so that one has

$$\left(\frac{\xi}{b}\right)^2 [K_0^2(\xi) + K_1^2(\xi)] \approx \frac{1}{b^2}. \tag{7.66}$$

The neutron-removal probability then becomes

$$\frac{dP_{-\mathrm{n}}(b)}{dE} \approx \frac{16\pi}{9}\left(\frac{Z_{\mathrm{T}}e}{\hbar v}\right)^2 \frac{1}{b^2}\frac{dB(\mathrm{E1})}{dE}. \tag{7.67}$$

Integrating both sides with respect to the excitation energy E leads to the neutron-removal probability (7.55) derived in the adiabatic approximation.

Even when the $B(\mathrm{E1})$ strength is concentrated in the low excitation energy region, ξ becomes larger than unity at a certain large impact parameter value. When $\xi \gg 1$, the asymptotic expansion of the modified Bessel function can be used to give

$$K_0^2(\xi) + K_1^2(\xi) \approx \frac{\pi}{\xi}\mathrm{e}^{-2\xi}. \tag{7.68}$$

Thus in the perturbation theory the neutron-removal probability $P_{-\mathrm{n}}(b)$ decreases exponentially with the impact parameter b beyond a certain b value. Because of this behaviour, the breakup cross section becomes finite as it should.

Moreover, a closed expression can be derived for the integrated cross section of the neutron removal. It is reasonable to assume that the nuclear process is confined to small impact parameters, and there exists a certain critical impact parameter, b_{min}, beyond which only the Coulomb breakup process occurs. The nuclear breakup cross section is ignored at this stage; our concern here is to evaluate the Coulomb contribution only. In the next section the nuclear and Coulomb contributions will be treated simultaneously.

By using the relations

$$2\pi \int_{x_0}^{\infty} dx\, x K_\mu^2(x) = \pi {x_0}^2 [K_{\mu-1}(x_0)K_{\mu+1}(x_0) - K_\mu^2(x_0)],$$

$$K_{\mu+1}(x) - K_{\mu-1}(x) = \frac{2\mu}{x} K_\mu(x), \tag{7.69}$$

the integrated cross section can be evaluated as

$$\begin{aligned}\frac{d\sigma}{dE} &= 2\pi \int_{b_{\rm min}}^{\infty} db\, b \frac{dP_{-\rm n}(b)}{dE} \\ &= \frac{16\pi}{9}\left(\frac{Z_{\rm T}e}{\hbar v}\right)^2 2\pi\xi_{\rm m} K_0(\xi_{\rm m}) K_1(\xi_{\rm m}) \frac{dB({\rm E1})}{dE},\end{aligned} \tag{7.70}$$

where $\xi_{\rm m}$ is defined as $\xi_{\rm m} = E^* b_{\rm min}/\hbar v$. The total neutron-removal cross section is given by integrating the right-hand side of Eq. (7.70) over the excitation energy E.

7.4 Coulomb breakup reaction of ^{11}Be

As an example for the Coulomb breakup process, the ^{11}Be+^{208}Pb reaction will now be discussed. As before, the g.s. of ^{11}Be will be described as a ^{10}Be core and an s-orbit halo neutron with binding energy $|\varepsilon_0| = 0.503$ MeV. The halo-neutron wave function is constructed with a Woods–Saxon potential as before.

The Coulomb breakup probability in the perturbation theory is expressed in Eq. (7.65) as the squared strength of the electric field [cf. Eq. (7.36)] multiplied by the dipole strength function of the projectile, $dB(\text{E1})/dE$. The squared strength of the electric field is proportional to the number of equivalent photons multiplied with the frequency,

$$\omega N(b,\omega) = \frac{Z_{\rm T}^2 \alpha}{\pi^2}\left(\frac{c}{v}\right)^2 \left(\frac{\xi}{b}\right)^2 \left[K_1^2(\xi) + \frac{1}{\gamma^2} K_0^2(\xi)\right], \tag{7.71}$$

This product, $\omega N(b,\omega)$, will be called the photon-energy distribution.

The photon-energy distribution seen by the projectile crossing the Coulomb field of the target is shown in Fig. 7.3. The left panel is the case of impact parameter $b = 10$ fm and the right panel is the case of $b = 50$ fm. Each panel shows the photon-energy distribution at three different incident energies. The dipole strength function, $dB(\text{E1})/dE$, of Eq. (7.9) in the Woods–Saxon potential model is shown for a reference.

The photon-energy distribution depends strongly on the impact parameter and incident energy. As the incident energy gets higher, the high-frequency component increases at the expense of the low-frequency component. For large impact parameters, the low-frequency component dominates. These features can be understood intuitively. According to Eq. (7.36), the square root of Eq. (7.71) is proportional to the Fourier transform of the electric field induced by the target nucleus.

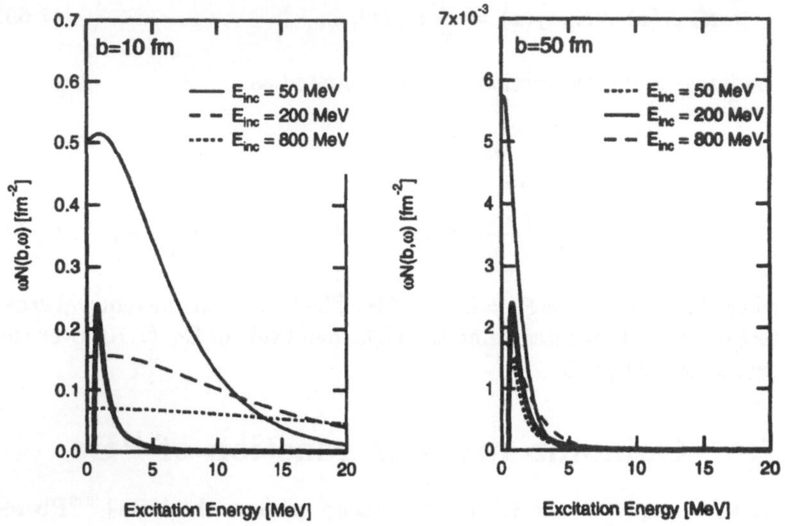

Figure 7.3: Photon energy distribution, Eq. (7.71), as a function of the excitation energy for the $^{11}Be+^{208}Pb$ collision. The left panel shows the distribution for impact parameter $b=10$ fm, and the right panel for $b=50$ fm. Each case is shown for three incident energies. The dipole strength function, Eq. (7.9), in the Woods–Saxon potential model is also shown for reference (thick line, arbitrary scale).

As the incident energy gets higher, the period of existence of the electric field becomes shorter, and that shifts the spectrum towards the high-frequency component. For larger impact parameters, the change of the electric field is less sudden, which shifts the spectrum towards the low-frequency component.

In the adiabatic approximation the neglect of the excitation energy amounts to neglecting the dependence of the photon-energy distribution on the excitation energy. Such an assumption may, of course, be acceptable only for events with small impact parameters. The correct photon-energy distribution decreases rapidly for larger impact parameter values, as is shown by the right-hand panel of Fig. 7.3.

The dynamics of the process and the performance of the approximations are best studied via the reaction probabilities. The probability of neutron removal is obtained by integrating the product of the photon-energy distribution and the dipole strength function over the excitation energy. In Fig. 7.4 the neutron-removal probability is plotted against the impact parameter at incident energies of $800\,A$ and $50\,A$ MeV.

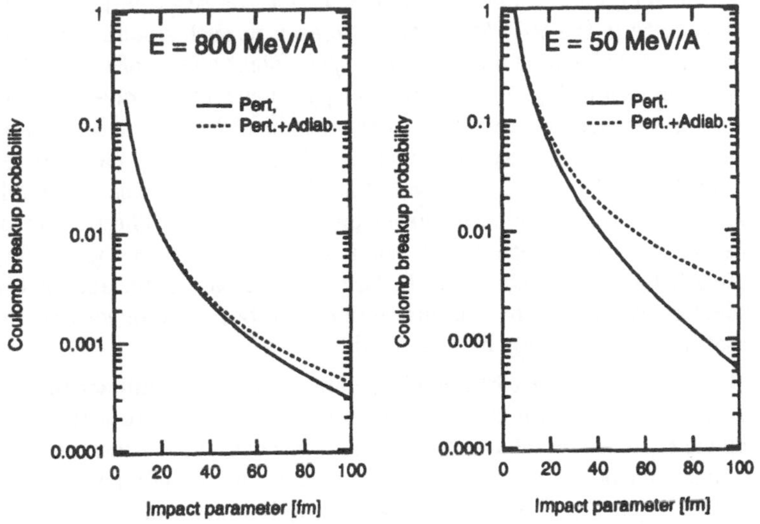

Figure 7.4: Neutron-removal probability in the ^{11}Be+^{208}Pb collision in the perturbation theory. The left panel is for the incident energy of 800 A MeV and the right panel is for 50 A MeV. The probability in the adiabatic approximation, which shows a simple b^{-2} dependence, is shown by dotted curves.

The solid curve shows the result in the perturbation theory. To allow for relativistic effects, the expression in the perturbation theory, Eq. (7.64), is modified as

$$\frac{dP_{-\mathrm{n}}(b)}{dE} = \frac{16\pi}{9}\left(\frac{Z_{\mathrm{T}}e}{\hbar v}\right)^2\left(\frac{\xi}{b}\right)^2\left[K_1^2(\xi) + \frac{1}{\gamma^2}K_0^2(\xi)\right]\frac{dB(\mathrm{E1})}{dE}, \quad (7.72)$$

where ξ is given by $\xi = E^*b/\gamma\hbar v$ [cf. Eq. (7.41)]. To obtain the neutron-removal probability, the above expression is integrated over the excitation energy. The strength distribution of the Woods–Saxon potential model is used. The probability in the adiabatic approximation, which is given by Eq. (7.55) and has a simple b^{-2} dependence, is also shown by dotted curves. The neutron-removal probability is larger at 50 A MeV than at 800 A MeV, reflecting the fact that the photon-energy distribution is larger at lower incident energies as is seen in Fig. 7.3. The target Coulomb field changes relatively slowly during a low-energy collision, which increases the number of equivalent low-frequency photons. Since the dipole strength of the halo nucleus is concentrated in low excitation energies, the Coulomb breakup probability is larger in collisions of lower incident energies.

In the very small impact parameter region the reaction probability exceeds unity in the 50 A MeV case. However, at such small impact parameters the collision is dominated by the violent effects of the nuclear force. As is seen in Fig. 7.5, this impact parameter region does not contribute to the neutron removal. In the region of $b < 20$ fm the dotted curve in the adiabatic approximation in Fig. 7.4 is very close to the perturbation calculation. However, the discrepancy increases towards larger impact parameters and lower incident energies. These observations are consistent with the previous considerations on the validity of the adiabatic approximation. In large impact-parameter and/or low incident-energy collisions the interaction time becomes longer and the excitation energy cannot be neglected.

The adiabatic approximation is reliable in the small impact parameter region, where both the nuclear and the Coulomb interactions are significant. In the adiabatic approximation both the nuclear and the Coulomb mechanisms can be taken into account, together with their interference [cf. Eq. (7.51) for the reaction probability and Eq. (7.52) for the neutron-removal probability]. Figure 7.5 shows such an example for the neutron-removal probability. For 800 A MeV, the phase-shift functions have been calculated from the nuclear density multiplied by the nucleon–nucleon profile function, whereas for 50 A MeV, from the op-

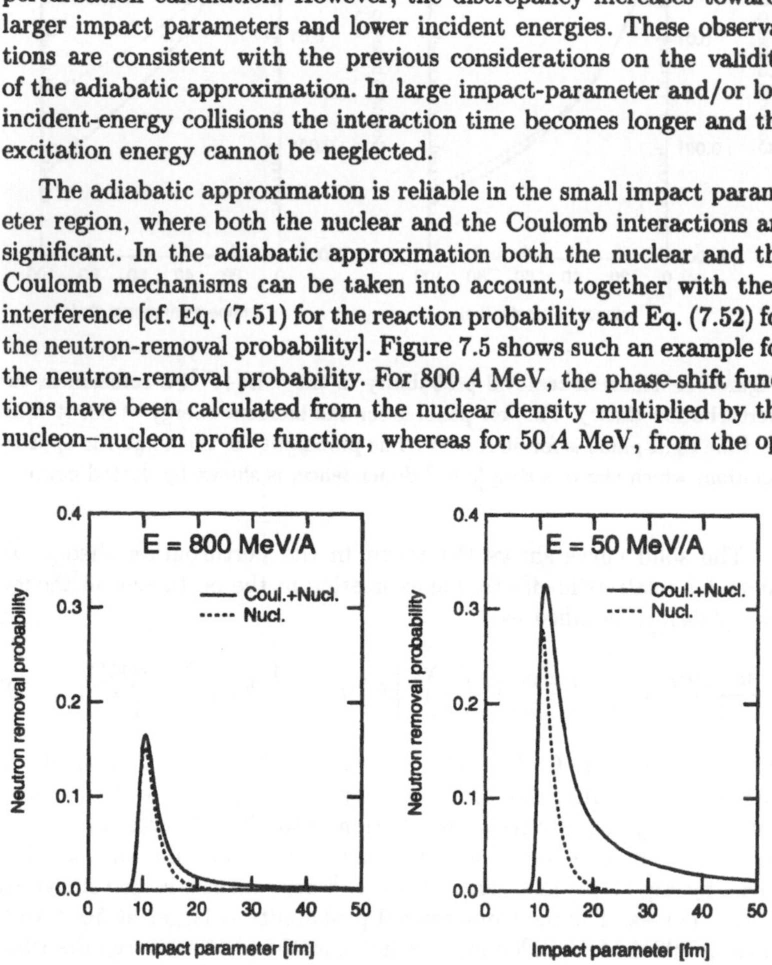

Figure 7.5: Neutron-removal probability, Eq. (7.52), calculated with the eikonal and adiabatic approximation for the collision of $^{11}Be+^{208}Pb$ at the incident energy of 800 A MeV (left panel) and 50 A MeV (right panel). The solid curves include the effects of both the nuclear and the Coulomb interactions, while the dotted curves include only nuclear interactions.

tical potential. In the high-energy case the neutron removal is induced predominantly by the nuclear interaction. For larger impact parameters, however, the nuclear breakup probability decreases rapidly. For $b > 20$ fm, only the Coulomb breakup mechanism contributes, and the perturbation theory is appropriate. At the lower incident energy the Coulomb contribution is much larger than at the higher energy, and even the nuclear contribution is somewhat larger.

In summary, there is no uniformly valid approach, but the eikonal and adiabatic approximation is suitable for $b < 20$ fm, while the perturbative approach is applicable to $b > 20$ fm. The cross section is an integral over b, and the the reaction probability in the integrand should be calculated accordingly.

Some cross sections calculated in this manner for the collision of the ^{11}Be projectile with a few target nuclei are shown in Fig. 7.6. It is seen that the neutron-removal cross section induced by the Coulomb field increases rapidly as a function of the target charge number. According to the perturbative theory, the neutron-removal cross section is proportional to the square of the target charge number [cf. Eq. (7.70)]. For reactions with light target nuclei, the Coulomb breakup contribution is negligible, but, for heavy target nuclei, its contribution is dominant. The cross section shows strong energy dependence in accord with the energy dependence of the probabilities seen in Figs. 7.4 and 7.5. For

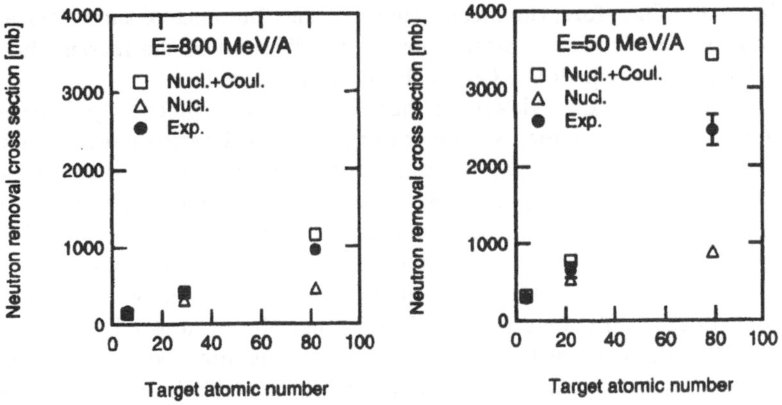

Figure 7.6: Neutron-removal cross section in collisions of ^{11}Be with various targets as a function of the target charge number at the incident energy of 800 A MeV (left) and 41 A MeV (right). The filled circles are the measured neutron-removal cross sections [75, 131]. The open circles are calculated nuclear and Coulomb breakup cross sections, whereas the open triangles include only the nuclear mechanism.

heavy targets, theory somewhat overestimates the data. This may be related to the bound-to-bound transitions in the ^{11}Be nucleus from the $\frac{1}{2}^+$ g.s. to $\frac{1}{2}^-$ excited state, which has a substantial dipole strength and is ignored in the present treatment, as is discussed in conjunction with the dipole strength function in Fig. 7.2.

7.5 Postacceleration phenomena

In the Coulomb breakup of ^{11}Li on ^{208}Pb measurements of the longitudinal momentum distribution of the two neutrons and the core nucleus, ^{9}Li, have revealed that the average longitudinal velocity of the core ^{9}Li is larger than that of the removed neutrons [149,165]. The difference has been explained with the following mechanism. In approaching the target nucleus, the projectile is decelerated by the target Coulomb field. Around the distance of closest approach the projectile is broken up by the Coulomb field. Then the core nucleus gets accelerated by the Coulomb field, but the neutrons keep their velocity which they had at the time of the breakup. Since, according to this scenario, the velocity difference is due to the acceleration after the breakup, the effect is called a postacceleration phenomenon.

If the breakup proceeds via a resonance state of long lifetime, the projectile as a whole may be accelerated by the target Coulomb field before it decays. In that case the postacceleration effect is expected to be small. Therefore, the measurement of the difference of the velocities should provide direct information on the lifetime of the intermediate state through which the Coulomb breakup proceeds.

There is another viewpoint, however, which predicts no postacceleration. In the arguments leading to the postacceleration mechanism it is assumed that the breakup occurs instantaneously somewhere at the point of the closest approach, which is not well-founded. Instantaneous transition is not a result of the models that describe the breakup but an interpretation attached to them. It is actually difficult to estimate how long it takes for the projectile to break up in a 'direct' process. If it is substantially long, postacceleration may be suppressed.

The difference of the longitudinal velocities may be estimated by using classical dynamics. Let the projectile–target relative velocity be v at the distance b (closest approach), where the projectile is assumed to be split up. Then energy conservation gives

$$\frac{1}{2}M_{\mathrm{CT}}v^2 + \frac{Z_{\mathrm{C}}Z_{\mathrm{T}}e^2}{b} = \frac{1}{2}M_{\mathrm{CT}}v_f^2, \tag{7.73}$$

where M_{CT} is the reduced mass of the core and the target and v_f is the final velocity of the core nucleus. The velocity difference between the core and the neutron can then be estimated to be

$$v_f - v \approx \frac{Z_C Z_T e^2}{M_{CT} v b}. \tag{7.74}$$

In the breakup experiments with ^{11}Li [149, 165] and ^{11}Be [148] the observed difference of the average velocities is consistent with the classical estimate (7.74). This is considered an evidence for the direct mechanism of the removal of the neutrons. However, a controversial measurement for the ^{11}Be breakup shows no difference between the velocities of the removed neutron and the core [166], which apparently indicates that the breakup time is substantially long to suppress the postacceleration effect. In Sect. 7.1 the appearance of the strong dipole strength at low excitation energies has been explained without assuming a resonant state in the exit channel. Therefore, the existence of a resonance that could account for the suppression of the postacceleration effect looks unlikely for ^{11}Be.

There have been several theoretical attempts to clarify the breakup mechanism and postacceleration effects. Neither the simple perturbation theory nor the eikonal approximation produces the difference of the average velocity. A description of Coulomb postacceleration requires a treatment beyond these approaches. Several attempts have been reported, including a direct method to solve the time-dependent Schrödinger equation with a time-dependent external field [167–171] and a higher-order perturbation treatment [172].

Part II

Structure of light exotic nuclei

Chapter 8

Introduction to Part II

8.1 Overview

The nuclear collisions discussed in Part I reveal that some light nuclei have very peculiar structures. Most attention has been focussed on halo structure. In the introduction to this book it has been shown that halo structure can be expected especially in the nuclei lying near the border-lines of stability against nucleon emission, i.e., the nucleon drip lines.

The subject of Part II is the description of all aspects of exotic structure in a unified approach. To start with, a selection of problems concerned with light exotic nuclei will be presented in this section, with special emphasis, again, on halo structure. The nature of halo structure emerging from these examples will provide guide-lines for their theoretical description (Sect. 8.2). This prepares the ground for introducing the theoretical approach adopted in this book for this purpose (Sect. 8.3).

In light nuclei situated in the vicinity of the drip line the separation energy of one or a few nucleons is small, typically 1 MeV or less. The binding of the bulk of the nucleus is, nevertheless, saturated normally. The large difference between the average binding energy and the binding of the last nucleons may cause non-uniformity in the density distribution as well. While the bulk nucleons form a normal-density core, the weakly bound nucleons stay mostly in a surface region of dilute density. The core may not be passive either. Owing to the outstanding stability of the α-particles, the core may be clustered into α and other clusters. This suggests that the mean field may not be well-defined, but at least it must be soft to deformation or must be prone to give way to clustering. As a consequence, the sequence of SM orbits is variable

Table 8.1: Neutron and proton separation energies ($|\varepsilon_n|$ and $|\varepsilon_p|$) in MeV. For Borromean nuclei $|\varepsilon_n|$ and $|\varepsilon_p|$ have been replaced by the two-nucleon separation energies, $|\varepsilon_{2n}|$ and $|\varepsilon_{2p}|$, respectively. The symbols for β-stable nuclei are underlined. The separation energies for particle-unstable nuclei are replaced by '—'. Values estimated from systematics are marked by asterisks. The data are taken from Ref. [175].

$Z=5$	$\|\varepsilon_n\|$	$Z=6$	$\|\varepsilon_n\|$	$N=8$	$\|\varepsilon_p\|$
$\underline{^{10}\mathrm{B}}$	8.44	$\underline{^{12}\mathrm{C}}$	18.72	$\underline{^{15}\mathrm{N}}$	10.21
$\underline{^{11}\mathrm{B}}$	11.45	$\underline{^{13}\mathrm{C}}$	4.95	$\underline{^{16}\mathrm{O}}$	12.13
^{12}B	3.37	^{14}C	8.18	^{17}F	0.60
^{13}B	4.88	^{15}C	1.22	^{18}Ne	3.92
^{14}B	0.97	^{16}C	4.25	^{19}Na	—
^{15}B	2.77	^{17}C	0.73	^{20}Mg	2.31
^{16}B	—	^{18}C	4.19	^{21}Al	—
^{17}B	1.39	^{19}C	0.16	^{22}Si	$\|-0.01\|^*$
^{18}B	—	^{20}C	3.34		
^{19}B	0.50*	^{21}C	—		
		^{22}C	1.12*		

from one nucleus to its neighbour.

As the separation energy of 1 MeV is smaller than, or of the same order as, the pairing energy ($\approx 12/\sqrt{A}$ MeV) [173,174], the pair correlation between the valence nucleons plays a crucial role in the binding near the drip line. This is exemplified by the data given in Table 8.1. The heaviest existing isotope of B ($Z=5$) is ^{19}B. The odd-N isotopes ^{16,18}B are, however, unstable to neutron emission, which shows the significance of the contribution of the pairing energy in the binding of the even-N isotopes. The neutron separation energy[1] is $|\varepsilon_n| = 0.97$ MeV for the heaviest odd-N B isotope, ^{14}B, but it becomes as large as 2.77 MeV when one more neutron is added. It is also due to the pairing effect that two or four more neutrons can be stuck to ^{15}B.

According to Chap. 1, halo structure of the projectile shows up in large cross sections and narrow fragment momentum distributions. The relevant separation energies of all halo nuclei are very small, and all empirical findings can be accounted for in a picture exploiting this fact. This picture consists of one or two halo neutrons orbiting around a passive core in an orbit of small separation energy, and that is the essence

[1]The interfragment energy, $E_{a_1+a_2} - E_{a_1} - E_{a_2}$, which is negative for a bound system, is denoted by ε.

of halo structure[2] [176]. It should be emphasized, however, that the small separation energy is only necessary but not sufficient to produce halos. In Table 8.1 ^{14}B is seen to have a pretty small neutron separation energy (0.97 MeV), but, in fact, the interaction cross sections of its collisions are quite normal (see Fig. 1.1), so its size cannot be too large. *A posteriori* one can find some explanation of why there is no halo. The small separation energy produces a halo configuration, but its weight may be quite small in the actual wave function. The ^{14}B is an odd–odd nucleus, and the halo configuration may easily mix with other configurations. But it would be hazardous to derive, from this example, an *a priori* criterion.

The isotope ^{17}B is conjectured to have a two-neutron halo [20]; its two-neutron separation energy, $|\varepsilon_{2n}|$, is 1.39 MeV, and the interaction cross section is significantly larger than that expected from Eq. (1.1) with the standard A-dependence. The $|\varepsilon_{2n}|$ value of ^{19}B is estimated to be only 0.50 MeV, so it is reasonable to imagine ^{19}B as four weakly bound neutrons surrounding a ^{15}B core. The interaction cross section is also enhanced [20] as is shown in Fig. 1.1. Should ^{19}B be portrayed as wearing a four-neutron halo, two two-neutron halos or just a neutron skin? To be consistent with the halo concept developed for one- and two-neutron halos [176], one should say that this is a four-neutron halo if it moves with an inextricable five-body motion, and it behaves as a neutron skin if the five-body motion can be decomposed into s.p. motions. Although this particular question is too subtle and the case is too complicated, this is the type of questions to be addressed by structure theory.

---

[2]The concept of a neutron halo is akin to, but not the same as, that of a neutron skin. The neutron skin is a neutron cover on the nuclear surface, which may arise from any large number of neutrons. Unlike the neutron halo, it is essentially a bulk phenomenon since it is not associated with a particular orbit that is separated from the rest by its extension or by its non-s.p. nature. It may exist without a clear separation of the orbits into core orbits and skin orbits. The skin thickness is defined as the difference between the neutron and proton rms radii. Nuclei with $N > Z$ along the stability region have usually very thin neutron skins if at all, and the skin thickness grows gradually towards the neutron drip line. This behaviour is related to that of the Fermi levels. In normal nuclei the proton and neutron Fermi levels are at about the same energy, while in excessively neutron-rich nuclei the neutron Fermi level may lie substantially higher. It is these nuclei that have thick neutrons skins. Thus in these nuclei the one-, two- etc. few-neutron separation energies are smaller than in normal nuclei (though are not necessarily as small as in a halo nucleus). Except for very light nuclei and apart from pairing fluctuations, the neutron separation energy grows gradually towards the line of stability. This contrasts sharply with an isotopic chain starting with a neutron-halo nucleus, in which the separation energy is small for the first one or two neutrons, but turns abruptly normal for the rest. In this book, the concept of skin thickness is used for halo nuclei as well, but halo nuclei are considered distinct from those wearing neutron skins.

By adding one proton to the B isotopes, one obtains the C isotopes with $Z = 6$, which can accommodate more neutrons. All isotopes up to the drip-line nucleus ^{22}C, except for the neutron-unstable ^{21}C, have been produced and studied. The $|\varepsilon_n|$ or $|\varepsilon_{2n}|$ values of the even-N C isotopes are fairly large, but decrease to 4.19, 3.34 and 1.12 MeV for ^{18,20,22}C, while those of the odd-N isotopes are much smaller, 1.22, 0.73 and 0.16 MeV for ^{15,17,19}C, respectively. The isotope ^{19}C has an extremely small separation energy,[3] and, indeed, it is a one-neutron-halo nucleus [15, 16]. The most probable value of the spin–parity of its g.s. is $\frac{1}{2}^+$ [15, 177]. If so, then its simplest possible SM configuration is $(0d_{5/2})^4(1s_{1/2})$. It will cost a lot of effort to understand why the still vacant $0d_{5/2}$-orbit is not occupied before the $1s_{1/2}$-orbit is filled.

The proton separation energy $|\varepsilon_p|$ also exhibits even–odd staggering. In the case of the $N = 8$ isotones (Table 8.1) the odd-Z nuclei ^{21}Al and ^{19}Na are particle-unstable, and the heaviest odd-Z particle-stable isotone is ^{17}F, next to doubly magic ^{16}O. For even Z, however, a few more protons can be added, to form ^{18}Ne, ^{20}Mg and ^{22}Si. Although ^{21}Al is unstable to proton emission, with one or two more neutrons, bound nuclei can again be produced: for ^{22,23}Al, $|\varepsilon_p| = 0.02$ and 0.13 MeV, respectively.

The correlation between $|\varepsilon_n|$, $|\varepsilon_p|$ and the interaction cross section [178] elucidates nuclear structure near the drip lines. But the extremities in the composition of light exotic nuclei are often beyond our imagination, and the explanation of the stability of some extreme structures in terms of the SM is feeble and is lagging behind empirical knowledge.

The maximum number, N_{max}, of neutrons to be accommodated in an isotope of proton number $Z = 3$ is 8, and the nucleus obtained is the typical halo nucleus: ^{11}Li. According to the SM, the last two neutrons should fill the $0p_{1/2}$ subshell, but a strong or dominant $(1s_{1/2})^2$ configuration has also been proposed [179–181]. For $Z = 4$, $N_{max} = 10$, and the configuration of the last two neutrons in ^{14}Be may be a mixture of $(1s_{1/2})^2$ and $(0d_{5/2})^2$ [20, 182]. By adding one more proton ($Z = 5$), one obtains $N_{max} = 14$, skipping 12. The six neutrons in the sd shell may prefer the configuration $(0d_{5/2})^4(1s_{1/2})^2$ to $(0d_{5/2})^6$. For $Z = 6$, $N_{max} = 16$, so there are eight neutrons in the sd shell etc. One can thus make first guesses, but common sense educated on the SM is usually not enough to explain the behaviour of exotic nuclei.

So far we have mostly been concerned with the stability of nuclei, i.e., with the existence of bound states. The structures formed can gen-

[3] Recent experiments imply a larger neutron separation energy $|\varepsilon_n|$ than that quoted in Table 8.1: 0.53±0.13 MeV (Coulomb dissociation experiment) [15] and 0.8±0.3 MeV (one-neutron knockout reaction) [177].

erally be visualized in a SM or in a cluster-model picture. The low-lying excited states are likely to have similar structures. But near the limits of stability against nucleon emission (i.e., near the drip lines) all excited states may be unbound. They are likely to decay according to their cluster patterns (cf. Table 12.1 later). Since the barriers keeping the components together are rather low, all but a few lowest-lying excited unbound states may be dissolved in the continuum. Contrary to a naive view, however, the possibility of decay into more than two fragments does not speed up decay. The resistance of the system to three-body decay looks as if there were an effective three-body barrier (cf. p. 273).

In conjunction with excited states, collective excitations should also be mentioned. Nuclear collectivity [174] comes from nuclear shape and its deformities. Deformation may be permanent, and then the associated collective motion is rotation, or dynamic, which implies vibration. Rotational and vibrational states involve excitations of a motion in orientation variables and in deformation variables, respectively. The microscopic picture of a permanently deformed nucleus involves a deformed potential well filled up with nucleons. The microscopic picture underlying vibration assumes a 'vacuum state' (i.e., a physical state with no vibration), which is excited to vibration by coherent particle-hole excitations. The rotational states can be arranged into sequences, 'rotational bands', each of which involves a well-defined deformed ('intrinsic') state, and the members of a band differ in the rate of their rotation. The vibrational states can be classified according to whether protons and neutrons move together (isoscalar modes) or against each other (isovector modes), whether spin-up and spin-down particles move together (no-spin mode) or against each other (spin vibrational mode) and according to multipolarity. Some particular vibrational modes exist in principle in all nuclei: those consisting of intershell particle-hole excitations, i.e., excitations in which particles are coherently lifted from one shell to another. These are the so-called giant multipole resonances. In discussing the collective modes of excitation, one should keep in mind, however, that really collective motion requires a sufficient number of particles, which may not be available in the very light nuclei of our concern.

Nevertheless, some exotic nuclei might be visualized as being permanently deformed. However, a nuclear state generated by a deformed potential well may grossly overlap with a two-cluster or with a multi-cluster state, and, indeed, the states regarded as deformed are usually the same as those regarded as clustered. Thus, for a light nucleus, the picture of permanent deformation is just a macroscopic visualization of cluster structure. (Similarly, a two-cluster picture seems to be viable for heavy nuclei that are believed to be permanently deformed [183].) Yet the 'collective' degrees of freedom to be excited in a light exotic nucleus are primarily not orientation variables but intercluster variables.

The behaviour of the standard multipole vibrational modes is influenced by special circumstances.

The vacuum state is mostly the g.s., which has an exotic structure. These nuclei are very asymmetric in protons and neutrons, which is bound to affect especially isovector modes. For neutron-halo nuclei, in particular, the vibrational motion of the neutron halo is expected to decouple from the rest, giving rise to a 'soft dipole mode' [11,142,143] (cf. Sect. 7.1). This is a special case of pygmy resonances, which are low-energy fragments of the giant dipole resonances, decoupled when there are low-l particle states near the s.p. threshold [147].

Another special circumstance is that most, if not all, of the particle states taking part in the particle-hole excitations that constitute the multipole modes are unbound. This makes the coupling of the multipole modes to the continuum very strong, which may largely smear out the multipole resonances. The soft dipole mode has been observed as a strength concentration [149–151], but its interpretation as a discrete state does not seem to hold [152–154].

Since the halo neutrons stay largely out of the core, the isobaric analogues of halo states[4] must contain very extensive proton orbits or must violate the isospin symmetry appreciably. Experiment has not yet reached the stage of such studies, but a theoretical study indicates that halo analogue states must exist and isospin mixing cannot be very strong in them (see Sect. 13.1.4).

8.2 Description of exotic structure

We have seen in Sect. 8.1 that the SM may be a good starting point to explore the configurations contained in the nuclear wave function even for nuclei near the drip lines. But the solution of the dynamical problem requires more. The spatial shape of the wave function for a system with weakly bound nucleons must be quite different from that of normal nuclei. When expanded over an HO basis, the wave function must contain many-$\hbar\omega$ components with substantial weights. Thus the solution of the dynamical problem requires huge HO bases or bases of other types that are better suited to the problem.

[4]The isobaric analogue of a state of a nucleus is a state in which a neutron is replaced by a proton or *vice versa* such that the space-spin part of the wave function is unchanged. Such replacements are generally allowed by the Pauli principle for nuclei with $N > Z$ and $N < Z$, respectively. In actual analogue pairs the spatial wave functions are bound to differ a little owing to the difference in the Coulomb force.

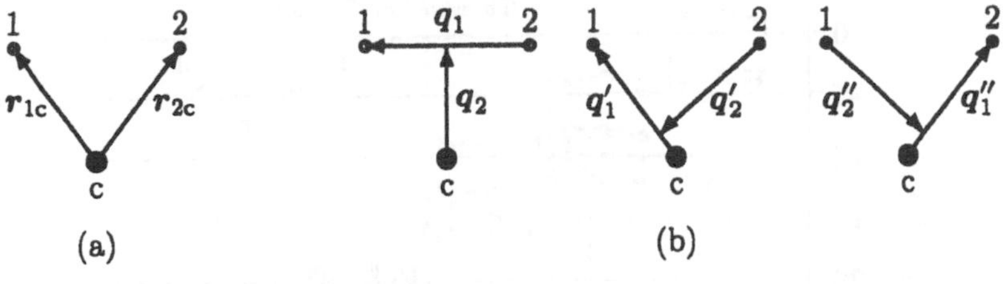

Figure 8.1: Intrinsic coordinates $\boldsymbol{r}_{ic} = \boldsymbol{r}_i - \boldsymbol{R}_c$ used in the cluster-orbital SM (a) and the Jacobi coordinates used in the Faddeev formalism and in some variational formalisms [cf. Eq. (B.10)] (b).

Moreover, the conventional nucleon–nucleon interactions used in the SM may not be appropriate for halo nucleons. The nucleons moving mostly outside the dense nuclear medium feel much less of Pauli effects than those confined inside, thus it seems unjustifiable to apply interactions deduced from the properties of normal nuclei [184] to halo nucleons. The interaction must be closer to the free nucleon–nucleon force and may be more attractive than the usual SM interactions.

The simplest SM-like approach to exotic nuclei is called the cluster-orbital SM [185, 186]. In this approach the core and the n valence nucleons are treated differently: the core is assumed to be inert, and only the valence nucleons are treated explicitly. The Hamiltonian[5] is thus

$$H = H_c + \sum_{i=1}^{n}\left[\frac{\boldsymbol{p}_i^2}{2\mu} + U(\boldsymbol{r}_{ic})\right] + \sum_{i<j}^{n}\left[V(\boldsymbol{r}_{ic} - \boldsymbol{r}_{jc}) + \frac{1}{m_c}\boldsymbol{p}_{ic}\cdot\boldsymbol{p}_{jc}\right], \quad (8.1)$$

where H_c is the Hamiltonian of the core, m_c is its mass, $\boldsymbol{r}_{ic}$ is the vector pointing from the core to particle i (see Fig. 8.1a), $\boldsymbol{p}_{ic}$ is the momentum conjugate to $\boldsymbol{r}_{ic}$, μ is the nucleon–core reduced mass, and the spin–isospin dependence of the potentials is suppressed for simplicity. The term $m_c^{-1}\boldsymbol{p}_{ic}\cdot\boldsymbol{p}_{jc}$ arises because of the use of the coordinates with respect to the core [187]. Since in the cluster-orbital SM translationally invariant coordinates are used, the excitation of the valence nucleons never gives rise to spurious c.m. excitations.[6]

[5]Throughout this book, the symbols $\prod_{i<j}^{n}$ and $\sum_{i<j}^{n}$ stand for $\prod_{i=1}^{n-1}\prod_{j=i+1}^{n}$ and $\sum_{i=1}^{n-1}\sum_{j=i+1}^{n}$, respectively.

[6]In the ordinary SM the nuclear states are built up from nucleon orbits generated by a potential fixed in space, and the dependence of the wave function on the total c.m. cannot be factored out. Each SM state can be expanded in the from $\sum_i \varphi_i(\text{c.m.})\Psi_i(\text{intrinsic})$, and the mixing of different c.m. states $\varphi_i(\text{c.m.})$ is obviously spurious. Such problems can be avoided by constructing the nuclear states from translationally invariant elements.

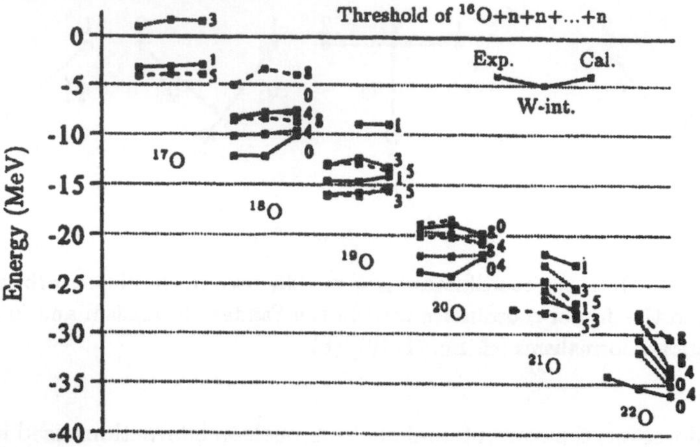

Figure 8.2: Low-lying energy levels of oxygen isotopes. The numbers on the right of each column stand for $2J$. 'Cal.' stands for cluster-orbital SM calculation with a modified Minnesota potential [190], 'W-int.' is the SM calculation with the so-called W interaction [191,192]. The data are taken from [193,194].

The interaction U may be determined from data on the core+one-nucleon system. The interaction V acting between the valence nucleons can be chosen to be an effective interaction reproducing the basic properties of the core+two-nucleon system. The Pauli principle is taken into account by choosing the valence s.p. orbits to be orthogonal to those assumed to be occupied in the core, which is an implicit way of taking into account the state of the core.

The cluster-orbital SM model has been applied, among others, to the ^{16}O+few-nucleon problems. The success of the cluster-orbital SM in this case (Fig. 8.2) is not surprising since ^{16}O as a core is duly assumed to be inert. It has also been applied, however, to ^{11}Li in a ^{9}Li+n+n model [188, 189], where ^{9}Li is described by the simple HOSM wave function. Although the halo nucleons must certainly be more active than those in the core, there is no guarantee that ^{9}Li is good as a passive core. The nucleus ^{9}Li has open proton and neutron shells. As a consequence, it has a low-lying excited state and its continuum states are also rather excitable. The nucleus ^{11}Li is distorted through the excitation of the ^{9}Li core by the halo neutrons, which may contribute to its binding appreciably.

The SM-like feature in this model is that the intrinsic coordinates are the nucleon–core relative coordinates (Fig. 8.1a). Therefore, the basis of the cluster-orbital SM can be constructed out of intrinsic s.p.

states. The cluster-orbital character of the model lies in the choice of the s.p. orbits. For two-neutron-halo nuclei the SM-like V-shaped configurations converge rather slowly, and so do the configurations written in the similar Y-shaped coordinates (Fig. 8.1b). The convergence is much faster with a combination of bases formulated with the Y- and T-shaped (Fig. 8.1b) systems of coordinates [189, 195]. This strongly suggests that both the correlations between the two valence neutrons and between each valence neutron and the core are important.

For core+n+n systems it is easy to incorporate such configurations, which leads to the general three-body models of two-neutron-halo nuclei [196]. The underlying three-body problem can be solved by the exact Faddeev formalism [196], by the hyperspherical method [196] or by variational methods [197]. The coordinates used by the Faddeev formalism and by some other variational methods[7] are shown in Fig. 8.1b. All three-body models can be regarded as macroscopic three-cluster models, where the three clusters are the core and the two nucleons as 'single-nucleon' clusters.

The need for treating two nucleons outside a core means that the building blocks of the description of these nuclei are not s.p. orbits. The basic entity on which the behaviour of these nuclei rests is not s.p. motion but three-particle motion. Even if the core plus one nucleon system could be described as a particle orbiting in a potential, this does not provide a suitable basis for the description of the nucleus that contains one or a few more nucleons. This seems to exclude the validity of a mean-field picture for these nuclei. The concept of a mean field is shaken by the apparent properties of light exotic nuclei, and the rest of this section is devoted to examining this question in some more detail.

A strongest argument against the mean-field picture may be the existence of clustering within the core. Although clustering might be approximated by permanent deformation, which is consistent with a mean field, when treated as clusters, no single mean field is assumed, nor is it generated by the model itself. A semirigid two-centred system can as well be described by analogy with molecules. The nuclear molecular model based on this analogy does not use a mean field; it is just a particular type of microscopic cluster model [198, 199]. In these models, the relative motion of the cluster centres is assumed to be restricted like in a molecule, and the non-clustered single nucleons are treated as

[7]The hyperspherical coordinates cannot be depicted in two or three dimensions. They include six coordinates: a hyperradius ρ defined by $\rho^2 = \mu^{-1}[\mu_{12}\mathbf{q}_1^2 + \mu_{(12)3}\mathbf{q}_2^2]$, where the factors μ_{12} and $\mu_{(12)3}$ are the reduced masses corresponding to the partitioning defined by $\{\mathbf{q}_1, \mathbf{q}_2\}$ and μ is an arbitrary constant, the solid angles of $\hat{\mathbf{q}}_1$ and $\hat{\mathbf{q}}_2$, and a hyperangle, γ, defined by $\rho \sin\gamma = \sqrt{\mu^{-1}\mu_{12}}q_1$ and $\rho\cos\gamma = \sqrt{\mu^{-1}\mu_{(12)3}}q_2$.

valence particles. These nucleons are given sufficient freedom to form 'chemical bonds' between the clusters.

A further defect of the mean-field concept in exotic nuclei is that some of the s.p. orbits which should contribute to the mean field may be unbound. In such a case the nucleus becomes bound as a result of a departure from the s.p. picture underlying the mean-field concept. Nevertheless, the Hartree–Fock method has been generalized to such cases [200]. All in all, despite all counterindications, Hartree–Fock theory can make qualitative predictions for light exotic nuclei [201, 202].

A s.p. model can, however, be formulated without reference to a mean field if one introduces some special constraint. One may assign the A nucleons any s.p. wave functions ϕ_i, $(i=1,\ldots,A)$ and construct a Slater determinant out of such functions. This Slater determinant can be used as a trial function, and the parametrization of ϕ_i may be chosen such that the optimization does not necessarily lead to a mean field. There is a widely used framework of this kind, which appears to be successful: the fermionic [203] or antisymmetrized molecular dynamics (AMD) [204]. In the AMD, the spatial parts of ϕ_i are Gaussians of fixed spread attached to variable centres $\boldsymbol{s}_i$: $\phi_i = \varphi_{\boldsymbol{s}_i}(\boldsymbol{r}_i) \sim \mathrm{e}^{-\frac{1}{2}\beta(\boldsymbol{r}_i-\boldsymbol{s}_i)^2}$. Such s.p. wave functions are called Gaussian wave packets [cf. Eqs. (9.65) and (9.66) later]. The optimization consists in finding an arrangement of the position vectors $\boldsymbol{s}_i$ with which the total energy reaches a minimum. The AMD produces a very characteristic clustered arrangement for each exotic nucleus [204] albeit it apparently includes no correlation.[8] Since clustering is not a model assumption but a result emerging from the AMD, this finding supports cluster models.

To sum up the foregoing considerations, one can say that a fair description of an exotic nucleus should allow for all possible correlations between the core and the valence nucleons as well as for the possibility of core excitation. Since in light nuclei with large neutron or proton excess clustering may dominate over the mean-field-governed structure, the most realistic picture is a multicluster picture. A model based on such a picture involves a few-body problem with composite bodies, i.e., clusters, in addition to some unclustered nucleons to be treated as single-nucleon clusters.

[8] Here the no-correlation case is identified with a Slater determinant. If it were identified with the product of s.p. wave functions, then antisymmetrization itself should also be viewed as a type of correlation. Antisymmetry (as well as parity projection) plays an important role in giving rise to clustering in the AMD. In more realistic models clustering is brought about by an interplay of dynamic correlations with antisymmetry, see Sect. 12.3 later.

8.3 Cluster approach with Gaussians

In Sect. 8.2 we have seen that the difficulty facing the theory of light exotic nuclei lies in that the independent-particle model is not applicable. These nuclei fail to develop a passive enough core and their valence nucleons do not move independently enough. For example, in Borromean nuclei the interaction of the halo nucleons with the core is not enough to bring about binding, so that the lowest-lying s.p. orbits with respect to the core are unbound. The slight contribution to the binding coming from the interaction between the valence nucleons makes the system bound. Therefore, the simplest picture of these nuclei involves at least three interacting bodies. But even the core may be quite soft, and the only relatively rigid ingredients are smaller clusters. This fact calls for a fully microscopic description of these nuclei, with little constraint on the nucleonic motion. The only constraint that seems justifiable is the assumption of the formation of α- and three-nucleon clusters.

In a microscopic scheme all nucleonic degrees of freedom are involved, and all are subjected to antisymmetrization. To derive the nuclear properties on a sound theoretical footing, one has to solve an equation of motion. In a non-relativistic quantum-mechanical description this is the time-independent Schrödinger equation for A nucleons:

$$H\Psi = E\Psi. \tag{8.2}$$

The potential V in the Hamiltonian $H = T + V$ is to consist of the interactions between all pairs of nucleons: $V = \sum V_{ij}$. In such a scheme the *cluster ansatz* means that the nuclear wave function contains the wave functions Φ_i of its subsystems i as *preset* factors, but the full wave function is subsequently antisymmetrized:

$$\Psi = \mathcal{A}\{\Phi_1\Phi_2 \ldots \phi_{\text{rel}}\}. \tag{8.3}$$

Thus the cluster intrinsic degrees of freedom are frozen, and the wave-function factor ϕ_{rel} describing the intercluster relative motion is to be determined. To this end, the Schrödinger equation (8.2) is solved with the cluster ansatz (8.3) variationally.

This is the main physically motivated simplification that we are going to adopt. It is also to be emphasized that even the nucleons frozen in a cluster are treated individually; they enter into interactions and particle exchanges with the others in exactly the same fashion as the unfrozen nucleons, but the intrinsic wave functions Φ_i of these clusters are fixed to those provided by a simple SM. The underlying philosophy is that the nucleons encapsulated in these clusters are much more passive than the others, and, therefore, the details of their intrinsic structures cannot affect the behaviour of the whole nucleus too much.

The cluster ansatz does, however, imply another model approximation: the nucleon–nucleon force V_{ij} cannot be chosen to be that acting between free nucleons; it has to be an *effective interaction*, which predicts the low-energy behaviour of the two-nucleon system correctly but is still compatible with a schematic description of the internal structures of the clusters. The use of an effective force is the other basic assumption underlying the scheme devised for the description of light exotic nuclei. This simplification comes mainly from physical considerations and is not an absolute technical necessity. In fact, we shall see that, for some of the lightest nuclei, where the cluster ansatz can be avoided or softened, the method has been used with realistic nucleon–nucleon interactions as well (see Sect. 13.9).

The implementation of a microscopic cluster model does involve considerable technical difficulties. This grouping of the degrees of freedom amounts to breaking with the s.p. description, and $\Phi_1\Phi_2\ldots\phi_{\text{rel}}$ in Eq. (8.3) is asymmetrical in the particle labels. Restoring it by an antisymmetrizing operator $\mathcal{A}$ implies explicit permutations of the labels, which is always extremely involved except in a s.p. scheme. This problem can be overcome by maintaining a relationship with a s.p. scheme. That can be achieved if the functional forms of the ingredients of the wave function are chosen in particular ways. The wave function should consist of building blocks of certain analytical forms. It should be expanded into terms of such convenient forms, each containing some parameters, and then the solution of the Schrödinger equation reduces to determining these parameters and the coefficients variationally.

The microscopic theory of light nuclei to be presented in this book is based on a variational approach. A variational approach is characterized, first, by the form of the trial function and the way it contains its variational parameters and, second, how the problem is solved. The trial function Ψ of our approach is built up from Gaussians of slightly generalized forms, which are called *correlated Gaussians* [205]. A correlated Gaussian of a set of intrinsic coordinates $\boldsymbol{x}_i$ looks like $\exp(-\frac{1}{2}\sum_{i,j}A_{ij}\boldsymbol{x}_i\cdot\boldsymbol{x}_j)$, where A is an appropriate matrix. The trial function consists of a great number of terms of the same form,

$$\Psi = \sum_k c_k\psi_k, \tag{8.4}$$

each containing a set of the width-like nonlinear parameters $(A_k)_{ij}$ and a coefficient c_k.

Since the $(A_k)_{ij}$ parameters of two such terms can be arbitrarily close to each other, the functions ψ_k are ***not orthogonal*** to each other, and one can well substitute one with another in an approximate expansion. Therefore, it is not important to find the very best set of the nonlinear parameters. It has proved useful to adopt an approximate optimization for them, which is a combination of a stochastic sampling with a trial-and-error selection procedure. The variational determination of the linear parameters then boils down to matrix diagonalizations. The functions ψ_k can then be considered elements of a nonorthogonal basis. The procedure is called a stochastic variational procedure, and the full name of the approach itself is the ***stochastic variational method with correlated Gaussian bases.*** It will sometimes be referred to by the acronym *SVM*.

What qualifies this approach to be dedicated to describing exotic nuclei is that it does not include the usual basic hypotheses of nuclear theory that are questionable for exotic nuclei. In this approach no mean field, no shell structure, no structureless core and no structureless clusters are invoked. All nucleonic degrees of freedom are treated explicitly. The basis states have the correct symmetries: ***exact antisymmetry*** with respect to permutation, ***translation invariance, definite angular momentum, parity and isospin*** ('projection before variation').

The strength of this approach is the treatment of the dynamics, i.e., the accuracy of the solution of the n-cluster problem. In its complete form the SVM with correlated Gaussian bases has been shown to be numerically exact. When each cluster is a single nucleon ($n = A$), the solution will thus give the exact solution of the A-nucleon problem. Although the full treatment of the A-particle dynamics is extremely challenging, it looks feasible for nuclei of mass-numbers $A \leq 8$, and it has indeed been implemented for a few cases up to $A = 6$ [206]. This provides a unique opportunity to test the cluster ansatz.

The structure theory of light exotic nuclei and its applications will be presented in the following scheme.

In Chap. 9 we shall introduce the correlated Gaussian formalism in its full generality. To make the physical content of the approach clear, we will build up the trial function gradually. To master the formalism as well as to relate it to the cluster models, we introduce a generating function for the correlated Gaussian basis function. This will be followed by showing how to evaluate matrix elements involved in the dynamical calculations and in various other physical quantities.

The SVM will be introduced in Chap. 10. The method can be generalized and has been applied to the description of discrete unbound states as well, and the methodology of that will also be discussed there.

The application of the approach to exotic nuclei requires that it be extended to systems of clusters, and discussing that requires an introduction to clustering and to cluster models. Clustering is a property of any nuclear states to some extent, and cluster models are nuclear models constructed to describe states that are clustered to a great extent.

These are the subjects of Chap. 11. The correlated Gaussian model is a full-fledged cluster model itself, and it is in principle understandable without reference to any other versions of the cluster model. The review of the cluster models in Chap. 11 will, however, help to understand the phenomena shown by exotic nuclei and the relationship of the correlated Gaussian approach to the other approaches commonly used in the literature. The basic version of the cluster model is what is called the resonating-group method (RGM), and the correlated Gaussian basis can be viewed as a particular representation of a very elaborate kind of RGM. The scheme of the analytic calculations of the correlated Gaussian matrix elements is closely related to another representation of the RGM, viz., the generator-coordinate method (GCM). The macroscopic models are mainly discussed since they are very popular in describing light exotic nuclei. The clue to understanding the macroscopic versions of the cluster models from the microscopic point of view is to know how the concept of intercluster relative motion can be introduced in the microscopic approach. That leads to the orthogonality-condition model (OCM), which relies on the solution of the problem of the norm operator, which, in turn, implies the underlying HO cluster model etc. This network of intertwining interdependences is expounded in Chap. 11.

Once the cluster models have been introduced, the correlated Gaussian approach can be made specific to multicluster systems as is required by the light exotic nuclei. That is the subject of Chap. 12. The correlated Gaussian approach will reduce to a cluster model as a result of constraining the relative motions of some groups of nucleons. The correlated Gaussian cluster model becomes then well-defined by giving the principles of constructing the state space and by choosing the effective nucleon–nucleon interaction.

The last and most voluminous chapter of Part II, Chap. 13, is devoted to concrete physical applications. The light exotic nuclei to be covered include ^{6}He, ^{6}Li, ^{8}He, ^{7}Li, ^{7}Be, ^{8}Li, ^{8}B, ^{9}Li, ^{9}C, ^{9}Be, ^{9}B, ^{10}Be, ^{11}Be, ^{10}Li and ^{11}Li. In addition, a few examples will also be given for the combination of structure calculations with reaction theory. These will include both Glauber analyses of the style of Part I and RGM-style generalizations of the structure model to low-energy collisions.

In Part I most of the discussion is focussed on halo nuclei, and halo nuclei play an important role in Part II as well. In a deductive treatment, however, halo structure should emerge as a result of structure calculations, without any special provision made for its appearance. The reader will yet be acquainted with some theoretical aspects of the halo phenomenon itself by reading App. C, where Borromean and Efimov states will be discussed.

Chapter 9

Correlated Gaussian approach

9.1 Preliminary notes

9.1.1 Motivation

The description of a multiparticle system involves a great number of degrees of freedom, and the wave function will depend on all corresponding variables. Having A particles, a theoretical description has to cope with $A-1$ intrinsic vectorial variables and A spin–isospin variables simultaneously. For s.p. problems, partial-wave expansions $\Psi(\boldsymbol{r}) = \sum_{lm} \frac{u_l(r)}{r} Y_{lm}(\hat{\boldsymbol{r}})$ are useful to express the function of $\boldsymbol{r}$ in terms of well-known functions Y_{lm} with unknown coefficients u_l, which depend only on the radial variable r. For bound states, these expansions can usually be truncated to just a few terms. The functions u_l can then be determined either by numerical solution of the (generally coupled) radial Schrödinger equations or by expanding them over a basis set and diagonalizing the Hamiltonian ('basis-set method').

A naive multiparticle analogue of Y_{lm} is a (coupled) product of Y_{lm} [cf. Eq. (9.20)], which are functions of the angles belonging to the intrinsic coordinates, to be denoted, say, by $\boldsymbol{x}_1, \ldots, \boldsymbol{x}_{A-1}$. An expansion in terms of such functions will contain unknown coefficients depending on the corresponding radial coordinates, $u(x_1, \ldots, x_{A-1})$, to be treated explicitly. A direct numerical solution of the (coupled) partial differential equations that could be derived for u is entirely unfeasible, but one may try to resort to some kind of variational solution. The first problem is how to reduce the continuous variables $x_1, \ldots, x_{A-1}$ to a finite number of numbers, which can still represent the full function. In principle, one could discretize each radial variable on a mesh with, say, p points. Alternatively, one may expand the function $u(x_1, \ldots, x_{A-1})$ in terms of $u_1(x_1) \cdots u_{A-1}(x_{A-1})$, with p functions taken for each u_i. One thus ends up either with p^{A-1} mesh points or with p^{A-1} basis functions for each combination of the angular momenta, and p^{A-1} soon becomes prohibitively large with increasing A.

All discretization methods but the Monte Carlo methods face this difficulty beyond a certain number of particles. The quantum Monte Carlo methods, which include the variational Monte Carlo method and the Green's function Monte Carlo method (GFMC), have proved to be most successful in that they are able to go beyond the four-nucleon problem [207]. The secret of the efficiency of the Monte Carlo methods is the use of an importance sampling[1] of the most relevant parts of the configuration space. (Now the configuration space is a tensor product [26] of the intrinsic-coordinate, spin and isospin spaces of the individual nucleons.) It is then natural to ask whether importance sampling could be applied to basis-set methods as well. In quantum Monte Carlo methods the importance sampling generates the most adequate configurations to be used in the integration. In a basis-set method, however, one looks for the most appropriate basis functions to approximate the eigenfunctions of the Hamiltonian. For such a method to be viable, one needs to have an easily tractable set of functions suitable for the construction of basis sets and a sampling technique. That is the essence of the SVM (with correlated Gaussian bases).

Most basis sets used to solve multiparticle problems are based on uncorrelated single-particle (s.p.) bases. With such bases, however, it is extremely difficult to describe the correlations between the particles, although correlations are of prime importance in nuclear physics. The most important types of correlation to be allowed for are the short-range correlation due to the repulsive core of the nucleon–nucleon force, clustering, α-clustering in particular, which is the most striking property of many light nuclei, and the correlation at large distances, which gains particular importance in halo nuclei. To describe the correlation between the nucleons, one should use appropriate coordinates.

The SVM with correlated Gaussian bases is not excessively complicated, yet we introduce it step by step, inductively, since it is important

[1] Sampling of the integration region in Monte Carlo integration with a probability density function biassed so as to maximize accuracy. Since the optimum density function is not known *a priori*, it may be explored by sampling strategies, e.g., by the Metropolis algorithm [208]. In a generalized sense, any approximation method using stochastic sampling strategies based on theoretical bias may be called importance sampling. In the sampling procedure for basis selection presented in this book the energy gain coming from the possible basis states is evaluated, which is not the case in the quantum Monte Carlo method.

to understand its essentials deeply. As a first step, in Sect. 9.1.2 we recall the theoretical foundation of the variational methods and introduce the one-dimensional analogue of the approach, together with the stochastic variational procedure. The presentation of the actual formalism is started in Sect. 9.1.3 with the introduction of some systems of coordinates.

9.1.2 Essentials

The bound-state variational methods are based on the following theorem [26]:

Theorem 1 *The square-integrable function Ψ_0 satisfies the Schrödinger equation $H\Psi_0 = E_0\Psi_0$ if and only if the energy functional*

$$E[\Psi] = \frac{\langle\Psi|H|\Psi\rangle}{\langle\Psi|\Psi\rangle} \tag{9.1}$$

is stationary, i.e. all variations $\delta E[\Psi]$ are zero, at $\Psi = \Psi_0$, and then the eigenvalue is $E_0 = E[\Psi_0]$.

An approximate solution can thus be obtained by choosing a suitable trial function Ψ with a number of parameters and choosing the parameters such that $E[\Psi]$ be stationary with respect to the variations of all these parameters.

The crucial element of a variational method is the setting up of the trial function. A flexible enough trial function would mostly contain an expansion over a number of 'basis' terms. Unlike a proper quantum-mechanical representation, the 'basis' need not be a proper basis, i.e., a complete set of orthonormal states, which spans a 'representation'. When we construct a trial function for a variational method, by representation we only mean an (approximate) expression of the wave function as a linear combination of some appropriate functions, and by basis states we mean these functions. In such a case we try to approximate an unknown function with as few terms as is sufficient to attain a certain accuracy. For this purpose, it is better to choose a few suitable elements from a set of functions of the same functional form, $\{g_a(x)\}$, defined by different values of a continuous parameter a. Obviously, such 'bases' will not be orthogonal; on the contrary, if the parametric function $g_a(x)$ is continuous with respect to its parameter a, the basis functions belonging to nearly equal values of a will also be nearly identical. This property is in fact an advantage because it helps to optimize the choice of the elements of the basis. For a s.p. problem, one of the most popular parametric functions is a Gaussian or, more correctly, a nodeless HO function, with the parameter being its width parameter. A set of such radial functions is shown in Fig. 9.1. The Gaussian is qualified for this role because any (square-integrable) wave function with angular momentum lm can be approximated, to any desired accuracy, by a linear combination of such functions [209]:

$$F_{lm}(\boldsymbol{r}) \approx \sum_{k=1}^{\mathcal{K}} c_k \mathrm{e}^{-\frac{1}{2}a_k r^2} \mathcal{Y}_{lm}(\boldsymbol{r}), \quad \text{with} \quad \mathcal{Y}_{lm}(\boldsymbol{r}) = r^l Y_{lm}(\hat{\boldsymbol{r}}), \tag{9.2}$$

where $Y_{lm}(\hat{\boldsymbol{r}})$ is a spherical harmonic [and $\mathcal{Y}_{lm}(\boldsymbol{r})$ is a solid spherical harmonic]. Since for a spherically symmetrical potential lm are good quantum numbers, the function $F_{lm}(\boldsymbol{r})$ is a very good trial function for a s.p. bound-state Ψ in such a potential. To understand the essentials of the method, let us stick to this case for a while.

The parameters of this trial function are the width parameters $\{a_k\}$ and the expansion coefficients $\{c_k\}$. For a one-dimensional case it would not be excessively difficult to vary all and find fully stationary solutions. But for this case it is not so important to vary $\{a_k\}$ at all, for it is easy to choose a set $\{a_k\}$ with which, at a slight expense in dimension, the variation of $\{c_k\}$ can produce equally good energies. The width of the density belonging to wave function $\mathrm{e}^{-\frac{1}{2}ar^2}$ is $a^{-1/2}$, and one should just guess what spatial range the solution will span, and choose an interval $[(a_{\max})^{-1/2}, (a_{\min})^{-1/2}]$ that embraces this range by one or two orders of magnitude from below and above. Then it is reasonable to divide this interval into bins with a number of a_k values systematically, e.g., by geometric progression. The dimension of the basis should somewhat

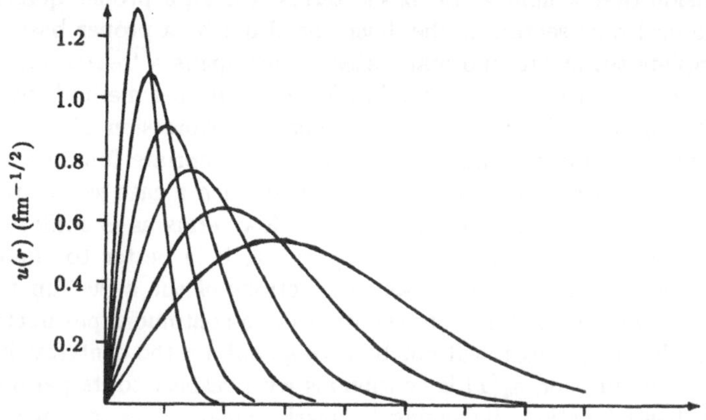

Figure 9.1: Radial Gaussian basis functions $u(r)=2(a^3/\pi)^{1/4} r \mathrm{e}^{-\frac{1}{2}ar^2}$ for $l=0$ and $a=\frac{1}{8},\frac{1}{4},\frac{1}{2},1,2,4\ \mathrm{fm}^{-2}$.

exceed the node number of the solutions one is interested in. A dimension $\mathcal{K}=10$ is typically enough, unless one wants to describe high-node (excited) states. One can imagine that the exact choice of the set $\{a_k\}$ is immaterial within broad limits.

Once the nonlinear parameters a_k have been fixed, the method reduces to a linear variational (or Ritz) method. In the Ritz method the general form of the trial function is

$$\Psi = \sum_{i=1}^{\mathcal{K}} c_i \psi_i, \tag{9.3}$$

and the stationarity condition leads to an algebraic eigenvalue problem of the following form:

$$\mathcal{H}c = E\mathcal{N}c. \tag{9.4}$$

Here $\mathcal{H}$ and $\mathcal{N}$ are, respectively, the (Hermitean) matrices of the Hamiltonian and of the overlap:

$$\mathcal{H}_{ij} = \langle\psi_i|H|\psi_j\rangle \quad \text{and} \quad \mathcal{N}_{ij} = \langle\psi_i|\psi_j\rangle \quad (i,j = 1,\ldots,\mathcal{K}). \tag{9.5}$$

The solution of Eq. (9.4) requires matrix diagonalization, which is numerically simple. There are $\mathcal{K}$ eigenvalues and eigenfunctions, which are considered approximate solutions to the Schrödinger equation.

The linear variational method is derivable from a *minimum principle* [26], which means that the lowest stationary point of the functional $E[\Psi]$ is a minimum. Moreover, the variational problem has the following convenient properties:

Theorem 2 *Hylleraas–Undheim theorem* [187]*: Let $E_1 \leq E_2 \leq \cdots \leq E_{\mathcal{K}}$ be the $\mathcal{K}$ energies provided by a $\mathcal{K}$-dimensional basis. When one more element is added to the basis, the resulting energies E'_i ($i = 1,\ldots,\mathcal{K}+1$) will be related to E_i as follows:*

$$E'_1 \leq E_1 \leq E'_2 \leq E_2 \leq \cdots \leq E'_{\mathcal{K}} \leq E_{\mathcal{K}} \leq E'_{\mathcal{K}+1}. \tag{9.6}$$

From the two theorems one may deduce the following corollaries:

1. If the basis is extended so that it fully spans the state space, the energies will converge to the exact energies. *A fortiori*, if the basis fully spans a well-defined subspace of the state space, the energies will converge to the exact energies in that subspace.

2. The ith approximate energy E_i is an upper bound of the ith exact energy ('mini-max theorem' [187]).

Corollary 1 says that the result usually improves by adding new elements to the basis. As a consequence, it may be convenient to construct a basis by including new elements one by one. (This will, of course, be an actual advantage only for a multidimensional problem, where p^{A-1} is too large for a basis dimension.) The utility of a new element can then be judged by seeing how much its inclusion has deepened the binding. There is no general recipe as to what particular values of a should be taken and in what sequence. The simplest thing to do is to choose $a^{-1/2}$ randomly, within an interval $[(a_{\max})^{-1/2}, (a_{\min})^{-1/2}]$, but admit the function only if its contribution deepens the binding substantially. The procedure can be continued until numerical convergence, which is reached when no element is found to deepen the energy appreciably. This stochastic sampling combined with the 'admittance test' (see p. 248 later) can be considered an approximate variation of the nonlinear parameters. Since each test step involves full diagonalization over a basis augmented with the new element to be tested, the variational procedure is exact with respect to the linear parameters.

Further details of the stochastic variational procedure will be discussed in Chap. 10.

9.1.3 Coordinates and correlations

The first piece of formalism comes with the introduction of a number of different kinds of coordinates, and that is the subject of this subsection. Once these formulae are at hand, it will be easy to comprehend the trial function itself, which is to be introduced in Sect. 9.2.

Let $\boldsymbol{r}_i$ $(i=1,\ldots,A)$ denote the s.p. coordinate vectors with respect to a centre fixed in space. These are suitable for formulating the dynamical problem if the c.m. of the system can be considered to be fixed in space, but for a light nucleus that would not be a good approximation. We shall rather introduce other types of coordinates, from which the dependence on the overall position of the system can be removed. A set of vectors that specify the relative positions of the particles with respect to each other but do not specify the location of the system in space are to be called *intrinsic coordinates*. Such are, e.g., the interparticle coordinates,

$$\boldsymbol{r}_{ij} = \boldsymbol{r}_i - \boldsymbol{r}_j \qquad (i = 1,\ldots,A-1,\ j = i+1,\ldots,A), \tag{9.7}$$

which will not be used too much since they are not independent of each other when $A>2$.

The standard system of coordinates we shall use is that of the Jacobi coordinates, which are defined as

$$\boldsymbol{x}_j = \sqrt{\frac{j}{j+1}}\left(\frac{1}{j}\sum_{i=1}^{j}\boldsymbol{r}_i - \boldsymbol{r}_{j+1}\right) \quad (j = 1,\ldots,A-1), \tag{9.8}$$

$$\boldsymbol{x}_A = \frac{1}{\sqrt{A}}\sum_{i=1}^{A}\boldsymbol{r}_i. \tag{9.9}$$

The coordinates defined in Eq. (9.8) are referred to as intrinsic Jacobi vectors. It is more common to use the vectors

$$\boldsymbol{y}_j = \frac{1}{j}\sum_{i=1}^{j}\boldsymbol{r}_i - \boldsymbol{r}_{j+1} \quad (j = 1,\ldots,A-1), \tag{9.10}$$

$$\boldsymbol{R}_{\text{c.m.}} \equiv \boldsymbol{y}_A = \frac{1}{A}\sum_{i=1}^{A}\boldsymbol{r}_i, \tag{9.11}$$

which are also called Jacobi vectors. The meaning of this set is more transparent: $\boldsymbol{y}_j$ are vectors connecting a particle with the c.m. of a subset of particles singled out in previous steps, while $\boldsymbol{R}_{\text{c.m.}}$ is the vector pointing to the c.m. of the whole system. But $\boldsymbol{x}_j$ differ from $\boldsymbol{y}_j$ just in constant factors, which are in principle arbitrary. The constant factors in the definition of $\boldsymbol{x}_j$ are chosen [210] so as to simplify the treatment of Gaussians.[2] To retain the meaning of the terms of Eqs. (9.8)–(9.11), the coefficients should be modified when the masses are non-equal [187]. Equations (9.8) and (9.9) can be rewritten as

$$\boldsymbol{x}_i = \sum_{k=1}^{A} U_{ik}\boldsymbol{r}_k \quad (i = 1,\ldots,A), \tag{9.12}$$

where the transformation matrix

$$U = \begin{pmatrix} \frac{1}{\sqrt{2}} & -\frac{1}{\sqrt{2}} & 0 & \cdots & 0 & 0 \\ \frac{1}{\sqrt{6}} & \frac{1}{\sqrt{6}} & -\sqrt{\frac{2}{3}} & \cdots & 0 & 0 \\ \vdots & \vdots & \vdots & \vdots & \vdots & \vdots \\ \frac{1}{\sqrt{A(A-1)}} & \frac{1}{\sqrt{A(A-1)}} & \frac{1}{\sqrt{A(A-1)}} & \cdots & \frac{1}{\sqrt{A(A-1)}} & -\sqrt{\frac{A-1}{A}} \\ \frac{1}{\sqrt{A}} & \frac{1}{\sqrt{A}} & \frac{1}{\sqrt{A}} & \cdots & \frac{1}{\sqrt{A}} & \frac{1}{\sqrt{A}} \end{pmatrix} \tag{9.13}$$

[2]Since $\sum_{i=1}^{A}\boldsymbol{r}_i^2 = \sum_{i=1}^{A}\boldsymbol{x}_i^2$, a product of s.p. Gaussians of the same parameter will transform into a product of Gaussians of such Jacobi coordinates *of the same parameter*:

$$\prod_{i=1}^{A}\left(\frac{\beta}{\pi}\right)^{3/2} \mathrm{e}^{-\frac{1}{2}\beta\boldsymbol{r}_i^2} = \prod_{i=1}^{A}\left(\frac{\beta}{\pi}\right)^{3/2} \mathrm{e}^{-\frac{1}{2}\beta\boldsymbol{x}_i^2}.$$

is seen to be orthogonal.[3]

Finally, we shall use the term *relative coordinates* for some generalizations of the intrinsic Jacobi coordinates. These are vectors connecting the c.m. of two subgroups of the A particles. The c.m. vector $\bar{\boldsymbol{\xi}}_j$ of a subgroup of a_j particles is defined in conformity with Eq. (9.9) as

$$\bar{\boldsymbol{\xi}}_j = \frac{1}{\sqrt{a_j}} \sum_i^{a_j} \boldsymbol{r}_i \tag{9.14}$$

and in conformity with Eq. (9.11) as

$$\bar{\boldsymbol{t}}_j = \frac{1}{a_j} \sum_i^{a_j} \boldsymbol{r}_i \tag{9.15}$$

[cf. Eq. (B.2)]. The relative vector between two subsystems is then defined, by analogy with Eq. (9.8), by

$$\boldsymbol{\rho} = \sqrt{\frac{a_1 a_2}{a_1 + a_2}} \left(\frac{1}{\sqrt{a_1}} \bar{\boldsymbol{\xi}}_1 - \frac{1}{\sqrt{a_2}} \bar{\boldsymbol{\xi}}_2 \right) \tag{9.16}$$

and, by analogy with Eq. (9.10), by

$$\boldsymbol{q} = \bar{\boldsymbol{t}}_1 - \bar{\boldsymbol{t}}_2. \tag{9.17}$$

One obtains a full set of relative vectors by taking a hierarchic succession of binary groupings of particles.[4] For example, for a system of three identical particles, there is just one legitimate set of relative vectors: the vector connecting two of the particles and the vector connecting the c.m. of the previous two with the third one ('T-pattern', with one particle sitting at the end of each arm of the letter T). Of course, when the particles are identical, two patterns differing only in a permutation of the particles are considered the same. For four identical particles, there are two patterns: a three-particle subsystem is connected with a fourth particle ('K-pattern') or two two-particle subsystems connected ('H-pattern'). As a more complete example, Fig. 9.2 shows all topologically different sets of independent relative coordinates for a system of six identical particles.

The coefficients in Eqs. (9.8) and (9.16) are chosen so that the relative coordinates transform into each other by orthogonal transformations, and, *a fortiori*, they can be constructed by subjecting a Jacobi

[3]Similar formulae, valid for the conventional Jacobi coordinates (9.10) and (9.11), are given in App. B.

[4]The binary groupings are formed iteratively, with a new group either fully including or fully excluding a group formed in previous steps.

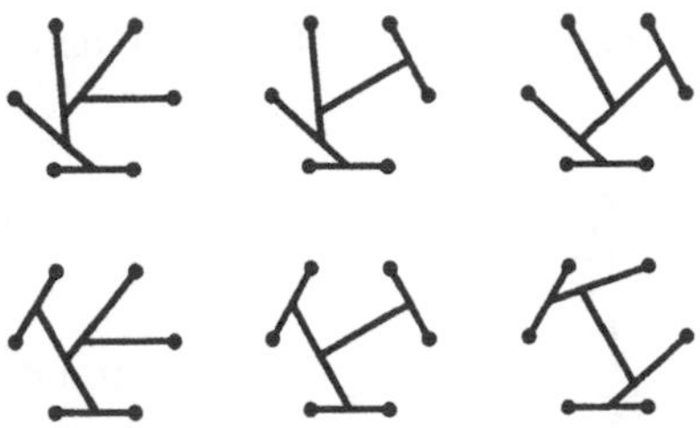

Figure 9.2: All topologically different sets of independent relative coordinates for a system of six identical particles. The lengths of the lines may be identified with the lengths of the vectors in parenthesis in Eq. (9.16).

set to such a transformation. Any relative coordinates $\boldsymbol{x}'$ can be obtained from the Jacobi coordinates $\boldsymbol{x}$ by an appropriate orthogonal transformation (rotation and/or reflection) $\mathcal{R}$:

$$\boldsymbol{x}' = \mathcal{R}\boldsymbol{x}. \tag{9.18}$$

It follows from the orthogonal-transformation property that the intrinsic kinetic energy (which is the total kinetic energy minus the c.m. kinetic energy) takes the form $\sum_{i=1}^{A-1} \boldsymbol{\pi}_i^2/2m = \sum_{i=1}^{A-1} \boldsymbol{\pi}_i'^2/2m$, where $\boldsymbol{\pi}_i$ and $\boldsymbol{\pi}_i'$ are, respectively, the momenta conjugate to $\boldsymbol{x}_i$ and $\boldsymbol{x}_i'$.

It is clear that each set of coordinates discussed so far emphasizes a particular type of correlation. Each type of correlation is then to be represented by trial-function terms given in terms of appropriate relative coordinates. For example, for four particles, pair correlation and asymptotic partitioning into two pairs are described most economically with the H-shaped set of vectors. In this way each coordinate set may be associated with a particular *'arrangement'* of the particles. To have good asymptotics everywhere, it is economical to include terms with all possible relative vectors. This will be formulated in the next section.

9.2 Variational trial function

9.2.1 Formulation in terms of relative coordinates

For an A-particle problem, one should introduce a set of intrinsic Jacobi coordinates $\boldsymbol{x}_1, \ldots, \boldsymbol{x}_{A-1}$ to be regarded as an $(A-1)\times 1$ (one-column) matrix $\boldsymbol{x}$ of the Cartesian vectors. We exclude the vector pointing to the total c.m. from the outset to emphasize that the dynamics of the problem can be solved without reference to the c.m.

As a direct generalization of Eq. (9.2), the expansion may take the form

$$\begin{aligned} F_{LM_L}(\boldsymbol{x}) &\approx \sum_{k_1,\dots,k_{A-1}} \sum_l c_{k_1,\dots,k_{A-1}l} \prod_{i=1}^{A-1} \mathrm{e}^{-\frac{1}{2}a_{k_i}\boldsymbol{x}_i^2} \theta_{lLM_L}(\boldsymbol{x}) \\ &= \sum_{\{kl\}} c_{kl} \exp\left(-\tfrac{1}{2}\tilde{\boldsymbol{x}}A_k\boldsymbol{x}\right) \theta_{lLM_L}(\boldsymbol{x}), \end{aligned} \tag{9.19}$$

with

$$\theta_{lLM_L}(\boldsymbol{x}) = \left[\mathcal{Y}_{l_1}(\boldsymbol{x}_1)\cdots\mathcal{Y}_{l_{A-1}}(\boldsymbol{x}_{A-1})\right]_{LM_L}. \tag{9.20}$$

The second line of Eq. (9.19) gives an identical reformulation of the first to facilitate later generalizations. The subscript $k \equiv \{k_1,\dots,k_{A-1}\}$ labels the different sets of product Gaussians, which are written in the second line as the exponential function of a diagonal quadratic form. To resolve that notation, we should mention that the tilde marks matrix transposition, and the quadratic form $\tilde{\boldsymbol{x}}A_k\boldsymbol{x}$ involves scalar products of the Cartesian vectors:

$$\tilde{\boldsymbol{x}}A_k\boldsymbol{x} = \sum_{i=1}^{A-1} \boldsymbol{x}_i \cdot \left[\sum_{j=1}^{A-1} (A_k)_{ij}\boldsymbol{x}_j\right] \equiv \sum_{i=1}^{A-1}\sum_{j=1}^{A-1} (A_k)_{ij}\, \boldsymbol{x}_i \cdot \boldsymbol{x}_j. \tag{9.21}$$

As the first line of Eq. (9.19) shows, here A_k is to be taken diagonal:

$$(A_k)_{ij} = a_{k_i}\delta_{ij}. \tag{9.22}$$

This restriction will be lifted later on [cf. Eq. (9.26)].

As is expressed in Eq. (9.20), the angular factor $\theta_{lLM_L}(\boldsymbol{x})$ may be imagined as composed of the solid spherical harmonics for the individual Jacobi vectors; then the label l may denote a set of orbital angular momentum quantum numbers. The symbol $[\cdots]$ means vector coupling[5] in an order, which is, in principle, arbitrary but, once chosen, it should be adhered to for all basis elements. Thus l implicitly includes not only $l_1,\dots,l_{A-1}$ but also the angular momentum quantum numbers of the intermediate coupling steps. The angular factors $[Y_{l_1}\cdots Y_{l_{A-1}}]_{LM_L}$ of all possible functions θ_{lLM_L} constructed in this way form a complete set. It is most natural to take the form

$$\begin{aligned} \theta_{lLM_L}(\boldsymbol{x}) = &\left[\cdots[[\mathcal{Y}_{l_1}(\boldsymbol{x}_1)\mathcal{Y}_{l_2}(\boldsymbol{x}_2)]_{L_{12}}\mathcal{Y}_{l_3}(\boldsymbol{x}_3)]_{L_{123}}\cdots\mathcal{Y}_{l_{A-1}}(\boldsymbol{x}_{A-1})\right]_{LM_L}, \\ &\text{with } l = \{l_1,\dots,l_{A-1},L_{12},L_{123},\dots\}. \end{aligned} \tag{9.23}$$

The dimensional explosion of p^{A-1} basis functions for each l mentioned at the beginning of Sect. 9.1.1 would occur if the summations

[5]The symbol $[\cdots]_{LM}$ will mostly be used for coupling two angular momenta.

over $\{kl\} = \{k_1, \ldots, k_{A-1}, l_1, \ldots\}$ were to be treated independently. But that would only be so if the motions in the A–1 degrees of freedom were independent as in an independent-particle model. That is, however, not the case, so that

$$\sum_{\{kl\}} \equiv \sum_{\{k_1 \cdots k_{A-1} l_1 \cdots\}} \neq \sum_{k_1} \cdots \sum_{k_{A-1}} \sum_{l_1} \cdots \tag{9.24}$$

The approximation in Eq. (9.19) may well be accurate even if the single sum on the left-hand side of Eq. (9.24) contains by far fewer terms. The number of terms needed is often substantially reduced owing just to the correlations that blur the independent-particle picture. Furthermore, for higher l values it is enough to have fewer $c_{lk} \neq 0$.

A particular angular-momentum set belonging to a particular arrangement will be referred to as a *'configuration'* or a 'channel'. Since each arrangement implies a particular partitioning, the asymptotics of the function can be well approximated without including terms with very high orbital angular momenta just in that particular partition. The convergence of the asymptotics in a particular channel in terms of another set of coordinates is in general slow, and the computational cost of using configurations with high orbital angular momenta is quite high [211–214].

A basis function of the form $\exp(-\frac{1}{2}\sum_{i=1}^{A-1} a'_{k'_i} x'^2_i)$ is expressed in terms of a specific set of relative coordinates $\{x'_i\}$, thus it emphasizes some specific correlations. In particular, the asymptotics in the partition conforming to these coordinates is easily reproduced, while it is very costly to describe any other asymptotics with this form. However, correlations of quite different types and, in particular, the asymptotics in several partitions, may be important. It is therefore desirable to include several arrangements in the expansion:

$$\begin{aligned} F_{LM_L}(\boldsymbol{x}) \quad &\approx \quad \sum_{\{kl\}} c_{kl} \exp\left(-\tfrac{1}{2}\tilde{\boldsymbol{x}} A_k \boldsymbol{x}\right) \theta_{lLM_L}(\boldsymbol{x}) \\ &+ \quad \sum_{\{k'l'\}} c'_{k'l'} \exp\left(-\tfrac{1}{2}\tilde{\boldsymbol{x}}' A'_{k'} \boldsymbol{x}'\right) \theta_{l'LM_L}(\boldsymbol{x}') \\ &+ \quad \cdots, \end{aligned} \tag{9.25}$$

where $(A'_{k'})_{ij} = a'_{k'_i}\delta_{ij}$. These sums may overlap with each other substantially, but, since even the terms within each sum are nonorthogonal to each other, this does not necessarily pose extra difficulties.

This scheme becomes simpler by two straightforward generalizations, which will be introduced in the next subsection.

9.2.2 Formulation without reference to relative coordinates

Correlated Gaussians

The first generalization makes it possible to treat all possible relative coordinate systems, and even their mixtures, simultaneously. All these may be contained in a single sum as in Eq. (9.19) if we allow A_k to be arbitrary positive-definite symmetric matrices, or, equivalently, if they are constructed via an orthogonal transformation by a matrix G from a diagonal matrix with positive elements:

$$A_k = \tilde{G}A_k^{(0)}G, \text{ where } (A_k^{(0)})_{ij} = a_{k_i}\delta_{ij}, \quad a_{k_i} > 0. \tag{9.26}$$

The relationship between Eqs. (9.25) and (9.26) may be understood as follows. Consider the terms of the second arrangement in the expansion (9.25). The relative coordinates $\boldsymbol{x}'$ can be obtained from the Jacobi coordinates $\boldsymbol{x}$ by an orthogonal transformation $\mathcal{R}$ as in Eq. (9.18), thus the quadratic form $\tilde{\boldsymbol{x}}'A'_{k'}\boldsymbol{x}'$ is reduced to the form of $\tilde{\boldsymbol{x}}A_k\boldsymbol{x}$ with $A_k = \tilde{\mathcal{R}}A'_k\mathcal{R}$. With the choice $G = \mathcal{R}$ and $A_k^{(0)} = A'_k$, the second exponential $\exp\left(-\frac{1}{2}\tilde{\boldsymbol{x}}'A'_{k'}\boldsymbol{x}'\right)$ can in this way be cast into the form of the first. The angular factor $\theta_{l'LM_L}(\boldsymbol{x}')$ is a polynomial[6] of $\boldsymbol{x}'$, and it can obviously be expressed in terms of a combination of another set of similar polynomials, $\theta_{lLM_L}(\boldsymbol{x})$, with various l.

The matrix G is an $(A-1)\times(A-1)$ orthogonal matrix, which contains $\frac{1}{2}(A-1)(A-2)$ parameters, while $A_k^{(0)}$ contains $A-1$ parameters. The matrix A_k is thus specified by $\frac{1}{2}(A-1)A$ parameters. Although no restriction on the parameters of G is in principle necessary, it is not advisable to have too many free or variational parameters. It is most reasonable to start the construction of the basis by selecting all discrete G that generate the arrangements that are expected, on physical grounds, to play any role in the system considered, and include more general transformations later only if they are needed [cf. Eq. (9.25)].

The quadratic form $\tilde{\boldsymbol{x}}A_k\boldsymbol{x}$ in the exponent is very flexible to express some important correlations. They especially emphasize clustering and intercluster relative motion, which is very useful in the description of light nuclei. The long-range correlations implied by the intercluster relative motion are observable in the wave-function tails stretching into the (closed) disintegration channels, and the shapes, angular momenta and relative sizes of these tails are mostly influenced by the values of the various separation energies.

It is instructive to show that the form $\exp(-\frac{1}{2}\tilde{\boldsymbol{x}}A_k\boldsymbol{x})$ can express two-particle correlations as well. The two-particle correlations for the

[6]The solid spherical harmonics are also polynomials; see pp. 105–106 of Ref. [187].

A-particle system are best represented by a correlation factor which depends on the interparticle coordinates (9.7): $F=\prod_{i<j}^{A} f_{ij}(\boldsymbol{r}_{ij})$ [215–219]. The calculations involving such correlation factors are, however, fairly involved for more than three particles, and the matrix elements have to be calculated by Monte Carlo integration. To make the scheme simpler, the correlation factors are sometimes expressed as linear combinations of Gaussians $\exp(-\alpha_{ij}\boldsymbol{r}_{ij}^2)$. The A-particle basis function then contains a product of these Gaussians:

$$\prod_{i<j}^{A} \exp(-\alpha_{ij}^{(k)}\boldsymbol{r}_{ij}^2) = \exp\left(-\sum_{i<j}^{A} \alpha_{ij}^{(k)}\boldsymbol{r}_{ij}^2\right). \tag{9.27}$$

Correlation factors of this form are widely used in variational calculations (see, e.g., in Refs. [219,220]).

How is the function (9.27) related to a Gaussian $\exp(-\frac{1}{2}\tilde{\boldsymbol{x}}A_k\boldsymbol{x})$? In the special case of Gaussians of the same widths, i.e., all $\alpha_{ij}^{(k)} = \alpha$, these two forms are identical.[7] The pool of Gaussian widths used in the present approach may contain the equal values and, with that, the uniform two-particle correlations, but are much more general. It is useful to relate the two forms for the case of unequal Gaussian widths as well. It will now be shown that they are equivalent in general. As is mentioned above, A_k has $\frac{1}{2}(A-1)A$ independent parameters, and the number of $\alpha_{ij}^{(k)}$ in Eq. (9.27) is the same. Therefore, one can construct A_k in terms of $\alpha_{ij}^{(k)}$, and transform $\prod_{i<j}^{A}\exp(-\alpha_{ij}^{(k)}\boldsymbol{r}_{ij}^2)$ into $\tilde{\boldsymbol{x}}A_k\boldsymbol{x}$:

$$\prod_{i<j}^{A} \exp(-\alpha_{ij}^{(k)}\boldsymbol{r}_{ij}^2) = \mathrm{e}^{-\frac{1}{2}\tilde{\boldsymbol{x}}A_k\boldsymbol{x}}. \tag{9.28}$$

To find the relationship between A_k and $\alpha_{ij}^{(k)}$, one should first write

$$\sum_{i<j}^{A} \alpha_{ij}^{(k)}\boldsymbol{r}_{ij}^2 = \sum_{i=1}^{A}\left(\sum_{l=1}^{i-1}\alpha_{li}^{(k)} + \sum_{l=i+1}^{A}\alpha_{il}^{(k)}\right)\boldsymbol{r}_i^2 - 2\sum_{i<j}^{A}\alpha_{ij}^{(k)}\boldsymbol{r}_i\cdot\boldsymbol{r}_j. \tag{9.29}$$

Then A_k can be expressed as

$$A_k = \tilde{U}^{-1}B_kU^{-1}, \tag{9.30}$$

[7] The Gaussian exponents are equal: $-\sum_{i<j}^{A}\frac{\alpha}{A}\boldsymbol{r}_{ij}^2 = -\sum_{i}^{A-1}\alpha\boldsymbol{x}_i^2$, where $\{\boldsymbol{x}_i\}$ is a set of intrinsic Jacobi coordinates.

with

$$(B_k)_{ij} = \begin{cases} 2\left(\sum_{l=1}^{i-1} \alpha_{li}^{(k)} + \sum_{l=i+1}^{A} \alpha_{il}^{(k)}\right) & \text{if } i = j, \\ -2\alpha_{ij}^{(k)} & \text{if } i < j, \\ -2\alpha_{ji}^{(k)} & \text{if } i > j, \end{cases} \tag{9.31}$$

where U is defined in Eq. (9.13). The matrix U of Eq. (9.13) is orthogonal, so that $U^{-1} = \tilde{U}$ and thus $A_k = UB_k\tilde{U}$, but the form (9.30) for A_k is valid even if U is not orthogonal, like, e.g., in the case of the conventional Jacobi coordinates (9.10)–(9.11). The redundancy of the vector set $\{r_{ij}\}$ thus does no harm; the structure of the matrix B_k ensures that the c.m. term gives no contribution, i.e., the elements of the Ath row and Ath column of $\tilde{U}^{-1}B_kU^{-1}$ vanish and the remaining $(A-1)\times(A-1)$ matrix defines A_k through $\{\alpha_{ij}^{(k)}\}$.

The correspondence between $\alpha_{ij}^{(k)}$ and A_k comes in handy in the optimization of the parameters. In principle, one can choose various combinations of the parameters for optimization. One of the possibilities is to single out different systems of relative coordinates (each defining a particular $\mathcal{R}$) and try to find the most appropriate $A_k^{(0)}$ parameters, i.e., Gaussian widths. These parameters can be singled out from an interval which corresponds to the extension of the wave function. One has to use different matrices $\mathcal{R}$ to incorporate different correlations. A drawback of this choice is that one has to optimize continuous variables ($A_k^{(0)}$) and discrete variables (systems of relative coordinates) at the same time. Another possibility, the optimization of G and $A_k^{(0)}$, is inconvenient; it contains too many and too involved parameters, whose appropriate values are difficult to find. In practice, the correspondence between $\alpha_{ij}^{(k)}$ and A_k offers the best possibility: the pair correlation functions can be best optimized via $\alpha_{ij}^{(k)}$ since their meaning is plausible.

Although the positiveness of all $\alpha_{ij}^{(k)}$ guarantees that A_k is positive definite, the positive definite nature of A_k does not necessarily imply that all parameters $\alpha_{ij}^{(k)}$ have to be positive.[8] In practice, however, the

[8]To show that $\alpha_{ij}^{(k)}$ can indeed be negative, we give an example. Let us rewrite an H-type Gaussian trial function into a sum of pair-correlation terms:

$$\begin{aligned} &\exp\left[-a_1(\boldsymbol{r}_1-\boldsymbol{r}_2)^2 - a_2(\boldsymbol{r}_3-\boldsymbol{r}_4)^2 - 4a_3\left(\frac{\boldsymbol{r}_1+\boldsymbol{r}_2}{2} - \frac{\boldsymbol{r}_3+\boldsymbol{r}_4}{2}\right)^2\right], \\ &= \exp[-(a_1-a_3)(\boldsymbol{r}_1-\boldsymbol{r}_2)^2 - (a_2-a_3)(\boldsymbol{r}_3-\boldsymbol{r}_4)^2 - a_3(\boldsymbol{r}_1-\boldsymbol{r}_3)^2 \\ &\quad -a_3(\boldsymbol{r}_1-\boldsymbol{r}_4)^2 - a_3(\boldsymbol{r}_2-\boldsymbol{r}_3)^2 - a_3(\boldsymbol{r}_2-\boldsymbol{r}_4)^2]. \end{aligned}$$

The parameters a_1, a_2 and a_3 must be positive in order to ensure the finite norm of the trial function, but $\alpha_{12} = a_1 - a_3$ as well as $\alpha_{34} = a_2 - a_3$ can either be negative

inclusion of negative $\alpha_{ij}^{(k)}$ does not offer appreciable advantages, so it is more convenient to use merely positive $\alpha_{ij}^{(k)}$ for simplicity.

With the transformation (9.26), a basis function belonging to any particular arrangement can be simply expressed in terms of the coordinates of any other arrangement. We can anticipate a similar relationship between the corresponding matrix elements since the sequence of the integrations and the transformation (9.26) can always be reversed. Consequently, it is enough to calculate the formulae for the matrix elements between one type of basis states; the matrix elements between other states can be obtained from that by simple transformations. Thus this generalization greatly improves the utility of the basis. A basis whose spatial building block is $\exp\left(-\frac{1}{2}\tilde{x}A_k x\right)\theta_{lLM_L}(x)$ is called a *correlated Gaussian basis.*

Global-vector representation

The second generalization that we may adopt is to allow the function $\theta_{lLM_L}(x)$ not to carry definite values of the intermediate angular momenta. This is justifiable because these are not good quantum numbers in most systems one is interested in. Then $\theta_{lLM_L}(x)$ may be chosen to be any functions, of all Jacobi vectors, that carry the appropriate total orbital angular momentum LM_L. The most convenient choice is

$$\theta_{lLM_L}(x) \equiv \theta_{lKLM_L}(x) = v_l^{2K}\mathcal{Y}_{LM_L}(v_l), \tag{9.32}$$

with

$$v_l = \sum_{i=1}^{A-1} u_{li}x_i \equiv \tilde{u}_l x. \tag{9.33}$$

The vector v_l is sometimes called 'global', to indicate that it is a combination of all intrinsic coordinates. The values of the coefficients $u_{l1}, \ldots, u_{l(A-1)}$, which may be arranged in an $(A-1)\times 1$ matrix u_l, are parameters of the basis just as a_{k_i} are. The use of this $\theta_{lLM_L}(x)$, instead of Eq. (9.23), is a simplification indeed since the calculation of matrix elements containing Eq. (9.23) becomes very complicated as the number of nucleons increases. It is interesting to compare Eq. (9.32)

---

or positive. The above equation can be derived by using

$$\left(\frac{1}{k}\sum_{i=1}^{k}\boldsymbol{r}_i - \frac{1}{N-k}\sum_{j=k+1}^{N}\boldsymbol{r}_j\right)^2 = -\frac{1}{k^2}\sum_{i<j}^{k}(\boldsymbol{r}_i-\boldsymbol{r}_j)^2$$

$$+\frac{1}{k(N-k)}\sum_{i=1}^{k}\sum_{j=k+1}^{N}(\boldsymbol{r}_i-\boldsymbol{r}_j)^2 - \frac{1}{(N-k)^2}\sum_{k+1\le i<j}^{N}(\boldsymbol{r}_i-\boldsymbol{r}_j)^2.$$

and Eq. (9.23) from another point of view as well. Equation (9.23) has a transparent form only when expressed in terms of the set of relative coordinates in which it has been constructed; in other coordinates it will look very complicated. Equation (9.32), on the contrary, will have exactly the same form in any set of relative coordinates; only the coefficients u_l have to undergo a transformation. All in all, Eq. (9.32) is much simpler than Eq. (9.23).

As was mentioned before Eq. (9.23), the functions of the form of Eq. (9.20) with a fixed coupling scheme form a complete set of functions in the state space of the angular motion of an A-particle system, and it is important to see whether a trial function with angular part (9.32) spans the same function space. To this end, we write down the formula for the composition of a solid spherical harmonic of $v = u_1 x_1 + u_2 x_2$ out of those of x_1 and x_2 [221]:

$$v^{2K}\mathcal{Y}_{LM}(v) = \frac{(2K+L)!}{B_{KL}} \sum_{\substack{k_1 l_1, k_2, l_2 \geq 0 \\ 2k_1+l_1+2k_2+l_2=2K+L}} u_1^{2k_1+l_1} u_2^{2k_2+l_2}$$
$$\times \frac{B_{k_1 l_1} B_{k_2 l_2}}{(2k_1+l_1)!(2k_2+l_2)!} \sqrt{\frac{(2l_1+1)(2l_2+1)}{4\pi(2L+1)}} \langle l_1 0 l_2 0 | L0 \rangle$$
$$\times x_1^{2k_1} x_2^{2k_2} \left[\mathcal{Y}_{l_1}(x_1)\mathcal{Y}_{l_2}(x_2)\right]_{LM}, \tag{9.34}$$

where the symbol $\langle j_1 m_1 j_2 m_2 | jm \rangle$ denotes a Clebsch–Gordan coefficient and B_{kl} is

$$B_{kl} = \frac{4\pi(2k+l)!}{2^k k!(2k+2l+1)!!}. \tag{9.35}$$

First, it is to be observed that, if $K = 0$, then the triangular inequality for l_1, l_2 and L cannot be satisfied unless $L = l_1 + l_2$ and all $k_1 = k_2 = 0$. Thus, a function $\theta_{lKLM_L}(x)$ with $K = 0$ only includes partial-wave angular momenta $l_1, \ldots, l_{A-1}$ of the stretched coupling $l_1 + l_2 + \cdots = L$, and $K > 0$ is actually needed to include contributions of higher angular momenta of the individual relative motions.

Second, since $\langle l_1 0 l_2 0 | L0 \rangle = 0$ unless $l_1 + l_2 + L$ is even, the parity of the function $v^{2K}\mathcal{Y}_{LM}(v)$ can only be 'natural', i.e., $(-1)^L$. Thus the expansion (9.34) can only be inverted if $l_1 + l_2 + L =$ is even. The invertibility means that coefficients c_i can be found with a suitable set of vectors $\{v_i\}$ [$i = 1, \ldots, n$, where n is the number of terms on the right-hand side of Eq. (9.34)] such that

$$x_1^{2k_1} x_2^{2k_2} \left[\mathcal{Y}_{l_1}(x_1)\mathcal{Y}_{l_2}(x_2)\right]_{LM} = \sum_{i=1}^{n} c_i v_i^{2q} \mathcal{Y}_{LM}(v_i)$$
$$(2q - 2k_1 - 2k_2 = l_1 + l_2 - L). \tag{9.36}$$

Similarly, a (9.34)-type expansion of Eq. (9.32) would consist of vector-coupled products with natural parity in each coupling step. Since the complete set $\{[Y_{l_1} \cdots Y_{l_{A-1}}]_{LM_L}\}$ includes elements which carry unnatural values of the total and/or intermediate parities, a set of functions of the form of $v^{2K}\mathcal{Y}_{LM}$ cannot be complete. Therefore, Eq. (9.23) can only be expressed as a combination of terms (9.32) multiplied by scalar terms if each of these parity values is natural [187]. To get a function $\theta_{lLM_L}(\boldsymbol{x})$ of unnatural parity or to include unnatural parity in the intermediate coupling, one has to modify Eq. (9.32) in some way. For example, one way of constructing a trial function with parity $(-1)^{L+1}$ is through

$$\theta_{lLM_L}(\boldsymbol{x}) = [v_l^{2K}\mathcal{Y}_L(\boldsymbol{v}_l)\mathcal{Y}_1(\boldsymbol{v}_l')]_{LM_L}, \tag{9.37}$$

where $\boldsymbol{v}_l$ and $\boldsymbol{v}_l'$ consist of two independent sets of $A-1$ numbers.

To sum up, one has the choice of making the basis complete either by putting together elements formulated in different relative sets and transforming them via Eq. (9.18) or constructing correlated Gaussians in terms of a Jacobi set or combine these two procedures. The formulation in relative coordinates is tailored for describing cluster correlations, but the correlated Gaussians are more general; they are suitable for describing any correlations, including two-particle correlations.

In this context one should mention a fact, which has been implicit so far: namely, that $\theta_{lLM_L}(\boldsymbol{x})$ may not be the only angle-dependent factor in a term of Eq. (9.19). Let us assume that a term of Eq. (9.19), expressed in terms of the Jacobi vectors $\boldsymbol{x}$, describes correlations represented by a set of relative coordinates $\boldsymbol{x}'$; this assumes that it results from a term $\exp(-\frac{1}{2}\tilde{\boldsymbol{x}}'A'_{k'}\boldsymbol{x}')\theta_{l'LM_L}(\boldsymbol{x}')$ of Eq. (9.25) subjected to a transformation (9.18). Then, according to Eq. (9.26), A_k will have off-diagonal elements $(A_k)_{ij}$, so that $\exp(-\frac{1}{2}\tilde{\boldsymbol{x}}A_k\boldsymbol{x})$ will contain a factor $\exp[-(A_k)_{ij}\boldsymbol{x}_i\cdot\boldsymbol{x}_j]$. This factor obviously depends on the relative angle of $\boldsymbol{x}_i$ and $\boldsymbol{x}_j$, so much so that its angular dependence can be expressed as an infinite multipole sum with the formula

$$\mathrm{e}^{a\boldsymbol{x}\cdot\boldsymbol{y}} = 4\pi\sum_{lm} i_l(axy)Y_{lm}(\hat{\boldsymbol{x}})Y_{lm}^*(\hat{\boldsymbol{y}}), \tag{9.38}$$

where i_l is a modified spherical Bessel function of the first kind [43]. Consequently, $\exp(-\frac{1}{2}\tilde{\boldsymbol{x}}A_k\boldsymbol{x})$ will contain all partial waves lm.

9.2.3 Full form

For this scheme to be applicable in actual physical problems, one has to include the intrinsic degrees of freedom of the particles and to impose the proper permutation symmetry. In nuclear physics this actually means the inclusion of spin and isospin and taking care of antisymmetrization. The spins and the isospins of the nucleons are described by the respective eigenfunctions $\chi_{\frac{1}{2}m_s}(i)$ and $\eta_{\frac{1}{2}m_t}(i)$, which are coupled successively to total spin SM_S and isospin TM_T:

$$\chi_{SM_S} = \left[\chi_{\frac{1}{2}}\chi_{\frac{1}{2}}\chi_{\frac{1}{2}}\cdots\right]_{SM_S}, \qquad \eta_{TM_T} = \left[\eta_{\frac{1}{2}}\eta_{\frac{1}{2}}\eta_{\frac{1}{2}}\cdots\right]_{TM_T}. \tag{9.39}$$

For brevity, the intermediate quantum numbers are suppressed. The antisymmetry is effected by the antisymmetrization operator $\mathcal{A}$ defined by (cf. App. D)

$$\mathcal{A} = \frac{1}{\sqrt{A!}} \sum_{P}^{A!} \text{sign}(P)P, \tag{9.40}$$

where the sum runs over all permutations of the A nucleon labels and sign(P) stands for the sign of the permutation P.

A general element of the basis will thus be written as

$$\begin{aligned} \psi_i &\equiv \psi_{il_i(L_iS_i)JMTM_T}(\boldsymbol{x}) \\ &= \mathcal{A}\left\{ \mathrm{e}^{-\frac{1}{2}\tilde{\boldsymbol{x}}A_i\boldsymbol{x}} \left[\theta_{l_iL_i}(\boldsymbol{x})\chi_{S_i}\right]_{JM}\eta_{TM_T} \right\}. \end{aligned} \tag{9.41}$$

As was mentioned above, the trial function becomes more flexible if the elements of A_i are not constrained to be the same for each value of the set $\{l_iL_iS_i\}$, thus A_i may carry the same subscript as $\{lLS\}$.

The trial function is then expressed as a linear combination of the basis elements (9.41) as is given in Eq. (9.3). The variational parameters are the coefficients $\tilde{c} = (c_1, \ldots, c_K)$ and, in each term (9.41), at most $\frac{1}{2}(A-1)A$ parameters of A_i. These can be either a_{i_j} $(j = 1, \ldots, A-1)$ and the $\frac{1}{2}(A-1)(A-2)$ parameters of G or the $\frac{1}{2}(A-1)A$ pair-correlation parameters α_{ij}. Furthermore, the $A-1$ components u_{li} of the vector $\boldsymbol{v}_l$ introduced in Eq. (9.32) may have to be varied also. For a satisfactory approximation, one needs a great number of terms, and it would indeed be a formidable task to vary all these parameters to find their values that make the energy functional fully stationary. Especially the nonlinear parameters would be very difficult to treat, and that is what can be avoided by the stochastic variational procedure. One can now imagine that a p^{A-1} grid would really produce grossly overlapping functions. Furthermore, it is conceivable that there are infinite numbers of equally good parameter sets, and it would be useless to find the very best among them even if it were possible. The optimum procedure is probably that which has been worked out: proper variation of the linear parameters and stochastic sampling combined with an admittance test for the values of the nonlinear parameters.

In this section we have seen how to set up a correlated Gaussian trial function. The utility of a trial function depends on three circumstances:

(i) whether it is flexible enough to accommodate itself to all correlations to be described, (ii) whether the calculation of the Hamiltonian matrix elements is feasible and (iii) whether the variational procedure implied is efficient enough. For the time being, we have argued for the flexibility of the trial function and indicated that the variational procedure may be feasible. That these statements really hold will be demonstrated in Chap. 10. The calculation of the matrix elements is presented in Sects. 9.3 and 9.4. By deriving a generating function for a basis element, in Sect. 9.3 we shall prepare the ground for calculating the matrix elements (Sect. 9.4) of not only the Hamiltonian but also of any operators of practical interest. The physical quantities of our interest will be presented in Sect. 9.5.

9.3 Generating function

The trial function constructed in Sect. 9.2 contains the essentials of the general version of the approach. Nevertheless, the formalism should be elaborated further before proceeding to more concrete physical ingredients. This will give insight into the relation of the correlated Gaussian formalism to the generator-coordinate formalism of the cluster model, and will open the way to introducing cluster ansätze in Sect. 12.1. At the same time, this will pave the way before the evaluation of the matrix elements in Sect. 9.4.

First, in Sect. 9.3.1 the notion of generating functions is introduced. Then, in Sect. 9.3.2 the generating function of the correlated Gaussians will be presented. In Sect. 9.3.3 we shall define Slater determinants of Gaussian wave packets and show how the c.m. wave function factors out. In Sect. 9.3.4 the relationship between such a Slater determinant and a correlated Gaussian will be established.

9.3.1 Definition

A function $g(s,x)$ is called the generating function of the function set $\{f_n(x)\}$ if the coefficients of its Taylor expansion in s are the functions $f_n(x)$:

$$g(s,x) = \sum_{n=0}^{\infty} \frac{1}{n!} f_n(x) s^n. \tag{9.42}$$

Then $f_n(x)$ is expressible in terms of the s-derivatives of $g(s,x)$:

$$f_n(x) = \left. \frac{\partial^n g(s,x)}{\partial s^n} \right|_{s=0}. \tag{9.43}$$

The functional form of the generating function is often much simpler than that of the function set generated. The function $g(s,x)$ is therefore most useful in evaluating integrals whose integrand contains $f_n(x)$ provided that the sequence of integration with respect to x and the differentiation with respect to s can be exchanged. For instance, a matrix element between $f_n(x)$ and $f_{n'}(x)$ can be expressed as

$$\int dx f_n(x) O f_{n'}(x) = \frac{\partial^n}{\partial s^n}\frac{\partial^{n'}}{\partial s'^{n'}}\left[\int dx g(s,x) O g(s',x)\right]\Bigg|_{s=0,s'=0}. \quad (9.44)$$

As a simple example for $f_n(x)$, we consider the normalized eigenfunctions $\psi_n(x)$ of a one-dimensional HO:

$$\psi_n(x) = \left(\frac{\beta}{\pi}\right)^{1/4}\frac{1}{\sqrt{2^n n!}}H_n\left(\sqrt{\beta}x\right)e^{-\frac{1}{2}\beta x^2}, \quad (9.45)$$

where H_n is a Hermite polynomial and the width constant β is defined as $\beta = m\omega/\hbar$, with m being the mass and $\hbar\omega$ the oscillator quantum. The function (9.45) can be generated via the relationship

$$g(s,x) \equiv \left(\frac{\beta}{\pi}\right)^{1/4} e^{-\frac{1}{2}\beta(x-s)^2+\frac{1}{4}\beta s^2} = \sum_{n=0}^{\infty}\frac{1}{\sqrt{2^n n!}}\left(\sqrt{\beta}s\right)^n \psi_n(x). \quad (9.46)$$

This generating function[9] is a Gaussian function of x displaced by s from the origin. The products of such functions can always be cast into

[9]The function $g(s,x)$ is a kind of 'coherent state' of HO eigenfunctions, a term borrowed from optics. The probability of finding an HO state with n quanta in it is proportional to $[u_n(s)]^2$ with $u_n(s) = \sqrt{\beta^n/(2^n n!)}s^n$. To get the overall norm of these relative probabilities, one should sum up these terms, to obtain $\sum_{n=0}^{\infty}[u_n(s)]^2 = \exp(\frac{1}{2}\beta s^2)$, thus the probability of finding the nth HO state in $g(s,x)$ is $P_n = (n!)^{-1}\lambda^n\exp(-\lambda)$, where $\lambda = \frac{1}{2}\beta s^2$. A coherent state is thus a superposition of HO states in which the weights follow a Poisson distribution of parameter λ. For a Poisson distribution the parameter λ is the average and variance at the same time. (See Ref. [222] for the generalized coherent state method and its applications.)

The function $g(s,x)$ is often used as the kernel of a transformation called the Bargmann transformation [223]. The Bargmann transform $\Psi(s)$ of a square-integrable function $\psi(x)$ is defined by

$$\Psi(s) = \int_{-\infty}^{+\infty} dx\, g(s,x)\psi(x).$$

By extending s to a complex variable, the Bargmann transformation can be made invertible:

$$\psi(x) = \int d\mu(s)\, g^*(s,x)\Psi(s),$$

where the Bargmann measure $d\mu(s)$ is defined by

$$d\mu(s) = \frac{\beta}{2\pi}e^{-\frac{\beta}{2}ss^*}d\,\mathrm{Re}\,s\; d\,\mathrm{Im}\,s = \frac{\beta}{2\pi}e^{-\frac{\beta}{2}\rho^2}\rho\, d\rho\, d\theta,$$

the form $e^{-\frac{1}{2}Ax^2+Bx}$, whose integral over the full axis is very simple:

$$\int_{-\infty}^{+\infty} dx e^{-\frac{1}{2}Ax^2+Bx} = \left(\frac{2\pi}{A}\right)^{1/2} \exp\left(\frac{1}{2}\frac{B^2}{A}\right). \tag{9.47}$$

Furthermore, the monomials times $e^{-\frac{1}{2}Ax^2+Bx}$ are also easy to integrate; these formulae can be obtained from Eq. (9.47) by differentiating both sides with respect to A and/or B repeatedly. Thus Eq. (9.46) gives a handy tool for treating one-dimensional HO wave functions.

To generalize the relationship (9.46) to the eigenfunctions $\psi_{nlm}(\boldsymbol{x})$ of the isotropic three-dimensional HO, one may use spherical coordinates. Then an analogy with Eq. (9.46) survives if the variables x, s are replaced by vectors and the expansion is complemented by an expansion over spherical harmonics:[10]

$$g(\boldsymbol{s},\boldsymbol{x}) \equiv g^{(\beta)}(\boldsymbol{s},\boldsymbol{x})$$
$$\equiv \left(\frac{\beta}{\pi}\right)^{3/4} e^{-\frac{1}{2}\beta(\boldsymbol{x}-\boldsymbol{s})^2+\frac{1}{4}\beta\boldsymbol{s}^2} = \left(\frac{\beta}{\pi}\right)^{3/4} e^{-\frac{1}{2}\beta\boldsymbol{x}^2+\beta\boldsymbol{x}\cdot\boldsymbol{s}-\frac{1}{4}\beta\boldsymbol{s}^2} \tag{9.48}$$
$$= \sum_{nlm} \sqrt{\frac{4\pi}{2^n n!(2n+2l+1)!!}} \left(\frac{\beta}{2}\right)^{2n+l} s^{2n+l} Y_{lm}^*(\hat{s})\psi_{nlm}(\boldsymbol{x}). \tag{9.49}$$

with $s = \rho e^{i\theta}$ and the integration extends over the whole complex s-plane. The invertibility is due to the following properties of the kernel $g(s,x)$:

$$(\dagger) \quad \int_{-\infty}^{+\infty} dx\, g(s,x)g^*(s',x) = e^{\frac{\beta}{2}ss'^*}, \qquad \int d\mu(s)\, g^*(s,x)g(s,x') = \delta(x-x').$$

The Bargmann transform of an HO wave function, $\psi_n(x)$, is just $u_n(s)$, an order-n monomial of s. It forms an orthonormal set with respect to the Bargmann measure:

$$\int d\mu(s)\, u_n^*(s)u_{n'}(s) = \delta_{nn'}.$$

(This set of monomials is, of course, complete.) This orthonormality relation helps to prove the closure relation for the HO eigenfunctions: $\sum_{n=0}^{\infty} \psi_n(x)\psi_n^*(x') = \delta(x-x')$ by simply substituting Eq. (9.46) into the second equation of (†) twice.

An important application of the Bargmann transformation is the evaluation of resonating-group kernels (cf. Sect. 11.5). For this the Bargmann transformation must be generalized to three dimensions as shown in Eqs. (9.48) and (9.49). See Sect. 11.7.2 for more details and Refs. [224, 225] for applications.

[10]It is more common to define the generating function of the HO eigenfunctions differently. For a one-dimensional HO [26], this definition is $g(t,x) = \exp(tb_x^\dagger)\psi_0(x) = (\beta/\pi)^{1/4}\exp(-\frac{1}{2}\beta x^2 + \sqrt{2\beta}xt - \frac{1}{2}t^2) = \sum_{n=0}^{\infty}\left(1/\sqrt{n!}\right)t^n\psi_n(x)$, where $b_x^\dagger$ is defined in Eq. (I.3); for a three-dimensional HO [187], similarly, $g(\boldsymbol{t},\boldsymbol{x}) = (\beta/\pi)^{3/4}\exp(-\frac{1}{2}\beta\boldsymbol{x}^2 + \sqrt{2\beta}\boldsymbol{x}\cdot\boldsymbol{t} - \frac{1}{2}\boldsymbol{t}^2)$. One can obtain Eqs. (9.46) and (9.48) from these by substituting $t = \sqrt{\beta/2}s$, $\boldsymbol{t} = \sqrt{\beta/2}\boldsymbol{s}$.

In evaluating integrals containing $g(\boldsymbol{s}, \boldsymbol{x})$, it is useful to write down a three-dimensional version of the formula (9.47):

$$\int d\boldsymbol{x}\, \mathrm{e}^{-\frac{1}{2}A\boldsymbol{x}^2+\boldsymbol{B}\cdot\boldsymbol{x}} = \left(\frac{2\pi}{A}\right)^{3/2} \exp\left(\frac{1}{2}\frac{\boldsymbol{B}^2}{A}\right), \tag{9.50}$$

where $\boldsymbol{x}$ and $\boldsymbol{B}$ are three-dimensional Cartesian vectors.

It follows from Eq. (9.49) that an eigenfunction $\psi_{nlm}(\boldsymbol{x})$ can be expressed as

$$\psi_{nlm}(\boldsymbol{x}) = \sqrt{\frac{2^{3n+l}n!(2n+2l+1)!!}{4\pi[(2n+l)!]^2\beta^{2n+l}}} \times \left[\frac{d^{2n+l}}{ds^{2n+l}}\int d\hat{\boldsymbol{s}} Y_{lm}(\hat{\boldsymbol{s}}) g(\boldsymbol{s}, \boldsymbol{x})\right]\Bigg|_{s=0}. \tag{9.51}$$

For completeness, we write down an explicit form of the HO eigenfunctions here: $\psi_{nlm}(\boldsymbol{x}) = R_{nl}(x) Y_{lm}(\hat{\boldsymbol{x}})$, where

$$R_{nl}(x) = \left(\frac{\beta^3}{\pi}\right)^{1/4} \left[\frac{2^{l-n+2}(2n+2l+1)!!}{n!}\right]^{1/2} \left(\sqrt{\beta}x\right)^l \mathrm{e}^{-\frac{1}{2}\beta x^2} \times \sum_{k=0}^{n} \frac{(-1)^{n+k} n! 2^k}{(n-k)!k!(2k+2l+1)!!} \left(\sqrt{\beta}x\right)^{2k}. \tag{9.52}$$

The phase convention[11] in Eq. (9.52) ensures that $R_{nl}(x)$ is positive for x greater than the outermost nodal point.

The functions $\psi_{0lm}(\boldsymbol{x})$ are, apart from constant factors, equal to the 'Gaussian' terms of the trial function (9.2) of the one-particle problem, and are also the basic ingredients of Eq. (9.19) for the multiparticle problem. The relationship (9.51) shows that it can be generated by displaced Gaussians or Gaussian wave packets. These considerations can be generalized to multiparticle functions, and that is done next for correlated Gaussians.

9.3.2 Generating a correlated Gaussian

Let us consider a correlated Gaussian of the form of Eq. (9.41) with the angle dependence given by Eq. (9.32), and, for simplicity, suppress the label i of the basis element as well as all labels implicit in i. In the following deliberations it is enough to manipulate with the spatial part

[11] This convention is used in Ref. [173]. Another convention [226,227] often used, which differs from the adopted formula by a factor of $(-1)^n$, makes $R_{nl}(x)$ positive in the vicinity of the origin.

of the function behind the antisymmetrizer, to be denoted by f_{KLM_L}. The notation implied can be summarized as follows:

$$\psi_i = [\psi_{L;S;TM_T}(\boldsymbol{x})]_{JM} = \mathcal{A}\{[f_{KL}(u, \boldsymbol{x}, A)\chi_S]_{JM}\, \eta_{TM_T}\}, \tag{9.53}$$

where the uncoupled function reads [cf. Eq. (9.41)]

$$\psi_{LM_L;SM_S;TM_T}(\boldsymbol{x}) = \mathcal{A}\{f_{KLM_L}(u, \boldsymbol{x}, A)\chi_{SM_S}\eta_{TM_T}\}, \tag{9.54}$$

with

$$f_{KLM_L}(u, \boldsymbol{x}, A) = \theta_{KLM_L}(u, \boldsymbol{x})e^{-\frac{1}{2}\tilde{\boldsymbol{x}}A\boldsymbol{x}}, \tag{9.55}$$

$$\theta_{KLM_L}(u, \boldsymbol{x}) = v^{2K+L}Y_{LM_L}(\hat{\boldsymbol{v}}), \tag{9.56}$$

$$\boldsymbol{v} = \sum_{i=1}^{A-1} u_i\boldsymbol{x}_i = \tilde{u}\boldsymbol{x}, \qquad \tilde{u} = (u_1, \ldots, u_{A-1}). \tag{9.57}$$

Note that u is a one-column matrix of $A-1$ real numbers, whereas $\boldsymbol{x}$ is a one-column matrix of $A-1$ Cartesian vectors, the Jacobi vectors.

By using the well-known identity

$$(\boldsymbol{v}\cdot\boldsymbol{e})^k = v^k \sum_{\substack{n,l\geq 0\\ 2n+l=k}} B_{nl} \sum_{m=-l}^{l} Y_{lm}(\hat{\boldsymbol{v}})Y_{lm}^*(\hat{\boldsymbol{e}}), \quad \text{with } |\boldsymbol{e}| = 1, \tag{9.58}$$

where B_{nl} was defined in Eq. (9.35), one can see that $f_{KLM_L}(u, \boldsymbol{x}, A)$ can be derived from the generating function

$$g(\boldsymbol{\varsigma}; \boldsymbol{x}, A) = \exp\left(-\frac{1}{2}\tilde{\boldsymbol{x}}A\boldsymbol{x} + \tilde{\boldsymbol{\varsigma}}\boldsymbol{x}\right) \quad \text{with } \boldsymbol{\varsigma} = \alpha\boldsymbol{e}u, \ |\boldsymbol{e}| = 1. \tag{9.59}$$

(The parameter of the generating function is the length α of the vector $\alpha\boldsymbol{e}$.) The expression of $f_{KLM_L}(u, \boldsymbol{x}, A)$ is as follows:

$$f_{KLM_L}(u, \boldsymbol{x}, A) = \frac{1}{B_{KL}}\int d\hat{\boldsymbol{e}}\, Y_{LM_L}(\hat{\boldsymbol{e}})\left[\frac{d^{2K+L}}{d\alpha^{2K+L}}g(\alpha\boldsymbol{e}u; \boldsymbol{x}, A)\right]\Bigg|_{\alpha=0}. \tag{9.60}$$

Remember that $\boldsymbol{v}$ contains the physical coordinates $\boldsymbol{x}$, but $\boldsymbol{e}$ is just an angle parameter. Notice the formal similarity and difference between Eq. (9.59) and the single-variable generating function $g(\boldsymbol{s}, \boldsymbol{x})$ of Eq. (9.48).

When the function $\theta_{LM_L}(\boldsymbol{x})$ of the type of Eq. (9.37) is needed, the generating function g of Eq. (9.59) must be modified to include another factor $e^{\alpha'\boldsymbol{v}'\cdot\boldsymbol{e}'}$. Since the ensuing derivation for this modified case follows exactly the same pattern, it suffices to present the derivation for $\theta_{LM_L}(\boldsymbol{x})$ of Eq. (9.56). The generating of $\theta_{LM_L}(\boldsymbol{x})$ of Eq. (9.23) requires more variables [187]. A sketchy derivation for that case is presented in App. G.2.

The generating function (9.59) is remarkably simple. The integrals involving Eq. (9.59) are especially easy to evaluate owing to the $3n$-dimensional generalization of the integration formulae (9.47) and (9.50):

$$\mathcal{I} \equiv \int d\boldsymbol{x}\, \mathrm{e}^{-\frac{1}{2}\tilde{\boldsymbol{x}}A\boldsymbol{x}+\tilde{\boldsymbol{B}}\boldsymbol{x}} = \left[\frac{(2\pi)^n}{\det A}\right]^{3/2} \mathrm{e}^{\frac{1}{2}\tilde{\boldsymbol{B}}A^{-1}\boldsymbol{B}}, \tag{9.61}$$

where $\int d\boldsymbol{x} \equiv \int d\boldsymbol{x}_1 \cdots \int d\boldsymbol{x}_n$. The integrals of expressions consisting of monomials times $\exp(-\frac{1}{2}\tilde{\boldsymbol{x}}A\boldsymbol{x}+\tilde{\boldsymbol{B}}\boldsymbol{x})$ are also easy to calculate as was outlined after Eq. (9.47). Let us mention a few special cases here:

$$\int d\boldsymbol{x}\boldsymbol{x}_i\, \mathrm{e}^{-\frac{1}{2}\tilde{\boldsymbol{x}}A\boldsymbol{x}+\tilde{\boldsymbol{B}}\boldsymbol{x}} = (A^{-1}\boldsymbol{B})_i\mathcal{I}, \tag{9.62}$$

$$\int d\boldsymbol{x}\, (\boldsymbol{x}_i \cdot \boldsymbol{x}_j)\mathrm{e}^{-\frac{1}{2}\tilde{\boldsymbol{x}}A\boldsymbol{x}+\tilde{\boldsymbol{B}}\boldsymbol{x}} = \left[3(A^{-1})_{ij} + (A^{-1}\boldsymbol{B})_i \cdot (A^{-1}\boldsymbol{B})_j\right]\mathcal{I}, \tag{9.63}$$

$$\int d\boldsymbol{x}\, [\boldsymbol{x}_i\boldsymbol{x}_j]_{\kappa\mu}\mathrm{e}^{-\frac{1}{2}\tilde{\boldsymbol{x}}A\boldsymbol{x}+\tilde{\boldsymbol{B}}\boldsymbol{x}} = \left[(A^{-1}\boldsymbol{B})_i(A^{-1}\boldsymbol{B})_j\right]_{\kappa\mu}\mathcal{I} \quad (\kappa = 1, 2), \tag{9.64}$$

where $[\boldsymbol{ab}]_{\kappa\mu}$ denotes $\sum_{m_1 m_2}\langle 1m_1 1m_2|\kappa\mu\rangle a_{m_1} b_{m_2}$, with the spherical 'components' a_i of a vector $\boldsymbol{a}$ being $a_1 = -\frac{1}{\sqrt{2}}(a_x + ia_y)$, $a_0 = a_z$, $a_{-1} = \frac{1}{\sqrt{2}}(a_x - ia_y)$ (see footnotes on pp. 202 and 221).

We shall elaborate further on Eq. (9.59) to get an even more useful form of the generating function, which shows how the correlated Gaussian basis is related to a s.p. basis. Establishing such a relationship would be useful not only to evaluate matrix elements but also to include clusters. This relationship will be obtained in Sect. 9.3.4, but before that, in Sect. 9.3.3, a Slater determinant of Gaussian wave packets will be introduced as a prerequisite.

9.3.3 Gaussian wave packets

Let the motion of the ith nucleon with mass m, spin projection m_s^i and isospin projection m_t^i be described by the s.p. Gaussian wave packet (or shifted Gaussian) centred around $\boldsymbol{s}_i$,

$$\hat{\varphi}_{\boldsymbol{s}_i m_s^i m_t^i}(\boldsymbol{r}_i) = \varphi_{\boldsymbol{s}_i}(\boldsymbol{r}_i)\chi_{\frac{1}{2}m_s^i}\eta_{\frac{1}{2}m_t^i}, \tag{9.65}$$

with

$$\varphi_{\boldsymbol{s}_i}(\boldsymbol{r}_i) \equiv \varphi_{\boldsymbol{s}_i}^{(\beta)}(\boldsymbol{r}_i) = \left(\frac{\beta}{\pi}\right)^{3/4} \mathrm{e}^{-\frac{1}{2}\beta(\boldsymbol{r}_i-\boldsymbol{s}_i)^2} \quad \text{and} \quad \beta = \frac{m\omega}{\hbar}, \tag{9.66}$$

where $\boldsymbol{r}_i$ is the position vector of the nucleon, $\chi_{\frac{1}{2}m_s^i}$ and $\eta_{\frac{1}{2}m_t^i}$ are its spin and isospin functions. The angular frequency ω may be taken an arbitrary constant; it will just provide an overall scale for the spatial extension of the basis states and may be chosen for convenience.[12] The parameter $\boldsymbol{s}_i$ is a 'generator coordinate', i.e., a continuous parameter of a wave function, over which a physical wave function can be expanded [141, 228]. A Slater determinant of these Gaussian packets can be written as

$$\phi_\kappa(\boldsymbol{s}_1,\ldots,\boldsymbol{s}_A) = \mathcal{A}\left\{\prod_{i=1}^{A} \hat{\varphi}_{\boldsymbol{s}_i m_s^i m_t^i}(\boldsymbol{r}_i)\right\}, \tag{9.67}$$

where $\kappa = (m_s^1 m_t^1, \ldots, m_s^A m_t^A)$ is the set of the spin–isospin quantum numbers of the nucleons.

A spin–isospin coupled multiparticle Gaussian wave packet can be expressed as a linear combination of ϕ_κ of Eq. (9.67):

$$\begin{aligned}\Phi_{SM_STM_T}(\boldsymbol{s}_1,\ldots,\boldsymbol{s}_A) &= \mathcal{A}\{\varphi_{\boldsymbol{s}_1}(\boldsymbol{r}_1)\cdots\varphi_{\boldsymbol{s}_A}(\boldsymbol{r}_A)\chi_{SM_S}\eta_{TM_T}\} \\ &= \sum_\kappa c_\kappa \phi_\kappa(\boldsymbol{s}_1,\ldots,\boldsymbol{s}_A),\end{aligned} \tag{9.68}$$

where c_κ is a product of the Clebsch–Gordan coefficients which couple the spin and isospin as defined in Eq. (9.39). The physical meaning of the state (9.68) is transparent. It describes an ensemble of particles, each tied with a spring to a particular centre pinned down in space. After elimination of the c.m. motion, such a set of states could be used as a representation of any nuclear states, and states of this form are used in the model called AMD [204] (cf. Sect. 8.2). This is a model for nuclei and nuclear collisions. To identify such a state with a nuclear state, one has to project it to definite linear[13] and angular momentum and parity. By letting some particle centres coincide, one gets a state in which the particles are clustered. Such states are widely used as basis states in cluster models (see Sect. 11.7). In the present approach, however, Eq. (9.68) is used just as a tool to generate the correlated Gaussians.

To factor out the c.m. dependence from the wave function (9.68), we transform the s.p. coordinates into the Jacobi coordinates. Similarly to Eq. (9.12), the s.p. generator coordinates are transformed to Jacobi

[12] For the multicluster approximation to be introduced in Sect. 12.1, the value of β is no longer arbitrary; it will have to be appropriate, as a s.p. HO parameter, for the description of the intracluster motion.

[13] Projecting a state to definite (linear) momentum $\boldsymbol{P}_A$ factors out the dependence on the c.m. coordinate $\boldsymbol{x}_A$ as $\exp[-(i/\hbar)\boldsymbol{P}_A\cdot\boldsymbol{x}_A/\sqrt{A}]$, which makes it possible to eliminate the c.m. motion immediately.

generator coordinates as

$$\boldsymbol{S}_i = \sum_{k=1}^{A} U_{ik}\boldsymbol{s}_k \quad (i = 1, \ldots, A). \tag{9.69}$$

The product of Gaussians in Eq. (9.68) is $\exp[-\frac{1}{2}\beta\sum_{i=1}^{A}(\boldsymbol{r}_i - \boldsymbol{s}_i)^2]$. Using Eqs. (9.12) and (9.69), the quadratic expression $\sum_{i=1}^{A}(\boldsymbol{r}_i - \boldsymbol{s}_i)^2$ can be rewritten as

$$\sum_{i=1}^{A}(\boldsymbol{r}_i - \boldsymbol{s}_i)^2 = \sum_{i,j,k=1}^{A}(\boldsymbol{x}_j - \boldsymbol{S}_j)U_{ji}\tilde{U}_{ik}(\boldsymbol{x}_k - \boldsymbol{S}_k) = \sum_{i=1}^{A}(\boldsymbol{x}_i - \boldsymbol{S}_i)^2. \tag{9.70}$$

Thus the product of the Gaussian s.p. wave packets can be written as a product of Gaussians depending on the Jacobi coordinates and Jacobi generator coordinates:

$$\prod_{i=1}^{A}\varphi_{\boldsymbol{s}_i}(\boldsymbol{r}_i) = \prod_{i=1}^{A}\varphi_{\boldsymbol{S}_i}(\boldsymbol{x}_i). \tag{9.71}$$

This transformation is so simple owing to the sophisticated definition of the lengths of the Jacobi vectors in Eqs. (9.8) and (9.9) (cf. footnote on p. 199). By using Eq. (9.71) and noting that the last factor of the product depends only on the c.m. coordinate, which is symmetric under the exchange of nucleons, the A-nucleon wave function can be rewritten as

$$\Phi_{SM_STM_T}(\boldsymbol{s}_1, \ldots, \boldsymbol{s}_A) = \Psi_{SM_STM_T}(\boldsymbol{S}_1, \ldots, \boldsymbol{S}_{A-1})\varphi_{\boldsymbol{S}_A}(\boldsymbol{x}_A), \tag{9.72}$$

which defines the intrinsic function that depends solely on the intrinsic coordinates (and contains, as parameters, the intrinsic Jacobi generator coordinates):

$$\Psi_{SM_STM_T}(\boldsymbol{S}_1, \ldots, \boldsymbol{S}_{A-1}) = \mathcal{A}\{\varphi_{\boldsymbol{S}_1}(\boldsymbol{x}_1)\ldots\varphi_{\boldsymbol{S}_{A-1}}(\boldsymbol{x}_{A-1})\chi_{SM_S}\eta_{TM_T}\}. \tag{9.73}$$

Note that $\varphi_{\boldsymbol{S}_i}(\boldsymbol{x}_i) = \mathrm{e}^{-\frac{1}{4}\beta \boldsymbol{S}_i^2}g(\boldsymbol{S}_i, \boldsymbol{x}_i)$, where $g(\boldsymbol{S}_i, \boldsymbol{x}_i)$ is the s.p. generating function given in Eq. (9.48). Therefore, the multiparticle function $\Psi_{SM_STM_T}$ could be used as a generating function of an element of an uncorrelated Gaussian basis for a multiparticle system. In the following it will be shown that it has to undergo an integral transformation in order to yield the generating function (9.59) of a correlated Gaussian.

9.3.4 Correlated Gaussian from single-particle states

The generating function (9.59) can be related to the product of the Gaussians centred around ($\boldsymbol{S}_i$, $i = 1, \ldots, A-1$) through an integral transformation [187,206]:

$$g(\varsigma; \boldsymbol{x}, A) = \left[(4\pi\beta)^{\frac{1}{2}(A-1)}\det C\right]^{-3/2} \exp\left(-\frac{1}{2\beta^2}\tilde{\varsigma}C^{-1}\varsigma\right)$$
$$\times \int d\boldsymbol{S} \exp\left[-\frac{1}{2}\tilde{\boldsymbol{S}}A(\beta C)^{-1}\boldsymbol{S} + \tilde{\varsigma}(\beta C)^{-1}\boldsymbol{S}\right] \prod_{i=1}^{A-1} \varphi_{\boldsymbol{S}_i}(\boldsymbol{x}_i), \quad (9.74)$$

where $\boldsymbol{S}$ stands for the set of generator coordinate vectors $\boldsymbol{S}_1, \ldots, \boldsymbol{S}_{A-1}$ supposed to be arranged in a column and

$$\beta C = I - \beta^{-1}A. \quad (9.75)$$

The quantity C is an $(A-1)\times(A-1)$ symmetric matrix, and I is the $(A-1)\times(A-1)$ unit matrix.

As a first step in verifying Eq. (9.74), the product $\prod_{i=1}^{A-1}\varphi_{\boldsymbol{S}_i}(\boldsymbol{x}_i)$ can be cast into the form of a single exponential function by using the matrix product of one-row and one-column matrices:

$$\prod_{i=1}^{A-1} \varphi_{\boldsymbol{S}_i}(\boldsymbol{x}_i) = \left(\frac{\beta}{\pi}\right)^{\frac{3}{4}(A-1)} \mathrm{e}^{-\frac{1}{2}\beta(\tilde{\boldsymbol{x}}\boldsymbol{x} - 2\tilde{\boldsymbol{x}}\boldsymbol{S} + \tilde{\boldsymbol{S}}\boldsymbol{S})}. \quad (9.76)$$

The verification may then proceed as follows:

$$\int d\boldsymbol{S}\, \mathrm{e}^{-\frac{1}{2}\tilde{\boldsymbol{S}}A(\beta C)^{-1}\boldsymbol{S} + \tilde{\varsigma}(\beta C)^{-1}\boldsymbol{S}} \prod_{i=1}^{A-1} \varphi_{\boldsymbol{S}_i}(\boldsymbol{x}_i)$$
$$= \left(\frac{\beta}{\pi}\right)^{\frac{3}{4}(A-1)} \int d\boldsymbol{S}\, \mathrm{e}^{-\frac{1}{2}\tilde{\boldsymbol{S}}A(\beta C)^{-1}\boldsymbol{S} + \tilde{\varsigma}(\beta C)^{-1}\boldsymbol{S}} \mathrm{e}^{-\frac{1}{2}\beta(\tilde{\boldsymbol{x}}\boldsymbol{x} - 2\tilde{\boldsymbol{x}}\boldsymbol{S} + \tilde{\boldsymbol{S}}\boldsymbol{S})} \quad (9.77)$$
$$= \left(\frac{\beta}{\pi}\right)^{\frac{3}{4}(A-1)} \int d\boldsymbol{S}\, \mathrm{e}^{-\frac{1}{2}\tilde{\boldsymbol{S}}[A(\beta C)^{-1} + \beta I]\boldsymbol{S} + [\tilde{\varsigma}(\beta C)^{-1} + \beta\tilde{\boldsymbol{x}}]\boldsymbol{S}} \mathrm{e}^{-\frac{1}{2}\beta\tilde{\boldsymbol{x}}\boldsymbol{x}} \quad (9.78)$$
$$= \left(\frac{\beta}{\pi}\right)^{\frac{3}{4}(A-1)} \int d\boldsymbol{S}\, \mathrm{e}^{-\frac{1}{2}\tilde{\boldsymbol{S}}C^{-1}\boldsymbol{S} + [\tilde{\varsigma}(\beta C)^{-1} + \beta\tilde{\boldsymbol{x}}]\boldsymbol{S}} \mathrm{e}^{-\frac{1}{2}\beta\tilde{\boldsymbol{x}}\boldsymbol{x}} \quad (9.79)$$
$$= \left(\frac{\beta}{\pi}\right)^{\frac{3}{4}(A-1)} \left[\frac{(2\pi)^{A-1}}{\det(C^{-1})}\right]^{3/2}$$
$$\times \mathrm{e}^{\frac{1}{2}[\tilde{\varsigma}(\beta C)^{-1} + \beta\tilde{\boldsymbol{x}}]C[(\beta C)^{-1}\varsigma + \beta\boldsymbol{x}]} \mathrm{e}^{-\frac{1}{2}\beta\tilde{\boldsymbol{x}}\boldsymbol{x}} \quad (9.80)$$
$$= \left[(4\pi\beta)^{\frac{1}{2}(A-1)} \det C\right]^{3/2} \mathrm{e}^{-\frac{1}{2}\tilde{\boldsymbol{x}}A\boldsymbol{x}} \exp\left(\frac{1}{2\beta^2}\tilde{\varsigma}C^{-1}\varsigma + \tilde{\varsigma}\boldsymbol{x}\right) \quad (9.81)$$

$$= \left[(4\pi\beta)^{\frac{1}{2}(A-1)} \det C\right]^{3/2} \exp\left(\frac{1}{2\beta^2}\tilde{\varsigma}C^{-1}\varsigma\right) g(\varsigma, \boldsymbol{x}, A). \tag{9.82}$$

In line (9.77) Eq. (9.76) was substituted. In line (9.78) only the integrand was rearranged. In line (9.79) Eq. (9.75) was used. In line (9.80) the integration over $\boldsymbol{S}$ was carried out with the help of Eq. (9.61). In line (9.81) a simple rearrangement was performed. In line (9.82) Eq. (9.59) was substituted. After rearrangement, this gives Eq. (9.74); Q.E.D.

Substituting Eq. (9.74) into Eq. (9.60) yields

$$f_{KLM_L}(u, \boldsymbol{x}, A) = \frac{1}{B_{KL}} \left[(4\pi\beta)^{\frac{1}{2}(A-1)} \det C\right]^{-3/2} \int d\hat{\boldsymbol{e}}\, Y_{LM_L}(\hat{\boldsymbol{e}})$$
$$\times \left\{ \frac{d^{2K+L}}{d\alpha^{2K+L}} \exp\left(-\frac{1}{2\beta^2}\tilde{\varsigma}C^{-1}\varsigma\right) \right.$$
$$\left. \times \int d\boldsymbol{S} \exp\left[-\frac{1}{2}\tilde{\boldsymbol{S}}A(\beta C)^{-1}\boldsymbol{S} + \tilde{\varsigma}(\beta C)^{-1}\boldsymbol{S}\right] \prod_{i=1}^{A-1} \varphi_{\boldsymbol{S}_i}(\boldsymbol{x}_i) \right\}\Bigg|_{\alpha=0}. \tag{9.83}$$

[Remember that $\varsigma = \alpha \boldsymbol{e} u$, with $|\boldsymbol{e}| = 1$, cf. Eq. (9.59).] Multiplying both sides by $\chi_{SM_S}\eta_{TM_T}$ and then applying the antisymmetrization operator to both sides leads to

$$\psi_{LM_L;SM_S;TM_T}(\boldsymbol{x})$$
$$= \frac{1}{B_{KL}} \left[(4\pi\beta)^{\frac{1}{2}(A-1)} \det C\right]^{-3/2} \int d\hat{\boldsymbol{e}}\, Y_{LM_L}(\hat{\boldsymbol{e}})$$
$$\times \left\{ \frac{d^{2K+L}}{d\alpha^{2K+L}} \exp\left(-\frac{1}{2\beta^2}\tilde{\varsigma}C^{-1}\varsigma\right) \right.$$
$$\times \int d\boldsymbol{S} \exp\left[-\frac{1}{2}\tilde{\boldsymbol{S}}A(\beta C)^{-1}\boldsymbol{S} + \tilde{\varsigma}(\beta C)^{-1}\boldsymbol{S}\right]$$
$$\left. \times \Psi_{SM_STM_T}(\boldsymbol{S}_1, \ldots, \boldsymbol{S}_{A-1}) \right\}\Bigg|_{\alpha=0}, \tag{9.84}$$

where Eq. (9.73) has been used. The function on the left-hand side differs from an element $\psi_{(LS)JMTM_T}$ of the correlated Gaussian basis merely in angular-momentum coupling [cf. Eq. (9.53)], thus the meaning of Eq. (9.84) is that a basis element can be generated from an integral transform of the intrinsic generator-coordinate state given in Eq. (9.73).

The integration over $\boldsymbol{S}$ in Eq. (9.84) may look formidable but is actually a simple case for the application of Eq. (9.61). Equation (9.84) is the basic formula for the calculation of matrix elements involving correlated Gaussians, and this is the point of link to cluster models. The matrix elements are to be dealt with immediately in Sect. 9.4, while the link to cluster models will be discussed later, in Sect. 12.1.

9.4 Evaluation of matrix elements

The first step in evaluating the matrix elements will be to uncouple the spin and orbital angular momenta (Sect. 9.4.1). The matrix elements of any interest are translation invariant. These will be calculated through their relationship to (translation-noninvariant) matrix elements between Slater determinants of Gaussian wave packets (Sect. 9.4.2). Then it would be logical to present explicit formulae for the matrix elements between Slater determinants. There is, however, no completely general formula for these; they are to be derived for each particular system one by one. Their functional forms are, however, the same for any system, and these imply generic forms for the translation-invariant matrix elements as well. These two types of matrix elements will be presented in Sects. 9.4.3 and 9.4.4, respectively.

9.4.1 Uncoupling

In Sect. 9.3 a generating-function relationship, (9.84), has been derived for a function $\psi_{LM_L;SM_S;TM_T}(\boldsymbol{x})$ of Eq. (9.54), which differs from a correlated Gaussian basis element, Eq. (9.53), only in angular-momentum coupling. In evaluating the matrix elements between two functions $\psi_{(LS)JMTM_T}$ of Eq. (9.53), the uncoupled functions $\psi_{LM_L;SM_S;TM_T}(\boldsymbol{x})$ are more convenient to use. Matrix elements of uncoupled functions will be obtained with the help of the Wigner–Eckart theorem (cf. e.g., Refs. [173, 226, 227]).

The Wigner–Eckart theorem states that a matrix element of a μ component of a rank κ spherical tensor operator[14] $O_{\kappa\mu}$ between states coupled, respectively, to angular momenta JM and $J'M'$ depends on μ,

[14]The $2\kappa+1$ operators $\{O_{\kappa\mu}\}$ ($\mu=-\kappa, -\kappa+1, \ldots, \kappa$) are called the μ-components of a spherical tensor operator O_κ of rank κ (κ is an integer or half integer) if, under rotations, they transform like the components of an angular-momentum eigenfunction. An alternative definition is given by their commutation relations with the components of the (dimensionless) angular momentum operator $J_0=J_z, J_\pm=J_x\pm iJ_y$:

$$[J_0, O_{\kappa\mu}] = \mu O_{\kappa\mu}, \qquad [J_\pm, O_{\kappa\mu}] = \sqrt{(\kappa\mp\mu)(\kappa\pm\mu+1)}O_{\kappa\,\mu\pm1}.$$

A third equivalent definition is provided by the property that, when acting on a state, a rank-κ spherical tensor adds to it angular momentum κ. The addition may be understood as follows: if $|\psi_{J'M'}\rangle$ is an eigenvector of the operators J^2, J_z with quantum numbers $J'M'$, then $\sum_{M'\mu}\langle J'M'\kappa\mu|JM\rangle O_{\kappa\mu}\psi_{J'M'}$ will be an angular-momentum eigenvector belonging to quantum numbers JM.

M and M' just via a factor $\langle J'M'\kappa\mu|JM\rangle$. This is usually expressed as an implicit definition for the reduced matrix elements, $\langle \alpha J||O_\kappa||\alpha' J'\rangle$, which, according to the theorem, do not depend on the projections[15]

$$\langle \alpha JM|O_{\kappa\mu}|\alpha' J'M'\rangle = (2J+1)^{-1/2}\langle J'M'\kappa\mu|JM\rangle\langle \alpha J||O_\kappa||\alpha' J'\rangle. \tag{9.85}$$

The labels α and α' stand for the other labels of the states involved.

The transition probability from the state $\alpha' J'$ to the state αJ due to the operator O_κ depends, among others, on the magnitude of the matrix element $\langle \alpha JM|O_{\kappa\mu}|\alpha' J'M'\rangle$. This probability is conveniently expressed in terms of the reduced transition probability defined by

$$\begin{aligned} B(\kappa, \alpha' J' \to \alpha J) &= \frac{1}{2J'+1}\sum_{M'\mu M} |\langle \alpha JM|O_{\kappa\mu}|\alpha' J'M'\rangle|^2 \\ &= \frac{1}{2J'+1}|\langle \alpha J||O_\kappa||\alpha' J'\rangle|^2. \end{aligned} \tag{9.86}$$

The reduced matrix element of the tensorial product,

$$O_{\kappa\mu} \equiv [O_{\kappa_1} O_{\kappa_2}]_{\kappa\mu}\,, \tag{9.87}$$

of two operators, O_{κ_1} and O_{κ_2}, acting in tensor-product state spaces can be expressed in terms of products of matrix elements reduced in the two state spaces. When the two spaces are the ordinary space and the spin space, this formula reads (see, e.g., [226])

$$\begin{aligned} &\langle (LS)J||[O_{\kappa_1}(\text{space})O_{\kappa_2}(\text{spin})]_\kappa||(L'S')J'\rangle \\ &\quad = \sqrt{(2J+1)(2J'+1)(2\kappa+1)}\begin{Bmatrix} L' & S' & J' \\ \kappa_1 & \kappa_2 & \kappa \\ L & S & J \end{Bmatrix} \\ &\quad \times \langle L||O_{\kappa_1}(\text{space})||L'\rangle\langle S||O_{\kappa_2}(\text{spin})||S'\rangle. \end{aligned} \tag{9.88}$$

Here $\{\cdots\}$ is a $9j$ symbol [cf. Eq. (G.22)]. This composition formula and the Wigner–Eckart theorem (9.85) itself can be used to uncouple the space and spin parts, while it is useful to keep the isospin part intact:

$$\langle \psi_{(LS)JMTM_T}|[O_{\kappa_1}(\text{space})O_{\kappa_2}(\text{spin})]_{\kappa\mu}O_{\kappa_3\mu_3}(\text{isospin})|\psi_{(L'S')J'M'T'M'_T}\rangle$$

---

[15] This convention is used, e.g., by Bohr and Mottelson [173]. Lawson's convention [227] is different; it is given by $\langle \alpha J||O_\kappa||\alpha' J'\rangle_{\text{Lawson}} = (2J+1)^{-1/2}\langle \alpha J||O_\kappa||\alpha' J'\rangle$. There is a third convention, followed, e.g., by Messiah [26], by deShalit and Talmi [226] and by Brussaard and Glaudemans [184]: this differs from the convention adopted by a phase factor $(-1)^{2\kappa}$.

$$= \sqrt{(2L+1)(2S+1)(2\kappa+1)(2J'+1)} \begin{Bmatrix} L' & S' & J' \\ \kappa_1 & \kappa_2 & \kappa \\ L & S & J \end{Bmatrix}$$
$$\times \frac{\langle J'M'\kappa\mu|JM\rangle}{\langle L'M'_L\kappa_1\mu_1|LM_L\rangle\langle S'M'_S\kappa_2\mu_2|SM_S\rangle}$$
$$\times \langle \psi_{LM_L;SM_S;TM_T}|O_{\kappa_1\mu_1}(\text{space})O_{\kappa_2\mu_2}(\text{spin})O_{\kappa_3\mu_3}(\text{isospin})$$
$$\times |\psi_{L'M'_L;S'M'_S;T'M'_T}\rangle, \tag{9.89}$$

where the uncoupled basis functions $\psi_{LM_L;SM_S;TM_T}$ tally with the uncoupled operator

$$O = O_{\kappa_1\mu_1}(\text{space})O_{\kappa_2\mu_2}(\text{spin})O_{\kappa_3\mu_3}(\text{isospin}). \tag{9.90}$$

The function $\psi_{LM_L;SM_S;TM_T}$ is expressed in Eq. (9.84) as a multiple derivative of an integral transform of the generator-coordinate function $\Psi_{SM_STM_T}(\boldsymbol{S}_1,\ldots,\boldsymbol{S}_{A-1})$ given in Eq. (9.73). The function (9.73) is a Slater determinant of Gaussian wave packets, from which the c.m. factor has been removed. Equation (9.84) is the starting point for the evaluation of any matrix elements. Let us assume that O is a translation-invariant operator. Then one has

$$\langle \psi_{LM_L;SM_S;TM_T}|O|\psi_{L'M'_L;S'M'_S;T'M'_T}\rangle$$
$$= \frac{1}{B_{KL}B_{K'L'}}\left[(4\pi\beta)^{A-1}\det C\right]^{-3/2}\int d\hat{\boldsymbol{e}}\int d\hat{\boldsymbol{e}}'\, Y^*_{LM}(\hat{\boldsymbol{e}})Y_{L'M'}(\hat{\boldsymbol{e}}')$$
$$\times \Bigg[\frac{\partial^{2K+L+2K'+L'}}{\partial\alpha^{2K+L}\partial\alpha'^{2K'+L'}}e^{-\frac{1}{2}\tilde{X}\mathcal{C}X}\int d\boldsymbol{T}\, e^{-\frac{1}{2}\tilde{\boldsymbol{T}}\mathcal{Q}\boldsymbol{T}+\tilde{X}\boldsymbol{T}}$$
$$\times \langle \Psi_{SM_STM_T}(\boldsymbol{S}_1,\ldots,\boldsymbol{S}_{A-1})|O|\Psi_{S'M'_ST'M'_T}(\boldsymbol{S}'_1,\ldots,\boldsymbol{S}'_{A-1})\rangle\Bigg]\Bigg|_{\substack{\alpha=0\\ \alpha'=0}}, \tag{9.91}$$

where

$$\tilde{\boldsymbol{T}} \equiv (\boldsymbol{T}_1,\ldots,\boldsymbol{T}_{2A-2}) = (\boldsymbol{S}_1,\ldots,\boldsymbol{S}_{A-1},\boldsymbol{S}'_1,\ldots,\boldsymbol{S}'_{A-1}), \tag{9.92}$$

and the $(2A-2)\times(2A-2)$ matrices, $\mathcal{C}$ and $\mathcal{Q}$ and the one-coulumn matrix of $2A-2$ vectors $\boldsymbol{X}$ are defined conventionally [206] as

$$\mathcal{C} = \begin{pmatrix} C & 0 \\ 0 & C' \end{pmatrix}, \qquad \mathcal{Q} = \begin{pmatrix} C^{-1}-\beta I & 0 \\ 0 & C'^{-1}-\beta I \end{pmatrix},$$
$$\boldsymbol{X} = \begin{pmatrix} \alpha\beta^{-1}C^{-1}u\boldsymbol{e} \\ \alpha'\beta^{-1}C'^{-1}u'\boldsymbol{e} \end{pmatrix}, \tag{9.93}$$

where C is defined in Eq. (9.75) and C' is the same, but with A replaced by A'. Equation (9.91) is pretty involved, but it can be evaluated fairly systematically.

9.4.2 Including the centre of mass

In evaluating $\langle\Psi_{SM_STM_T}(\boldsymbol{S}_1,\ldots,\boldsymbol{S}_{A-1})|O|\Psi_{S'M'_ST'M'_T}(\boldsymbol{S'}_1,\ldots,\boldsymbol{S'}_{A-1})\rangle$ the difficulty is that it is expressed in terms of Jacobi coordinates, which are asymmetrical in particle labels, thus keeping track of antisymmetrization is very complicated in these coordinates. However, by returning to s.p. coordinates, one can get around this hitch. The s.p. coordinates can be brought back into the wave function owing to the factorization property of the Slater determinants of Gaussian wave packets, Eq. (9.72). The operator O is supposed to be translation-invariant, but for any observable of physical interest, there should exist a corresponding observable defined in the laboratory frame. Conversely, an observable defined in the laboratory frame may only have significance for the nucleus itself if its c.m. dependence can be separated from its dependence on the intrinsic variables.

The c.m. motion itself is a s.p. Gaussian wave packet, and, to facilitate the treatment of the c.m., we first discuss some matrix elements of s.p. wave packets (9.65). These matrix elements are also building blocks of the intrinsic many-particle matrix elements (see Sect. 9.4.3), and, by discussing them here, we are making preparations for the calculation of intrinsic matrix elements as well.

The operators involved in the dynamical solution of the problem are the unit operator, the kinetic-energy operator, $T_1=(-\hbar^2/2m)\boldsymbol{\nabla}^2_{\boldsymbol{r}_1}$, and the potential-energy operator. The potential may include a spin-orbit term, which contains the (dimensionless) relative orbital angular momentum,

$$\boldsymbol{l}_{12}=(\boldsymbol{r}_1-\boldsymbol{r}_2)\times(\boldsymbol{p}_1-\boldsymbol{p}_2)\frac{1}{2\hbar}=-\tfrac{1}{2}i(\boldsymbol{r}_1-\boldsymbol{r}_2)\times(\boldsymbol{\nabla}_1-\boldsymbol{\nabla}_2). \quad (9.94)$$

The spatial factor of a potential term can always be expressed as an integral, over $\boldsymbol{r}$, of $\delta(\boldsymbol{r}_1-\boldsymbol{r}_2-\boldsymbol{r})$; e.g., $V(\boldsymbol{r}_1-\boldsymbol{r}_2)=\int d\boldsymbol{r}\,V(\boldsymbol{r})\delta(\boldsymbol{r}_1-\boldsymbol{r}_2-\boldsymbol{r})$. Here we give formulae for the matrix elements, between Gaussian wave packets, of the unit operator, of T_1, of $\delta(\boldsymbol{r}_1-\boldsymbol{r}_2-\boldsymbol{r})$, of $\delta(\boldsymbol{r}_1-\boldsymbol{r}_2-\boldsymbol{r})\boldsymbol{l}_{12}$ and, as a simple example for operators not involved in the Hamiltonian, for $\boldsymbol{r}_1^2$:

$$\langle\varphi_{\boldsymbol{s}_1}|\varphi_{\boldsymbol{s}_2}\rangle=\exp\left[-\frac{\beta}{4}(\boldsymbol{s}_1-\boldsymbol{s}_2)^2\right], \quad (9.95)$$

$$\langle\varphi_{\boldsymbol{s}_1}|T_1|\varphi_{\boldsymbol{s}_2}\rangle=\frac{\hbar^2}{2m}\frac{\beta}{2}\left[3-\frac{\beta}{2}(\boldsymbol{s}_1-\boldsymbol{s}_2)^2\right]\langle\varphi_{\boldsymbol{s}_1}|\varphi_{\boldsymbol{s}_2}\rangle, \quad (9.96)$$

$$\langle\varphi_{\boldsymbol{s}_1}|\boldsymbol{r}_1^2|\varphi_{\boldsymbol{s}_2}\rangle=\frac{1}{2\beta}\left[3+\frac{\beta}{2}(\boldsymbol{s}_1+\boldsymbol{s}_2)^2\right]\langle\varphi_{\boldsymbol{s}_1}|\varphi_{\boldsymbol{s}_2}\rangle, \quad (9.97)$$

$$\langle\varphi_{\boldsymbol{s}_1}\varphi_{\boldsymbol{s}_2}|\delta(\boldsymbol{r}_1-\boldsymbol{r}_2-\boldsymbol{r})|\varphi_{\boldsymbol{s}_3}\varphi_{\boldsymbol{s}_4}\rangle$$

$$= \left(\frac{\beta}{2\pi}\right)^{3/2} \exp\left\{-\frac{\beta}{2}\left[\boldsymbol{r} - \frac{1}{2}(\boldsymbol{s}_1 + \boldsymbol{s}_3 - \boldsymbol{s}_2 - \boldsymbol{s}_4)\right]^2\right\}$$
$$\times \langle \varphi_{\boldsymbol{s}_1}|\varphi_{\boldsymbol{s}_3}\rangle\langle \varphi_{\boldsymbol{s}_2}|\varphi_{\boldsymbol{s}_4}\rangle, \tag{9.98}$$
$$\langle \varphi_{\boldsymbol{s}_1}\varphi_{\boldsymbol{s}_2}|\delta(\boldsymbol{r}_1 - \boldsymbol{r}_2 - \boldsymbol{r})\boldsymbol{l}_{12}|\varphi_{\boldsymbol{s}_3}\varphi_{\boldsymbol{s}_4}\rangle$$
$$= -\frac{1}{2}i\beta\boldsymbol{r} \times (\boldsymbol{s}_3 - \boldsymbol{s}_4)\langle \varphi_{\boldsymbol{s}_1}\varphi_{\boldsymbol{s}_2}|\delta(\boldsymbol{r}_1 - \boldsymbol{r}_2 - \boldsymbol{r})|\varphi_{\boldsymbol{s}_3}\varphi_{\boldsymbol{s}_4}\rangle. \tag{9.99}$$

To verify Eqs. (9.95) and (9.98), one may use Eq. (9.50). One way of deriving Eq. (9.96) is by employing

$$\nabla^2_{\boldsymbol{r}_1}\varphi_{\boldsymbol{s}_2}(\boldsymbol{r}_1) = \nabla^2_{\boldsymbol{s}_2}\varphi_{\boldsymbol{s}_2}(\boldsymbol{r}_1). \tag{9.100}$$

Eq. (9.97) can be proven, e.g., by using

$$\langle \varphi_{\boldsymbol{s}_1}|\boldsymbol{r}_1^2|\varphi_{\boldsymbol{s}_2}\rangle = -\frac{\partial}{\partial\alpha}\langle \varphi_{\boldsymbol{s}_1}|e^{-\alpha \boldsymbol{r}_1^2}|\varphi_{\boldsymbol{s}_2}\rangle\bigg|_{\alpha=0}, \tag{9.101}$$

and the matrix element of the Gaussian can be calculated via Eq. (9.50). The derivation of Eq. (9.99) is even more straightforward.

To come closer to the c.m. problem, we separate the c.m. term of the A-nucleon Hamiltonian written in the laboratory system:

$$H = \sum_{i=1}^{A}\frac{\boldsymbol{p}_i^2}{2m} + \sum_{i<j}^{A} V_{ij} = \left(\sum_{i=1}^{A-1}\frac{\boldsymbol{\pi}_i^2}{2m} + \sum_{i<j}^{A} V_{ij}\right) + \frac{\boldsymbol{\pi}_A^2}{2m} \equiv H_{\text{rel}} + T_{\text{c.m.}}, \tag{9.102}$$

where $\boldsymbol{\pi}_i$ is the momentum canonically conjugate to the Jacobi coordinate $\boldsymbol{x}_i$. Then, because of Eq. (9.72), the matrix elements of the Hamiltonian can be written as

$$\langle \Phi_{SM_STM_T}(\boldsymbol{s}_1,\ldots,\boldsymbol{s}_A)|H|\Phi_{S'M'_ST'M'_T}(\boldsymbol{s}'_1,\ldots,\boldsymbol{s}'_A)\rangle$$
$$= \langle \Psi_{SM_STM_T}(\boldsymbol{S}_1,\ldots,\boldsymbol{S}_{A-1})|H_{\text{rel}}|\Psi_{S'M'_ST'M'_T}(\boldsymbol{S}'_1,\ldots,\boldsymbol{S}'_{A-1})\rangle$$
$$\times\langle \varphi_{\boldsymbol{S}_A}|\varphi_{\boldsymbol{S}'_A}\rangle$$
$$+\langle \Psi_{SM_STM_T}(\boldsymbol{S}_1,\ldots,\boldsymbol{S}_{A-1})|\Psi_{S'M'_ST'M'_T}(\boldsymbol{S}'_1,\ldots,\boldsymbol{S}'_{A-1})\rangle$$
$$\times\langle \varphi_{\boldsymbol{S}_A}|T_{\text{c.m.}}|\varphi_{\boldsymbol{S}'_A}\rangle. \tag{9.103}$$

To express the matrix element of H_{rel}, let us integrate the matrix element of H, i.e., Eq. (9.103), over $\boldsymbol{S}_A$. On the right-hand side of Eq. (9.103) the $\boldsymbol{S}_A$ integration gives $\int d\boldsymbol{S}_A\,\langle \varphi_{\boldsymbol{S}_A}|\varphi_{\boldsymbol{S}'_A}\rangle = (4\pi/\beta)^{3/2}$ and $\int d\boldsymbol{S}_A\,\langle \varphi_{\boldsymbol{S}_A}|T_{\text{c.m.}}|\varphi_{\boldsymbol{S}'_A}\rangle = 0$. The former can be obtained from Eq. (9.95) by using Eq. (9.50), while the latter can be verified by using Eq. (9.96) and the integration formula

$$\int d\boldsymbol{S}_A\left[-\tfrac{1}{4}\beta(\boldsymbol{S}_A-\boldsymbol{S}'_A)^2\right]^n \mathrm{e}^{-\frac{1}{4}\beta(\boldsymbol{S}_A-\boldsymbol{S}'_A)^2} = \left(-\frac{1}{2}\right)^n (2n+1)!!\left(\frac{4\pi}{\beta}\right)^{3/2}, \tag{9.104}$$

which derives from Eq. (9.50) by repeated differentiation in terms of the coefficient in the exponent. Summarizing these results leads to the desired expression for the matrix elements of $H_{\rm rel}$:

$$\begin{aligned}&\langle\Psi_{SM_STM_T}(\boldsymbol{S}_1,\ldots,\boldsymbol{S}_{A-1})|H_{\rm rel}|\Psi_{S'M'_ST'M'_T}(\boldsymbol{S}'_1,\ldots,\boldsymbol{S}'_{A-1})\rangle\\ &=\left(\frac{\beta}{4\pi}\right)^{3/2}\int d\boldsymbol{S}_A\langle\Phi_{SM_STM_T}(\boldsymbol{s}_1,\ldots,\boldsymbol{s}_A)|H|\Phi_{S'M'_ST'M'_T}(\boldsymbol{s}'_1,\ldots,\boldsymbol{s}'_A)\rangle.\end{aligned} \tag{9.105}$$

To understand the physical meaning of the above derivation, it is useful to reinterpret the integration over $\boldsymbol{S}_A$ as a projection to total momentum zero. Once this is recognized, it becomes obvious that the integral in the second term on the right-hand side of Eq. (9.103) vanishes since the bra is an eigenfunction of the kinetic energy belonging to eigenvalue zero.

The relationship between the integration over $\boldsymbol{S}_A$ and a momentum projection is non-trivial. The projection operator that projects any (translation-noninvariant) state onto a definite total momentum value $\boldsymbol{p}$ is [229]

$$\mathcal{P}_{\boldsymbol{p}} = \frac{1}{(2\pi\hbar)^3}\int d\boldsymbol{a}\,\mathrm{e}^{-\frac{i}{\hbar}\boldsymbol{p}\cdot\boldsymbol{a}}\mathcal{T}_{\boldsymbol{a}}, \tag{9.106}$$

where $\mathcal{T}_{\boldsymbol{a}}$ is an operator that translates the system of particles by the vector $\boldsymbol{a}$.[16] The projection of $\Phi_{SM_STM_T}(\boldsymbol{s}_1,\ldots,\boldsymbol{s}_A)$ onto momentum zero can be evaluated as

$$\begin{aligned}&\mathcal{P}_0\Phi_{SM_STM_T}(\boldsymbol{s}_1,\ldots,\boldsymbol{s}_A)\\ &\quad=\frac{1}{(2\pi\hbar)^3}\int d\boldsymbol{a}\,\Phi_{SM_STM_T}(\boldsymbol{s}_1-\boldsymbol{a},\ldots,\boldsymbol{s}_A-\boldsymbol{a})\\ &\quad=\frac{1}{(2\pi\hbar)^3}\int d\boldsymbol{a}\,\Psi_{SM_STM_T}(\boldsymbol{S}_1,\ldots,\boldsymbol{S}_{A-1})\varphi_{\boldsymbol{S}_A-\sqrt{A}\boldsymbol{a}}(\boldsymbol{x}_A)\end{aligned}$$

[16]The operator $\mathcal{T}_{\boldsymbol{a}}$ acting on the elements Φ of an A-particle state space is defined as $\mathcal{T}_{\boldsymbol{a}}\Phi(\boldsymbol{r}_1,\ldots,\boldsymbol{r}_A)=\Phi(\boldsymbol{r}_1+\boldsymbol{a},\ldots,\boldsymbol{r}_A+\boldsymbol{a})$, and it may be expressed as a product of the s.p. translation operators: $\mathcal{T}_{\boldsymbol{a}}=\prod_{j=1}^A\exp(i\boldsymbol{a}\cdot\hat{\boldsymbol{p}}_j/\hbar)=\exp(i\boldsymbol{a}\cdot\hat{\boldsymbol{P}}/\hbar)$, where $\hat{\boldsymbol{p}}_j$ is the momentum operator of particle j and $\hat{\boldsymbol{P}}=\sum_{j=1}^A\hat{\boldsymbol{p}}_j$ is the total momentum operator. It is then clear that the projection operator $\mathcal{P}_{\boldsymbol{p}}$ can be witten as $\mathcal{P}_{\boldsymbol{p}}=\delta(\hat{\boldsymbol{P}}-\boldsymbol{p})$. From this form it is easy to see that $\mathcal{P}_{\boldsymbol{p}}\mathcal{P}_{\boldsymbol{p}'}=\delta(\boldsymbol{p}-\boldsymbol{p}')\mathcal{P}_{\boldsymbol{p}}$ and $[T_{\rm c.m.},\mathcal{P}_{\boldsymbol{p}}]=0$. It is useful to note that the effect of the s.p. translation operator on a Gaussan wave packet (9.66) can be written as $\exp(i\boldsymbol{a}\cdot\hat{\boldsymbol{p}}_j/\hbar)\varphi_{\boldsymbol{s}_j}(\boldsymbol{r}_j)=\varphi_{\boldsymbol{s}_j-\boldsymbol{a}}(\boldsymbol{r}_j)$. This property is used in Eq. (9.107).

$$= \frac{1}{(2\pi\hbar)^3}\frac{1}{A^{3/2}}\int d\boldsymbol{S}_A\, \Psi_{SM_STM_T}(\boldsymbol{S}_1,\ldots,\boldsymbol{S}_{A-1})\varphi_{\boldsymbol{S}_A}(\boldsymbol{x}_A)$$
$$= \frac{1}{(2\pi\hbar)^3}\frac{1}{A^{3/2}}\int d\boldsymbol{S}_A\, \Phi_{SM_STM_T}(\boldsymbol{s}_1,\ldots,\boldsymbol{s}_A), \qquad (9.107)$$

where the second equality is obtained by using the separability of the c.m. motion [Eq. (9.72)] and the property that the replacement of $\boldsymbol{s}_i \to \boldsymbol{s}_i - \boldsymbol{a}$ is equivalent to changing just the c.m generator coordinate, and the third step is a change of the integration variable, $\boldsymbol{a} \to \boldsymbol{S}_A$. The function (9.107) is essentially the same as is involved in Eq. (9.105), and it shows that the integration over $\boldsymbol{S}_A$ produces a state with total momentum zero.

Since momentum projecting implies gaining translation invariance [cf. footnote after Eq. (9.68)], a procedure based on momentum projecting is suitable for the removal of the c.m. motion in general. Therefore, the application of Eq. (9.105) is not limited to the case of identical Gaussians. It can be used to remove the c.m. motion for systems of clusters of different size parameters as well (cf. Sect. 11.5.2).

9.4.3 Generic forms of determinantal matrix elements

With Eqs. (9.95)–(9.99) at hand, we are ready to derive generic forms for the matrix elements of the most important operators between the corresponding Slater determinants $\phi_\kappa(\boldsymbol{s}_1,\ldots,\boldsymbol{s}_A)$ of Eq. (9.67). The rules of calculating these quantities are well-known [230–232]. Here the generic forms are given without proof, but, for the sake of completeness, some details of their derivations are given in App. E. The operators of our present concern are the unit, the kinetic energy, the squared radius and the potential operators. The potential V_{ij} may depend on the internucleon displacement, on the relative momentum, on the relative orbital angular momentum and on the spins and isospins as well; it may be central or non-central. For the spin–isospin coupled states $\Phi_{SM_STM_T}(\boldsymbol{s}_1,\ldots,\boldsymbol{s}_A)$ of Eq. (9.68) these formulae are as follows:

$$\langle\Phi_{SM_STM_T}(\boldsymbol{s}_1,\ldots,\boldsymbol{s}_A)|\Phi_{SM_STM_T}(\boldsymbol{s}'_1,\ldots,\boldsymbol{s}'_A)\rangle$$
$$= \sum_i C_i^{(o)} \mathrm{e}^{-\frac{1}{2}\tilde{\boldsymbol{t}}A_i^{(o)}\boldsymbol{t}}, \qquad (9.108)$$

$$\langle\Phi_{SM_STM_T}(\boldsymbol{s}_1,\ldots,\boldsymbol{s}_A)|\sum_{i=1}^{A}\frac{\boldsymbol{p}_i^2}{2m}|\Phi_{SM_STM_T}(\boldsymbol{s}'_1,\ldots,\boldsymbol{s}'_A)\rangle$$
$$= \frac{\hbar^2}{2m}\frac{\beta}{2}\sum_i C_i^{(o)}\left(3A - \tilde{\boldsymbol{t}}A_i^{(o)}\boldsymbol{t}\right)\mathrm{e}^{-\frac{1}{2}\tilde{\boldsymbol{t}}A_i^{(o)}\boldsymbol{t}}, \qquad (9.109)$$

$$\langle \Phi_{SM_STM_T}(\boldsymbol{s}_1,\ldots,\boldsymbol{s}_A)|\sum_{i=1}^{A} \boldsymbol{r}_i^2|\Phi_{SM_STM_T}(\boldsymbol{s}'_1,\ldots,\boldsymbol{s}'_A)\rangle$$
$$= \frac{1}{2\beta}\sum_i C_i^{(\mathrm{o})}\left(3A - \tilde{\boldsymbol{t}}A_i^{(\mathrm{o})}\boldsymbol{t} + \beta\tilde{\boldsymbol{t}}\boldsymbol{t}\right)\mathrm{e}^{-\frac{1}{2}\tilde{\boldsymbol{t}}A_i^{(\mathrm{o})}\boldsymbol{t}}, \tag{9.110}$$

$$\langle \Phi_{SM_STM_T}(\boldsymbol{s}_1,\ldots,\boldsymbol{s}_A)|\sum_{i<j}^{A} V_{ij}|\Phi_{S'M'_ST'M'_T}(\boldsymbol{s}'_1,\ldots,\boldsymbol{s}'_A)\rangle$$
$$= \left(\frac{\beta}{2\pi}\right)^{3/2}\int d\boldsymbol{r}\, V(r)\mathrm{e}^{-\frac{1}{2}\beta r^2}\sum_i C_i^{(\mathrm{p})}P_i(\boldsymbol{t},\boldsymbol{r})\mathrm{e}^{-\frac{1}{2}\tilde{\boldsymbol{t}}A_i^{(\mathrm{p})}\boldsymbol{t}+\boldsymbol{d}_i\cdot\boldsymbol{r}}, \tag{9.111}$$

where $\tilde{\boldsymbol{t}}$ is a one-row matrix of $2A$ Cartesian vectors,

$$\tilde{\boldsymbol{t}} \equiv (\boldsymbol{t}_1,\ldots,\boldsymbol{t}_{2A}) = (\boldsymbol{s}_1,\ldots,\boldsymbol{s}_A,\boldsymbol{s}'_1,\ldots,\boldsymbol{s}'_A), \tag{9.112}$$

$A_i^{(\mathrm{o})}$ and $A_i^{(\mathrm{p})}$ are $2A\times 2A$ real symmetric matrices, $C_i^{(\mathrm{o})}$ are combination coefficients, $V(r)$ is the radial form factor of the interaction, $P_i(\boldsymbol{t},\boldsymbol{r})$ are polynomials of $\boldsymbol{t}$ and $\boldsymbol{r}$, and $\boldsymbol{d}_i$ are linear combinations of $\boldsymbol{t}_j$, so that $\boldsymbol{d}_i\cdot\boldsymbol{r}$ has the form $\boldsymbol{d}_i\cdot\boldsymbol{r} = \sum_{j=1}^{2A}\mathcal{D}_{(i)j}\boldsymbol{t}_j\cdot\boldsymbol{r}$. The polynomials reduce to unity $[P_i(\boldsymbol{t},\boldsymbol{r}) = 1]$ for angular-momentum independent forces, and they are very simple for the spin-orbit (as well as for tensor and other standard angular-momentum dependent) interactions (see App. E). The constants in $C_i^{(\mathrm{o})}$, $A_i^{(\mathrm{o})}$, $C_i^{(\mathrm{p})}$ and $A_i^{(\mathrm{p})}$ etc. are most conveniently calculated for each particular system by computer algebra.

9.4.4 Translation-invariant matrix elements

We have now presented matrix elements between states containing the c.m. motion. It remains to show how to remove the c.m. motion from Eqs. (9.108)–(9.111).

Each operator of our present interest can be written as a sum of an intrinsic term and of a c.m. term via a Jacobi transformation. Then each matrix element will be a sum of two terms, like Eq. (9.103) for the Hamiltonian, and each term is factored, just as Eq. (9.103), into an overlap times a matrix element. According to Eq. (9.95), the overlap $\langle\varphi_{\boldsymbol{S}_A}|\varphi_{\boldsymbol{S}'_A}\rangle$ in the first term is

$$\langle\varphi_{\boldsymbol{S}_A}|\varphi_{\boldsymbol{S}'_A}\rangle = \exp\left[-\frac{\beta}{4}(\boldsymbol{S}_A - \boldsymbol{S}'_A)^2\right], \tag{9.113}$$

which contributes to the exponents of Eqs. (9.108)–(9.111) with the same term. The Jacobi transformation will cause these exponents to separate into intrinsic and c.m. terms:

$$\tilde{\boldsymbol{t}}A_i^{(k)}\boldsymbol{t} = \tilde{\boldsymbol{T}}B_i^{(k)}\boldsymbol{T} + \tfrac{1}{2}\beta(\boldsymbol{S}_A - \boldsymbol{S}'_A)^2 \quad (k = \mathrm{o},\mathrm{p}). \tag{9.114}$$

Here $\boldsymbol{T}$ is the matrix of vectors introduced in Eq. (9.92), and $B_i^{(k)}$ is the $(2A-2)\times(2A-2)$ symmetric matrix defined by dropping the Ath and $(2A)$th rows and columns of the matrix $\mathcal{U}A_i^{(k)}\tilde{\mathcal{U}}$, where

$$\mathcal{U} = \begin{pmatrix} U & 0 \\ 0 & U \end{pmatrix}. \tag{9.115}$$

It is clear from Eq. (9.103) that the polynomials $P_i(\boldsymbol{t},\boldsymbol{r})$ and the vectors $\boldsymbol{d}_i$ defined in Eq. (9.111) can only depend on the relative generator coordinates. The dependence of the matrix elements on the c.m. variables, $\boldsymbol{S}_A$ and $\boldsymbol{S}'_A$, can thus be factored out and eliminated via Eq. (9.105). Then the matrix elements can be expressed in terms of the Jacobi generator coordinates:

$$\langle\Psi_{SM_STM_T}(\boldsymbol{S}_1,\ldots,\boldsymbol{S}_{A-1})|\Psi_{SM_STM_T}(\boldsymbol{S}'_1,\ldots,\boldsymbol{S}'_{A-1})\rangle$$
$$= \sum_i C_i^{(o)} \mathrm{e}^{-\frac{1}{2}\tilde{\boldsymbol{T}}B_i^{(o)}\boldsymbol{T}}, \tag{9.116}$$

$$\langle\Psi_{SM_STM_T}(\boldsymbol{S}_1,\ldots,\boldsymbol{S}_{A-1})|\sum_{i=1}^{A-1}\frac{\boldsymbol{\pi}_i^2}{2m}|\Psi_{SM_STM_T}(\boldsymbol{S}'_1,\ldots,\boldsymbol{S}'_{A-1})\rangle$$
$$= \frac{\hbar^2}{2m}\frac{\beta}{2}\sum_i C_i^{(o)}\left[3(A-1) - \tilde{\boldsymbol{T}}B_i^{(o)}\boldsymbol{T}\right]\mathrm{e}^{-\frac{1}{2}\tilde{\boldsymbol{T}}B_i^{(o)}\boldsymbol{T}}, \tag{9.117}$$

$$\langle\Psi_{SM_STM_T}(\boldsymbol{S}_1,\ldots,\boldsymbol{S}_{A-1})|\sum_{i=1}^{A-1}\boldsymbol{x}_i^2|\Psi_{S'M'_ST'M'_T}(\boldsymbol{S}'_1,\ldots,\boldsymbol{S}'_{A-1})\rangle$$
$$= \frac{1}{2\beta}\sum_i C_i^{(o)}\left[3(A-1) - \tilde{\boldsymbol{T}}B_i^{(o)}\boldsymbol{T} + \beta\tilde{\boldsymbol{T}}\boldsymbol{T}\right]\mathrm{e}^{-\frac{1}{2}\tilde{\boldsymbol{T}}B_i^{(o)}\boldsymbol{T}}, \tag{9.118}$$

$$\langle\Psi_{SM_STM_T}(\boldsymbol{S}_1,\ldots,\boldsymbol{S}_{A-1})|\sum_{i<j}^{A}V_{ij}|\Psi_{S'M'_ST'M'_T}(\boldsymbol{S}'_1,\ldots,\boldsymbol{S}'_{A-1})\rangle$$
$$= \left(\frac{\beta}{2\pi}\right)^{3/2}\int d\boldsymbol{r}\, V(r)\mathrm{e}^{-\frac{1}{2}\beta r^2}\sum_i C_i^{(p)}P_i(\boldsymbol{T},\boldsymbol{r})\mathrm{e}^{-\frac{1}{2}\tilde{\boldsymbol{T}}B_i^{(p)}\boldsymbol{T}+\boldsymbol{D}_i\cdot\boldsymbol{r}}. \tag{9.119}$$

Here the vector $\boldsymbol{D}_i$ is given by $\sum_{j=1}^{2A-2}\widehat{D}_{(i)j}\boldsymbol{T}_j$, where $\widehat{D}_{(i)j}$ are formed out of the elements of the row vector $\tilde{\mathcal{D}}_{(i)}\tilde{\mathcal{U}}$ by omitting its Ath and $(2A)$th columns.

Equations (9.116)–(9.119) show that the $\boldsymbol{T}$-dependence of the matrix elements is rather simple and the integration over $\boldsymbol{T}$ contained in Eq. (9.91) is to be performed analytically with the formulae given in Eqs. (9.61)–(9.64). The variables α, α', $\boldsymbol{e}$ and $\boldsymbol{e}'$ enter only through the vector $\boldsymbol{X}$, and the operations on these variables contained in Eq. (9.91)

$$M = \mathfrak{C}\{\mathfrak{P}[\mathfrak{G}(\mathfrak{T}\{\mathfrak{E}[\mathfrak{D}(m)]\})]\}$$

Operation	$\mathfrak{C}$	$\mathfrak{P}$ $\mathfrak{G}$ $\mathfrak{T}$	$\mathfrak{E}$	$\mathfrak{D}$	m
Location	Eq. (9.89)	App. F	Sect. 9.4.4	App. E	Eqs. (9.95)–(9.99)

Explanation

- m: elementary matrix elements of the type of Eqs. (9.95)–(9.99)
- $\mathfrak{D}$: (Gothic D) operation that, acting on m, creates determinantal matrix elements $\langle\Psi_{SM_STM_T}(s_1,\ldots,s_A)|O|\Psi_{S'M'_ST'M'_T}(s'_1,\ldots,s'_A)\rangle$ (9.108)–(9.111)
- $\mathfrak{E}$: (Gothic E) operation that eliminates the c.m. motion [e.g., Eq. (9.105)]
- $\mathfrak{T}$: (Gothic T) integral transformation which involves integration over $\boldsymbol{T}$ [Eq. (9.91)];
- $\mathfrak{G}$: (Gothic G) operation that generates matrix elements belonging to a certain number of excitation quanta [differentiation and setting the variables to 0; cf. Eq. (9.91)]
- $\mathfrak{P}$: (Gothic P) operation that provides for orbital angular momentum projection [cf. Eq. (9.91)]
- $\mathfrak{C}$: (Gothic C) operation that takes care of angular-momentum and isospin coupling [as in Eq. (9.89)]

Figure 9.3: Schematic layout of the steps of evaluation of the matrix elements $M = \mathfrak{C}\Big\{\mathfrak{P}\Big[\mathfrak{G}\Big(\mathfrak{T}\{\mathfrak{E}[\mathfrak{D}(m)]\}\Big)\Big]\Big\}$ defined in Eq. (9.89).

are also to be performed analytically. The procedure is elaborated further in App. F, but details are only given for the simplest matrix element, the overlap.

The fairly complex procedure of evaluating matrix elements between correlated Gaussians involves several steps and some details are relegated to the appendices. The logic and the whereabouts of each step of calculations are summarized in Fig. 9.3.

Concluding this subsection, we mention that, although the formalism has been elaborated with the use of a particular set of relative coordinates, a Jacobi set, that was only for simplicity's sake. It should be emphasized that matrix elements between basis functions with any sets of relative coordinates can be evaluated in this unified framework without extra coordinate transformations. This is so because of a form invariance of the correlated Gaussians with respect to transformations on the relative coordinates. Between any two sets of relative vectors, $\boldsymbol{x}$

and x', there exists a transformation matrix T, such that $x = Tx'$. By form invariance we mean the property of the correlated Gaussians that the acting of T on the matrices A and u,

$$A' = \tilde{T}AT, \qquad u' = Tu, \tag{9.120}$$

leaves its form entirely unaltered [cf. Eq. (9.55)]:

$$f_{KLM}(u, A, x) = f_{KLM}(u', A', x'). \tag{9.121}$$

Another important property is that in a cluster-model framework some of the generator coordinates are set to coincide (see Sect. 12.1); this will reduce the dimension of T but will not change the structure, and thus the calculation, of any of the matrix elements.

9.5 Physical quantities

In Sect. 9.4 some generic matrix elements have been evaluated between correlated Gaussian states. Now the matrix elements of the operators of some of the important physical quantities will be derived. Some derivations will be carried through explicitly, while only hints will be given in the case of some others. In the course of this elaboration most of the important observables will be defined.

All operators of our concern have tensorial forms, which can be decoupled according to Eq. (9.90). In Sects. 9.5.1 and 9.5.2 we give examples for s.p. and two-particle physical observables, respectively. By physical observables we mean permutation-invariant observables.[17] The properties that characterize the relative motion of subsystems, and hence are not permutation-invariant, will be discussed in Sect. 11.3 later.

9.5.1 One-body operators

Among the important s.p. (or one-body) operators of a nucleus are those whose expectation values characterize its size: the squared radii of its proton, neutron and matter distributions.[18]

[17]The eigenfunctions of coordinate operators are Dirac-δ's, which are not elements of the Hilbert space. Following Ref. [26], we will yet regard them as observables, although, in an orthodox sense, they are not observables.

[18]The radii considered throughout this book are radii of point-particle distributions. The charge density is obtained by folding the point-proton distribution with the intrinsic charge distribution of the proton. It can be shown that the corresponding squared radii are additive: $\langle r^2_{\text{charge}}\rangle = \langle r^2_{\text{point}}\rangle + \langle r^2_{\text{intrinsic}}\rangle$, where $\langle r^2_{\text{intrinsic}}\rangle = 0.74\ \text{fm}^2$.

To define quantities for protons, neutrons and nucleons simultaneously, we introduce three projection operators P_i as follows:

$$P_i^{\rm p} = \tfrac{1}{2}[1-(\tau_i)_z], \quad P_i^{\rm n} = \tfrac{1}{2}[1+(\tau_i)_z], \quad P_i^{\rm N} = P_i^{\rm p} + P_i^{\rm n} = 1, \tag{9.122}$$

where $(\tau_i)_z$ is the Pauli operator of the isospin z-projection of particle i.

The operators for the point-proton, point-neutron and point-matter square radii are, with superscript p, n and N, respectively,

$$O_{00} = \frac{1}{n}\sum_{i=1}^{A} \bar{r}_i^2 P_i, \tag{9.123}$$

where n is the number of non-zero terms in the sum (Z, N or A for $P_i = P_i^{\rm p}$, $P_i^{\rm n}$ and $P_i^{\rm N}$, respectively), $\bar{\boldsymbol{r}}_i$ is the vector connecting the c.m. with particle i,

$$\bar{\boldsymbol{r}}_i = \boldsymbol{r}_i - \boldsymbol{R}_{\rm c.m.}, \tag{9.124}$$

and $\boldsymbol{R}_{\rm c.m.}$ is the vector pointing to the c.m., defined in Eq. (9.11). As another important operator, the electric 2^L-pole operator should be mentioned, which enters into the formulae of the electric 2^L-pole moments and transition rates:

$$O_{LM} = \sum_{i=1}^{A} \bar{r}_i^L Y_{LM}(\hat{\bar{\boldsymbol{r}}}_i) P_i^{\rm p}. \tag{9.125}$$

The longitudinal electron scattering form factors are matrix elements of the operator

$$O_{LM} = \sum_{i=1}^{A} j_L(k\bar{r}_i) Y_{LM}(\hat{\bar{\boldsymbol{r}}}_i) P_i^{\rm p}, \tag{9.126}$$

where k is the magnitude of momentum transfer [cf. Eq. (2.34)] and $j_L(z)$ is a spherical Bessel function of the first kind [43].

Underlying these operators are the nuclear densities, which are less directly accessible to experiment. The s.p. density distribution in a state Ψ is defined as

$$\rho(\boldsymbol{r}) = \langle\Psi|\sum_i \delta(\bar{\boldsymbol{r}}_i - \boldsymbol{r}) P_i|\Psi\rangle. \tag{9.127}$$

This is normalized such that $n \equiv \int d\boldsymbol{r}\rho(\boldsymbol{r})$ is equal to Z, N or A for $P_i = P_i^{\rm p}$, $P_i^{\rm n}$ and $P_i^{\rm N}$, respectively. Owing to the full (anti)symmetry of the state Ψ, this can be rewritten identically as

$$\rho(\boldsymbol{r}) = n\langle\Psi|\delta(\bar{\boldsymbol{r}}_1 - \boldsymbol{r}) P_1|\Psi\rangle. \tag{9.128}$$

Here the bra-ket implies integration over an independent set of the $A-1$ vectorial variables of the ordinary space and an 'integration' (actually a summation) over all the A spin–isospin coordinates, ξ. Let us choose the spatial variables such that they contain $\bar{\boldsymbol{r}}_1$, and let us denote the other $A-2$ coordinates by η. Writing these explicitly yields

$$\begin{aligned}\rho(\boldsymbol{r}) &= n\int d\bar{\boldsymbol{r}}_1\int d\xi\int d\eta\, \Psi^*(\bar{\boldsymbol{r}}_1,\xi,\eta)P_1\delta(\bar{\boldsymbol{r}}_1-\boldsymbol{r})\Psi(\bar{\boldsymbol{r}}_1,\xi,\eta)\\ &= n\int d\xi\int d\eta\, \Psi^*(\boldsymbol{r},\xi,\eta)P_1\Psi(\boldsymbol{r},\xi,\eta). \end{aligned}\tag{9.129}$$

This form is useful to relate $\rho(\boldsymbol{r})$ to the s.p. density matrix $\rho[\boldsymbol{r},\boldsymbol{r}']$ (see, e.g., [233])

$$\rho[\boldsymbol{r},\boldsymbol{r}'] = n\int d\xi\int d\eta\, \Psi^*(\boldsymbol{r}',\xi,\eta)P_1\Psi(\boldsymbol{r},\xi,\eta). \tag{9.130}$$

Obviously, $\rho(\boldsymbol{r}) = \rho[\boldsymbol{r},\boldsymbol{r}]$.

By analogy, one can define the momentum density distribution as

$$\varrho(\boldsymbol{k}) = \langle\Psi|\sum_i \delta(\bar{\boldsymbol{k}}_i-\boldsymbol{k})P_i|\Psi\rangle = n\langle\Psi|\delta(\bar{\boldsymbol{k}}_1-\boldsymbol{k})P_1|\Psi\rangle, \tag{9.131}$$

where $\hbar\bar{\boldsymbol{k}}_i$ is the momentum canonically conjugate to $\bar{\boldsymbol{r}}_i$. To arrive at a more transparent formula, insert $\int d\boldsymbol{K}\,|\boldsymbol{K}\rangle\langle\boldsymbol{K}|$ and $\int d\boldsymbol{K}'\,|\boldsymbol{K}'\rangle\langle\boldsymbol{K}'|$ in front of and behind $\delta(\bar{\boldsymbol{k}}_1-\boldsymbol{k})P_1$, respectively, where $|\boldsymbol{K}\rangle$ is a state in the space spanned by $\bar{\boldsymbol{r}}_i$, viz. an eigenstate of $\bar{\boldsymbol{k}}_1$:

$$\begin{aligned}\varrho(\boldsymbol{k}) &= n\int d\boldsymbol{K}\int d\boldsymbol{K}'\,\langle\Psi|\boldsymbol{K}\rangle\langle\boldsymbol{K}|\delta(\bar{\boldsymbol{k}}_1-\boldsymbol{k})P_1|\boldsymbol{K}'\rangle\langle\boldsymbol{K}'|\Psi\rangle\\ &= n\langle\Psi|\boldsymbol{k}\rangle P_1\langle\boldsymbol{k}|\Psi\rangle. \end{aligned}\tag{9.132}$$

To see the meaning of this formula, one has to insert further unit operators:

$$\begin{aligned}\varrho(\boldsymbol{k}) &= n\int d\boldsymbol{r}'\int d\xi\int d\eta\int d\boldsymbol{r}\,\langle\Psi|\boldsymbol{r}'\xi\eta\rangle\langle\boldsymbol{r}'|\boldsymbol{k}\rangle P_1\langle\boldsymbol{k}|\boldsymbol{r}\rangle\langle\boldsymbol{r}\xi\eta|\Psi\rangle\\ &= \int d\boldsymbol{r}'\int d\boldsymbol{r}\,\frac{\mathrm{e}^{i\boldsymbol{k}\cdot\boldsymbol{r}'}}{(2\pi)^{3/2}}\rho[\boldsymbol{r},\boldsymbol{r}']\frac{\mathrm{e}^{-i\boldsymbol{k}\cdot\boldsymbol{r}}}{(2\pi)^{3/2}}. \end{aligned}\tag{9.133}$$

Thus the momentum distribution is a double Fourier transform of the one-particle density matrix.

The momentum structure of a nucleus is in fact more often characterized by the longitudinal form factor, which, for an elastic transition, is the Fourier transform of the ordinary s.p. density [apart from a factor of $(2\pi)^{-3/2}$]:

$$F(\boldsymbol{k}) = \int d\boldsymbol{r}\,\mathrm{e}^{i\boldsymbol{k}\cdot\boldsymbol{r}}\rho(\boldsymbol{r}). \tag{9.134}$$

If $\rho(\boldsymbol{r})$ is the proton s.p. distribution, $F(\boldsymbol{k})$ is the longitudinal charge (or Coulomb) form factor[19] [233]. In the one-photon-exchange (i.e., Born-type) and no-recoil approximation to the electron elastic scattering cross section $F(\boldsymbol{k})$ can be extracted from the angular distribution data. Using the well-known multipole expansion formula,

$$e^{i\boldsymbol{k}\cdot\boldsymbol{r}} = 4\pi \sum_{lm} i^l Y^*_{lm}(\hat{\boldsymbol{k}}) j_l(kr) Y_{lm}(\hat{\boldsymbol{r}}), \tag{9.135}$$

we see that the expectation value of the operator in Eq. (9.126) yields just a multipole of Eq. (9.134):

$$\langle\Psi|O_{LM}|\Psi\rangle = \int d\boldsymbol{r}\, j_L(kr) Y_{LM}(\hat{\boldsymbol{r}}) \rho(\boldsymbol{r}) = \frac{1}{4\pi i^L} \int d\hat{\boldsymbol{k}}\, Y_{LM}(\hat{\boldsymbol{k}}) F(\boldsymbol{k}). \tag{9.136}$$

Although the form factor is informative of the momentum content of the wave function, yet it is not to be identified with a momentum distribution. The difference between the two quantities is most plausible for a s.p. system. The momentum density distribution described by the wave function $\psi(\boldsymbol{r})$ is $(2\pi)^{-3}|\int d\boldsymbol{r} e^{-i\boldsymbol{k}\cdot\boldsymbol{r}} \psi(\boldsymbol{r})|^2$, while the form factor is $\int d\boldsymbol{r} e^{i\boldsymbol{k}\cdot\boldsymbol{r}} |\psi(\boldsymbol{r})|^2$.

All operators involved in the above considerations can be expressed in terms of the density operators defined by

$$\hat{\rho}(\boldsymbol{r}) = \sum_{i=1}^{A} \delta(\bar{\boldsymbol{r}}_i - \boldsymbol{r}) P_i \tag{9.137}$$

as follows:

$$\begin{aligned} O_{\kappa\mu} &\equiv \sum_{i=1}^{A} f(\bar{r}_i) Y_{\kappa\mu}(\hat{\bar{\boldsymbol{r}}}_i) P_i \\ &= \int d\boldsymbol{r}\, f(r) Y_{\kappa\mu}(\hat{\boldsymbol{r}}) \sum_{i=1}^{A} \delta(\bar{\boldsymbol{r}}_i - \boldsymbol{r}) P_i \\ &= \int d\boldsymbol{r}\, f(r) Y_{\kappa\mu}(\hat{\boldsymbol{r}}) \hat{\rho}(\boldsymbol{r}). \end{aligned} \tag{9.138}$$

The projector P_i may be generalized to project out the spin-up or spin-down state together with, or without, the isospin z-projection. By choosing appropriate forms for the function $f(r)$ and for the projector P_i, it is possible to generate almost all operators of physical interest.

[19] The longitudinal form factor is sometimes [234] defined as $Z^{-1}F(\boldsymbol{k})$. It is called longitudinal because, in the description of electron scattering, it appears as a result of the photon component parallel to the momentum transfer $\boldsymbol{k}$. The contributions of the other photon components are called transverse components.

The basic matrix element needed is therefore that of $\hat{\rho}(\boldsymbol{r})$. Relying on Eq. (9.84), we consider matrix elements of $\hat{\rho}(\boldsymbol{r})$ between Gaussian wave packets [cf. Eq. (9.127)]:

$$\rho(\boldsymbol{r}) = \langle \Psi_{SM_STM_T}(\boldsymbol{S}_1,\ldots,\boldsymbol{S}_{A-1})|\hat{\rho}(\boldsymbol{r})|\Psi_{S'M'_ST'M'_T}(\boldsymbol{S}'_1,\ldots,\boldsymbol{S}'_{A-1})\rangle \tag{9.139}$$

$$= \sum_{i=1}^{A} \frac{1}{(2\pi)^3} \int d\boldsymbol{k}\, \langle \Psi_{SM_STM_T}(\boldsymbol{S}_1,\ldots,\boldsymbol{S}_{A-1})| \times e^{i\boldsymbol{k}\cdot(\tilde{\boldsymbol{r}}_i-\boldsymbol{r})} P_i |\Psi_{S'M'_ST'M'_T}(\boldsymbol{S}'_1,\ldots,\boldsymbol{S}'_{A-1})\rangle, \tag{9.140}$$

where the δ-function has been expanded in terms of plane waves. The factoring of the wave function (9.72) into intrinsic and c.m. parts enables one to evaluate the matrix element in Eq. (9.140) in terms of s.p. coordinates. To show this, let us write down a corresponding matrix element between wave functions of the type of Eq. (9.72):

$$\begin{aligned}\langle \Phi_{SM_STM_T}(\boldsymbol{s}_1,\ldots,\boldsymbol{s}_A)|e^{i\boldsymbol{k}\cdot(\boldsymbol{r}_i-\boldsymbol{r})}P_i|\Phi_{S'M'_ST'M'_T}(\boldsymbol{s}'_1,\ldots,\boldsymbol{s}'_A)\rangle \\ = \langle \Psi_{SM_STM_T}(\boldsymbol{S}_1,\ldots,\boldsymbol{S}_{A-1})| \\ \times e^{i\boldsymbol{k}\cdot(\boldsymbol{r}_i-\boldsymbol{R}_{\text{c.m.}}-\boldsymbol{r})}P_i|\Psi_{S'M'_ST'M'_T}(\boldsymbol{S}'_1,\ldots,\boldsymbol{S}'_{A-1})\rangle \\ \times\langle \varphi_{\boldsymbol{S}_A}|e^{i\boldsymbol{k}\cdot\boldsymbol{R}_{\text{c.m.}}}|\varphi_{\boldsymbol{S}'_A}\rangle.\end{aligned} \tag{9.141}$$

The c.m. variables $\boldsymbol{S}_A$ and $\boldsymbol{S}'_A$ can be chosen to be equal and, for that matter, to be zero ($\sum_{i=1}^{A}\boldsymbol{s}_i = \sum_{i=1}^{A}\boldsymbol{s}'_i = 0$), without any loss of generality, and, from now on, we will do so. Then, with the use of Eq. (9.50), the last factor of the matrix element can be shown to be $\exp\left(-\boldsymbol{k}^2/4A\beta\right)$. One thus gets

$$\begin{aligned}\rho(\boldsymbol{r}) &= \sum_{i=1}^{A} \frac{1}{(2\pi)^3}\int d\boldsymbol{k}\, \exp\left(\frac{\boldsymbol{k}^2}{4A\beta}\right) \\ &\quad\times\langle \Phi_{SM_STM_T}(\boldsymbol{s}_1,\ldots,\boldsymbol{s}_A)|e^{i\boldsymbol{k}\cdot(\boldsymbol{r}_i-\boldsymbol{r})}P_i|\Phi_{S'M'_ST'M'_T}(\boldsymbol{s}'_1,\ldots,\boldsymbol{s}'_A)\rangle \\ &\equiv \langle \Phi_{SM_STM_T}(\boldsymbol{s}_1,\ldots,\boldsymbol{s}_A)|\sum_{i=1}^{A} o_i|\Phi_{S'M'_ST'M'_T}(\boldsymbol{s}'_1,\ldots,\boldsymbol{s}'_A)\rangle,\end{aligned} \tag{9.142}$$

where $\sum_i o_i$ is a density operator corrected for the c.m. 'correlation':

$$o_i = \frac{1}{(2\pi)^3}\int d\boldsymbol{k}\, \exp\left(\frac{\boldsymbol{k}^2}{4A\beta}\right) e^{i\boldsymbol{k}\cdot(\boldsymbol{r}_i-\boldsymbol{r})}P_i. \tag{9.143}$$

Equation (9.142) may be interpreted as saying that the c.m.-free form of the density matrix element is the matrix element of a redefined one-body operator, (9.143), between A-nucleon wave functions which include the c.m. degree of freedom. Note, however, that Eq. (9.142) is a very formal expression, which is not quite correct mathematically. It is easy to see that the integral in Eq. (9.143) diverges, thus o_i is ill-defined. The matter is that the sequence of integrations over the physical coordinates and over $\boldsymbol{k}$ must not have been reversed; when performed in the original sequence, the integrals become convergent owing to the fact that $\langle \cdots | e^{i\boldsymbol{k}\cdot(\boldsymbol{r}_i-\boldsymbol{r})} P_i | \cdots \rangle$ in the second line of Eq. (9.142) goes to zero for large $|\boldsymbol{k}|$ rapidly. All in all, in Eq. (9.142) the order of integration should be as indicated in the first two lines.

It is now straightforward to evaluate a matrix element of Eq. (9.143) along the lines of Sect. 9.4, with the correct sequence of integrations:

$$
\begin{aligned}
&\langle \varphi_{s_1}\chi_{\frac{1}{2}m_s^1}\eta_{\frac{1}{2}m_t^1}|o_1|\varphi_{s_2}\chi_{\frac{1}{2}m_s^2}\eta_{\frac{1}{2}m_t^2}\rangle \\
&\equiv \frac{1}{(2\pi)^3}\int d\boldsymbol{k}\, \exp\left(\frac{\boldsymbol{k}^2}{4A\beta}\right)\langle \varphi_{s_1}|e^{i\boldsymbol{k}\cdot(\boldsymbol{r}_1-\boldsymbol{r})}|\varphi_{s_2}\rangle\langle \chi_{\frac{1}{2}m_s^1}\eta_{\frac{1}{2}m_t^1}|P_1|\chi_{\frac{1}{2}m_s^2}\eta_{\frac{1}{2}m_t^2}\rangle \\
&= \left(\frac{A}{A-1}\frac{\beta}{\pi}\right)^{3/2} \exp\left[-\frac{A}{A-1}\beta\left(\boldsymbol{r}-\frac{\boldsymbol{s}_1+\boldsymbol{s}_2}{2}\right)^2\right]\langle \varphi_{s_1}|\varphi_{s_2}\rangle \\
&\times \langle \chi_{\frac{1}{2}m_s^1}\eta_{\frac{1}{2}m_t^1}|P_1|\chi_{\frac{1}{2}m_s^2}\eta_{\frac{1}{2}m_t^2}\rangle,
\end{aligned}
\tag{9.144}
$$

where the second equality may be verified by using Eq. (9.50). •

In the point-nucleon approximation the magnetic moment and magnetic dipole transition rate are matrix elements of the magnetic dipole operator

$$
\boldsymbol{\mu} = \mu_N \sum_{i=1}^{A}\left(g_{l_i}\boldsymbol{l}_i + g_{s_i}\frac{1}{2}\boldsymbol{\sigma}_i\right),
\tag{9.145}
$$

where $\mu_N = e\hbar/2mc$ is the nuclear magneton (e and m are the charge and mass of the proton, c is the speed of light), and the g-factors are

$$
g_{l_i} = P_i^{\mathrm{p}}, \qquad g_{s_i} = 5.586P_i^{\mathrm{p}} - 3.826P_i^{\mathrm{n}}.
\tag{9.146}
$$

The operator of Eq. (9.145) contains both isoscalar and isovector terms [cf. Eq. (13.4)], which are vectors either in the ordinary or in the spin space. The treatment of the spin term is simple; problems seem to arise only with the $\boldsymbol{l}_i$ term. Strictly speaking, the angular momentum $\boldsymbol{l}_i$ is to be defined relative to the c.m. of the nucleus. It can, however, be replaced effectively by the s.p. angular momentum provided that in the wave functions (9.68), which contain the c.m. motion, $\sum_{i=1}^{A} \boldsymbol{s}_i = 0$ is chosen. Then they have the properties $\langle \boldsymbol{R}_{\mathrm{c.m.}}\rangle = 0$, $\langle \boldsymbol{\pi}_A\rangle = 0$ and $\langle \boldsymbol{R}_{\mathrm{c.m.}} \times \boldsymbol{\pi}_A\rangle = 0$, where $\langle\cdots\rangle$ denotes an expectation value. Therefore, any matrix element of $\boldsymbol{\mu}$ between generator-coordinate functions of the type of Eq. (9.73) reduces to ones which involve the s.p. operators $\boldsymbol{l}_i = -i\boldsymbol{r}_i \times \boldsymbol{\nabla}_i$. A s.p. matrix element of $\boldsymbol{l}_1$ can be evaluated, e.g., through the following steps:

$$\langle\varphi_{s_1}|\boldsymbol{r}_1\times\boldsymbol{\nabla}_1|\varphi_{s_2}\rangle = -\beta\langle\varphi_{s_1}|\boldsymbol{r}_1\times(\boldsymbol{r}_1-\boldsymbol{s}_2)|\varphi_{s_2}\rangle = \beta\langle\varphi_{s_1}|\boldsymbol{r}_1|\varphi_{s_2}\rangle\times\boldsymbol{s}_2,$$
$$\langle\varphi_{s_1}|\boldsymbol{r}_1|\varphi_{s_2}\rangle = \langle\varphi_{s_1}|\boldsymbol{r}_1-\boldsymbol{s}_1|\varphi_{s_2}\rangle + \boldsymbol{s}_1\langle\varphi_{s_1}|\varphi_{s_2}\rangle$$
$$= \beta^{-1}\boldsymbol{\nabla}_{s_1}\langle\varphi_{s_1}|\varphi_{s_2}\rangle + \boldsymbol{s}_1\langle\varphi_{s_1}|\varphi_{s_2}\rangle = \frac{\boldsymbol{s}_1+\boldsymbol{s}_2}{2}\langle\varphi_{s_1}|\varphi_{s_2}\rangle, \quad (9.147)$$

where Eq. (9.95) is also used, and the result is

$$\langle\varphi_{s_1}|\boldsymbol{l}_1|\varphi_{s_2}\rangle = -\frac{1}{2}i\beta(\boldsymbol{s}_1\times\boldsymbol{s}_2)\langle\varphi_{s_1}|\varphi_{s_2}\rangle. \quad (9.148)$$

The transverse electron scattering form factors are matrix elements of the current density operator $\hat{\boldsymbol{j}}(\boldsymbol{r})$, which consists of the convection and magnetization currents:

$$\hat{\boldsymbol{j}}(\boldsymbol{r}) = \hat{\boldsymbol{j}}^{\mathrm{c}}(\boldsymbol{r}) + \hat{\boldsymbol{j}}^{\mathrm{m}}(\boldsymbol{r}), \quad (9.149)$$
$$\hat{\boldsymbol{j}}^{\mathrm{c}}(\boldsymbol{r}) = \frac{e}{2m}\sum_{i=1}^{A}\left[\bar{\boldsymbol{p}}_i\delta(\bar{\boldsymbol{r}}_i-\boldsymbol{r}) + \delta(\bar{\boldsymbol{r}}_i-\boldsymbol{r})\bar{\boldsymbol{p}}_i\right]P_i^{\mathrm{p}}, \quad (9.150)$$
$$\hat{\boldsymbol{j}}^{\mathrm{m}}(\boldsymbol{r}) = \boldsymbol{\nabla}\times\left[\frac{e\hbar}{2m}\sum_{i=1}^{A}\delta(\bar{\boldsymbol{r}}_i-\boldsymbol{r})g_{s_i}\frac{1}{2}\boldsymbol{\sigma}_i\right]. \quad (9.151)$$

Here $\bar{\boldsymbol{p}}_i$ is the momentum of the nucleon in the c.m. system and $\boldsymbol{\nabla}$ is understood to act on the variable $\boldsymbol{r}$. The momenta $\bar{\boldsymbol{p}}_i$ can be expressed in terms of $\boldsymbol{p}_i$, the momenta in the laboratory frame, with the trick used in Eq. (9.142). For a matrix element

$$\boldsymbol{j}(\boldsymbol{r}) = \boldsymbol{j}^{\mathrm{c}}(\boldsymbol{r}) + \boldsymbol{\nabla}\times\boldsymbol{j}^{\mathrm{m}}(\boldsymbol{r}) \quad (9.152)$$

of the current $\hat{\boldsymbol{j}}^{\mathrm{c}}(\boldsymbol{r}) + \hat{\boldsymbol{j}}^{\mathrm{m}}(\boldsymbol{r})$ [Eq. (9.149)], one can write

$$\boldsymbol{j}^k(\boldsymbol{r}) = \langle\Psi_{SM_STM_T}(\boldsymbol{S}_1,\ldots,\boldsymbol{S}_{A-1})|\hat{\boldsymbol{j}}^k(\boldsymbol{r})|\Psi_{S'M_S'T'M_T'}(\boldsymbol{S}_1',\ldots,\boldsymbol{S}_{A-1}')\rangle$$
$$= \langle\Phi_{SM_STM_T}(\boldsymbol{s}_1,\ldots,\boldsymbol{s}_A)|\sum_{i=1}^{A}\boldsymbol{o}_i^k|\Phi_{S'M_S'T'M_T'}(\boldsymbol{s}_1',\ldots,\boldsymbol{s}_A')\rangle$$
$$(k = \mathrm{c},\ \mathrm{m}), \quad (9.153)$$

where the 'effective operators' $\boldsymbol{o}_i^k$ [cf. Eq. (9.143)] contain the ordinary momenta $\boldsymbol{p}_i = -i\hbar\boldsymbol{\nabla}_{\boldsymbol{r}_i}$:

$$\boldsymbol{o}_i^{\mathrm{c}} = -\frac{1}{(2\pi)^3}ic\mu_N\int d\boldsymbol{k}\,\exp\left(\frac{k^2}{4A\beta}\right)\left[e^{i\boldsymbol{k}\cdot(\boldsymbol{r}_i-\boldsymbol{r})}\boldsymbol{\nabla}_{\boldsymbol{r}_i} + \boldsymbol{\nabla}_{\boldsymbol{r}_i}e^{i\boldsymbol{k}\cdot(\boldsymbol{r}_i-\boldsymbol{r})}\right]P_i^{\mathrm{p}},$$
$$\boldsymbol{o}_i^{\mathrm{m}} = \frac{1}{(2\pi)^3}c\mu_N\int d\boldsymbol{k}\,\exp\left(\frac{k^2}{4A\beta}\right)e^{i\boldsymbol{k}\cdot(\boldsymbol{r}_i-\boldsymbol{r})}g_{s_i}\frac{1}{2}\boldsymbol{\sigma}_i. \quad (9.154)$$

For the convection current, the s.p. matrix elements can be obtained, e.g., by the use of

$$\begin{aligned}\langle\varphi_{s_1}|e^{i\boldsymbol{k}\cdot(\boldsymbol{r}_1-\boldsymbol{r})}(-i\hbar)\nabla_{\boldsymbol{r}_1}|\varphi_{s_2}\rangle &= i\hbar\nabla_{s_2}\langle\varphi_{s_1}|e^{i\boldsymbol{k}\cdot(\boldsymbol{r}_1-\boldsymbol{r})}|\varphi_{s_2}\rangle\\ &= \frac{1}{2}i\hbar[\beta(\boldsymbol{s}_1-\boldsymbol{s}_2)-i\boldsymbol{k}]\langle\varphi_{s_1}|e^{i\boldsymbol{k}\cdot(\boldsymbol{r}_1-\boldsymbol{r})}|\varphi_{s_2}\rangle, \end{aligned}\tag{9.155}$$

$$\begin{aligned}\langle\varphi_{s_1}|-i\hbar\nabla_{\boldsymbol{r}_1}e^{i\boldsymbol{k}\cdot(\boldsymbol{r}_1-\boldsymbol{r})}|\varphi_{s_2}\rangle &= \langle\varphi_{s_2}|e^{-i\boldsymbol{k}\cdot(\boldsymbol{r}_1-\boldsymbol{r})}(-i\hbar)\nabla_{\boldsymbol{r}_1}|\varphi_{s_1}\rangle^*\\ &= -\frac{1}{2}i\hbar[\beta(\boldsymbol{s}_2-\boldsymbol{s}_1)-i\boldsymbol{k}]\langle\varphi_{s_1}|e^{i\boldsymbol{k}\cdot(\boldsymbol{r}_1-\boldsymbol{r})}|\varphi_{s_2}\rangle, \end{aligned}\tag{9.156}$$

where the actual form of $\langle\varphi_{s_1}|e^{i\boldsymbol{k}\cdot(\boldsymbol{r}_1-\boldsymbol{r})}|\varphi_{s_2}\rangle$,

$$\langle\varphi_{s_1}|e^{i\boldsymbol{k}\cdot(\boldsymbol{r}_1-\boldsymbol{r})}|\varphi_{s_2}\rangle = \left(\frac{\pi}{\beta}\right)^{3/2}\exp\left[-\frac{k^2}{4\beta}+i\boldsymbol{k}\cdot\left(\boldsymbol{r}-\frac{\boldsymbol{s}_1+\boldsymbol{s}_2}{2}\right)\right]\langle\varphi_{s_1}|\varphi_{s_2}\rangle, \tag{9.157}$$

has been used, which can be obtained with the help of Eq. (9.50), and in Eq. (9.156) the hermiticity of $-i\hbar\nabla$ has been exploited. One can see that, upon summing up the two terms, the $\boldsymbol{k}$-dependence cancels out from the coefficient of $\langle\varphi_{s_1}|\varphi_{s_2}\rangle$. Thus the $\boldsymbol{k}$-dependence of the matrix element of the convection current is the same as that of the density, the second line of Eq. (9.144). Therefore, the third line of Eq. (9.144) is readily applicable, and for the matrix element of the convection current we have

$$\begin{aligned}&\langle\varphi_{s_1}\chi_{\frac{1}{2}m_s^1}\eta_{\frac{1}{2}m_t^1}|\boldsymbol{o}_1^{\mathrm{c}}|\varphi_{s_2}\chi_{\frac{1}{2}m_s^2}\eta_{\frac{1}{2}m_t^2}\rangle\\ &= ic\mu_N\beta\left(\frac{A}{A-1}\frac{\beta}{\pi}\right)^{3/2}(\boldsymbol{s}_1-\boldsymbol{s}_2)\exp\left[-\frac{A}{A-1}\beta\left(\boldsymbol{r}-\frac{\boldsymbol{s}_1+\boldsymbol{s}_2}{2}\right)^2\right]\\ &\quad\times\langle\varphi_{s_1}|\varphi_{s_2}\rangle\delta_{m_s^1m_s^2}\left(\frac{1}{2}-m_t^1\right)\delta_{m_t^1m_t^2}. \end{aligned}\tag{9.158}$$

The s.p. matrix element of the magnetization current is then trivially

$$\begin{aligned}&\langle\varphi_{s_1}\chi_{\frac{1}{2}m_s^1}\eta_{\frac{1}{2}m_t^1}|\boldsymbol{o}_1^{\mathrm{m}}|\varphi_{s_2}\chi_{\frac{1}{2}m_s^2}\eta_{\frac{1}{2}m_t^2}\rangle\\ &= \frac{1}{2}c\mu_N\left(\frac{A}{A-1}\frac{\beta}{\pi}\right)^{3/2}\exp\left[-\frac{A}{A-1}\beta\left(\boldsymbol{r}-\frac{\boldsymbol{s}_1+\boldsymbol{s}_2}{2}\right)^2\right]\\ &\quad\times\langle\varphi_{s_1}|\varphi_{s_2}\rangle\langle\chi_{\frac{1}{2}m_s^1}|\boldsymbol{\sigma}_1|\chi_{\frac{1}{2}m_s^2}\rangle\langle\eta_{\frac{1}{2}m_t^1}|g_{s1}|\eta_{\frac{1}{2}m_t^2}\rangle. \end{aligned}\tag{9.159}$$

To conclude the discussion of the s.p. operators of our interest, we should mention two operators, which are not observables as they are not Hermitean [26], but are important from the point of view of nuclear structure: those of the β-decay. The Fermi transition and the Gamow–Teller transition take place via the transition operators $o_i = (\tau_i)_\pm$ and

$o_i = \sigma_i(\tau_i)_\pm$, respectively. Since these operators do not depend on the spatial variables, the evaluation of their matrix elements does not pose any difficulties.

9.5.2 Two-body operators

Two-body (or two-particle) operators may be treated completely analogously to the one-body operators shown in Sect. 9.5.1. Derivation will only be carried through for the two-particle density,[20]

$$\begin{aligned}\rho(\boldsymbol{r},\boldsymbol{r}') = \langle \Psi_{SM_STM_T}(\boldsymbol{S}_1,\ldots,\boldsymbol{S}_{A-1})| \\ \times\hat{\rho}(\boldsymbol{r},\boldsymbol{r}')|\Psi_{S'M'_ST'M'_T}(\boldsymbol{S}'_1,\ldots,\boldsymbol{S}'_{A-1})\rangle,\end{aligned} \tag{9.160}$$

where

$$\hat{\rho}(\boldsymbol{r},\boldsymbol{r}') = \sum_{i\neq j}^{A} \delta(\bar{\boldsymbol{r}}_i - \boldsymbol{r})\delta(\bar{\boldsymbol{r}}_j - \boldsymbol{r}')P_{ij}. \tag{9.161}$$

The projector P_{ij} may be a product of P_i and P_j, with the same or with different superscripts, or it may depend on some joint property (e.g., it may project onto the two-particle spin triplet state). The density $\rho(\boldsymbol{r},\boldsymbol{r}')$ may also be called the *pair correlation function* since it shows the probability of finding a pair of particles at $\boldsymbol{r}$, $\boldsymbol{r}'$. The two-particle density operator $\hat{\rho}(\boldsymbol{r},\boldsymbol{r}')$ is as fundamental in expressing the matrix elements of two-particle operators as the s.p. density $\hat{\rho}(\boldsymbol{r})$ of Eq. (9.137) is for matrix elements of s.p. operators. Any two-particle operator can be expressed in terms of $\hat{\rho}(\boldsymbol{r},\boldsymbol{r}')$. For example, when $O = \sum_{i<j} f(\boldsymbol{r}_i,\boldsymbol{r}_j)P_{ij}$, one can write [cf. Eq. (9.138)]

$$O = \frac{1}{2}\int d\boldsymbol{r}\int d\boldsymbol{r}'\, f(\boldsymbol{r},\boldsymbol{r}')\hat{\rho}(\boldsymbol{r},\boldsymbol{r}'). \tag{9.162}$$

The δ-functions can again be expanded, as in Eq. (9.140), and a matrix element between generator-coordinate functions that involve the c.m. motion can be separated into c.m. and intrinsic factors as in Eq. (9.141). Once $\sum_{i=1}^{A}\boldsymbol{s}_i = \sum_{i=1}^{A}\boldsymbol{s}'_i = 0$ is imposed, an intrinsic matrix element of

[20] The two-particle density $\rho(\boldsymbol{r},\boldsymbol{r}')$ is not to be confused with the s.p. density matrix $\rho[\boldsymbol{r},\boldsymbol{r}']$ of Eq. (9.130). To see the difference more explicitly, one can write, for the point mass densities,

$$\begin{aligned}\rho[\boldsymbol{r},\boldsymbol{r}'] &= A\int d\boldsymbol{r}_2\cdots\int d\boldsymbol{r}_A\Psi^*(\boldsymbol{r},\boldsymbol{r}_2,\ldots,\boldsymbol{r}_A)\Psi(\boldsymbol{r}',\boldsymbol{r}_2,\ldots,\boldsymbol{r}_A),\\ \rho(\boldsymbol{r},\boldsymbol{r}') &= A\int d\boldsymbol{r}_3\cdots\int d\boldsymbol{r}_A|\Psi(\boldsymbol{r},\boldsymbol{r}',\boldsymbol{r}_3,\ldots,\boldsymbol{r}_A)|^2.\end{aligned}$$

the density operator can be expressed, similarly to Eq. (9.142), as

$$\begin{aligned}\rho(\boldsymbol{r},\boldsymbol{r}') = \sum_{i\neq j}^{A} \frac{1}{(2\pi)^6} \int d\boldsymbol{k} \int d\boldsymbol{k}' \exp\left[\frac{(\boldsymbol{k}+\boldsymbol{k}')^2}{4A\beta}\right] \\ \times \langle \Phi_{SM_STM_T}(\boldsymbol{s}_1,\ldots,\boldsymbol{s}_A)| \\ \times e^{i\boldsymbol{k}\cdot(\boldsymbol{r}_i-\boldsymbol{r})} e^{i\boldsymbol{k}'\cdot(\boldsymbol{r}_j-\boldsymbol{r}')} P_{ij} |\Phi_{S'M_S'T'M_T'}(\boldsymbol{s}_1',\ldots,\boldsymbol{s}_A')\rangle \\ \equiv \langle \Phi_{SM_STM_T}(\boldsymbol{s}_1,\ldots,\boldsymbol{s}_A)| \sum_{i\neq j}^{A} o_{ij} |\Phi_{S'M_S'T'M_T'}(\boldsymbol{s}_1',\ldots,\boldsymbol{s}_A')\rangle,\end{aligned} \tag{9.163}$$

with the operator o_{ij} defined by

$$o_{ij} = \frac{1}{(2\pi)^6} \int d\boldsymbol{k} \int d\boldsymbol{k}' \exp\left[\frac{(\boldsymbol{k}+\boldsymbol{k}')^2}{4A\beta}\right] e^{i\boldsymbol{k}\cdot(\boldsymbol{r}_i-\boldsymbol{r})} e^{i\boldsymbol{k}'\cdot(\boldsymbol{r}_j-\boldsymbol{r}')} P_{ij}. \tag{9.164}$$

The two-particle matrix element will be

$$\begin{aligned}&\langle \varphi_{\boldsymbol{s}_1}\chi_{\frac12 m_s^1}\eta_{\frac12 m_t^1}\varphi_{\boldsymbol{s}_2}\chi_{\frac12 m_s^2}\eta_{\frac12 m_t^2}|o_{12}|\varphi_{\boldsymbol{s}_3}\chi_{\frac12 m_s^3}\eta_{\frac12 m_t^3}\varphi_{\boldsymbol{s}_4}\chi_{\frac12 m_s^4}\eta_{\frac12 m_t^4}\rangle \\ &= \left(\frac{A}{A-2}\frac{\beta^2}{\pi^2}\right)^{3/2} \exp\left[-\frac{A}{A-2}\frac{\beta}{2}\left(\boldsymbol{r}+\boldsymbol{r}' - \frac{\boldsymbol{s}_1+\boldsymbol{s}_3+\boldsymbol{s}_2+\boldsymbol{s}_4}{2}\right)^2\right] \\ &\times \exp\left[-\frac{\beta}{2}\left(\boldsymbol{r}-\boldsymbol{r}' - \frac{\boldsymbol{s}_1+\boldsymbol{s}_3-\boldsymbol{s}_2-\boldsymbol{s}_4}{2}\right)^2\right] \langle\varphi_{\boldsymbol{s}_1}|\varphi_{\boldsymbol{s}_3}\rangle\langle\varphi_{\boldsymbol{s}_2}|\varphi_{\boldsymbol{s}_4}\rangle \\ &\times \langle \chi_{\frac12 m_s^1}\eta_{\frac12 m_t^1}\chi_{\frac12 m_s^2}\eta_{\frac12 m_t^2}|P_{12}|\chi_{\frac12 m_s^3}\eta_{\frac12 m_t^3}\chi_{\frac12 m_s^4}\eta_{\frac12 m_t^4}\rangle.\end{aligned} \tag{9.165}$$

Here, again, Eq. (9.50) was used.

As an example, let us consider an operator o_{12} which only depends on the interparticle displacements, $\boldsymbol{r}_{ij} = \boldsymbol{r}_i - \boldsymbol{r}_j$. Its matrix elements can be derived from those of a δ-function, $\delta(\boldsymbol{r}_1-\boldsymbol{r}_2-\boldsymbol{r})$, as is given in Eq. (9.98). Since now $f(\boldsymbol{r},\boldsymbol{r}')$ only depends on the combination $\boldsymbol{r}-\boldsymbol{r}'$, it is useful to change the variables in Eq. (9.162) such as $\{\boldsymbol{r},\boldsymbol{r}'\} \to \{\boldsymbol{r}-\boldsymbol{r}', \frac12(\boldsymbol{r}+\boldsymbol{r}')\}$, and it is useful to integrate Eq. (9.165) over $\boldsymbol{r}_+ = \frac12(\boldsymbol{r}+\boldsymbol{r}')$. The result is

$$\begin{aligned}&\int d\boldsymbol{r}_+ \langle \varphi_{\boldsymbol{s}_1}\chi_{\frac12 m_s^1}\eta_{\frac12 m_t^1}\varphi_{\boldsymbol{s}_2}\chi_{\frac12 m_s^2}\eta_{\frac12 m_t^2}|o_{12}|\varphi_{\boldsymbol{s}_3}\chi_{\frac12 m_s^3}\eta_{\frac12 m_t^3}\varphi_{\boldsymbol{s}_4}\chi_{\frac12 m_s^4}\eta_{\frac12 m_t^4}\rangle \\ &= \left(\frac{\beta}{2\pi}\right)^{3/2} \exp\left[-\frac{\beta}{2}\left(\boldsymbol{r}-\boldsymbol{r}' - \frac{\boldsymbol{s}_1+\boldsymbol{s}_3-\boldsymbol{s}_2-\boldsymbol{s}_4}{2}\right)^2\right] \langle\varphi_{\boldsymbol{s}_1}|\varphi_{\boldsymbol{s}_3}\rangle\langle\varphi_{\boldsymbol{s}_2}|\varphi_{\boldsymbol{s}_4}\rangle \\ &\times \langle \chi_{\frac12 m_s^1}\eta_{\frac12 m_t^1}\chi_{\frac12 m_s^2}\eta_{\frac12 m_t^2}|P_{12}|\chi_{\frac12 m_s^3}\eta_{\frac12 m_t^3}\chi_{\frac12 m_s^4}\eta_{\frac12 m_t^4}\rangle,\end{aligned} \tag{9.166}$$

which is essentially the same formula as Eq. (9.98) as it indeed should be.

To evaluate the matrix elements of a general many-body operator is cumbersome, but handy *ad hoc* methods may be devised for operators of particular forms. As an example, the special case of a product operator will be treated in App. E.4.

Chapter 10

Variational procedure

10.1 Basis optimization

The description of light exotic nuclei involves solving few-body and few-cluster problems variationally. The general scheme of the construction of the Hamiltonian and overlap matrices in the correlated Gaussian bases has already been discussed. Since this comprises the full formalism for the treatment of some systems (the few-body systems), it is useful to turn now to the methods of the solution of dynamical problems. Part II is mainly concerned with bound states, so the main objective of this chapter is to discuss the methodology of the bound-state variational problem associated with correlated Gaussian trial functions. This is the subject of Sects. 10.1–10.3. Since, however, light exotic nuclei often have very few bound states or none, discrete unbound states have to be treated as well. A generalization of the bound-state methods to such unbound states will be introduced in Sects. 10.4 and 10.5.

The most conventional variational method for the solution of quantum-mechanical bound-state problems is the diagonalization of the Hamiltonian in a state space spanned by some appropriate functions $\psi_1, \ldots, \psi_K$. One sets up a basis in a well-defined way, e.g., by using a complete set of states that contain no variable parameters, and then obtains the energy by diagonalization. A typical example of this direct approach is the nuclear SM. The applicability of the direct approach is limited by the infeasibility of treating too large matrices.

Another method is basis optimization, which is devised to avoid the problem of huge basis dimensions in the direct approach. In this method the basis itself is set up on variational grounds. One constructs a basis that looks suitable for reproducing the wave function. The construction may proceed as a selection of basis elements from a large pool of functions. The selection strategy has to be tailored for the state sought, which is, of course, not known *a priori*. What still helps to define the pool of functions is that one can envisage *a priori* the extension of the state. The ***stochastic variational method*** (SVM[1]) ***is a basis optimization method*** in which the suitable basis states are singled out from a pool of functions by a ***trial-and-error procedure***. The SVM was originally proposed in Ref. [235] and was later refined [236] and successfully applied to multicluster descriptions of light exotic nuclei [213, 214]. It was subsequently generalized and refined further to be applicable to diverse quantum-mechanical few-body systems emerging in any branches of microphysics [187, 206, 237, 238].

The correlated Gaussian basis functions presented in Chap. 9 depend on a number of continuous nonlinear parameters and on discrete parameters (e.g., angular momentum quantum numbers). These parameters define the shape of a basis function and determine how well the subspace spanned by the basis contains the true eigenfunction. For instance, the basis function ψ_i of Eq. (9.41) contains $\frac{1}{2}(A-1)A$ pair-correlation parameters α_{jk} (or, equivalently, the elements of A_i) and the discrete labels $l_i L_i S_i$, including the intermediate quantum numbers of angular momentum coupling. Since there are usually more than one, say $\mathcal{K}$, basis functions, the trial function also contains the coefficients $(c_1, \ldots, c_{\mathcal{K}})$ as linear variational parameters. The energy expectation value (9.1) to be minimized will be a function of all these nonlinear, linear and discrete parameters.

The variational construction of the basis, i.e., the optimization of the nonlinear and discrete parameters, has to go along with the variational determination of the linear parameters, i.e., with the diagonalization. Any variation of the nonlinear parameters should be followed by a diagonalization of Eq. (9.4). Each such step yields partially optimized values of the coefficients c_k and of the energy functional. In this way a full optimization of the parameters would involve re-calculations of the matrix elements and re-diagonalizations a great number of times, which requires extremely long computer time.

To have a crude guess of how much work the optimization requires, let us consider a simple A-nucleon system restricted to a single configuration defined by a set of quantum numbers. When a combination of $\mathcal{K}$ correlated Gaussians of spherical symmetry is needed, one faces an optimization problem with $\mathcal{K}$ linear parameters and at least $\frac{1}{2}\mathcal{K}(A-1)A$ nonlinear parameters. This number increases quadratically with the number of particles. By taking $A=4$ and $\mathcal{K}=200$ as a typical case, one ends up with 1200 nonlinear parameters.

[1]Except for this chapter, the acronym SVM is more often used for the specific version of the stochastic variational method that involves correlated Gaussian bases.

Another serious problem involving the minimization of a function is the omnipresence of local minima, whose number tends to increase rapidly with the size of the problem.

There are many different methods for the optimization of a function, and these optimizations can be divided into two categories: the deterministic and the stochastic optimizations.

A deterministic minimization procedure moves downwards the slope of the function according to a well-defined strategy. There exist a great number of elaborate algorithms (conjugate gradient, Powell (direction set) algorithms etc. [239]). Since they are deterministic, from the same starting values they always lead to the same (local or global) minimum. The drawback of these methods is that they are generally time-consuming and tend to converge to whichever local minimum they first encounter. There is no good recipe for choosing the starting point and for escaping from a local minimum.

Stochastic optimizations address the problem of finding the global minimum by making random steps [240, 241]. In principle, they are not affected by the existence of any number of undesired local minima. A trivial (but very laborious) stochastic optimization would be, e.g., the repetition of deterministic searches from a great number of random starting points and adopting the minimum of the local minima. Numerous more sophisticated strategies have been developed: e.g., simulated annealing [242], genetic algorithms [243] etc. But the simplest stochastic optimization is just picking up random points in the parameter space and adopting that which yields the smallest value. This may not sound very economical, but, wrapped into sophisticated strategies, it may turn out to be most efficient.

To anticipate the content of Sect. 10.2, we now sum up the essence of the SVM for basis construction. The basis is built up element by element, by trial-and-error steps. In each step several random states are tested by adding it to the basis obtained in the previous step, but only one new element is adopted: the one whose inclusion lowers the g.s. energy E_0 most. The basis is enlarged until 'convergence' is reached, whose criterion controls the accuracy. Since the basis optimization is not simultaneous for all elements and the number of trials in each step is finite, one cannot claim that a basis constructed in this way is the best of any bases of that particular dimension. The procedure may not even produce a local energy minimum with respect to parameter variations. Yet it is never bogged down in local minima because the function $E_0(\mathcal{K})$ is monotonous (see Theorem 2 of Sect. 9.1.2). The great number of trials involved in the whole procedure ensures with great probability that the procedure approaches the exact g.s. energy.

To make the optimization feasible, there seems to be one viable alternative of the SVM: to assume a regular pattern for the change of the parameters from one element to the other, and optimize the parameters involved in the pattern. Each pattern can be associated with a "grid" in the parameter space. The grid is given by some *ad hoc* rule and each point of the grid defines a basis function. The grid can be defined by some simple functions (e.g., a geometric progression [244,245], random tempering [220,246,247] and Chebyshev grid [197]), and the parameters of these are the parameters to be optimized. The number of parameters in the functions that define the grid must be much smaller than in the original problem. The grid methods have the drawback that, for complicated cases, they are much less selective than the SVM, so that a good enough grid-pattern basis may be much larger than an SVM basis of similar quality.

The correlated Gaussian basis contains functions of the same analytical form but of different values of their continuous parameters. Such functions are especially suited both for the SVM and for grid-pattern bases. If all parameters of two functions (nearly) coincide, the linear independence is lost and numerical inaccuracies may in principle occur. In practice, however, the selection criteria and the grid constants, respectively, can always be chosen so as to avoid this pathological case.

10.2 Stochastic optimization

10.2.1 Random basis and sorting

The way stochastic optimization works is best illustrated by a simple example: the problem of the α-particle with the Malfliet–Tjon potential MT–V [248] described in terms of correlated Gaussians. (This potential is given by $\sum_{i=1,2} V_i \frac{1}{r} e^{-\mu_i r}$, with V_1=1458.05 MeV fm, V_2=−578.09 MeV fm, μ_1=3.11 fm^{-1}, μ_2=1.55 fm^{-1}.) The spin–isospin part of the α-particle wave function is taken to be totally antisymmetric, so that its orbital part becomes totally symmetric. There are a number of different methods to calculate the g.s. energy for this case, and they all agree very well with each other, giving −31.36 MeV for the g.s. energy.

First a basis $\{\psi_k\}$ is generated by taking different random values for the nonlinear parameters. Random $\{\alpha_{ij}^{(k)}\}$ sets [cf. Eq. (9.28) and p. 205] are taken as follows:

$$\psi_k = \exp\left[-\sum_{i<j} \alpha_{ij}^{(k)} (\boldsymbol{r}_i - \boldsymbol{r}_j)^2\right], \quad 0 < \frac{1}{\sqrt{\alpha_{ij}^{(k)}}} \leq 8.5 \quad \text{fm}. \qquad (10.1)$$

The functions obtained in this way are included in the basis one by one, and the g.s. energy is calculated in each step. The convergence of the energy as a function of the basis dimension $\mathcal{K}$ is shown in Fig. 10.1. Surprisingly enough, such a completely random basis gives quite a good g.s. energy with $\mathcal{K} = 200$. It is seen that different random bases lead to the same g.s. energy. It is instructive to see what happens if an uncorrelated basis is used in a similar procedure. Therefore, the above calculation has been repeated with s.p.-like states

$$\psi_k = \exp\left[-\frac{1}{2}\sum_{i=1}^{4} \alpha_i^{(k)}(\boldsymbol{r}_i - \boldsymbol{R}_{\rm c.m.})^2\right], \tag{10.2}$$

which contain four parameters $\alpha_1^{(k)}, \ldots, \alpha_4^{(k)}$. The energy curve obtained with this basis is also displayed in Fig. 10.1. Its convergence is extremely slow and unconvincing.

Such test calculations show that a random correlated basis may be much more economical than an uncorrelated basis. Nevertheless, one may wonder whether each of the $\mathcal{K}$ states in a random basis is really important. What happens if some of them are omitted? Is it possible

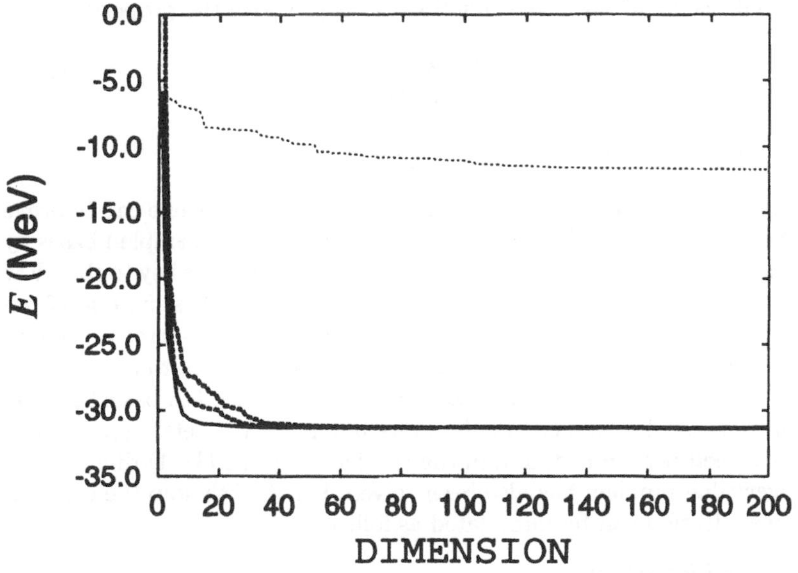

Figure 10.1: Convergence of the α-particle energy on random basis sets: two random sets of the correlated functions of Eq. (10.1) (thick dotted lines); reordered basis (solid line); s.p.-like basis of Eq. (10.2) (thin dotted line). Exact energy: -31.36 MeV; energy with the $\mathcal{K} = 200$ fully random basis: -31.34 MeV.

to reduce the basis by including only a minimal set of functions? To explore this possibility, we have included the same $\mathcal{K}$ basis states one by one in a different order with the following prescription:

1. The first state, $\psi^{(1)}$, is that which gives the smallest energy expectation value.

2. Each of the remaining $\mathcal{K}-1$ states ψ_k is tested in a two-dimensional calculation with the basis $\{\psi^{(1)}, \psi_k\}$, and $\psi^{(2)}$ is identified with ψ_k that produces the lowest g.s. energy.

3. This reordering is carried all the way through.

The energies calculated by these reordered bases are also shown in Fig. 10.1. To approximate the energy obtained with the full basis within 0.1 MeV, one needs a minimal set of just 50 basis states. Thus it has been confirmed that the randomly generated basis states are not equally important, so some of them can be omitted. Actually, none of them is indispensable; if any one of them is omitted, some others will compensate for the loss. This is owing to a 'denseness' property of the pool of correlated Gaussian functions $\{\psi\}$: for any $\psi_k \in \{\psi\}$ and for any $\varepsilon(>0)$, there always exist functions $\psi' \in \{\psi\}$ with the property

$$\left| \frac{|\langle\psi'|\psi_k\rangle|^2}{\langle\psi'|\psi'\rangle\langle\psi_k|\psi_k\rangle} - 1 \right| < \varepsilon. \tag{10.3}$$

10.2.2 Trial-and-error search

The example presented in Sect. 10.2.1 has shown that one can improve the convergence of the energy produced by randomly sampled bases by selection of the basis states by an 'admittance test'. The key to the SVM is a gradual construction of the basis by choosing the best function from among many random trials in each step. The trial-and-error procedure used in each step of the SVM is called *competitive selection.*

In the kth step the parameter set A_k defining the kth basis function ψ_k is to be determined. In the preceding steps the sets $A_1, \ldots, A_{k-1}$ have been determined, and in the $(k-1)$th step the $(k-1)$-dimensional eigenvalue problem has also been solved. For the kth step the competitive selection can be formulated as follows:

Competitive selection

s1. Different parameter sets $\{A_k^{(1)}, \ldots, A_k^{(n)}\}$ are generated randomly.

s2. The n k-dimensional eigenvalue problems obtained by substituting these n parameter sets, consecutively, for A_k are solved, and the corresponding energies $\{E_k^{(1)}, \ldots, E_k^{(n)}\}$ are determined.

Table 10.1: Convergence of the energy, in MeV, of the α-particle with the Malfliet–Tjon potential. The exact energy is −31.36 MeV.

Basis | $\mathcal{K}$	5	10	15	20	25	30
Random 1	−15.61	−23.04	−24.73	−27.69	−27.99	−28.86
Random 2	−9.97	−14.64	−16.89	−22.18	−26.21	−28.78
Selected 1	−26.75	−30.11	−30.78	−31.02	−31.16	−31.23
Selected 2	−26.12	−30.55	−30.97	−31.14	−31.21	−31.27
Refinement	−30.39	−30.96	−31.15	−31.25	−31.29	−31.32

s3. The parameter set $A_k^{(i)}$ that produces the lowest of the energies $\{E_k^{(1)},\ldots,E_k^{(n)}\}$ is chosen to be A_k, which defines ψ_k.

This procedure may be continued by increasing k to $k+1$.

The essential motivation behind this strategy is the need to sample different parameter sets as quickly as possible. Obviously, when enlarging the basis by one element, it is not necessary to recompute the whole Hamiltonian matrix. Nor is it necessary to perform a full diagonalization whenever a new parameter set is generated since the diagonalization of a k-dimensional matrix of the structure of

$$\begin{pmatrix} E_1^{(k-1)} & 0 & \cdots & 0 & \langle\psi_1|H|\psi_k\rangle \\ 0 & E_2^{(k-1)} & \cdots & 0 & \langle\psi_2|H|\psi_k\rangle \\ \vdots & \vdots & \ddots & \vdots & \vdots \\ 0 & 0 & \cdots & E_{k-1}^{(k-1)} & \langle\psi_{k-1}|H|\psi_k\rangle \\ \langle\psi_k|H|\psi_1\rangle & \langle\psi_k|H|\psi_2\rangle & \cdots & \langle\psi_k|H|\psi_{k-1}\rangle & \langle\psi_k|H|\psi_k\rangle \end{pmatrix} \tag{10.4}$$

is extremely simple [187]. The competitive selection substantially improves the convergence, and its performance can be optimized by choosing the number of trials suitably ($n \sim$10–100).

Table 10.1 compares the g.s. energies calculated with different random and selected bases. (The case with 'refinement' will be discussed in Sect. 10.2.3.) The results with the random bases are far from convergence, but those with the selected bases are fairly near, and the last value is quite close to the exact value in spite of the small basis dimension. Two sets are shown to emphasize that the energy converges to the same value, independently of the random starting point. The fulfilment of this agreeable property can be checked in any problem. If it is found to be satisfied, it strongly suggests that the limit is actually the exact value on the model space defined by the intrinsic limits of the basis (angular-momentum quantum numbers, range of Gaussian parameters).

10.2.3 Refining

It is important to produce an acceptable basis with as low dimension as possible. Since the optimization achieved via the competitive selection is very limited, there is room for various improvements. The main limitation is that, when the kth function is adopted, the previously chosen functions are kept necessarily fixed. One may nevertheless try to make up for the lack of simultaneous optimization by reconsidering the previously adopted elements. The procedure introduced for this purpose is called the refinement cycle. It may be implemented once a $\mathcal{K}$-dimensional basis has been constructed via competitive selection. The refinement cycle goes over all basis elements $i=1,\ldots,\mathcal{K}$. The step for the ith element is the following:

Refinement cycle

r1. As possible replacements for A_i, new random parameter sets $(A_i^{(1)},\ldots,A_i^{(n)})$ are generated.

r2. The parameters A_i are replaced by the n new candidates consecutively to obtain n new g.s. energies $(E_{\mathcal{K}}^{(1)},\ldots,E_{\mathcal{K}}^{(n)})$.

r3. If the best of the new energies is lower than the original one, the old parameters A_i are replaced with the new ones; otherwise the original ones are kept.

Note that, again, there is no need for full diagonalization when a new parameter set is tested. One may repeat the refining procedure until the improvement obtained is no longer significant.

The energy resulting from a refinement procedure is also shown in Table 10.1. The cycle was repeated several times until no further gain was obtained. The resulting energy is then virtually independent of which of the two sets is used as a starting point. When the dimension is so low, the refinement increases the accuracy to a great extent. When the refinement has converged, further improvement can only be achieved by enlarging the basis. Another example for the refinement will be shown in the next section.

For nuclear problems the SVM may have to be modified owing to the presence of non-central interaction terms. These couple the different spin–isospin and partial-wave channels, requiring a more complex approach. A basis function of label k is then characterized by a quantum number set g_k, in addition to the parameter set A_k. The number m of quantum-number combinations that come into play may be large, and the corresponding subspaces may not be orthogonal to each other. Therefore, the g-sets also have to be tested—systematically for small m

values and stochastically for large ones [237]. The competitive selection is then to be modified accordingly. Up to the $(k-1)$th step, the parameters $A_1, \ldots, A_{k-1}$ and the corresponding discrete labels $g_1, \cdots, g_{k-1}$ have to have been determined. The kth step of the competitive selection, with the systematic examination of the g-sets, will go as follows:

Modified competitive selection

s1. Different parameter sets $\{A_k^{(1)}, \ldots, A_k^{(n)}\}$ are generated randomly. This opens an n-element cycle for $i=1, \ldots, n$:

s2–1. For each set $A_k^{(i)}$, the m k-dimensional eigenvalue problems obtained by combination of $A_k^{(i)}$ with each of the quantum number sets $\{g_{[1]}, \cdots, g_{[m]}\}$ are solved, and the corresponding energies, $\{E_k^{(i1)}, \ldots, E_k^{(im)}\}$, are calculated.

s2–2. The $g_{[j]}$ set that produces the lowest of $\{E_k^{(i1)}, \ldots, E_k^{(im)}\}$ is adopted as $g_k^{(i)}$ to accompany the ith parameter set $A_k^{(i)}$ for the kth basis state.

By repeating the steps s2 for $i, \ldots, n$, energies $\{E_k^{(1)}, \ldots, E_k^{(n)}\}$ are produced.

s3. The parameter set $A_k^{(i)}$ and quantum number set $g_k^{(i)}$ whose combination produces the lowest of the energies $\{E_k^{(1)}, \ldots, E_k^{(n)}\}$ is chosen to be A_k and g_k, which define ψ_k.

The refinement cycle is devised to be unbiassed: new random parameters are generated irrespective of their previous values. This certainly helps to avoid the procedure's being trapped in local energy minima, but does not help to hit the precise position of a minimum. If the energy cannot be improved any more by refinement, one can be reasonably confident that one is close to the global minimum or to a minimum that is almost as satisfactory as the global minimum, and it makes sense to find this minimum itself. This can be done by seeking improvement in the vicinity of the point resulting from the refinement. This step may be called fine tuning.

In practice, the fine-tuning step can be implemented by choosing new random parameters $\{A_i^{(1)}, ..., A_i^{(n)}\}$ from the interval $[pA_i, p^{-1}A_i]$ sandwiching the result of the previous step A_i; the parameter p, which determines the size of the interval, may be chosen to be, e.g., 0.7–0.8. The trial-and-error search thus proceeds in exactly the same way as in the case of the refinement, except that the search interval is more limited.

A combination of the three steps may also be useful when it is important to reach high accuracy with a strictly limited basis size. Then refining and fine tuning may be introduced at an earlier stage, and the fine tuning may be followed by further basis enlargements. If fine tuning is applied, then the search strategy will be efficient if each fine-tuning cycle is preceded by a refinement cycle, and the last step is a fine-tuning cycle.

The SVM with the correlated Gaussians has been thoroughly tested for quite different few-body problems against other powerful methods [187]. The well-established high-accuracy methods existing in atomic and molecular physics offer especially stringent tests. The most accurate results for some of the Coulombic 3–6-particle systems [238, 249–251] have been reproduced or improved. The method proved to be very useful and powerful also in subnuclear and solid-state physics [252–254].

10.2.4 Description of excited states

Variational methods are designed primarily to describe g.s. But any variational method that involves diagonalization does produce excited states as well. The SVM is not excepted, and it can be made accurate for any particular excited state.

In the SVM the basis is usually constructed so as to belong to definite angular momentum, parity, isospin etc., whichever observable commutes with the Hamiltonian.[2] The lowest-lying state in each such quantum-number set can be counted as the g.s. in that subspace. But there are light nuclei that have bound excited states with the same angular momentum, parity and isospin as those of the g.s. (e.g., excited 0^+ states in ^{10}Be or ^{12}C), which have to be described with the same basis as the g.s. All states obey the mini-max theorem (Sect. 9.1.2). This says that a diagonalization of the Hamiltonian over a $\mathcal{K}$-dimensional basis provides an upper bound not only for the g.s. but also for each of the $\mathcal{K}-1$ excited states. It may happen that the basis set found for the g.s. is fairly good for some excited states (of the same quantum numbers) as well. But by optimizing the g.s., the energies of the excited states will not converge at the same rate.

When the aim is to describe an excited state, one may try to minimize the upper bound of that particular excited state. The procedure to be followed is very similar to that presented in Sect. 10.2, but now the basis selection is governed by another eigenvalue. To get an estimate

[2]The SVM works in general with mixed bases as well. For example, one may choose not to apply angular-momentum coupling. The minimum principle will still ensure that the resulting g.s. may converge to the exact g.s., which carries pure angular momentum (except for cases of angular-momentum degeneracy), but the 'trial-and-error angular-momentum coupling' involved is costly in computing time.

for the ith eigenvalue, we have to start at least with an i-dimensional basis, but, in practice, it is better to start with a basis in which all lower eigenvalues ($k=1,\ldots,i-1$) are already 'stable' to some extent. The diagonalization procedure ensures that the ith eigenstate is orthogonal to all the lower ones, and the ith state will be affected through this orthogonality. Higher-lying states pick up errors from the lower-lying states in this way. There are three procedures tested for finding optimized SVM bases for excited states.

The simplest approach is to minimize $\sum_{i=1}^{N} w_i^2(\epsilon - E_i)^2$, where E_i are energies of the first few states to be optimized and ϵ is an estimated lower bound of all (exact) eigenvalues. This form causes the optimization procedure to pull all energies $E_1,\ldots,E_N$ downwards, without regard to whether they are positive or negative. The relative accuracies of the states can be controlled by the weight factor w_i, and it is advisable to set the accuracy of the g.s. to be highest. The disadvantage of this simple prescription is that the convergence is slow. Since the states are orthogonal to each other mostly owing to their spatial parts, they may have very different spatial shapes, and so their simultaneous optimization may not be easy. Even a level 'crossing', i.e., a change in the sequence of two states of different natures, may occur.

A simple alternative is to construct a basis of a certain size by optimizing the description of the g.s. and then add states optimizing the first excited state etc. In each of these steps competitive selection and refinement may be combined as described in Sect. 10.2.

A third possibility is to set up a basis with competitive selection and refinement, just as in the previous case, and make a refinement cycle for the excited state to be optimized. The advantage of this procedure is that the dimension of the basis does not increase. The optimization of the excited state, however, leads to loss of accuracy for the g.s.

These alternatives are exemplified in Fig. 10.2 by a calculation of the g.s. and the first excited state of the α-particle produced by the Malfleit–Tjon potential (cf. p. 246). The 0^+ excited state of the α-particle is actually unbound, so the bound state is an artefact of this simple potential. The g.s. energy is -31.36 MeV, and the energy of the first excited state is -8.50 MeV. (In the figure the energy of the excited state is shifted by -22 MeV.) The difference between the spatial structures of the two states is clearly indicated by the fact that their rms radii are 1.4 fm and 4.6 fm, respectively.

In Fig. 10.2a the basis was constructed by competitive selection, minimizing $w_1^2(\epsilon - E_1)^2 + w_2^2(\epsilon - E_2)^2$ with $w_1=3$, $w_2=1$ and $\epsilon=-32$ MeV. With the basis enlarged, both energies converge towards their exact values: they are about 0.05 MeV above their exact values at $\mathcal{K}=100$, but that can be easily improved by adding more states or by

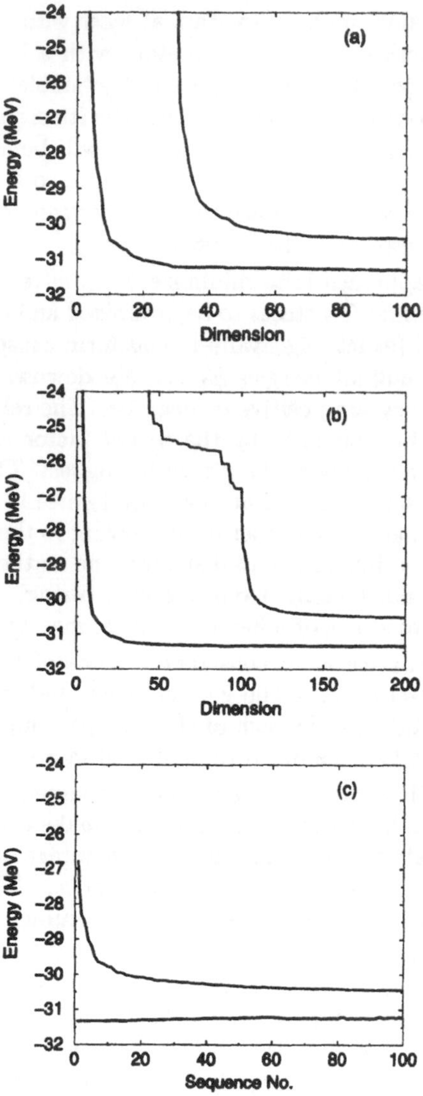

Figure 10.2: Comparison of the energies produced by three procedures to optimize the basis for the g.s. and first excited state of the α-particle. The energy of the excited state is shifted by -22 MeV. In case (a) the optimized functional is $w_1^2(\epsilon - E_1)^2 + w_2^2(\epsilon - E_2)^2$. In case (b) the first 100 elements of the basis optimize E_1, and the second 100 optimize E_2. In case (c) the 100-dimensional basis of panel (b) is refined by optimizing E_2, and the energies E_1 and E_2 are plotted versus the sequence number of the basis state being refined. See text for details.

refinement. In Fig. 10.2b the g.s. was optimized with $\mathcal{K}=100$ ($w_1=1$, $w_2=0$) and then another hundred states were added to optimize the first excited state ($w_1=0$, $w_2=1$). The energy of the excited state is reproduced very well. Figure 10.2c shows a refinement for the excited state starting with a $\mathcal{K}=100$ basis obtained by minimizing the g.s. energy. The refinement cycle has been carried through once. The figure shows that the g.s. energy slightly oscillates.

10.3 Short-range and long-range behaviour

As a full illustration for the performance of the SVM combined with the correlated Gaussian basis, a challenging example is chosen: the hypernucleus[3] ${}_{\Lambda\Lambda}^{\ 4}$H, which is a p+n+Λ+Λ four-particle system. The structure problem of a light exotic nucleus may pose two main difficulties: the treatment of short-range correlations and halo-like dilute substructures. In ${}_{\Lambda\Lambda}^{\ 4}$H the distribution of the Λ's extends to large distances in comparison with that of the nucleons, and there is a strong two-body correlation between them. Thus this example challenges the SVM in the most critical points. The system ${}_{\Lambda\Lambda}^{\ 4}$H has not yet been observed experimentally, but its existence is supported by theoretical calculations [255, 256].

For the p–n interaction, the Minnesota force to be given in Table 12.2 has been employed. It is suited for this role as it reproduces the energy and the size of the deuteron (see Table 12.3). For the Λ–nucleon force the effective potential A of Ref. [256] is used. This reproduces the binding energies of the lightest Λ-hypernuclei (${}_{\Lambda}^{3}$H, ${}_{\Lambda}^{4}$H, ${}_{\Lambda}^{4}$He, ${}_{\Lambda}^{4}$H* and ${}_{\Lambda}^{4}$He*) fairly well. (Here $*$ denotes the first excited state.) The Λ–Λ potential is constructed [257] so as to be phase-equivalent to the Nijmegen potential D [258], and in the ${}^1S_0$ channel it has a strong central repulsion whose strength at $r=0$ is about 4.4 GeV. Neither the Λ–nucleon nor the Λ–Λ potential is attractive enough to produce a two-body bound state. The three-body system p+n+Λ, however, forms a very weakly bound state. The calculated Λ separation energy is just 0.18 MeV, which is in good agreement with the experimental value of 0.13±0.05 MeV. The energy of ${}_{\Lambda\Lambda}^{\ 4}$H was calculated for spin–isospin states $(S,T)=(0,0)$, (0,1), (1,0) and (1,1) [256], and it is found that ${}_{\Lambda\Lambda}^{\ 4}$H forms a bound state only for $(S,T)=(1,0)$. The resultant binding is very weak: the separation energy of one and two Λ-particles is 0.23 MeV and 0.41 MeV, respectively. It is the short-range repulsion between the two Λ-particles that causes the short-range correlation, and

[3] A hypernucleus is a nucleus which contains one or more hyperons (e.g., Λ-, Σ-particles).

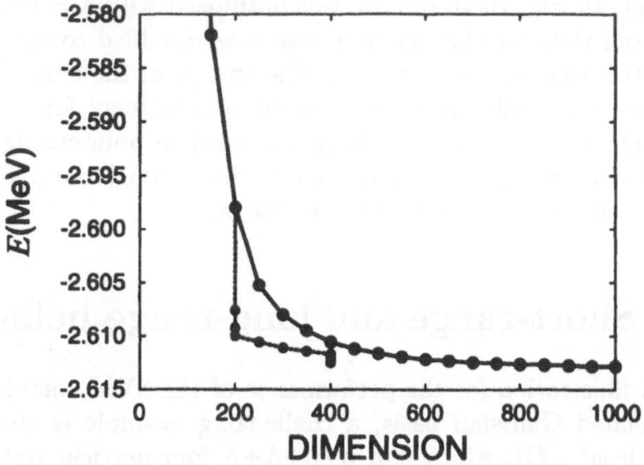

Figure 10.3: Convergence of the $_{\Lambda\Lambda}^{\ 4}$H energy with pure competitive selection (full line, $E = -2.61286$ MeV with $\mathcal{K} = 1000$) and with competitive selection combined with refinements (dashed line, $E = -2.61251$ MeV with $\mathcal{K} = 400$).

it is the very weak binding that causes the spatially extended density distribution.

Figure 10.3 shows the convergence of the $_{\Lambda\Lambda}^{\ 4}$H energy as a function of the basis dimension. The basis was set up by the competitive selection with $n = 75$ trials in each step. The basis (10.1) was used, such that the total orbital angular momentum was $L = 0$, and the number of spin–isospin subspaces was $m = 3$. The correlation parameters were allowed to be in the interval $0 < 1/\sqrt{\alpha_{ij}^{(k)}} \leq 80$ fm [cf. Eq. (10.1)]. To predict the energy within an accuracy of 10 keV, a basis size of $\mathcal{K} = 400$ was needed. Refinement was also applied; three refinement cycles were performed at $\mathcal{K} = 200$ as well as at 400. One can see that the first refinement cycles brings the energy down close to that reached by the competitive selection around $\mathcal{K} = 400$.

Although the 'convergence' of $E(\mathcal{K})$ is convincing, the limit cannot ever be attained. Therefore, one should check whether the wave function is also close to convergence. An overall measure of the wave function's convergence may be the convergence of the rms radius (Fig. 10.4) or of the virial ratio $\langle\Psi|W|\Psi\rangle/(2\langle\Psi|T|\Psi\rangle)$ or, more precisely (Fig. 10.5), of

$$\eta = \left|\frac{\langle\Psi|W|\Psi\rangle}{2\langle\Psi|T|\Psi\rangle} - 1\right|, \quad \text{with} \quad W = \sum_{i=1}^{A} \mathbf{r}_i \cdot \frac{\partial V}{\partial \mathbf{r}_i}, \tag{10.5}$$

where T and V are the kinetic and potential energies and A is the number of particles. According to the virial theorem [187], the quantity

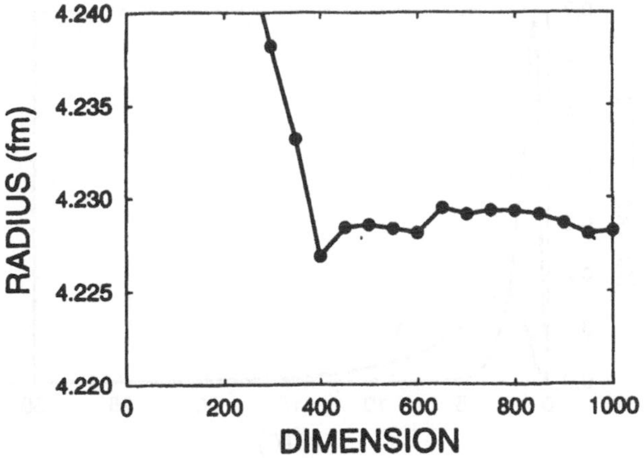

Figure 10.4: Convergence of the rms radius of ${}_{\Lambda\Lambda}^{4}$H as a function of the basis dimension.

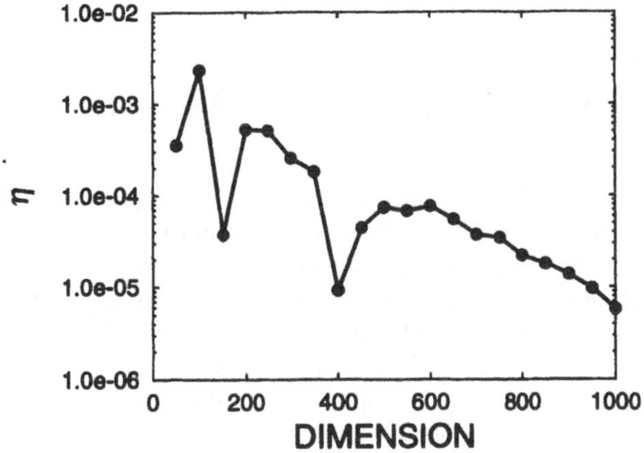

Figure 10.5: Departure of the virial ratio from unity for ${}_{\Lambda\Lambda}^{4}$H as a function of the basis dimension.

η must vanish for the true eigenstate Ψ of the Hamiltonian. In Figs. 10.4 and 10.5 wave functions obtained with competitive selection were used. The figures show that the SVM description is indeed accurate.

The density distribution and the correlation function between the particles provide us with more detailed information on the structure of the system. Figure 10.6 shows the nucleon and Λ density distributions

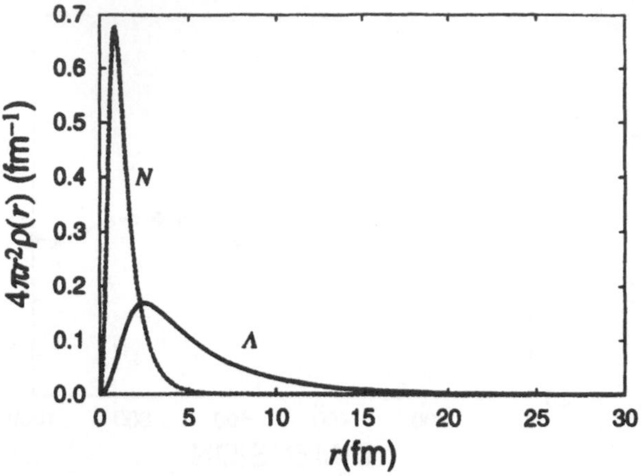

Figure 10.6: Density distributions of nucleons (dashed line) and Λ-particles (solid line) as a function of r, the distance from the c.m. of the pn subsystem (normalized to unity).

defined by

$$\rho_B(\mathbf{r}) = N_B^{-1}\langle\Psi|\sum_i \delta(\boldsymbol{r}_i - \boldsymbol{R}^{N}_{c.m.} - \boldsymbol{r})P_i^{B}|\Psi\rangle, \quad B = N, \Lambda, \qquad (10.6)$$

which differ from Eq. (9.127) in the following respects: (i) the vector $\boldsymbol{R}_{c.m.}$ is replaced by $\boldsymbol{R}^{N}_{c.m.}$, the c.m. coordinate of the nucleonic subsystem; (ii) the projector P_i^B now selects the nucleons ($P_i^B = P_i^N$) or the Λ-particles ($P_i^B = P_i^\Lambda$); (iii) owing to the factor N_B^{-1}, where N_B is the number of particles B, the density is now normalized to unity. The Λ distribution extends much further out than the nucleon distribution. The nucleons stay mostly in a sphere of radius of 3 fm, while the probability of finding the Λ-particles in that sphere is only about 30%. The ${}^{4}_{\Lambda\Lambda}$H nucleus thus exhibits a Λ-particle-halo structure. The exponential tail of the Λ distribution is well described up to long distances in spite of the basis functions' Gaussian falloff.

The function

$$C(\mathbf{r}) = \langle\Psi|\sum_{i<j} \delta(\mathbf{r}_i - \mathbf{r}_j - \mathbf{r})P_i^\Lambda P_j^\Lambda|\Psi\rangle \qquad (10.7)$$

shown in Fig. 10.7 gives information on the correlation between the constituent Λ-particles. The peak of $r^2C(\mathbf{r})$ appears at about 3 fm, and beyond that almost 80% of the probability is distributed over a broad range. This means that the two Λ-particles are outside each other's

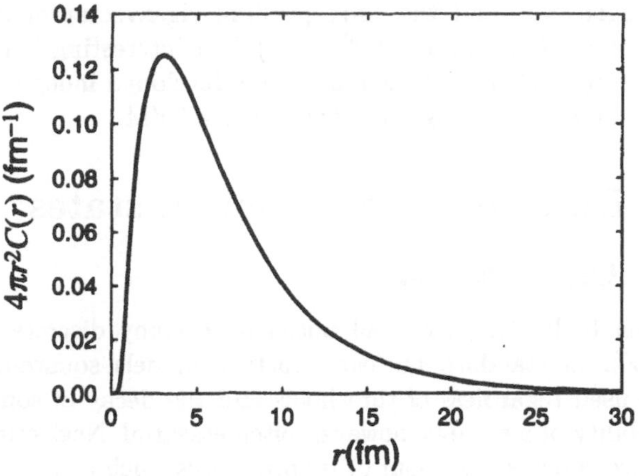

Figure 10.7: Correlation function of two Λ-particles as a function of their distance r.

interaction range (of ∼1 fm) almost all the time. The correlation function has vanishingly small amplitudes at short distances, as it should. This point is still worth emphasizing since the basis has been put together from Gaussians that peak right at zero interparticle distances. The point is not just that the correlated Gaussians are flexible; what is more is that the SVM is capable of picking up suitable functions from a suitable pool of correlated Gaussians.

The demonstration of the performance of the SVM should be concluded by stating that the SVM is very practical for the purpose of setting up a good basis set. That the basis it provides is not strictly optimal is not a serious disadvantage. But its convergence can be improved by some fine-tuning if one can afford to recompute the matrices many times. Such a fine-tuning can be applied to the parameters of the new basis function, which has passed the 'admittance test', and it can involve a deterministic method [239].

It should be mentioned that the SVM requires high numerical accuracy for the calculation of the energy in each step. If the energy were calculated by Monte Carlo integration, which has an inherent statistical uncertainty, the selection procedure would never work. Even when the energy is calculated by a deterministic method, such as a diagonalization, the accuracy of the numerical calculations has to be high. The analytic evaluation of the matrix elements is thus an essential prerequisite in the SVM.

An importance-sampling algorithm called the stochastic diagonalization method [259] has been constructed for the computation of the smallest eigenvalue and the corresponding eigenvector of extremely large matrices (of order up to $10^{35}\times10^{35}$). It is interesting to note that, though the algorithm used there has been developed independently, it includes procedures analogous to those of the SVM.

10.4 Description of unbound states

10.4.1 Classification

In addition to bound states, all nuclei have many discrete unbound states [260]. In standard nuclear structure models square-integrable bases are used regardless of thresholds and the decay of some states. The instability of a state is, however, often essential. Neglecting it may lead to false conclusions about other properties, such as partial widths or distributions of electromagnetic strengths. The problem is especially acute for light exotic nuclei, which often have just one bound state or none. Without consideration of unbound states or disregarding their instability, the description would be incomplete and unreliable.

Unbound states may be put on the same footing as bound states by the concept of the S-matrix pole (see, e.g., [42]). The relevant theory requires some rigorous mathematics, but, in giving a concise summary, we shall keep the level of rigour low. In the simple-minded time-independent scattering theory the S-matrix is a function of the energy or of the wave number [$S=S(E)$ or $S=S(k)$], and its element S_{ij} connects incoming channel i with outgoing channel j. The elements of the row S_{ij} ($j=1,\dots,n$) of the S-matrix are parameters in the asymptotic form of the wave function describing scattering from channel i to all channels.

For a collision in which the outgoing channels also contain just two fragments, the Schrödinger equation can be reduced to a set of 'coupled two-body Schrödinger equations' [cf. Eq. (A.7)]. Let us denote the incoming channel by i and let us assume for simplicity that the interaction is of finite range and that all channels are s-wave channels. For such a system, the asymptotic forms of the radial channel wave functions u look like

$$\begin{aligned} u_i(r) &= k_i^{-1/2}(\mathrm{e}^{-ik_ir} - S_{ii}\mathrm{e}^{ik_ir}), \\ u_j(r) &= -k_j^{-1/2}S_{ij}\mathrm{e}^{ik_jr} \qquad (j \neq i), \end{aligned} \tag{10.8}$$

where k_j are linked to E by $k_j \sim (E - E_1^{(j)} - E_2^{(j)})^{1/2}$, with $E_\alpha^{(j)}$ being the energy of fragment α in channel j.

As a mathematical concept, the Schrödinger equation can be extended to any complex E and $k \sim \sqrt{E}$, which implies an analytic continuation of S as a function of E or k to the entire complex plane.

The elements of the S-matrix will be analytic functions of E or k except for the essential singularities[4] at the threshold energies and for their poles [42]. At a pole of an S_{ij}, the wave function u_j, as defined in Eq. (10.8), blows up, but this does not imply that the problem becomes meaningless. To illuminate the meaning of a pole, it is useful to renormalize the wave function so as to keep it finite at the pole as well. Let us assume that there is a pole in $S_{ii}(k)$ at $k=\bar{k}$, and let us choose the overall norm of the wave function such that the asymptotic forms of the channel wave functions u may look like

$$\begin{aligned} u_i(r) &\sim k_i^{-1/2}(S_{ii}^{-1}\mathrm{e}^{-ik_ir} - \mathrm{e}^{ik_ir}), \\ u_j(r) &\sim -k_j^{-1/2}S_{ii}^{-1}S_{ij}\mathrm{e}^{ik_jr} \qquad (j \neq i). \end{aligned} \tag{10.9}$$

The channel wave numbers k_j corresponding to the pole position $\bar{k}$ will be denoted by $\bar{k}_j$. Then, at $k=\bar{k}$, we have

$$\begin{aligned} u_i(r) &\sim -\bar{k}_i^{-1/2}\mathrm{e}^{i\bar{k}_ir}, \\ u_j(r) &\sim -\bar{k}_j^{-1/2} \lim_{\substack{k_i\to\bar{k}_i \\ k_j\to\bar{k}_j}} (S_{ii}^{-1}S_{ij})\mathrm{e}^{i\bar{k}_jr} \qquad (j \neq i). \end{aligned} \tag{10.10}$$

We see that the amplitude of the incoming wave vanishes and in any other channel ($j\neq i$) there is either no outgoing wave at all or $|S_{ij}|\to\infty$ in the same order as $|S_{ii}|$, i.e., S_{ij} must also have a pole there. The former case happens trivially when there is no coupling between channel j and the rest, but otherwise it is just accidental [42]; it may be caused by a coincidence of the pole with a recoupling caused by dynamic effects, which is also a kind of singular event. It seems thus justifiable to say that, if S_{ii} has a pole at a particular wave number, in general S_{ij} have poles of the same order for all j. By similar arguments one can convince oneself that a pole of any particular S_{ij} ($j\neq i$) generally implies poles in S_{ij} for all j.

The fact that u_i at the pole contains no incoming wave at all actually means that the wave function is independent of i. Indeed, in the limit $k\to\bar{k}$ the same wave function can be reached whichever channel contains the incoming wave for $k\neq\bar{k}$. It thus follows that, in general, *all elements of the S-matrix have poles at the same energy.* The converse is also true: if the wave function obeys outgoing boundary conditions in

[4]An n-channel S-matrix would be a 2^n-valued function since the square root in each $k_j \sim (E-E_1^{(j)}-E_2^{(j)})^{1/2}$ can take either sign. To make it single-valued, a 2^n-sheet Riemann surface is defined. The sheets join along the n branch cuts, which follow the positive real energy axis, with one cut starting at each threshold energy. The essential singularities are the starting points of the cuts (the so-called branch points) at the thresholds.

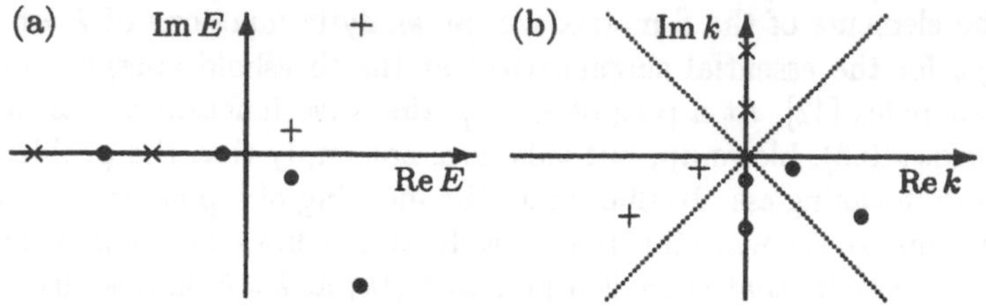

Figure 10.8: Typical positions of the bound-state (×), resonance (•), antiresonance (+) and virtual-state (∘) poles in the complex energy plane (a) and in the complex momentum plane (b). In the k-plane the physical resonances lie between the Re k axis and the dotted lines.

all channels (except, possibly, for channels with vanishing asymptotic wave functions), the S-matrix has a pole there.

It should be added that not only the positions but also the orders of the poles of all S-matrix elements are the same. One can easily verify that, if $S(E)$ has a first-order pole at $E = \bar{E} \equiv (\hbar^2/2m)\bar{k}^2$, so does $S(k)$ at $k = \bar{k}$ and *vice versa*.[5] Higher-order poles can be regarded as resulting from the coincidence of first-order poles (degeneracy) [261].

The distinctive feature of the wave function at the pole is that it contains no incoming wave. Such a state cannot be an actual physical state (remember: it belongs to a complex energy!), but can be considered an approximation to a (time-dependent) state that has been created in a collision, but has lived so long that it has forgotten in which channel it was born. Such a state may provide a precise realization of a compound resonance according to Bohr's concept.

The bound-state asymptotics $e^{-\kappa r}$ can be rewritten as $e^{-\kappa r} = e^{ikr}$, with $k = i\kappa$, which is like Eq. (10.10). Indeed, for a bound state the energies are negative and the wave numbers are positive imaginary in all channels: $k_j = i\kappa_j$, where $\kappa_j > 0$, and there is no incoming wave in any channel. From Eq. (10.10) it follows that *at any bound-state energy the S-matrix has a pole.*

For simplicity, let us deal with single-channel problems and omit channel labels. The positions of the S-matrix poles of the important type are shown schematically in Fig. 10.8. The S-matrix is unitary. For a single-channel problem the unitarity reads as $|S(E)|^2 = 1$, so $S(E)$ can be parametrized as $S(E) = e^{2i\delta}$, where δ is the (real) phase shift. The total (elastic) cross section for a partial wave l will be [26]

$$\sigma_l = \frac{\pi}{k^2}(2l+1)|1-S|^2. \tag{10.11}$$

[5] At the vicinity of a pole, $S(E) \sim (E-\bar{E})^{-1} \sim [(k+\bar{k})(k-\bar{k})]^{-1} \sim (k-\bar{k})^{-1}$.

The unitarity relation can be continued analytically as

$$[S(E^*)]^* S(E) = 1 \tag{10.12}$$

[262, 263]. From this it follows that, if $S(E)$ has a first-order pole at $E_R - \frac{i}{2}\Gamma$, it is bound to have a zero[6] at $E_R + \frac{i}{2}\Gamma$. (The energies E_R and Γ are real and positive.) Let us assume that the pole is close enough to the real energy axis for the effect of the pole to dominate. Then $S(E)$ can be written near E_R as

$$S(E) \approx e^{2i\phi} \frac{E - E_R - \frac{i}{2}\Gamma}{E - E_R + \frac{i}{2}\Gamma}, \tag{10.13}$$

where ϕ is a constant phase factor. This is the Breit–Wigner formula for the S-matrix. The Breit–Wigner formula is simplified when the 'background phase' ϕ can be taken to be zero. Then the cross section (10.11) reduces to the well-known form:

$$\sigma_l(E) \approx \frac{\pi}{k^2}(2l+1)\frac{\Gamma^2}{(E - E_R)^2 + \frac{1}{4}\Gamma^2}, \tag{10.14}$$

which shows a clear-cut peak of width Γ at energy E_R. Thus *a first-order pole of $S(E)$ or $S(k)$ that lies in the fourth quadrant of the* complex E- and k-plane (i.e., $E = E_R - \frac{i}{2}\Gamma$, $k = \kappa - i\gamma$, with $E_R, \Gamma, \kappa, \gamma > 0$, see Fig. 10.8) *gives rise to a Breit-Wigner resonance*, and that is why resonances are associated with S-matrix poles. This association has proved very satisfactory even though it is not impossible that a Breit–Wigner-like peak might be produced without an S-matrix pole. It may happen that $\kappa < \gamma$, so that $E_R = \hbar^2(\kappa^2 - \gamma^2)/(2m)$ becomes negative. That is unphysical of a resonance, and a pole of this kind is called an 'unphysical resonance'. The k plane is thus divided into physical and unphysical regions by the lines $\mathrm{Im}\, k = \pm \mathrm{Re}\, k$. Whether physical or not, a resonance pole will only affect $S(E)$ for real energies if Γ is sufficiently small.

---

[6]Suppose that $S(E)$ is given by $a(E)/(E - E_R + \frac{i}{2}\Gamma)$ near the pole, with $a(E_R - \frac{i}{2}\Gamma) \neq 0$. Then the unitarity relation (10.12) entails $[a(E^*)]^* a(E)/[(E - E_R - \frac{i}{2}\Gamma)(E - E_R + \frac{i}{2}\Gamma)] = 1$, which implies $a(E_R + \frac{i}{2}\Gamma) = 0$. That is, $S(E)$ is given by $[(E - E_R - \frac{i}{2}\Gamma)/(E - E_R + \frac{i}{2}\Gamma)]b(E)$ with $[b(E^*)]^* b(E) = 1$.

In Fig. 10.8a there is a pole at $E_R + \frac{i}{2}\Gamma$ corresponding to the pole at $E_R - \frac{i}{2}\Gamma$, which apparently contradicts $S(E_R + \frac{i}{2}\Gamma) = 0$. This contradiction can be resolved by noting that $S(E)$ is a double-valued function, owing to the sign of $k \sim \pm\sqrt{E}$. It can be made single-valued by doubling the complex E-plane into two Riemann sheets defined by $\mathrm{Im}\, k \geq 0$ and $\mathrm{Im}\, k < 0$, respectively. At $E_R + \frac{i}{2}\Gamma$ the S-matrix does indeed have a zero on the Riemann sheet with $\mathrm{Im}\, k > 0$ and does indeed have a pole on the Riemann sheet with $\mathrm{Im}\, k < 0$.

The wave function belonging to a resonance pole is called a Gamow wave function. The tail of a Gamow wave function oscillates with exponentially increasing amplitudes: $e^{ikr} = e^{i(\kappa - i\gamma)r} = e^{i\kappa r}e^{\gamma r}$. Such a 'state' is obviously not a proper quantum-mechanical state. While looking pretty unphysical, this behaviour is a natural consequence of what is expected from a resonance state. A resonance state should decay; and $E_0 = E_R - \frac{i}{2}\Gamma$ implies an exponential decay[7] of lifetime $\hbar/\Gamma$. But to be stationary at the same time, it has to oscillate with a spatially increasing amplitude.

There is one more notable kind of S-matrix pole: that which is called a 'virtual' or 'antibound' state. A virtual state belongs to negative real energies like bound states, but negative imaginary wave numbers: $k = -i\kappa$, with $\kappa > 0$ (see Fig. 10.8). Obviously, $\kappa > 0$ implies a wave-function tail of $e^{\kappa r}$, which shows that a virtual 'state' is farthest from a proper state. It is not a 'state', yet its position may be a notable property of a quantum-mechanical problem. A virtual state (or an unphysical resonance) near the threshold causes the phase shift to behave in a particular way [262].

It is also to be noted that if a certain radial wave function $u^{(k)}(r)$ satisfies the Schrödinger equation with (complex) energy $E = E_R - \frac{i}{2}\Gamma$ and asymptotic form e^{ikr}, then so does $u^*(r)$, with energy E^* and asymptotic form e^{-ik^*r}. When E_R, $\Gamma > 0$, the state at E^* corresponds to a quasistationary state, which may be considered to describe capture, and it is sometimes called an antiresonance. Thus the poles of $S(k)$ off the imaginary k-axis occur pairwise, symmetrically to this axis [cf. Fig. 10.8]. The poles on the imaginary k-axis also occur pairwise, but there is no symmetry between them. In fact, any bound state has a virtual-state counterpart. (A virtual state, however, may form a pair with another virtual state.)

10.4.2 Localization of resonances

The standard techniques to find the poles of the S-matrix in a theoretical model work well mostly for simple interactions and systems. When there are only two-body channels, a pole can be found by solving the scattering problem at many complex energies and by searching for the singularity [264]. Because of Eq. (10.10), finding the S-matrix poles is equivalent to solving an eigenvalue problem with outgoing boundary condition. The problem of resonances or any other S-matrix poles is thus a generalization of bound-state problems. A direct numerical

[7] This can be verified by substituting $E_R - \frac{i}{2}\Gamma$ for the energy E in the time factor of a stationary state $\psi = e^{-\frac{i}{\hbar}Et}\psi_0$ and considering the time-dependence of the probability $|\psi|^2$.

solution to this eigenvalue problem is only feasible for two-body problems [265]. A few-body or few-cluster problem would require variational techniques [266], but these are well-established and simple enough only for bound states. Therefore, several attempts have been made to trace back the unbound-state problems to bound-state problems.

There are two methods of this type widely used in atomic and nuclear physics. In the first method, called the real stabilization method [267,268], the Schrödinger equation is solved in a box, i.e., on a square-integrable (L^2) basis, which makes all solutions look like bound states. From among the bound-state-looking discrete states, those corresponding to the resonances are singled out by exploiting their stability against changes of the box size. Typical trajectories of resonance poles as functions of the basis dimension (which is related to the box size) are shown in Fig. 10.9. The method turns out to be suitable only for narrow resonances with decay into two-body channels.[8]

In the second method, called the *complex scaling method* (CSM) [273, 274], the coordinate r is rotated into the complex plane by the transformation $r \to e^{i\theta} r$ with $0 < \theta < \frac{\pi}{2}$, which may transform the resonance wave functions into square-integrable functions, while leaving the pole position intact. The condition for the Gamow wave function to become square-integrable can be found out by replacing r with $e^{i\theta}r$ in the asymptotic form e^{ikr}:

$$\exp\left(ike^{i\theta}r\right) = \exp\left[i|k|e^{i(\theta-|\arg(k)|)}r\right] \quad (k = |k|e^{-i|\arg(k)|}). \quad (10.15)$$

This is square-integrable provided $\theta > |\arg(k)|$. From this it follows that a resonance can be found as a bound-state-like solution to the transformed problem provided its wave number obeys $|\arg(k)| < \theta$.

The transformation (or 'scaling') also rotates the energy continua[9]

[8] An alternative way of locating resonances is to search for stability of solutions against a (real) spatial scaling of the problem [269]. Another procedure is based on calculating the level density of the continuum, which closely follows the pattern of resonances. The continuum level density may be calculated approximately by smoothing a discrete level density obtained by an L^2 discretization [270]. The R-matrix microscopic scattering formalism also uses L^2 discretization and is an efficient tool for localizing resonances [271].

[9] To see the transformation of the energy continua of the scattering problem, one should look at the transformation of the asymptotic form (10.8) upon $r' = re^{i\theta}$ and introduce a new wave number, $k_j' = k_j e^{i\theta}$. The asymptotic form will then look like

$$u_j(r) \sim k_j^{-1/2}[\delta_{ij}e^{-ik_jr'} - S_{ij}'(k_j)e^{ik_jr'}] \sim k_j'^{-1/2}\{\delta_{ij}e^{-ik_j'r} - S_{ij}'[k_j(k_j')]e^{ik_j'r}\},$$

where S' is the transformed S-matrix. The relationship $k_j' = k_j e^{i\theta}$ implies $E_j' = E_j e^{2i\theta}$ for the energies with respect to the nearest threshold from below. The asymptotic form of the wave function of the rotated problem at k_j' is thus similar to that of the original problem at k_j. Indeed, the continua are formed by the positive real

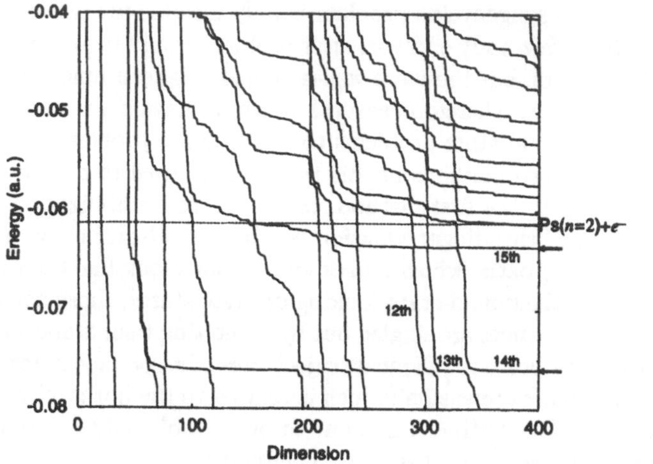

Figure 10.9: Typical pole trajectories as functions of the basis dimension controlling the box size. (0^+ resonances of the $Ps^- = e^- + e^+ + e^-$ system with the SVM with correlated Gaussians. Energy measured from the three-particle threshold.) Each trajectory has a plateau around the resonance energy, and at any dimension there is an eigenvalue near the resonance. The length of the plateau shows staggering due to the randomness of the basis. Some trajectories are labelled by their sequence numbers. [272]

from the real energy axis by angles -2θ. A typical pattern of a rotated energy plane is shown in Fig. 10.10.

The transformed problem can also be solved on a square-integrable basis, and the solution looks again like closing the system into a box. It involves the solution of a complex eigenvalue problem. All eigenenergies will of course be discrete. Some will belong to bound states and resonances E_i that lie close enough to the real energy axis ($|\arg E_i| < 2\theta$), and the rest will represent discretized points of the rotated continua. The eigenenergies that represent resonances can be distinguished from the continuum points by being stable against the choice of θ, while the discretized continuum is rotated with varying θ [cf. Fig. 10.10]. This method has the advantage that it can be applied to complicated problems as well, even with three- or more-body final states [275]. Its limitation is that it works well for narrow enough resonances only; broad resonances, let alone virtual states, cannot be located in this way. An

values of E'_j. These are related to the 'physical' energies as $E_j = E'_j e^{-2i\theta}$. It is therefore plausible that the continua get rotated by -2θ. [The mathematical meaning of the continua is that they are cuts of the Riemann surface of $S(E)$, cf. footnote on p. 261].

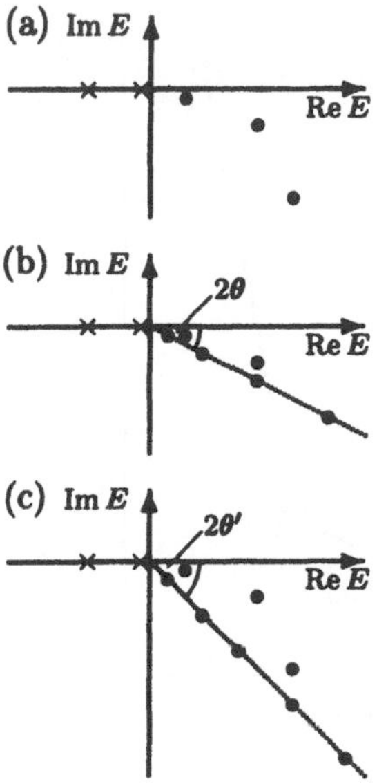

Figure 10.10: Bound-state (×) and resonance (•) poles of a single-channel problem in the complex energy plane without rotation (a), the eigenvalues obtained by diagonalization after rotation by θ (b), and by θ' (c). The discretized continuum eigenvalues are denoted by ∘. The third resonance pole cannot be found in panel (b) since θ is not large enough.

example for the behaviour of a resonance pole with respect to the varying of θ is shown in Fig. 10.11. The basis set used may give a good approximation for the rotated wave function of the resonance but, of course, cannot be complete. That is why the complex energy depends slightly on θ. The point where the rate of change shows a minimum is considered the position of the resonance.

A third bound-state type method [276] is that of the *analytic continuation in a coupling constant* (ACCC). This approach draws on the intuitive observation that any unbound state can be made bound by making the interaction sufficiently more attractive. The unbound-state problem is thus first transformed into a bound-state problem by artificially changing a strength parameter (the 'coupling constant') in the potential. The bound-state problem is solved for a number of artificial values of this strength parameter, and the result is extrapolated to its physically meaningful value.

It is not trivial that such an extrapolation can be made since a resonance state looks so different from a bound state, and the pole trajectory produced by varying the potential strength often suffers a

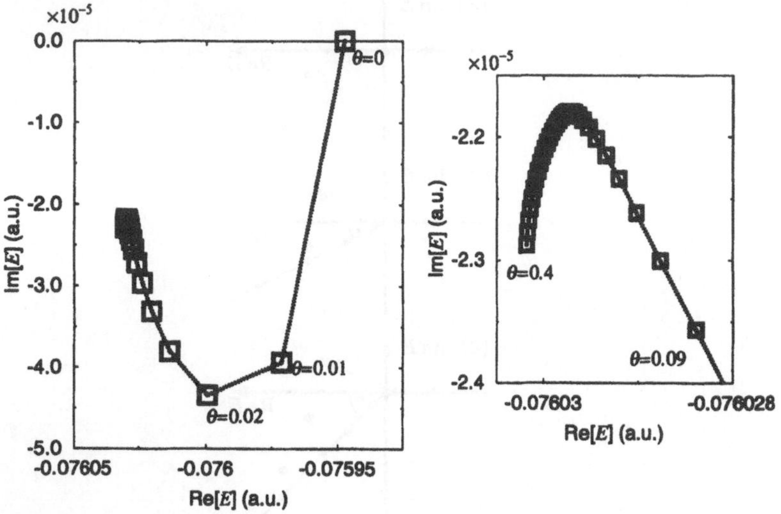

Figure 10.11: Pole trajectory in the complex energy plane of one of the resonances of Ps^- shown in Fig. 10.9 as a function of the rotation angle θ, incremented by steps of 0.01. The vicinity of the stationary point is magnified on the right. [272]

fracture at the threshold (Fig. 10.12). That an extrapolation is, nevertheless, possible is indicated by two observations. First, in the region in which the potential is attractive a resonance wave function looks exactly like a bound-state wave function and, second, the apparently so different tail of the resonance wave function is actually the analytic continuation of the bound-state tail. Moreover, as will be seen in Sect. 10.5, the fracture of the trajectory (which may or may not be at the threshold) is a trivial effect. The complex wave number of the pole, k, depends on the real potential strength λ as $k(\lambda)=f(\sqrt{\lambda-\lambda_0})$, where λ_0 is a constant [see Eq. (10.16) later]. This trajectory is bound to suffer a fracture and a bifurcation where λ passes through λ_0. The fracture can be explained without assuming that $f(x)$ is ill-behaved, and there is empirical evidence that $f(x)$ is indeed a smooth function, which can be extrapolated.

This extrapolation is actually an analytic continuation of the real bound-state eigenfunction to the complex wave-number region. The extrapolation is usually made through Padé approximations. The coefficients of the Padé approximant are determined by the real eigenvalues of the Hamiltonian for several different potential strengths. Thus only real matrix elements of the Hamiltonian are needed, and these need not be calculated repeatedly since the set of potential matrices to be used in the bound-state problems differ from each other just in the value of a strength. This is contrary to the real stabilization method or the CSM. Moreover, there is no obvious limit for the position of the pole to be found; even virtual states are amenable. The extrapolation requires very accurate bound-state energies, which the SVM is able to produce.[10] Although the ACCC method is much less wide-spread than the other two, some of the more interesting cases for unbound states in the correlated Gaussian approach have been produced by the ACCC method. Therefore, in Sect. 10.5 we shall illustrate its performance.

10.5 Analytic continuation in the coupling constant

10.5.1 Pole trajectories

To extrapolate the bound-state results to the positive-energy region, one has to know the qualitative behaviour of the poles with respect to changes of the potential strength. To discuss this, it is useful to write the Hamiltonian of the system as $H(\lambda)=H_1+\lambda H_2$, where H_2 is an attractive interaction. Let us consider a state that is unbound for the 'physical' value of λ but assume that there exist some larger real values of λ for which this state is bound. When λ is decreased, the bound state approaches the threshold and may become a resonance or a virtual state. The value of λ for which $E(\lambda)=0$ will be denoted by λ_0.

As an illustration for the behaviour of the poles as functions of the potential strength, let us consider a problem with a square-well potential, with $H_1=\boldsymbol{p}^2/2m$, $H_2=V(r)$, where $V(r)=-V_0\Theta(a-r)$. The Schrödinger equation can then be solved analytically, which makes it possible to locate the S-matrix poles easily. The examples to be shown are for $l=0$ and 1.

In the uppermost panel of Fig. 10.12 the poles that lie close to the origin are displayed. Seen are two bound states, one virtual state and a pair of resonant-antiresonant states in the s-wave and one bound state, one virtual state and a pair of resonant-antiresonant states in the p-wave. The middle and lower panels follow the poles when the potential is made less attractive. This is the path that can only be followed through exactly for model problems like that under study. For realistic problems the pole trajectory will be pursued by extrapolation.

[10]The closer the energy is to the threshold, the more difficult it is to achieve accuracy; in this respect the SVM is not an exception.

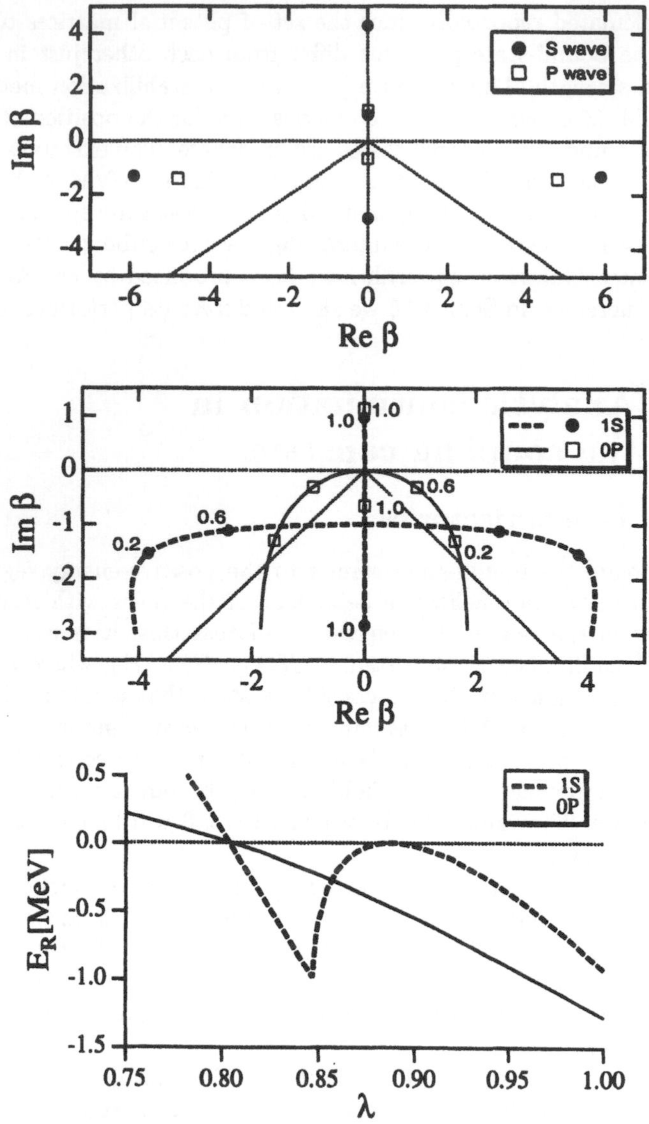

Figure 10.12: S-matrix poles in a square-well potential of radius a and depth $-\lambda V_0$. The parameters [260, 277] are $\hbar^2/2m = 4$ MeV fm^2, $a = 2$ fm and $V_0 = 25$ MeV for $l = 0$ and 12.25 MeV for $l = 1$. Uppermost panel: s- and p-wave poles in the complex $\beta = ka$-plane; $\lambda = 1$. The oblique lines mark $\text{Im}\,\beta = \pm\,\text{Re}\,\beta$. Middle panel: s- and p-wave pole trajectories in the β-plane, with some λ values displayed. The oblique line as above. Lowest panel: (real parts of) the energies of the 1s and 0p states as functions of λ.

The middle panel displays the trajectories of one pair of poles for each l obtained by decreasing λ. By decreasing λ, the s-wave bound-state and virtual-state poles approach each other along the imaginary axis and merge at some negative imaginary value. By further decreasing λ, the poles move, symmetrically, perpendicularly to the imaginary axis into the fourth and third quadrants, respectively. This is typical for all s-wave resonances that decay into two fragments. The non-s-wave poles behave differently. The bound-state p-wave pole meets its virtual-state partner at the origin, and both leave the origin, symmetrically with respect to the imaginary axis and tangentially with respect to the real axis. In Fig. 10.12 the p-wave resonance pole passes through the physical region and eventually enters the unphysical region (cf. p. 263). The lowest panel shows the real parts of the 1s and 0p pole energies (E_R) as functions of λ. The two curves behave very differently near the threshold. By decreasing λ, the s-state energy touches the $E_R=0$ axis, and turns negative again—reflecting the fact that the pole moves, through the origin, to the negative imaginary k-axis—and it shows a kink where $\mathrm{Re}\,k$ becomes non-zero. On the other hand, the passage of the p-wave state through the threshold is smooth.

The p-wave differs from the s-wave significantly owing to the r^{-2}-like centrifugal barrier, which may give rise to long-lived, i.e., narrow resonances. A Coulomb barrier with its r^{-1}-shape tail acts even more effectively as a confining force, but the pole trajectories in such cases will qualitatively look the same as the p-wave example of Fig. 10.12. It

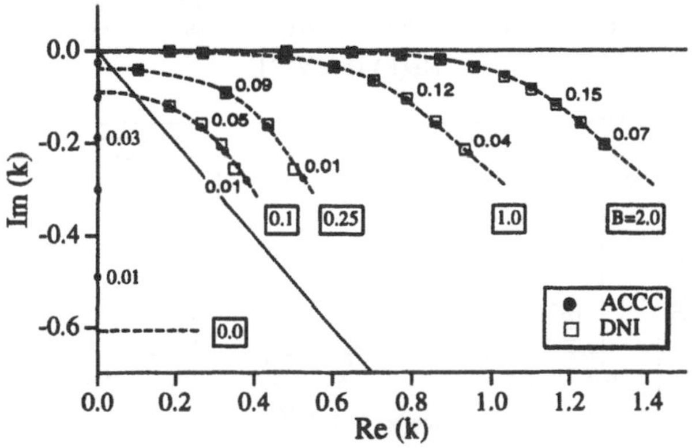

Figure 10.13: Pole trajectories in the two-range Gaussian potential $V(r)=-8\lambda\exp[-(r/2.5)^2]+B\exp[-(r/5)^2]$ ($\hbar=c=m=1$). The trajectories are obtained by varying λ, and the points marked belong to equidistant λ. The oblique line marks $\mathrm{Im}\,k=-\mathrm{Re}\,k$. ACCC results are compared with those with direct numerical integration (DNI) [265]. The boxed numbers are B values.

is then natural that with finite-range barriers the trajectories behave in an intermediate way. Some s-wave resonance trajectories are compared in Fig. 10.13 for the case of a Gaussian well plus a Gaussian barrier with five barrier heights B (including $B=0$). The with-barrier case is seen to differ from the no-barrier case in the position of the branching-off of the trajectory. The trajectories for $B=1$ and 2 leave the imaginary k-axis near the origin, entering almost immediately the region of physical resonances in a way apparently indistinguishable from the trajectories of higher partial waves.

When the resonance decays predominantly in one step into more than two fragments, there is again a qualitative change in the pole trajectory. This is illustrated in Fig. 10.14, where two pure s-wave three-body resonance trajectories are shown: one with a barrier and the other without. There are no bound states (nor resonances) in the two-body subsystems in either case, so the system decays to three-body final states, indeed. It is seen that the branching-off point is (close to) the origin, irrespective of whether the potential has a barrier or not.

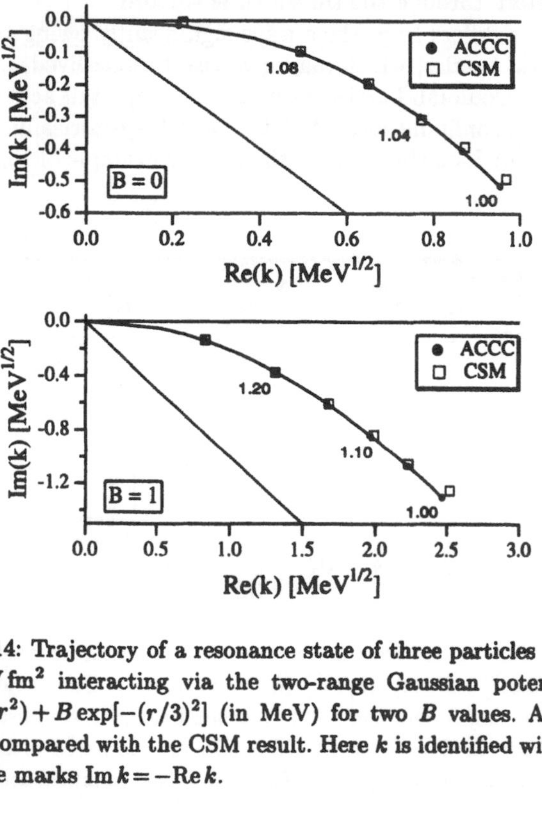

Figure 10.14: Trajectory of a resonance state of three particles with $\hbar^2/m=$ 41.47 MeV fm^2 interacting via the two-range Gaussian potential $V(r)=-80\exp(-r^2)+B\exp[-(r/3)^2]$ (in MeV) for two B values. ACCC trajectories are compared with the CSM result. Here k is identified with $E^{1/2}$. The oblique line marks $\operatorname{Im}k=-\operatorname{Re}k$.

Thus the behaviour of a three-body s-wave resonance is similar to a two-body case with finite barrier. This is consistent with the usual interpretation that the three-body system has an effective barrier against disintegration even for zero orbital angular momentum (see, e.g., p. 69 of Ref. [187]).

10.5.2 Analytic continuation of pole trajectories

The examples have shown that the bound-state pole trajectory along the imaginary k-axis obtained by decreasing the strength λ of the attractive potential bifurcates at one point. The bifurcated trajectory is perpendicular to the imaginary k-axis. This kind of bifurcation is produced most simply by a function $i\sqrt{\lambda-\lambda_0}$, where λ_0 is the λ value belonging to the bifurcation. For s-wave poles it has been proved that $k(\lambda) \sim i\sqrt{\lambda-\lambda_0}$, indeed [260, 276]. It has also been proved that, for $l>0$, $k(\lambda)$ is an analytic function in the vicinity of the origin. It is just natural to assume that it ceases to be analytic just at the bifurcation. For $\lambda>\lambda_0$ the pole is on the imaginary axis, while for $\lambda<\lambda_0$ its trajectory is perpendicular to it. The simplest possible parametrization that bears out this behaviour in the vicinity of $\lambda \simeq \lambda_0$ is

$$k(\lambda) + i\kappa_0 \sim i\sqrt{\lambda - \lambda_0} \quad \text{for } \lambda \simeq \lambda_0, \tag{10.16}$$

where $-i\kappa_0$ ($\kappa_0 \geq 0$) is the position of the bifurcation point. It has been shown that this parametrization works well [278]. As was discussed in Sect. 10.5.1, $\kappa_0>0$ for two-body decay into an s-wave without a strong barrier and $\kappa_0 = 0$ (exactly or approximately) otherwise. It is thus convenient to introduce the variable

$$x = \sqrt{\lambda - \lambda_0}, \quad \text{with } k(\lambda_0) \text{ the bifurcation point}, \tag{10.17}$$

and extrapolate $k(\lambda) \equiv k[\lambda(x)]$, which is assumed to be an analytic function in the bound-state region, to the region $\lambda<\lambda_0$. The behaviour (10.16) of $k(\lambda)$ allows one to perform the extrapolation with a Padé approximant of the form

$$k(\lambda) = i\,\frac{c_0 + c_1 x + c_2 x^2 + \cdots + c_M x^M}{1 + d_1 x + d_2 x^2 + \cdots + d_N x^N} - i\kappa_0. \tag{10.18}$$

The resulting value of c_0 should be zero; c_0 may be kept as a parameter for numerical check. Since for the no-barrier case λ_0 is not known *a priori*, it has to be determined iteratively, from the stability of the extrapolated trajectory.

The $M+N+1$ coefficients of the $[M,N]$ Padé approximant (10.18) are calculated in the bound-state region, and they are therefore real. If

$\lambda < \lambda_0$, then x will be imaginary and $k(\lambda)$ may become complex. The bound-state problem has to be solved for various values of the coupling constant λ ($> \lambda_0$) and c_i and d_i have to be determined by a fitting procedure along with λ_0. For a reliable approximation, one has to solve the bound-state problem to high precision (typically 4 or more digits) and consistently from deep to weak binding up to the threshold, and that demands high accuracy from the variational calculation. The SVM is capable of such a precision, and, for not more than three clusters, even grid-type Gaussian bases are suitable. An SVM basis can produce higher precision with the same dimension than a grid basis, but the optimum SVM basis is not the same for a deep-lying state as for a near-threshold state. Basis optimization for each λ value may cost more computing time than using one larger (grid) basis for all λ's.

As a test for the ACCC, Fig. 10.13 shows a comparison of the ACCC with the direct numerical integration. For a three-body model a direct calculation is not possible. Therefore, for the case of Fig. 10.14, the ACCC results are compared with the CSM. The tests show the ACCC method to be satisfactory, and similar agreement with the CSM can be obtained for realistic three-cluster calculations [278,279]. The examples show that the method is indeed applicable to any poles whose trajectory can be traced back to the bound-state region, and the accuracy required by the description of the important unbound states of light exotic nuclei is often attainable. Since overlapping resonances become well-localized bound states when the binding is strengthened, their extrapolation from bound states does not pose any extra problems.

The ACCC method has been introduced for the case of a single decay channel, but that is not a strong restriction. In a multichannel case the main point to be observed is that extrapolation has to be done from below the lowest-lying threshold since the variational energies are eigenenergies only in the true bound-state region. If possible, it is preferable to choose the coupling constant λ such that it shifts the resonance without shifting the thresholds. In general, however, that is not possible. Then due care is to be taken of the energies of the subsystems as well to ensure that all calculations used in the extrapolation be genuine bound-state calculations. Such a problem will be discussed in Sect. 13.4.2. No problem should arise from the multilevel structure of the Riemann surface of multichannel problems.[11]

---

[11]In an n-channel problem $S(E)$ has 2^n Riemann sheets (cf. footnote on p. 261), of which all but two are remote from the physical energy axis. There are 2^{n-1} poles belonging to each resonance [280]. When the coupling is weak, each of them is located on a different sheet. The scattering observables are affected only by one pole, that lying on a sheet adjacent to the physical energies ('the resonance pole'), apart from exceptional cases [264]. It is the resonance poles that are 'continuations' of the bound-state poles, so the extrapolation should follow these poles.

As applied so far, the ACCC method is a mere extrapolation of the energy eigenvalues from the bound-state region to the unbound-state region. But the energy extrapolation works because the unbound-state wave function is an analytic continuation of the bound-state wave function. If that is so, however, the wave function itself could be extrapolated at every point of the configuration space, which would actually amount to a numerical analytic continuation. Since the counterpart of this procedure for the CSM (the 'back-rotation' of the wave function) is very unstable [281], it would be interesting to see how this procedure works for the ACCC method. A more practical approach would be, however, to make extrapolations for the matrix elements of observables, but no examples are available as yet.

Chapter 11

Cluster models

11.1 Preliminary notes

As was stated at the beginning of Chap. 9, the complete version of the correlated Gaussian basis becomes too complicated when applied to systems as large as most of the exotic nuclei of current interest. To make this assertion plausible, it perhaps suffices to mention the numbers of topologically independent systems of relative coordinates (cf. Fig. 9.2) for a few important cases. For $A = 7$, 8, 11 and 16 identical particles this number is 11, 23, 207 and 10905, respectively. It is thus obvious that even stepping from $A = 7$ to 8 is a great leap; but, tackling, e.g., ^{11}Li or, let alone, ^{16}O, is beyond our reach in the foreseeable future. To simplify the basis, one can omit some of the spin or isospin configurations (e.g., spin 1 for like nucleon pairs), or one may include fewer orbital angular momenta. One may as well truncate the Gaussian expansion. For the relative motion of fragments between which the breakup energy is high, one may retain just a single term. For instance, a partitioning of ^{6}He into d+d+(nn) is not likely to influence the neutron-halo structure because the asymptotic part of the relative wave function of these three subsystems is damped rapidly. This may be expressed by saying that this partitioning splits up a 'stable' subsystem in the nucleus, an α-cluster. On the other hand, for short relative distances a basis function for this arrangement overlaps with others very much, thus can be omitted with confidence. This reasoning leads naturally to the conclusion that it is safe to build up the wave function from blocks in which the stable subsystems are frozen or, at least, their intrinsic state spaces are thoroughly truncated.

These rather technical considerations can also be said in more physical terms: light exotic nuclei tend to behave as composed of well-defined subsystems called 'clusters'. The most well-known examples for clustering are α-decaying states [282], but the generality of clustering phenomena is shown by the fact that α-particles, tritons and helions can be picked up or knocked out from a number of nuclei, especially light nuclei, with substantial probability. This suggests that the states of the clusters can be used as building blocks in describing the nuclear states concerned [199]. Furthermore, one can view any colliding systems as composed of two clusters, the target and the projectile.

The subject of clustering appears here in two contexts: as a measurable property of the nuclei and as a cornerstone of their description. In this chapter both lines will be followed through. The cluster fragmentation properties of exotic nuclei are both characteristic and accessible to experiments. It was seen in Part I that the most important reaction channel for halo nuclei is a breakup, which takes place predominantly as a direct reaction. On the other hand, the cluster model seems, at least in principle, general enough to reproduce all specific properties of the exotic nuclei. The subject of this chapter is nuclear clustering and cluster models.

We set out in Sect. 11.2 by discussing the phenomenon of clustering and continue in Sect. 11.3 by presenting the formalism of its description in a general model-independent approach. Then in Sect. 11.4 we introduce the common ansatz of the cluster models and discuss the intercluster relative motion. In Sect. 11.5 the basic version of the cluster models, dubbed the 'resonating-group method'[1] (RGM), will be elaborated. Some of the basic concepts can be best elucidated via an HO approximation to the RGM, and that is the subject of Sect. 11.6. The next step is to show a more practical formulation of cluster models, the generator-coordinate method (GCM), which also incorporates the multicentre SM (Sect. 11.7). This is the method that is related to the correlated Gaussian approach most closely. In Sect. 11.8 the equation of motion of macroscopic cluster models will be derived formally from the microscopic picture, which provides a link to the most popular phenomenological models of light exotic nuclei. Since the Pauli principle is formulated in these models as an orthogonality condition, the archetype of these models is called the orthogonality-condition model (OCM). Finally, in Sect. 11.8.2, we shall further elaborate on the relationship between microscopic and macroscopic models. The incorporation of the cluster hypothesis into the correlated Gaussian formalism will be the subject of Chap. 12.

Although the subject of cluster models is rich in physics, this order

[1]This term refers to an oscillatory motion of particles grouped into clusters. The actual meaning is narrower than the literal meaning, and the wording tastes somewhat historical.

of presentation is rather formal because of economic, aesthetic and didactic considerations. We shall, however, not neglect physical aspects. In what sense are the clusters building blocks? In what sense can a system of nucleons be described as a system of clusters? How is it possible that the Pauli principle does not destroy nuclear clusters? How are cluster models related to other microscopic models? Answers to all these intriguing questions will be unfolded from the formalism. In most papers on cluster models these issues remain implicit and sometimes semantic mismatches deteriorate understanding.

The cluster model transcends the confines of nuclear physics. It can be readily applied to other systems consisting of two or more composite particles. For example, two atoms may be treated as two clusters, each consisting of a nucleus and electrons. The adiabatic treatment of molecules [26] can be surpassed following the cluster model (see Sect. 11.8.2) or the interaction between two atoms can be derived from the Coulomb interaction and the atomic wave functions following the derivation of the interaction between nuclear clusters (see Sect. 11.7.3), but taking into account 'cluster distortion' (see Sect. 11.7.4). The nucleon–nucleon force may be derived similarly from the dynamics of 'nucleonic clusters', consisting of quarks and gluons. The interaction between the constituents is governed by quantum chromodynamics, from which it is too difficult to derive the nuclear force. The phenomenological approaches imitating the nuclear cluster model are therefore very useful: some properties of the interaction can be derived and the behaviour of two-nucleon and other two-baryon systems can be elucidated.

For simplicity, the cluster models will be presented for a two-cluster case, and angular momenta and isospin will be treated rather implicitly. Our aim here is to elucidate the physics underlying the models rather than to provide readily applicable formulae. It is straightforward to generalize these formulae to multicluster models. Some of this formalism is explicitly rewritten for three-cluster systems in Sect. 11.8.2.

11.2 Basic concepts of clustering

The propensity of nuclei to behave as composite systems of subsystems that preserve their identity is called clustering. In quantum-mechanical terms one can say that a system behaves as a composite of two clusters if its wave function has a large component of the form of

$$\mathcal{A}_{12}\{\Phi_1(\xi_1)\Phi_2(\xi_2)\phi(\boldsymbol{q})\}, \tag{11.1}$$

where Φ_i are the intrinsic wave functions of cluster a_i, $\phi(\boldsymbol{q})$ is any function of the relative coordinate vector $\boldsymbol{q}$ defined in Eq. (9.17) and the operator $\mathcal{A}_{12}$ takes care of antisymmetry [cf. Eq. (D.7)]. That the wave function has such a component means that it is not orthogonal to a formula of the form (11.1). Thus a state may carry (11.1)-type clustering even though it does not contain a term of the form (11.1) explicitly.

At this stage we should qualify the meaning of the phrase that 'the clusters preserve their identity'. The identity of a cluster in a nucleus is never preserved in one particular sense: one can never pinpoint which nucleons form it. The antisymmetrizer in Eq. (11.1) erases the correspondence between nucleonic and cluster labels, and even the relative vectors connecting clusters are mixed up. In the observables of the united system their contributions cannot be disentangled. Once two clusters overlap, the Pauli principle will squeeze each of them out from the spatial and momentum regions occupied by the other. But all these statements are fully consistent with the state (11.1), and that is what is meant by the clusters' preserved identity.

Inside the nuclei, however, the push and pull by other nucleons distorts the clusters. If Φ_1 and Φ_2 are the g.s. of the two clusters as individual nuclei, the form (11.1) cannot ever be exact. Terms with cluster excited states will be mixed in. In the region of the configuration space of high density it is especially futile to hope to disentangle the two clusters. But, as will be seen in Sect. 11.7.4, in that region the Pauli projection makes any states look rather alike [283], and the distortion effects do not tend to make them very unlike either. In particular, the components $\mathcal{A}_{12}\{\Phi_1^{(i)}(\xi_1)\Phi_2^{(j)}(\xi_2)\phi_{ij}(\boldsymbol{q})\}$ (with $i,j>0$ denoting cluster excited states) will greatly overlap with $\mathcal{A}_{12}\{\Phi_1^{(0)}(\xi_1)\Phi_2^{(0)}(\xi_2)\phi_{00}(\boldsymbol{q})\}$ (where the zeros label cluster g.s.) [284]. These qualitative arguments help to understand why the cluster concept has led to a viable model. These statements can be made quantitative by introducing an amplitude which describes the distribution of a cluster with respect to the rest of the nucleus in much the same way as a s.p. spectroscopic amplitude function does [cf. Eqs. (11.11) and (11.28) later].

It is usual to claim that there is substantial clustering only on the nuclear surface. In microscopic terms this statement should be interpreted as saying that, in the regions of the configuration space where the density is large, there is a large overlap with other configurations. But this does not mean that the component (11.1) is necessarily negligible there. On the contrary, those regions may still contribute to the overlap with the state (11.1) substantially. Cluster models rely on this hypothesis, and the results bear it out. The cluster picture is blurred inside the nuclear medium, but any microscopic picture is blurred there. The concept of clusters inside the nucleus is as viable as the s.p. orbits, which are also subject to smear-out.

As a preparation for the quantitative description of clustering, let us now formulate the problem more precisely.

To be specific, let us consider clustering of a nucleus of intrinsic state Ψ into clusters $a_1, \ldots, a_n$, with mass numbers also denoted by $a_1, \ldots, a_n$ $(a_1 + \cdots + a_n = A)$ and with intrinsic wave functions $\Phi_1(\xi_1), \ldots, \Phi_n(\xi_n)$. The coordinates $\xi_1, \ldots, \xi_n$ are sets of cluster intrinsic coordinates [e.g., intrinsic Jacobi coordinates (see App. B), together with spin–isospin coordinates]. To complement the cluster intrinsic coordinates, one should introduce the cluster c.m. coordinates either as in Eq. (9.15) or as in Eq. (9.14):

$$\bar{\boldsymbol{t}}_j = \frac{1}{a_j} \sum_{i=a_1+\cdots+a_{j-1}+1}^{a_1+\cdots+a_j} \boldsymbol{r}_i,$$

$$\bar{\boldsymbol{\xi}}_j = \frac{1}{\sqrt{a_j}} \sum_{i=a_1+\cdots+a_{j-1}+1}^{a_1+\cdots+a_j} \boldsymbol{r}_i \qquad (j = 1, \ldots, n). \qquad (11.2)$$

Accordingly, the intercluster Jacobi coordinates can either be defined in terms of $\bar{\boldsymbol{t}}_j$,

$$\boldsymbol{q}_j = \frac{1}{\sum_{i=1}^{j} a_i} \sum_{i=1}^{j} a_i \bar{\boldsymbol{t}}_i - \bar{\boldsymbol{t}}_{j+1} \qquad (j = 1, \ldots, n-1), \qquad (11.3)$$

$$\boldsymbol{q}_n = \frac{1}{\sum_{i=1}^{n} a_i} \sum_{i=1}^{n} a_i \bar{\boldsymbol{t}}_i = \frac{1}{A} \sum_{i=1}^{A} \boldsymbol{r}_i \equiv \boldsymbol{R}_{\text{c.m.}}, \qquad (11.4)$$

or in terms of $\bar{\boldsymbol{\xi}}_j$:

$$\boldsymbol{\rho}_j = \sqrt{\frac{a_{j+1} \sum_{i=1}^{j} a_i}{\sum_{i=1}^{j+1} a_i}} \left[\frac{1}{\sum_{i=1}^{j} a_i} \sum_{i=1}^{j} \sqrt{a_i} \bar{\boldsymbol{\xi}}_i - \frac{1}{\sqrt{a_{j+1}}} \bar{\boldsymbol{\xi}}_{j+1} \right]$$
$$(j = 1, \ldots, n-1),$$

$$\boldsymbol{\rho}_n = \frac{1}{\sqrt{\sum_{i=1}^{n} a_i}} \sum_{i=1}^{n} \sqrt{a_i} \bar{\boldsymbol{\xi}}_i = \frac{1}{\sqrt{A}} \sum_{i=1}^{A} \boldsymbol{r}_i = \boldsymbol{x}_A. \qquad (11.5)$$

In describing the clustering and fragmentation properties and the intercluster relative motion, Eq. (11.3) will be used mostly. The formalism of these more conventional cluster Jacobi coordinates is presented in App. B. But in formulating the microscopic cluster models in calculable forms, we shall mostly resort to Eqs. (11.5), which are familiar from Chap. 9.

11.3 Theory of clustering

11.3.1 Cluster subspace

A nucleus can break up into fragments via radioactive decay or direct breakup reactions if the fragments are preformed in it as clusters with some probability. Therefore, clustering and fragmentation properties are closely related. The probability of clustering must be related to some sort of probability amplitude, which might be interpreted as a relative motion wave function. Associated with this, one can define momentum densities etc.

These properties differ somewhat from the bulk properties, which were the subject of Sect. 9.5. While the density and momentum density distributions introduced in Sect. 9.5 are bulk properties, i.e., they are formed by all particles with respect to their c.m., now we shall consider the pattern formed by the relative motion of groups of the constituents. For this reason no permutation-invariant observables[2] can be associated with the interfragment motion (apart from those involving only large interfragment distances[3]), and it is not trivial how to describe this motion without breaking the permutation symmetry.

Reaction theory has been formulated in Part I with disregard of antisymmetry. This is a good approximation as far as the target–projectile relative motion is concerned [cf., e.g., Eq. (3.48)] because of the high energy of the relative motion, but is questionable for the intercluster relative motion within a nucleus [cf. Eq. (4.1)], even if the exotic nucleus as a projectile is assumed to be fully clustered. In this section we will attempt to reconcile the two viewpoints. The intercluster relative wave function has to be reinterpreted as an amplitude function [39,40] associated with clustering, and the cross section has to be scaled by the norm square of this amplitude, which depends on the extent of the preformation of the fragments as clusters.

In Sect. 11.2 clustering has been defined as the property of the state that it has a component of the form of Eq. (11.1). The form (11.1), with well-defined cluster intrinsic wave functions, defines a subspace of the total state space. The nucleus is clustered to the extent that its wave

[2] An observable (a Hermitean operator with complete systems of eigenvectors) can represent directly measurable quantities if it obeys some elementary symmetry properties that the system itself has, e.g., invariance under rotation through 2π or permutation invariance. Such observables are often called ***physical*** observables [26].

[3] It will be shown that antisymmetrization causes finite-range effects. The quantities associated with the interfragment motion may be directly measurable if they can be defined with operators that only refer to the interfragment motion in an asymptotic region. An example is the relative energy, which is a difference between the total energy, which is a bulk property, and the subsystem energies, defined for the subsystems taken apart.

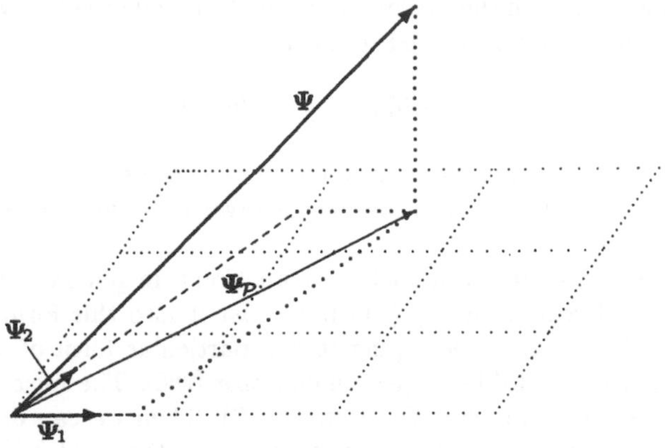

Figure 11.1: Projection $\Psi_{\mathcal{P}}$ of a vector Ψ onto a plane spanned by two (nonorthogonal) vectors, Ψ_1 and Ψ_2.

function belongs to this particular subspace. One can liken the wave function to a spatial vector and then the cluster subspace to a plane (see Fig. 11.1). The component $\Psi_{\mathcal{P}}$ of vector Ψ in the plane is defined by the prescription that $\Psi - \Psi_{\mathcal{P}}$ be perpendicular to the plane, i.e., $\Psi - \Psi_{\mathcal{P}} \perp \Psi_i$, where $\{\Psi_i\}$ $(i = 1, 2)$ is a (not necessarily orthogonal) complete set of vectors in the plane. Thus $\Psi_{\mathcal{P}}$ can always be expressed as $\Psi_{\mathcal{P}} = \phi_1 \Psi_1 + \phi_2 \Psi_2$ with suitable coefficients ϕ_1, ϕ_2 such that the vector Ψ be written as $\Psi = \Psi_{\mathcal{P}} +$ orthogonal terms. Similarly, it must be possible to write the nuclear wave function Ψ as a term belonging to the cluster subspace plus orthogonal terms. One should just find appropriate vectors corresponding to Ψ_i and coefficients corresponding to ϕ_i.

If there is some $a_1 + \cdots + a_n$ clustering in the nucleus, this can be written as

$$\Psi = \mathcal{A}_{1\cdots n} \left\{ \Phi_1(\xi_1) \cdots \Phi_n(\xi_n) \phi(q_1, \ldots, q_{n-1}) \right\} + \text{orthogonal terms.} \tag{11.6}$$

Here $\mathcal{A}_{1\cdots n}$ is an intercluster antisymmetrizer, $\phi(q_1, \cdots, q_{n-1})$ represents the intercluster relative motion, and the 'orthogonal terms' are orthogonal to the first term for any choice of ϕ. The effect of the operator $\mathcal{A}_{1\cdots n}$ is that, from each of the $a_1! \cdots a_n!$ permutations created by $\mathcal{A}_1 \cdots \mathcal{A}_n$ in the cluster intrinsic states, it creates $A!/(a_1! \cdots a_n!)$ permutations involving intercluster transpositions. Moreover, $\mathcal{A}_{1\cdots n}$ carries a normalization factor, such that (cf. App. D)

$$\mathcal{A} = \mathcal{A}_{1\cdots n} \mathcal{A}_1 \cdots \mathcal{A}_n. \tag{11.7}$$

We assume that both the nuclear wave function and the cluster intrinsic wave functions are normalized to unity:

$$\langle\Psi|\Psi\rangle = \langle\Phi_1|\Phi_1\rangle = \cdots = \langle\Phi_n|\Phi_n\rangle = 1, \tag{11.8}$$

where each bra-ket implies integration over the respective intrinsic variables. For simplicity, we first confine ourselves to binary clustering, i.e., $n=2$.

A microscopic nuclear model wave function is in general constructed in a form different from Eq. (11.6). To cast it into this form, one has to define the subspace belonging to the particular form of clustering concerned via a suitable projection operator [285]. The next step is to construct such a projector $\mathcal{P}$ (Sect. 11.3.2). Then we can discuss the various amplitudes that characterize clustering (Sect. 11.3.3). Finally, some methods for the calculation of these amplitudes will be sketched (Sect. 11.3.4).

11.3.2 Projection to the cluster subspace

The subspace of clustering into clusters a_1 and a_2 can be defined by the cluster intrinsic states $\Phi_1(\xi_1)$ and $\Phi_2(\xi_2)$ themselves. The projector should be built up from fully antisymmetrical states. To this end, one may define 'test states'

$$\Psi_{\boldsymbol{R}} = \mathcal{A}_{12}\{\Phi_1(\xi_1)\Phi_2(\xi_2)\delta(\boldsymbol{R}-\boldsymbol{q})\}, \tag{11.9}$$

which describe the clusters pinned down in the relative position specified by the vectorial parameter $\boldsymbol{R}$ ('parameter coordinate' [283]). It is these states that correspond to Ψ_1 and Ψ_2 in Fig. 11.1. The first term of Eq. (11.6), to be written as $\mathcal{P}\Psi$, can then be expressed as

$$\mathcal{P}\Psi = \int d\boldsymbol{R}\,\phi(\boldsymbol{R})\Psi_{\boldsymbol{R}}. \tag{11.10}$$

The cluster component of Ψ is characterized by the amplitude function $\phi(\boldsymbol{R})$, which corresponds to ϕ_1 and ϕ_2 in Fig. 11.1 and is yet to be constructed. It is, however, straightforward to construct another amplitude function, g, which is called the fragmentation amplitude or spectroscopic amplitude:

$$\begin{aligned} g(\boldsymbol{R}) &\equiv \langle\Psi_{\boldsymbol{R}}|\Psi\rangle = \langle\Psi_{\boldsymbol{R}}|\mathcal{P}|\Psi\rangle \\ &= \langle\mathcal{A}_{12}\{\Phi_1(\xi_1)\Phi_2(\xi_2)\delta(\boldsymbol{R}-\boldsymbol{q})\}|\Psi\rangle \qquad (11.11) \\ &= \left(\frac{A!}{a_1!a_2!}\right)^{1/2}\langle\Phi_1(\xi_1)\Phi_2(\xi_2)\delta(\boldsymbol{R}-\boldsymbol{q})|\Psi\rangle, \qquad (11.12) \end{aligned}$$

where the first equality holds because of $\Psi_{\boldsymbol{R}} = \mathcal{P}\Psi_{\boldsymbol{R}}$ and, in the last step, the antisymmetry of Ψ and

$$\mathcal{A}_{1\cdots n}\mathcal{A} = \left(\frac{A!}{a_1!\cdots a_n!}\right)^{1/2}\mathcal{A} \tag{11.13}$$

were used. While the permutation symmetry of Eq. (11.11) is explicit, that of Eq. (11.12) is hidden. It is the latter form that is involved in the theory of direct reactions in which nucleus A is broken into or is created from a_1+a_2.[4] Note furthermore that the δ-functions are inserted just to facilitate the bra-ket notation for overlaps of a function of A–1 variables with a function of $(a_1-1)+(a_2-1) = A-2$ variables. Instead of $g(\boldsymbol{R})$, one may as well use its Fourier transform, which is a momentum-space representative of the spectroscopic amplitude:

$$\begin{aligned}\tilde{g}(\boldsymbol{k}) &= (2\pi)^{-3/2}\int d\boldsymbol{R}\,\exp(-i\boldsymbol{k}\cdot\boldsymbol{R})g(\boldsymbol{R}) \\ &= (2\pi)^{-3/2}\langle\mathcal{A}_{12}\{\Phi_1(\xi_1)\Phi_2(\xi_2)\exp(-i\boldsymbol{k}\cdot\boldsymbol{q})\}|\Psi\rangle. \end{aligned} \tag{11.14}$$

Here we should stop for some discussion. The states $\Psi_{\boldsymbol{R}}$, with all possible values of the label $\boldsymbol{R}$, span the subspace in which the nucleus behaves like an ensemble of the clusters a_1, a_2 of states $\Phi_1(\xi_1), \Phi_2(\xi_2)$. It looks as if the set $\{\Psi_{\boldsymbol{R}}\}$ could be considered a basis in this subspace, like the analogous set $\{\delta(\boldsymbol{R}-\boldsymbol{q})\}$ in a s.p. problem. But there is an important difference. The function $\delta(\boldsymbol{R}-\boldsymbol{q})$ is an eigenfunction of an observable, the position vector $\boldsymbol{q}$, and hence $\{\delta(\boldsymbol{R}-\boldsymbol{q})\}$ is a complete set of orthonormal states. In the cluster case, however, there is no such observable [284]. (Note that in $\Psi_{\boldsymbol{R}}$ the coordinate $\boldsymbol{q}$ is mixed up by $\mathcal{A}_{12}$.) It is thus not surprising that the set $\{\Psi_{\boldsymbol{R}}\}$ is not orthonormal. It is interesting to see that the nonorthogonality arises from the identity of the nucleons:

$$\begin{aligned}\langle\Psi_{\boldsymbol{R}}|\Psi_{\boldsymbol{R}'}\rangle &\equiv N(\boldsymbol{R},\boldsymbol{R}') && (11.15)\\ &= \delta(\boldsymbol{R}-\boldsymbol{R}') + \text{ exchange terms,} && (11.16)\end{aligned}$$

and the exchange terms come from the antisymmetry of $\Psi_{\boldsymbol{R}}$, i.e., from the indistinguishability of the nucleons.[5]

---

[4]For example, in a reaction in which a projectile picks up particle a_1 from $A = a_1+a_2$, the structure of A is represented by $g(\boldsymbol{R})$ if the antisymmetrization between A and the projectile is neglected.

[5]As defined in Eq. (11.9), the test state $\Psi_{\boldsymbol{R}} \equiv \Psi_{\boldsymbol{R}}(\boldsymbol{q})$ will produce the property (11.16) for the overlap (11.15) only if the clusters are not identical. If a_1 and a_2 are identical clusters, the antisymmetrizer $\mathcal{A}_{12}$ can be written as a sum of two terms, each containing half of the permutations: $\mathcal{A}_{12} = \mathcal{A}'_{12}+\mathcal{A}'_{12}P^{12}$, where P^{12} exchanges the two identical clusters as a whole. The phase factor ϵ of P^{12} is $+1$

Nevertheless, $\{\Psi_R\}$ can still be considered a nonorthogonal 'basis', and Eq. (11.10) can be interpreted as an expansion over this basis. Then $\phi(\boldsymbol{R})$ is to be viewed as an expansion coefficient of continuous labels. The nonorthogonality of the basis $\{\Psi_R\}$, however, implies another major difference: the expansion coefficient $\phi(\boldsymbol{R})$ is not equal to the projection function $g(\boldsymbol{R})$. This is natural in view of the geometrical analogy: when Ψ_1 and Ψ_2 are two nonorthogonal basis vectors of a Euclidean space, $\Psi_{\mathcal{P}} = \phi_1\Psi_1 + \phi_2\Psi_2$, $g_1 = \Psi_1 \cdot \Psi_{\mathcal{P}} = \Psi_1 \cdot \Psi$ and $g_2 = \Psi_2 \cdot \Psi_{\mathcal{P}} = \Psi_2 \cdot \Psi$, then $g_1 = \phi_1$ and $g_2 = \phi_2$ are satisfied if and only if $\Psi_i \cdot \Psi_j = \delta_{ij}$ (see Fig. 11.2). By this analogy, $\phi(\boldsymbol{R})$ and $g(\boldsymbol{R})$, respectively, are called the contravariant and covariant components of clustering. In the geometrical case the covariant and the contravariant components are related by $g_i = \sum_j N_{ij}\phi_j$, where $N_{ij} = \Psi_i \cdot \Psi_j$. The corresponding formula is now obtained by substitution of Eq. (11.10) into Eq. (11.11) and by using the definition (11.15):

$$g(\boldsymbol{R}) = \int d\boldsymbol{R}'\, N(\boldsymbol{R}, \boldsymbol{R}')\phi(\boldsymbol{R}') = \mathcal{N}\phi(\boldsymbol{R}), \tag{11.17}$$

where the definition of the norm operator $\mathcal{N}$ is implicit. It is an integral operator whose kernel is $N(\boldsymbol{R}, \boldsymbol{R}')$, and its action on a function f is defined as

$$\mathcal{N}f(\boldsymbol{R}) \equiv (\mathcal{N}f)(\boldsymbol{R}) = \int d\boldsymbol{R}'\, N(\boldsymbol{R}, \boldsymbol{R}')f(\boldsymbol{R}'). \tag{11.18}$$

The norm operator is invertible in the subspace of physical interest,[6] so that

$$\phi(\boldsymbol{R}) = (\mathcal{N}^{-1}g)(\boldsymbol{R}) = \mathcal{N}^{-1}g(\boldsymbol{R}). \tag{11.19}$$

It will be seen later [cf. Eq. (11.53)] that the inverse of $\mathcal{N}$ is also an integral operator. With its kernel denoted by $N^{-1}(\boldsymbol{R}, \boldsymbol{R}')$, Eq. (11.19) takes the form

$$\phi(\boldsymbol{R}) = \int d\boldsymbol{R}'\, N^{-1}(\boldsymbol{R}, \boldsymbol{R}')g(\boldsymbol{R}'). \tag{11.20}$$

---

or -1 depending on whether the number of nucleons in the cluster is even (boson) or odd (fermion). The exchange of the two clusters induces a sign change $\boldsymbol{q} \to -\boldsymbol{q}$, so that $\mathcal{A}_{12}\{\Phi_1\Phi_2\delta(\boldsymbol{R}-\boldsymbol{q})\} = \Psi'_{\boldsymbol{R}}(\boldsymbol{q}) + \epsilon\Psi'_{\boldsymbol{R}}(-\boldsymbol{q}) = \Psi'_{\boldsymbol{R}}(\boldsymbol{q}) + \epsilon\Psi'_{-\boldsymbol{R}}(\boldsymbol{q})$, where $\Psi'_{\boldsymbol{R}} = \mathcal{A}'_{12}\{\Phi_1\Phi_2\delta(\boldsymbol{R}-\boldsymbol{q})\}$. Then $N(\boldsymbol{R}, \boldsymbol{R}')$ of Eq. (11.15) [and $H(\boldsymbol{R}, \boldsymbol{R}')$ of Eq. (11.43) as well] takes the form $F(\boldsymbol{R}, \boldsymbol{R}') + \epsilon F(\boldsymbol{R}, -\boldsymbol{R}')$. To avoid this trivial duplication, the definition of $\Psi_{\boldsymbol{R}}$ may be changed to $\Psi_{\boldsymbol{R}} = 2^{-1/2}\mathcal{A}_{12}\{\Phi_1\Phi_2\delta(\boldsymbol{R}-\boldsymbol{q})\}$. Two clusters count identical if not only their compositions but their states are also identical. For simplicity, the possibility that some clusters may be identical will be mostly disregarded.

[6] The invertibility problem will be discussed later, see Eq. (11.54).

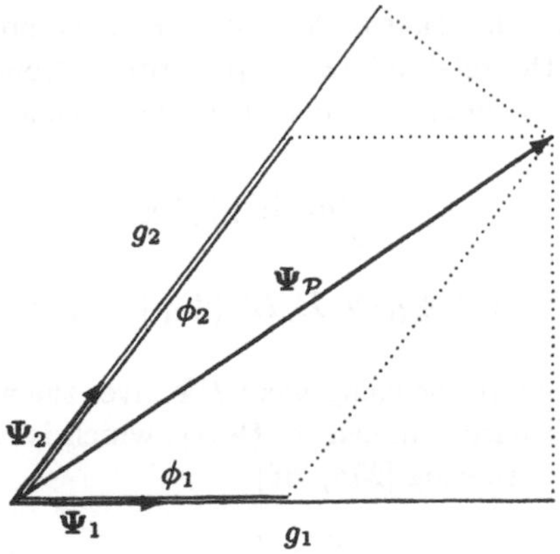

Figure 11.2: Covariant components g_i and contravariant components ϕ_i of a vector $\Psi_{\mathcal{P}}$ which lies in the plane spanned by the nonorthogonal unit vectors Ψ_1, Ψ_2.

The form (11.19)–(11.20) may be viewed as an alternative definition of the contravariant component, and this is a practicable formula. Whatever the form of Ψ, the components $g(\boldsymbol{R})$ and $\phi(\boldsymbol{R})$ can be calculated (see App. G.3). Once $\phi(\boldsymbol{R})$ has been constructed, $\mathcal{P}\Psi$ can be expressed explicitly. One has just to substitute Eqs. (11.20) and (11.11) into Eq. (11.10). With the use of the Dirac notation, one obtains

$$\mathcal{P}|\Psi\rangle = \int d\boldsymbol{R}\,|\Psi_{\boldsymbol{R}}\rangle\phi(\boldsymbol{R}) = \int d\boldsymbol{R}\int d\boldsymbol{R}'\,|\Psi_{\boldsymbol{R}}\rangle N^{-1}(\boldsymbol{R},\boldsymbol{R}')\langle\Psi_{\boldsymbol{R}'}|\Psi\rangle. \tag{11.21}$$

Since $|\Psi\rangle$ may be any state of the state space, Eq. (11.21) gives the formula for the projector itself [285] as[7]

$$\mathcal{P} = \int d\boldsymbol{R}\int d\boldsymbol{R}'\,|\Psi_{\boldsymbol{R}}\rangle N^{-1}(\boldsymbol{R},\boldsymbol{R}')\langle\Psi_{\boldsymbol{R}'}|. \tag{11.22}$$

[7]One can verify Eq. (11.22) more elegantly by reference to the property that any Hermitean operator $\mathcal{P}$ for which $\mathcal{P}^2 = \mathcal{P}$ holds is a projection operator [26]. Indeed,

$$\begin{aligned}
&\mathcal{P}^2|\Psi\rangle \\
&= \int d\boldsymbol{R}\int d\boldsymbol{R}'\int d\boldsymbol{R}''\int d\boldsymbol{R}'''\,|\Psi_{\boldsymbol{R}}\rangle N^{-1}(\boldsymbol{R},\boldsymbol{R}')\langle\Psi_{\boldsymbol{R}'}|\Psi_{\boldsymbol{R}''}\rangle N^{-1}(\boldsymbol{R}'',\boldsymbol{R}''')\langle\Psi_{\boldsymbol{R}'''}|\Psi\rangle \\
&= \int d\boldsymbol{R}\int d\boldsymbol{R}'\,|\Psi_{\boldsymbol{R}}\rangle N^{-1}(\boldsymbol{R},\boldsymbol{R}')\langle\Psi_{\boldsymbol{R}'}|\Psi\rangle = \mathcal{P}|\Psi\rangle,
\end{aligned}$$

where $\mathcal{N}^{-1}\mathcal{N}\mathcal{N}^{-1} = \mathcal{N}^{-1}$ has been used. To show that it projects just onto the cluster subspace $\{\Psi_{\boldsymbol{R}}\}$, one should recognize two obvious facts: first, that $\mathcal{P}\Psi \in \{\Psi_{\boldsymbol{R}}\}$ for any Ψ and, second, that $\mathcal{P}\Psi_{\boldsymbol{R}} = \Psi_{\boldsymbol{R}}$ implies that the projections span the whole cluster subspace $\{\Psi_{\boldsymbol{R}}\}$.

The projector has this unusual form owing to the nonorthogonality of $\{\Psi_{\boldsymbol{R}}\}$. Although the ket-bra $|\Psi_{\boldsymbol{R}}\rangle\langle\Psi_{\boldsymbol{R}}|$ is (proportional to) a projector, yet the projection operator to subspace $\{\Psi_{\boldsymbol{R}}\}$ cannot simply be

$$\Lambda = \int d\boldsymbol{R}\, |\Psi_{\boldsymbol{R}}\rangle\langle\Psi_{\boldsymbol{R}}|, \tag{11.23}$$

because $\Lambda^2 = \int d\boldsymbol{R} \int d\boldsymbol{R}' \, |\Psi_{\boldsymbol{R}}\rangle N(\boldsymbol{R},\boldsymbol{R}')\langle\Psi_{\boldsymbol{R}'}| \neq \Lambda$, due to the nonorthogonality (11.16).

The norm square of the component $\mathcal{P}\Psi$ gives the weight of the state in the cluster subspace spanned by $\{\Psi_{\boldsymbol{R}}\}$, which is called the amount or probability of clustering [284, 286]:

$$\mathfrak{P} \equiv \langle\mathcal{P}\Psi|\mathcal{P}\Psi\rangle = \langle\Psi|\mathcal{P}|\Psi\rangle = \int d\boldsymbol{R} \int d\boldsymbol{R}' \, \langle\Psi|\Psi_{\boldsymbol{R}}\rangle N^{-1}(\boldsymbol{R},\boldsymbol{R}')\langle\Psi_{\boldsymbol{R}'}|\Psi\rangle. \tag{11.24}$$

To understand this important result, let us invoke the geometrical analogy again. In geometry it is well-known that the length square of $\boldsymbol{\Psi}_{\mathcal{P}}$ is $\boldsymbol{\Psi}_{\mathcal{P}} \cdot \boldsymbol{\Psi}_{\mathcal{P}} = \sum_i \phi_i g_i$. From Eq. (11.24) one can recognize that the analogy holds for a Hilbert space of states as well. The amount of clustering can indeed be rewritten as

$$\begin{aligned} \mathfrak{P} &= \int d\boldsymbol{R}\, \phi^*(\boldsymbol{R}) g(\boldsymbol{R}) = \int d\boldsymbol{R}\, g^*(\boldsymbol{R}) \phi(\boldsymbol{R}) \\ &\equiv \langle\!\langle \phi | g \rangle\!\rangle = \langle\!\langle g | \phi \rangle\!\rangle = \langle\!\langle \phi | \mathcal{N} | \phi \rangle\!\rangle = \langle\!\langle g | \mathcal{N}^{-1} | g \rangle\!\rangle. \end{aligned} \tag{11.25}$$

In the second line a special bra-ket notation has been introduced for inner-product-like and matrix-element-like integrals involving integration over the parameter coordinate. These inner products will be assigned physical meaning in Sect. 11.4.2.

It is very important to understand that a state Ψ may have a component in the subspace spanned by $\{\Psi_{\boldsymbol{R}}\}$ even though its functional form is completely different. Indeed, any Ψ that is not orthogonal to all $\{\Psi_{\boldsymbol{R}}\}$ contains some amount of $a_1 + a_2$ clustering. Thus, e.g., a SM state may lie largely, or even completely [283, 287, 288], in the cluster subspace $\{\Psi_{\boldsymbol{R}}\}$. In Sect. 11.6 we shall see that a close correspondence exists between the SM and the cluster-model subspaces.

11.3.3 Amplitudes related to clustering

The structure information can be extracted from the reaction experiments by comparison of the measured cross sections with model calculations. In the transition amplitude of a direct breakup reaction the information on the structure of the disintegrating nucleus is represented either by the amplitude (11.11) or by (11.14). The amplitude (11.11) is the function that is to be substituted for the wave function $\phi_0(\boldsymbol{r}_1^{\mathrm{P}})$ of Eq. (4.1) in Part I. Thus it is these quantities that can be extracted from breakup experiments. The former gives information on the spatial distribution, while the latter on the momentum distribution, of the relative motion. Therefore, $|g(\boldsymbol{R})|^2$ and $|\tilde{g}(\boldsymbol{k})|^2$ are called the functions of density and momentum distribution, respectively. The common norm square of these amplitudes is called the spectroscopic factor:

$$S \equiv \langle \Psi | \Lambda | \Psi \rangle = \int d\boldsymbol{R}\, \langle \Psi | \Psi_{\boldsymbol{R}} \rangle \langle \Psi_{\boldsymbol{R}} | \Psi \rangle$$
$$= \int d\boldsymbol{R}\, |g(\boldsymbol{R})|^2 = \int d\boldsymbol{k}\, |\tilde{g}(\boldsymbol{k})|^2 = \langle\!\langle g | g \rangle\!\rangle. \quad (11.26)$$

To have a feeling for the meaning of this quantity, take $g_1^2 + g_2^2$ in the geometrical example. Although this quantity differs from the amount of clustering (11.25) and its meaning is less transparent, it is certainly characteristic of clustering. This is the quantity that represents the extent of preformation in direct reaction theories [289], and thus it directly affects the magnitude of the breakup cross section.

The relationship of the spectroscopic factor to the amount of clustering can be elucidated by trying to decompose a state into cluster components. In fact, the definition of the amount of clustering does not provide a complete set of orthonormal states into which any state is conveniently decomposed. For example, the complete orthonormal sets $\{\Phi_1^{(i)}(\xi_1)\}$ for cluster a_1 and $\{\Phi_2^{(j)}(\xi_2)\}$ for a_2 do not provide such a means. Although one can define subspaces $\{\Psi_{\boldsymbol{R}}^{(ij)}\}$, where $\Psi_{\boldsymbol{R}}^{(ij)} = \mathcal{A}_{12}\{\Phi_1^{(i)}(\xi_1)\Phi_2^{(j)}(\xi_2)\delta(\boldsymbol{R}-\boldsymbol{q})\}$, these do in general overlap with each other since the antisymmetrization destroys the orthogonality which holds for the subsystem states. Thus the amount $\mathfrak{P}_{ij}$ of each clustering ij is a sensible quantity, but the amounts of clustering belonging to them do not add up to unity. Such an example will be seen later in Table 11.2.

It is, however, possible to decompose Ψ with the help of the unit operator $1 = \sum_{ij} \int d\boldsymbol{R}\, |\Phi_1^{(i)}\Phi_2^{(j)}\delta(\boldsymbol{R}-\boldsymbol{q})\rangle\langle\Phi_1^{(i)}\Phi_2^{(j)}\delta(\boldsymbol{R}-\boldsymbol{q})|$:

$$1 = \langle \Psi | \Psi \rangle = \sum_{ij} \int d\boldsymbol{R}\, \langle \Psi | \Phi_1^{(i)}\Phi_2^{(j)}\delta(\boldsymbol{R}-\boldsymbol{q})\rangle \langle \Phi_1^{(i)}\Phi_2^{(j)}\delta(\boldsymbol{R}-\boldsymbol{q}) | \Psi \rangle$$
$$= \frac{a_1!a_2!}{A!} \sum_{ij} \int d\boldsymbol{R}\, \langle \Psi | \mathcal{A}_{12}\{\Phi_1^{(i)}\Phi_2^{(j)}\delta(\boldsymbol{R}-\boldsymbol{q})\}\rangle$$
$$\times \langle \mathcal{A}_{12}\{\Phi_1^{(i)}\Phi_2^{(j)}\delta(\boldsymbol{R}-\boldsymbol{q})\} | \Psi \rangle = \frac{a_1!a_2!}{A!} \sum_{ij} S_{ij}, \quad (11.27)$$

where Eq. (11.12) has been used and the spectroscopic factor S_{ij} is defined by analogy with Eq. (11.26). This decomposition shows that $(a_1!a_2!/A!)\sum_{ij} S_{ij}$ does carry some probability meaning, viz., that a non-antisymmetrized state $\Phi_1^{(i)}\Phi_2^{(j)}\phi(\boldsymbol{q})$ forms only a small fraction of the antisymmetrical state Ψ. Equation (11.27), however, is more commonly interpreted as a sum rule for the spectroscopic factor: $\sum_{ij} S_{ij} = A!/(a_1!a_2!)$. The spectroscopic factor itself may exceed unity, and has no direct probability meaning.

Similarly, calling the functions $|g|^2$ and $|\tilde{g}|^2$ distribution functions is imprecise. If they really were what they are called, their norm squares should be unity or they should be equal to the clustering probability $\mathfrak{P}$, none of which is the case. The matter is that since $\Psi_{\boldsymbol{R}}$ is not orthogonal to $\Psi_{\boldsymbol{R}'}$, in Eq. (11.26) the contributions of the different functions $\Psi_{\boldsymbol{R}}$ are counted with unequal weights, and a similar statement holds for their Fourier transform. One could define amplitudes of the required properties as follows:

$$G(\boldsymbol{R}) = \mathcal{N}^{1/2}\phi(\boldsymbol{R}) = \mathcal{N}^{-1/2}g(\boldsymbol{R}), \tag{11.28}$$

$$\tilde{G}(\boldsymbol{k}) = (2\pi)^{-3/2}\int d\boldsymbol{R}\, \exp(-i\boldsymbol{k}\cdot\boldsymbol{R})G(\boldsymbol{R}). \tag{11.29}$$

With the amplitude $G(\boldsymbol{R})$, the amount of clustering takes the simple form

$$\mathfrak{P} = \int d\boldsymbol{R}\, |G(\boldsymbol{R})|^2 = \langle\!\langle G|G\rangle\!\rangle. \tag{11.30}$$

Nevertheless, it is to be noted that G and $\tilde{G}$ are not quite unambiguous (see Sect. 11.4.2), and play no role in nuclear reaction theory [289,290]. On the other hand, for two-body disintegrations of light nuclei, the difference between g and G is very small [291]. Calling them distributions is just a convention.

Towards the asymptotic region $g(\boldsymbol{R})$ and $G(\boldsymbol{R})$ approach each other since the Pauli exchanges, which cause $\mathcal{N}$ to differ from unity [cf. Eq. (11.16) and footnote on p. 282], are short-ranged. This is important since the asymptotic region has some specific significance in the description of some phenomena. For example, clustering is the essential property of cluster-decaying (e.g., α-decaying) states [286]. For decay channels, the amplitudes $g(\boldsymbol{R})$ and $G(\boldsymbol{R})$ do not vanish asymptotically, and $g(\boldsymbol{R}) \simeq G(\boldsymbol{R})$ there carries unambiguous information on the partial decay width [286]. For a channel with angular momentum LM, the partial decay width Γ can be expressed in terms of the radial factors[8] $g_L(R) = R\int d\widehat{\boldsymbol{R}}\, Y_{LM}^*(\widehat{\boldsymbol{R}})g(\boldsymbol{R})$ or $G_L(R) = R\int d\widehat{\boldsymbol{R}}\, Y_{LM}^*(\widehat{\boldsymbol{R}})G(\boldsymbol{R})$.

[8]This relationship is formulated conventionally through the reduced-width amplitude $\gamma_L(r_c)$ borrowed from Wigner's R-matrix theory. For the derivation of these

The formulae for the amplitudes associated with clustering can be easily generalized to the case of n clusters by a formal reinterpretation of Eqs. (11.9)–(11.30). One can use the notation $\boldsymbol{R}$ for the set of $\{\boldsymbol{R}_1, \ldots, \boldsymbol{R}_{n-1}\}$ and $\boldsymbol{k}$ for the set of $\{\boldsymbol{k}_1, \ldots, \boldsymbol{k}_{n-1}\}$ and carry this change of interpretation through. Thus $\int d\boldsymbol{R}$ will actually stand for $\int d\boldsymbol{R}_1 \cdots \int d\boldsymbol{R}_{n-1}$ and so will $\langle\!\langle \cdots | \cdots | \cdots \rangle\!\rangle$ be. The scalar product $\boldsymbol{k}\cdot\boldsymbol{R}$ will in this way be read as $\boldsymbol{k}_1\cdot\boldsymbol{R}_1 + \cdots + \boldsymbol{k}_{n-1}\cdot\boldsymbol{R}_{n-1}$. For example, the test state

$$\Psi_{\boldsymbol{R}_1,\ldots,\boldsymbol{R}_{n-1}} = \mathcal{A}_{1\cdots n}\{\Phi_1(\xi_1)\cdots\Phi_n(\xi_n)\delta(\boldsymbol{R}_1-\boldsymbol{q}_1)\cdots\delta(\boldsymbol{R}_{n-1}-\boldsymbol{q}_{n-1})\} \tag{11.31}$$

will be written as $\Psi_{\boldsymbol{R}}$. The kernel $N(\boldsymbol{R}_1, \ldots, \boldsymbol{R}_{n-1}; \boldsymbol{R}'_1, \ldots, \boldsymbol{R}'_{n-1})$ will just be written as $N(\boldsymbol{R}, \boldsymbol{R}')$, so that $\mathcal{N}$ will be defined as before but understood to act on $n-1$ variables. Apart from this generalization of notation, the only change to be made is that the statistical factor $a_1!a_2!/A!$ in Eqs. (11.27) is to be replaced by $a_1!\cdots a_{n-1}!/A!$. In this way all formulae (11.9)–(11.30) introduced for the two-cluster system will be valid for the n-cluster system as well. To show the use of the generalized definition, let us write down the formulae (11.24), (11.25) and (11.30) for the amount of clustering in an extended notation:

$$\begin{aligned}\mathfrak{P} \equiv \langle\Psi|\mathcal{P}|\Psi\rangle &= \int d\boldsymbol{R}_1 \cdots \int d\boldsymbol{R}_{n-1} \int d\boldsymbol{R}'_1 \cdots \int d\boldsymbol{R}'_{n-1} \\ &\times \langle\Psi|\Psi_{\boldsymbol{R}_1,\ldots,\boldsymbol{R}_{n-1}}\rangle N^{-1}(\boldsymbol{R}_1, \ldots, \boldsymbol{R}_{n-1}; \boldsymbol{R}'_1, \ldots, \boldsymbol{R}'_{n-1}) \\ &\times \langle\Psi_{\boldsymbol{R}'_1,\ldots,\boldsymbol{R}'_{n-1}}|\Psi\rangle \\ &= \int d\boldsymbol{R}_1 \cdots \int d\boldsymbol{R}_{n-1}\, \phi^*(\boldsymbol{R}_1, \ldots, \boldsymbol{R}_{n-1}) g(\boldsymbol{R}_1, \ldots, \boldsymbol{R}_{n-1})\end{aligned} \tag{11.32}$$

---

formulae see Ref. [292]:

$$\gamma(r_c) = \left(\frac{\hbar^2}{2\mu_{12}r_c}\right)^{1/2} g_L(r_c) \approx \left(\frac{\hbar^2}{2\mu_{12}r_c}\right)^{1/2} G_L(r_c),$$

where μ_{12} is the reduced mass. (In fact, the reduced-width amplitude should contain the R-matrix eigenfunction in $g_L(r)$ instead of a resonance wave function, but for a narrow isolated resonance such a substitution is a good approximation [286,293].) In taking $g_L(r_c) \approx G_L(r_c)$ here, it has been assumed that r_c is greater than the range of the Pauli effects. The width is then $\Gamma = 2P_L(r_c)\gamma_L^2(r_c)$, where $P_L(r_c)$ stands for the penetrability of the barrier beyond r_c. Note that both $\gamma^2(r_c)$ and $P_L(r_c)$ depend on the choice of the channel radius r_c, but their product does not. Now, if there are more than one decay channels, the reduced widths defined consistently, e.g., taken at radii corresponding to the touching distances, are characteristic of the preformation of the clusters, i.e., the weight of the particular cluster configuration in the parent state. To make this more informative, the reduced width is often compared with the so-called Wigner limit [292] $\gamma_W^2(r_c) = 3\hbar^2/(2\mu_{12}r_c^2)$, which is the partial width of a s.p. state whose wave function is uniform up to r_c and zero beyond. The ratio of the reduced width $\gamma^2(r_c)$ to the Wigner limit is called the dimensionless reduced width [286]: $\theta^2(r_c) = \gamma^2(r_c)/\gamma_W^2(r_c)$.

$$= \int d\boldsymbol{R}_1 \cdots \int d\boldsymbol{R}_{n-1}\, g^*(\boldsymbol{R}_1,\ldots,\boldsymbol{R}_{n-1})\phi(\boldsymbol{R}_1,\ldots,\boldsymbol{R}_{n-1})$$
$$= \int d\boldsymbol{R}_1 \cdots \int d\boldsymbol{R}_{n-1}\, G^*(\boldsymbol{R}_1,\ldots,\boldsymbol{R}_{n-1})G(\boldsymbol{R}_1,\ldots,\boldsymbol{R}_{n-1})$$
$$= \langle\!\langle \phi|g\rangle\!\rangle = \langle\!\langle g|\phi\rangle\!\rangle = \langle\!\langle G|G\rangle\!\rangle. \qquad (11.33)$$

11.3.4 Calculation of the clustering properties

We have now defined the quantities characterizing the cluster composition and fragmentation, and it remains to give some indication for their calculation when the wave functions are given in the correlated Gaussian formalism. All quantities are expressible as overlaps of correlated Gaussian wave functions Ψ and wave functions of the type of $\Psi_{\boldsymbol{R}}$ [Eq. (11.9)] or, more generally, of $\Psi_{\boldsymbol{R}_1,\ldots,\boldsymbol{R}_{n-1}}$ [Eq. (11.31)]. In these functions the relative motion is described by δ-functions. The states Ψ are expressible in terms of Gaussian wave packets $\Phi_{SM_STM_T}(\boldsymbol{s}_1,\ldots,\boldsymbol{s}_A)$ of Eq. (9.68), and we should be able to express $\Psi_{\boldsymbol{R}_1,\ldots,\boldsymbol{R}_{n-1}}$ similarly; then all overlaps of our concern could be calculated in terms of overlaps of Slater determinants of shifted Gaussians, Eq. (9.108). This can be achieved by using one of the expressions [232] of the δ-function in terms of Gaussians, e.g.,

$$\delta(\boldsymbol{R}-\boldsymbol{q}) = (2\pi)^{-3}\left(\frac{\beta}{4\pi}\right)^{3/4}\int d\boldsymbol{k}\, \exp\left(-i\boldsymbol{k}\cdot\boldsymbol{R}+\frac{\boldsymbol{k}^2}{2\beta}\right)\int d\boldsymbol{S}\, e^{i\boldsymbol{k}\cdot\boldsymbol{S}}\varphi_{\boldsymbol{S}}(\boldsymbol{q}). \qquad (11.34)$$

This formula can be easily verified by performing the integrations as indicated with the use of Eq. (9.61).[9]

Equation (11.34) should be substituted into each of the δ-functions in $\Psi_{\boldsymbol{R}_1,\ldots,\boldsymbol{R}_{n-1}}$, which will give rise to a product of Gaussians $\varphi_{\boldsymbol{S}_i}(\boldsymbol{q}_i)$ of the relative variables $\boldsymbol{q}_i$ and of cluster intrinsic states. The standard way to proceed is shown in the context of calculating cluster-model matrix elements in Sect. 11.5.2 later. The calculation of the amplitudes $g(\boldsymbol{R}_1,\ldots,\boldsymbol{R}_{n-1})$ and $\phi(\boldsymbol{R}_1,\ldots,\boldsymbol{R}_{n-1})$ will be discussed with some more detail in App. G.3.

The treatment of $\mathcal{N}^{-1}$ also needs some comments. To invert $\mathcal{N}$, one should solve its eigenvalue problem and express it in terms of its eigenfunctions and eigenvalues. This treatment is generally applicable, and it will be discussed in Sects. 11.4.2 and 11.6.2.

[9] Equation (11.34) can be derived without knowing the final result as follows. Write the folding formula $\varphi_0(\boldsymbol{S}-\boldsymbol{q}) = \int d\boldsymbol{R}\,\varphi_0(\boldsymbol{S}-\boldsymbol{R})g(\boldsymbol{R})$ for $\varphi_0(\boldsymbol{S}-\boldsymbol{q}) \equiv \varphi_{\boldsymbol{S}}(\boldsymbol{q})$ and $g(\boldsymbol{R}) = \delta(\boldsymbol{q}-\boldsymbol{R})$. Since the Fourier transform of a folding is $(2\pi)^{3/2}$ times the product of the Fourier transforms of its 'factors' [26], the Fourier transform of $\delta(\boldsymbol{q}-\boldsymbol{R})$ as a function of $\boldsymbol{R}$ is $(2\pi)^{-3}$ times the ratio of those of $\varphi_{\boldsymbol{S}}(\boldsymbol{q})$ and of $\varphi_{\boldsymbol{S}}(0)$ (as functions of $\boldsymbol{S}$). Then Eq. (11.34) is obtained by an inverse Fourier transformation of this expression.

Here we mention a method of calculating $\mathfrak{P}$ which is especially useful if the nuclear state Ψ itself has been calculated in a microscopic cluster model. The projection $\mathcal{P}\Psi$ of Eq. (11.21) obviously has the property that it minimizes the norm of the 'orthogonal terms' in Eq. (11.6) (cf. Fig. 11.1), so that $\delta\langle\Psi-\mathcal{P}\Psi|\Psi-\mathcal{P}\Psi\rangle = 0$. This can be used as a variational prescription to find $\mathcal{P}\Psi$ itself [284]. It is useful to represent the cluster subspace $\{\Psi_{\boldsymbol{R}_1,\ldots,\boldsymbol{R}_{n-1}}\}$ in the same way as Ψ itself in the dynamical calculation [294]. If Ψ is constructed in the correlated Gaussian approach, $\mathcal{P}\Psi$ is to be expressed by the same correlated Gaussians as Ψ. This method is presented in App. G.3. Examples for results for the amount of clustering will be seen in their physical context in Sect. 11.7.4 and in Chap. 13.

11.4 Basic concepts of cluster models

11.4.1 Overview

The starting point of a standard cluster model is a wave-function ansatz: taking one term or a few terms of the form of Eq. (11.1). In these models all nucleons are treated explicitly, which is expressed by calling them microscopic cluster models. Owing to the scattering asymptotics, any microscopic model of low-energy nuclear collisions may be considered a microscopic cluster model. Our interest is, however, focussed on bound and resonant states that are cast into the mould of a cluster model by dynamics rather than by a trivial boundary condition.

It is not unrealistic to assume in many cases that a few cluster configurations, or just a single one, may be preponderant so that all other terms of the wave function are negligible. The relevance of the concept of clustering to nuclear motion is shown most spectacularly by the success of cluster models. They have proved to be extremely useful in explaining nuclear structure and reactions. Starting with complete sets of clusters states $\{\Phi_1^{(i)}\}$ and $\{\Phi_2^{(j)}\}$ and relative-motion states $\{\phi'_{ijk}\}$, by systematic orthogonalization one can set up a complete set, $\{\Psi_{ijk}\}$, of two-cluster states of the form of [295]

$$\Psi_{ijk} = \mathcal{A}_{12}\left\{\Phi_1^{(i)}(\xi_1)\Phi_2^{(j)}(\xi_2)\phi_{ijk}(\boldsymbol{q})\right\}. \qquad (11.35)$$

Based on this, the cluster model provides a representation of the nuclear wave function and a full microscopic theory of nuclei [283]. The cluster model is, however, not always the simplest of conceivable approaches. Therefore, it has been used mostly where it is the most realistic model and where it has no rivals. Apart from scattering states, these are the states of cluster-decaying resonance states and the bound states with small intercluster separation energies [296]. It is extremely difficult to treat such cases by the SM [286, 297]. Many light nuclei and most exotic nuclei fall into this category. The macroscopic, phenomenological and heuristic versions of the model are tractable for a broader range of systems. For example, molecular-orbital models [298] and algebraic cluster models [299, 300] are very useful for heavy-ion resonances, and some macroscopic two-cluster models may be viable alternatives of collective rotational models of the low-lying states of deformed nuclei [183] (cf. p. 183).

The formalisms of cluster models differ in the treatment of two technical difficulties that plague cluster models: antisymmetrization and the disentanglement of the c.m. motion. The actual forms of the cluster model can be viewed as different solutions to these problems. In models that rely on single mean fields, like the SM, none of these is a major difficulty (at least for light nuclei). In such models the nucleons are described as moving relative to the c.m. *of a core or of the whole system*, mostly assumed to be fixed, whereas in cluster models the *cluster* c.m. move with respect to each other. It is obvious that the cluster c.m. cannot be assumed to be fixed, but, for small systems, even the total c.m. is to be treated properly. Antisymmetrization is also much simpler in mean-field models. Owing to the orthogonality of the s.p. orbits, the Pauli principle may be observed by allowing single occupancy of each orbit and by including rather simple, and often not too large, exchange terms. This makes it appear as if antisymmetrization did not play a significant role. In microscopic cluster models, however, s.p. orbits are either not defined or are not orthogonal to each other, which makes explicit antisymmetrization both cumbersome and indispensable.

These considerations seem to suggest that no cluster model is meaningful without explicit antisymmetrization. Nevertheless, some macroscopic cluster models with no explicit antisymmetrization do very well, which is paradoxical. It would be wrong to infer from this that antisymmetrization is negligible. These models should not be derived from microscopic models by simply neglecting the exchange terms. They are related to the underlying microscopic scheme in a more sophisticated fashion.

11.4.2 Intercluster relative motion

In the pure form of the $a_1 + a_2$ two-cluster model of nucleus A it is assumed that the nuclear wave function Ψ can be written as an antisymmetrized product of cluster intrinsic wave functions $\Phi_i(\xi_i)$ and of a function $\phi(\boldsymbol{q})$ of the vector $\boldsymbol{q}$ connecting the two cluster c.m.:

$$\Psi = \mathcal{A}_{12} \left\{ \Phi_1(\xi_1)\Phi_2(\xi_2)\phi(\boldsymbol{q}) \right\} \tag{11.36}$$

[cf. Eq. (11.6)]. The angular momenta are suppressed for simplicity in most of this chapter. This provides an ansatz for an approximate solution of the dynamical problem,

$$H\Psi = E\Psi, \tag{11.37}$$

with

$$H = \sum_{j=1}^{A} T_j - T_{\mathrm{c.m.}} + \sum_{j<k}^{A} V_{jk}. \tag{11.38}$$

The intrinsic Hamiltonian H excludes the kinetic energy of the total c.m., $T_{\mathrm{c.m.}}$, so the problem is evidently translation-invariant. In an orthodox interpretation the functions Φ_i are identified with the g.s. wave functions of the isolated clusters a_i, i.e., they are the g.s. solutions of

$$H_i\Phi_i(\xi_i) = E_i\Phi_i(\xi_i). \tag{11.39}$$

Then, Φ_i being fixed, one can reduce the full Schrödinger equation, Eq. (11.37), to an equation for the relative-motion function, ϕ. Our aim with the following formal manipulation is to derive an equation for the relative motion and recast it into the form of a two-particle Schrödinger equation. The practical, essentially variational, solution of the problem will be discussed in Sects. 11.5 and 11.7.

To obtain an equation for the relative motion, one should disentangle $\phi(\boldsymbol{q})$ from the antisymmetrizer, which can be done, similarly to Eqs. (11.9) and (11.10), by inserting a dummy integration and a corresponding δ-function:

$$\begin{aligned}\Psi &= \int d\boldsymbol{R}\,\phi(\boldsymbol{R})\mathcal{A}_{12}\left\{\Phi_1(\xi_1)\Phi_2(\xi_2)\delta(\boldsymbol{R}-\boldsymbol{q})\right\}\\ &= \int d\boldsymbol{R}\,\phi(\boldsymbol{R})\Psi_{\boldsymbol{R}},\end{aligned} \tag{11.40}$$

where

$$\Psi_{\boldsymbol{R}} = \mathcal{A}_{12}\left\{\Phi_1(\xi_1)\Phi_2(\xi_2)\delta(\boldsymbol{R}-\boldsymbol{q})\right\}. \tag{11.41}$$

The second line may be interpreted as an expansion in terms of a basis $\{\Psi_{\boldsymbol{R}}\}$ with $\phi(\boldsymbol{R})$ an expansion coefficient.

As was mentioned in Sect. 11.3, no physical observables can be defined for the relative distance and for the relative momentum, nor is there a Hamiltonian of the relative motion. The role of the relative distance is to be played by the ***parameter coordinate*** $\boldsymbol{R}$, and we shall see in what sense it can do that. The vector $\boldsymbol{R}$ is a good candidate to play the role of a relative-distance vector since it is unaffected by the permutation operators. It is a continuous label of the 'basis' $\{\Psi_{\boldsymbol{R}}\}$, and it is a vector of spatial displacement associated with that of the two cluster c.m. It is not an observable though, nor is its value an eigenvalue of an observable. It is, nevertheless, possible to define a linear Hermitean operator $\boldsymbol{Q}$, whose eigenvalues are $\boldsymbol{R}$, and the system of its eigenfunctions, belonging to $\boldsymbol{R}$, is almost complete in the two-cluster subspace.[10] Apart from this stipulation on the completeness, $\boldsymbol{Q}$ has all the properties of an observable. Thus, in a utilitarian sense, $\boldsymbol{R}$ can be regarded as an eigenvalue of an observable.

To elaborate further on the meaning of the parameter coordinate $\boldsymbol{R}$, let us consider a space $\{F(\boldsymbol{R})\}$ of the functions of $\boldsymbol{R}$. One can define a Hermitean operator $\boldsymbol{\mathcal{R}}$ such that $\boldsymbol{\mathcal{R}}F(\boldsymbol{R}) = \boldsymbol{R}F(\boldsymbol{R})$ for any $F(\boldsymbol{R})$. Then any particular vector, say $\boldsymbol{R}'$, of the vector space $\{\boldsymbol{R}\}$ will be an eigenvalue, of $\boldsymbol{\mathcal{R}}$, belonging to eigenfunction $\delta(\boldsymbol{R}' - \boldsymbol{R})$. It is then useful to define $\{F(\boldsymbol{R})\}$ in full analogy with the state space of a s.p. problem. The Hermitean operators that act on the elements of the function space $\{F(\boldsymbol{R})\}$ and have complete sets of eigenfunctions are analogous to observables. Since $\boldsymbol{R}$ is the eigenvalue of such an operator, it may be regarded as a representation label, i.e., an argument of a 'wave function'. Then it is meaningful to define matrix elements with integration over $\boldsymbol{R}$ and define other operators, reminiscent of observables in the space representation, involving differentiation and integration with respect to $\boldsymbol{R}$. The operators and the matrix elements defined in the $\boldsymbol{R}$-space will be denoted by script capitals and by $\langle\!\langle \cdots | \cdots | \cdots \rangle\!\rangle$, respectively.

Projected onto the subspace spanned by $\{\Psi_{\boldsymbol{R}}\}$, the Schrödinger equation (11.37) reduces to

$$\mathcal{H}\phi(\boldsymbol{R}) = E\mathcal{N}\phi(\boldsymbol{R}), \tag{11.42}$$

where $\mathcal{H}$ and $\mathcal{N}$ are integral operators whose kernels are [cf. Eq. (11.15)]

$$H(\boldsymbol{R}, \boldsymbol{R}') = \langle \Psi_{\boldsymbol{R}} | H | \Psi_{\boldsymbol{R}'} \rangle, \tag{11.43}$$

$$N(\boldsymbol{R}, \boldsymbol{R}') = \langle \Psi_{\boldsymbol{R}} | \Psi_{\boldsymbol{R}'} \rangle. \tag{11.44}$$

[For the definition of an integral operator see Eq. (11.18).] The norm operator $\mathcal{N}$ appears in Eq. (11.42) because the basis $\{\Psi_{\boldsymbol{R}}\}$ is nonorthogonal. It is the norm operator that connects the covariant components $g(\boldsymbol{R})$ and contravariant components $\phi(\boldsymbol{R})$ of Ψ, defined as is usual for

[10] This operator may be given in its spectral form, $\boldsymbol{Q} = \int d\boldsymbol{R}\, |\Xi_{\boldsymbol{R}}\rangle \boldsymbol{R} \langle \Xi_{\boldsymbol{R}}|$, with $\Xi_{\boldsymbol{R}} = \mathcal{N}^{-1/2}\Psi_{\boldsymbol{R}}$, where $\mathcal{N}$ is defined in Eq. (11.17), and the set of its eigenfunctions is complete in the Pauli-allowed part of the two-cluster subspace [see Eq. (11.54)]. Note that here, in keeping with the convention used throughout, the quantities of the abstract Dirac formalism are not distinguished from their representatives in coordinate space. Thus, in the spirit of the Dirac formalism it would be more precise but less explicit to write $|\boldsymbol{R}\rangle$ instead of $|\Xi_{\boldsymbol{R}}\rangle$.

decompositions into nonorthogonal bases [cf. Eqs. (11.17) and (11.19)]:

$$g(\boldsymbol{R}) = \langle\Psi_{\boldsymbol{R}}|\Psi\rangle = \mathcal{N}\phi(\boldsymbol{R}), \tag{11.45}$$

$$\phi(\boldsymbol{R}) = \mathcal{N}^{-1}g(\boldsymbol{R}), \tag{11.46}$$

where it is implicitly assumed that $\phi(\boldsymbol{R})$ is in the subspace in which $\mathcal{N}$ is invertible.

It is natural that the optimum approximate solution to a problem in a restricted subspace is obtained just by projecting the basic equation of the full problem onto the subspace. It is nevertheless important to mention that, for bound states, Eq. (11.42) can in fact be derived from the variational principle, Theorem 1 on p. 195. The energy functional (9.1) can be now written as

$$E[\Psi] = \frac{\langle\!\langle\phi|\mathcal{H}|\phi\rangle\!\rangle}{\langle\!\langle\phi|\mathcal{N}|\phi\rangle\!\rangle}, \tag{11.47}$$

and its stationarity can be expressed by performing the variation:[11]

$$\frac{(\langle\!\langle\delta\phi|\mathcal{H}|\phi\rangle\!\rangle + \langle\!\langle\phi|\mathcal{H}|\delta\phi\rangle\!\rangle)\langle\!\langle\phi|\mathcal{N}|\phi\rangle\!\rangle - \langle\!\langle\phi|\mathcal{H}|\phi\rangle\!\rangle(\langle\!\langle\delta\phi|\mathcal{N}|\phi\rangle\!\rangle + \langle\!\langle\phi|\mathcal{N}|\delta\phi\rangle\!\rangle)}{\langle\!\langle\phi|\mathcal{N}|\phi\rangle\!\rangle^2}$$

$$= \langle\!\langle\phi|\mathcal{N}|\phi\rangle\!\rangle^{-1}(\langle\!\langle\delta\phi|\mathcal{H} - E\mathcal{N}|\phi\rangle\!\rangle + \langle\!\langle\phi|\mathcal{H} - E\mathcal{N}|\delta\phi\rangle\!\rangle) = 0, \tag{11.48}$$

where, in the second step, Eq. (11.47) has been used. The last expression can be zero if and only if Eq. (11.42) holds, and that is what we wanted to prove. For scattering states the variational foundation of Eq. (11.42) is not based on the energy functional, but the alternative variational schemes are all based on the wave function produced by the projection equation (11.42) [301].

It became possible to obtain an equation of motion for a relative variable at the expense of incorporating the Pauli effects in the kernels (11.43) and (11.44). The Dirac-δ's do not make the kernels singular; their role is just to extract one variable, specific to each permutation, from the integration implied by the matrix element. The norm operator itself carries the effects of the Pauli principle. It is thus plausible that it can be used to represent Pauli effects in approximate schemes. The norm operator is Hermitean, commutes with the relative-motion angular-momentum operators $\mathcal{L}^2$ and $\mathcal{L}_z$ defined in terms of $\mathcal{R}$ and $\nabla_{\boldsymbol{R}}$. The common eigenfunctions $\phi_{nlm}(\boldsymbol{R})$ of $\mathcal{N}$, $\mathcal{L}^2$ and $\mathcal{L}_z$, defined by

$$\mathcal{N}\phi_{nlm}(\boldsymbol{R}) = \nu_{nl}\phi_{nlm}(\boldsymbol{R}), \tag{11.49}$$

[11]The variation of a functional $L = \int dx\, F[f(x)]$ is given by $\delta L = \int dx\, (\partial F/\partial f)\delta f$. When $f(x)$ is complex and $L = \int dx\, F[f(x), f^*(x)]$, $\delta L = \int dx\, (\partial F/\partial f)\delta f + \int dx\, (\partial F/\partial f^*)(\delta f)^*$.

form a complete set in the L^2 Hilbert space of functions of one variable. It can be shown[12] that

$$\nu_{nl} \geq 0 \quad \text{and} \quad \lim_{n\to\infty} \nu_{nl} = 1. \tag{11.50}$$

It is convenient to express $\mathcal{N}$ and its functions in the form of a spectral representation, such as

$$\mathcal{N} = \sum_{nlm} |\phi_{nlm}\rangle\!\rangle \nu_{nl} \langle\!\langle \phi_{nlm}|, \tag{11.51}$$

$$\mathcal{N}^{\kappa} = \sum_{nlm} |\phi_{nlm}\rangle\!\rangle \nu_{nl}^{\kappa} \langle\!\langle \phi_{nlm}|, \qquad (\kappa > 0), \tag{11.52}$$

$$\mathcal{N}^{-\kappa} = \sum_{\substack{nlm \\ \nu_{nl}\neq 0}} |\phi_{nlm}\rangle\!\rangle \nu_{nl}^{-\kappa} \langle\!\langle \phi_{nlm}|, \qquad (\kappa > 0). \tag{11.53}$$

From Eq. (11.53) it is clear that $\mathcal{N}$ can only be inverted in the subspace spanned by the eigenfunctions ϕ_{nlm} with $\nu_{nl} \neq 0$, and the definition of the inverse implied by Eq. (11.53) is

$$\mathcal{N}\mathcal{N}^{-1} = \mathcal{N}^{-1}\mathcal{N} = \mathcal{I}, \tag{11.54}$$

where $\mathcal{I}$ is the projector to that subspace:

$$\mathcal{I} = \sum_{\substack{nlm \\ \nu_{nl}\neq 0}} |\phi_{nlm}\rangle\!\rangle \langle\!\langle \phi_{nlm}|. \tag{11.55}$$

---

[12]The former property is a direct consequence of the fact that $\langle\!\langle \phi_{nlm}|\mathcal{N}|\phi_{nlm}\rangle\!\rangle$ is the norm square of a physical state of the system, thus cannot be negative. The second property can be made plausible by the following considerations. The range of $\mathcal{K} = 1 - \mathcal{N}$ is comparable with the sum of the 'ranges' of the density distributions of Φ_1 and Φ_2, thus any L^2 function ϕ that is zero within this range is an approximate eigenfunction of $\mathcal{N}$ belonging to eigenvalue $\nu \simeq 1$. Since there is an infinite number of linearly independent functions of this property, there must be an accumulation point of the eigenvalues at value 1. This is a limiting value provided there is no other accumulation point. Any other accumulation point has to be produced by functions whose 'range' is finite. Any infinite orthogonal set of finite-range functions has to contain elements with unlimited node numbers. These correspond to infinitely large relative momenta. However, the exchange operator $\mathcal{K}$ is of finite range in the momentum space as well. Thus any eigenfunctions ϕ_{nlm} of $\mathcal{N}$ that carry large momenta have to belong to eigenvalues close to unity. Therefore, the assumed second accumulation point must coincide with the first one, and the value unity is indeed a limiting value of ν_{nl}. In fact, the eigenfunctions ϕ_{nlm} get progressively longer-range as well as more curly as n is increased. For the case that Φ_1 and Φ_2 are HOSM states of the same oscillator quantum, ϕ_{nlm} are also HO s.p. states of the same $\hbar\omega$. The last node of the HO states goes to infinity with n, both in the coordinate space and in the momentum space. The eigenvalues of $\mathcal{N}$ for problems involving more than two clusters have more complicated properties.

It is important to note that there may exist some non-vanishing functions $\phi_0(\boldsymbol{q})$ that, substituted into $\phi(\boldsymbol{q})$, make the wave function (11.36) vanish, so that $\langle\Psi|\Psi\rangle \equiv \langle\!\langle\phi_0|\mathcal{N}|\phi_0\rangle\!\rangle = 0$. To find such a $\phi_0(\boldsymbol{R})$, let us expand it in terms of the complete set of functions $\phi_{nlm}(\boldsymbol{R})$: $\phi_0(\boldsymbol{R}) = \sum_{nlm} c_{nlm}\phi_{nlm}(\boldsymbol{R})$. From $\langle\!\langle\phi_0|\mathcal{N}|\phi_0\rangle\!\rangle = 0$ it follows that the coefficients c_{nlm} satisfy

$$\sum_{nlm} \nu_{nl}|c_{nlm}|^2 = 0. \tag{11.56}$$

Considering the properties of ν_{nl} [see Eq. (11.50)], Eq. (11.56) entails $\nu_{nl} = 0$ for all non-zero c_{nlm}. Thus ϕ_0 must satisfy $\mathcal{N}\phi_0(\boldsymbol{R}) = 0$, i.e., $\mathcal{A}_{12}\{\Phi_1\Phi_2\phi_0(\boldsymbol{q})\} = 0$. This indicates that the total nuclear wave function vanishes because of the antisymmetrization, i.e., because $\phi_0(\boldsymbol{q})$ violates the Pauli principle. The eigenfunctions $\phi_0(\boldsymbol{R})$, of $\mathcal{N}$, that belong to eigenvalue zero are thus to be interpreted as ***Pauli-forbidden relative-motion states*** and are also called redundant states.

One should, however, not interpret the effect of $\mathcal{N}$ in Eq. (11.42) as removing Pauli-forbidden states. In fact, $\mathcal{A}_{12}\{\Phi_1\Phi_2\phi_0(\boldsymbol{q})\} = 0$ implies that $\langle\Psi_{\boldsymbol{R}}|H|\mathcal{A}_{12}\{\Phi_1\Phi_2\phi_0(\boldsymbol{q})\}\rangle = 0$, i.e., $\mathcal{H}\phi_0(\boldsymbol{R}) = 0$, which shows that ϕ_0 satisfies Eq. (11.42) regardless of the energy E. Moreover, Ψ is not altered by adding ϕ_0 to ϕ behind the antisymmetrizer, i.e., $\mathcal{A}_{12}\{\Phi_1\Phi_2\phi(\boldsymbol{q})\} = \mathcal{A}_{12}\{\Phi_1\Phi_2[\phi(\boldsymbol{q}) + c_0\phi_0(\boldsymbol{q})]\}$. Thus, if at least one Pauli-forbidden relative-motion state exists, the function ϕ, as defined in Eq. (11.36), is not unambiguous. To ensure that it is equal to the function $\phi(\boldsymbol{R})$ defined unambiguously in Eq. (11.46), one should impose the condition $\mathcal{I}\phi(\boldsymbol{R}) = \phi(\boldsymbol{R})$, i.e., $\langle\!\langle\phi_0|\phi\rangle\!\rangle = 0$ [302].

With Eq. (11.42) we managed to get an equation for the relative motion. As will be seen in Sect. 11.5, it serves well as the basic equation for a practicable microscopic model. We shall work on this equation further to arrive at a firm interpretation in terms of the cluster internal and relative modes of motion. This is crucial for a macroscopic interpretation of the model. As we saw in Part I, reaction theory cannot do without macroscopic concepts, and the elementary models of light exotic nuclei are all macroscopic. One has to know whether the relative-motion function $\phi(\boldsymbol{R})$ can be interpreted as an intercluster relative ***wave function***. The fact that ϕ and $\phi + c\phi_0$ give the same multiparticle function Ψ is a warning against its interpretation as a relative wave function. Further warning comes from its normalization: $\langle\Psi|\Psi\rangle = 1$ implies $\langle\!\langle\phi|\mathcal{N}|\phi\rangle\!\rangle = 1$, which is not like the normalization of a wave function. Indeed, Eq. (11.42) is not like a Schrödinger equation because of $\mathcal{N}$ on the right-hand side. To cure Eq. (11.42), one can perform a transformation as follows: insert $\mathcal{I} = \mathcal{N}^{-\kappa}\mathcal{N}^{\kappa}$ in Eq. (11.42) after $\mathcal{H}$ and multiply both sides by $\mathcal{N}^{\kappa-1}$. One thus obtains

$$\mathcal{H}_\kappa \psi_\kappa(\boldsymbol{R}) = E\psi_\kappa(\boldsymbol{R}), \tag{11.57}$$

with

$$\psi_\kappa(\boldsymbol{R}) = \mathcal{N}^\kappa \phi(\boldsymbol{R}), \qquad \mathcal{H}_\kappa = \mathcal{N}^{\kappa-1}\mathcal{H}\mathcal{N}^{-\kappa}. \tag{11.58}$$

If $\kappa \neq 0$, ψ_κ is free from the additive ambiguity characteristic of ϕ. Equation (11.57) has the form of a Schrödinger equation with a Hamiltonian, which is non-Hermitean in general because the Hermitean operators $\mathcal{H}$ and $\mathcal{N}$ do not commute. This is unreasonable for the intercluster Hamiltonian of a pure cluster model. There is, however, a distinguished value of κ, viz. $\kappa = \frac{1}{2}$, which implies that (i) the relative wave function is normalized as a wave function should be: $\langle\psi_{1/2}|\psi_{1/2}\rangle = 1$, (ii) the Hamiltonian becomes Hermitean. The equations

$$\psi(\boldsymbol{R}) \equiv \psi_{1/2}(\boldsymbol{R}), \qquad \tilde{\mathcal{H}} \equiv \mathcal{H}_{1/2}, \tag{11.59}$$

$$\tilde{\mathcal{H}}\psi(\boldsymbol{R}) = E\psi(\boldsymbol{R}), \tag{11.60}$$

have these properties.[13]

It should be noted that, although the foregoing derivation has disqualified both $g(\boldsymbol{R})$ and $\phi(\boldsymbol{R})$ as candidates for a relative wave function, it did not rule out their use in any other context. In fact, as we have seen in Sect. 9.5, the densities and momentum densities that can be extracted from experiment are those which belong to $g(\boldsymbol{R}) = \psi_1(\boldsymbol{R})$ and not those belonging to any other possible $\psi_\kappa(\boldsymbol{R})$. That is so because reaction theories contain just $g(\boldsymbol{R})$.[14] For light systems it makes little error to identify the two types of quantities since the eigenvalues of $\mathcal{N}$ are either (almost) zero or close to unity.

11.5 The resonating-group method

The RGM is essentially a practical implementation of the general cluster model discussed up to now. The ansatz (11.40) expresses that $\boldsymbol{\Psi}_{\boldsymbol{R}}$ is fixed and $\phi(\boldsymbol{R})$ is to be determined through the dynamical solution of the problem. Since, for a bound-state case, this solution can be derived

[13] The Schrödinger equation for the relative motion is still not quite unique. Subjected to a unitary transformation, it yields another equation of equally good formal properties. What distinguishes Eqs. (11.59) and (11.60) from all transformed equations (which imply similarly transformed observables) is the best correspondence with macroscopic observables [303].

[14] In fact, e.g., the theory of cluster decay and γ-induced direct breakup can be reformulated so that g is replaced by ψ, but those of stripping and pickup reactions cannot be. What prevents such a reformulation in the latter cases is that the transferred cluster is part of two different nuclei before and after the collision [289, 290].

from the variational principle stated on p. 195, the RGM is essentially a variational method. In this section we make the formalism more explicit and introduce what makes the RGM calculable: the description of the cluster internal states by the HOSM. Then we shall show how the matrix elements can actually be calculated.

11.5.1 Essentials

To display the structure of Eqs. (11.42) and (11.60), one should separate the intrinsic and intercluster terms of the microscopic Hamiltonian (11.38):

$$H = H_{a_1}(\xi_1) + H_{a_2}(\xi_2) + T_{12}(\boldsymbol{q}) + V_{12}(\xi_1, \xi_2, \boldsymbol{q}), \tag{11.61}$$

where[15] $T_{12}(\boldsymbol{q}) = -(\hbar^2/2\mu_{12})\nabla_{\boldsymbol{q}}^2$. Substituting Eqs. (11.41) and (11.61) into Eq. (11.43) and using $\mathcal{A}_{12}^2 = [A!/a_1!a_2!]^{1/2}\mathcal{A}_{12}$ [cf. Eq. (D.13)] and then Eq. (11.39), we obtain

$$H(\boldsymbol{R}, \boldsymbol{R}') = [E_1 + E_2 + T_{12}(\boldsymbol{R})]N(\boldsymbol{R}, \boldsymbol{R}') + V(\boldsymbol{R}, \boldsymbol{R}') \tag{11.62}$$

where $T_{12}(\boldsymbol{R}) = (-\hbar^2/2\mu_{12})\nabla_{\boldsymbol{R}}^2$ and

$$\begin{aligned} V(\boldsymbol{R}, \boldsymbol{R}') &= \sqrt{A!/a_1!a_2!}\langle \Phi_1\Phi_2\delta(\boldsymbol{R}-\boldsymbol{q})|V_{12}|\Psi_{\boldsymbol{R}'}\rangle \\ &= \langle \mathcal{A}_{12}\{\Phi_1\Phi_2\delta(\boldsymbol{R}-\boldsymbol{q})V_{12}\}|\Psi_{\boldsymbol{R}'}\rangle. \end{aligned} \tag{11.63}$$

Equation (11.62) is obtained by shifting the antisymmetrizer in the bra to the right through the *permutation-invariant* operator H, and then by allowing the terms of H defined in Eq. (11.61) to act on the left. One can, furthermore, separate the direct terms[16] of $N(\boldsymbol{R}, \boldsymbol{R}')$ and $V(\boldsymbol{R}, \boldsymbol{R}')$:

$$N(\boldsymbol{R}, \boldsymbol{R}') = \delta(\boldsymbol{R}-\boldsymbol{R}') - K(\boldsymbol{R}, \boldsymbol{R}'), \tag{11.64}$$

$$V(\boldsymbol{R}, \boldsymbol{R}') = V_{\rm D}(\boldsymbol{R})\delta(\boldsymbol{R}-\boldsymbol{R}') - V_{\rm EX}(\boldsymbol{R}, \boldsymbol{R}') \tag{11.65}$$

[15] The potential $V_{12}(\xi_1, \xi_2, \boldsymbol{q})$ acts between clusters; it is not to be mixed up with the nucleon–nucleon potential V_{jk} introduced in Eq. (11.38).

[16] When the two clusters are identical, the terms created by the permutations will be equal pairwise. That renders $V(\boldsymbol{R}, \boldsymbol{R}')$ to have the form $F(\boldsymbol{R}, \boldsymbol{R}') + \epsilon F(\boldsymbol{R}, -\boldsymbol{R}')$, where $\epsilon = 1$ for bosonic and -1 for fermionic clusters (cf. footnote on p. 286). For bosons with even partial waves and for fermions with odd partial waves, the angular-momentum projections of the two terms will be identical, which takes away the extra normalization factor $2^{-1/2}$ in the test state of identical clusters, so the definition of the direct potential $V_{\rm D}(\boldsymbol{R})$ will remain meaningful. (For opposite-parity partial waves, the two terms cancel each other, signalling the Pauli-forbiddenness of these partial waves.) We shall see in Chap. 12, that the nucleon–nucleon potential may contain the operator P_{ij}^{r}, which exchanges the spatial coordinates of the nucleons. The direct potential term is defined such that P_{ij}^{r} is replaced by the product $-P_{ij}^{\sigma}P_{ij}^{\tau}$ of spin- and isospin-exchange operators before the exchange term is detached, which is an identity for antisymmetrical states. The spin–isospin dependence of V is now suppressed.

[cf. Eq. (11.16)]. The exchange terms, as integral-equation kernels, are of short range and they are well-behaved. With Eqs. (11.62), (11.64) and (11.65), the equation of motion (11.42) takes the form

$$[T_{12}(\boldsymbol{R}) + V_{\rm D}(\boldsymbol{R}) - \varepsilon]\phi(\boldsymbol{R}) + \int d\boldsymbol{R}' \Big\{[\varepsilon - T_{12}(\boldsymbol{R})]K(\boldsymbol{R},\boldsymbol{R}') - V_{\rm EX}(\boldsymbol{R},\boldsymbol{R}')\Big\}\phi(\boldsymbol{R}') = 0, \qquad (\varepsilon = E - E_1 - E_2), \tag{11.66}$$

where ε is the energy of the relative motion.

This is the equation of motion of the RGM, but the less explicit Eq. (11.42) will also be referred to as such. Equation (11.66) is an integrodifferential equation for a function which depends on the intercluster displacement. Although, as discussed in Sect. 11.4.2, it should not be interpreted as a Schrödinger equation, the integral operator is sometimes regarded as an explicitly energy-dependent nonlocal potential.[17] The local potential can be expressed in terms of cluster densities. Restricting ourselves to a Wigner-type (i.e., no-exchange) central nucleon–nucleon interaction, we can write (see explanation below)

$$\begin{aligned}
&V_{\rm D}(\boldsymbol{R})\delta(\boldsymbol{R}-\boldsymbol{R}')\\
&\quad= \langle \Phi_1\Phi_2\delta(\boldsymbol{R}-\boldsymbol{q})| \sum_{i\in a_1,\, j\in a_2} V(\boldsymbol{r}_j-\boldsymbol{r}_i)|\Phi_1\Phi_2\delta(\boldsymbol{R}'-\boldsymbol{q})\rangle\\
&\quad= \delta(\boldsymbol{R}-\boldsymbol{R}')\langle \Phi_1\Phi_2\delta(\boldsymbol{R}-\boldsymbol{q})|\\
&\qquad\times \sum_{i\in a_1,\, j\in a_2} V[(\boldsymbol{r}_j-\bar{\boldsymbol{t}}_2)-(\boldsymbol{r}_i-\bar{\boldsymbol{t}}_1)-\boldsymbol{q}]|\Phi_1\Phi_2\rangle\\
&\quad= \delta(\boldsymbol{R}-\boldsymbol{R}')\int d\boldsymbol{\zeta}\int d\boldsymbol{\eta}\, V(\boldsymbol{\zeta}-\boldsymbol{\eta}-\boldsymbol{R})\\
&\qquad\times \sum_{i\in a_1}\langle\Phi_1|\delta(\boldsymbol{r}_i-\bar{\boldsymbol{t}}_1-\boldsymbol{\eta})|\Phi_1\rangle \sum_{j\in a_2}\langle\Phi_2|\delta(\boldsymbol{r}_j-\bar{\boldsymbol{t}}_2-\boldsymbol{\zeta})|\Phi_2\rangle\\
&\quad= \delta(\boldsymbol{R}-\boldsymbol{R}')\int d\boldsymbol{\zeta}\int d\boldsymbol{\eta}\, \rho_1(\boldsymbol{\eta})V(\boldsymbol{\zeta}-\boldsymbol{\eta}-\boldsymbol{R})\rho_2(\boldsymbol{\zeta}).
\end{aligned} \tag{11.67}$$

In the first step the identity $\delta(\boldsymbol{R}-\boldsymbol{q})\delta(\boldsymbol{R}'-\boldsymbol{q}) = \delta(\boldsymbol{R}-\boldsymbol{R}')\delta(\boldsymbol{R}-\boldsymbol{q})$ is used and $0 = \bar{\boldsymbol{t}}_1 - \bar{\boldsymbol{t}}_2 - \boldsymbol{q}$ [cf. Eq. (9.17)] is inserted. In the second step two dummy delta functions and the corresponding integrations over the vectors $\boldsymbol{\zeta}$ and $\boldsymbol{\eta}$ are introduced, whereby the potential can be brought out from the matrix element and the matrix element factorizes. In the last step the two factors are identified with the densities of the two clusters according to the general definition given in Eq. (9.127). To see that the two matrix elements correspond to Eq. (9.127), one has

[17] A quantum-mechanical potential $\mathcal{V}$ is said to be nonlocal if it is given by an integral operator: $(\mathcal{V}\psi)(\boldsymbol{r}) = \int d\boldsymbol{r}'\, V(\boldsymbol{r},\boldsymbol{r}')\psi(\boldsymbol{r}')$, where $V(\boldsymbol{r},\boldsymbol{r}') \not\propto \delta(\boldsymbol{r}-\boldsymbol{r}')$.

to recognize that $r_i - \bar{t}_1$ and $r_j - \bar{t}_2$, the position vectors with respect to the cluster c.m., are the same as $\bar{r}_i$ in Eq. (9.127) for the whole nucleus. Equation (11.67) is just the familiar double-folding formula [cf. Eq. (5.2) of Part I] often used to generate phenomenological nucleus–nucleus interactions.

The spin- and isospin-dependent terms of the exchange forces contribute when the nucleonic spins or isospins add up to non-zero values for both clusters. For example, when both clusters have neutron excess, the isospin-dependent interaction term $V(r_j - r_i)\tau_i \cdot \tau_j$ (where $\frac{1}{2}\tau$ are isospin vector operators) gives rise to folding terms containing the differences between the neutron and proton densities.

It is tempting to omit the nonlocal terms in Eq. (11.66), but the exchange terms are very large unless the relative momentum of the two fragments is large (which may be an acceptable approximation for scattering at tens of MeV per nucleon and beyond, but not for bound-state problems).

We now turn to the relative-motion Schrödinger equation (11.60). With the definition given in Eqs. (11.59) and (11.58), the following alternative formulae are obtained:

$$\tilde{\mathcal{H}} = (E_1 + E_2)\mathcal{I} + \mathcal{N}^{-1/2}\mathcal{T}\mathcal{N}^{1/2} + \mathcal{N}^{-1/2}\mathcal{V}\mathcal{N}^{-1/2} \tag{11.68}$$

$$= (E_1 + E_2)\mathcal{I} + \mathcal{N}^{1/2}\mathcal{T}\mathcal{N}^{-1/2} + \mathcal{N}^{-1/2}\mathcal{V}^\dagger\mathcal{N}^{-1/2}, \tag{11.69}$$

where $\mathcal{T}$ and $\mathcal{V}$ are integral operators with kernels $T_{12}(\boldsymbol{R})\delta(\boldsymbol{R}-\boldsymbol{R}')$ and $V(\boldsymbol{R},\boldsymbol{R}')$ of Eq. (11.63), respectively. The first formula is just Eq. (11.62) rewritten into operator form and multiplied by $\mathcal{N}^{-1/2}$ from both sides, yielding $\tilde{\mathcal{H}} = \mathcal{N}^{-1/2}\mathcal{H}\mathcal{N}^{-1/2}$. The second formula comes from an analogous derivation, with the left–right directions reversed. In particular, in Eq. (11.43) the antisymmetrizer in the ket is shifted to the left through H and then the terms of H are made to act on the right. Note that both in Eqs. (11.68) and (11.69) the last two terms are non-Hermitean, and the respective terms are each other's Hermitean conjugates. Then it follows from the equality of Eqs. (11.68) and (11.69) that the sums of the last two terms are Hermitean and identical:

$$\begin{aligned}\mathcal{T}+\tilde{\mathcal{V}} &\equiv \mathcal{N}^{-1/2}\mathcal{T}\mathcal{N}^{1/2}+\mathcal{N}^{-1/2}\mathcal{V}\mathcal{N}^{-1/2}\\ &= \mathcal{N}^{1/2}\mathcal{T}\mathcal{N}^{-1/2}+\mathcal{N}^{-1/2}\mathcal{V}^\dagger\mathcal{N}^{-1/2}.\end{aligned} \tag{11.70}$$

With this, the Schrödinger equation for the relative motion reads

$$(\mathcal{T} + \tilde{\mathcal{V}})\psi = \varepsilon\psi. \tag{11.71}$$

The formula (11.70) may be viewed as an implicit expression for the potential $\tilde{\mathcal{V}}$. This shows that the intercluster potential gains contribution not only from the nucleon–nucleon potential but from the nucleonic kinetic energy as well, but the separation of the two contributions is not unambiguous.

In practice the g.s. wave functions Φ_i of the isolated clusters a_i have to be identified by approximate solutions of Eq. (11.39): $\Phi_i \equiv \Phi_i^{(0)}$, where

$$H_i \Phi_i^{(0)}(\xi_i) \simeq E_i^{(0)} \Phi_i^{(0)}(\xi_i). \tag{11.72}$$

To make the kernels in Eq. (11.66) calculable, the clusters are mostly described by the HOSM. The term RGM is usually used for this version of the cluster model. Once the SM is adopted for Φ_i, Eq. (11.62) is not exact any more, thus Eqs. (11.42) and (11.66) are no longer equivalent. Equation (11.66) is an approximation to Eq. (11.42), but it is not necessarily less realistic than Eq. (11.42). The advantage of Eq. (11.66) over Eq. (11.42) is that the asymptotic behaviour of its solution is determined by $\varepsilon = E - E_1 - E_2 (= E - E_1^{(0)} - E_2^{(0)})$, which may be set correct whether or not the approximate energy $E_1^{(0)} + E_2^{(0)}$ is correct. Therefore, for scattering problems it is more convenient to use Eq. (11.66).

The g.s. of the isolated clusters imply size parameters that produce the minimum energies and/or the correct radii for the isolated clusters. It may as well be meaningful to minimize the energy of the composite system by varying the cluster size parameters. That improves the model but makes the associated picture less transparent, although the departure from the value that yields the energy minimum for the isolated subsystems is not unphysical; it may be attributed to distortions due to the presence of the other cluster.

The model becomes even more transparent if the oscillators of the two clusters belong to the same quanta $\hbar\omega$. In such a case the norm kernel eigenfunctions $\phi_{nlm}(\boldsymbol{R})$ will also be HO eigenfunctions belonging to the same $\hbar\omega$. Furthermore, it is in this case that, for the first few values of n, l, the eigenvalues ν_{nl} are zero, which implies that exactly Pauli-forbidden relative-motion states $\phi_{nlm}(\boldsymbol{R})$ exist. When the two oscillator quanta are not equal but are close to each other, the ν_{nl} values that become zero in the equal-$\hbar\omega$ limit will be very small. The corresponding eigenfunctions will be strongly suppressed in the low-energy relative-motion states, and that is why these relative-motion components are called 'almost forbidden states'.[18] These questions will be discussed in more detail in Sect. 11.6.

[18] To comply with the Pauli principle, the energy expectation value in an almost forbidden state is high [304]. They are thus usually embedded in the continuum, where they appear as spurious [305] or possibly non-spurious [306] 'Pauli resonances'.

11.5.2 Matrix elements

The kernels $N(\boldsymbol{R},\boldsymbol{R}')$ and $H(\boldsymbol{R},\boldsymbol{R}')$ of the integral operators involved in the RGM equation (11.42) are dubbed the RGM kernels. To calculate them is not simple because the RGM is formulated in terms of intrinsic and relative coordinates, in which antisymmetrization is extremely involved. To make them calculable, the RGM wave function is somehow expressed in terms of s.p. wave functions, i.e., Slater determinants. The calculation of the RGM kernels is, in some sense, a prototype of the calculation of the matrix elements of the correlated Gaussians. In fact, the sample calculation given in App. H for the correlated Gaussian formalism runs parallel with the forthcoming derivation. Since we shall now deal with HO wave functions explicitly, it is more convenient to use the relative variable $\boldsymbol{\rho}$ of Eq. (9.16) or (11.5), rather than $\boldsymbol{q}$ of Eq. (11.3).

For the time being, let us confine ourselves to two clusters whose intrinsic motions are described by HO g.s. of equal oscillator quanta $\hbar\omega$. All systematic methods to evaluate the kernels do somehow involve shifted-Gaussian relative-motion states, such as given here:

$$\Psi(\boldsymbol{S}) = \mathcal{A}_{12}\left\{\Phi_1(\xi_1)\Phi_2(\xi_2)\varphi_{\boldsymbol{S}}(\boldsymbol{\rho})\right\}. \tag{11.73}$$

For simplicity, the angular momenta have again been suppressed. The shifted Gaussians $\varphi_{\boldsymbol{S}}$ are defined by Eq. (9.66). They are qualified for this role by the following properties: (i) they are related to the Dirac-δ involved in the RGM basis state $\Psi_{\boldsymbol{R}}$ by a variety of handy integral transformations [232, 307], (ii) a shifted-Gaussian cluster-model state appended with appropriate c.m. wave functions is essentially a Slater determinant, (iii) a function can be approximated by a linear combination of shifted Gaussians with arbitrary accuracy [308], and thus the shifted Gaussians lend themselves to very flexible nonorthogonal bases, (iv) the shifted-Gaussian states are essentially the generating functions of the HO s.p. states (cf. Sect. 9.3).

The cluster-model states with shifted Gaussians in the relative motion, (11.73), are the basic ingredients of the generator-coordinate formalism. This formalism may just serve as a technique to evaluate the RGM kernels or, alternatively, may provide new physical bases, which can be substantiated independently. The technique for calculating the RGM kernels is based on properties (i) and (ii), and is called the *generator-coordinate technique*. From among the bases related to shifted Gaussians, three should be mentioned. The representation of the relative motion by shifted Gaussians [property (iii)] is called the GCM [232]. Owing to property (iv), it may be useful to take the relative motion function a combination of HO eigenfunctions with different node numbers, and that is often called the algebraic version of the

RGM [309]. Finally, it is again property (iv) that enables one to use correlated Gaussians to represent the relative motions. In this section we confine ourselves to the generator-coordinate technique; the GCM is the subject of Sect. 11.7, and the correlated Gaussian cluster model, which is a central subject of this book, will be presented in Chap. 12.

To each transformation used in the generator-coordinate technique there corresponds a particular integral formula for the Dirac-δ, and the transformation consists in substituting this formula for the δ that is included in the RGM basis state $\Psi_{\boldsymbol{R}}$. For instance, one can obtain a double integral transformation [232] by applying Eq. (11.34). For an a_1+a_2 two-cluster case, $\boldsymbol{\rho}=\sqrt{\mu}\boldsymbol{q}$, where $\mu=a_1a_2/(a_1+a_2)$ [cf. Eqs. (9.16), (9.17)]. Let us introduce a corresponding parameter coordinate, $\underline{\boldsymbol{R}} = \sqrt{\mu}\boldsymbol{R}$, and substitute

$$\delta(\boldsymbol{R}-\boldsymbol{q}) = \mu^{3/2}\delta(\underline{\boldsymbol{R}}-\boldsymbol{\rho}), \tag{11.74}$$

into Eq. (11.41). Applying Eq. (11.34) to $\delta(\underline{\boldsymbol{R}}-\boldsymbol{\rho})$ and inserting Eq. (11.41) in the kernels, one gets

$$\begin{aligned}
&\langle\Psi_{\boldsymbol{R}}|O|\Psi_{\boldsymbol{R}'}\rangle \\
&= (2\pi)^{-6}\mu^3\left(\frac{\beta}{4\pi}\right)^{3/2}\int d\boldsymbol{k}\int d\boldsymbol{k}'\,\exp\left[i\sqrt{\mu}(\boldsymbol{k}\cdot\boldsymbol{R}-\boldsymbol{k}'\cdot\boldsymbol{R}')+\frac{\boldsymbol{k}^2+\boldsymbol{k}'^2}{2\beta}\right] \\
&\times\int d\boldsymbol{S}\int d\boldsymbol{S}'\,\exp[-i(\boldsymbol{k}\cdot\boldsymbol{S}-\boldsymbol{k}'\cdot\boldsymbol{S}')]\langle\mathcal{A}_{12}\{\Phi_1(\xi_1)\Phi_2(\xi_2)\varphi_{\boldsymbol{S}}(\boldsymbol{\rho})\} \\
&\times|O|\mathcal{A}_{12}\{\Phi_1(\xi_1)\Phi_2(\xi_2)\varphi_{\boldsymbol{S}'}(\boldsymbol{\rho})\}\rangle.
\end{aligned} \tag{11.75}$$

The aim is now to relate the state $\mathcal{A}_{12}\{\Phi_1\Phi_2\varphi_{\boldsymbol{S}}(\boldsymbol{\rho})\}$ to a Slater determinant. One can multiply Eq. (11.75) by $1=\langle\varphi_{\boldsymbol{S}_A}(\boldsymbol{x}_A)|\varphi_{\boldsymbol{S}_A}(\boldsymbol{x}_A)\rangle$, where $\varphi_{\boldsymbol{S}_A}(\boldsymbol{x}_A)$ is the c.m. wave packet. Then, it is easy to show that the product of the c.m. and relative Gaussians is a product of cluster c.m. Gaussians pinned down at certain points $\boldsymbol{s}_1$, $\boldsymbol{s}_2$ [cf. Eqs. (9.71) and (9.72)]:

$$\varphi_{\boldsymbol{S}_A}(\boldsymbol{x}_A)\varphi_{\boldsymbol{S}}(\boldsymbol{\rho}) = \varphi_{\boldsymbol{s}_1}(\bar{\boldsymbol{\xi}}_1)\varphi_{\boldsymbol{s}_2}(\bar{\boldsymbol{\xi}}_2), \tag{11.76}$$

with the cluster centres being at

$$\boldsymbol{s}_1 = \frac{1}{\sqrt{A}}(\sqrt{a_1}\boldsymbol{S}_A+\sqrt{a_2}\boldsymbol{S}), \qquad \boldsymbol{s}_2 = \frac{1}{\sqrt{A}}(\sqrt{a_2}\boldsymbol{S}_A-\sqrt{a_1}\boldsymbol{S}). \tag{11.77}$$

This definition is an inverted version of Eqs. (11.5) applied to two fragments, with the physical coordinates replaced by generator coordinates:

$$\boldsymbol{S} = \frac{1}{\sqrt{A}}(\sqrt{a_2}\boldsymbol{s}_1-\sqrt{a_1}\boldsymbol{s}_2), \qquad \boldsymbol{S}_A = \frac{1}{\sqrt{A}}(\sqrt{a_1}\boldsymbol{s}_1+\sqrt{a_2}\boldsymbol{s}_2). \tag{11.78}$$

A cluster c.m. Gaussian multiplied by the g.s. cluster intrinsic wave function will be equal to an ordinary HOSM state centred at position $\boldsymbol{s}_i$:

$$\Phi^{\rm SM}_{i\boldsymbol{s}_i} = \Phi_i \varphi_{\boldsymbol{s}_i}(\bar{\boldsymbol{\xi}}_i). \tag{11.79}$$

This statement is a special case of a Bethe–Rose theorem, which says that, so long as there are no intershell excitations, *the c.m.-dependence of an HOSM wave function can be factored out as a g.s. HO wave function* [310, 311]. Using these three tricks, one can obtain a matrix element between Slater determinants:

$$\langle \mathcal{A}_{12}\{\Phi_1(\xi_1)\Phi_2(\xi_2)\varphi_{\boldsymbol{S}}(\boldsymbol{\rho})\}|O|\mathcal{A}_{12}\{\Phi_1(\xi_1)\Phi_2(\xi_2)\varphi_{\boldsymbol{S}'}(\boldsymbol{\rho})\}\rangle \tag{11.80}$$

$$= \langle \mathcal{A}_{12}\{\Phi_1(\xi_1)\Phi_2(\xi_2)\varphi_{\boldsymbol{S}}(\boldsymbol{\rho})\varphi_{\boldsymbol{S}_A}(\boldsymbol{x}_A)\} \times |O|\mathcal{A}_{12}\{\Phi_1(\xi_1)\Phi_2(\xi_2)\varphi_{\boldsymbol{S}'}(\boldsymbol{\rho})\varphi_{\boldsymbol{S}_A}(\boldsymbol{x}_A)\}\rangle \tag{11.81}$$

$$= \langle \mathcal{A}_{12}\{\Phi_1(\xi_1)\Phi_2(\xi_2)\varphi_{\boldsymbol{s}_1}(\bar{\boldsymbol{\xi}}_1)\varphi_{\boldsymbol{s}_2}(\bar{\boldsymbol{\xi}}_2)\} \times |O|\mathcal{A}_{12}\{\Phi_1(\xi_1)\Phi_2(\xi_2)\varphi_{\boldsymbol{s}'_1}(\bar{\boldsymbol{\xi}}_1)\varphi_{\boldsymbol{s}'_2}(\bar{\boldsymbol{\xi}}_2)\}\rangle \tag{11.82}$$

$$= \langle \mathcal{A}_{12}\{\Phi^{\rm SM(0)}_{1\boldsymbol{s}_1}(\boldsymbol{r}_1,\ldots,\boldsymbol{r}_{a_1})\Phi^{\rm SM(0)}_{2\boldsymbol{s}_2}(\boldsymbol{r}_{a_1+1},\ldots,\boldsymbol{r}_A)\} \times |O|\mathcal{A}_{12}\{\Phi^{\rm SM(0)}_{1\boldsymbol{s}'_1}(\boldsymbol{r}_1,\ldots,\boldsymbol{r}_{a_1})\Phi^{\rm SM(0)}_{2\boldsymbol{s}'_2}(\boldsymbol{r}_{a_1+1},\ldots,\boldsymbol{r}_A)\}\rangle, \tag{11.83}$$

with [cf. Eqs. (11.77)]

$$\boldsymbol{s}'_1 = \frac{1}{\sqrt{A}}(\sqrt{a_1}\boldsymbol{S}_A + \sqrt{a_2}\boldsymbol{S}'), \qquad \boldsymbol{s}'_2 = \frac{1}{\sqrt{A}}(\sqrt{a_2}\boldsymbol{S}_A - \sqrt{a_1}\boldsymbol{S}'). \tag{11.84}$$

The function $\mathcal{A}_{12}\{\Phi^{\rm SM(0)}_{1\boldsymbol{s}_1}\Phi^{\rm SM(0)}_{2\boldsymbol{s}_2}\}$ is a Slater determinant, like Eq. (9.67) for the few-body problem. The evaluation of such a matrix element is straightforward and is the subject of App. E.

In a general cluster model, however, it is often necessary to allow the cluster intrinsic HO quanta $\hbar\omega$ to be different [307]. To include that case, we use a superscript on the symbol of the shifted Gaussian that displays the Gaussian (or HO) parameter [cf. Eq. (9.66)]: $\varphi^{(\beta)}_{\boldsymbol{S}} \equiv \varphi_{\boldsymbol{S}}$. A most straightforward treatment of this case is to use the integral formula

$$\left(\frac{4\pi b}{\beta_1\beta_2}\right)^{3/4} \varphi^{(b)}_{\boldsymbol{S}}(\boldsymbol{\rho}) = \int d\boldsymbol{S}_A\, \varphi^{(\beta_1)}_{\boldsymbol{s}_1}(\bar{\boldsymbol{\xi}}_1)\varphi^{(\beta_2)}_{\boldsymbol{s}_2}(\bar{\boldsymbol{\xi}}_2) \tag{11.85}$$

$$= \int d\boldsymbol{x}_A\, \varphi^{(\beta_1)}_{\boldsymbol{s}_1}(\bar{\boldsymbol{\xi}}_1)\varphi^{(\beta_2)}_{\boldsymbol{s}_2}(\bar{\boldsymbol{\xi}}_2), \quad \text{with } b = \frac{A\beta_1\beta_2}{a_1\beta_1 + a_2\beta_2}. \tag{11.86}$$

Equation (11.85) can be verified by performing a Gaussian integration (9.50) and the laborious substitution of Eq. (11.77) for $\boldsymbol{s}_1$ and $\boldsymbol{s}_2$; Eq. (11.86) is then a trivial consequence of the symmetry of $\varphi_{\boldsymbol{s}}(\boldsymbol{r})$ in $\boldsymbol{s}$ and $\boldsymbol{r}$. A matrix element between two-centre Slater determinants with different HO quanta should be integrated over S_A with the help of Eq. (11.86). The derivation to come follows the pattern of Eqs. (11.80)–(11.83) in backward direction. But now the centres in the bra and in the ket should be independent of each other, i.e., Eq. (11.84) is to be replaced by

$$s_1' = \frac{1}{\sqrt{A}}(\sqrt{a_1}S_A' - \sqrt{a_2}S'), \qquad s_2' = \frac{1}{\sqrt{A}}(\sqrt{a_2}S_A' + \sqrt{a_1}S'). \tag{11.87}$$

One can thus write

$$\int dS_A \, \langle \mathcal{A}_{12}\{\Phi_{1s_1}^{\mathrm{SM}(0)}(r_1,\ldots,r_{a_1})\Phi_{2s_2}^{\mathrm{SM}(0)}(r_{a_1+1},\ldots,r_A)\} \times |O|\mathcal{A}_{12}\{\Phi_{1s_1'}^{\mathrm{SM}(0)}(r_1,\ldots,r_{a_1})\Phi_{2s_2'}^{\mathrm{SM}(0)}(r_{a_1+1},\ldots,r_A)\}\rangle \tag{11.88}$$

$$= \int dS_A \, \langle \mathcal{A}_{12}\{\Phi_1(\xi_1)\Phi_2(\xi_2)\varphi_{s_1}^{(\beta_1)}(\bar{\xi}_1)\varphi_{s_2}^{(\beta_2)}(\bar{\xi}_2)\} \times |O|\mathcal{A}_{12}\{\Phi_1(\xi_1)\Phi_2(\xi_2)\varphi_{s_1'}^{(\beta_1)}(\bar{\xi}_1)\varphi_{s_2'}^{(\beta_2)}(\bar{\xi}_2)\}\rangle \tag{11.89}$$

$$= \left(\frac{4\pi b}{\beta_1\beta_2}\right)^{3/4} \langle \mathcal{A}_{12}\{\Phi_1(\xi_1)\Phi_2(\xi_2)\varphi_S^{(b)}(\rho)\} \times |O|\mathcal{A}_{12}\{\Phi_1(\xi_1)\Phi_2(\xi_2)\varphi_{s_1'}^{(\beta_1)}(\bar{\xi}_1)\varphi_{s_2'}^{(\beta_2)}(\bar{\xi}_2)\}\rangle \tag{11.90}$$

$$= \left(\frac{4\pi b}{\beta_1\beta_2}\right)^{3/2} \langle \mathcal{A}_{12}\{\Phi_1(\xi_1)\Phi_2(\xi_2)\varphi_S^{(b)}(\rho)\} \times |O|\mathcal{A}_{12}\{\Phi_1(\xi_1)\Phi_2(\xi_2)\varphi_{S'}^{(b)}(\rho)\}\rangle. \tag{11.91}$$

In stepping from Eq. (11.89) to (11.90), Eq. (11.85) was used, and then, from Eq. (11.90) to (11.91), the analogous integration, Eq. (11.86), was performed. In keeping with this, up to Eq. (11.90), the notation $\langle\cdots|\cdots|\cdots\rangle$ includes $\int dx_A$, but in Eq. (11.91) it does not. The integration over the c.m. generator coordinate S_A is tantamount to projecting the bra onto total momentum zero [cf. discussion after Eq. (9.105)]:

$$\int dS_A \, \Phi_{S_A} = \int dS_A \, \exp(-ik\cdot\sqrt{A}S_A)\mathcal{T}_{\sqrt{A}S_A}\Phi_0\Big|_{k=0}, \tag{11.92}$$

where $\mathcal{T}_{S_A}$ is a shift operator and Φ_{S_A} is any wave function centred at S_A. The right-hand side is indeed a constant times the momentum projector defined in Eq. (9.106). That is why the c.m. motion can be eliminated by integration over either S_A or S_A'.

Equation (11.75) expresses concisely how to relate an RGM kernel to a matrix element between shifted Gaussian wave packets. Equations (11.80)–(11.83) and (11.88)–(11.91), on the other hand, relate the matrix elements of shifted-Gaussian cluster-model states to those of Slater determinants. With this, we have shown a method for evaluating RGM kernels, and any matrix elements between RGM states can be traced back to those of Slater determinants similarly. Once the kernels are evaluated, it is straightforward to solve the RGM equation (11.42). In Sect. 11.7 we shall show that the matrix elements between two-cluster Gaussian wave packets, the GCM kernels, can also be used in solving a reformulated dynamical problem.

So far we have not cared about angular momenta. We show now how to treat the orbital angular momentum for the case of spin-zero clusters. One may introduce test states with definite orbital angular momenta as

$$\Psi_R^{(LM)} = \mathcal{A}_{12}\left\{\Phi_1(\xi_1)\Phi_2(\xi_2)\frac{\delta(R-q)}{q}Y_{LM}(\widehat{\boldsymbol{q}})\right\}. \tag{11.93}$$

By using the identity

$$\delta(\boldsymbol{R}-\boldsymbol{q}) = \sum_{LM}\frac{\delta(R-q)}{Rq}Y_{LM}(\widehat{\boldsymbol{q}})Y_{LM}^*(\widehat{\boldsymbol{R}}), \tag{11.94}$$

the test state $\Psi_{\boldsymbol{R}}$ of Eq. (11.9) can be expanded in terms of $\Psi_R^{(LM)}$ as

$$\Psi_{\boldsymbol{R}} = \sum_{LM}\frac{\Psi_R^{(LM)}}{R}Y_{LM}^*(\widehat{\boldsymbol{R}}), \tag{11.95}$$

and the function $\phi(\boldsymbol{R})$ can also be expanded:

$$\phi(\boldsymbol{R}) = \sum_{LM}\frac{\phi_{LM}(R)}{R}Y_{LM}(\widehat{\boldsymbol{R}}). \tag{11.96}$$

By using the orthonormality relation for the spherical harmonics, the RGM expansion (11.40) can be modified as

$$\begin{aligned}\Psi &= \int d\boldsymbol{R}\,\phi(\boldsymbol{R})\Psi_{\boldsymbol{R}} = \sum_{LM}\sum_{L'M'}\int d\boldsymbol{R}\,\frac{\phi_{LM}(R)}{R}Y_{LM}(\widehat{\boldsymbol{R}})\frac{\Psi_R^{(L'M')}}{R}Y_{L'M'}^*(\widehat{\boldsymbol{R}})\\ &= \sum_{LM}\int_0^\infty dR\,\phi_{LM}(R)\Psi_R^{(LM)}.\end{aligned} \tag{11.97}$$

The kernels in the radial RGM equation will thus be just

$$\langle\Psi_R^{(LM)}|O|\Psi_{R'}^{(L'M')}\rangle = RR'\int d\widehat{\boldsymbol{R}}\int d\widehat{\boldsymbol{R}}'\,Y_{LM}(\widehat{\boldsymbol{R}})\langle\Psi_{\boldsymbol{R}}|O|\Psi_{\boldsymbol{R}'}\rangle Y_{L'M'}^*(\widehat{\boldsymbol{R}}'), \tag{11.98}$$

which are easy to calculate from Eq. (11.75). Since the operators O in the RGM equation are scalar with respect to rotation, the kernels vanish unless $L = L'$ and $M = M'$, and, in addition, they are independent of M. Thus they can be calculated by taking the average over M, i.e., by

$$\begin{aligned}\langle\Psi_R^{(LM)}|O|\Psi_{R'}^{(LM)}\rangle &= \frac{1}{2L+1}\sum_M\langle\Psi_R^{(LM)}|O|\Psi_{R'}^{(LM)}\rangle \\ &= \frac{1}{4\pi}RR'\int d\widehat{\boldsymbol{R}}\int d\widehat{\boldsymbol{R}}'\, P_L\left(\frac{\boldsymbol{R}\cdot\boldsymbol{R}'}{RR'}\right)\langle\Psi_{\boldsymbol{R}}|O|\Psi_{\boldsymbol{R}'}\rangle,\end{aligned} \tag{11.99}$$

where P_L is a Legendre polynomial. Since all ingredients of the RGM equation are independent of M, so will be $\phi(\boldsymbol{R})$ itself. Be that enough for an introduction to the treatment of the angular momenta in the RGM.

11.6 The harmonic-oscillator cluster model

11.6.1 The model

As was said in Sect. 11.5, a specific property of the RGM is that the cluster intrinsic states are described by the HOSM. It is tempting to examine what happens if one puts an HO eigenfunction also for the intercluster function in Eq. (11.36):

$$\Psi = \mathcal{A}_{12}\left\{\Phi_1(\xi_1)\Phi_2(\xi_2)\psi_{nlm}(\boldsymbol{q})\right\}. \tag{11.100}$$

The model obtained in this way will become especially transparent if one takes all $\hbar\omega$ to be the same. This is what is called the HO cluster model [283, 312, 313]. Although not a dynamical model, it is very illuminating since it is closely related to the HOSM. *A priori* it looks as meaningful as the HOSM; after all, it is as justifiable to identify the intercluster potential with an HO potential as the s.p. potential. The tail of the intercluster motion in the HO cluster model is as incorrect as the tail of the s.p. motion in the HOSM. Furthermore, the requirement of the identity of the intercluster and intracluster $\hbar\omega$ scales the spatial extensions of these modes of motion just in accord with the SM.

To come up to this point, let us consider two non-interacting particles of equal mass confined by the same HO. The Hamiltonian has the $\{\boldsymbol{r}_1, \boldsymbol{r}_2\} \to \{\boldsymbol{x}_1, \boldsymbol{x}_2\}$ transformation property $H_{\text{HO}}(\boldsymbol{r}_1) + H_{\text{HO}}(\boldsymbol{r}_2) = H_{\text{HO}}(\boldsymbol{x}_1) + H_{\text{HO}}(\boldsymbol{x}_2)$ [cf. Eqs. (9.8) and (9.9)]. This implies the Talmi–Moshinsky transformation [210, 313] for the eigenfunctions:

$$\begin{aligned}&[\psi_{n_1l_1}(\boldsymbol{r}_1)\psi_{n_2l_2}(\boldsymbol{r}_2)]_{LM} \\ &\quad= \sum_{n\lambda N\Lambda}\langle n\lambda N\Lambda; L|n_1l_1n_2l_2; L\rangle[\psi_{n\lambda}(\boldsymbol{x}_1)\psi_{N\Lambda}(\boldsymbol{x}_2)]_{LM},\end{aligned} \tag{11.101}$$

where the coefficients $\langle n\lambda N\Lambda; L|n_1l_1n_2l_2; L\rangle$ can be given analytically. With consecutive transformations like (11.101), any multiparticle HO state given in terms of any relative coordinates can be rewritten into any other relative or s.p. coordinates. As a consequence, all HO cluster-model states can be rewritten exactly into any other HO cluster model or into a SM and *vice versa*, with a finite number of terms. The transformation (11.101) conserves the number of oscillator quanta, which implies that a state that carries $Q\,\hbar\omega$ can be expanded in terms of states of $Q\,\hbar\omega$ in any other arrangement. For basis states that carry the minimum number of quanta allowed by the Pauli principle or not many more, this means that they will be simple in other HO models as well.

How do the HOSM and cluster models compare? Let us consider the HOSM of ^{20}Ne in the sd-shell: the 0s, 0p and 1s0d shells are occupied by 4, 12 and 4 nucleons, respectively, and the state carries $(4\times0+12\times1+4\times2)\,\hbar\omega=20\,\hbar\omega$ excitation quanta [i.e. $(20+20\times\frac{3}{2})\,\hbar\omega$ energy]. A viable alternative model for ^{20}Ne is an ^{16}O+α cluster model [307], with g.s. clusters and relative-motion functions $\phi(\boldsymbol{R})$ being HO eigenfunctions. The ^{16}O+α model corresponding to this SM configuration must also carry the same number of excitation quanta. It thus follows that the corresponding ^{16}O+α states must have $8\hbar\omega$ for the relative motion, i.e., $2N+L=8$. This finding says that the $(N,L)=(4,0)$, (3,2), (2,4), (1,6), (0,8) cluster-model states are degenerate, and, rewritten into a one-centre form, each state lies in the sd-subspace. Moreover, since in the HOSM these states are the lowest-lying Pauli-allowed states, for the HO cluster model, the ^{16}O-α relative motion with $2N+L<8$ is Pauli-forbidden. The HOSM provides general guidance for the Pauli-allowed cluster-model states. For the quantum numbers N,L for the a_1–a_2 intercluster relative motion to be Pauli-allowed, it is necessary that

$$2N+L \geq \sum_{i=1}^{A}(2n_i+l_i) - \sum_{i=1}^{a_1}(2n_i+l_i) - \sum_{i=1}^{a_2}(2n_i+l_i) \qquad (11.102)$$

be satisfied, where the summations on the right-hand side are for the lowest-lying SM orbits in each nucleus, with single occupancy in each spin–isopin projection. The condition (11.102) is called the Wildermuth rule.[19]

The correspondence between the HOSM and the HO cluster model can be made more explicit by reference to the common symmetry of

[19]The condition (11.102) is sufficient for core+α systems, but not in general. For example, the ^{16}O+^{16}O, ^{40}Ca+^{16}O and ^{40}Ca+^{40}Ca systems require $2N+L$ values that are higher by 4, 4 and 20, respectively, than what is required by Eq. (11.102).

these models, which comes from the symmetry of the HO Hamiltonian itself.

The three-dimensional HO Hamiltonian is invariant under the unitary transformations of the dynamical variables $\{\boldsymbol{r}, \boldsymbol{p}\}$ which rotate the HO quanta in the three-dimensional space [313, 314]. These transformations span the U(3) (three-dimensional unitary) group (for details, see App. I). Naturally, the HO Hamiltonian is also invariant with respect to the transformations of the subgroups of U(3), of which notable are the group of the three-dimensional special unitary transformations, SU(3), and its subgroup, the group of three-dimensional spatial rotations, SO(3). The Hamiltonian of A or $A-1$ independent oscillators is also invariant with respect to all these transformations.

The symmetry properties of a Hamiltonian always have an imprint on the level structure and the degeneracies. The levels of a multiparticle HOSM are highly degenerate, and the degeneracy is split by the residual interactions according to the symmetry properties of the eigenfunctions. It is therefore important to find a systematic way of classifying the eigenfunctions according to their transformation properties. The set of eigenfunctions belonging to any particular level transform into itself when subjected to any transformation belonging to the symmetry group. This behaviour is usually expressed by saying that such a set spans a corresponding representation of the group. One can usually split the set of degenerate states into smaller sets that are not mixed by the transformations. The smallest possible sets are called the irreducible representations. These may be further split into smaller sets which are irreducible representations of the subgroups.

The irreducible representations of the SU(3) are labelled by the quantum numbers $(\lambda\mu)$ defined in App. I, and hence so are the HOSM as well as the HO cluster-model states. Since the SO(3) group is a subgroup of SU(3), the states with definite SU(3) quantum numbers can be chosen such that they are also eigenstates of $\boldsymbol{L}^2 = L_x^2 + L_y^2 + L_z^2$ and of L_z, where $\boldsymbol{L}$ is the total orbital angular momentum. Thus the model states bear angular-momentum labels LM and, possibly, further labels as well (see App. I).

The SU(3) labels of the states of a composite system can be put together from those of its subsystems, in close analogy with angular-momentum coupling. The $2N + L = 8$ states of the single relative-motion oscillator span the irreducible representation $(\lambda\mu) = (N0) = (80)$, which couple with representations (00) for the two 'SU(3)-scalar', i.e., isotropic (see App. I) clusters, ^{16}O and α, adding up to $(\lambda\mu) = (80)$. With this, we have determined the subset, of the sd-shell state space of ^{20}Ne, that describes two-cluster configurations. It so happens that the g.s. of ^{20}Ne in the SU(3) SM is just a $|(\lambda\mu)LM\rangle = |(80)00\rangle$ state and there is a single $|(80)00\rangle$ state in the lowest SM subspace. This fact can be interpreted as an indication that the g.s. of ^{20}Ne is a good cluster-model state [315]. The residual interaction removes the degeneracy between the members of the (80) multiplet. In agreement with experiment, they form a rotational-like level sequence, which can be interpreted as a 'cluster band'.

It is instructive to apply alternative HO models for trivial cases. The closed-shell 0p HO model space for ^{16}O, which carries 12 $\hbar\omega$ excitation quanta, is one-dimensional, and forms the SU(3)-scalar representation $(\lambda\mu) = (00)$. The 12 $\hbar\omega$ ^{12}C+α model [287] has $2N + L = 4$ in the relative-motion subspace, so that $(N, L) = (2, 0)$, $(1, 2)$ and $(0, 4)$ are allowed. Let us couple these states to construct ^{16}O states. To get the g.s., which has $(L, M) = (0, 0)$, the $L = 0$, 2 and 4 relative-motion states should be coupled, respectively, with the 0^+, 2^+ or 4^+ spin–parity member of the $(\lambda\mu) = (04)$ g.s. band of ^{12}C. Since there is just one Slater determinant belonging to the closed-shell configuration $(0s)^4(0p)^{12}$ in ^{16}O, the SM g.s. coincides with all three cluster-model states. It is the Pauli projection implicit in these three apparently different states of ^{16}O that makes them coincide,[20] and the Pauli projection happens to imply angular-momentum and SU(3) projections as well.

Another famous example for the coincidence of states in different models is the HO cluster models of ^{6}Li. Both the α+d and the h+t model g.s. carry $2\hbar\omega$, and, since all clusters are SU(3)-scalar, their SU(3) quantum numbers are equal to those of the relative-motion HO, i.e., $(\lambda\mu) = (20)$. In the cluster-model picture the $(\lambda\mu) = (20)$ subspace can be sliced up to $(N, L) = (1, 0)$ and $(0, 2)$ subspaces, in which the states are defined unambiguously by their quantum numbers L, M. Thus the state $|(\lambda\mu)LM\rangle = |(20)LM\rangle$ with clusterization α+d coincides with the state $|(20)LM\rangle$ with clusterization h+t. Similar statements can be said about the α+(2n) and t+t configurations of ^{6}He.

These coincidences are all examples from very simple nuclei, and from the 0 $\hbar\omega$ subspace, where the HO cluster-model space is the same as the ordinary SM space. At higher energies and for larger nuclei the overlap between the two models becomes insignificant and the trivial coincidences have less and less to do with reality. The SM and the cluster model perform very differently also in improving the description of the states that coincide in the simplest approach; for weakly bound states the cluster model is definitely superior. It is also fair to say that the cluster model performs extremely well even for states that are *par excellence* 'SM states', like those in ^{16}O with s.p. excitation [287].

[20]The paradoxes caused by the Pauli principle are discussed in more detail in Ref. [290].

11.6.2 Eigenvalue problem of the norm operator

The HO cluster model has one more important aspect: it gives the clue to the solution of the eigenvalue problem of the norm operator with clusters described by SM's of the same $\hbar\omega$. The eigenfunctions are the HO eigenfuntions $\psi_{nlm}(\boldsymbol{R})$ of Eq. (9.51) belonging to the same $\hbar\omega$. This general statement will be proved for the special case of two SU(3)-scalar clusters. More precisely, it will be shown that *the common complete set of eigenfunctions of $\mathcal{N}$ and the angular momentum $\mathcal{L}^2$, $\mathcal{L}_z$* (cf. p. 297) *are the HO eigenfunctions ψ_{nlm}* belonging to the same $\hbar\omega$. Since $\{\psi_{nlm}\}$ is a complete set of eigenfunctions of $\mathcal{L}^2$ and $\mathcal{L}_z$, what we have to show is that they are eigenfunctions of $\mathcal{N}$ as well.

To prove this, let us consider $\mathcal{N}\psi_{nlm}(\boldsymbol{R})$ for a two-cluster problem with SU(3)-scalar clusters:

$$\begin{aligned}\mathcal{N}\psi_{nlm}(\boldsymbol{R}) &= \langle \mathcal{A}_{12}\{\Phi_1(\xi_1)\Phi_2(\xi_2)\delta(\boldsymbol{R}-\boldsymbol{q})\}|\mathcal{A}_{12}\{\Phi_1(\xi_1)\Phi_2(\xi_2)\psi_{nlm}(\boldsymbol{q})\}\rangle \\ &= \sum_{n'l'm'} \langle \mathcal{A}_{12}\{\Phi_1(\xi_1)\Phi_2(\xi_2)\psi_{n'l'm'}(\boldsymbol{q})\} \\ &\qquad\qquad \times |\mathcal{A}_{12}\{\Phi_1(\xi_1)\Phi_2(\xi_2)\psi_{nlm}(\boldsymbol{q})\}\rangle\psi_{n'l'm'}(\boldsymbol{R}) \\ &= \nu_{nl}\psi_{nlm}(\boldsymbol{R}). \end{aligned} \tag{11.103}$$

In the first step the definitions of the integral operator, Eq. (11.18), and of the kernel, Eq. (11.44), have been used. In the second step the closure relation $\sum_{nlm}\psi^*_{nlm}(\boldsymbol{q})\psi_{nlm}(\boldsymbol{R}) = \delta(\boldsymbol{R}-\boldsymbol{q})$ has been used. The third step holds because $\mathcal{A}_{12}$ acts on the nucleonic labels and does not change the number of HO quanta. The norm square of the HO cluster-model state is denoted by ν_{nl}; it is independent of m because of the Wigner–Eckart theorem. It follows from the equality of the two extreme sides of Eq. (11.103) that ν_{nl} is an eigenvalue of $\mathcal{N}$. With this, it has been shown that any HO function $\psi_{nlm}(\boldsymbol{R})$ is an eigenfunction of $\mathcal{N}$. Since $\{\psi_{nlm}\}$ form a complete set of functions, there are no other common eigenfunctions of $\mathcal{L}^2$, $\mathcal{L}_z$ and $\mathcal{N}$. This completes the proof.

From Eq. (11.103) it follows that the eigenvalue ν_{nl} is

$$\begin{aligned}\nu_{nl} = \binom{A}{a_1}^{1/2} &\langle \Phi_1(\xi_1)\Phi_2(\xi_2)\psi_{nlm}(\boldsymbol{q}) \\ &\times|\mathcal{A}_{12}|\Phi_1(\xi_1)\Phi_2(\xi_2)\psi_{nlm}(\boldsymbol{q})\rangle. \end{aligned} \tag{11.104}$$

The antisymmetrizer $\mathcal{A}_{12}$ is SU(3)-scalar, which means that it does not change the SU(3) transformation properties of the function it acts on. (See App. I.) According to the SU(3) Wigner–Eckart theorem (I.19) [cf. the Wigner–Eckart theorem (9.85) for the SO(3) group], a matrix element of an SU(3)-scalar operator between states belonging to the same irreducible representation $(\lambda\mu)$ depends only on the representation label; it is independent of the subgroup labels. Since now both states are labelled by $(\lambda\mu)=(2n+l,0)$, the eigenvalue ν_{nl} depends only on $Q=2n+l$, and for a fixed value of $2n+l$, it does not depend on l.

With this we have given the solution to the eigenvalue problem for systems of two SU(3)-scalar [i.e., $(\lambda\mu)=(00)$] clusters. If one or both of the cluster states Φ_i are non-scalar [i.e., have $(\lambda_i\mu_i)\neq(00)$] and/or there are more than two clusters, the eigenfunctions are still HO states with definite SU(3) quantum numbers, but some of the degeneracy of the eigenvalues is split [232].

Since the generating function $g(\boldsymbol{S},\boldsymbol{\rho})$ of the HO functions, given in Eq. (9.48), is essentially a shifted Gaussian, i.e., $g(\boldsymbol{S},\boldsymbol{\rho})=\mathrm{e}^{\frac{1}{4}\beta S^2}\varphi_{\boldsymbol{S}}(\boldsymbol{\rho})$, one can generate the matrix elements of HO functions from the matrix elements of $\varphi_{\boldsymbol{S}}(\boldsymbol{\rho})$. For a two-cluster case,

$$\int d\boldsymbol{R}\int d\boldsymbol{R}' g(\boldsymbol{S},\sqrt{\mu}\boldsymbol{R})\mathcal{N}(\boldsymbol{R},\boldsymbol{R}')g(\boldsymbol{S}',\sqrt{\mu}\boldsymbol{R}')=\mathrm{e}^{\frac{1}{4}\beta(S^2+S'^2)}\langle\Psi(\boldsymbol{S})|\Psi(\boldsymbol{S}')\rangle, \tag{11.105}$$

with $\Psi(\boldsymbol{S})$ defined in Eq. (11.73). The overlap $\langle\Psi(\boldsymbol{S})|\Psi(\boldsymbol{S}')\rangle$ can be evaluated as shown in Eqs. (11.80)–(11.83), and $\langle\psi_{nlm}|\mathcal{N}|\psi_{n'l'm'}\rangle$ can be calculated as in the generating relationship (9.51) or with the Bargmann transformation technique mentioned in a footnote on p. 213 [224,225]. The generating-function approach can be used whenever the cluster intrinsic states are described by the HOSM with equal quanta. In more general cases the overlap $\langle\Psi(\boldsymbol{S})|\Psi(\boldsymbol{S}')\rangle$ may still be calculated analytically, but the SU(3) classification does not hold, and the eigenvalue problem has to be solved by diagonalization on a wider basis.

Equation (11.104) shows again that $\nu_{nl}=0$ if and only if the antisymmetrized state $\mathcal{A}_{12}\{\Phi_1\Phi_2\psi_{nlm}\}$ vanishes. In the dynamical equation (11.60) the Pauli-forbidden components get discarded, but that is not the only effect of $\mathcal{N}$. It rescales the relative-motion components according to the corresponding eigenvalues. (Since some ν_{nl} may be larger than 1, some components may be amplified by the Pauli effects.) When the HO quanta of the clusters are not the same or Φ_i are not HO states, there are no strictly vanishing eigenvalues, but the scaling may still strongly suppress some 'almost Pauli-forbidden' states. The Pauli effects on the relative motion will be further discussed in Sect. 11.8.

11.7 The generator-coordinate method and the two-centre shell model

11.7.1 Generator-coordinate method

In Sect. 11.5.2 the RGM kernels are expressed in terms of shifted-Gaussian relative-motion states. Let us now examine the same expressions from the point of view of the wave function itself. With Eq. (11.74) inserted in Eq. (11.41) and the integral representation (11.34) used for $\delta(\underline{\boldsymbol{R}} - \boldsymbol{\rho})$, Eq. (11.40) takes the form

$$\Psi = \int d\boldsymbol{R}\, \phi(\boldsymbol{R}) \mathcal{A}_{12}\{\Phi_1\Phi_2\delta(\boldsymbol{R}-\boldsymbol{q})\} = \mu^{3/2}\int d\boldsymbol{R}\, \phi(\boldsymbol{R}) \mathcal{A}_{12}\{\Phi_1\Phi_2\delta(\underline{\boldsymbol{R}}-\boldsymbol{\rho})\}$$

$$= \int d\underline{\boldsymbol{R}}\, \phi(\mu^{-1/2}\underline{\boldsymbol{R}}) \mathcal{A}_{12}\{\Phi_1\Phi_2\delta(\underline{\boldsymbol{R}} - \boldsymbol{\rho})\} = \int d\boldsymbol{S}\, f(\boldsymbol{S})\Psi(\boldsymbol{S}), \qquad (11.106)$$

where

$$\Psi(\boldsymbol{S}) = \mathcal{A}_{12}\{\Phi_1(\xi_1)\Phi_2(\xi_2)\varphi_{\boldsymbol{S}}(\boldsymbol{\rho})\}, \qquad (11.107)$$

$$f(\boldsymbol{S}) = (2\pi)^{-3}\left(\frac{\beta}{4\pi}\right)^{3/4} \int d\boldsymbol{k} \int d\boldsymbol{R}\, \phi(\mu^{-1/2}\boldsymbol{R}) \exp\left[i\boldsymbol{k}\cdot(\boldsymbol{S}-\boldsymbol{R}) + \frac{\boldsymbol{k}^2}{2\beta}\right]. \qquad (11.108)$$

The function $\Psi(\boldsymbol{S})$ has the form of Eq. (11.73), but now the two cluster oscillator constants are not necessarily the same. Note that the integration over $\boldsymbol{S}$ has been put first, and such a reshuffling of the integrations may not be legitimate mathematically; no wonder that the expansion in Eq. (11.106) has pathological mathematical properties. This may be conjectured from the fact that the $\boldsymbol{k}$-integration diverges apparently. Yet Eq. (11.106) is conventionally interpreted as an alternative ansatz for the wave function. By substituting the ansatz Eq. (11.106) into the Schrödinger equation and projecting the Schrödinger equation onto $\langle\Psi(\boldsymbol{S})|$, we have

$$\int d\boldsymbol{S}'\, \langle\Psi(\boldsymbol{S})|H|\Psi(\boldsymbol{S}')\rangle f(\boldsymbol{S}') = E\int d\boldsymbol{S}'\, \langle\Psi(\boldsymbol{S})|\Psi(\boldsymbol{S}')\rangle f(\boldsymbol{S}'). \qquad (11.109)$$

This equation can be derived from the variational principle, Theorem 1 on p. 195, in complete analogy with the derivation given in Eqs. (11.47)–(11.48).

Any continuous expansion parameter of a wave function, like $\boldsymbol{S}$, is called a generator coordinate. The dynamical equation, of the type of Eq. (11.109), that is implied by a wave-function ansatz containing an expansion over generator coordinates is called a Hill–Wheeler equation, and its use in the solution of the Schrödinger equation is called a GCM in a general sense. Since the kinetic energy does not act on the generator-coordinate $\boldsymbol{S}$, the Hill–Wheeler equation contains no differential operator; it is an integral equation. In the cluster model the term GCM has a more specific meaning: it means a generator-coordinate scheme in which the generator coordinates are the relative distance vectors between the HO centres of the clusters, like in Eq. (11.106).[21] Since there are GCM 'basis' states that almost completely overlap with each other, i.e., $\lim_{S' \to S} \Psi(S') = \Psi(S)$, the expression (11.106) is highly pathological,[22] and the Hill–Wheeler amplitude $f(S)$ is not a function; it is a rather ill-behaved distribution in general.

Yet Eq. (11.107) has a very plausible meaning: it is a two-cluster state whose relative motion is a Gaussian peaking at a relative displacement S. Even a single state of this form is a good starting point for describing a two-cluster system, and it is conceivable that just a few functions of this kind, belonging to different vectors S, could combine into a realistic wave function. The trouble originates, of course, from the continuous set. All mathematical inconveniences disappear as soon as Eq. (11.106) is discretized:

$$\Psi = \sum_k f(S_k)\Psi(S_k), \tag{11.110}$$

$$\sum_{k'} \langle \Psi(S_k)|H|\Psi(S_{k'})\rangle f(S_{k'}) = E \sum_{k'} \langle \Psi(S_k)|\Psi(S_{k'})\rangle f(S_{k'}) \quad \text{for all } k. \tag{11.111}$$

This is a set of homogeneous linear equations: a generalized matrix eigenvalue problem. A discrete expansion (11.110) is well-behaved provided $\det\left(\langle \Psi(S_k)|\Psi(S_{k'})\rangle\right) \neq 0$. Owing to the favourable properties of the shifted Gaussians, it is easy to choose an appropriate shifted-Gaussian basis for any bound-state two-cluster problem. Since the basis states in Eq. (11.110) can be chosen to fit the problem very well, Eq. (11.111) provides remarkably accurate solutions to the Schrödinger equation of the relative motion with very few terms. This may also be understood by noting that while a single RGM basis function Ψ_R describes an unrealistic situation of two clusters pinned down sharply at point R, in the generator-coordinate basis function $\Psi(S)$ the two clusters perform a relative motion of reasonable span. Moreover, it is easy to make a realistic guess for the GCM basis: for any partial wave the interval that S should span is given by the expected size of the system, while the density of the S points should allow a good overlap between neighbouring Gaussians.

[21]Since the width parameter of the Gaussians representing the relative motion in the correlated Gaussian method is also a continuous parameter, the correlated Gaussian method is also a GCM in a general sense.

[22]One has the feeling that the closer S' is to S, the more $f(S')\Psi(S')$ should cancel $f(S)\Psi(S)$ or, rather, that all $f(S)$ should vanish, while $\int dS\, f(S)\Psi(S) \neq 0$. This property is reminiscent of the Dirac delta. Indeed, when Eq. (11.106) is discretized, the finer the mesh the larger the staggering is in $f(S)$, so that $f(S)$ does not converge with the mesh made finer.

It is often said that *the RGM and the GCM are equivalent.* The equivalence is in fact twofold.

The dynamical aspect of the equivalence can be shown by deriving, from Eq. (11.42), the Hill–Wheeler equations (11.109) and (11.111), respectively, by the ansätze

$$\phi = \int d\boldsymbol{S}\, f(\boldsymbol{S})\varphi_{\boldsymbol{S}} \quad \text{and} \quad \phi = \sum_k f(\boldsymbol{S}_k)\varphi_{\boldsymbol{S}_k}. \tag{11.112}$$

The variational principle ensures that $f(\boldsymbol{S})$, the solution of the Hill-Wheeler equation, provides, through Eq. (11.112), the solution ϕ of the RGM equation (11.42) because the two bases, $\{\Psi_{\boldsymbol{R}}\}$ and $\{\Psi(\boldsymbol{S})\}$, span the same function space. For a discretized Eq. (11.112), the converse is also true: Equation (11.112) can be rewritten into the set of linear equations

$$\langle\!\langle \varphi_{\boldsymbol{S}_j}|\phi\rangle\!\rangle = \sum_k f(\boldsymbol{S}_k)\langle\!\langle \varphi_{\boldsymbol{S}_j}|\varphi_{\boldsymbol{S}_k}\rangle\!\rangle, \tag{11.113}$$

which can be solved provided $\det(\langle\!\langle \varphi_{\boldsymbol{S}_j}|\varphi_{\boldsymbol{S}_k}\rangle\!\rangle) \neq 0$, so ϕ also implies $\{f(\boldsymbol{S}_k)\}$ unambiguously in general. This aspect of the equivalence is made complete by the fact that the scattering boundary conditions can also be formulated in the GCM [301, 316, 317].

The technical aspect of the equivalence is expressed by the relationship

$$\langle \Psi(\boldsymbol{S})|O|\Psi(\boldsymbol{S}')\rangle = \int d\boldsymbol{R}\int d\boldsymbol{R}'\, \varphi_{\boldsymbol{S}}(\sqrt{\mu}\boldsymbol{R})\langle \Psi_{\boldsymbol{R}}|O|\Psi_{\boldsymbol{R}'}\rangle \varphi_{\boldsymbol{S}'}(\sqrt{\mu}\boldsymbol{R}'). \tag{11.114}$$

The integral in terms of $\boldsymbol{R}$ can be interpreted as a folding of the RGM kernel as a function of $\boldsymbol{R}$ with a Gaussian function of $\boldsymbol{R}$, and a similar statement holds for $\boldsymbol{R}'$.

The RGM kernel can be unfolded from the Gaussians just like the Dirac-δ itself [cf. Eq. (11.34) and the attached footnote] through a pair of Fourier transformations and inverse Fourier transformations. The result has already been given in Eq. (11.75) [note that, according to Eq. (11.107), the matrix element on the right-hand side of Eq. (11.75) is nothing but $\langle \Psi(\boldsymbol{S})|O|\Psi(\boldsymbol{S}')\rangle$]. The equivalence of the RGM and GCM formulated in Eqs. (11.114) and (11.75) means that their kernels are mutually expressible in terms of each other.

Note that the relationship between the GCM and the RGM for more than two clusters depends on the treatment of the angular momentum. To elucidate this point, let us briefly summarize the standard treatment of the angular momentum in the GCM for a two-cluster case. Still disregarding the cluster spins, we can write [cf. Eq. (11.97)]

$$\begin{aligned}\Psi &= \int d\boldsymbol{S}\, f(\boldsymbol{S})\Psi(\boldsymbol{S}) \\ &= \sum_{LM}\sum_{L'M'}\int d\boldsymbol{S}\, \frac{f_{LM}(S)}{S} Y_{LM}(\widehat{\boldsymbol{S}}) \frac{\Psi_{L'M'}(S)}{S} Y^*_{L'M'}(\widehat{\boldsymbol{S}}) \\ &= \sum_{LM}\int_0^\infty dS\, f_{LM}(S)\Psi_{LM}(S),\end{aligned} \tag{11.115}$$

with

$$\begin{aligned}\Psi_{LM}(S) &= S\int d\widehat{\boldsymbol{S}}\, Y_{LM}(\widehat{\boldsymbol{S}})\Psi(\boldsymbol{S}) \\ &= S\int d\widehat{\boldsymbol{S}}\, Y_{LM}(\widehat{\boldsymbol{S}})\mathcal{A}_{12}\left\{\Phi_1\Phi_2\left(\frac{\beta}{\pi}\right)^{3/4} e^{-\frac{1}{2}\beta(\boldsymbol{\rho}-\boldsymbol{S})^2}\right\} \\ &= 4\pi S\mathcal{A}_{12}\left\{\Phi_1\Phi_2\left(\frac{\beta}{\pi}\right)^{3/4} e^{-\frac{1}{2}\beta(\rho^2+S^2)} i_L(\beta\rho S) Y_{LM}(\hat{\rho})\right\},\end{aligned} \tag{11.116}$$

where the multipole expansion (9.38) of the exponential function has been used. With Eq. (11.115), the Hill–Wheeler equation (11.111) will decouple into partial-wave equations as is allowed by the dynamics implied by the model assumptions.

This treatment of the angular momentum can be generalized to more than two clusters. To illustrate this case, let us write down formulae for a three-cluster model in which all clusters are characterized by the same β and the total orbital angular momentum quantum numbers LM are good quantum numbers. The ansatz takes the form [cf. Eq. (11.115)]

$$\begin{aligned}\Psi &= \int d\boldsymbol{S}_1 \int d\boldsymbol{S}_2\, f(\boldsymbol{S}_1,\boldsymbol{S}_2)\Psi(\boldsymbol{S}_1,\boldsymbol{S}_2) \\ &= \sum_{l_1 l_2 LM}\sum_{l_1' l_2' L'M'} \int d\boldsymbol{S}_1 \int d\boldsymbol{S}_2\, \frac{f_{[l_1 l_2]LM}(S_1,S_2)}{S_1 S_2} [Y_{l_1}(\widehat{\boldsymbol{S}}_1) Y_{l_2}(\widehat{\boldsymbol{S}}_2)]_{LM} \\ &\quad\times \frac{\Psi_{[l_1' l_2']L'M'}(S_1,S_2)}{S_1 S_2}[Y^*_{l_1'}(\widehat{\boldsymbol{S}}_1) Y^*_{l_2'}(\widehat{\boldsymbol{S}}_2)]_{L'M'} \\ &= \sum_{l_1 l_2 LM}\int_0^\infty dS_1 \int_0^\infty dS_2\, f_{[l_1 l_2]LM}(S_1,S_2)\Psi_{[l_1 l_2]LM}(S_1,S_2),\end{aligned} \tag{11.117}$$

where

$$\begin{aligned}\Psi_{(l_1 l_2)LM}(S_1,S_2) &= S_1 S_2 \int d\widehat{\boldsymbol{S}}_1 \int d\widehat{\boldsymbol{S}}_2\, [Y_{l_1}(\widehat{\boldsymbol{S}}_1) Y_{l_2}(\widehat{\boldsymbol{S}}_2)]_{LM}\Psi(\boldsymbol{S}_1,\boldsymbol{S}_2), \\ \Psi(\boldsymbol{S}_1,\boldsymbol{S}_2) &= \mathcal{A}_{123}\,\{\Phi_1\Phi_2\Phi_3\varphi_{\boldsymbol{S}_1}(\boldsymbol{\rho}_1)\varphi_{\boldsymbol{S}_2}(\boldsymbol{\rho}_2)\}.\end{aligned} \tag{11.118}$$

This framework does, however, have a serious drawback: the summations over l_1, l_2 are infinite,[23] and, in practice, for large values of S_1, S_2, one has to include a great number of terms. It is therefore preferable to use angular momentum projection rather than coupling:

$$\Psi_{LM} = \mathcal{P}_{LM} \int dS_1 \int dS_2 \, f(S_1, S_2) \Psi(S_1, S_2). \tag{11.119}$$

With this, the wave function will no longer involve l_1 and l_2, and the equivalence with the RGM becomes less complete. Equation (11.119) produces pure L, which can then be mixed if required by the interaction. The form (11.119) can be made even more convenient for models that are constructed so as to have a symmetry axis in an intrinsic frame. Such a frame can be spanned by appropriately constraining S_1 and S_2. In such a frame the projection K of the angular momentum onto the symmetry axis is a good quantum number, and the correct angular momentum in the laboratory frame can be obtained [318], as in the collective model [141], by applying the operator

$$\mathcal{P}^L_{MK} = \frac{2L+1}{8\pi^2} \int d\widehat{S} \, D^{L^*}_{MK}(\widehat{S}) \mathcal{R}(\widehat{S}), \tag{11.120}$$

where[24] D^L_{MK} is a rotation matrix [54], $\mathcal{R}$ is the rotation operator acting on the spatial variables and $\widehat{S}$ is a solid angle defined usually by a triplet of Euler angles. This solid angle gives an alignment of the frame S_1, S_2 in space.

The GCM basis function $\Psi(S_1, S_2, \ldots)$ boils down to Eq. (9.72) of the Gaussian wave packet formalism when $n = A$, i.e., each cluster contains just one nucleon and the cluster relative coordinates are the Jacobi set. As will be seen later, this correspondence will help to generalize the correlated Gaussian model to composite clusters.

The physics involved in the GCM will be discussed further in Sect. 11.7.3. Before coming to that point, we shall present a reformulation of the GCM problem, which is somewhat more involved, but is more satisfactory both as a representation of the RGM and as a technique to calculate RGM matrix elements.

---

[23]To the description of the scattering of $(a_1a_2)+a_3$ this does not apply since in that case l_1 belongs to the nucleus (a_1a_2), so that it is naturally restricted.

[24]This operator is not a genuine projector. It projects onto total orbital angular momentum LK, but then rotates the state into an L^2, L_z eigenstate of quantum numbers LM, while leaving all other quantum numbers intact.

11.7.2 The method of complex generator coordinates

In the RGM kernels (11.43) and (11.44) the intercluster relative variable is involved in Dirac-δ's, and the key idea in the GCM representation is the recognition that the δ-function can be expressed in terms of Gaussian wave packets [cf. Eq. (11.34)]. The δ-function appears in a similar position also in the quantities related to clustering: in the spectroscopic amplitude (11.11) and in the amount of clustering (11.24) and (11.32). In Eq. (11.75) the RGM kernel is expressed as an integral transform of a GCM kernel, which is a matrix element between two GCM basis states (11.107). The function $\phi(\boldsymbol{R})$ can be expanded in terms of HO eigenfunctions as well, and then the RGM is represented by HO cluster model states (11.100). Each version of the RGM requires specific matrix elements.

The aim of this subsection is to show how to treat these matrix elements in a transparent and unified way via the Bargmann transformation [223] (cf. footnote on p. 212). It will be seen that this formalism can be classified as using another formula [Eq. (11.126)] for the Dirac-δ. At the same time, we obtain an alternative formulation of the cluster model.

The Bargmann transform of the RGM kernel for an operator O is defined by

$$\mathfrak{O}(\boldsymbol{S}, \boldsymbol{S}'^*) = \langle \Xi(\boldsymbol{S}) | O | \Xi(\boldsymbol{S}') \rangle, \tag{11.121}$$

where in the state

$$\Xi(\boldsymbol{S}) = \mathcal{A}_{12} \left\{ \Phi_1(\xi_1)\Phi_2(\xi_2) g^*(\boldsymbol{S}, \boldsymbol{\rho}) \right\} \tag{11.122}$$

the relative motion is represented by the generating function (9.48) but $\boldsymbol{S}$ is a complex vector now. For simplicity, a common β value is used in the bra and in the ket although a generalization is straightforward [225]. Equation (9.48) shows that the generating function is related to the shifted Gaussian as

$$g(\boldsymbol{S}, \boldsymbol{\rho}) = \left(\frac{\beta}{\pi}\right)^{3/4} \mathrm{e}^{-\frac{1}{2}\beta\rho^2 + \beta\boldsymbol{\rho}\cdot\boldsymbol{S} - \frac{1}{4}\beta\boldsymbol{S}^2} = \mathrm{e}^{\frac{1}{4}\beta\boldsymbol{S}^2} \varphi_{\boldsymbol{S}}(\boldsymbol{\rho}), \tag{11.123}$$

so the Bargmann transform can be expressed in terms of the GCM kernel $\langle \Psi(\boldsymbol{S}) | O | \Psi(\boldsymbol{S}') \rangle$ as [cf. Eq. (11.105)]

$$\mathfrak{O}(\boldsymbol{S}, \boldsymbol{S}'^*) = \mathrm{e}^{\frac{1}{4}\beta(\boldsymbol{S}^2 + \boldsymbol{S}'^{*2})} \langle \Psi(\boldsymbol{S}) | O | \Psi(\boldsymbol{S}') \rangle, \tag{11.124}$$

where

$$\Psi(\boldsymbol{S}) = \mathcal{A}_{12} \left\{ \Phi_1(\xi_1)\Phi_2(\xi_2) \varphi^*_{\boldsymbol{S}}(\boldsymbol{\rho}) \right\}. \tag{11.125}$$

The three-dimensional complex vectors $\boldsymbol{S}$ and $\boldsymbol{S}'$ may be called the complex generator coordinates. A representation of the wave function by functions $\Psi(\boldsymbol{S})$ of Eq. (11.125) or their use as a means to calculate the RGM kernels is called the complex generator-coordinate method (CGCM). The utility of this method has been recognized by several authors [319–322]. More on the CGCM can be found in Refs. [141,232, 323].

One of the most important properties of $g(\boldsymbol{S},\boldsymbol{\rho})$ is that it satisfies the relation (see footnote on p. 212)

$$\int d\mu(\boldsymbol{S})\, g(\boldsymbol{S},\boldsymbol{\rho})g^*(\boldsymbol{S},\boldsymbol{R}) = \delta(\boldsymbol{R}-\boldsymbol{\rho}), \tag{11.126}$$

$$\int d\boldsymbol{R}\, g(\boldsymbol{S},\boldsymbol{R})g^*(\boldsymbol{S}',\boldsymbol{R}) = \mathrm{e}^{\frac{\beta}{2}\boldsymbol{S}\cdot\boldsymbol{S}'^*}, \tag{11.127}$$

where the Bargmann measure is defined by

$$d\mu(\boldsymbol{S}) = \left(\frac{\beta}{2\pi}\right)^3 \prod_{i=x,y,z} \mathrm{e}^{-\frac{\beta}{2}S_iS_i^*}\, d\,\mathrm{Re}\, S_i\; d\,\mathrm{Im}\, S_i. \tag{11.128}$$

The function space, spanned by functions of $\boldsymbol{S}$, in which the scalar product is defined with respect to the Bargmann measure (11.128), is called the Bargmann space. It is useful to denote the real and imaginary parts of $\boldsymbol{S}$ such that $\boldsymbol{S} \equiv \boldsymbol{X} + \frac{i}{\hbar\beta}\boldsymbol{P}$, and with that

$$d\mu(\boldsymbol{S}) = \frac{1}{(2\pi\hbar)^3}\mathrm{e}^{-\frac{\beta}{2}\boldsymbol{X}^2 - \frac{1}{2\hbar^2\beta}\boldsymbol{P}^2}\, d\boldsymbol{X}d\boldsymbol{P}. \tag{11.129}$$

Equation (11.126) emphasizes that it is an alternative of Eq. (11.34). It follows from this property that the RGM kernel is related to the Bargmann transform as

$$\langle\Psi_{\boldsymbol{R}}|O|\Psi_{\boldsymbol{R}'}\rangle = \int d\mu(\boldsymbol{S}) \int d\mu(\boldsymbol{S}')\, g^*(\boldsymbol{S},\boldsymbol{R})\mathfrak{O}(\boldsymbol{S},\boldsymbol{S}'^*)g(\boldsymbol{S}',\boldsymbol{R}'), \tag{11.130}$$

which is an analogue of Eq. (11.75). The matrix element $\mathfrak{O}(\boldsymbol{S},\boldsymbol{S}'^*)$ is calculated in the standard way, and the integrations implied by the Bargmann measure are also fairly standard. While in Eq. (11.75) the order of integrations must be kept as prescribed lest divergence may occur, there is no need for such stipulations here.

The complex analogue of the GCM expansion (11.106) and (11.108) is formulated as

$$\Psi = \int d\mu(\boldsymbol{S})\, f(\boldsymbol{S})\Xi(\boldsymbol{S}), \tag{11.131}$$

$$f(\boldsymbol{S}) = \int d\boldsymbol{R}\, \phi(\boldsymbol{R})g(\boldsymbol{S},\boldsymbol{R}). \tag{11.132}$$

Equation (11.131) is easily proved by substituting Eq. (11.132) into Eq. (11.131) and using Eq. (11.126). The GCM weight function $f(\boldsymbol{S})$ is nothing but the Bargmann transform of $\phi(\boldsymbol{R})$. Conversely, $\phi(\boldsymbol{R})$ is obtained from $f(\boldsymbol{S})$ by

$$\phi(\boldsymbol{R}) = \int d\mu(\boldsymbol{S})\, f(\boldsymbol{S}) g^*(\boldsymbol{S}, \boldsymbol{R}). \tag{11.133}$$

With the Bargmann measure, the expansion has no pathological mathematical behaviour, and the equivalence between the RGM and the CGCM is clearly established.

Equation (11.130), which expresses the relationship between the RGM and CGCM kernels, plays a central role in the application of the Bargmann transform technique to the cluster model. Its direct applications may be somewhat tedious, but there are some more convenient indirect methods to calculate the RGM kernels from the CGCM kernels. Some useful formulae of this type are tabulated in Ref. [225].

The extending of $g(\boldsymbol{S}, \boldsymbol{\rho})$ to complex variables $\boldsymbol{S}$ is also useful in generating HO functions. Equation (9.49) can be generalized as

$$g(\boldsymbol{S}, \boldsymbol{R}) = \sum_{\alpha} \psi_{\alpha}^*(\boldsymbol{R}) P_{\alpha}(\boldsymbol{S}), \tag{11.134}$$

where α is any convenient label for the three-dimensional HO functions. The function $P_{\alpha}(\boldsymbol{S})$ is the Bargmann transform of the HO function $\psi_{\alpha}(\boldsymbol{R})$, i.e., $P_{\alpha}(\boldsymbol{S}) = \int d\boldsymbol{R}\, \psi_{\alpha}(\boldsymbol{R}) g(\boldsymbol{S}, \boldsymbol{R})$. In the Cartesian coordinate representation α means $\{n_x n_y n_z\}$ and $P_{\alpha}(\boldsymbol{S})$ is given by

$$P_{n_x n_y n_z}(\boldsymbol{S}) = u_{n_x}(S_x) u_{n_y}(S_y) u_{n_z}(S_z), \tag{11.135}$$

with

$$u_m(S_i) = \sqrt{\frac{1}{m!}\left(\frac{\beta}{2}\right)^m} S_i^m. \tag{11.136}$$

In the polar coordinate representation α stands for $\{nlm\}$ and $P_{\alpha}(\boldsymbol{S})$ reads

$$P_{nlm}(\boldsymbol{S}) = \sqrt{\frac{4\pi}{2^n n!(2n+2l+1)!!}\left(\frac{\beta}{2}\right)^{2n+l}} \boldsymbol{S}^{2n} \mathcal{Y}_{lm}(\boldsymbol{S}), \tag{11.137}$$

where $\mathcal{Y}_{lm}$ is a solid spherical harmonic [see Eq. (9.2)]. Substituting Eq. (11.134) into Eqs. (11.126) and (11.127) leads to orthonormality and closure relations for $P_{\alpha}(\boldsymbol{S})$ with respect to the Bargmann measure:

$$\int d\mu(\boldsymbol{S})\, P_{\alpha}(\boldsymbol{S}) P_{\alpha'}^*(\boldsymbol{S}) = \delta_{\alpha\alpha'}, \tag{11.138}$$

$$\sum_{\alpha} P_{\alpha}(\boldsymbol{S}) P_{\alpha}^*(\boldsymbol{S}') = e^{\frac{\beta}{2}\boldsymbol{S}\cdot\boldsymbol{S}'^*}. \tag{11.139}$$

It follows from Eqs. (11.138) and (11.139) that for any analytic function, $\psi(\boldsymbol{S})$,

$$\int d\mu(\boldsymbol{S}')\, e^{\frac{\beta}{2}\boldsymbol{S}\cdot\boldsymbol{S}'^{*}}\, \psi(\boldsymbol{S}') = \psi(\boldsymbol{S}), \tag{11.140}$$

which is easily proved by expanding $\psi(\boldsymbol{S})$ in terms of $P_\alpha(\boldsymbol{S})$ and using the orthonormality relations. Thus the function $e^{\frac{\beta}{2}\boldsymbol{S}\cdot\boldsymbol{S}'^{*}}$ behaves in the Bargmann space as a Dirac-δ.

The matrix element of the operator O between basis states (11.100) is related to the Bargmann transform (11.121) as

$$\begin{aligned}\mathfrak{O}(\boldsymbol{S}, \boldsymbol{S}'^{*}) = \sum_{\alpha\alpha'} & P_\alpha(\boldsymbol{S}) P^{*}_{\alpha'}(\boldsymbol{S}') \langle \mathcal{A}_{12}\{\Phi_1(\xi_1)\Phi_2(\xi_2)\psi_\alpha(\boldsymbol{\rho})\} \\ & \times | O | \mathcal{A}_{12}\{\Phi_1(\xi_1)\Phi_2(\xi_2)\psi_{\alpha'}(\boldsymbol{\rho})\}\rangle, \end{aligned} \tag{11.141}$$

which can be proven with the use of Eq. (11.134). The matrix elements between HO basis states can be easily obtained by expanding the Bargmann transform in terms of the polynomials $P_\alpha(\boldsymbol{S})$ and $P^{*}_{\alpha'}(\boldsymbol{S}')$. This technique has been widely used for the evaluation of the RGM kernels in the HOSM basis [224, 286, 324, 325].

To sum up the results of this subsection, we can say that the matrix elements between shifted Gaussian or HO basis states can be formulated in a unified manner in terms of the Bargmann transformation. All these matrix elements can be derived from the CGCM kernel (11.124), which can be calculated from the same procedure as that presented in Sect. 11.5.2. The CGCM does not suffer from the pathological mathematical behaviour inherent in the GCM with real variables.

11.7.3 Two-centre shell model

The GCM may be viewed as a shifted-Gaussian representation of the RGM. But it can be assigned a richer physical content by considering its basis function as a two-centre SM wave function. To this end, the cluster intrinsic wave functions should be complemented with cluster c.m. wave functions, which can be brought in through Eq. (11.76) for the case of identical oscillators and through Eq. (11.85) for any case:

$$\begin{aligned}\Psi_{\boldsymbol{S}}\varphi_{\boldsymbol{S}_A}(\boldsymbol{x}_A) &= \mathcal{A}_{12}\{\Phi_1(\xi_1)\Phi_2(\xi_2)\varphi_{\boldsymbol{S}}(\boldsymbol{\rho})\varphi_{\boldsymbol{S}_A}(\boldsymbol{x}_A)\} \\ &= \mathcal{A}_{12}\{\Phi_1(\xi_1)\Phi_2(\xi_2)\varphi_{\boldsymbol{s}_1}(\bar{\boldsymbol{\xi}}_1)\varphi_{\boldsymbol{s}_2}(\bar{\boldsymbol{\xi}}_2)\} \\ &= \mathcal{A}_{12}\{\Phi^{\mathrm{SM}(0)}_{1\boldsymbol{s}_1}(\boldsymbol{r}_1,\ldots,\boldsymbol{r}_{a_1})\Phi^{\mathrm{SM}(0)}_{2\boldsymbol{s}_2}(\boldsymbol{r}_{a_1+1},\ldots,\boldsymbol{r}_A)\} \\ &\qquad \text{for identical HO,}\end{aligned} \tag{11.142}$$

$$\begin{aligned}\Psi_{\boldsymbol{S}} &= \mathcal{A}_{12}\{\Phi_1(\xi_1)\Phi_2(\xi_2)\varphi^{(b)}_{\boldsymbol{S}}(\boldsymbol{\rho})\} \qquad (11.143)\\ &= \left(\frac{\beta_1\beta_2}{4\pi b}\right)^{3/4}\int d\boldsymbol{S}_A\, \mathcal{A}_{12}\{\Phi_1(\xi_1)\Phi_2(\xi_2)\varphi_{\boldsymbol{s}_1}(\bar{\boldsymbol{\xi}}_1)\varphi_{\boldsymbol{s}_2}(\bar{\boldsymbol{\xi}}_2)\}\end{aligned}$$

$$= \left(\frac{\beta_1\beta_2}{4\pi b}\right)^{3/4}\int d\boldsymbol{S}_A\,\mathcal{A}_{12}\{\Phi^{\mathrm{SM}(0)}_{1s_1}(\boldsymbol{r}_1,\ldots,\boldsymbol{r}_{a_1})$$
$$\times\Phi^{\mathrm{SM}(0)}_{2s_2}(\boldsymbol{r}_{a_1+1},\ldots,\boldsymbol{r}_A)\}\ \text{for general HO.} \tag{11.144}$$

A generator-coordinate basis state is thus identical to a two-centre SM state. This is a convenient property, which assigns the GCM meaning even when the GCM basis contains just one element. This is a viable approximation especially for a multicentre case,

$$\Phi_{s_1,\ldots,s_n}(\boldsymbol{r}_1,\ldots,\boldsymbol{r}_A)$$
$$= \mathcal{A}_{1\cdots n}\{\Phi^{\mathrm{SM}(0)}_{1s_1}(\boldsymbol{r}_1,\ldots,\boldsymbol{r}_{a_1})\ldots\Phi^{\mathrm{SM}(0)}_{ns_n}(\boldsymbol{r}_{a_1+\cdots+a_{n-1}+1},\ldots,\boldsymbol{r}_A)\}, \tag{11.145}$$

where the relative position of the centres provides ample freedom for a reasonable description of the nucleus. Note that the product of the cluster c.m. wave functions implicit in Eq. (11.145) can again be rewritten as a product of relative Gaussians times a Gaussian of the total c.m. [cf. Eq. (9.71)]: $\prod_{i=1}^n \varphi_{s_i}(\bar{\boldsymbol{\xi}}_i)=\prod_{i=1}^n \varphi_{\boldsymbol{S}_i}(\boldsymbol{\rho}_i)$, where the nth vectors belong to the motion of the total c.m.: $\boldsymbol{S}_n\equiv\boldsymbol{S}_A$, $\boldsymbol{\rho}_n\equiv\boldsymbol{x}_A$. With this, one can, again, factor out the c.m. function: $\Phi_{s_1,\ldots,s_n} = \varphi_{\boldsymbol{S}_A}(\boldsymbol{x}_A)\Psi_{\boldsymbol{S}_1,\ldots,\boldsymbol{S}_{n-1}}$, where

$$\Psi_{\boldsymbol{S}_1,\ldots,\boldsymbol{S}_{n-1}}(\xi_1,\ldots,\xi_n,\boldsymbol{\rho}_1,\ldots,\boldsymbol{\rho}_{n-1})$$
$$= \mathcal{A}_{1\cdots n}\{\Phi_1(\xi_1)\cdots\Phi_n(\xi_n)\varphi_{\boldsymbol{S}_1}(\boldsymbol{\rho}_1)\cdots\varphi_{\boldsymbol{S}_{n-1}}(\boldsymbol{\rho}_{n-1})\}. \tag{11.146}$$

A multicentre SM state is to be optimized with respect to the relative position of the centres. Such a multicentre SM is one of the original formulations of the cluster model, called Brink's cluster model [231]. This model is primarily used as a means to build up light self-conjugate ($N = Z$) nuclei with even N and even Z from α-clusters. The refinements of this model, which use just a few GCM terms, are useful to display a molecular-like picture for light nuclei [198].

The two-centre SM looks like a model for a molecule. Nevertheless, there are more differences than analogies, and it is illuminating to discuss these differences. For a molecule, $\boldsymbol{S}$ is the displacement between the two nuclei and, as such, it is a dynamical variable, in addition to $\boldsymbol{\rho}$, which is the displacement between the c.m. of the two electron clouds.

The two-centre function $\Psi_{\boldsymbol{S}}(\xi_1,\xi_2,\boldsymbol{\rho})$ defined in Eq. (11.143) is reminiscent of a model wave function of the electron cloud of a diatomic molecule, with the nuclei taken at rest at relative displacement $\boldsymbol{S}$. In this model the full wave function of the molecule, including the wave function of the nucleus as well, is

$$\Psi_{\boldsymbol{S}'}(\boldsymbol{S},\xi_1,\xi_2,\boldsymbol{\rho}) = \delta(\boldsymbol{S}-\boldsymbol{S}')\Psi_{\boldsymbol{S}'}(\xi_1,\xi_2,\boldsymbol{\rho}). \tag{11.147}$$

The overlap between two model wave functions (11.147) differing in the relative displacement of the nuclei, is, of course,

$$\int dS \int d\xi_1 \int d\xi_2 \int d\rho\, \Psi^*_{S'}(S,\xi_1,\xi_2,\rho)\Psi_{S''}(S,\xi_1,\xi_2,\rho) = \delta(S'-S''). \tag{11.148}$$

In an adiabatic approximation the multielectron state $\Psi_{S'}(\xi_1,\xi_2,\rho)$, in which S' is a parameter, can be determined by solving the electronic problem in the 'external' field of the two nuclei. (The nuclei are taken non-identical here for simplicity.) In this approximation the full wave function can be written as a linear combination of functions $\Psi_{S'}(S,\xi_1,\xi_2,\rho)$, which looks like the GCM ansatz (11.106), but, because of the orthogonality (11.148), reduces to a simple product:

$$\Psi = \int dS'\, f(S')\Psi_{S'}(S,\xi_1,\xi_2,\rho) = f(S)\Psi_S(\xi_1,\xi_2,\rho). \tag{11.149}$$

As a consequence, $f(S)$ can play the role of the internuclear wave function.

The fact that the electronic wave function $\Psi_S(\xi_1,\xi_2,\rho)$ still contains S shows that the separation of the nuclear and electronic motions is not complete. The Hamiltonian of the full problem is

$$H_{\text{mol.}} = T_S + H_{\text{el.}}, \tag{11.150}$$

where T_S is the kinetic energy of the internuclear motion and $H_{\text{el.}}$ depends on the electronic degrees of freedom as well. By substitution of Eq. (11.149) into the Schrödinger equation corresponding to Eq. (11.150), a Schrödinger equation can be derived for the nuclear relative-motion wave function $f(S)$:

$$(T_S + V_{\text{intermol.}})f(S) = Ef(S), \tag{11.151}$$

with a potential

$$V_{\text{intermol.}} = \frac{\langle\Psi_S|H_{\text{el.}}|\Psi_S\rangle}{\langle\Psi_S|\Psi_S\rangle} - E_1^{(0)} - E_2^{(0)} + \text{minor corrections}. \tag{11.152}$$

In the nuclear two-cluster problem, however, there are no point-like heavy nuclei in the centres of the clusters, thus there is no dynamical variable S. The functions $\Psi_S(\xi_1,\xi_2,\rho)$ obey no approximate orthogonality. On the contrary, for large S, S', their overlap depends only on $S-S'$, so that they are large for $S \simeq S'$ even in the limit $S,S' \to \infty$. No Schrödinger equation can be derived for $f(S)$ as there is no relative kinetic energy for the variable S. This is a major difference spoiling the analogy between Eq. (11.149) and Eqs. (11.106), (11.107). (Remember the pathological mathematical behaviour of $f(\boldsymbol{S})$ of the GCM!) Since in the GCM the relative-motion equation is not a Schrödinger equation, the analogue of Eq. (11.152), the so-called energy surface,

$$E[\boldsymbol{S}] = \frac{\langle \Psi(\boldsymbol{S})|H|\Psi(\boldsymbol{S})\rangle}{\langle \Psi(\boldsymbol{S})|\Psi(\boldsymbol{S})\rangle}, \tag{11.153}$$

cannot be interpreted as a potential. Although the Hill–Wheeler expansion (11.106) is a very useful ingredient of the GCM approach, the GCM amplitude $f(\boldsymbol{S})$ as well as the GCM kernels $\langle \Psi(\boldsymbol{S})|H|\Psi(\boldsymbol{S}')\rangle$, $\langle \Psi(\boldsymbol{S})|\Psi(\boldsymbol{S}')\rangle$ in Eq. (11.109) should not be endowed with direct physical meaning. The relationship between the intercluster potential and the energy surface (11.153) will be discussed in Sect. 11.8.2.

As a last remark, we should mention how the two-centre SM incorporates the one-centre SM as a limiting case. The integration domain in Eq. (11.109) does contain $\boldsymbol{S}=0$ and in Eq. (11.111) one may choose $\boldsymbol{S}=0$ as one of the mesh points. Yet it should also be clear that, apart from trivial exceptions, $\Psi(0) \equiv 0$ over the whole configuration space because of multiple occupancies of the s.p. states caused by the $\boldsymbol{S}=0$ constraint. The energy surface $E[\boldsymbol{S}]$ of Eq. (11.153) at $\boldsymbol{S}=0$ is of zero over zero form, thus it may be meaningful to take the $\boldsymbol{S} \to 0$ limit. One should expand both the numerator and the denominator around $\boldsymbol{S}=0$. The order-Q terms in the expansion of $\Psi(\boldsymbol{S})$ contain altogether Q partial differentiations with respect to the components of $\boldsymbol{S}$. It follows from Eqs. (9.46) and (9.51) that, with each differentiation, the number of HO quanta attached to the relative motion is increased by 1. The first non-vanishing term in $\lim_{\boldsymbol{S}\to 0} \Psi(\boldsymbol{S})$ will contain at least as many quanta as is required by the Wildermuth condition (11.102). (All terms that contain fewer quanta will be eliminated by the antisymmetrizer.) In this way the non-vanishing relative motion will belong to the SU(3) representation $(Q0)$, where $Q=2N+L$ satisfies the inequality (11.102). For SU(3)-scalar clusters the state $\Psi(\boldsymbol{S})$ will belong to the irreducible representation $(\lambda\mu) = (Q0)$ in the limit, and, if the energy surface is angular-momentum projected, the angular momentum will be well-defined.

11.7.4 Cluster distortion

In this last subsection of the microscopic cluster model, it is appropriate to pay some attention to the cluster intrinsic states. Up till now we have just hinted that they have something to do with the g.s. of the separate clusters. Let us assert now that $\Phi_i(\xi_i)$ are taken to approximate the g.s. of independent clusters, and raise the questions what this means, whether it is satisfactory and how to improve on it.

Do we really think that the clusters can survive intact while penetrating each other? They are already linked by the correlation implied by the elimination of the c.m. motion, but what mixes them up internally is the Pauli principle. The density of a product of cluster SM states, $\Phi^{\rm SM}_{1s_1}(\boldsymbol{r}_1,\ldots,\boldsymbol{r}_{a_1})\Phi^{\rm SM}_{2s_2}(\boldsymbol{r}_{a_1+1},\ldots,\boldsymbol{r}_A)$, is the sum of those of $\Phi^{\rm SM}_{1s_1}$ and $\Phi^{\rm SM}_{2s_2}$. The density of a corresponding state from which the cluster c.m. has been eliminated, $\Phi_1(\xi_1)\Phi_2(\xi_2)\varphi^{(b)}_{\boldsymbol{S}}(\boldsymbol{q})$ [with b defined in Eq. (11.86)] is almost equal to the sum of those of $\Phi_1(\xi_1)$ and $\Phi_2(\xi_2)$ (there is a minor c.m. correction), but for $\mathcal{A}_{12}\{\Phi_1(\xi_1)\Phi_2(\xi_2)\varphi^{(b)}_{\boldsymbol{S}}(\boldsymbol{q})\}$ that is not the case. From the region of overlap the two densities will tend to expel each other. This is sometimes expressed by saying that the clusters undergo *Pauli distortion* [283]. But one can expect that the interaction between them is also a major source of disturbance, which is called the *dynamic or specific* [283] *distortion*.

How can we allow for these effects? We can modify

$$\Psi = \mathcal{A}_{12}\{\Phi_1(\xi_1)\Phi_2(\xi_2)\phi(\boldsymbol{q})\} \equiv \mathcal{A}_{12}\{\Phi^{(0)}_1(\xi_1)\Phi^{(0)}_2(\xi_2)\phi(\boldsymbol{q})\} \quad (11.154)$$

in either of the following ways:

(a) the cluster sizes may be made variable;
(b) the clusters may be allowed to be excited;
(c) some internal degree of freedom, which may be considered frozen in the free cluster, may be activated;
(d) another clusterization is admixed:

$$\begin{aligned}
&\text{(a)}\quad \Psi = \mathcal{A}_{12}\{\Phi^{(0')}_1(\xi_1)\Phi^{(0')}_2(\xi_2)\phi(\boldsymbol{q})\},\\
&\text{(b)}\quad \Psi = \sum_{i,j=0,1,\ldots} \mathcal{A}_{12}\{\Phi^{(i)}_1(\xi_1)\Phi^{(j)}_2(\xi_2)\phi_{ij}(\boldsymbol{q})\},\\
&\text{(c)}\quad \Psi = \mathcal{A}_{12}\left\{\mathcal{A}_{1'1''}\{\Phi'^{(0)}_1(\xi'_1)\Phi''^{(0)}_1(\xi''_1)\phi'(\boldsymbol{q}_1)\}\Phi^{(0)}_2(\xi_2)\phi(\boldsymbol{q})\right\},\\
&\text{(d)}\quad \Psi = \sum_{(ij)=(12),(1'2'),\ldots} \mathcal{A}_{ij}\{\Phi^{(0)}_i(\xi_i)\Phi^{(0)}_j(\xi_j)\phi_{ij}(\boldsymbol{q}_{ij})\},
\end{aligned} \quad (11.155)$$

where

(a) $\Phi^{(0')}_k$ denotes a SM g.s. with a size parameter β'_k different from β_k in $\Phi^{(0)}_k$;
(b) $\Phi^{(i)}_1$ and $\Phi^{(j)}_2$ denote different states of clusters a_1 and a_2 and ϕ_{ij} the corresponding relative-motion functions;
(c) $\{\xi'_1, \xi''_1, \boldsymbol{q}_1\} = \xi_1$ give a partitioning within cluster a_1, and their functions describe an inner clustering within cluster a_1;
(d) the labels (ij) denote groupings into different cluster pairs.

It is obvious that these wave function forms contain some additional freedom for the clusters to undergo distortions. These forms are to be taken as trial functions, and the variational principle will take care that the additional degrees of freedom be used to lower the energy and hence letting the clusters distort. Form (b) is distinguished by providing an expansion in terms of a complete set. The other forms provide a framework which is rather heuristic than systematic. But since it is difficult to use a nearly complete basis, the heuristic forms may be equally useful.

In Chap. 13 we shall see examples for the application of methods (b)–(d) and their combinations to light exotic nuclei. To give a qualitative feeling for the significance and properties of the distortion effects, we now discuss an example for ^{6}Li here. Let ^{6}Li be described in an α+d GCM called the breathing cluster model [326]:

$$\Psi = \sum_{i=0}^{N_\alpha-1} \sum_{j=0}^{N_d-1} \mathcal{A}_{\alpha d}\{\Phi_\alpha^{(i)}(\xi_\alpha)\Phi_d^{(j)}(\xi_d)\phi_{ij}(\boldsymbol{q})\} \tag{11.156}$$

or

$$\Psi = \sum_{i=1}^{N'_\alpha} \sum_{j=1}^{N'_d} \sum_{k} \mathcal{A}_{\alpha d}\{\Phi_\alpha^{(\beta_i)}(\xi_\alpha)\Phi_d^{(\beta_j)}(\xi_d)\varphi_{\boldsymbol{S}_k}^{(b_{ij})}(\boldsymbol{q})\}. \tag{11.157}$$

Here $\Phi_\alpha^{(\beta)}$ and $\Phi_d^{(\beta)}$ are intrinsic SM g.s.[25] in an HO well of size parameter β, the sets of which are optimized for the free clusters, and b_{ij} is defined as b in Eq. (11.86). The two formulae are equivalent if $N'_\alpha = N_\alpha$ and $N'_d = N_d$. Convergence was attained with $N_\alpha = N'_\alpha = 3$ and $N_d = N'_d = 5$. Since for the p–n motion five optimized β values produce numerically exact energies and wave functions, this model can as well be considered an α+p+n model described in a cluster Jacobi system of $\boldsymbol{q}_1 = \boldsymbol{r}_p - \boldsymbol{r}_n$, $\boldsymbol{q}_2 = \frac{1}{2}(\boldsymbol{r}_p + \boldsymbol{r}_n) - \boldsymbol{r}_\alpha$. The model is called the breathing model since the cluster excited states in Eq. (11.156) are the breathing modes. Although these allow for spherically symmetric distortions only, combined with Pauli exchanges, they extend the state space towards deformed clusters as well. That Pauli exchanges intermingle spherical and deformed configurations is illustrated as follows.

Let us consider a GCM Slater determinant state Φ_{1100} of ^{6}Li of the form of Eqs. (9.67) and (9.68) and use the identities $\mathcal{A}\{\Phi(1,\dots,6)\} \equiv P_{15}P_{26}\mathcal{A}\{\Phi(1,\dots,6)\} \equiv \mathcal{A}\{P_{15}P_{26}\Phi(1,\dots,6)\}$:

[25] The coordinates ξ_α, ξ_d are to be understood to contain well-defined sets of s.p. coordinates, e.g., coordinates 1, 2, 3, 4 and 5, 6, respectively, which implies corresponding precise definitions of $\boldsymbol{q}$ and of the intrinsic and intercluster antisymmetrizers $\mathcal{A}_\alpha$, $\mathcal{A}_d$ and $\mathcal{A}_{\alpha d}$.

$$
\begin{aligned}
\Phi_{1100} &= \mathcal{A}\left\{\prod_{i=1}^{6}\hat{\varphi}_{s_i m_i^s m_i^t}(\boldsymbol{r}_i)\right\}\\
&= \mathcal{A}_{(1234)(56)}\left\{\mathcal{A}_{1234}\left\{\hat{\varphi}_{s_1\frac{1}{2}\frac{1}{2}}(\boldsymbol{r}_1)\hat{\varphi}_{s_2\frac{1}{2}-\frac{1}{2}}(\boldsymbol{r}_2)\hat{\varphi}_{s_3-\frac{1}{2}\frac{1}{2}}(\boldsymbol{r}_3)\hat{\varphi}_{s_4-\frac{1}{2}-\frac{1}{2}}(\boldsymbol{r}_4)\right\}\right.\\
&\qquad\left.\times\mathcal{A}_{56}\left\{\hat{\varphi}_{s_5\frac{1}{2}\frac{1}{2}}(\boldsymbol{r}_5)\hat{\varphi}_{s_6\frac{1}{2}-\frac{1}{2}}(\boldsymbol{r}_6)\right\}\right\}\\
&= \mathcal{A}_{(1234)(56)}\left\{\mathcal{A}_{1234}\left\{\hat{\varphi}_{s_5\frac{1}{2}\frac{1}{2}}(\boldsymbol{r}_1)\hat{\varphi}_{s_6\frac{1}{2}-\frac{1}{2}}(\boldsymbol{r}_2)\hat{\varphi}_{s_3-\frac{1}{2}\frac{1}{2}}(\boldsymbol{r}_3)\hat{\varphi}_{s_4-\frac{1}{2}-\frac{1}{2}}(\boldsymbol{r}_4)\right\}\right.\\
&\qquad\left.\times\mathcal{A}_{56}\left\{\hat{\varphi}_{s_1\frac{1}{2}\frac{1}{2}}(\boldsymbol{r}_5)\hat{\varphi}_{s_2\frac{1}{2}-\frac{1}{2}}(\boldsymbol{r}_6)\right\}\right\}.
\end{aligned}
\tag{11.158}
$$

In these equations the full antisymmetrizer has been written as a product as in Eq. (D.6). Now $s_1, \ldots, s_4$ can be chosen to coincide approximately or exactly, and the same applies to s_5 and s_6. If so, then the second expression can be interpreted as describing two spherical subsystems displaced (see Fig. 11.3a), while in the third expression one of the two overlapping subsystems is elongated (see Fig. 11.3b). From the identity of the two expressions we learn that configurations involving purely spherical clusters cannot be distinguished from some that involve non-spherical ones. Consequently, allowing for spherical distortions enriches the subspace with deformed clusters as well.

One can evaluate the distortion effects through their contribution to the energy and through the amounts of clustering $\mathfrak{P}_{ij}$ belonging to $\{\Psi_{\boldsymbol{R}}^{(ij)}\}$, where $\Psi_{\boldsymbol{R}}^{(ij)} = \mathcal{A}_{\alpha\mathrm{d}}\{\Phi_\alpha^{(i)}\Phi_\mathrm{d}^{(j)}\delta(\boldsymbol{R}-\boldsymbol{q})\}$.

Table 11.1 compares different versions of the cluster model calculated with the same simple central effective force (Volkov 2, with Majorana parameter $M = 0.6$ [327]). The first and second rows contain results of ordinary GCM calculations with equal and unequal size parameters; the third to fifth rows come from calculations in which the clusters are described as superpositions of HOSM states of different size

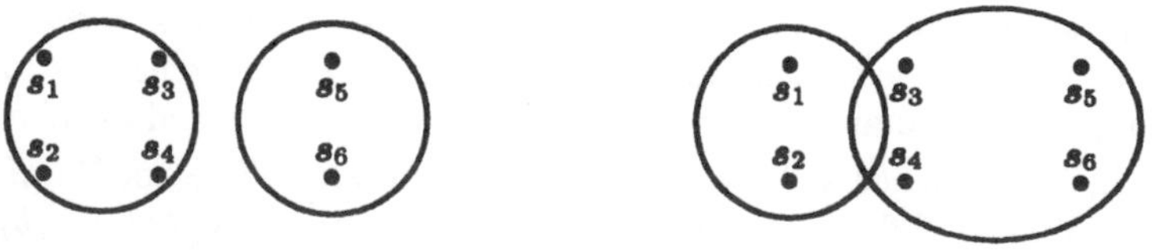

(a) (b)

Figure 11.3: Pauli exchanges intermingle spherically symmetric and deformed cluster configurations: the two configurations, given precisely in Eq. (11.158), are indistinguishable.

Table 11.1: Total energies E and α+d relative energies $\varepsilon = E - E_\alpha - E_d$ in six models [328], with single-configuration cluster states, frozen multiconfiguration cluster states and breathing cluster states, alternatively, all optimized. [See Eqs. (11.156) and (11.157).]

Model					E	ε
N'_α	N'_d	N_α	N_d	Cluster states (β's in fm^{-2})	(MeV)	(MeV)
1	1	1	1	$\beta_\alpha = \beta_d$=0.469	−27.48	−2.49
1	1	1	1	β_α = 0.528, β_d=0.164	−28.12	−2.45
3	5	1	1	frozen α, d	−28.80	0.37[a]
3	5	3	1	breathing α, frozen d	−29.12	0.05[a]
3	5	1	5	frozen α, breathing d	−30.38	−1.21
3	5	3	5	breathing α, d	−30.44	−1.27

[a]Unbound state.

parameters, but in the two-cluster calculation one or both of the clusters are frozen in their g.s. [model (11.156) with i and/or j restricted to 0]; the last row results from the full breathing model (11.156). It is seen that the distortion of the deuteron is the most significant single effect. The multiconfiguration cluster g.s. deepen the internal bindings of the clusters at the expense of the intercluster binding, and ε becomes positive. However, by switching on the distortion, not only the overall binding becomes deeper, but the intercluster binding is also restored. Thus it is fair to say that the distortion effects seem to restore the validity of the cluster model.

The weight $\mathfrak{P}_{ij}$ of each ij pair of states in the g.s. of ^{6}Li can be seen in Table 11.2, along with the corresponding spectroscopic factors. The model used here is a breathing model with parameters similar to those used in Table 11.1. In accord with the moderate energy gain coming from the distortion effects, the weight of the g.s. clusters, $\mathfrak{P}_{00} = 0.92$, is not much smaller than unity. Yet some of the pairs of excited states are admixed substantially. Owing to the nonorthogonality of the clusterizations (cf. p. 289), the sum of their amounts appreciably exceeds unity. The admixtures tend to decrease with growing excitation energy, but there is a departure from monotonicity for states involving a state of the deuteron close to zero energy. The spectroscopic factors are mostly close to the amounts of clustering, but one exception $[(ij) = (10)]$ is a warning against mixing them up.

In summary, the contribution of the distortion enhances the binding to some extent, but our picture of the nucleus remains essentially unaltered. The weights of some cluster excited states are substantial in

Table 11.2: Amounts $\mathfrak{P}_{ij}$ [Eq. (11.24)] and spectroscopic factors S_{ij} [Eq. (11.26)] of $\alpha_i + d_j$ clustering in the model g.s. of ^{6}Li, where α_0 and d_0 label the cluster g.s. and α_i and d_j, with $i, j \geq 1$ denote the cluster breathing modes [284]. The energies E_{d_j} and E_{α_i} are given in MeV.

E_{d_j}		-2.22	1.66	10.82	52.32	713.68
E_{α_i}						
−28.30	$\mathfrak{P}_{0j}$	0.92	0.13	0.39	0.073	0.00027
	S_{0j}	0.93	0.12	0.27	0.048	0.00026
−6.33	$\mathfrak{P}_{1j}$	0.21	0.0016	0.013	0.0023	0.000013
	S_{1j}	0.040	0.0015	0.0097	0.0019	0.000013
275.82	$\mathfrak{P}_{2j}$	0.00022	0.000024	0.00014	0.0000020	0.00000042
	S_{2j}	0.00012	0.000024	0.00012	0.0000018	0.00000040

the g.s. wave function, yet the amount of g.s.+g.s. clustering is hardly smaller than unity. One should assess the significance of this statement in view of the fact that one of the clusters in this model is the deuteron, which we will declare to be too distortable to be included in the correlated Gaussian cluster model [see Sect. 12.2.2]. The coexistence of the large admixture of the excited states and the almost complete g.s. clustering can be understood by noting that the excited two-cluster configurations ($\{\Psi_{\boldsymbol{R}}^{(ij)}\}$ with $i, j > 0$) highly overlap with each other as well as with the g.s. configuration $\{\Psi_{\boldsymbol{R}}^{(00)}\}$. Without antisymmetrization, all these configurations $\{\Psi_{\boldsymbol{R}}^{(ij)}\}$ would be orthogonal to each other, so we have to attribute this overlap to the Pauli effects. One can say that *Pauli effects* reduce the state space available for distortion, and thereby *diminish distortion.* Thus the Pauli principle does not only cause difficulties but also simplifies the structure of nuclei, and, in this respect, it is an essential component in the success of the cluster models.

11.8 The orthogonality-condition model

It was mentioned on p. 303 that the exchange terms in the kernels (11.64) and (11.65) are non-negligible. How can this statement be reconciled with the fact that the collision of any two nuclei is described in the optical model with local potentials remarkably well? How should we comment on the fact that macroscopic potential models do seem to work fairly well for bound states [329], including cases of exotic nuclei [196] (cf. Sect. 8.2)?

11.8.1 The nonlocality problem

To find the clue to this paradox, let us consider the simple special case of a large cluster, described by a single HO configuration, interacting with a single (the Ath) nucleon. Let us neglect the c.m. motion; then the 'intercluster' coordinate will be equal to a s.p. coordinate: $\boldsymbol{q} = \boldsymbol{r}_A$, and the total wave function will reduce to a SM Slater determinant $\Phi = \mathcal{A}\{\psi_{000}(\boldsymbol{r}_1)\cdots\psi_{n_{A-1}l_{A-1}m_{A-1}}(\boldsymbol{r}_{A-1})\phi(\boldsymbol{r}_A)\}$, where the spin–isospin labels have been suppressed. The relative-motion function $\phi(\boldsymbol{r}_A)$, which can be assumed to have good angular momentum, can be expanded in terms of the HO states: $\phi=\sum_{n=0}^{\infty} c_{nlm}\psi_{nlm}$. The terms of ϕ containing ψ_{nlm} with $\{nlm\} = \{000\}, \ldots, \{n_{A-1}l_{A-1}m_{A-1}\}$ can be omitted since they give rise to determinants with two identical columns, which vanish. (For simplicity, we take the case when the core spins and isospins are paired off.) All other components will be intact. Thus the sole effect of the Pauli principle on the relative-motion function is that the Pauli-forbidden components are projected out.

It is instructive to formulate this observation as follows. Let $\Phi_{\boldsymbol{R}}$ be $\Phi_{\boldsymbol{R}}=\mathcal{A}\{\psi_{000}(\boldsymbol{r}_1)\cdots\psi_{n_{A-1}l_{A-1}m_{A-1}}(\boldsymbol{r}_{A-1})\delta(\boldsymbol{R}-\boldsymbol{r}_A)\}$, which defines $N(\boldsymbol{R},\boldsymbol{R}')=\langle\Phi_{\boldsymbol{R}}|\Phi_{\boldsymbol{R}'}\rangle$ and a corresponding operator $\mathcal{N}$. Then

$$\mathcal{N}\psi_{nlm}(\boldsymbol{R})=\begin{cases}\psi_{nlm}(\boldsymbol{R}) & \text{if } \psi_{nlm} \text{ is unoccupied in } \Phi,\\ 0 & \text{if } \psi_{nlm} \text{ is occupied in } \Phi.\end{cases} \tag{11.159}$$

This equation can be reinterpreted as the eigenvalue equation of $\mathcal{N}$, Eq. (11.49), and the eigenvalues are

$$\nu_{nl}=\begin{cases}1 \text{ if } \psi_{nlm} \text{ is unoccupied in } \Phi,\\ 0 \text{ if } \psi_{nlm} \text{ is occupied in } \Phi.\end{cases} \tag{11.160}$$

Equation (11.159) expresses that, when one of the clusters is a heavy core and the other is a nucleon, the norm operator reduces to a projector to Pauli-allowed states: $\mathcal{N}=\mathcal{I}$ [cf. Eq. (11.55)]. Thus Eqs. (11.60) and (11.71) can be rewritten as

$$\mathcal{I}\mathcal{H}\mathcal{I}\psi(\boldsymbol{R}) = E\psi(\boldsymbol{R}), \qquad \mathcal{I}(\mathcal{T}+\mathcal{V})\mathcal{I}\psi = \varepsilon\psi. \tag{11.161}$$

So far our considerations have been exact for the system of a large core plus a single particle. We know that $\phi_{nlm} = \psi_{nlm}$ is valid for any two clusters described by the HOSM, and so is the property that the Pauli-forbidden relative-motion states are eigenstates belonging to $\nu_{nl}=0$. But $\mathcal{N}=\mathcal{I}$ does not hold in a more general case, which implies that Eq. (11.159) has to be modified, too. From the spectral decompositions (11.51)–(11.53) it is clear that any power of the norm operator always contains this projector $\mathcal{I}$ as a factor: $\mathcal{N}^{\kappa} = \mathcal{I}\mathcal{N}^{\kappa}\mathcal{I}$. Thus

Eqs. (11.60) and (11.71) can be rewritten as

$$\mathcal{I}\tilde{\mathcal{H}}\mathcal{I}\psi(\boldsymbol{R}) = E\psi(\boldsymbol{R}), \qquad \mathcal{I}(\mathcal{T}+\tilde{\mathcal{V}})\mathcal{I}\psi = \varepsilon\psi, \tag{11.162}$$

where $\tilde{\mathcal{H}} = \mathcal{N}^{-1/2}\mathcal{H}\mathcal{N}^{-1/2}$ has been used. Note that Eqs. (11.161) and (11.162) are satisfied by the Pauli-forbidden states at $\varepsilon = 0$, which is meaningless, thus $\varepsilon = 0$ should be excluded.

We see that in general the effect of $\mathcal{N}$ does not reduce to a projection, yet we argue that the projection implicit in it is by far the most important of its effects. The appearance of $\mathcal{I}$ is a source of strong nonlocality in the Hamiltonian, and plays a dominant role in Eq. (11.161). Its effect is that it constrains the inner nodes of the wave function, thus washing out the details of the dynamics. It is therefore tempting to believe that this is perhaps the only important nonlocality in the relative Hamiltonian in Eq. (11.71), so, in between two operators $\mathcal{I}$, the potential $\tilde{\mathcal{V}}$ may be approximated by a simple local potential $\widetilde{\mathcal{V}}$. Understood in this way, Eq. (11.162) is the equation of motion of an approach called the orthogonality-condition model (OCM) [302, 304]. In the OCM Eq. (11.162) is the starting point: one forgets all microscopic details, e.g., whether the two fragments can be represented by the same $\hbar\omega$. It is a basically macroscopic model since its equation of motion is an equation for the relative degree of freedom. But it incorporates microscopic effects through the projector $\mathcal{I}$, which depends on the intracluster states.

An even more useful form of the OCM equation can be obtained by omitting the second projector from $\mathcal{I}(\mathcal{T}+\tilde{\mathcal{V}})\mathcal{I}\psi=\varepsilon\psi$:

$$\mathcal{I}(\mathcal{T}+\tilde{\mathcal{V}})\psi = \varepsilon\psi. \tag{11.163}$$

Since the left-hand side is in the Pauli-allowed subspace, the equality can only hold if ψ on the right-hand side is also in that subspace. Then it is redundant to put $\mathcal{I}$ in front of ψ on the left-hand side, which justifies the omission. The two equations are not equivalent only at $\varepsilon = 0$, but that is beneficial. While a redundant state is a solution of Eq. (11.162) at $\varepsilon=0$, it is not a solution of Eq. (11.163).

A more practicable version of the OCM equation (11.163) may be obtained by putting the nonlocality artificially into an extra potential term:

$$[\mathcal{T}+\widetilde{\mathcal{V}}+\lambda(1-\mathcal{I})]\psi = \varepsilon\psi, \tag{11.164}$$

where $1-\mathcal{I}$ projects to the Pauli-forbidden states and λ is a large positive energy.[26] To understand the relationship between Eqs. (11.163)

[26] For a multicluster system there are an infinite number of Pauli-forbidden states, so it is not simpler to solve Eq. (11.164) than Eq. (11.162). A practical method to eliminate Pauli-forbidden states in a complex system has been presented in Ref. [330].

and (11.164), one should imagine that, with $\lambda = 0$, the lowest-lying states have the same quantum numbers as the Pauli-forbidden states, thus the Pauli-forbidden states are strongly mixed into them. The role of $\lambda(1 - \mathcal{I})$ is to enhance the energy of such states very much. When λ is large, the variational principle underlying the Schrödinger equation [cf. the Theorem 1 in Sect. 9.1.2] will cause the weight of the Pauli-forbidden components in the lowest-lying states to be thoroughly reduced. In the limit $\lambda \to \infty$ they are completely eliminated, but it is simpler, and—for practical purposes—enough, to have λ large (e.g., 10 GeV). Note that this prescription also produces an OCM equation that is meaningful at $\varepsilon = 0$ as well.

There exist local potentials $\widetilde{\mathcal{V}}$ with which the OCM produces a good approximation to the RGM, so that it describes bound states, scattering states [304,331,332] as well as resonances [333] with remarkable success. It is possible to derive a potential $\widetilde{\mathcal{V}}$ from the RGM, which is indeed almost local [334,335]. It is found that, in most cases, this local potential $\widetilde{\mathcal{V}}$ is very close to the folding-model potential. It is reasonable to choose it to have the shape of the folding potential and rescale its strength. The OCM can be generalized to more than two clusters [336] and to more than one channel [287]; for such complicated cases it is much more feasible than the RGM.

The effect of the Pauli projection is that the sequence of states starts with $n > 0$, which requires deep enough potentials $\widetilde{\mathcal{V}}$. The potential may be determined phenomenologically [304] or microscopically [337]. The OCM potentials have Gaussian-like shapes, and, near the bottom, they do not differ very much from the oscillator consistent with the HOSM. Therefore, the Pauli-forbidden relative-motion states are very similar to the lowest-lying states in the OCM potentials used in an ordinary Schrödinger equation

$$(\mathcal{T} + \widetilde{\mathcal{V}})\psi = \varepsilon\psi. \qquad (11.165)$$

Consequently, it is reasonable to replace Eq. (11.162) or (11.164) by Eq. (11.165), with the extra prescription that the lowest-lying states that do not obey Eq. (11.102) are to be discarded [338]. (This prescription cannot be generalized to multicluster systems.) The model thus obtained is the local-potential cluster model, which is as extreme among cluster models as is the extreme s.p. model among the SM.

One can thus say that the microscopic cluster model does support the local intercluster potentials of the folding type. Yet in deriving the local potentials one need not assume that the exchange terms in Eq. (11.65) are negligible. The clue is that the potential appears in the expression $\mathcal{N}^{-1/2}\mathcal{V}\mathcal{N}^{-1/2} \equiv \mathcal{I}\mathcal{N}^{-1/2}\mathcal{V}\mathcal{N}^{-1/2}\mathcal{I}$. The exchange terms of this expression are proportional to $\mathcal{I}$, and it is reasonable to neglect the remaining nonlocality.

11.8.2 Local intercluster potential

We have seen that the two-cluster problem reduces to a Schrödinger equation for the relative motion, $(\mathcal{T}+\tilde{\mathcal{V}}-\varepsilon)\psi=0$ [cf. Eq. (11.71)] with a nonlocal potential $\tilde{\mathcal{V}}$. It was also indicated that this potential can be approximated by local potentials $\widetilde{\mathcal{V}}$. More precisely, it is possible to construct approximately equivalent local potentials. The term equivalence may be used in two senses. *Phase-equivalence* means that two potentials produce the same asymptotic wave functions, i.e., the same phase shifts and bound-state energies. *Trivial equivalence* means that the two potentials produce the same wave functions as well. Any equivalence may hold for a certain partial wave and it may only be valid for a certain energy. Nonlocal optical potentials have actually been introduced in an attempt to eliminate the energy dependence of local optical potentials [339, 340].

A very plausible semiclassical (WKB) prescription [341] to determine a local potential which is approximately phase-equivalent to $\tilde{\mathcal{V}}$ may be started [342] with an implicit definition of a local momentum $\boldsymbol{p}$. In the WKB approximation to the RGM problem the relative energy ε can be expressed in terms of the Wigner transform of the RGM kernels $H(\boldsymbol{R},\boldsymbol{R}')$ and $N(\boldsymbol{R},\boldsymbol{R}')$:

$$\varepsilon = \frac{H^{(W)}[\boldsymbol{R},\boldsymbol{p}(\boldsymbol{R})]}{N^{(W)}[\boldsymbol{R},\boldsymbol{p}(\boldsymbol{R})]} - (E_1^{(0)} + E_2^{(0)}), \tag{11.166}$$

where the Wigner transform of a function $O(\boldsymbol{R},\boldsymbol{R}')$ is defined by

$$O^{(W)}(\boldsymbol{R},\boldsymbol{p}) = \int d\boldsymbol{s}\, \exp\left(\tfrac{i}{\hbar}\boldsymbol{s}\cdot\boldsymbol{p}\right) O\left(\boldsymbol{R}-\tfrac{1}{2}\boldsymbol{s}, \boldsymbol{R}+\tfrac{1}{2}\boldsymbol{s}\right), \tag{11.167}$$

and Eq. (11.166) defines the local momentum $\boldsymbol{p}(\boldsymbol{R})$ for a given ε. The local potential $\widetilde{\mathcal{V}}(\boldsymbol{R})$ approximately equivalent to $\tilde{\mathcal{V}}$ is then obtained from another implicit equation:

$$\frac{1}{2\mu}[\boldsymbol{p}(\boldsymbol{R})]^2 + \widetilde{\mathcal{V}}(\boldsymbol{R}) = \varepsilon, \tag{11.168}$$

where μ is the reduced mass.

Equivalent local potentials may be constructed for the RGM equation (11.66) similarly. This equation contains the local potential $V_{\rm D}(\boldsymbol{R})$ and a nonlocal potential $\mathcal{G}$, whose kernel is

$$G(\boldsymbol{R},\boldsymbol{R}') = [\varepsilon - T_{12}(\boldsymbol{R})]K(\boldsymbol{R},\boldsymbol{R}') - V_{\rm EX}(\boldsymbol{R},\boldsymbol{R}'). \tag{11.169}$$

The effect of $\mathcal{G}$ on the function ϕ can be written as

$$\mathcal{G}\phi(\boldsymbol{R}) \equiv (\mathcal{G}\phi)(\boldsymbol{R}) = \int d\boldsymbol{R}'\, G(\boldsymbol{R},\boldsymbol{R}')\phi(\boldsymbol{R}'). \tag{11.170}$$

Equation (11.170) lends itself to constructing a trivially-equivalent local potential $\mathcal{V}^{\mathrm{T}}(\boldsymbol{R})$:

$$\mathcal{V}^{\mathrm{T}}(\boldsymbol{R}) = V_{\mathrm{D}}(\boldsymbol{R}) + \frac{1}{\phi(\boldsymbol{R})} \int d\boldsymbol{R}'\, G(\boldsymbol{R},\boldsymbol{R}')\phi(\boldsymbol{R}'), \qquad (11.171)$$

which can be verified by direct substitution into Eq. (11.66). This $\mathcal{V}^{\mathrm{T}}(\boldsymbol{R})$ is seen to have inconvenient properties: it is singular at the nodes of $\phi(\boldsymbol{R})$ and it depends on the energy through $\phi(\boldsymbol{R})$.

The semiclassical prescription to determine a (phase-equivalent) local potential again starts with an implicit definition of a local momentum $\boldsymbol{p}(\boldsymbol{R})$. This local momentum is substituted into the momentum of the Wigner transform $G^{(W)}$ of $G(\boldsymbol{R},\boldsymbol{R}')$. The equivalent local potential is thus determined by the following pair of transcendental equations:

$$\begin{aligned} \mathcal{V}'(\boldsymbol{R}) &= V_{\mathrm{D}}(\boldsymbol{R}) + G^{(W)}[\boldsymbol{R},\boldsymbol{p}(\boldsymbol{R})], \\ \frac{1}{2\mu}[\boldsymbol{p}(\boldsymbol{R})]^2 &+ \mathcal{V}'(\boldsymbol{R}) = \varepsilon. \end{aligned} \qquad (11.172)$$

It has been shown that the two prescriptions, Eqs. (11.166)–(11.168) and (11.172), do actually lead to the same potential, $\mathcal{V}'(\boldsymbol{R}) = \widetilde{\mathcal{V}}(\boldsymbol{R})$ [342], and, consequently, the same wave function. The solutions of Eqs. (11.71) and (11.66) differ in the interior region ($\psi = \mathcal{N}^{1/2}\phi$), and the solution belonging to the local-potential problem is much closer to ψ than to ϕ [342]. Equation (11.172) was used to convert a nonlocal nucleon–nucleon potential, derived from quark exchanges between nucleons, into a local potential [343].

An alternative method [337] is based on the expression of the energy surface (11.153) in terms of the OCM:

$$\frac{\langle\Psi(S)|H|\Psi(S)\rangle}{\langle\Psi(S)|\Psi(S)\rangle} - E_1^{(0)} - E_2^{(0)} = \frac{\langle\!\langle\varphi_S|\mathcal{N}^{1/2}(\mathcal{T}+\widetilde{\mathcal{V}})\mathcal{N}^{1/2}|\varphi_S\rangle\!\rangle}{\langle\Psi(S)|\Psi(S)\rangle}. \qquad (11.173)$$

This equality should hold for any $\boldsymbol{S}$, and should be inverted to determine $\widetilde{\mathcal{V}}(\boldsymbol{R})$. The inversion is done in practice by parametrizing $\widetilde{\mathcal{V}}(\boldsymbol{R})$ and fitting its parameters to the energy surface for a physically important range of S. The former procedure produces potentials for the local-potential cluster model, while the latter for the OCM.

It should be noted that the derivation of an OCM potential or a local potential from the RGM is somewhat similar to finding local potentials equivalent to a nonlocal potential. There may be different definitions for the equivalence. The equivalence can only be approximate unless angular-momentum dependence (and hence parity dependence) is allowed. Correspondingly, both prescriptions (11.168) and (11.173) have different versions depending on whether or not the Wigner transforms or the energy surfaces are projected to good parity or angular-momentum. The parity dependence is strongest for identical clusters: for them the equivalent potential is only meaningful for even partial waves. The odd-even staggering is also pretty strong for systems of nearly equal clusters.

All derivations produce deep potentials in the sense that, when put into a Schrödinger equation, they accommodate the Pauli-forbidden relative-motion states. From the OCM the forbidden states are, of course, automatically excluded, but from the local-potential model the Pauli-forbidden states have to be eliminated by an extra prescription. Both models reproduce the results of the underlying RGM, and show the viability of macroscopic models. It may of course be noticeable that the local-potential model is more approximate. Figure 11.4 compares the RGM relative wave function of the g.s. of ^{7}Li in an α+t model with its approximants in the OCM and in the local-potential model. The OCM potential, $V(r) = -67.87 \exp[-(r/2.740 \text{ fm})^2]$ MeV, has been deduced from the nucleon–nucleon force through the angular-momentum projected version of Eq. (11.173) [344]. In the local-potential model the

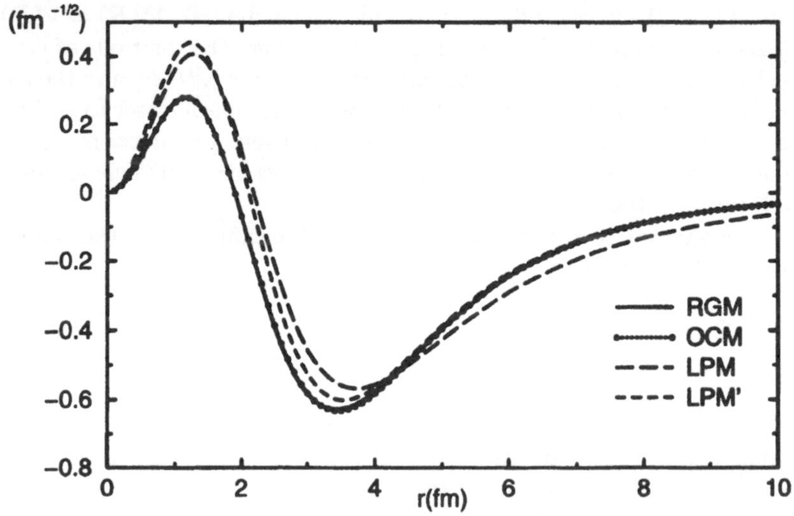

Figure 11.4: RGM relative wave function $u_1(r) = r \int d\hat{r}\, Y^*_{1m}(\hat{r})\psi(\boldsymbol{r})$, calculated with the Volkov 2 force [327], of the g.s. of ^{7}Li=α+t, compared with corresponding OCM and local-potential-model (LPM) results. The OCM potential is deduced from the RGM. The potential in the LPM is the same as that of the OCM; the LPM$'$ potential is the same multiplied by 1.074 to produce the same relative energy ε.

same potential is used or is readjusted to give the same energy. One can see that the OCM agrees with the RGM within the line width, and the local-potential model also produces qualitatively similar functions. Note that this potential has been optimized for the OCM, and for the local-potential model more suitable potentials may exist.

In phenomenological macroscopic cluster models Pauli effects are often taken into account by repulsive-core local potentials, which are shallow in the sense that they support no Pauli-forbidden states [345]. Although no shallow potentials can be derived from a microscopic framework, shallow potentials may still produce the same phase shift as the well-founded deep potentials. There is a systematic way, called the supersymmetric transformation [346], of constructing phase-equivalent potentials that produce one node fewer; the repulsive core of these potentials is of r^{-2} form. From the viewpoint of nuclear structure, such phase-equivalent potentials are in principle not satisfactory since the wave functions they yield may differ substantially from the wave function that is well founded microscopically. In a wave function normalized to unity even the tail region may be affected through the normalization. Nevertheless, it is not so easy to pinpoint empirically the superiority of the deep potentials in low-energy problems [347, 348]. The explanation of this ambiguity is that the repulsive-core treatment is close to a semiclassical treatment of the Pauli effects. Semiclassically, Pauli exclusion amounts to exclusion from a certain domain of the phase space [349]. This domain is bounded by a curve that corresponds to full expulsion from an inner spatial zone for low-momentum motion and, correspondingly, full exclusion from a low-momentum zone for overlapping fragments.

The OCM and the local-potential model perform well for the states that correspond to $(\lambda\mu)=(2n+l,0)$ in the SU(3) model. The local potential produces the states with the same quantum numbers as the HO potential, but splits them up in the orbital angular momentum. For Gaussian-shaped potentials, which come from the folding procedure, it so happens that the L-sequence obtained in this way follows the $L(L+1)$ rule, like a rotational band. In nuclei near shell closure there are several such cluster bands amenable to this description. Not only the level sequence but also the electromagnetic transition rates (with no or little effective charges [184]) and decay properties are well reproduced simultaneously. The most convincing cases are ^{20}Ne=^{16}O+α [329], ^{44}Ti=^{40}Ca+α [350, 351] and ^{212}Po=^{208}Pb+α [352, 353]. Alpha-decay [354] and other [355] cluster-decay rates can be predicted for a virtually complete range of cluster-decay cases. Furthermore, heavy deformed nuclei, which belong to the classical realm of the rotational model, can also be described as two-cluster systems [183] (cf. p. 183).

11.9 Microscopic versus macroscopic approach

In Sect. 11.8 we have seen that, in some approximation, a macroscopically meaningful and tractable equation of motion can be derived from the microscopic equation of motion (11.60) or (11.71). In this section a more rigorous relationship will be established between the two approaches. The significance of this issue for us is that, while this book follows the microscopic approach, the macroscopic approach to light exotic nuclei is also very common. First we shall start with the correspondence between the microscopic and macroscopic observables.

11.9.1 Observables

Let us consider a two-cluster system with clusters represented by states $\Phi_1^{(0)}$ and $\Phi_2^{(0)}$ with no inner excitations, carrying $Q_1^{(0)}$ and $Q_2^{(0)}$ excitation quanta of the same $\hbar\omega$, respectively. The rigorous relationship to be derived applies to this case and to observables $\hat{O}$ of the type of

$$\hat{O}+\hat{O}_{\text{c.m.}}(\boldsymbol{R}_{\text{c.m.}}) = \hat{O}_1(\xi_1)+\hat{O}_2(\xi_2)+\hat{O}_{12}(\boldsymbol{q})+\hat{O}_{\text{c.m.}}(\boldsymbol{R}_{\text{c.m.}}). \quad (11.174)$$

This includes all spatial observables that are linear or quadratic in the s.p. coordinates and/or momenta. These observables are separable in this sense even for an n-cluster case as is given for $\sum_i \boldsymbol{r}_i^2$ in Eq. (B.19) and for $T=(2m)^{-1}\sum_i \boldsymbol{p}_i^2$ in Eq. (B.20). To familiarize the reader with these involved formulae, we give them for the $A=a_1+a_2$ two-cluster case here:

$$\begin{aligned}
\sum_{i=1}^{A} \boldsymbol{r}_i^2 &= \sum_{k=1}^{a_1-1} \frac{k}{k+1}\boldsymbol{t}_k^2 + \sum_{k=1}^{a_2-1} \frac{k}{k+1}\boldsymbol{t}_{a_1+k}^2 + \frac{a_1a_2}{A}\boldsymbol{q}_1^2 + A\boldsymbol{q}_2^2 \\
&= \frac{1}{2}(\boldsymbol{r}_1-\boldsymbol{r}_2)^2 + \frac{2}{3}\left[\frac{1}{2}(\boldsymbol{r}_1+\boldsymbol{r}_2)-\boldsymbol{r}_3\right]^2 + \cdots \\
&\quad + \frac{a_1-1}{a_1}\left[\frac{1}{a_1-1}(\boldsymbol{r}_1+\cdots+\boldsymbol{r}_{a_1-1})-\boldsymbol{r}_{a_1}\right]^2 \\
&\quad + \frac{1}{2}(\boldsymbol{r}_{a_1+1}-\boldsymbol{r}_{a_1+2})^2 + \frac{2}{3}\left[\frac{1}{2}(\boldsymbol{r}_{a_1+1}+\boldsymbol{r}_{a_1+2})-\boldsymbol{r}_{a_1+3}\right]^2 + \cdots \\
&\quad + \frac{a_2-1}{a_2}\left[\frac{1}{a_2-1}(\boldsymbol{r}_{a_1+1}+\cdots+\boldsymbol{r}_{A-1})-\boldsymbol{r}_A\right]^2 \\
&\quad + \frac{a_1a_2}{A}(\bar{\boldsymbol{t}}_1-\bar{\boldsymbol{t}}_2)^2 + A\boldsymbol{R}_{\text{c.m.}}^2, \qquad (11.175)
\end{aligned}$$

$$\sum_{i=1}^{A} \frac{\boldsymbol{p}_i^2}{2m} = \sum_{k=1}^{a_1-1} \frac{\boldsymbol{\pi}_k^2}{2\frac{k}{k+1}m} + \sum_{k=1}^{a_2-1} \frac{\boldsymbol{\pi}_{a_1+k}^2}{2\frac{k}{k+1}m} + \frac{\boldsymbol{\varpi}_1^2}{2\frac{a_1a_2}{A}m} + \frac{\boldsymbol{\varpi}_2^2}{2Am}, \qquad (11.176)$$

where ϖ_1 and ϖ_2 are the relative and c.m. momenta canonically conjugate to $\bar{q}\equiv\bar{t}_1-\bar{t}_2$ and $\boldsymbol{R}_{\text{c.m.}}$, respectively. As a simple example for a non-scalar, non-isoscalar operator, we can take the electric dipole operator $e\sum_i \bar{r}_i Y_{1\mu}(\hat{\bar{r}}_i)P_i^{\text{p}} = e\sqrt{3/4\pi}\sum_i(\bar{\boldsymbol{r}}_i)_{1\mu}P_i^{\text{p}}$, where e is the electric charge unit and $(\bar{\boldsymbol{r}}_i)_{1\mu}$ are the spherical tensor components of $\bar{\boldsymbol{r}}_i = \boldsymbol{r}_i - \boldsymbol{R}_{\text{c.m.}}$ [cf. Eq. (9.124)]. One can see without explicitly writing down the result that this operator is also expressible in the required form: the Cartesian vectors $\boldsymbol{r}_i$ can be expressed in terms of intrinsic, relative and c.m. vectors $\boldsymbol{t}_j$ as $\boldsymbol{r} = T^{-1}\boldsymbol{t}$ [cf. Eq. (B.7)], where T^{-1} is a block diagonal matrix, with the blocks given by Eq. (B.5). But true enough, there are important operators, like the two-body interaction, which do not satisfy the condition (11.174), and they are outside the scope of the theorem to come.

From now on, let us deal with $\hat{O}=\hat{O}_1(\xi_1)+\hat{O}_2(\xi_2)+\hat{O}_{12}(\boldsymbol{q})$, without the c.m. term, and, to avoid unimportant complications, assume that $\hat{O}$ is rotation-invariant. It will be shown that a matrix element of this $\hat{O}$ between $\Psi=\mathcal{A}_{12}\{\Phi_1^{(0)}\Phi_2^{(0)}\phi(\boldsymbol{q})\}$ and $\Psi'=\mathcal{A}_{12}\{\Phi_1^{(0)}\Phi_2^{(0)}\phi'(\boldsymbol{q})\}$ can be rewritten as [287, 356]

$$\langle\Psi|\hat{O}|\Psi'\rangle = \sum_{ij(\neq 0)}\left(\frac{\nu_<}{\nu_>}\right)^{1/2} \langle\!\langle\psi|\phi_i\rangle\!\rangle\langle\!\langle\phi_i|\mathcal{O}|\phi_j\rangle\!\rangle\langle\!\langle\phi_j|\psi'\rangle\!\rangle, \qquad (11.177)$$

$$\psi = \mathcal{N}^{1/2}\phi, \quad \psi' = \mathcal{N}^{1/2}\phi', \qquad (11.178)$$

where $\sum_{ij(\neq 0)}$ means summation over all i and j for which $\nu_i \neq 0$ and $\nu_j \neq 0$, ϕ_i, ϕ_j are the norm-kernel eigenfunctions, $\nu_< = \min\{\nu_i, \nu_j\}$, $\nu_> = \max\{\nu_i, \nu_j\}$, and the kernel of the macroscopic operator $\mathcal{O}$ is

$$O(\boldsymbol{R},\boldsymbol{R}') = \langle\Phi_1^{(0)}\Phi_2^{(0)}\delta(\boldsymbol{R}-\boldsymbol{q})|\hat{O}|\Phi_1^{(0)}\Phi_2^{(0)}\delta(\boldsymbol{R}'-\boldsymbol{q})\rangle. \qquad (11.179)$$

Equation (11.177) can be proved as follows. The first step is to express the matrix element in terms of the eigenfunctions of the norm kernel. To this end, we write $\langle\Psi|\hat{O}|\Psi'\rangle$ as $\langle\Phi_1^{(0)}\Phi_2^{(0)}\phi|\mathcal{A}_{12}\hat{O}\mathcal{A}_{12}|\Phi_1^{(0)}\Phi_2^{(0)}\phi'\rangle$, insert two unit operators $1=\sum_i|\phi_i\rangle\!\rangle\langle\!\langle\phi_i|$ in it, then substitute $\phi = \mathcal{N}^{-1/2}\psi$ and $\phi'=\mathcal{N}^{-1/2}\psi'$ and use $\mathcal{N}\phi_i=\nu_i\phi_i$:

$$\begin{aligned}\langle\Psi|\hat{O}|\Psi'\rangle &= \langle\Phi_1^{(0)}\Phi_2^{(0)}\phi|\mathcal{A}_{12}\hat{O}\mathcal{A}_{12}|\Phi_1^{(0)}\Phi_2^{(0)}\phi'\rangle \\ &= \sum_{ij}\langle\!\langle\phi|\phi_i\rangle\!\rangle\langle\Phi_1^{(0)}\Phi_2^{(0)}\phi_i|\mathcal{A}_{12}\hat{O}\mathcal{A}_{12}|\Phi_1^{(0)}\Phi_2^{(0)}\phi_j\rangle\langle\!\langle\phi_j|\phi'\rangle\!\rangle \\ &= \sum_{ij(\neq 0)}(\nu_i\nu_j)^{-1/2}\langle\!\langle\psi|\phi_i\rangle\!\rangle\langle\Phi_1^{(0)}\Phi_2^{(0)}\phi_i|\mathcal{A}_{12}\hat{O}\mathcal{A}_{12}|\Phi_1^{(0)}\Phi_2^{(0)}\phi_j\rangle\langle\!\langle\phi_j|\psi'\rangle\!\rangle.\end{aligned} \qquad (11.180)$$

The next step is to disentangle $\hat{O}$ from the antisymmetrizers. To carry this out, it is useful to shift one of the antisymmetrizers through

$\hat{O}$ (as $[\mathcal{A}_{12},\hat{O}]=0$) and put another unit operator, this time $\mathbf{1}=\sum_{\kappa\lambda k}|\Phi_1^{(\kappa)}\Phi_2^{(\lambda)}\phi_k\rangle\langle\Phi_1^{(\kappa)}\Phi_2^{(\lambda)}\phi_k|$, between $\mathcal{A}_{12}^2$ and $\hat{O}$. The number of oscillator quanta carried by ϕ_i and ϕ_j, respectively, are to be denoted by Q_i and Q_j. The position of the antisymmetrizer $\mathcal{A}_{12}^2$ should be chosen so as to be adjacent to the eigenfunction that carries fewer excitation quanta, i.e., adjacent to ϕ_i if $Q_i<Q_j$ and to ϕ_j if $Q_i>Q_j$. (For $Q_i=Q_j$, the choice of the side is indifferent.) Let us assume that $Q_i\leq Q_j$. Then, as will be explained below, the following derivation holds:

$$\begin{aligned}
&\langle\Phi_1^{(0)}\Phi_2^{(0)}\phi_i|\mathcal{A}_{12}\hat{O}\mathcal{A}_{12}|\Phi_1^{(0)}\Phi_2^{(0)}\phi_j\rangle=\langle\Phi_1^{(0)}\Phi_2^{(0)}\phi_i|\mathcal{A}_{12}^2\hat{O}|\Phi_1^{(0)}\Phi_2^{(0)}\phi_j\rangle\\
&=\sum_{\kappa\lambda k}\langle\Phi_1^{(0)}\Phi_2^{(0)}\phi_i|\mathcal{A}_{12}^2|\Phi_1^{(\kappa)}\Phi_2^{(\lambda)}\phi_k\rangle\\
&\times\Big(\langle\Phi_1^{(\kappa)}|\hat{O}_1|\Phi_1^{(0)}\rangle\delta_{\lambda 0}\delta_{kj}+\langle\Phi_2^{(\lambda)}|\hat{O}_2|\Phi_2^{(0)}\rangle\delta_{\kappa 0}\delta_{kj}+\langle\phi_k|\hat{O}_{12}|\phi_j\rangle\delta_{\kappa 0}\delta_{\lambda 0}\Big)
\end{aligned}\tag{11.181}$$

$$\begin{aligned}
&=\sum_{\kappa}\langle\Phi_1^{(0)}\Phi_2^{(0)}\phi_i|\mathcal{A}_{12}^2|\Phi_1^{(\kappa)}\Phi_2^{(0)}\phi_j\rangle\langle\Phi_1^{(\kappa)}|\hat{O}_1|\Phi_1^{(0)}\rangle\\
&+\sum_{\lambda}\langle\Phi_1^{(0)}\Phi_2^{(0)}\phi_i|\mathcal{A}_{12}^2|\Phi_1^{(0)}\Phi_2^{(\lambda)}\phi_j\rangle\langle\Phi_2^{(\lambda)}|\hat{O}_2|\Phi_2^{(0)}\rangle\\
&+\langle\Phi_1^{(0)}\Phi_2^{(0)}\phi_i|\mathcal{A}_{12}^2|\Phi_1^{(0)}\Phi_2^{(0)}\phi_i\rangle\langle\phi_i|\hat{O}_{12}|\phi_j\rangle
\end{aligned}\tag{11.182}$$

$$=\nu_i\left[\left(\langle\Phi_1^{(0)}|\hat{O}_1|\Phi_1^{(0)}\rangle+\langle\Phi_2^{(0)}|\hat{O}_2|\Phi_2^{(0)}\rangle\right)\delta_{ij}+\langle\phi_i|\hat{O}_{12}|\phi_j\rangle\right]\tag{11.183}$$

$$=\nu_i\langle\Phi_1^{(0)}\Phi_2^{(0)}\phi_i|\hat{O}|\Phi_1^{(0)}\Phi_2^{(0)}\phi_j\rangle\equiv\nu_i\langle\!\langle\phi_i|\mathcal{O}|\phi_j\rangle\!\rangle.\tag{11.184}$$

As an explanation for this derivation, first it should be noted that any matrix element of $\mathcal{A}_{12}^2$ may only be non-zero if the total number of HO quanta in the bra is equal to that in the ket. That is why the summations reduce to one term in the third term of Eq. (11.182). For the first and second terms of Eq. (11.183), we should use $Q_1^{(0)}\leq Q_1^{(\kappa)}$ and $Q_2^{(0)}\leq Q_2^{(\lambda)}$ in addition, which hold because $\Phi_i^{(0)}$ are the g.s. For the first term to be non-zero, $Q_1^{(0)}\leq Q_1^{(\kappa)}$ and $Q_1^{(0)}+Q_i=Q_1^{(\kappa)}+Q_j$ should hold simultaneously, which implies $Q_i\geq Q_j$, which is consistent with the assumption $Q_i\leq Q_j$ only if $Q_i=Q_j$. For the second term one obtains similarly that it is only non-zero if $Q_i=Q_j$. Once $Q_i=Q_j$ has been ensured, the sums over κ and λ reduce to one term, with $\kappa=0$ and $\lambda=0$, respectively, which causes $\langle\Phi_1^{(0)}\Phi_2^{(0)}\phi_i|\mathcal{A}_{12}^2|\Phi_1^{(0)}\Phi_2^{(0)}\phi_i\rangle=\nu_i$ to factor out from all three terms in Eq. (11.183). Finally, the three terms can be put together in one as in Eq. (11.184).

The last step to arrive at Eq. (11.177) is simply the substitution of Eq. (11.184) into Eq. (11.180) and taking also the case $Q_i>Q_j$ into consideration. Alternatively, putting Eq. (11.183) into Eq. (11.180), one gets an even more explicit expression:

$$\langle\Psi|\hat{O}|\Psi'\rangle = \langle\psi|\psi'\rangle\left(\langle\Phi_1^{(0)}|\hat{O}_1|\Phi_1^{(0)}\rangle + \langle\Phi_2^{(0)}|\hat{O}_2|\Phi_2^{(0)}\rangle\right) + \sum_{ij(\neq 0)}\left(\frac{\nu_<}{\nu_>}\right)^{1/2}\langle\psi|\phi_i\rangle\langle\phi_i|\hat{O}_{12}|\phi_j\rangle\langle\phi_j|\psi'\rangle. \quad (11.185)$$

The last term would reduce to $\langle\psi|\hat{O}_{12}|\psi'\rangle$ if all non-zero eigenvalues ν_i were the same. This shows that the function ψ would completely behave like a macroscopic relative wave function only in this case; otherwise the terms of the matrix element belonging to different values of ν_i, ν_j are weighted with $(\nu_</\nu_>)^{1/2}$. All $\nu_i(\neq 0)$ are never strictly the same, but for a pair of light clusters of not very different sizes this may not be a bad approximation.

11.9.2 The fishbone model

As a most important application, we derive the dynamical equation of motion for the relative motion of two clusters, with the use of Eq. (11.177). The condition of validity of Eq. (11.177) is that the operator O have the property (11.174). While Eq. (11.174) is fulfilled by the kinetic energy exactly, it would only be fulfilled by the internucleon potential if it were linear or quadratic (like an HO potential). Nevertheless, there is reason to believe [357] that Eq. (11.177) provides the optimum macroscopic equation, the best possible equation to replace the OCM. The macroscopic model to be derived is called the fishbone model.

For H to satisfy the conditions of theorem (11.177), let us replace the potential $V_{12}(\xi_1, \xi_2, \boldsymbol{q})$ in Eq. (11.61) by $V_{12}(\xi_1, \xi_2, \boldsymbol{q}) \approx V_{12}(\boldsymbol{q})$. It seems natural to identify the approximate potential with the folding potential (11.67), with which the kernel of the potential $\mathcal{V} \approx \mathcal{V}_D$ will look like $V_{12}(\boldsymbol{R}, \boldsymbol{R}') = V_D(\boldsymbol{R})\delta(\boldsymbol{R} - \boldsymbol{R}')$ [cf. Eq. (11.63)]. The kernel of the kinetic energy $\mathcal{T}$ is $T_{12}(\boldsymbol{R})\delta(\boldsymbol{R} - \boldsymbol{R}')$ [cf. definition below Eq. (11.69)]. Let us now apply the theorem to $\langle\Psi_n|H - E|\Psi\rangle = 0$, where $\Psi = \mathcal{A}_{12}\{\Phi_1^{(0)}\Phi_2^{(0)}\phi(\boldsymbol{q})\}$ is the model wave function and $\Psi_n = \mathcal{A}_{12}\{\Phi_1^{(0)}\Phi_2^{(0)}\phi_n(\boldsymbol{q})\}$, with ϕ_n an eigenfunction of the norm operator:

$$\sum_{ij(\neq 0)}\left(\frac{\nu_<}{\nu_>}\right)^{1/2}\langle\psi_n|\phi_i\rangle\langle\phi_i|\mathcal{T} + \mathcal{V}_D - \varepsilon|\phi_j\rangle\langle\phi_j|\psi\rangle = 0, \quad (11.186)$$

where $\psi_n = \mathcal{N}^{1/2}\phi_n$. The substitution of V_D for V_{12} can be substantiated by the observation that V_D would appear if, instead of the neglect of the ξ_1, ξ_2-dependence of V_{12}, the exchange terms were neglected in the last term within the round brackets in Eq. (11.181).

Multiplying Eq. (11.186) by $|\psi_n\rangle\!\rangle$, adding it up for all n and using $\sum_n |\psi_n\rangle\!\rangle\langle\!\langle\psi_n| = 1$, one obtains

$$\sum_{ij(\neq 0)} \left(\frac{\nu_<}{\nu_>}\right)^{1/2} |\phi_i\rangle\!\rangle\langle\!\langle\phi_i|\mathcal{T} + \mathcal{V}_{\rm D} - \varepsilon|\phi_j\rangle\!\rangle\langle\!\langle\phi_j|\psi\rangle\!\rangle = 0. \qquad (11.187)$$

This is the basic equation of the fishbone model. It is more conventional to add as well as subtract $(\mathcal{T} + \mathcal{V}_{\rm D} - \varepsilon)|\psi\rangle\!\rangle$:

$$(\mathcal{T}+\mathcal{V}_{\rm D}-\varepsilon)|\psi\rangle\!\rangle - \sum_{ij} |\phi_i\rangle\!\rangle\langle\!\langle\phi_i|\mathcal{T}+\mathcal{V}_{\rm D}-\varepsilon|\phi_j\rangle\!\rangle M_{ij}\langle\!\langle\phi_j|\psi\rangle\!\rangle = 0, \quad (11.188)$$

where[27]

$$M_{ij} = \begin{cases} 1 & \text{if at least one of } \nu_i \text{ and } \nu_j \text{ is zero,} \\ 1 - \left(\frac{\nu_<}{\nu_>}\right)^{1/2} & \text{otherwise.} \end{cases} \qquad (11.189)$$

For large values of both i and j, the nonlocal terms go to zero. Neglecting all nonlocal terms but those belonging to Pauli-forbidden states, one obtains an equation which can be reduced to the OCM equation (11.163). The role of the other terms is to correct for the Pauli effects that are neglected in the OCM. This is the most explicit way of displaying Pauli effects, and that is why it was useful to present it.

The fishbone model can be generalized to more than two clusters [358] and is valid for clusters of unequal size parameters or more complicated intrinsic structures in a reasonable approximation [357, 358].

11.9.3 Three-cluster system

To follow up the correspondence between microscopic and macroscopic approaches, let us proceed to the case of three clusters. We shall show that the macroscopic potential between a_1 and a_2 derivable from a macroscopic three-cluster model is not exactly the same as that derivable from an a_1+a_2 two-cluster model, but it may be a good approximation to neglect the three-body effects. Finally we shall, however, show a quantity for which three-body effects may not be negligible.

Let us take a simplest possible three-cluster model, that of a_0+1+2, in which a_0 is a composite cluster, but 1 and 2 are single nucleons. The

[27] Subjected to the transformation $\phi = \mathcal{N}^{-1/2}\psi$, Eq. (11.187) takes a similar form. The transform $\bar{M}$ of the matrix M has a 'fishbone structure', i.e., $\bar{M}_{ij} = \bar{M}_{jj}$ for $i < j$ and $\bar{M}_{ij} = \bar{M}_{ii}$ for $i > j$; hence the name of the model.

expansion corresponding to Eq. (11.40) will be

$$\Psi = \int d\boldsymbol{R}_1 \int d\boldsymbol{R}_2\, \phi(\boldsymbol{R}_1, \boldsymbol{R}_2)\Psi_{\boldsymbol{R}_1,\boldsymbol{R}_2}, \tag{11.190}$$

where

$$\Psi_{\boldsymbol{R}_1,\boldsymbol{R}_2} = \mathcal{A}_{012}\left\{\Phi_0(\xi_0)\Phi_1(\xi_1)\Phi_2(\xi_2)\delta(\boldsymbol{R}_1 - \boldsymbol{q}_{12})\delta(\boldsymbol{R}_2 - \boldsymbol{q}_{0(12)})\right\}. \tag{11.191}$$

Here Φ_1 and Φ_2 are the nucleonic spin–isospin states, $\boldsymbol{q}_{12} = \boldsymbol{r}_1 - \boldsymbol{r}_2$, $\boldsymbol{q}_{0(12)} = \frac{1}{2}(\boldsymbol{r}_1 + \boldsymbol{r}_2) - \bar{\boldsymbol{t}}_0$ [cf. Eq. (11.3)] and $\mathcal{A}_{012}$ is the intercluster antisymmetrizer, $\mathcal{A}_{012} \equiv \mathcal{A}_{\{0\}\{1\}\{2\}}$, as is defined in Eq. (D.7). The other subscript combinations to be used for $\boldsymbol{q}$ (e.g., $\boldsymbol{q}_{01}$) are to be interpreted similarly. The equation corresponding to Eq. (11.42) looks exactly like Eq. (11.42), but the kernels have two pairs of variables:

$$H(\boldsymbol{R}_1, \boldsymbol{R}_2; \boldsymbol{R}_1', \boldsymbol{R}_2') = \langle \Psi_{\boldsymbol{R}_1,\boldsymbol{R}_2} | H | \Psi_{\boldsymbol{R}_1',\boldsymbol{R}_2'} \rangle, \tag{11.192}$$

$$N(\boldsymbol{R}_1, \boldsymbol{R}_2; \boldsymbol{R}_1', \boldsymbol{R}_2') = \langle \Psi_{\boldsymbol{R}_1,\boldsymbol{R}_2} | \Psi_{\boldsymbol{R}_1',\boldsymbol{R}_2'} \rangle. \tag{11.193}$$

One can obtain a three-particle Schrödinger equation of the type of Eq. (11.60) with a Hermitean Hamiltonian similarly to the two-particle case:

$$\tilde{\mathcal{H}}\psi(\boldsymbol{R}_1, \boldsymbol{R}_2) = E\psi(\boldsymbol{R}_1, \boldsymbol{R}_2), \tag{11.194}$$

where $\psi(\boldsymbol{R}_1, \boldsymbol{R}_2) = \mathcal{N}^{1/2}\phi(\boldsymbol{R}_1, \boldsymbol{R}_2)$ and $\tilde{\mathcal{H}} = \mathcal{N}^{-1/2}\mathcal{H}\mathcal{N}^{-1/2}$. It remains to relate the Hamiltonian $\tilde{\mathcal{H}}$ to those of the two-body subsystems. To this end, let us write the microscopic Hamiltonian like Eq. (11.61):

$$\begin{aligned} H = {} & H_0(\xi_0) + T_{12}(\boldsymbol{q}_{12}) + T_{0(12)}(\boldsymbol{q}_{0(12)}) \\ & + V_{01}(\xi_0, \boldsymbol{q}_{01}) + V_{02}(\xi_0, \boldsymbol{q}_{02}) + V_{12}(\boldsymbol{q}_{12}). \end{aligned} \tag{11.195}$$

Substituting Eqs. (11.191) and (11.195) into Eq. (11.192) yields

$$\begin{aligned} & H(\boldsymbol{R}_1, \boldsymbol{R}_2; \boldsymbol{R}_1', \boldsymbol{R}_2') = [E_0 + T_{12}(\boldsymbol{R}_1) + T_{0(12)}(\boldsymbol{R}_2) + V_{12}(\boldsymbol{R}_1)] \\ & \times N(\boldsymbol{R}_1, \boldsymbol{R}_2; \boldsymbol{R}_1', \boldsymbol{R}_2') + V_{01}(\boldsymbol{R}_1, \boldsymbol{R}_2; \boldsymbol{R}_1', \boldsymbol{R}_2') + V_{02}(\boldsymbol{R}_1, \boldsymbol{R}_2; \boldsymbol{R}_1', \boldsymbol{R}_2'), \end{aligned} \tag{11.196}$$

where $E_0 = \langle \Phi_0 | H_0 | \Phi_0 \rangle$ and

$$\begin{aligned} & V_{0i}(\boldsymbol{R}_1, \boldsymbol{R}_2; \boldsymbol{R}_1', \boldsymbol{R}_2') \\ & = \left[\frac{(a_0+2)!}{a_0!1!1!}\right]^{1/2} \langle \Phi_0\Phi_1\Phi_2\delta(\boldsymbol{R}_1 - \boldsymbol{q}_{12})\delta(\boldsymbol{R}_2 - \boldsymbol{q}_{0(12)}) | V_{0i} | \Psi_{\boldsymbol{R}_1',\boldsymbol{R}_2'} \rangle \\ & = \langle \mathcal{A}_{012}\{\Phi_0\Phi_1\Phi_2\delta(\boldsymbol{R}_1 - \boldsymbol{q}_{12})\delta(\boldsymbol{R}_2 - \boldsymbol{q}_{0(12)})V_{0i}\} | \Psi_{\boldsymbol{R}_1',\boldsymbol{R}_2'} \rangle \\ & \qquad (i = 1,\ 2). \end{aligned} \tag{11.197}$$

Thus $\tilde{\mathcal{H}}$ can be expressed as

$$\tilde{\mathcal{H}} = E_0\mathcal{I} + \mathcal{N}^{-1/2}(\mathcal{T}_{12}+\mathcal{T}_{0(12)}+\mathcal{V}_{12})\mathcal{N}^{1/2} + \mathcal{N}^{-1/2}(\mathcal{V}_{01}+\mathcal{V}_{02})\mathcal{N}^{-1/2}, \tag{11.198}$$

which corresponds to Eq. (11.68).

The terms of $\tilde{\mathcal{H}}$ are all three-body operators, and there seems to be no obvious way of reducing them to sums of two-body terms, unless approximations are introduced. To achieve such a reduction, one should neglect the exchanges between 2 and the rest of the system in $V_{01}(\boldsymbol{R}_1, \boldsymbol{R}_2; \boldsymbol{R}_1', \boldsymbol{R}_2')$. A neglect of wave function terms (belonging to some particular exchanges) has to be accompanied by a renormalization, which is automatic in the definition of the antisymmetrizers: $\langle\mathcal{A}_{012}\{\cdots\}|\mathcal{A}_{012}\{\cdots\}\rangle \approx \langle\mathcal{A}_{01}\{\cdots\}|\mathcal{A}_{01}\{\cdots\}\rangle$. Once these exchanges are neglected, one can integrate over ξ_2 and, after a coordinate transformation from $\{\boldsymbol{q}_{12}, \boldsymbol{q}_{0(12)}\}$ to $\{\boldsymbol{q}_{01}, \boldsymbol{q}_{02}\}$, whose Jacobian is unity, one can also integrate over $\boldsymbol{q}_{02}$:

$$\begin{aligned} V_{01}(\boldsymbol{R}_1, \boldsymbol{R}_2; \boldsymbol{R}_1', \boldsymbol{R}_2') \\ &\approx \langle\mathcal{A}_{01}\{\Phi_0\Phi_1\Phi_2\delta(\boldsymbol{R}_1 - \boldsymbol{q}_{12})\delta(\boldsymbol{R}_2 - \boldsymbol{q}_{0(12)})V_{01}\} \\ &\quad\times|\mathcal{A}_{01}\{\Phi_0\Phi_1\Phi_2\delta(\boldsymbol{R}_1' - \boldsymbol{q}_{12})\delta(\boldsymbol{R}_2' - \boldsymbol{q}_{0(12)})\}\rangle \\ &= \delta(\boldsymbol{R}_1 - \boldsymbol{R}_1')\langle\mathcal{A}_{01}\{\Phi_0\Phi_1\delta(\boldsymbol{R}_2 + \tfrac{1}{2}\boldsymbol{R}_1 - \boldsymbol{q}_{01})V_{01}\} \\ &\quad\times|\mathcal{A}_{01}\{\Phi_0\Phi_1\delta(\boldsymbol{R}_2' + \tfrac{1}{2}\boldsymbol{R}_1' - \boldsymbol{q}_{01})\}\rangle \\ &\equiv \delta(\boldsymbol{R}_1 - \boldsymbol{R}_1')V_{01}(\boldsymbol{R}_2 + \tfrac{1}{2}\boldsymbol{R}_1; \boldsymbol{R}_2' + \tfrac{1}{2}\boldsymbol{R}_1'). \end{aligned} \tag{11.199}$$

The kernel $V_{01}(\boldsymbol{R}; \boldsymbol{R}')$ is nothing but Eq. (11.63) specified to the system a_0+1. It is a term of the Hamiltonian kernel of the two-cluster problem. By neglecting certain exchange terms similarly, one can also obtain

$$\begin{aligned} V_{02}(\boldsymbol{R}_1, \boldsymbol{R}_2; \boldsymbol{R}_1', \boldsymbol{R}_2') &\approx \delta(\boldsymbol{R}_1 - \boldsymbol{R}_1')V_{02}(\boldsymbol{R}_2 - \tfrac{1}{2}\boldsymbol{R}_1; \boldsymbol{R}_2' - \tfrac{1}{2}\boldsymbol{R}_1'), \\ N(\boldsymbol{R}_1, \boldsymbol{R}_2; \boldsymbol{R}_1', \boldsymbol{R}_1') &\approx \delta(\boldsymbol{R}_1 - \boldsymbol{R}_1')N_{01}(\boldsymbol{R}_2 + \tfrac{1}{2}\boldsymbol{R}_1; \boldsymbol{R}_2' + \tfrac{1}{2}\boldsymbol{R}_1') \\ &\approx \delta(\boldsymbol{R}_1 - \boldsymbol{R}_1')N_{02}(\boldsymbol{R}_2 - \tfrac{1}{2}\boldsymbol{R}_1; \boldsymbol{R}_2' - \tfrac{1}{2}\boldsymbol{R}_1'). \end{aligned} \tag{11.200}$$

Here the kernels $N_{0i}(\boldsymbol{R}; \boldsymbol{R}')$ $(i = 1,\ 2)$ are two-cluster analogues of $N(\boldsymbol{R}_1, \boldsymbol{R}_2; \boldsymbol{R}_1', \boldsymbol{R}_2')$. Since this decomposition can only be achieved at the expense of neglecting exchange terms, the three-body nature of the microscopically deduced Hamiltonian is obviously seen to be a consequence of the Pauli principle.

Such exchange terms are always very important in microscopic cluster-model calculations, so these approximations seem to be very rough, at least for small values of the arguments. Detailed analyses for several three-cluster systems [358–360] have revealed that the three-body potential term of $\mathcal{H}$ is indeed strong, but that of $\tilde{\mathcal{H}}$ is very weak. More precisely, it was shown that the matrix elements of the residual term $\mathcal{V}_{\rm res.}$ of $\tilde{\mathcal{H}}$, written in the form

$$\tilde{\mathcal{H}} = E_0\mathcal{I}+\mathcal{T}_{12}+\mathcal{T}_{0(12)}+\mathcal{V}_{12}+\mathcal{N}_{01}^{-1/2}\mathcal{V}_{01}\mathcal{N}_{01}^{-1/2}+\mathcal{N}_{02}^{-1/2}\mathcal{V}_{02}\mathcal{N}_{02}^{-1/2}+\mathcal{V}_{\rm res.}, \tag{11.201}$$

is virtually negligible. Here $\mathcal{N}_{01}^{-1/2}\mathcal{V}_{01}\mathcal{N}_{01}^{-1/2}$ and $\mathcal{N}_{02}^{-1/2}\mathcal{V}_{02}\mathcal{N}_{02}^{-1/2}$ are the potentials appearing in the two-body Hamiltonians (11.68), with $\mathcal{N}_{01}$ and $\mathcal{N}_{02}$ being the operators whose kernels are N_{01} and N_{02}. This finding shows that the three-body approach to three-cluster systems is reasonable, but it is also a warning that this approach contains an approximation in addition to those involved in the macroscopic two-cluster model.

The reduction of the three-body term as a result of the transformation from $\{\phi, \mathcal{H}\}$ to $\{\psi, \tilde{\mathcal{H}}\}$ can be made plausible by studying the limiting case of an HO cluster model in which the potential is $V_{\rm HO}(\boldsymbol{q}_{12})+V_{\rm HO}(\boldsymbol{q}_{0(12)})$. Then the Hamiltonians $\mathcal{H}$ and $\tilde{\mathcal{H}}$ are replaced by

$$\begin{aligned}\mathcal{H}_{\rm HO} &= (E_0 + \mathcal{T}_{12} + \mathcal{V}_{12} + \mathcal{T}_{0(12)} + \mathcal{V}_{0(12)})\mathcal{N} \\ &= \mathcal{N}(E_0 + \mathcal{T}_{12} + \mathcal{V}_{12} + \mathcal{T}_{0(12)} + \mathcal{V}_{0(12)}),\end{aligned} \tag{11.202}$$

$$\begin{aligned}\tilde{\mathcal{H}}_{\rm HO} &= (E_0 + \mathcal{T}_{12} + \mathcal{V}_{12} + \mathcal{T}_{0(12)} + \mathcal{V}_{0(12)})\mathcal{I} \\ &= \mathcal{I}(E_0 + \mathcal{T}_{12} + \mathcal{V}_{12} + \mathcal{T}_{0(12)} + \mathcal{V}_{0(12)}).\end{aligned} \tag{11.203}$$

The operator $\mathcal{T}_{12} + \mathcal{V}_{12} + \mathcal{T}_{0(12)} + \mathcal{V}_{0(12)}$ in Eq. (11.202) has a common complete set of eigenstates with $\mathcal{N}$, and that is why it commutes with $\mathcal{N}$ and Eq. (11.203) is so simple. It is clear that $\tilde{\mathcal{H}}_{\rm HO}$ contains just local two-body terms (apart from the projector $\mathcal{I}$), while $\mathcal{H}_{\rm HO}$ contains the large three-body Pauli terms implied by $\mathcal{N}$. As is shown by the numerical tests mentioned earlier [358–360], the realistic models tend to mimic this behaviour.

With this, the correspondence between the microscopic and macroscopic models seems to be settled. Within the limits of the validity of the microscopic cluster model (with no distortion), it is justifiable to replace the microscopic model by a corresponding OCM-type macroscopic model. The internal structure of the cluster(s) affects the dynamics of the problem, in addition, through the rescaling of the components of the wave function by $(\nu_</\nu_>)^{1/2}$. We will now show, however, that the cluster substructure does show up in another way as well when the system is split up into clusters.

The substructure enters into the disintegration process through the amplitude defined in Eq. (11.45). For the fragmentation of a_0+1+2 into a_0 plus a bound state,

$$\Phi_{12} = \mathcal{A}_{12}\{\Phi_1\Phi_2\phi_{12}\}, \tag{11.204}$$

of 1+2, this amplitude reads

$$g_{0(12)}(\boldsymbol{R}) = \langle \mathcal{A}_{0(12)}\{\Phi_0\Phi_{12}\delta(\boldsymbol{R}-\boldsymbol{q}_{0(12)})\}|\Psi\rangle. \tag{11.205}$$

Here the internucleon antisymmetrizer $\mathcal{A}_{12}$ plays the role of an intracluster antisymmetrizer, and $\mathcal{A}_{0(12)} \equiv \mathcal{A}_{\{0\}\{12\}}$ is the intercluster antisymmetrizer between a_0 and (12), such that $\mathcal{A}_{012} = \mathcal{A}_{0(12)}\mathcal{A}_{12}$. In a macroscopic model, whose normalized wave function is $\psi(\boldsymbol{R}_1, \boldsymbol{R}_2)$, one would naively replace Eq. (11.205) by

$$g_{0(12)}(\boldsymbol{R}) = \int d\boldsymbol{R}_1\, \phi_{12}^*(\boldsymbol{R}_1)\psi(\boldsymbol{R}_1, \boldsymbol{R}_2). \tag{11.206}$$

But, in fact, substitution of Eqs. (11.204), (11.190) and (11.191) into Eq. (11.205) results in

$$\begin{aligned} g_{0(12)}(\boldsymbol{R}) &= \int d\boldsymbol{R}_1 \int d\boldsymbol{R}_1' \int d\boldsymbol{R}_2'\, \phi_{12}^*(\boldsymbol{R}_1) \\ &\quad \times \langle \mathcal{A}_{0(12)}\mathcal{A}_{12}\{\Phi_0\Phi_1\Phi_2\delta(\boldsymbol{R}_1-\boldsymbol{q}_{12})\delta(\boldsymbol{R}-\boldsymbol{q}_{0(12)})\} \\ &\quad \times |\mathcal{A}_{012}\{\Phi_0\Phi_1\Phi_2\delta(\boldsymbol{R}_1'-\boldsymbol{q}_{12})\delta(\boldsymbol{R}_2'-\boldsymbol{q}_{0(12)})\}\rangle\phi(\boldsymbol{R}_1', \boldsymbol{R}_2') \\ &= \int d\boldsymbol{R}_1\, \phi_{12}^*(\boldsymbol{R}_1)\mathcal{N}\phi(\boldsymbol{R}_1, \boldsymbol{R}) \\ &= \int d\boldsymbol{R}_1\, \phi_{12}^*(\boldsymbol{R}_1)\mathcal{N}^{1/2}\psi(\boldsymbol{R}_1, \boldsymbol{R}). \end{aligned} \tag{11.207}$$

In the second step $\mathcal{A}_{0(12)}\mathcal{A}_{12} = \mathcal{A}$ and Eq. (11.193) have been used, and in the third step $\mathcal{N}^{-1/2}\psi$ has been substituted for ϕ. Thus Eq. (11.206) is not correct.

As another spectroscopic amplitude, $g_{(02)1}$ is examined. The wave functions involved are

$$\Psi_{02} = \mathcal{A}_{02}\{\Phi_0(\xi_0)\Phi_2(\xi_2)\phi_{02}(\boldsymbol{q}_{02})\}, \tag{11.208}$$

$$\Psi = \mathcal{A}_{012}\{\Phi_0(\xi_0)\Phi_1(\xi_1)\Phi_2(\xi_2)\phi'(\boldsymbol{q}_{02}, \boldsymbol{q}_{(02)1})\}. \tag{11.209}$$

Here $\phi'(\boldsymbol{q}_{02}, \boldsymbol{q}_{(02)1}) \equiv \phi(\boldsymbol{q}_{12}, \boldsymbol{q}_{0(12)})$, so that this Ψ is identical to that given in Eqs. (11.190) and (11.191). It is convenient to introduce the kernel

$$\begin{aligned} N'(\boldsymbol{R}_1, \boldsymbol{R}_2; \boldsymbol{R}_1', \boldsymbol{R}_2') &= \langle \mathcal{A}_{012}\{\Phi_0\Phi_1\Phi_2\delta(\boldsymbol{R}_1-\boldsymbol{q}_{02})\delta(\boldsymbol{R}_2-\boldsymbol{q}_{(02)1})\} \\ &\quad \times |\mathcal{A}_{012}\{\Phi_0\Phi_1\Phi_2\delta(\boldsymbol{R}_1'-\boldsymbol{q}_{02})\delta(\boldsymbol{R}_2'-\boldsymbol{q}_{(02)1})\}\rangle, \end{aligned} \tag{11.210}$$

which defines an operator $\mathcal{N}'$. The spectroscopic amplitude is then calculated following the pattern of Eq. (11.207):

$$g_{(02)1}(\boldsymbol{R}) = \int d\boldsymbol{R}_1 \int d\boldsymbol{R}_2' \int d\boldsymbol{R}'\, \phi_{02}^*(\boldsymbol{R}_1)N'(\boldsymbol{R}_1, \boldsymbol{R}; \boldsymbol{R}_1', \boldsymbol{R}_2')\phi'(\boldsymbol{R}_1', \boldsymbol{R}_2')$$

$$= \int d\boldsymbol{R}_1\, \phi_{02}^*(\boldsymbol{R}_1) \mathcal{N}' \phi'(\boldsymbol{R}_1, \boldsymbol{R})$$

$$= \int d\boldsymbol{R}_1\, \psi_{02}^*(\boldsymbol{R}_1) \mathcal{N}_{02}^{-1/2} \mathcal{N}'^{1/2} \psi'(\boldsymbol{R}_1, \boldsymbol{R}), \tag{11.211}$$

where $\psi_{02} \equiv \mathcal{N}_{02}^{1/2} \phi_{02}$ and $\psi' \equiv \mathcal{N}'^{1/2} \phi'$. Note that the transformation $\{q_{12}, q_{0(12)}\} \rightarrow \{q_{02}, q_{(02)1}\}$ applied to the eigenvalue equation of $\mathcal{N}$ shows that the eigenvalues of $\mathcal{N}'$ are identical to those of $\mathcal{N}$ and the eigenfunctions of $\mathcal{N}'$ are related to those of $\mathcal{N}$ via a Talmi–Moshinsky transformation [210].

For the ^{6}Li$\rightarrow\alpha$+d spectroscopic factor the inclusion of $\mathcal{N}^{1/2}$ in Eq. (11.207) results in an increase by a factor of 35% [361]. This is a consequence of the facts that the first non-zero eigenvalue ν, that which departs from unity most, is $\frac{13}{8}$, and the corresponding eigenfunction has a great overlap with ψ. The increase of the ^{5}He+p spectroscopic factor is less spectacular since the effect of $\mathcal{N}'^{1/2}$ is partly compensated for by $\mathcal{N}_{02}^{-1/2}$. (The 'leading' eigenvalue of $\mathcal{N}_{02}$ is $\frac{5}{4}$.) The overall conclusion is that the three-cluster Pauli effects in the spectroscopic amplitudes are difficult to assess, and the safest policy is to use the correct formulae (11.207) and (11.211), no matter whether we work in a microscopic or in a macroscopic model.

All in all, it seems justifiable to reduce the Schrödinger equation of a microscopic problem to a macroscopic Schrödinger equation with two-body forces provided the Pauli projection is done correctly. But the signature of cluster substructure is bound to reappear in the spectroscopic amplitudes. That can also be treated in the macroscopic framework with the use of norm operators borrowed from the microscopic approach.

Chapter 12

Cluster model in the correlated Gaussian approach

12.1 Multicluster approximation

In Chap. 11 the conventional formulations of cluster models have been introduced, and the physical aspects of clustering have been reviewed. In Sect. 11.5 we saw that in the practicable forms of the microscopic cluster models the cluster intrinsic motion is described by the HOSM. The versions of the microscopic cluster models are specific mainly in the way in which they represent the intercluster relative motion. We reviewed the shifted-Gaussian representation, called the GCM, and mentioned the HO representation, called the algebraic version of the RGM. In this chapter the correlated Gaussian representation of the relative motion will be discussed, which is called the *correlated Gaussian cluster model*. It will be derived as an approximation to the correlated Gaussian approach to the multinucleon problem introduced in Chap. 9. This derivation is the subject of the present section. In Sect. 12.2 the physical aspects and practical details of the correlated Gaussian cluster model will be presented. Some more general issues concerning clustering and its description will be analysed with the tool of correlated Gaussians in Sect. 12.3.

In a cluster model the treatment of cluster intrinsic motions is simpler than that of cluster relative motions. At the beginning of Chap. 11 we have indicated how to reduce the correlated Gaussian formulation of the few-nucleon problem to a cluster model. One should omit some of the spin–isospin configurations (e.g., spin 1 for like nucleon pairs) and, for the relative motions to be reduced to cluster intrinsic motions, discard some orbital angular momenta and truncate the Gaussian expansion.

In order to localize the parts of the wave function that should be truncated, one should cast the nuclear wave function into the form

$$\Psi = \sum_{i_1 \cdots i_n} C_{i_1 \cdots i_n} \mathcal{A}_{1 \cdots n} \Big\{ \Phi_1^{(i_1)}(\xi_1) \cdots \Phi_n^{(i_n)}(\xi_n) \phi_{i_1 \cdots i_n}(\rho_1, \ldots, \rho_{n-1}) \Big\}, \tag{12.1}$$

where $\{\Phi_j^{(i_j)}(\xi_j)\}$ are complete sets of internally antisymmetrical intrinsic states $(i_j = 0, 1, \ldots)$ of the n subsystems $(j = 1, \ldots, n)$, the vectors $\{\rho_1, \ldots, \rho_{n-1}\}$ are Jacobi coordinates [cf. Eq. (11.5)] connecting the subsystems. Equation (12.1) consists of terms similar to the first term of Eq. (11.6). Any square-integrable function Ψ can always be expanded in this way, but the expansion may not be unambiguous.[1]

The expansion Eq. (9.3) in terms of the correlated Gaussians (9.41) takes the form (12.1) if some arrangements are discarded. The arrangements to be retained are those which conform to the cluster pattern to be described. For example, if the α-particle is to be described as d+d, only the H-shaped arrangement should be included, while a t+p model only requires the K-shaped arrangement. Since a basis containing a single arrangement may be complete, this is not an approximation yet.

Then Eq. (12.1) should be suitably truncated by introducing a model for each $\Phi_j^{(i_j)}(\xi_j)$. With this truncation, the subsystems are treated as clusters or, in other words, a cluster-model ansatz has been introduced. The models for the clusters are usually chosen so as to correspond with the isolated nuclei.

As was seen in Sect. 11.5, in the usual cluster models the cluster intrinsic wave functions are described by the translation-invariant HOSM. The most important property of this model is that, to any translation-invariant SM state $\Phi_j^{(i_j)}(\xi_j)$ of cluster j there corresponds a translation-noninvariant counterpart $\Phi_{js_j}^{\mathrm{SM}(i_j)}$, which is fixed to a centre

---

[1]The functions $\{\Phi_1^{(i_1)}(\xi_1) \cdots \Phi_n^{(i_n)}(\xi_n) \phi_{i_1 \cdots i_n}(\rho_1, \ldots, \rho_{n-1})\}$ can always be chosen so as to form a complete set provided all belong to the same clusterization. Then the expansion

$$\Psi = \sum_{i_1 \cdots i_n} C_{i_1 \cdots i_n} \Phi_1^{(i_1)}(\xi_1) \cdots \Phi_n^{(i_n)}(\xi_n) \phi_{i_1 \cdots i_n}(\rho_1, \ldots, \rho_{n-1})$$

is unambigous. By letting $\mathcal{A}_{1 \cdots n}$ act on both sides, the expansion remains valid, but the function set may become linearly dependent, which renders the expansion ambiguous. Further 'overcompleteness' and ambiguity is introduced by including terms belonging to different clusterizations.

with (normalized) cluster generator coordinate $\boldsymbol{s}_j$ [cf. Eq. (11.79)]:

$$\Phi^{\mathrm{SM}(i_j)}_{j\boldsymbol{s}_j}(\boldsymbol{r}_{a_1+\cdots+a_{j-1}+1},\ldots,\boldsymbol{r}_{a_1+\cdots+a_j}) = \Phi^{(i_j)}_j(\xi_j)\varphi_{\boldsymbol{s}_j}(\bar{\boldsymbol{\xi}}_j). \qquad (12.2)$$

If $\Phi^{(i_j)}_j$ is a lowest-lying configuration, $\Phi^{\mathrm{SM}(i_j)}_{j\boldsymbol{s}_j}$ is a Slater determinant.

The cluster ansatz should be introduced so as to maintain the analytic scheme of the correlated Gaussian formalism. This is ensured if it is introduced into the generating formula (9.84) of the correlated Gaussian state $\psi_{LM_L;SM_S;TM_T}$. In this, $\psi_{LM_L;SM_S;TM_T}$ is expressed in terms of a state $\Psi_{SM_STM_T}(\boldsymbol{S}_1,\ldots,\boldsymbol{S}_{A-1})$, where $\boldsymbol{S}_1,\ldots,\boldsymbol{S}_{A-1}$ are independent Jacobi generator coordinates. The state $\Psi_{SM_STM_T}(\boldsymbol{S}_1,\ldots,\boldsymbol{S}_{A-1})$ is reminiscent of a multicentre GCM basis state (11.146). We shall immediately see that the only difference is that in a GCM state more than one nucleon may sit at each centre. From $\Psi_{SM_STM_T}(\boldsymbol{S}_1,\ldots,\boldsymbol{S}_{A-1})$ one can obtain a similar function by letting some $\boldsymbol{s}_i$ coincide in Eq. (9.68). Such a state will generate a correlated Gaussian cluster-model state.

If the generator coordinates $\boldsymbol{s}_1,\ldots,\boldsymbol{s}_A$ are set equal to each other for groups of nucleons such that

$$\begin{aligned}
&\boldsymbol{s}_1=\cdots=\boldsymbol{s}_{a_1}\equiv\underline{\boldsymbol{s}}_{[1]}\equiv a_1^{-1/2}\boldsymbol{s}_{[1]}, \quad \boldsymbol{s}_{a_1+1}=\cdots=\boldsymbol{s}_{a_1+a_2}\equiv\underline{\boldsymbol{s}}_{[2]}\equiv a_2^{-1/2}\boldsymbol{s}_{[2]},\\
&\ldots,\boldsymbol{s}_{A-a_n+1}=\cdots=\boldsymbol{s}_A\equiv\underline{\boldsymbol{s}}_{[n]}\equiv a_n^{-1/2}\boldsymbol{s}_{[n]}, \qquad (12.3)
\end{aligned}$$

the generator-coordinate state (9.68) reduces to that describing n clusters. The vectors $\underline{\boldsymbol{s}}_{[i]}$ and $\boldsymbol{s}_{[i]}$ are the conventional and normalized cluster generator coordinates, respectively. The spin–isospin states are assumed to be coupled to conform to this cluster arrangement. From the cluster-model GCM wave function thus obtained,

$$\begin{aligned}
&\widehat{\Phi}_{SM_STM_T}(\boldsymbol{s}_{[1]},\ldots,\boldsymbol{s}_{[n]})\\
&\quad=\Phi_{SM_STM_T}(\underline{\boldsymbol{s}}_{[1]},\ldots,\underline{\boldsymbol{s}}_{[1]},\underline{\boldsymbol{s}}_{[2]},\ldots,\underline{\boldsymbol{s}}_{[2]},\ldots,\underline{\boldsymbol{s}}_{[n]},\ldots,\underline{\boldsymbol{s}}_{[n]}), \qquad (12.4)
\end{aligned}$$

the same c.m. wave function factors out as in Eq. (9.72):

$$\begin{aligned}
&\widehat{\Phi}_{SM_STM_T}(\boldsymbol{s}_{[1]},\ldots,\boldsymbol{s}_{[n]})\\
&=\mathcal{A}\Big\{\varphi_{\underline{\boldsymbol{s}}_{[1]}}(\boldsymbol{r}_1)\ldots\varphi_{\underline{\boldsymbol{s}}_{[1]}}(\boldsymbol{r}_{a_1})\ldots\varphi_{\underline{\boldsymbol{s}}_{[n]}}(\boldsymbol{r}_{A-a_n+1})\ldots\varphi_{\underline{\boldsymbol{s}}_{[n]}}(\boldsymbol{r}_A)\chi_{SM_S}\eta_{TM_T}\Big\}\\
&=\mathcal{A}_{1\ldots n}\Big\{\Big[\mathcal{A}_1\left\{\varphi_{\underline{\boldsymbol{s}}_{[1]}}(\boldsymbol{r}_1)\cdots\varphi_{\underline{\boldsymbol{s}}_{[1]}}(\boldsymbol{r}_{a_1})\chi_{S_1}\eta_{T_1}\right\}\cdots\\
&\qquad\times\mathcal{A}_n\left\{\varphi_{\underline{\boldsymbol{s}}_{[n]}}(\boldsymbol{r}_{A-a_n+1})\cdots\varphi_{\underline{\boldsymbol{s}}_{[n]}}(\boldsymbol{r}_A)\chi_{S_n}\eta_{T_n}\right\}\Big]_{SM_STM_T}\Big\}\\
&=\mathcal{A}_{1\ldots n}\Big\{\Big[\Phi^{(0)}_1(\xi_1)\varphi_{\boldsymbol{s}_{[1]}}(\bar{\boldsymbol{\xi}}_1)\cdots\Phi^{(0)}_n(\xi_n)\varphi_{\boldsymbol{s}_{[n]}}(\bar{\boldsymbol{\xi}}_n)\Big]_{SM_STM_T}\Big\}\\
&=\widehat{\Psi}_{SM_STM_T}(\boldsymbol{S}_{[1]},\ldots,\boldsymbol{S}_{[n-1]})\varphi_{\boldsymbol{S}_A}(\boldsymbol{x}_A). \qquad (12.5)
\end{aligned}$$

Here $\chi_{S_i M_{S_i}}$, $\eta_{T_i M_{T_i}}$ are cluster spin and isospin states and $[\cdots]_{SM_STM_T}$ couples them. In the second step the full antisymmetrizer is decomposed into a product of subsystem antisymmetrizers and intercluster antisymmetrizers [cf. Eq. (D.6) and the footnote on p. 504]. The functions $\Phi_i^{(0)}(\xi_i) \equiv \Phi_i(\xi_i)$ are the intrinsic HO g.s. of clusters i, while $\varphi_{s_{[j]}}(\bar{\boldsymbol{\xi}}_j)$ are the same as those which appeared in Eq. (12.2). The vectors $\boldsymbol{S}_{[1]}, ..., \boldsymbol{S}_{[n-1]}, \boldsymbol{S}_{[n]}(= \boldsymbol{S}_A)$ are intercluster generator coordinates analogous to the physical coordinates $\boldsymbol{\rho}_1, ..., \boldsymbol{\rho}_{n-1}, \boldsymbol{\rho}_n(= \boldsymbol{x}_A)$, and

$$\widehat{\Psi}_{SM_STM_T}(\boldsymbol{S}_{[1]}, ..., \boldsymbol{S}_{[n-1]}) \\ = \mathcal{A}_{1\cdots n}\left\{\left[\Phi_1(\xi_1)\cdots\Phi_n(\xi_n)\varphi_{\boldsymbol{S}_{[1]}}(\boldsymbol{\rho}_1)\cdots\varphi_{\boldsymbol{S}_{[n-1]}}(\boldsymbol{\rho}_{n-1})\right]_{SM_STM_T}\right\}. \tag{12.6}$$

This function is indeed very similar to the multicentre GCM basis function (11.146).

This completes the proof that by forcing some $\boldsymbol{s}_i$ to coincide in Eq. (9.68), a generator-coordinate state of Gaussian wave packets reduces to a cluster-model state. The product $\prod_{i=1}^{n-1}\varphi_{\boldsymbol{S}_{[i]}}(\boldsymbol{\rho}_i)$ in Eq. (12.6) is completely analogous to $\prod_{i=1}^{A-1}\varphi_{\boldsymbol{S}_i}(\boldsymbol{x}_i)$ in Eq. (9.74). Therefore, Eq. (12.6) can be used in a generating function just as well as Eq. (9.73); but it will generate a correlated Gaussian with an $(n-1)\times(n-1)$ dimensional, rather than an $(A-1)\times(A-1)$-dimensional, quadratic form in the exponent. When $\widehat{\Psi}_{SM_STM_T}(\boldsymbol{S}_{[1]}, ..., \boldsymbol{S}_{[n-1]})$ is substituted into Eq. (9.84), the cluster intrinsic wave functions will not be affected by the transformation. Thus in this way the transformation (9.84) yields a cluster-model basis function, in which the cluster intrinsic motion is described in the standard way, but the intercluster relative motion is represented by a correlated Gaussian function.

The reduction of the multinucleon problem to a cluster model is depicted pictorially in Fig. 12.1. This shows the reduction of the six arrangements shown in Fig. 9.2 to two when the six-nucleon problem is approximated by an α+N+N three-cluster model. The reduction consists in (i) omitting two of the six arrangements, (ii) reducing the number of α-configurations to one, with the same HO quantum attached to each intrinsic coordinate. The latter step makes the four arrangements pairwise indistinguishable. These ideas are elaborated in App. H.

One should notice two restrictive elements in this formalism.

First, the cluster HO parameters should be identical. This looks necessary for the c.m. wave function to factor out [cf. Eq. (9.72)]. However, as was shown in Eqs. (11.88)–(11.91), it is straightforward to overcome this restriction: if not factored out, the c.m. motion may be eliminated by integration over $\boldsymbol{S}_A$, i.e., by projecting the state to c.m. momentum

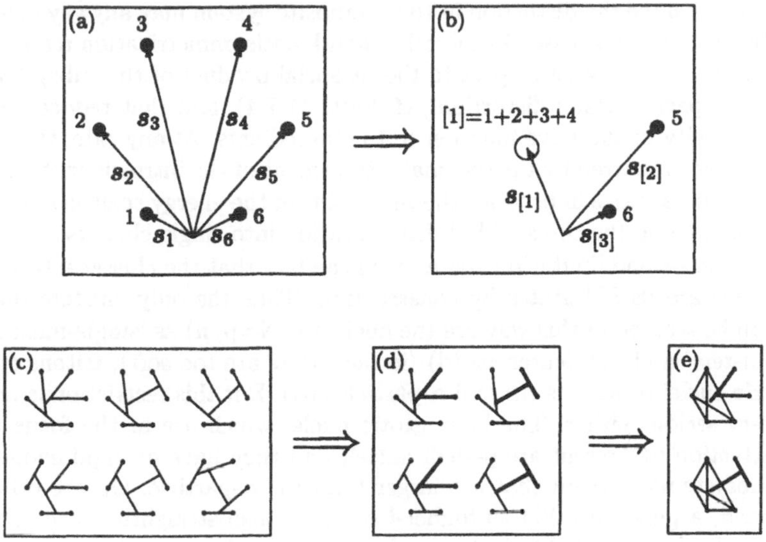

Figure 12.1: Pictorial reduction of the six-nucleon problem to an α+N+N cluster model in terms of generator coordinates (first row) and dynamical coordinates (second row). Panel (a): nucleonic generator coordinates $s_1, \ldots, s_6$ as in Eq. (9.67). Panel (b): cluster generator coordinates $s_{[1]}, s_{[2]}, s_{[3]}$ as in Eq. (12.4). Panel (c): the six arrangements shown also in Fig. 9.2. Panel (d): these are reduced to four in which the intrinsic coordinates of the α-cluster appear. Thin lines: intracluster coordinates, thick lines intercluster coordinates. Panel (e): the two arrangements in each column of panel (d) coincide when the HO quanta for the intrinsic motion of the α-cluster are the same, and the wave function can be expressed in terms of the Gaussians of the internucleonic relative distances (cf. footnote on p. 205).

zero [cf. also Eq. (9.107)]. On the other hand, the few-body Gaussian formalism itself is exempt from such a restriction, thus, if necessary, the above derivation can be repeated so as to avoid it. One should just start from Eq. (9.41), and the parameters of A_i should be allowed to describe clusters of different sizes. Thus the size parameter constraint is not an actual limitation.

But even the present form is satisfactory; although it is not a good approximation for the α-particle and the triton as independent nuclei, it may be quite acceptable in cluster models, where the clusters are immersed in each other. This statement may be valid owing to the cluster distortions discussed in Sect. 11.7.4. On the one hand, it may be a futile effort to set the cluster sizes very precisely to those of the separated systems since the distortion in the composite system may anyway modify the optimum sizes. On the other hand, antisymmetrization restricts the state space with respect to the tensorial product of the subsystem state spaces ('Pauli distortion', cf. Sect. 11.7.4), and that reduces the sensitivity of the wave function to the ingredients. At any rate, the inaccuracy incurred by the schematic treatment of the intrinsic motion is not critical since in practice the zero point of the energy scale is mostly chosen to be the threshold of disintegration into single clusters.

The second limitation comes from the fact that the cluster intrinsic states are 0s SM states by construction. Thus the only clusters that can be treated in this way are the nucleons (N=p, n) as 'single-nucleon clusters' and the deuterons (d) (in fact, they are too soft), tritons (t), ^{3}He nuclei or helions (h) and α-particles (α). But this restriction is not very serious either. The light exotic nuclei which are in the focus of attention at present are p-shell nuclei, and they have no rigid enough subsystems in them that are larger than the α-particle. On the other hand, a generalization to higher-l orbits is also straightforward. The standard recipe of the cluster model would use the generating-function property of the shifted Gaussians presented in Sect. 9.3.1. As is shown by Eq. (9.51), a p-wave orbit can be easily produced as a derivative of a slightly changed shifted Gaussian with respect to the coordinate of its centre. We get a p-orbit by putting the p-wave particles to different centres $\boldsymbol{s}$, projecting to $l = 1$ and differentiating with respect to $\boldsymbol{s}$. Then one can put this on top of the s-wave orbits by letting $\boldsymbol{s}$ coincide with the generator coordinate belonging to the s-wave particles. In practice, these operations have to be performed directly on the matrix elements, which can be done analytically. But the correlated Gaussian approach offers a more straightforward alternative: rather than confining more than four nucleons to one centre, one should treat some relative motions explicitly as Gaussian intercluster orbits.

To sum up the result of this section, we can say that a multicluster scheme can be incorporated into the correlated Gaussian approach in a straightforward way. A correlated Gaussian cluster-model state with s-wave clusters can be generated by Slater determinants of Gaussian wave packets by equating the vectors pointing to the centres of some of the wave packets. Since all matrix elements are to be evaluated via the generating functions, this is enough for the implementation of the multicluster approximation.

12.2 Model space and interactions

Section 12.1 was concerned with the general aspects of incorporating cluster ansätze in the Gaussian formalism. Now we shall introduce the physical ingredients of the model as applied to light exotic nuclei.

The model is composed of cluster-model subspaces, and it will help to understand its physical content in every particular case if one can decompose its wave function into shell-model subspaces as well. The method of this is presented in Sect. 12.2.1. The light nuclei will be surveyed according to their clustering properties in Sect. 12.2.2. Section 12.2.3 will expose the problem of the interaction to be used in the microscopic descriptions of light nuclei and gives a preview of the performance of the Minnesota interaction used in the examples to be presented later.

12.2.1 Characteristics of the state space

A multicluster wave function has the form of Eq. (12.1). As an exact expansion, it is preferable to take all terms to represent the samc clusterization, but, as soon as it is interpreted as a model ansatz, it may be useful to admit terms belonging to different clusterizations.[2] Thus the function space is in general spanned by a direct sum over a number of clusterizations. The subspace for each clusterization is spanned by $\{\Psi_{\boldsymbol{R}_1,\ldots,\boldsymbol{R}_{n-1}}\}$ defined in Eq. (11.41) or (11.31). The weight of each cluster component is characterized by the particular amount of clustering introduced in Sect. 11.3. It is equal to the expectation value $\mathfrak{P} = \langle\Psi|\mathcal{P}|\Psi\rangle$ [Eq. (11.24) or (11.32)], where $\mathcal{P}$ is the projector onto that subspace, Eq. (11.22). In practice, it is useful to analyse the subspaces further according to angular momenta.

The amount of clustering can be calculated for any nuclear state Φ given in microscopic terms, e.g., for a SM state. Conversely, one may ask what is the SM content in a cluster-model state (or in any state given in microscopic terms). For that, similarly, one needs to define a SM space and construct a projector onto that particular subspace of the full state space. From the SM point of view the state space is naturally divided into subspaces of definite numbers of quanta $\hbar\omega$ belonging to an HO Hamiltonian H_{HO}. The composition of a state is therefore to be analysed through the weight of the $Q\hbar\omega$ subspace for each Q. For that purpose one should define a projector that projects onto the $Q\hbar\omega$ subspace. A projector for Q quanta carried by protons, neutrons or

[2]This is in accord with the presentation of the correlated Gaussian formalism itself. The trial function was first written in Eq. (9.19) as an (exact) expansion in one particular arrangement, but then, in Eq. (9.25), it was reinterpreted as an approximate form, with more arrangements included. Note, however, that in the basis in Eq. (9.19) A_k is a diagonal matrix, while $\phi_{i_1\cdots i_n}(\boldsymbol{\rho}_1,\ldots,\boldsymbol{\rho}_{n-1})$ describing the relative motion in Eq. (12.1) may be an expansion over a fully correlated basis.

nucleons may be written in the form [362]

$$\mathcal{P}_Q = \frac{1}{2\pi}\int_0^{2\pi} d\theta \exp\left\{ i\theta \left(\sum_{i=1}^{A} P_i[(\hbar\omega)^{-1} H_{\rm HO}(i) - \tfrac{3}{2}] - Q \right) \right\}, \quad (12.7)$$

where P_i is one of $P_i^{\rm p}$, $P_i^{\rm n}$ and $P_i^{\rm N}$ defined in Eq. (9.122). To understand why this $\mathcal{P}_Q$ is a projector to the $Q\hbar\omega$ subspace, let it act on $\Phi = \sum_q c_q \Phi_q$, where Φ_q is an eigenfunction of $\sum_i P_i[(\hbar\omega)^{-1} H_{\rm HO}(i) - \frac{3}{2}]$ belonging to eigenvalue q:

$$\exp\left\{ i\theta \left(\sum_{i=1}^{A} P_i[(\hbar\omega)^{-1} H_{\rm HO}(i) - \tfrac{3}{2}] - Q \right) \right\} \Phi = \sum_q \exp[i\theta(q-Q)] c_q \Phi_q. \quad (12.8)$$

Then the integral vanishes unless $q = Q$ because an integral over a multiple of the period of an exponential function of imaginary argument vanishes. When $q = Q$, then $\mathcal{P}_Q \sum_q c_q \Phi_q = c_Q \Phi_Q$, which is indeed the projection required. It is to be noted that the operator (12.7) counts the excitation quanta belonging to the c.m. excitation as well, but in the correlated Gaussian formalism, just as in the usual microscopic cluster models, no such excitations appear. With Eq. (12.7), the amount of Q-quantum component in Φ is

$$\mathfrak{P}_Q = \langle \Phi | \mathcal{P}_Q | \Phi \rangle. \quad (12.9)$$

12.2.2 Clustering in light nuclei

As is revealed by phenomenological probing and by theoretical structure calculations, clustering is substantial in light nuclei. It is found that clustering into fragments of strong internal binding is especially strong. What should be strong in fact is the internal binding *with respect to the intercluster binding* (i.e., the binding energy with respect to the energy of separation into clusters). So the condition for prominent clustering can as well be formulated by saying that the intercluster separation energy should be small. In other words, strong clustering into a particular set of clusters can be expected in states that lie close below the threshold of disintegration into the particular clusters. Correspondingly, the dominant clustering in any state is that whose threshold lies closest above the energy of the state itself. These conditions of clustering are often referred to as the threshold rule [296] .

The best examples for clustered nuclei are ^{8}Be=α+α, ^{7}Li=α+t, ^{7}Be=α+h, ^{9}Be=α+α+n, ^{6}Li=α+p+n, ^{12}C=α+α+α. The deuteron only behaves like a fairly (but not completely) well-behaved cluster when no other open-shell clusters are present, like in ^{6}Li. The triton and the helion are decent clusters provided no protons and neutrons, respectively, are available with which to form an α-cluster. The α-cluster is expected to be the most perfect cluster owing to the large energy needed to break it up. This shows that the prominent clustering in light nuclei is mainly caused by the saturation of the 0s shell.

When applied to a broader range of light nuclei, these rules predict the cluster structures shown in Table 12.1 and in Fig. 12.2.

The properties of these light nuclei fluctuate between broad limits, and this fluctuation seems to be governed by the fluctuation of the energies of disintegration into their building blocks, the clusters. Therefore, this staggering can be attributed to the stepwise progression in the formation of clusters from one nucleus to the next. If so, then this tendency is most likely to be accounted for just by the cluster model with building blocks α, t, h, n, p.

For a multicluster system with single-nucleon 'clusters' most characteristic is the behaviour of the single nucleons, but the other details of the cluster structure may also have some observable effect. For example, in the behaviour of ^{11}Li the effect of the two halo-neutrons is of paramount importance, and the α+t system is likely to be more compact than in ^{7}Li. This implies stronger Pauli distortion, which is taken

Table 12.1: Clustering in the ground states of light nuclear systems. The nuclei marked by * are unbound.

Nucleus	Clustering
^{5}He*, ^{5}Li*	α+n, α+p
^{6}He, ^{6}Be*	α+n+n, α+p+p
^{6}Li	α+n+p
^{7}He*	α+n+n+n
^{7}Li, ^{7}Be	α+t, α+h
^{8}He	α+n+n+n+n
^{8}Li, ^{8}B	α+t+n, α+h+p
^{8}Be*	α+α
^{9}Li, ^{9}C	α+t+n+n, α+h+p+p
^{9}Be, ^{9}B*	α+α+n, α+α+p
^{10}Li*	α+t+n+n+n
^{10}Be, ^{10}C	α+α+n+n, α+α+p+p
^{10}B	α+α+n+p
^{11}Li, ^{11}O*	α+t+n+n+n+n, α+h+p+p+p+p
^{11}Be, ^{11}N	α+α+n+n+n, α+α+p+p+p
^{11}B, ^{11}C	α+α+t, α+α+h
^{12}C	α+α+α

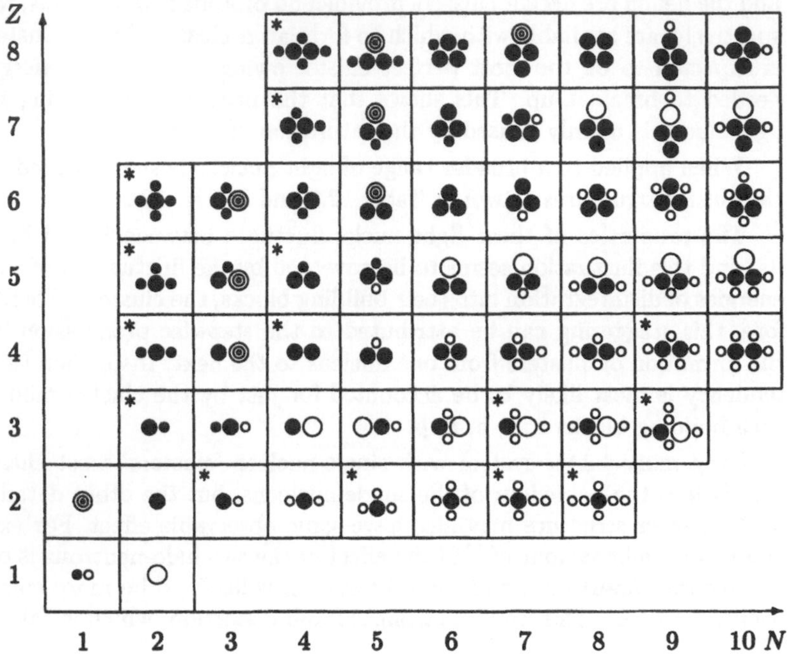

Figure 12.2: Lowest section of the nuclear chart with the nuclides symbolized by their composition out of clusters α (●), t (○), h (◎), p (•) and n (∘). The nuclides that are unbound with respect to nucleon or cluster emission are marked by ∗. (Cf. Fig. 1 on p. 2.)

into account by a multicluster model, and, possibly, stronger dynamic distortion, without necessarily implying that the clusters are dissolved. The cluster ansatz (11.10) does accommodate both well-separated and overlapping clusters, and the more the clusters overlap the more their state overlaps with a low-lying SM state. The cluster-model state may thus assimilate to a SM state in the limit of deep intercluster binding, as is shown in the example of ^{12}C. But clustering still shows up in the behaviour of the wave function near the nuclear surface, and, therefore, it is expected that the cluster model is more realistic than the SM just in describing the more prominent surface effects.

12.2.3 Effective force

The logic of theoretical physics is 'reductionistic': it requires that composite systems be described via interactions, between their 'elementary' constituents, which describe all their relevant subsystems. For composite nuclei the elementary constituents are the nucleons and the relevant subsystems are the few-nucleon systems. The nucleon–nucleon force used in the description should reproduce the properties of the few-nucleon systems. The nucleon–nucleon force is constrained by symmetry considerations and by the qualitative knowledge of the underlying quark-gluon dynamics, which can be visualized in terms of meson exchanges. The parameters of the force are determined by fitting to two- and three-nucleon data. This procedure is not unambiguous, but results in interactions that give realistic descriptions of two-nucleon and three-nucleon data. These so-called 'realistic' interactions (cf. Sect. 13.9.1 later) are still useful for the description of composite systems since their predictions depend weakly on the particular form of the interaction adopted. The realistic forces have very complicated functional forms [cf. Eq. (13.26)]; they should also contain either three-nucleon terms or short-range nonlocality (or both), which is not surprising since the nucleons themselves are composite.

So far as light nuclei are described in a few-body approach exactly, it is inevitable to use the realistic forces and cope with all the difficulties caused by their complicated forms. Once, however, the function space is truncated, it looks not only reasonable but also unavoidable to use forces that tally with the truncated state space. Such forces are called effective forces [363]. Using effective forces is reasonable since they may produce almost the same dynamics with far less complications, and necessary because, in a truncated state space, the realistic forces may fail to produce any reasonable results. For example, combined with a thoroughly truncated HO basis, any realistic force will fail to bind the deuteron. The effective forces should nevertheless be established well at a phenomenological level. The effective forces are of course ambiguous, but that property is shared by the most realistic potentials as well.

In a cluster model the cluster relative motion is treated accurately at the expense of a schematic treatment of cluster internal structure. It is clear, nevertheless, that the assumed cluster internal structure must be consistent in some sense with the nucleon–nucleon force. More precisely, the g.s. $\Phi_i(\xi_i)$ in the model of the separate cluster i should be a fair approximation to the solution of the Schrödinger equation (11.39). The model state Φ_i is an acceptable approximation if both the energy $E_i \equiv \langle\Phi_i|H_i|\Phi_i\rangle$ and the rms radius $\langle r^2\rangle^{1/2}$ are realistic. The cluster sizes always affect the intercluster dynamics, so the effective force to be used for the composite system ought to produce approximately correct radii in consistent models of the separate clusters. The cluster energy is less important; it gains some importance if the state space allows the clusters to break up. Then the relative position of the threshold energies depends on the cluster energies. For example, a model of ^{6}Li may be appreciably affected by the relative positions of the α+d and t+h thresholds, which come into play especially in the positive-energy regime. That the chosen size parameter really gives the precise energy minimum is only important in certain comparisons (e.g., when the distortion effects are assessed by comparing no-distortion and with-distortion calculations [364]). Therefore, for α-, t and h clusters, whose sizes are not very different, it is acceptable to adopt a common size parameter, which is approximately consistent with the force and with the known cluster radii.

The interaction should, at the same time, describe the relative motion of nucleons that belong to different clusters realistically. Since in the model there is virtually no restriction on the relative motion of such pairs of nucleons, this requirement is fulfilled by an interaction that is suited for free pairs of nucleons.

Thus the interaction has to describe the intrinsic motion of the nucleons within a cluster in a restricted model, and has to be realistic for a pair of free nucleons as well. To find such an interaction is an overconstrained problem. Therefore, these two requirements cannot be fulfilled exactly at the same time, but there may exist good compromises. It is clear that the force to be used must be an effective force. It should not contain the tensor force because that is inconsistent with the single 0s configuration of the cluster g.s. It is, however, possible to mock up the effect of the tensor force by a pure central interaction to some extent.[3]

It facilitates the construction of effective interactions that some parameters have selective effects. For simple systems, the potential matrix elements depend only on certain combinations of some exchange parameters [366]. Similarly, the spin-orbit term has no effect on the intracluster motion. Its only effect is a splitting between partial waves of the same orbital angular momentum and different total angular momenta. The central term should describe the average of the phase shifts in these partial waves, and the spin-orbit term enters only into reproducing their splitting.

Most nucleon–nucleon forces used in cluster models have been devised to reproduce bulk nuclear properties in Hartree-Fock-like mean-field models or have a microscopic foundation as G-matrix forces [327, 367, 368]. No wonder these are not very suitable for describing loosely bound light systems. From among the effective forces available, the

[3]It is possible to find central forces that reproduce the most important low-energy nucleon–nucleon phase shifts and, at the same time, describe d, t, h and α as clusters in a 0s HOSM realistically. But it is hardly possible to find an effective force that describes t, h and α as clusters in which only the 0s orbit is populated (0s clusters), and yet reproduces the deuteron properties and the nucleon–nucleon phase shifts with a tensor force [365].

Minnesota force [369] looks to be the most suitable, because its parameters were determined so as to reproduce the most important nucleon–nucleon phase shifts. Together with a spin-orbit force [370] constructed independently, it has the form

$$\begin{aligned} V_{ij} &= [V_1 f_1(r) + P^{t}_{ij} V_2 f_2(r) + P^{s}_{ij} V_3 f_3(r)]\,[P^{e}_{ij} + (u-1)P^{o}_{ij}] \\ &\quad + V_4 f_4(r) \boldsymbol{l}_{ij}\cdot\tfrac{1}{2}(\boldsymbol{\sigma}_i + \boldsymbol{\sigma}_j) \qquad (12.10) \\ &= \sum_{k=1}^{3} V_k f_k(r)(W_k + M_k P^{r}_{ij} + B_k P^{\sigma}_{ij} - H_k P^{\tau}_{ij}) \\ &\quad + V_4 f_4(r)(W_4 + M_4 P^{r}_{ij} + B_4 P^{\sigma}_{ij} - H_4 P^{\tau}_{ij}) \boldsymbol{l}_{ij}\cdot\tfrac{1}{2}(\boldsymbol{\sigma}_i + \boldsymbol{\sigma}_j), \qquad (12.11) \end{aligned}$$

where $r = |\boldsymbol{r}_{ij}|$, $\boldsymbol{r}_{ij} = \boldsymbol{r}_i - \boldsymbol{r}_j$, $f_k(r) = \exp(-\mu_k r^2)$, and the spin-triplet, spin-singlet, space-even and space-odd projection operators are

$$P^{t}_{ij} = \tfrac{1}{2}(1+P^{\sigma}_{ij}),\quad P^{s}_{ij} = \tfrac{1}{2}(1-P^{\sigma}_{ij}),\quad P^{e}_{ij} = \tfrac{1}{2}(1+P^{r}_{ij}),\quad P^{o}_{ij} = \tfrac{1}{2}(1-P^{r}_{ij}), \qquad (12.12)$$

with P^{σ}_{ij}, P^{τ}_{ij} and P^{r}_{ij} being spin-exchange, isospin-exchange and spatial-coordinate-exchange operators, respectively; the relative orbital angular momentum $\boldsymbol{l}_{ij}$ is defined as $\boldsymbol{l}_{ij} = -\frac{1}{2} i \boldsymbol{r}_{ij} \times (\boldsymbol{\nabla}_i - \boldsymbol{\nabla}_j)$ [cf. Eq. (9.94)]. The force parameters are given in Table 12.2.

The performance of the Minnesota force in reproducing the basic ingredients of the cluster model is given in Table 12.3 and Figs. 12.3 and 12.4. Table 12.3 contains its predictions for the bound states of d, t, h and α in two versions of the model for the cluster intrinsic states, and these are compared with exact calculations. The first model is of standard use in cluster models; in this the g.s. is described by a 0s HO

Table 12.2: The Minnesota nucleon–nucleon potential [369,370] parametrized as in Eq. (12.11). The value of u is to be specified for each application.

k	V_k (MeV)	μ_k (fm^{-2})	W_k	M_k	B_k	H_k
1	200.0	1.487	$\frac{1}{2}u$	$1-\frac{1}{2}u$	0	0
2	−178.0	0.639	$\frac{1}{4}u$	$\frac{1}{2}-\frac{1}{4}u$	$\frac{1}{4}u$	$\frac{1}{2}-\frac{1}{4}u$
3	−91.85	0.465	$\frac{1}{4}u$	$\frac{1}{2}-\frac{1}{4}u$	$-\frac{1}{4}u$	$-\frac{1}{2}+\frac{1}{4}u$
4[a]	−1182.2	3.0	1	0	0	0
4[b]	−50.0	1.0	1	0	0	0
4[c]	−629.4	2.0	1	0	0	0

[a]Spin-orbit term, useful for α+few-nucleon systems.
[b]Spin-orbit term, useful for $\alpha+\{^{\rm t}_{\rm h}\}$+N(+N).
[c]Spin-orbit term, useful for $\alpha+\alpha$+N+N.

Table 12.3: Predictions of the Minnesota force for the energies E (in MeV) and for the point rms radii $\langle r^2\rangle^{1/2}$ (in fm) of the isolated clusters in a single HO model with an optimized parameter β, in a five-state breathing model and in an exact calculation.

Cluster		Single-HO model	Breathing model	Exact	Experiment
d	E	−0.130	−2.196	−2.201	−2.225
	$\langle r^2\rangle^{1/2}$	1.327	1.920	1.916	1.96
t	E	−4.558	−6.023	−8.383	−8.482
	$\langle r^2\rangle^{1/2}$	1.485	1.693	1.705	1.57
α	E	−25.584	−26.483	−30.765	−28.296
				−29.937[a]	
	$\langle r^2\rangle^{1/2}$	1.357	1.404	1.405	1.47

[a]With Coulomb force.

state ('single-HO model') with optimized size. In the second, the nuclear state is taken to be a linear combination of a few such states of different size parameters, optimized simultaneously ('breathing model'). This improvement has been discussed in Sect. 11.7.4 and will be discussed later in Sect. 13.1 in the context of cluster models. It suffices to say here that if the multicluster model included deuteron clusters as such, then it would be very important to use the breathing model for it.[4] Otherwise even the results of the single-HO model are qualitatively correct, and certainly satisfactory for the multicluster model. One can see that the exact calculations reproduce the experimental energies extremely well, but even the single-HO model yields quite acceptable values.

The predictions for the S-wave low-energy nucleon–nucleon phase shifts are shown in Fig. 12.3. The $^3S_1$ phase shift is surprisingly accurate and the $^1S_0$ is also reasonable, given the fact that the central Minnesota interaction is constructed so as to simulate the effect of the tensor force.

The quality of the force can also be tested through the scattering lengths and effective ranges, which are the parameters of the phase shift in the vicinity of wave number $k=0$ [26]:

$$k \cot \delta = -\frac{1}{a} + \frac{1}{2} r k^2 \quad (l = 0). \tag{12.13}$$

For S-wave p+n scattering these quantities are compared with the empirical values in Table 12.4. The agreement is again reasonable.

[4]It is the deuteron whose description is improved by the breathing most thoroughly; it can easily be made virtually exact within the 0s approximation.

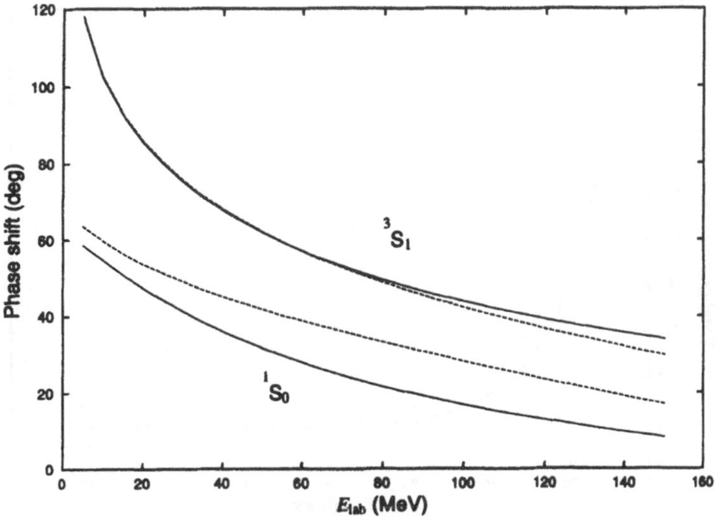

Figure 12.3: S-wave proton–neutron phase shifts as functions of the energy E_{lab} in the laboratory frame, calculated (solid lines) with the Minnesota force [369] with the parameters given in Table 12.2, compared with experiment (dashed lines [50]).

The results summarized in Fig. 12.3 and in Table 12.4 indicate that the Minnesota force is too weak in the $^1S_0$ partial wave. That partial wave implies $T=1$, and the same component is responsible for the underbinding of some of the $J^\pi=0^+$ states in the even-A oxygen isotopes (see Fig. 8.2). From this one can conclude that the Minnesota interaction is almost perfect in the $T=0$ channel but is a little weaker than it should be in $T=1$ channel.

Not surprisingly, the accuracy of the predictions of the Minnesota potential for higher partial waves is much less good. To give a feeling

Table 12.4: S-wave p+n scattering lengths a and effective ranges r produced by the Minnesota force compared with the data [371].

	a (fm)		r (fm)	
	Theory	Experiment	Theory	Experiment
$^3S_1$	5.427	5.424±0.004	1.758	1.759±0.005
$^1S_0$	−16.80	−23.748±0.010	2.885	2.75±0.05

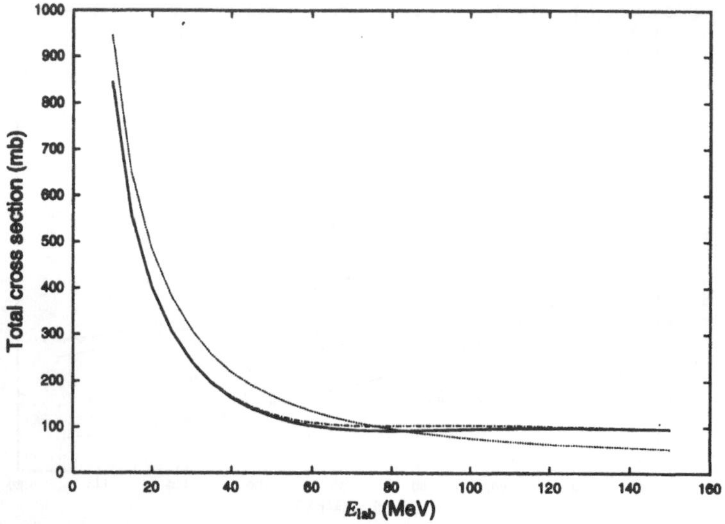

Figure 12.4: Total elastic cross section of the proton–neutron elastic scattering. The solid and dot-dashed lines are predictions with the Minnesota potential including a spin-orbit term [370] given as the first set in Table 12.2, with $u = 0.94$ (solid line) or with 1.0 (dashed line). The dotted line is experiment [50].

for the overall quality of the performance of the force, in Fig. 12.4 the total (elastic) cross section is shown for low-energy proton–neutron scattering. This is a sum of terms σ_{lS}^{JT} of the type of Eq. (10.11) for all total angular momenta J and for its $2J+1$ possible projections and is averaged over all projections of the orbital and spin angular momenta and for the total isospin[5] T:

$$\begin{aligned}\sigma &= \frac{1}{2}\sum_{lSJT}\frac{2J+1}{(2l+1)(2S+1)}\sigma_{lS}^{JT}\\ &= \frac{1}{2}\sum_{lSJT}\frac{2J+1}{(2l+1)(2S+1)}\frac{\pi}{k^2}(2l+1)\left|1-e^{2i\delta_{lS}^{JT}}\right|^2\\ &= \frac{2\pi}{k^2}\sum_{lSJT}\frac{2J+1}{2S+1}\sin^2\delta_{lS}^{JT}. \qquad (12.14)\end{aligned}$$

The agreement is still fair.

On these grounds, the Minnesota force is expected to give a reason-

[5]The p+n system contains $T=0$ and $T=1$ components with equal weight, so the averaging amounts to summing over T and dividing by 2.

ably realistic description for the low-energy states of systems of a few α-, t and h clusters and nucleons.

Since in the 0s models of the clusters no results depend on the exchange parameter u, this quantity is not well-established phenomenologically. Its role is to set the relative weights of the contributions coming from two-nucleon relative states of even and odd reflection symmetries. What is known is that it should not depart very much from its Serber value, $u=1$. Since in multicluster models all results do depend on it, it can serve as an adjustable parameter. The usual philosophy is to fit it to the properties of the two-cluster subsystems. In this way one may assign the the most important subsystem energy its exact value or more than one subsystem energies their nearly exact values, and there remains no free parameter for the composite system to be described.

The spin-orbit parameters shown in Table 12.2 are not very well-established either. None of them is fine-tuned for the two-nucleon system. They were used alternatively so as to get more realistic results.

12.3 Cluster correlations

12.3.1 Correlated versus uncorrelated description

If the motion of a multiparticle system is fully uncorrelated, its wave function is a product of s.p. wave functions. For a nuclear wave function to be meaningful, it should be antisymmetrical, and antisymmetrization is already a departure from the product form. The correlation implied may be called Pauli correlation. Removal of the c.m. motion also implies a departure from the product form, and the correlation involved is called the c.m. correlation. Correlations are induced by all other projections as well: e.g., by parity and angular-momentum projections. For identical fermions, a motion is usually called uncorrelated if it is described by a Slater determinant, i.e., the only correlation included is Pauli correlation. The variational solution to the dynamical equations of the problem of A identical fermions in the subspace of uncorrelated states is nothing but the Hartree–Fock model.

In Chap. 9 clustering and any other correspondence between the motions of the individual particles have been referred to as correlations. In the correlated Gaussian approach even the basis states may be strongly correlated, and the main merit of this approach is its flexibility to adapt itself to cluster correlations. In Chap. 11, however, we have seen that the uncorrelated SM [288] as well as the multicentre SM and the AMD [204] (which is an extreme case of a multicentre SM: each centre accommodates just one nucleon) may produce clustering. These examples show that clustering may actually exist without correlations, at least without dynamic correlations. It is therefore important to see in what respect a truly correlated model is superior to uncorrelated models in the description of clustering.

The models to be considered now include some which may be regarded as versions of the AMD and some as their improvements. The model wave functions are Slater determinants of Gaussian s.p. wave packets [cf. Eq. (9.67)],

$$\Phi = \mathcal{A}\left\{\prod_{i=1}^{A} \hat{\varphi}^{(\beta_i)}_{s_i m_s^i m_t^i}(\boldsymbol{r}_i)\right\}, \tag{12.15}$$

or their superpositions. In Fig. 10.1 we have seen that the energy convergence with a basis consisting of essentially uncorrelated states[6] may be rather slow, but here the basis parameters are optimized. In all models the central term of the Minnesota interaction is used with $u=1$ with no Coulomb force, the spins are coupled to $S=\frac{1}{2}$, to 0 and to 1 for t, α and ^{6}Li, respectively, and the parity is projected to +1 by adding up the state and its mirror image before variation.

In model 1 there is just one term, in which all Gaussians are set to have equal widths, and the common β parameter is optimized along with the positions of the Gaussian centres. This is equivalent to the spin- and parity-projected version of the AMD. In model 2 the β values are also allowed to be different and are optimized together with the Gaussian centres. Model 3 is a many-dimensional calculation. Models 1′, 2′ and 3′ are the same as models 1, 2 and 3, respectively, but the orbital angular momentum is projected to $L = 0$ after variation. In the many-dimensional calculation the β_i and $\boldsymbol{s}_i$ values are optimized stochastically.

From a basis state containing wave packets of non-equal β's it is impossible to factor out the c.m. motion. The intrinsic energy belonging to any basis state can be calculated by omitting the c.m. kinetic energy from the Hamiltonian [cf. Eq. (9.102)]. In this way, the contribution of the c.m. motion to the diagonal matrix elements (i.e., the expectation value of the c.m. kinetic energy) has been eliminated. What remains in the off-diagonal elements may be called a c.m. fluctuation effect. In the $\mathcal{K}$-dimensional case these are still present, but the minimization of the intrinsic Hamiltonian over the $\mathcal{K}A$-variable trial function will minimize the effect of the c.m. fluctuation as well. Thus the c.m. motion is expected to give a minor contribution even in that case [372].

Table 12.5 shows the results. The most conspicuous thing to be observed is how little binding the AMD provides.[7] It is interesting that

[6]Those basis states contain the c.m. correlation only.

[7]Most of the missing binding energy in an uncorrelated model comes from the

Table 12.5: Energies (in MeV) of Gaussian wave packet models for a few light nuclei with the Minnesota force (without Coulomb and spin-orbit interactions) and numerically exact results calculated with correlated Gaussians. The symbol $\mathcal{K}$ is the basis dimension, 1β and β's denote calculations in which a uniform β or individual β's have been optimized, L's and $L=0$ mean, respectively, no orbital angular momentum projection and orbital angular momentum projection to $L=0$ after variation.

	No. 1 $\mathcal{K}$=1 1β L's	No. 1′ $\mathcal{K}$=1 1β $L=0$	No. 2 $\mathcal{K}$=1 β's L's	No. 2′ $\mathcal{K}$=1 β's $L=0$	No. 3 $\mathcal{K}$=100 β's L's	No. 3′ $\mathcal{K}$=100 β's $L=0$	Exact $L=0$
t	−5.12	−5.77	−5.12	−5.77	−7.53	−7.94	−8.38
α	−25.72	−25.73	−25.72	−25.73	−27.96	−28.32	−30.76
^{6}Li	−9.47	−12.41	−19.58	−24.09	−25.12	−28.25	−36.17

for the single-configuration calculations the possibility of several β may not give any additional flexibility. All β values converge to 0.36 fm^{-2} for t and to 0.50 fm^{-2} for α, while for ^{6}Li four of the β's converge to 0.5 fm^{-2} and the rest to the lowest value it is allowed to take. By letting the Gaussian centres $\boldsymbol{s}_i$ variable, the nucleons gain more freedom, which improves the energy (cf. Table 12.3), and the energy gets better when the admixtures from higher angular momenta are removed. The effect of angular-momentum projection is most appreciable for ^{6}Li since, without projection, the two loose nucleons stray far, picking up high angular momenta. Wherever the ^{6}Li binding is less deep than the α binding, the α-cluster and the two nucleons tend to stay apart as far as they are allowed by the basis states.

The dimensions of the correlated Gaussian bases producing the exact results are 30, 50 and 200, respectively. With the uncorrelated basis states, however, the energy converges rather slowly even though their parameters are optimized, which clearly shows the superiority of the correlated bases.

One of the lessons to learn is that correlated and uncorrelated models behave very differently. In fact, the approaches that include correlations form a class distinct from the class of those excluding correlations. Uncorrelated models are one step more 'effective' than the correlated model presented in this book. The interactions useful in uncorrelated models are farther from those describing nucleon–nucleon data, and

neglect of short-range correlations. The AMD must perform better with forces of weaker repulsive cores and with readjusted forces.

their parameters may mock up the effects of correlations. Such 'effective' theories can be formally derived from '*ab initio*' theories by truncating the state space by projecting off the correlated part of the state space. Accordingly, the predictions of such models are likely to be restricted to s.p. properties.

The other lesson to learn is that, to get an (almost) exact result, it is advisable to use basis states that are correlated in compliance with the correlations involved in the physical problem itself.

12.3.2 Clustering in A-nucleon calculations

Section 12.3.1 was concerned with an aspect of the ***description*** of light nuclei. Therefore, approximate model calculations have had to be contrasted with exact ones. Once (virtually) exact calculations can be performed, the ***phenomena themselves*** can be scrutinized.

Our aim now is to demonstrate that nucleons tend to be clustered in a plausible sense. To this end we have to use a theoretical framework in which no bias for clustering is introduced. That is what exact calculations are useful for. However, the term 'exact' always have to be qualified. Unlike the cases of Sect. 12.3.1 and Sect. 13.9.1 later, the calculations to be considered now are exact in a restricted sense, but the restrictions introduced are totally unbiassed for or against clustering.

The calculations to be shown are '*ab initio*' calculations in the sense that they are A-body calculations for A-nucleon systems, but the Minnesota effective force given in Table 12.2 is used again, with $u = 1$, with the Coulomb and without a spin-orbit interaction since there is no need to produce the exact energy. From among the clustered nuclei portrayed in Fig. 12.2, two of the most simple and spectacular cases, ^{8}Be and ^{6}He, are considered.

The Minnesota potential reproduces the energies and sizes of the d, t, h and α-clusters (cf. Table 12.3), so if the bulk properties of ^{8}Be and ^{6}He are also described reasonably well, the clustering to be found is likely to be realistic, too.

The trial function is a combination of A-body correlated Gaussians (9.41). The only restriction to be introduced is that the angular and spin–isospin functions are fixed to be the simplest single functions. The total orbital angular momentum is restricted to $L=0$ and the angular part $\theta_{lL}(\boldsymbol{x})$ is chosen to be Eq. (9.32) with $K=0$. Note, however, that this does not exclude high values of the angular momenta l, which belong to the relative motions of (groups of) nucleons; any l values will get mixed in if required to minimize the energy. Half of all protons (and neutrons) are put into spin-up (↑) states and the other half into spin-down (↓) states. Each basis function contains $\frac{1}{2}(A-1)A$ pair-correlation parameters α_{ij}. These nonlinear parameters are determined by the SVM and the basis dimension is increased one by one to reach fair accuracy.

Figure 12.5 shows the convergence of the α, ^{6}He and ^{8}Be energies versus the basis dimension. The energy of the α-particle is very close to the experimental value. The energy of ^{6}He converges slowly, and a careful optimization makes ^{6}He bound with respect to the lowest α+n+n threshold. The two-neutron separation energy is found to be 0.1 MeV, which is smaller than the empirical value (0.976 MeV). The g.s. of ^{8}Be is found to be by more than 2 MeV lower than the $\alpha+\alpha$ threshold. It should in fact be 0.092 MeV above the threshold. Although this is not realistic, a bound ^{8}Be is more suited for the present purpose: in this way we shall have a well-behaved square-integrable wave function, and clustering is likely to be suppressed rather than exaggerated.

A most natural pictorial way to represent cluster structure is by means of density plots. That is, however, trivial only in models with projection after variation. In such models the wave function before angular-momentum and parity projection is like one in a body-fixed system for collective models, and the density plot, whether realistic or not, does display clustering [373]. But any (nearly) exact theory should work with properly projected basis states, and the resulting wave function should also have the correct parity and angular momentum. In particular, all zero-spin states are spherically symmetric.

Nevertheless, if clustering has any relevance to reality, it must be possible to display it, independent of the way of the description. One way is to define an intrinsic frame by diagonalizing the classical moment

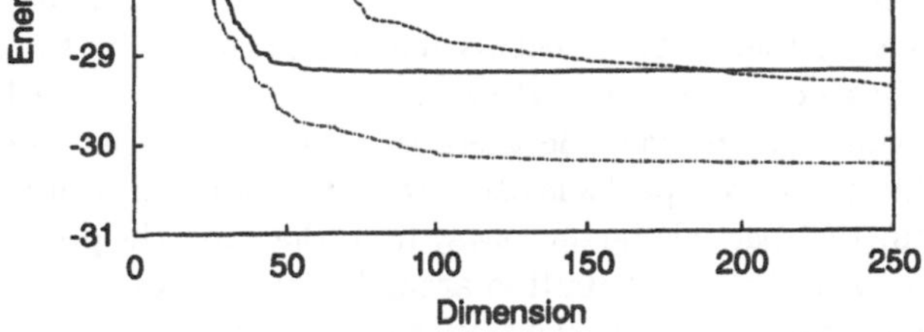

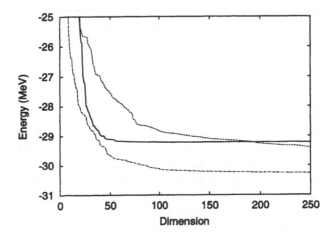

Figure 12.5: Energy of the α-particle (solid line), of ^{6}He (dashed line), and half of the energy of ^{8}Be (dot-dashed line) as functions of the basis dimension.

of inertia matrix, which may be constructed via a stochastic sampling of the configuration space [374]. The intrinsic shape of the ^{8}Be g.s. obtained in this way exhibits an $\alpha+\alpha$ cluster structure.

In the following we show an alternative way of displaying cluster structure, through two-particle densities, which are also called pair correlation functions (cf. Sect. 9.5.2). Let us introduce the notation [cf. Eqs. (9.160) and (9.161)]

$$P(\boldsymbol{r}\gamma, \boldsymbol{r}_0\gamma_0) = \langle\Psi| \sum_{i\neq j} \delta(\boldsymbol{r}_i - \boldsymbol{R}_{\rm c.m.} - \boldsymbol{r})\delta(\boldsymbol{r}_j - \boldsymbol{R}_{\rm c.m.} - \boldsymbol{r}_0)P_i^{\gamma} P_j^{\gamma_0}|\Psi\rangle, \tag{12.16}$$

where P_i^{γ} projects out a nucleon specified by γ which stands for either the spin label, up or down, or isospin label, p or n, or both [cf. Eq. (9.122)]. The label γ will be omitted from $P(\boldsymbol{r}\gamma, \boldsymbol{r}_0\gamma_0)$ if it is to denote any nucleons of any spin projections. Let $\boldsymbol{r}_0$ be a fixed vector, pointing to a 'test particle' positioned appropriately and let $\boldsymbol{r}$ be the variable of P. The function $P(\boldsymbol{r}\gamma, \boldsymbol{r}_0\gamma_0)$ shows how the fixed test particle γ_0 sees the density distribution of particles γ. All $P(\boldsymbol{r}\gamma, \boldsymbol{r}_0\gamma_0)$ functions to be shown are plotted in the xy-plane ($z=0$), with the test particle put in the same plane. The functions for ^{8}Be are defined in the c.m. frame, i.e., $\boldsymbol{R}_{\rm c.m.}=0$, but those to be shown for ^{6}He, are defined in the system in which the c.m. of the protons is at rest: $\boldsymbol{R}^{\rm p}_{\rm c.m.}=0$.

In Fig. 12.6 some pair correlation functions of ^{8}Be are plotted. All are seen from the viewpoint of a p$\downarrow$ test particle displaced from the centre by 1.91 fm, which is one half of the average distance between the two $\downarrow$ protons.

If there is α-clustering, the other $\downarrow$ proton has to be in the other cluster, and its distribution should map out the density of that cluster. Figure 12.6a bears out this expectation. The distribution is centred on the other half of the xy-plane, and is slightly elongated to the direction of the test proton. This does look like an α-cluster, slightly distorted. Figures 12.6b and c show the distribution of the two p$\uparrow$, and the four neutrons, respectively, and they fully map out the two-cluster system. The highest peaks in Figs. 12.6b and c appear almost at the same positions as in Fig. 12.6a. Their asymmetry is caused by the fact that the nucleons near the test particle are correlated with the fixed test particle, while on the far side there is no fixed particle. The ratio of the heights of the two peaks is also nearly the same in both panels. By moving the test particle farther away from the c.m., the position of the highest peak and the distribution around it are hardly changed.

These results show that the nucleons are distributed such that two protons and two neutrons tend to be close to each other. In order to see whether it is justifiable to call these clusters α-clusters, in Fig. 12.7 we compare the pair correlation function $P(\boldsymbol{r}, \boldsymbol{r}_0\,\mathrm{p}\downarrow)$ with the density

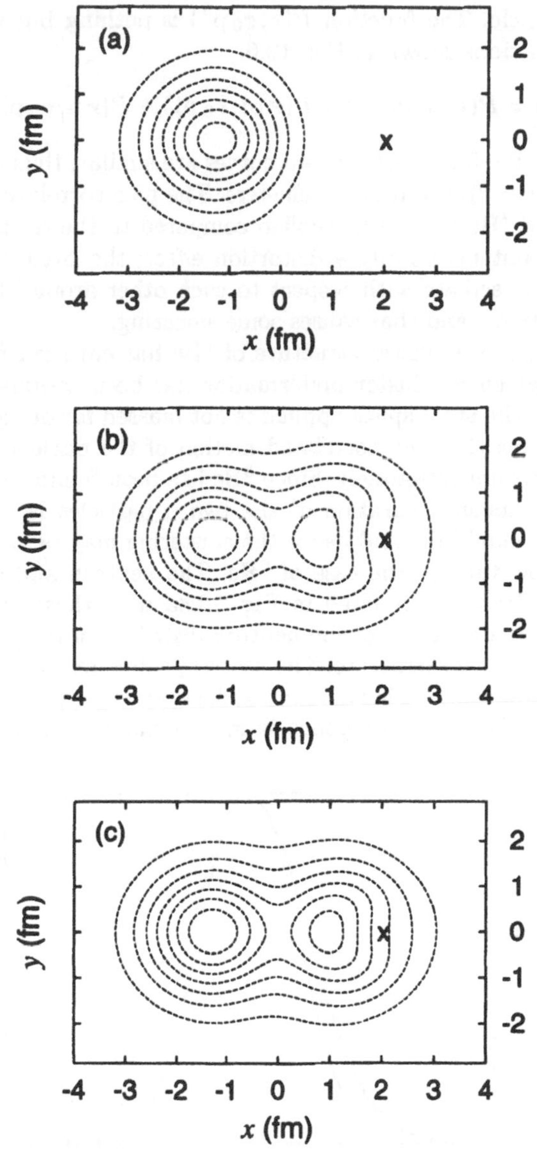

Figure 12.6: Pair correlation functions of ^{8}Be plotted as contour maps. The position of the test particle p$\downarrow$, marked by $\times$, is at $(x, y) = (1.91\,\mathrm{fm}, 0)$. The panels show $P(\boldsymbol{r}\,\mathrm{p}\downarrow, \boldsymbol{r}_0\,\mathrm{p}\downarrow)$ (a), $P(\boldsymbol{r}\,\mathrm{p}\uparrow, \boldsymbol{r}_0\,\mathrm{p}\downarrow)$ (b) and $P(\boldsymbol{r}\,\mathrm{n}, \boldsymbol{r}_0\,\mathrm{p}\downarrow)$ (c). The contours represent peaks, and the difference between any two neighbouring contour levels is the same. The units in panel (c) are twice as large as elsewhere.

of the α-particle. The function $P(\boldsymbol{r}, \boldsymbol{r}_0\,\mathrm{p}\downarrow)$ is nothing but the sum of the three functions shown in Fig. 12.6:

$$P(\boldsymbol{r}, \boldsymbol{r}_0\,\mathrm{p}\downarrow) = P(\boldsymbol{r}\,\mathrm{p}\downarrow, \boldsymbol{r}_0\,\mathrm{p}\downarrow) + P(\boldsymbol{r}\,\mathrm{p}\uparrow, \boldsymbol{r}_0\,\mathrm{p}\downarrow) + P(\boldsymbol{r}\,\mathrm{n}, \boldsymbol{r}_0\,\mathrm{p}\downarrow). \quad (12.17)$$

As expected, the falloffs of the two curves are similar, thus it is justifiable to say that ^{8}Be contains α-clusters. The pair correlation function $P(\boldsymbol{r}, \boldsymbol{r}_0\,\mathrm{p}\downarrow)$ of ^{8}Be is slightly swollen compared to the α-particle density. This is not necessarily a distortion effect; the α-clusters in ^{8}Be are bound to oscillate with respect to each other around their mean relative positions, and that causes some smearing.

All in all, the α-cluster structure of ^{8}Be has emerged from a calculation in which no cluster preformation has been assumed, and the truncation of the state space applied is not biassed for or against clustering. It seems that the correlated motion of the nucleons has been taken into account adequately. Since ^{8}Be has been found to be bound, the resultant mean separation of the two α-particles is likely to be smaller than should be, and hence the clustering may be underrated.

Let us now turn to the case of ^{6}He. The neutron and proton s.p. densities of ^{6}He are compared in Fig. 12.8a with those of ^{4}He. One can clearly see a long tail of the neutron distribution in ^{6}He, which is characteristic of halo structure. The central proton densities of ^{4}He and ^{6}He differ significantly, which, however, does not contradict the α+n+n structure for ^{6}He. This difference is mostly due to the oscillation of

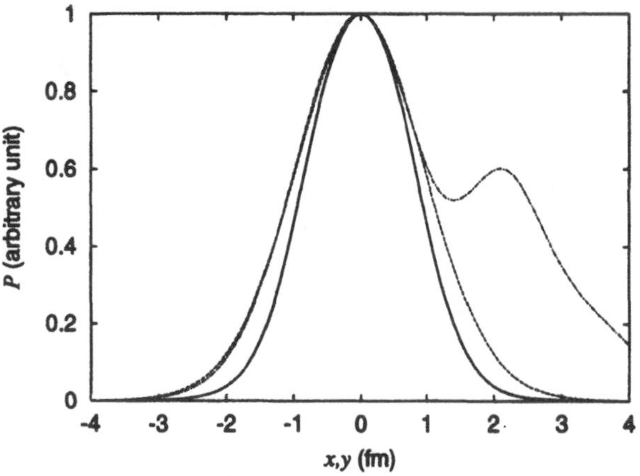

Figure 12.7: Pair correlation function $P(\boldsymbol{r}, \boldsymbol{r}_0\,\mathrm{p}\downarrow)$ of ^{8}Be with the origin put at the peak position. The dot-dashed curve shows P for $\boldsymbol{r} = (x, 0, 0)$ and the dashed curve for $\boldsymbol{r} = (0, y, 0)$. The solid curve is the α-particle density, whose maximum is scaled to be equal to the peak value of P.

the c.m. of the protons with respect to the total c.m. In Fig. 12.8b the densities with respect to the c.m. of the protons are plotted, and the difference is greatly reduced, indeed. That is why the correlation functions to be shown are also defined with respect to the proton c.m.

If the test particle is a neutron put far (e.g., 4 fm) from the proton c.m., it can be regarded as a halo neutron. Since the two halo-neutrons have opposite spin projections, $P(\boldsymbol{r}\,\mathrm{n}\downarrow, \boldsymbol{r}_0\,\mathrm{n}\downarrow)$ will map the density distribution of a core neutron and, indirectly, of the core (Fig. 12.9a).

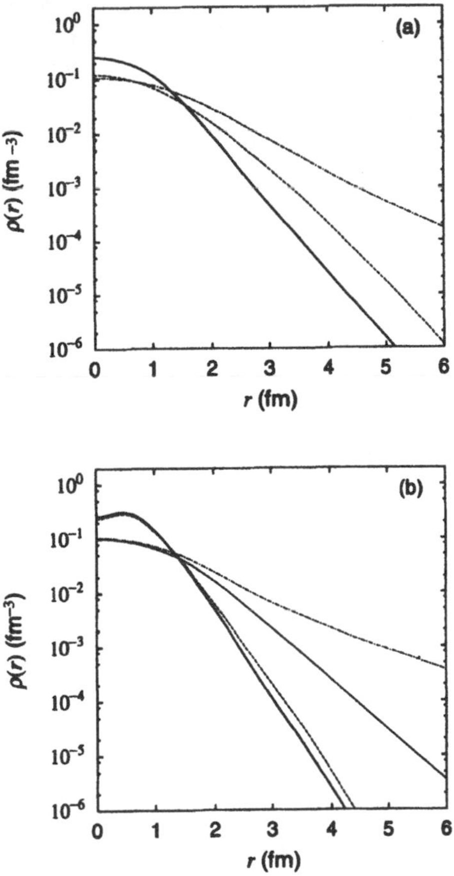

Figure 12.8: Proton and neutron densities of ^{4}He and ^{6}He with respect to the total c.m. (a) and to the proton c.m. (b). Thick solid curve: proton density of ^{4}He; thin solid curve: neutron density of ^{4}He; dashed curve: proton density of ^{6}He; dot-dashed curve: neutron density of ^{6}He. In (a) the two solid curves coincide.

The correlation of this test particle with the protons, e.g., $P(\boldsymbol{r}\,\mathrm{p}, \boldsymbol{r}_0\,\mathrm{n}\!\downarrow)$ is also very similar (not shown). This finding underpins the 'α-cluster plus halo' picture of ^{6}He.

The correlation function $P(\boldsymbol{r}\,\mathrm{n}\!\uparrow, \boldsymbol{r}_0\,\mathrm{n}\!\downarrow)$ (Fig. 12.9b) contains the contribution of two neutrons: the other halo neutron and the $\uparrow$ neutron

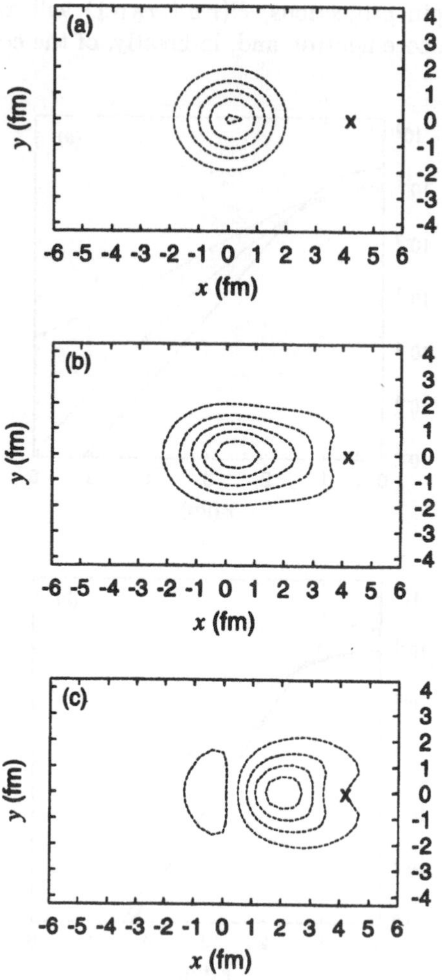

Figure 12.9: Pair correlation functions of ^{6}He plotted as contour maps. The position of the test particle $\mathrm{n}\!\downarrow$, marked by $\times$, is at $(x, y) = (4\,\mathrm{fm}, 0)$. The panels show $P(\boldsymbol{r}\,\mathrm{n}\!\downarrow, \boldsymbol{r}_0\,\mathrm{n}\!\downarrow)$ (a), $P(\boldsymbol{r}\,\mathrm{n}\!\uparrow, \boldsymbol{r}_0\,\mathrm{n}\!\downarrow)$ (b) and $P'(\boldsymbol{r}\,\mathrm{n}\!\uparrow, \boldsymbol{r}_0\,\mathrm{n}\!\downarrow)$ [Eq. (12.18)] (c). The contours show peaks, and the difference between any two neighbouring contour levels is the same.

in the core. To see the contribution of the halo neutron separately, we have to subtract the contribution of the core. This is done approximately in the function

$$P'(\boldsymbol{r}\,\mathrm{n}\!\uparrow, \boldsymbol{r}_0\,\mathrm{n}\!\downarrow) = \frac{P_{^6\mathrm{He}}(\boldsymbol{r}\,\mathrm{n}\!\uparrow, \boldsymbol{r}_0\,\mathrm{n}\!\downarrow)}{\rho_{^6\mathrm{He}}(\boldsymbol{r}_0\,\mathrm{n}\!\downarrow)} - \rho_{^4\mathrm{He}}(\boldsymbol{r}\,\mathrm{n}\!\uparrow), \tag{12.18}$$

whose contour map is plotted in Fig 12.9c. The function $P'(\boldsymbol{r}\,\mathrm{n}\!\uparrow, \boldsymbol{r}_0\,\mathrm{n}\!\downarrow)$ is normalized such that $\int d\boldsymbol{r}\, P'(\boldsymbol{r}\,\mathrm{n}\!\uparrow, \boldsymbol{r}_0\,\mathrm{n}\!\downarrow) = 1$. It indeed takes small values near the c.m. of the protons, thus it must be approximately equal to the distribution of the $\uparrow$ halo neutron viewed from the test particle. The halo neutron mostly resides between the test particle and the α-particle.

The probability distribution of the second particle, of course, depends on the position of the test particle. Figure 12.10 is similar to Fig. 12.9, but the test particle is put 2.4 fm away from the proton c.m. In this case the test particle may belong to the halo as well as to the α-cluster. That is why contours in Figs. 12.10a and b are distorted with respect to Figs. 12.9a and b. To identify the halo contribution to the contours in Fig. 12.10b, the definition of P' has had to be modified by subtracting the contribution that comes from the test neutron belonging to the α-particle. A non-unique but simplest plausible prescription for that is the following. The probability that the test particle belongs to α can be estimated to be $p = \rho_{^4\mathrm{He}}(\boldsymbol{r}_0\,\mathrm{n}\!\downarrow)/\rho_{^6\mathrm{He}}(\boldsymbol{r}_0\,\mathrm{n}\!\downarrow)$. When $|\boldsymbol{r}_0| = 2.4$ fm, p is 0.55. The halo contribution to the pair correlation function is then characterized approximately by the function

$$P''(\boldsymbol{r}\,\mathrm{n}\!\uparrow, \boldsymbol{r}_0\,\mathrm{n}\!\downarrow) = \frac{P_{^6\mathrm{He}}(\boldsymbol{r}\,\mathrm{n}\!\uparrow, \boldsymbol{r}_0\,\mathrm{n}\!\downarrow)}{\rho_{^6\mathrm{He}}(\boldsymbol{r}_0\,\mathrm{n}\!\downarrow)} - p\frac{P_{^4\mathrm{He}}(\boldsymbol{r}\,\mathrm{n}\!\uparrow, \boldsymbol{r}_0\,\mathrm{n}\!\downarrow)}{\rho_{^4\mathrm{He}}(\boldsymbol{r}_0\,\mathrm{n}\!\downarrow)} - (1-p)\rho_{^4\mathrm{He}}(\boldsymbol{r}\,\mathrm{n}\!\uparrow). \tag{12.19}$$

The contribution of the case that α includes the test particle is subtracted in the second term on the right-hand side, and the contribution of the case that α does not include it is subtracted in the third term. When p is close to zero, Eq. (12.19) reduces to Eq. (12.18). When $|\boldsymbol{r}_0|$ is small and p is close to unity, the first and second terms in Eq. (12.19) approximately cancel near the c.m. of the protons. This function is again normalized to unity when integrated over $\boldsymbol{r}$. The resultant contour map is drawn in Fig. 12.10c. One can see two peaks: One is near the test particle and the other is on the opposite side of the α-particle. The former peak is about twice as high as the latter one. The distribution near the c.m. of the protons is again very small.

The results for ^{6}He can be summarized by saying that in a calculation without model assumptions the α-cluster and the halo are discernible as separate entities. As a general conclusion, one can assert that a cluster picture emerges as an output even from calculations in

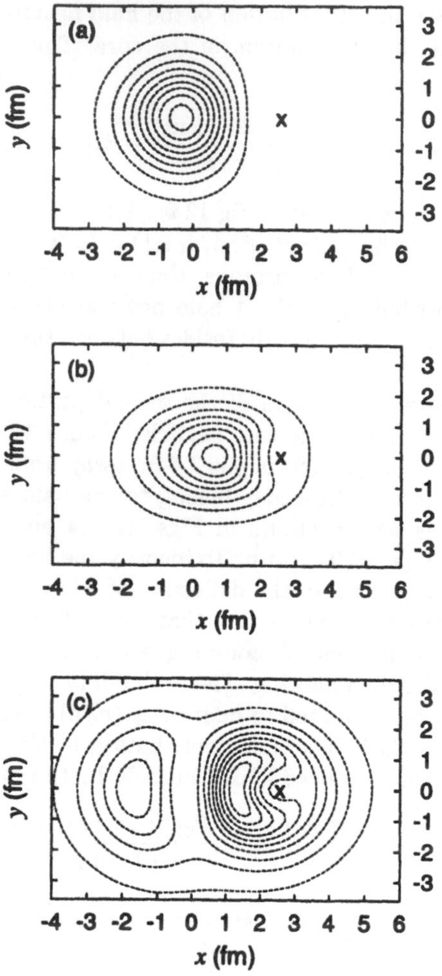

Figure 12.10: Pair correlation functions of ^{6}He plotted as contour maps. The position of the test particle n$\downarrow$, marked by $\times$, is at $(x,y) = (2.4\,\text{fm}, 0)$. The panels show $P(\boldsymbol{r}\,\text{n}\!\downarrow, \boldsymbol{r}_0\,\text{n}\!\downarrow)$ (a), $P(\boldsymbol{r}\,\text{n}\!\uparrow, \boldsymbol{r}_0\,\text{n}\!\downarrow)$ (b) and $P''(\boldsymbol{r}\,\text{n}\!\uparrow, \boldsymbol{r}_0\,\text{n}\!\downarrow)$ [Eq. (12.19)] (c). The contours show peaks, and the difference between any two neighbouring contour levels is the same.

which there are no cluster assumptions as input. It should be mentioned that the nuclear states scrutinized satisfy the conditions of the threshold rule (cf. p. 358), and the examples shown may be taken as illustrations for its validity. Based on this evidence, one can venture to apply the cluster models to such nuclei as is to be done in Chap. 13.

Chapter 13

Application to exotic nuclei

13.1 The structure of ^{6}He and ^{6}Li

13.1.1 Exposition

The six-nucleon systems have some unique properties that make them prime targets for close investigation.

The nucleus ^{6}He is the lightest object showing halo structure. Although it does not seem to have a very large halo, it may be a most perfect example of such a nucleus in many respects. Owing to the shell closure at $A = 4$, it is expected to behave as a tightly bound α-particle with two loose neutrons, i.e., as α+n+n. This property may make it a most perfect bound three-cluster system. Since neither the dineutron nor ^{5}He is bound, removal of any of the three clusters releases the binding of the rest as well. As was mentioned in the main Introduction, three-body systems of this property are called Borromean. A Borromean system has no two-body breakup threshold, which would introduce two-body correlations into its bound-state motion as well. The intercluster motion in ^{6}He is thus expected to be an exceptionally pure three-body motion. The ^{6}He g.s. may be viewed as a lighter and simpler analogue of the ^{11}Li g.s., which has a more pronounced neutron halo, but a much less inert core.

Because of its relative simplicity, ^{6}He is a good testing ground for the basic assumptions on the structure and description of halo nuclei. In particular, one may be able to probe the inertness of the α-cluster. The t+t clustering may be a competitor of α+N+N clustering, which could break the hypothetic core+n+n structure. Since the correlated Gaussian calculation can be performed exactly for $A = 6$, the utmost test, i.e., comparison with the exact solution, can also be carried through. These aspects will be the subject of Sect. 13.1.3.

The nucleus ^{6}Li is important because its electromagnetic properties are exceptionally well-known empirically, so they offer a testing ground for theory. The isobaric analogue states of the g.s. of ^{6}He in ^{6}Li and ^{6}Be offer examples for halo analogue states, and thus tests of charge symmetry of nuclei in extreme conditions. On the other hand, the g.s. of the three isobars may serve as a litmus test for the validity of the simulation of the tensor interaction by a central interaction. In the g.s. of ^{6}He and ^{6}Be (and in their analogue in ^{6}Li) the two nucleons are likely to be in a spin-singlet state, whereas in ^{6}Li they are overwhelmingly in a spin-triplet state. The tensor force has no contribution to binding in the former, but it has a substantial contribution in the latter. Therefore, it is expected that they cannot be described by exactly the same central force, and the discrepancy is a measure of the inaccuracy of the simulation. These aspects will be expounded in Sect. 13.1.4.

A further peculiarity of the α+N+N system is that it shows a number of resonances that are likely to decay directly to three-body channels, and their description is a tough challenge for theory. The tools presented in Sect. 10.4 are available, and they will be used here.

Because of these reasons, the six-nucleon systems had been extensively studied before. From among the stages that are in a direct line towards the SVM, we should first mention the implementation of a fully dynamical microscopic α+N+N model [375]. The next step was the description of the α-cluster distortion [284, 326, 366, 376–379]. The halo nature of ^{6}He was studied in Refs. [157, 185, 186, 380–382]. Spin-orbit and tensor forces have been introduced to improve the description of the intercluster motion in Refs. [365,383]. The (3N)+(3N) contribution was included in Ref. [384] and three-body resonances were described in a microscopic model in Ref. [152].

There have been several bouts in the SVM description of the six-nucleon problem [206,212,213,236,385,386], too. Reference [236] is concentrated on the performance of the SVM basis optimization, Ref. [212] is devoted to the description of the properties of the He isotopes, Ref. [213] gives a consistent treatment for the six-nucleon systems in the α+N+N model, Ref. [206] contains exact results for ^{6}He and ^{6}Li and in Refs. [385, 386] the α+N+N model is tested against various improvements. In Sect. 13.1.2 these approaches will be reviewed with an emphasis on the dynamical aspects (state spaces, energies). The test of the α+N+N model will be presented in Sect. 13.1.3. The predictions for the observables of the six-nucleon systems will be summarized in Sect. 13.1.4.

13.1.2 State spaces

Six-nucleon bound states have been described in eight different versions of the correlated Gaussian approach. These differ in the purpose of the investigation, in the choice of the state space and in the u parameter of the Minnesota force [cf. Eq. (12.10)]. These characteristics of the calculations are summarized in Table 13.1.

To describe these models, we should first specify the state spaces more precisely. The state space is built up in each model from all different relative-coordinate arrangements pertinent to the cluster composition assumed. All arrangements included for ^{6}He are shown schematically in Fig. 13.1. Thus, e.g., in α+N+N models the state space is defined by two Jacobi sets. It is convenient to refer to them as 'Y' and 'T' arrangements, respectively. In the subsystems t+p and h+n of models 6 and 7 the t and h clusters always couple to p and n, respectively, as is shown in Fig. 13.1.

Table 13.1: Summary of the work done on the six-nucleon bound states in eight versions of the SVM with correlated Gaussians. The symbol β stands for the HO width parameter for the α-particle. The symbol $*$ expresses that breathing distortion has been included; for α- and three-nucleon clusters this involves 3 and 4 states, respectively.

Mod.	Nuclei	State space	Force[a]		
	involved		u	S.o.	Coul.
1.	^{6}He, ^{8}He	α+N+N; $\beta = 0.606$ fm^{-2} [b]	1.14	no	no
2.	^{6}He, ^{8}He	α+N+N; $\beta = 0.54$ fm^{-2} [c]	1.15	no	no
3.	^{6}He, ^{6}Be	α+N+N; $\beta = 0.52$ fm^{-2} [c]	1.02[d]	yes	yes
	^{6}Li	α+N+N; $\beta = 0.52$ fm^{-2} [c]	0.92[e]	yes	yes
4.	^{6}He, ^{6}Li	α^*+N+N	0.98[d]	yes	yes
5.	^{6}He, ^{6}Li	{α^*+N+N; 3N+3N}	0.98[d]	yes	yes
6.	^{6}He, ^{6}Li	$\left\{\begin{pmatrix} t^*+p \\ h^*+n \end{pmatrix}+N+N\right\}$	0.98[d]	yes	yes
7.	^{6}He, ^{6}Li	$\left\{\begin{pmatrix} t^*+p \\ h^*+n \end{pmatrix}+N+N;\ 3N+3N\right\}$	0.98[d]	yes	yes
8.	^{6}He, ^{6}Li	exact	1.00	no	yes

[a]For u, see Eq. (12.10). S.o.: spin-orbit force. Coul.: Coulomb force.
[b]For deepest binding of free α.
[c]For realistic radius of free α.
[d]For $T = 1$ states.
[e]For $T = 0$ states.

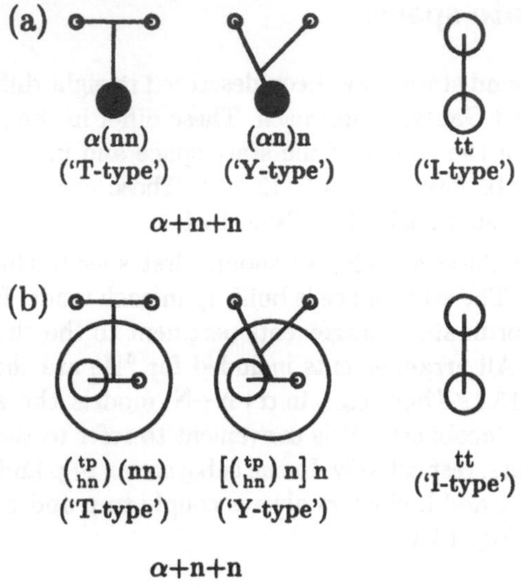

Figure 13.1: Relative vectors for two models of ^{6}He: {α+N+N; t+t} (a) and {(3N+N)+N+N; t+t} (b). Notation: α: ●; three-nucleon clusters: ◯; nucleon: ○.

Most of the calculations involve bases with $\theta_{\{l\}LM_L}(\boldsymbol{x})$ of the type of Eq. (9.23), so each basis state used has definite values of intermediate angular momenta $S, (l_1, l_2)L$. Such state spaces are conveniently characterized by the symbol Y or T, with the values of $S, (l_1, l_2)L$ in the subscript. The basis states of α+N+N form used in models 1–3 are specified in Table 13.2. The α+N+N sectors of the bases used for models 4–7 are similar to that in model 3. The 3N+3N motion in models 5 and 7 and the 3N+3N motion in models 6 and 7 have $l = 0$. In the exact calculation (model 8) all channels are included that have been found to contribute appreciably. Note that the forces have been slightly changed from one version of the model to another and from one isospin to the other, thus comparison of the energies obtained is only meaningful within one set of calculations, divided by horizontal lines.

Apart from the randomly selected Gaussians of the basis, the parameters of the model are the size parameter β of the α-particle and the mixing constant of the Minnesota force, u. The α size, to which the results are rather insensitive, has either been set to attain the energy minimum in Eq. (11.39) or to produce a realistic α size. The mixing parameter u controls the strength of the force in the odd partial waves relative to the even partial waves [cf. Eq. (12.10)], and so it affects the

Table 13.2: Bases used in models 1–3 of Table 13.1 for the six-nucleon systems. The energies, measured from the α+N+N threshold, are also shown. Incompatible results are separated by horizontal lines.

	Nucleus,	Subspace		Energy (MeV)	
	$J(T)$	T	Arrangement$_{S,(l_1,l_2)L}$	Theory	Exp.
1.	^{6}He, 0(1)	1	$T_{0,(0,0)0}$	−0.395	−0.975
	^{6}He, 0(1)	1	$Y_{0,(0,0)0}$	−0.387	
	^{6}He, 0(1)	1	$Y_{0,(1,1)0}$	−0.435	
	^{6}He, 0(1)	1	$T_{0,(0,0)0}, Y_{0,(0,0)0}$	−0.944	
	^{6}He, 0(1)	1	$T_{0,(0,0)0}, Y_{0,(1,1)0}$	−1.015	
	^{6}He, 0(1)	1	$Y_{0,(0,0)0}, Y_{0,(1,1)0}$	−0.605	
	^{6}He, 0(1)	1	$T_{0,(0,0)0}, Y_{0,(0,0)0}, Y_{0,(1,1)0}$	−1.016	
2.	^{6}He, 0(1)	1	$Y_{0,(0,0)0}$	−0.382	−0.975
	^{6}He, 0(1)	1	$Y_{0,(0,0)0}, Y_{0,(1,1)0}$	−0.601	
	^{6}He, 0(1)	1	$Y_{0,(0,0)0}, Y_{0,(1,1)0}, Y_{0,(2,2)0}$	−0.823	
	^{6}He, 0(1)	1	$Y_{0,(0,0)0}, Y_{0,(1,1)0}, Y_{0,(2,2)0},$ $Y_{0,(3,3)0}$	−0.923	
	^{6}He, 0(1)	1	$T_{0,(0,0)0}, Y_{0,(0,0)0}, Y_{0,(1,1)0}$	−0.990	
	^{6}He, 0(1)	1	$T_{0,(0,0)0}, T_{0,(2,2)0}, Y_{0,(0,0)0},$ $Y_{0,(1,1)0}, Y_{0,(2,2)0}, Y_{0,(3,3)0}$	−0.994	
3.	^{6}He, 0(1)	1	$T_{0,(0,0)0}, Y_{0,(0,0)0}, Y_{0,(1,1)0},$ $T_{1,(1,1)1}, Y_{1,(1,1)1}$	−0.970	−0.975
	^{6}Li, 0(1)	1	$T_{0,(0,0)0}, Y_{0,(0,0)0}, Y_{0,(1,1)0},$ $T_{1,(1,1)1}, Y_{1,(1,1)1}$	−0.083	−0.137
		0	$Y_{0,(0,0)0}, T_{0,(1,1)0}, Y_{0,(1,1)0},$ $Y_{1,(1,1)1}$		
	^{6}Li, 1(0)	0	$T_{1,(0,0)0}, Y_{1,(0,0)0}, Y_{1,(1,1)0},$ $Y_{1,(1,1)1}, T_{0,(1,1)1}, Y_{0,(1,1)1}$	−3.666	−3.700
		1	$Y_{1,(0,0)0}, T_{1,(1,1)0}, Y_{1,(1,1)0},$ $T_{1,(1,1)1}, Y_{1,(1,1)1}, Y_{0,(1,1)1}$		

results strongly. Its adopted value depends on whether the spin-orbit force is included or not because, when not included, it has to mock up the effects of the spin-orbit coupling. In model 1 it has been set so as to produce the right α+n phase shift in the $p_{3/2}$ wave, and was found to give a realistic value for the binding of ^{6}He, in accord with the SM picture of two $p_{3/2}$ valence neutrons. In models 2 and 3 it was changed a little to give more accurate binding for the g.s. of ^{6}He and, in model 3, for the g.s. of ^{6}Li separately. In models 4–7 u was set so as to reproduce the $p_{3/2}$ α+n phase shift in a single-channel model (force A in Table 13.6 later and in an $\{\alpha+n; t+t\}$ two-channel model (force B) with slightly different spin-orbit terms [384]. In the exact calculation $u=1$ was used. As was mentioned in Sect. 13.1.1, for the g.s. of ^{6}Li, an adjustment of the parameter is necessary because of the simulation of the effect of the tensor force by a central force.

In an α+N model consistent with model 4 apart from the lack of breathing, the subsystems ^{5}He and ^{5}Li were also examined. These nuclei are not bound, and the ACCC method was used to treat their states, with the coupling strength λ taken proportional to u. The results, compared with some others extracted from RGM calculations made for complex energies [cf. Sect. 10.4.1] are shown in Table 13.3. The agreement with the other theoretical results is satisfactory, which confirms the accuracy of the ACCC method. The discrepancy between theory and the standard experimental value [387] in the position of the $\frac{1}{2}^-$ state is caused by the fact that the parameters of a broad resonance are poorly defined empirically. The unique definition is that based on the S-matrix pole (last column of Table 13.3), and if that definition is used in the analyses [388], there is no discrepancy.

The $\frac{1}{2}^+$ state deserves special attention since bound-state-like models predict the existence of a low-lying $\frac{1}{2}^+$ state in ^{5}He [390, 391]. Table 13.3 shows that this state is pushed far off the vicinity of the real energy axis. Figure 13.2 displays the extrapolation from the bound-state region by varying u. The trajectory for the ^{5}He system passes through the virtual-state region and the unphysical resonance region and finally shows up in the physical region before reaching its physical

Table 13.3: Resonance energies E_R and widths Γ (in MeV) of ^{5}He and ^{5}Li with respect to the two-body threshold.

	ACCC		Scattering RGM [389]		Exp. [387]		S-matrix analysis [389]	
J^π	E_R	Γ	E_R	Γ	E_R	Γ	E_R	Γ
^{5}He								
$\frac{3}{2}^-$	0.77	0.64	0.76	0.63	0.89	0.60	0.80	0.65
$\frac{1}{2}^-$	1.98	5.4	1.89	5.20	5±1	4±1	2.07	5.57
$\frac{1}{2}^+$	12	190						
^{5}Li								
$\frac{3}{2}^-$	1.63	1.25	1.67	1.33	1.97	~1.5	1.69	1.23
$\frac{1}{2}^-$	3.0	6.4	2.70	6.25	7–12	5±2	3.18	6.60
$\frac{1}{2}^+$	42	197						

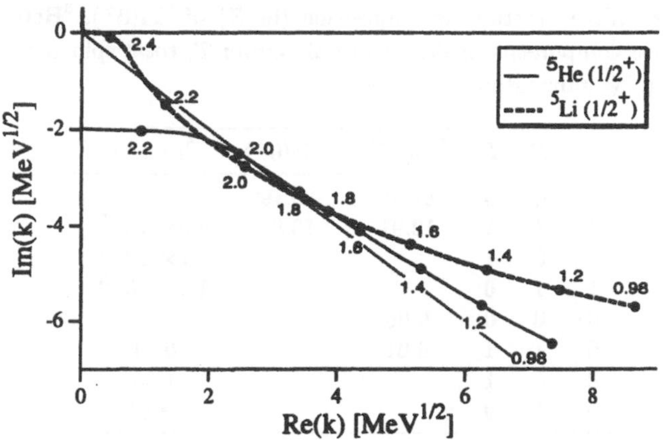

Figure 13.2: Trajectories of the $\frac{1}{2}^+$ states in ^{5}He and ^{5}Li as functions of the u parameter. Here k is identified with $E^{1/2}$. The dotted line marks Im k=−Re k.

value, $u=0.98$. This trajectory is reminiscent of the s-wave trajectory in a potential problem with no barrier (see Fig. 10.13). The point $u=0.98$ is very close to the borderline separating the physical region from the unphysical region. The width of the state is therefore predicted to be much larger than its energy. Such a state does not show up in the phase shift, so it cannot be called a proper resonance. The case of ^{5}Li is similar to that of ^{5}He, except that the trajectory resembles the pattern of the finite-barrier case of Fig. 10.13. It is thus fair to say that the existence of a low-lying $\frac{1}{2}^+$ resonance in ^{5}He and ^{5}Li is ruled out.

One can learn the significance of the individual subspaces of the state space from Table 13.2. One should remember, however, that these subspaces grossly overlap, so the energy gain coming from the inclusion of a subspace always depends on what state space it is added to. Having this in mind, one can infer that in ^{6}He the sequence of importance of the subspaces is $Y_{0,(1,1)0}$, $T_{0,(0,0)0}$, $Y_{0,(0,0)0}$ etc. The results with more extensive state spaces show that the direct sum of these three subspaces is practically satisfactory. The sequence of importance is in contrast with ^{6}Li, in which α+d clustering, which fits into the mould of the $T_{1,(0,0)1}$ subspace, is much stronger than ^{5}He+p clustering which can be associated with the $Y_{0,(1,1)0}$ pattern [392]. It is observable (see, e.g., sector 2 of Table 13.2) that, when an arrangement is missing, it can be substituted by configurations with higher orbital angular momenta in another arrangement, but it would be difficult to get as good an energy as with all arrangements because the convergence of the partial-wave

Table 13.4: Wave function decomposition (in %) of ^{6}Li(0^+), ^{6}He(0^+) and ^{6}Li(1^+) into components of definite total isospin T, total spin S and total orbital angular momentum L in model 3.

T	S	L	^{6}Li(0^+)	^{6}He(0^+)	^{6}Li(1^+)
1	0	0	87.02	86.19	
1	1	1	12.91	13.81	4.0×10^{-5}
1	0	1			1.1×10^{-4}
1	1	0			1.4×10^{-3}
0	0	0	0.06		
0	1	1	0.01		0.34
0	0	1			1.65
0	1	0			98.01

expansions is slow.

The angular-momentum composition of the three states investigated [^{6}He(g.s., 0^+), ^{6}Li(g.s., 1^+), ^{6}Li(analogue state, 0^+)] are shown in Table 13.4. The L=2 admixture is missing because there is no tensor force. The results agree with those of other microscopic calculations [365,384] very well, but are at variance with some of the phenomenological three-body models. In particular, for the $(S,L)=(1,1)$ component of ^{6}He and its analogue, one of the three-body models [197] tends to give considerably lower values than the microscopic models or other three-body calculations [393]. This may be understood by knowing that the arrangement in which the $(S,L)=(1,1)$ component naturally appears is Y-type, and the bases in Refs. [197] are of pure T-type. Another slight discrepancy is that in some macroscopic models the g.s. of ^{6}Li contains higher amounts of $(S,L)=(0,1)$ component: 4% [394], 5% [395], 3% [393]. No simple explanation is known for this disagreement.

13.1.3 Test of the approach

As presented in Chap. 12, the correlated Gaussian approach for ^{6}He is an α+n+n model. The validity of this model may be questioned based on the observation that, with interactions that reproduce the properties of the two-cluster subsystems, all α+n+n models underbind ^{6}He (with respect to the three-body threshold) by at least 0.2 MeV. This can be interpreted as a distortion effect (cf. Sect. 11.7.4). The breathing distortion has proved insufficient, but the inclusion of the t+t clusterization has been shown to be sufficient to produce the correct binding [384]. The question is whether the t+t clusterization is necessary as well.

Table 13.5: Binding energies and point matter rms radii of the α-particle in four models involved in the models of ^{6}He given in Table 13.1.

Model	1	4, 5	6, 7	8	Experiment
E (MeV)	−24.687	−25.595	−26.549	−29.940	−28.296
r_m (fm)	1.36	1.41	1.42	1.41	1.43[a]

[a] The empirical rms charge radius of the α-particle is 1.671 ± 0.014 fm [396, 397]. By unfolding the proton charge distribution following the recipe given in a footnote on p. 231, one obtains 1.43 fm for the point-proton distribution.

Since it goes beyond the α+n+n model, necessity to include it would upset the core+n+n picture even for the simplest halo nucleus.

Models 5–8 of Table 13.1 have been constructed to test whether it is really necessary to abandon the core+n+n model. The properties of the model α-particles consistent with those in models 1–8 are summarized in Table 13.5 and Fig. 13.3. It is seen that the HO model deviates from the exact result appreciably. It is remarkable that the inclusion of the breathing mode leaves the HO g.s. almost unaltered, while the {t+p; h+n} model approximates the exact result very well.

The results for ^{6}He are summarized in Table 13.6. In keeping with

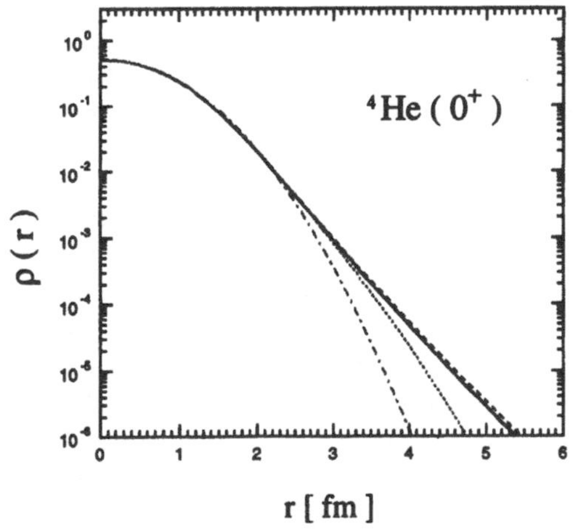

Figure 13.3: Density distribution in the α-particle g.s. in the four models of Tables 13.1, 13.3 and 13.5. Model 1 (single β): dot-dashed line; model 4, 5 (breathing): dotted line; model 6, 7 (t^*+p; h^*+n): dashed line; model 8 (exact four-body solution): solid line.

Table 13.6: ^{6}He energies E, α+n+n intercluster energies ε (in MeV) and rms radii (in fm) of the point matter, proton and neutron distributions (r_m, r_p and r_n, respectively) in models 1–8 of Table 13.1. Force A: $u = 0.98$, $V_4 = -591.1$ MeV; force B: $u = 0.92$, $V_4 = -691.1$ MeV. Model 1′ is the same as model 1 but u is adjusted to produce the correct ε. In model 3 $u = 1.02$. (Experiment: Ref. [387]; for other data, see Table 13.9 later.)

Mod.	E		ε		r_m	r_p	r_n
	Force A	Force B	Force A	Force B			
1.	−25.349	−24.909	−0.660	−0.220	2.52[a]	1.83[a]	2.81[a]
4.	−26.333	−25.914	−0.738	−0.319	2.49[a]	1.82[a]	2.77[a]
5.	−27.013	−26.557	−1.418	−0.961	2.42[b]	1.81[b]	2.68[b]
6.	−28.127	−27.749	−1.578	−1.200	2.34[b]	1.75[b]	2.59[b]
7.	−28.222	−27.839	−1.673	−1.290	2.33[b]	1.76[b]	2.57[b]
1′.	−25.684		−0.995		2.42[a]	1.77[a]	2.69[a]
3.	−24.955		−0.970		2.51[a]	1.87[a]	2.78[a]
8.	−30.07		−0.13				
Exp.	−29.271		−0.975		2.48	2.21	2.61
					2.33	1.72	2.59

[a]Radii calculated with force A. [b]Radii calculated with force B.

the case of the free α-particle, the breakup of the α-cluster deepens the energy much more than the breathing. At the same time there is still an overall underbinding with respect to the six-body threshold. This can be attributed to the treatment of the t and h clusters. Even though they are also allowed to breathe, the corresponding free clusters are still underbound by some 2.4 MeV with respect to the exact energy (see Table 13.7 later). Since ^{6}He itself is underbound by 3.4 MeV with respect to the exact binding (see Table 13.7), one can say that t and h are the main sources of discrepancy. On the other hand, since the force parameters have been chosen so as to reproduce ε in less complete state spaces, there is an overbinding with respect to the α+n+n threshold.

In comparing the effects of the t+t arrangement and the {t+p, h+n} breakup, one can observe that, while E is much deeper when the latter is included, ε is almost the same. This suggests that the improvement in both cases mainly comes from the treatment of the α-cluster. This can be understood by returning, for a moment, to the free α-particle. The state space $\left\{\binom{t^*+p}{h^*+n}+N+N\right\}$ does, of course, also allow the α-cluster in ^{6}He to be excited, and, with the inclusion of t+p and h+n,

Table 13.7: Model and exact energies in MeV with the Minnesota force (with Coulomb but without spin-orbit term and $u=1$.

	E_t	E_α	$E_{^6\mathrm{He}}$	ε
Model 7	-4.56^a -6.02^b	-26.549	-26.630	-0.081
Exact	-8.380	-29.937	-30.07	-0.133
Experiment	-8.481	-28.296	-29.271	-0.975

[a]With single β, as in t+t. [b]With four β, as in α=t+n.

not only the binding energy has become more realistic, but even more the excitation energy of the first excited state. The g.s. energy has been shifted from -25.595 to -26.549 MeV, while the excitation energy from 34.04 to 20.90 MeV, close to the empirical value of 20.21 MeV. Since this excited state almost coincides with the t+p and h+n thresholds (20.53 and 21.27 MeV in the model and 19.82 and 20.58 MeV empirically), the t+p and h+n clusterings must indeed be overwhelming in it. One can thus say that there is a great deal to improve on the conventional treatment of the α-cluster, and the {t+p; h+n}-like treatment appears to be adequate. The agreement between models 5 and 6 shows that the inclusion of t+t has a very similar effect. This is corroborated by the fact that model 7, in which both {t+p; h+n} and t+t are present, improves on model 6 ($\{(\substack{t^*+p\\h^*+n})+N+N\}$) very little. Correspondingly, the amount of t+t clustering increases appreciably between models 4 and 5, but less for the rest: it is 0.49, 0.54, 0.55, 0.57 for models 4, 5, 6 and 7, respectively. This indicates that the two ways of breaking up the α-cluster opens up much the same segment of the state space.

Similar extensions of the state space for the g.s. of ^{6}Li (Table 13.2) have similar but less drastic effects, showing that the inclusion of the {t+p; h+n} and t+h degrees of freedom are approximately equivalent. The difference between the behaviours of ^{6}He and ^{6}Li can be understood as follows. The two nucleons in ^{6}Li are more tightly bound and thus their overlap with the tail of the core is smaller, which makes the {t+p; h+n} effect smaller. On the other hand, the overlap of the t+h configuration with the α+N+N configuration is larger, which diminishes the t+h effect.

With these two improvements on the α+N+N model, the arsenal of conventional cluster models has been exhausted. The ultimate test of the cluster models is, however, a comparison with the exact results. Such a comparison is made in Table 13.7. In these calculations the force is compatible with that in the exact calculations ($u = 1$ and no spin-orbit force). It is for this reason that the total binding energy is less realistic than the ones shown before. The difference between the model and exact results is nevertheless not very large even in the total binding, but the agreement of the model with the exact result in the intercluster energy is almost perfect. In other words, the missing part of the total binding is just what is missing from the binding of the α-particle as well, and that, in turn, is just what is missing from the triton. This is very satisfactory since the purpose of the cluster model is just to describe the cluster relative motion as perfectly as possible, without perfecting the cluster intrinsic motion. The example shown is probably the only one in which the cluster model is compared with fully consistent exact calculations, and it is probably the most ample justification of the cluster model up to now.

13.1.4 Observables

In the previous subsection exact calculations have been used to justify the $\{\binom{t^*+p}{h^*+n}+n+n;\ t+t\}$ model and that model has been used to justify the $\{\binom{t^*+p}{h^*+n}+n+n\}$ scheme. The agreement of this with the $\{\alpha^*+n+n;\ t+t\}$ model justifies the $\{\binom{t^*+p}{h^*+n}+n+n\}$ model, but the link towards simpler models looks to be cut. For, in looking solely at the energies, we find discrepancy between models $\{\alpha^*+n+n;\ t+t\}$ and α+n+n. In Table 13.6, however, a most important representative of the other observables, the rms radii, are also included. These do depend on the model slightly, but they show remarkable correlation with the energy ε, especially the neutron radius, which carries the halo property. Case $1'$ is just the simple α+n+n model with u readjusted to give the correct intercluster energy. Its prediction for the radii agrees well with the more sophisticated model 5, which produces a similar value for ε. On these grounds, one can consider model α+n+n also realistic provided ε is accurate enough. (Model 3 also reproduces ε, but predicts somewhat different radii. This may be attributed to the different size parameters chosen for the α-core; the size used in model 3 is realistic for the free α-particle.)

Table 13.6 shows that the SVM cluster model predicts the neutron rms radius 0.9 fm larger than the proton radius. From the ingredients of the model it is obvious that this is due entirely to the two single neutrons, thus it is a genuine two-body halo. The empirical rms radii quoted there have been extracted from total cross sections [10, 68] via the Glauber theory (Part I). The empirical values depend very much on the type of the reaction model with which they have been extracted. Two empirical estimates for the difference between the neutron and proton rms radii of ^{6}He ('thickness of the neutron skin') are 0.4 fm [10] and 0.87 fm [68]. The theoretical results ($r_n - r_p = 0.87$ fm in model 2 and 0.91 fm in model 3) are close to 0.87 fm [68] and strongly support the existence of a thick neutron skin in ^{6}He. This will be corroborated by a new analysis on p. 470.

It is most interesting to know what structure the model will predict for the isobaric analogue of the ^{6}He g.s. in ^{6}Li, although the properties of the analogue state are difficult to access experimentally. That this state must have peculiar properties is known from the anomaly found in the inelastic proton [398] and pion [399] scattering transitions leading to this state. It is extremely loosely bound (−0.137 MeV, counted from the three-body threshold), or may be considered unbound as it lies above the α+d threshold. It can still be handled as a bound state because it can only decay via its $T=0$ contamination through a parity-violating transition.[1] Based on the well-known tendency of the growth of the halo with the lessening of binding, one may think that such a loose binding might give rise to an even much larger halo.

The calculations have been performed in model 3 with the model subspace given in the third section of Table 13.2. With the parameters fixed to ^{6}He, the energy predicted for the isobaric analogue state is very close (−0.083 MeV) to the empirical value (−0.137 MeV). By being even closer to zero, it may slightly overestimate the size of the halo. Note that the model contains all minute subspaces of the alien isospin $T=0$ as well in order to give a realistic estimate for the relative behaviours of the two states. But the resulting proton and neutron radii (2.77 and 2.70 fm) are almost equal and are very close to the neutron radius of ^{6}He (2.78 fm). Thus, in this state, ^{6}Li does look very much as if one of the halo neutrons of ^{6}He had been replaced by a proton, which is the very concept of an analogue state. It is also justifiable to say that the neutron–proton halo in the analogue state is very much like the neutron halo in the parent state. Table 13.4 shows that the admixture of $T=0$ in the analogue state is merely 0.07%. Although this impurity is an order of magnitude larger than that of the g.s. [173], it is still low, indeed.

One of the most challenging problems that arise for core+n+n-type Borromean nuclei is the description of the β-delayed deuteron emission. The β-transition may not only lead to a bound state (the g.s.) of the final nucleus but also to the continuum, and the lowest-lying threshold is that of the core plus deuteron channel. In such a case the two-neutron halo transfigures into a deuteron while emitting the electron and the antineutrino. The probability of this process depends sensitively on the

[1]The spin–parity–isospin of this state is $J^\pi(T)=0^+(1)$, while that of α and d are $J^\pi(T)=0^+(0)$ and $1^+(0)$, respectively. Indeed, the decay ^{6}Li$[0^+(1)] \to \alpha[0^+(0)] +$ d$[1^+(0)]$ not only requires isospin violation, but also $l=1$ relative partial wave, which implies parity −1, and hence parity violation.

overlap between the wave function of the neutron-halo state [with one nucleon spin and isospin reversed (Gamow–Teller transition)] and the wave function of the core plus deuteron system moving apart. In particular, recent theoretical analyses [381,382] have revealed an extreme sensitivity to the halo wave function up to distances as large as 15 fm. Thus this process is a stringent test of the reliability of the description of the halo. The probability per time, $[dW(E)/dE]dE$, of the emission of a deuteron in the energy interval $(E, E+dE)$, following the β-decay of ^{6}He, is proportional to the square of the overlap (see Ref. [381] for details)

$$\begin{aligned} o(E) &= \langle \mathcal{A}_{\alpha\mathrm{d}}\{\Phi_\alpha(\xi_\alpha)\Phi_\mathrm{d}(\xi_\mathrm{d})\varphi_{\alpha\mathrm{d}}(\mathbf{r}_{\alpha\mathrm{d}})\}|T_-|\Psi_{^6\mathrm{He}}\rangle \\ &= \int dr \int dR\, F_\mathrm{d}^{(0)}(r)\chi_E^{(0)}(R)g_{\alpha(\mathrm{nn})}^{(00)}(r,R), \end{aligned} \tag{13.1}$$

where the three radial functions, all belonging to orbital angular momentum 0, are as follows: $\frac{1}{r}F_\mathrm{d}^{(0)}(r)$ is the p–n radial relative wave function in the deuteron, $\frac{1}{R}\chi_E^{(0)}(R)$ belongs to the α–d motion in the final state and $\frac{1}{rR}g_{\alpha(\mathrm{nn})}^{(00)}(r,R)$ is the radial factor of the two-particle spectroscopic amplitude [cf. Eq. (11.11)]. The deuteron wave function used is fully consistent with the model, but the relative-motion function, which is a scattering wave function to be normalized asymptotically, has been generated by an α–d local potential. This is a good approximation because the integral in $o(E)$ picks up large contributions only from the asymptotic region. Owing obviously to the satisfactory description of the halo, the calculated probability dW/dE, shown in Fig. 13.4, reproduces the experimental results [400] reasonably well. The quality of agreement is very similar to that of Ref. [382], in which the final state is described in a more sophisticated manner.

The accuracy of the wave functions of the g.s. of ^{6}He and ^{6}Li with respect to each other has been tested in model 3 by calculating the rate of the Gamow–Teller β-transition (see Sect. 9.5) between them. For the $\log ft$ value[2] related to this quantity the model predicts $\log ft = 2.910$,

[2]The 'ft value' is a conventional measure of β-decay life time. The quantity t involved is just the half life in seconds and f is a universal dimensionless correction factor containing the trivial dependence of t^{-1} on the nuclear charge and size, on the decay energy and multipolarity. The ft value for an allowed β-transition can be expressed with the reduced transition probability B_{fi} as $ft = 2\pi^3\hbar^7 \ln 2/(g_\mathrm{F}^2 m_e^5 c^4 B_{fi})$, where m_e is the electron mass and g_F is the Fermi coupling constant, $g_\mathrm{F} = 0.896196 \times 10^{-4}$ MeV fm^3 [401]. [The dimensionless coupling constant G is often defined as $G = (m^2c/\hbar^3)g_\mathrm{F}$ with the nucleon mass m.] The ft value (in seconds), which can be measured experimentally, is thus related to the reduced transition probability by $ft = 5969/B_{fi}$. It is useful to compare the β-decay transition rates of any transitions in terms of their $\log ft$ values since $\log ft$ depends exclusively on the nuclear matrix elements. The reduced

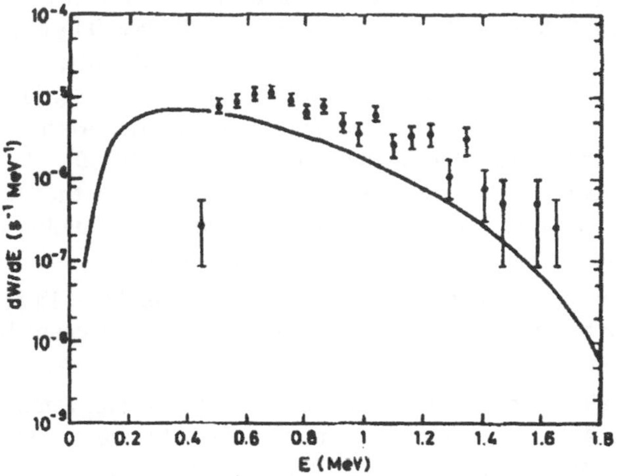

Figure 13.4: Transition probability dW/dE per time and energy unit (in s^{-1} MeV^{-1}) in the c.m. frame for the emission of deuterons following the β-decay of ^{6}He as a function of the c.m. energy E (MeV). The experimental points are from Ref. [400].

while the experimental value is 2.910±0.002 [387]. The agreement is perfect.

The form factors contain detailed information on the densities and currents of the nucleus in a form in which these enter into the description of electron scattering, viz. in a Fourier-transformed form [cf. Sect. 9.5.1]. Electron scattering experiments have only been performed with stable nuclei[3], so one has to resort to stable nuclei for comparison. The nucleus ^{6}Li is not too exotic, and yet it is useful to involve it in the group of nuclei studied because on ^{6}Li a nearly complete comparison with electron scattering data is possible. From the degree of agreement, one can assess the performance of the cluster model with correlated Gaussians for the exotic nuclei as well. The following results are thus to be considered as representing the over-all performance of the model.

Calculations have been performed for the elastic charge (E0) and

transition probability B_{fi} is expressed in terms of the reduced matrix elements of the Fermi and Gamow–Teller transition operators mentioned in Sect. 9.5.1 as $B_{fi} = B(\mathrm{F}, i \to f) + (g_{\mathrm{GT}}/g_{\mathrm{F}})^2 B(\mathrm{GT}, i \to f)$ [cf. Eq. (9.86)], where the ratio of the Gamow–Teller and Fermi coupling constants is $g_{\mathrm{GT}}/g_{\mathrm{F}} = -1.267\pm0.0035$ [401]. For precise details, see, e.g., Ref. [83].

[3]In the second phase of the RIKEN (Wako, near Tokyo, Japan) 'radioactive ion beam factory' project collisions of electrons with unstable nuclei will be performed.

magnetic (M1) form factors and the inelastic M1 form factor to the analogue of the ^{6}He g.s., $J^\pi(T)=0^+(1)$, $E_x=3.563$ MeV. The elastic charge form factor is the Fourier transform of the s.p. density, [Eq. (9.134)], which is the g.s. expectation value of the Dirac delta [Eq. (9.127)]. The magnetic elastic dipole form factor is, similarly, the Fourier transform of the expectation value of the dipole operator (9.145). The inelastic M1 form factor is the same as the elastic one, except that the g.s. expectation value in it is to be replaced by the corresponding transition matrix element between the two states concerned.

The results are compared with experiment in Figs. 13.5, 13.6 and 13.7. The finite size of the nucleons was taken into account by using a phenomenological nucleon form factor [403].

The calculated charge form factor is in very good agreement with experiment [402] up to the momentum transfer value where the data seem to show a diffraction minimum. Since, for low momentum transfer, the charge form factor is determined by the charge distribution, the theory should reproduce r_p well. Model 3 gives $r_p=r_n=2.44$ fm, which is indeed in good agreement with the empirical charge radius of 2.56–2.57 fm [396,402,406] (after the correction for the finite proton size). The discrepancy for large q is common to all realistic calculations using effective interactions. There are two mechanisms proposed to be responsible for this discrepancy: short-range correlation [366,376,377] and quadrupole contribution (arising from the $L=2$ wave-function component due to

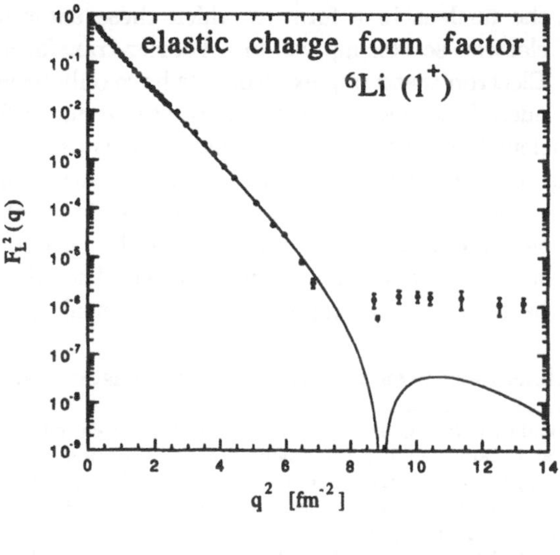

Figure 13.5: Elastic charge from factor for ^{6}Li. Data are taken from Ref. [402].

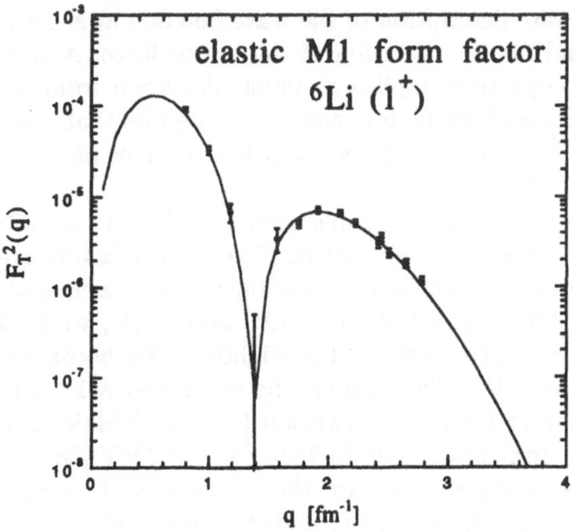

Figure 13.6: Elastic M1 from factor for ^{6}Li. Data are taken from Ref. [404].

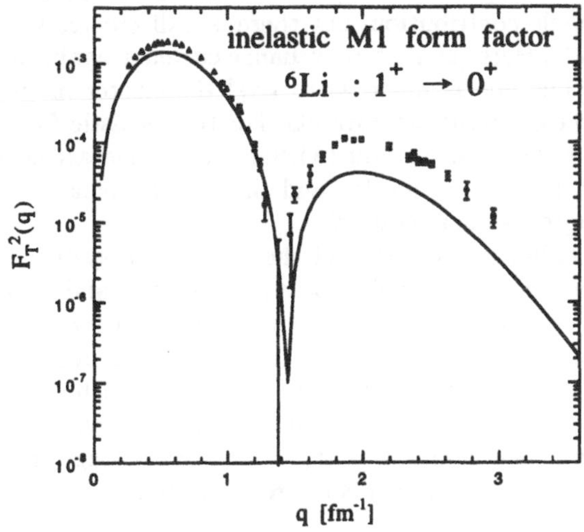

Figure 13.7: Inelastic M1 form factor for the transition to the 3.563 MeV 0^+ state of ^{6}Li. Data are taken from Ref. [405].

the tensor force) to the experimental charge form factor [407]. (The contribution of meson-exchange currents is non-negligible at large momentum transfers but far less than the discrepancy. [408]) These call for a more accurate description of the wave function at short internucleon distances and for the inclusion of the tensor force. A 0s HO description of the α-particle implies a similar deviation from the measured α-particle charge form factor, and it may well be that the discrepancy observed in ^{6}Li is due to the schematic treatment of the α-particle in ^{6}Li [376, 377, 409].

Both elastic and inelastic magnetic form factors are difficult to reproduce by theoretical calculations. These form factors are, nevertheless, in good agreement with experiment. The calculated elastic M1 form factor reproduces the data [404] excellently, while the inelastic form factor is correct in shape but slightly undershoots the data [405]. In keeping with this discrepancy, the calculated M1 transition width corresponding to the same transition (6.6×10^{-6} MeV) is 20% smaller than the experimental value, $(8.2\pm0.2)\times10^{-6}$ MeV [387].

It is interesting to compare these results with recent variational Monte Carlo calculations using realistic forces [408]. The accuracy of these calculations in energy is not too high, but the non-central terms and the short-range repulsion of the force are realistic, and all possible correlations are allowed for to some extent. The high-momentum region of the charge form factor is reproduced owing, undoubtedly, to the quadrupole contribution, but there is a discrepancy in the region of 2–2.5 fm^{-1}. Here the meson-exchange currents significantly improve the results but not as much as to rival the cluster model, which includes no meson-exchange currents. For the inelastic form factor, the variational Monte Carlo result (without the meson-exchange currents) agrees with the cluster model, and the meson-exchange currents improve the agreement substantially.

To conclude the discussion of the six-nucleon system, one can say that ^{6}He has been confirmed to be a halo nucleus, and almost all properties of the g.s. and the g.s. analogue have been very well reproduced. The SVM with the correlated Gaussian basis has proved to be a powerful framework for the description of light nuclei, which is flexible enough to describe halo properties as well. Further improvement is obtained by breaking up the α-cluster, but, at the present stage of experimental precision, the α+n+n model is entirely satisfactory.

13.2 The structure of ^{8}He

A simultaneous description of ^{8}He and ^{6}He with the same parameters poses further challenge to the correlated Gaussian cluster model. The nucleus ^{8}He is an interesting case because it has four loosely bound neutrons, and the behaviour of these is not easy to predict by using simple rules of thumb. It is a Borromean nucleus, but, when a neutron is removed, only one more is emitted as ^{6}He is stable. The separation energy of the last two neutrons is more (2.137 MeV) than that of ^{6}He (0.975 MeV), but the binding energy per excess neutron, 0.778, is indeed pretty low. This suggests that these neutrons are bound together and to the core rather loosely. There is thus no foregone conclusion as to whether ^{8}He supports a four-nucleon halo or, on the contrary, no halo at all. Current experiments are not entirely conclusive, but seem to indicate that ^{8}He has a neutron skin, which is thicker by 0.09 fm [10] or 0.06 fm [68] than in ^{6}He.

Apart from direct interest in the behaviour of this particular nucleus, the case of ^{8}He deserves attention from the point of view of understanding the working mechanism of the correlated Gaussian cluster model. It is a five-cluster system, α+n+n+n+n, but a rather simple one at that. It is therefore instructive to have a closer look at how the binding comes about from the contributions of the individual subspaces. This will be illustrated in the following by results for the energy and for the radii. To probe into the binding mechanism further, we shall also show model predictions for the distribution of the last two neutrons.

For the α+n+n+n+n system, there are nine non-equivalent arrangements; these are shown in Fig. 13.8. It is *a priori* clear, however, that some of these are in fact almost identical. For example, both $\{\alpha[(\text{nn})\text{n}]\}\text{n}$ and $(\alpha\text{n})[(\text{nn})\text{n}]$ are tailored for representing three neutrons closely coordinated, so none of them seems to be important but it is likely that they overlap substantially. Such intuitive considerations corroborated by numerical experiments may guide one in singling out the most important arrangements.

A review of the results is given in Table 13.8. The first and second segments of the table have been calculated with the parameter sets adopted in models 1 and 2 for ^{6}He, respectively. Therefore, they are denoted in the same way. In all configurations considered the neutron spins are coupled to zero pairwise; in setting up these pairs the pattern of the relative coordinates has been followed. All orbital angular momenta have been taken zero except those marked by subscripts 1, which means l-value 1. These subscripts occur pairwise, and the corresponding orbital angular momenta are coupled to 0.

The calculations with model version 1 show that quite different arrangements produce surprisingly similar results. The good energy obtained with the SM-like arrangement $\{[(\alpha\text{n})\text{n}]\text{n}\}\text{n}$ with pure $l=0$ shows the enormous flexibility of the correlated Gaussians. To appreciate this, one should recall that the closest relative of this arrangement in the SM would carry an excitation of 4 $\hbar\omega$ over the valence shell! The fact that the α+(4n)-like configuration $\alpha\{[(\text{nn})\text{n}]\text{n}\}$ performs so moderately shows that, in the 'all $l=0$ limit', the neutrons are more closely coordinated with the α cluster than with each other. At first sight, it looks surprising that the two α+(2n)+(2n)-like configurations do not work equally well; α[(nn)(nn)] is less satisfactory obviously because it is closer to an α+(4n)-type configuration like α\{[(nn)n]n\}. The arrangement (αn)[(nn)n] is a representative of α+(3n)+n-like formations, and, not surprisingly, is energetically disfavoured with $l=0$ for similar reasons. The α+(2n)+n+n-like arrangement yields a deep binding, which suggests that a configuration with two closely co-ordinated and two loosely co-ordinated neutrons, which is consistent with a two-neutron halo, has a large weight. The best result is, however, furnished by the $\{[(\alpha n)_1 n]_1 n\}n$ configuration, which is not entirely unexpected since it contains, as it were, the best single ^{6}He configuration, Y_{11}. The multi-configuration calculations again show that different arrangements can substitute for each other. It is clear that the basis should be constructed out of SM-like, α+(2n)+n+n-like and α+(2n)+(2n)-like arrangements. The basis adopted in model 2 is better in that it contains a number of $l=1$ configurations and shows numerical energy convergence.

Both models overbind ^{8}He slightly, but that is consistent with the overbinding of ^{6}He. The rms radii calculated in model 2 are shown in Table 13.9, compared with those of ^{6}He (cf. Table 13.6) and of the α-particle. The neutron skin of ^{8}He is thick but not as much as that of ^{6}He. It is interesting that, when the binding energy is artificially shifted by modifying the u parameter of the force, the radii are hardly changed. This suggests that the energy change, and hence the energy itself, is smeared all over the s.p. degrees of freedom. Thus, in this respect, ^{8}He

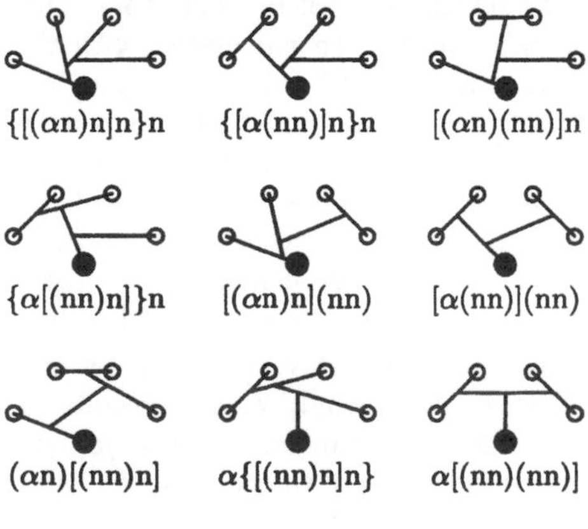

Figure 13.8: All topologically different sets of relative vectors for a system of five particles of which four are identical, with labelling tailored for ^{8}He=α+n+n+n+n.

Table 13.8: Bases used in models 1 and 2 of Table 13.1 for ^{8}He and the resulting energies.

Model	Subspace[a]	Energy[b] (MeV)
1.	$\{[(\alpha n)n]n\}n$	−2.51
	$\{[(\alpha n)_1 n]_1 n\}n$	−3.03
	$\{[\alpha(nn)]n\}n$	−2.96
	$[\alpha(nn)](nn)$	−2.68
	$\alpha[(nn)(nn)]$	−1.17
	$(\alpha n)[(nn)n]$	unbound
	$\alpha\{[(nn)n]n\}$	−1.37
	$\{[\alpha(nn)]n\}n+[\alpha(nn)](nn)$	−3.29
	$\{[\alpha(nn)]n\}n+\alpha\{[(nn)n]n\}$	−3.03
	$\{[\alpha(nn)]n\}n+[\alpha(nn)](nn)+\{[(\alpha n)n]n\}n$	−3.39
	$\{[\alpha(nn)]n\}n+[\alpha(nn)](nn)+\alpha\{[(nn)n]n\}$	−3.32
2.	$\{[(\alpha n)n]n\}n$	−2.49
	$\{[(\alpha n)n]n\}n$, $[(\alpha n)n](nn)$	−3.07
	$\{[(\alpha n)n]n\}n$, $[(\alpha n)n](nn)$, $[\alpha(nn)](nn)$	−3.23
	$\{[(\alpha n)n]n\}n$, $[(\alpha n)n](nn)$, $[\alpha(nn)](nn)$, $\{[\alpha(nn)]n\}n$	−3.28
	$\{[(\alpha n)n]n\}n$, $[(\alpha n)n](nn)$, $[\alpha(nn)](nn)$, $\{[\alpha(nn)]n\}n$, $\{[(\alpha n)_1 n]_1 n\}n$	−3.29
	$\{[(\alpha n)n]n\}n$, $[(\alpha n)n](nn)$, $[\alpha(nn)](nn)$, $\{[\alpha(nn)]n\}n$, $\{[(\alpha n)_1 n]_1 n\}n$, $[(\alpha n)_1 n]_1(nn)$	−3.30
Experiment		−3.112

[a]The neutron spins are coupled to 0 pairwise, and all orbital angular momenta are zero except those denoted by subscript 1, which are 1; these occur pairwise and are coupled to 0 in the following step.
[b]Measured from the α+n+n+n+n disintegration threshold.

is not like a compactified ^{6}He core surrounded by a two-neutron halo; rather, it has a four-neutron skin, which is denser than a usual halo. This observation is in full accord with what the relationships between the interaction cross sections and the two- and four-neutron-removal cross sections show (cf. Sect. 4.2.2). Recall that the cross section data are consistent with Eq. (4.18), which assumes an α-core, but are inconsistent with Eq. (4.17), which assumes a ^{6}He core.

To look into the structure of the neutron skins, it is instructive to compare the spectroscopic amplitudes of the two-neutron removal from ^{6}He and ^{8}He. The radial spectroscopic amplitudes plotted in Fig. 13.9 are those belonging to $l=0$ in the relative motion for n–n ($\boldsymbol{r}$) as well as for α–(nn) ($\boldsymbol{R}$):

$$g^{(0)}(r,R) = rR\int d\hat{\boldsymbol{r}}Y^*_{00}(\hat{\boldsymbol{r}})\int d\hat{\boldsymbol{R}}Y^*_{00}(\hat{\boldsymbol{R}})g(\boldsymbol{r},\boldsymbol{R}) \qquad (13.2)$$

where $g(\boldsymbol{r},\boldsymbol{R})$ is defined in Eq. (11.11). This component gives about 95% of the norm square of the full amplitude, and approximately (cf. Sect 11.3) represents the distribution of two neutrons with respect to the core α and ^{6}He. The similarity between the two functions is remarkable, and the ^{6}He amplitude is very similar to that of the macroscopic three-body models as well [196, 381]. They show a peak in a dineutron-like configuration, in which the two neutrons are much closer to each other than to the core, and another peak in an elongated configuration, with the neutrons farther from each other than from the core. The amplitude of ^{8}He is more compact as if the extra two neutrons were orbiting within the halo neutrons of ^{6}He, thus blurring the difference between the core and the halo.

It is instructive to compare the contour plot in Fig. 13.9a with those of Figs. 12.9c and 12.10c. The difference between the models of ^{6}He cannot be very significant. The model shown in Chap. 12 is restricted in angular momentum but is more complete in describing the spatial distribution of the nucleons within that. The two types of figures reflect different facets of the halo-neutron distribution. Figure 13.9a provides the *amplitude* of the halo-neutron distribution in T-type cluster Jacobi coordinates, while the figures in Chap. 12 give the distribution of a particle as is seen by another from a fixed position. Another difference is that the spectroscopic amplitude in Figure 13.9a is integrated over angles, while the pair correlation function is not. From the peak at

Table 13.9: Theoretical rms radii (in fm) of the point matter, proton and neutron distributions calculated in model version 2 (cf. Table 13.1) for ^{4}He, ^{6}He and ^{8}He, compared with empirical values [10, 68].

Nucleus	r_m		r_p		r_n	
	Th.	Exp.	Th.	Exp.	Th.	Exp.
^{4}He(0^+)	1.42	1.35 [10] 1.42 [68]	1.42	1.35 [10] 1.43 [68]	1.42	1.35 [10] 1.43 [68]
^{6}He(0^+)	2.46	2.48 [10] 2.33 [68]	1.80	2.21 [10] 1.72 [68]	2.67	2.61 [10] 2.59 [68]
^{8}He(0^+)	2.40	2.52 [10] 2.49 [68]	1.71	2.15 [10] 1.76 [68]	2.53	2.64 [10] 2.69 [68]

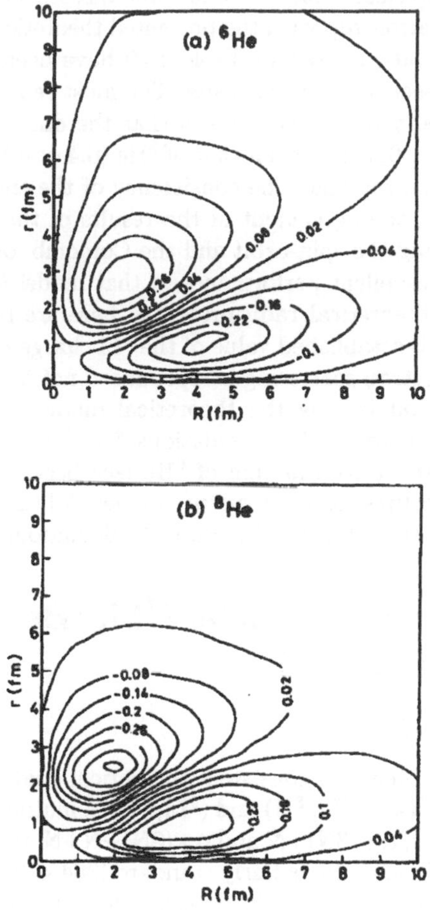

Figure 13.9: Radial two-neutron-removal spectroscopic amplitudes $g^{(0)}$ of ^{6}He (a) and ^{8}He (b). The values of g are given in fm^{-1}.

$(R, r) \sim (4, 1)$ in Fig. 13.9a it should follow that, if there is a halo neutron at 4 fm from the core centre, then there is likely to be another neutron nearby, at a distance of 1 fm. Figure 12.9c bears out this expectation: the second neutron does show some concentration there. Figure 13.9a also shows that, if the two neutrons are brought closer to the centre, say to 2.5 fm, then the distribution with respect to each other shows two distinct peaks: one very near and another farther away. Figure 12.10c shows these qualitative features. Thus the two types of figures seem to accord.

As regards empirical radii, there is some discord between them as well as in their relationship with the presented theoretical results. Both sets of empirical data quoted in Table 13.9 have been deduced from interaction cross section measurements. The most remarkable discrepancy between theory and these data is that the data put the skin of ^{8}He thicker (0.49 or 0.93 fm) than that of ^{6}He (0.4 or 0.87 fm). There is no easy explanation for this. The consistency of the energies obtained for the two nuclei, the agreement of the results of the present model with that containing the spin-orbit and the Coulomb force (Sect. 13.1) and, in turn, the excellent performance of that model for ^{6}Li strongly suggest that the theoretical radii presented here are realistic. At the same time, the well-established value of the ^{6}Li charge radius (2.44 fm) extracted from electron scattering is reproduced not by the interaction cross-section method but by the theoretical model (see Table 13.6). Finally, the agreement of the calculations for the p+^{6}He scattering performed using the wave function of ^{6}He (see Sect. 13.10 later) also supports the correctness of the structure model. All these indicate that this time it is not sure that the theoretical calculations err.

13.3 The mirror nuclei (^{7}Li, ^{7}Be), (^{8}Li, ^{8}B) and (^{9}Li, ^{9}C)

13.3.1 Exposition

This section is devoted to a consistent theoretical study of the chain of mirror pairs (^{7}Li, ^{7}Be), (^{8}Li, ^{8}B) and (^{9}Li, ^{9}C) in multicluster models of the type of α+(3N), α+(3N)+N and α+(3N)+N+N, respectively. Such a description of ^{7}Li and ^{7}Be is fairly standard [364,410–413], but it has to be included as a doorstep to the description of the more complicated nuclei, which currently attract considerable attention [10,414–419]. The study of the chain of nuclei ^{7,8,9}Li also paves the way for a proper description of ^{11}Li.

The treatment of the six nuclei under study is unified in that the same cluster intrinsic states are assumed and the same effective two-nucleon force is applied to all of them. In addition, the state spaces of the larger nuclei are built up from those of the smaller ones, and, for the mirror pairs, the same (mirrored) state space is used. The internal states of α, t and h are 0s HO intrinsic states of a common width parameter β=0.52 fm^{-2}. This yields nearly the correct value for the summed radii of the free α and t or h clusters and minimizes the sum of their energies. The u parameter of the Minnesota effective interaction has been chosen to be u=1. This value has been found to give overall agreement between the experimental and model g.s. energies of ^{7}Be,

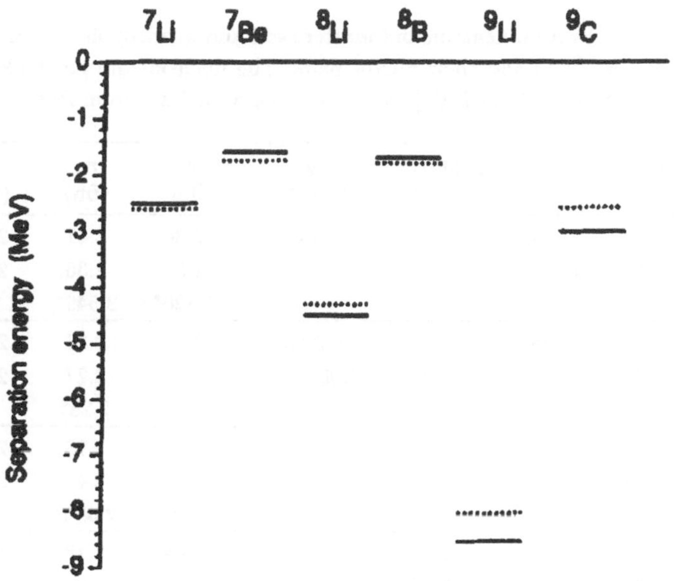

Figure 13.10: Experimental (solid line) and theoretical (dotted line) energies of the two-, three- and four-cluster systems (^{7}Li, ^{7}Be), (^{8}Li, ^{8}B) and (^{9}Li, ^{9}C) with respect to the threshold of disintegration into their constituent clusters.

^{8}B, ^{7}Li and ^{8}Li. For the spin-orbit force, the second of those given in Table 12.2 has been used, which gives the correct spacing between the $\frac{3}{2}^-$ and $\frac{1}{2}^-$ states of ^{7}Li and ^{7}Be.

The calculations performed include the g.s. energies, the point rms radii of the proton, neutron and matter distributions, the magnetic and quadrupole moments, nucleon density distributions and the one- and two-nucleon spectroscopic amplitudes. Here some of the most informative results have been selected for presentation.

For reference in the discussions of the individual nuclei, the results for all these nuclei are summarized in Fig. 13.10 and Table 13.10. Figure 13.10 depicts the g.s. energies and Table 13.10 shows the moments and radii.

13.3.2 The structure of ^{7}Li and ^{7}Be

The bound states of ^{7}Li and ^{7}Be of spin–parities $\frac{3}{2}^-$ and $\frac{1}{2}^-$ are obviously to be interpreted in the cluster model as spin-orbit doublets belonging to intercluster motion of $l = 1$. It is enough to take 7-dimensional bases to get a satisfactory description of these states. The

Table 13.10: Proton, neutron and matter radii and quadrupole and magnetic moments of ^{7,8,9}Li and their mirror pairs. The nucleons are point-like. The empirical radii are from Ref. [10], while those with * are from Ref. [420].

Nucleus		μ [387] (μ_N)	Q ($e\,\text{fm}^2$)	r_m (fm)	r_p (fm)	r_n (fm)
^{7}Be ($\frac{3}{2}^-$)	theor.	−1.27	−6.11	2.36	2.41	2.31
^{7}Be ($\frac{3}{2}^-$)	exp.	—	—	2.31	2.36	2.25
				2.420*	2.549*	2.237*
^{7}Li ($\frac{3}{2}^-$)	theor.	3.15	−3.65	2.33	2.27	2.38
^{7}Li ($\frac{3}{2}^-$)	exp.	3.26	−4.00±0.06[a]	2.33	2.27	2.38
					2.25[e]	
^{8}B (2^+)	theor.	1.42	6.65	2.56	2.73	2.24
^{8}B (2^+)	exp.	1.04	6.83±0.21[b]	2.38	2.45	2.27
				2.507*	2.680*	2.190*
^{8}Li (2^+)	theor.	1.17	2.23	2.44	2.18	2.58
^{8}Li (2^+)	exp.	1.65	3.27±0.06[b]	2.37	2.26	2.44
			3.11±0.05[c]			
			2.4±0.2[d]			
^{9}C ($\frac{3}{2}^-$)	theor.	−1.50	−5.04	2.50	2.64	2.16
^{9}C ($\frac{3}{2}^-$)	exp.	$\|\mu\|$=1.39	—	2.42	—	—
^{9}Li ($\frac{3}{2}^-$)	theor.	3.43	−2.74	2.39	2.10	2.52
^{9}Li ($\frac{3}{2}^-$)	exp.	3.44	−2.74[c]	2.32	2.18	2.39

[a]Ref. [421]; [b]Ref. [134]; [c]Ref. [422]; [d]Ref. [387]; [e]Ref. [396,406]

resulting energies and some other static properties are compared with experiment in Fig. 13.10 and in Table 13.10, respectively. The description may be deemed satisfactory. Although it would be easy to make the models of these nuclei more realistic, that was not the aim of this study. This model is sophisticated enough to base on it realistic models of the heavier nuclei and simple enough to make those models feasible.

13.3.3 The structure of ^{8}Li and ^{8}B

The nucleus ^{8}B has recently attracted attention owing to the proposal that it may have a proton halo [414,415] and to its role in the production of high-energy neutrinos in the sun. In the p–p chain of solar nucleosynthesis ^{8}B is produced by the reaction ^{7}Be(p,γ)^{8}B, and the β^+-decay of ^{8}B is responsible for the production of most of the high-energy neutrinos. To understand its structure, various three-cluster models have been developed [416–418]. Together with ^{8}Li, they are among the very few light odd-odd nuclei that are stable against nucleon emission.

For ^{8}Li there are three independent arrangements: (αt)n, (αn)t and (tn)α, and for ^{8}B the 'conjugate' arrangements. The g.s. have spin–parity 2^+. The first arrangement, with $l_1=1$, contains the g.s. doublet of ^{7}Li (^{7}Be), and therefore is intuitively the most important one. The ^{5}He+t-like and ^{5}Li+h-like partitions, which lie, respectively, by 0.89 MeV and 1.97 MeV above the three-body threshold, are also included, and so are the even higher lying α(tn) and α(hp) arrangements. As the even orbital angular momenta are unfavoured for the two-cluster relative motions, the lowest physically important values adding up to positive parity are $l_1=l_2=1$. (In test calculations even partial waves were also included and were found to give minor improvements. This is in accord with the remoteness of the $\frac{1}{2}^+$ states of the subsystems α+n, α+p from the physical region, cf. Sect. 13.1.2.) Both cluster-spin couplings ($S=0,1$) were allowed. The configurations that were found important and hence included are listed in Table 13.11. The choice of the state space was substantiated by a detailed analysis similar to that shown in Table 13.2 for the six-nucleon systems and by tests to assess the importance of higher-l configurations. The energy changed less than 0.01 MeV, which shows that the model space practically contains the exact cluster-model g.s.

In Table 13.11 the decomposition of the state into subspaces is given through the amounts of clustering, defined in Eq. (11.33). Since the sub-

Table 13.11: Cluster decomposition of the g.s. of ^{8}Li and ^{8}B into the subspaces of the bases used.

Configuration						Amount of clustering	
Arrangement		Angular momenta					
^{8}Li	^{8}B	l_1	l_2	L	S	^{8}Li	^{8}B
(αt)n	(αh)p	1	1	1	1	0.95	0.96
		1	1	2	0	0.01	0.01
		1	1	2	1	0.02	0.01
(αn)t	(αp)h	1	1	1	1	0.95	0.95
		1	1	2	0	0.02	0.01
		1	1	2	1	0.01	0.01
(tn)α	(hp)α	1	1	1	1	0.94	0.94
		1	1	2	0	0.01	0.01
		1	1	2	1	0.01	0.01

spaces are not orthogonal, the amounts of clustering over the subspaces do not usually add up to unity. One finds that in each arrangement the component with angular momentum label $(l_1 l_2)LS = (11)11$ has large and almost equal weights. Although the other components have small weights, their contributions to the binding are non-negligible. The (SL) composition irrespective of clustering is exhibited in Table 13.12.

The g.s. energies in Fig. 13.10 show that ^{8}B is slightly overbound, while ^{8}Li is slightly underbound. This difference may arise from the fact that the triton and the helion are not equally well represented by the SM wave function of fixed (and too small) size, for the helion is, in reality, appreciably larger than the triton.

As is shown in Table 13.10, ^{8}B has the thickest skin (a proton skin of 0.49 fm) of the $A = 8, 9$ nuclei considered. This skin thickness is not as large as is typical for a neutron-halo nucleus, the very small one-proton separation energy, $|\varepsilon| = 0.14$ MeV, yet indicates that it may originate from a one-proton halo. To see whether the nucleus consists of a proton orbiting faraway around a core-like ^{7}Be, we should examine the spectroscopic amplitude for one-proton removal. This function is defined as

$$g^{(1)}(r) = r \int d\hat{\boldsymbol{r}} Y^*_{1m}(\hat{\boldsymbol{r}}) g(\boldsymbol{r}) \tag{13.3}$$

[cf. Eq. (13.2)], with $g(\boldsymbol{r})$ given in Eq. (11.11). The function $g^{(1)}(r)$, shown in Fig. 13.11, does indeed have a prominent tail. This is a realistic estimate since the model intercluster energy is $E_{^8\mathrm{B}} - E_{^7\mathrm{Be}} = -0.18$ MeV, which tallies well with the experimental value of -0.14 MeV. It is instructive, however, to compare the tail of this amplitude with that of the ^{10}Be+n wave function, Fig. 1.2, which can be viewed as an approximate ^{10}Be+n amplitude of ^{11}Be. One can see that the tail of the ^{7}Be+p amplitude is appreciably smaller than that of the ^{10}Be+n amplitude. The difference is caused partly by the Coulomb and centrifugal barriers, which reduce the tail of the ^{8}B amplitude, and partly by components that are orthogonal to the core+N configuration (which

Table 13.12: Decomposition (in %) of the g.s. of ^{8}Li and ^{8}B into components of definite total spin S and total orbital angular momentum L.

S	L	^{8}Li	^{8}B
1	1	96.0	96.5
0	2	1.6	1.6
1	2	2.4	1.9

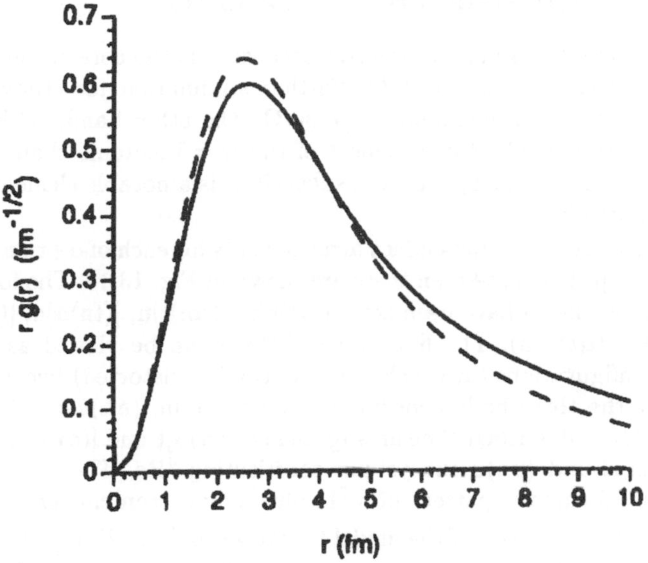

Figure 13.11: Spectroscopic amplitudes for one-proton removal from ^{8}B (solid line) and for one-neutron removal from ^{8}Li (dashed line).

are neglected in the ^{10}Be+n amplitude).

The experiments quoted in Sect. 6.3.2 have left open the question of the existence of a proton halo in ^{8}B. Theory seems to decide it with a compromise: there is a proton halo, but it is not very pronounced.

In electromagnetic moments there is some degree of agreement between theory and experiment (Table 13.10). The least satisfactory cases are the ^{8}Li and ^{8}B magnetic moments. One source of the discrepancy is the oversimplified internal wave functions taken for t and h. In this model the triton and helion spins contribute to the magnetic moments as single nucleons, i.e., with $2.79\,\mu_N$ and $-1.91\,\mu_N$, respectively, whereas the magnetic moments of the isolated triton and helion are $2.98\,\mu_N$ and $-2.13\,\mu_N$, respectively. Since the magnetic moments of ^{7}Li and ^{9}Li are reproduced better, it looks likely that the interaction of the spin-unsaturated t and h with the extra unpaired nucleon causes further trouble. To reproduce the magnetic moments of odd-odd nuclei is always very difficult. For the disagreement in the quadrupole moment of ^{8}Li, one can only say that the theoretical value quoted here is in close agreement with another theoretical estimate [418], while the experiments disagree.

13.3.4 The structure of ^{9}Li and ^{9}C

The nucleus ^{9}Li is of great interest because it is the core of the famous two-neutron-halo nucleus ^{11}Li. Further stimulation for study comes from recent experiments on ^{9}C [419]. On the other hand, ^{9}C looks to be the last particle-stable element of the $N=3$ isotonic chain. Owing to the closure of the $p_{3/2}$ proton subshell, it is a notable element of the proton drip line.

There are nine independent arrangements for each of α+t+n+n and α+h+p+p. For α+t+n+n these are shown in Fig. 13.12. The following six arrangements have been kept: [(αt)n]n, [(αn)t]n, [(tn)α]n, [(nn)α]t, [(αn)n]t, (αt)(nn). The first three of these can be viewed as ^{8}Li+n-type configurations (in which ^{8}Li is, of course, distorted) because they include the three-body coordinate systems (αt)n, (nα)t and (tn)α, in which ^{8}Li is described. The arrangements [(nn)α]t and [(αn)n]t contain the (nn)α and (αn)n subsystems constituting ^{6}He. The arrangement (αt)(nn) forms the pattern of a ^{7}Li plus a dineutron cluster.

In the full version of the model for the g.s. of ^{9}Li ($J^\pi = \frac{3}{2}^-$) the total orbital angular momentum and spin have been restricted to $L=1$ and $S=\frac{1}{2}$, respectively. The only orbital angular momenta included are: $(l_1, l_2, l_3) = (1,0,0)$, $(0,1,0)$, $(0,0,1)$ and $(1,1,1)$, except that $l=0$ has been excluded for the α+t, α+n and t+n motions for reasons explained in Sect. 13.3.3 in conjunction with ^{8}Li and ^{8}B. The intermediate quantum numbers in the coupling of l_i and s_i have been allowed to take values of 0 or 1. A great number of test calculations show that no significant improvement can be achieved by enlarging the model space any further. The g.s. energies in the full model are -8.05 and -2.62 MeV for ^{9}Li and ^{9}C, respectively.

The wave function in this model space is a combination of dozens of

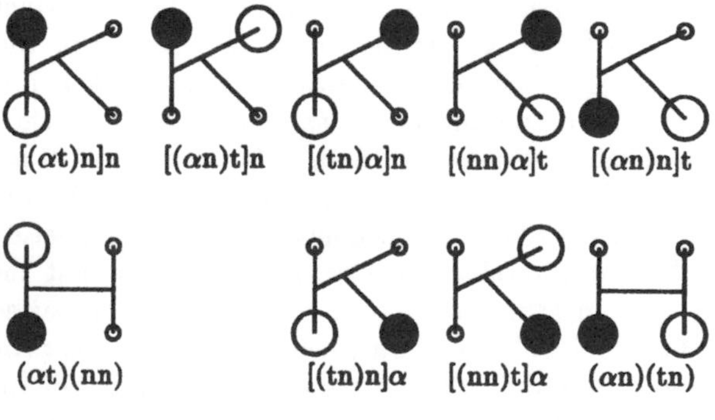

Figure 13.12: All independent arrangements of the system α+t+n+n. Notation: α: ●; t: ○; neutron: ∘. The first six arrangements are included.

Table 13.13: Cluster decomposition of the g.s. of ^{9}Li and ^{9}C into the subspaces of the simplified bases. $L=1$, $S=\frac{1}{2}$, $J=\frac{3}{2}$.

Configuration							Amount of clustering	
Arrangement		Angular momenta						
^{9}Li	^{9}C	l_1	l_2	l_{12}	l_3	s_{12}	^{9}Li	^{9}C
$[(\alpha t)n]n$	$[(\alpha h)p]p$	1	1	1	1	1	0.89	0.84
		1	0	1	0	0	0.19	0.20
		1	0	1	0	1	0.09	0.12
$[(\alpha n)t]n$	$[(\alpha p)h]p$	1	1	1	1	1	0.87	0.86
		1	0	1	0	0	0.11	0.11
		1	0	1	0	1	0.07	0.06
$[(tn)\alpha]n$	$[(hp)\alpha]p$	1	1	1	1	1	0.80	0.82
		1	0	1	0	0	0.13	0.14
$(nn)(\alpha t)$	$(pp)(\alpha h)$	0	0	0	1	0	0.40	0.44
$[(nn)\alpha]t$	$[(pp)\alpha]h$	0	0	0	1	0	0.34	0.22
		0	1	0	1	0	0.21	0.13
$[(\alpha n)n]t$	$[(\alpha p)p]h$	1	0	1	0	0	0.43	0.29
		1	1	0	1	0	0.23	0.17

different configurations. A reduction has been achieved by discarding all configurations with weights (amounts of clustering) less than 3%, whereby a simplified model space has been defined. This truncation goes with a loss of binding of 0.31 MeV and 0.13 MeV for ^{9}Li and ^{9}C, respectively, but the other physical quantities remain largely unaffected. The cluster composition of the wave function in the simplified model space is given in Table 13.13.

Figure 13.10 indicates that ^{9}Li and ^{9}C are underbound by some 0.5 and 0.4 MeV, respectively, by a force that reproduces the g.s. energies of ^{8}Li and ^{8}B more or less satisfactorily. The approximative nature of the solution of the four-cluster problem is only responsible for a minor part of this discrepancy. Allowance for the correct cluster sizes and dynamic cluster distortions could somewhat improve the energy because they deepen the energies of the multicluster system slightly more than those of the isolated clusters. But much of the (minor!) discrepancy may be inherent in the use of a simplified interaction.

The scanty experimental data for the rms radii and electromagnetic moments for the $A=9$ nuclei may be said to be satisfactorily reproduced (Table 13.10). The interaction cross section of ^{9}C on carbon target at 800 A MeV has recently been measured [18] and found to be slightly larger than that of ^{9}Li, thus it seems that the empirical radii will bear out the tendency of the theoretical prediction. The skin thickness of ^{9}C is 0.48 fm, which is not too large for a nucleus that has twice as many protons as neutrons.

13.3.5 Magnetic moments of mirror nuclei

The magnetic moment deserves due attention since it is a sensitive measure of the correctness of some components of the wave function. It is insensitive to fine details of the spatial parts. (The fine details affect the form factors, but the multipole moments are related to the zero momentum-transfer limits of the form factors [234].) But the magnetic moment is very sensitive to the weights and phases of the components belonging to different L, S.

As far as the magnetic moments of mirror nuclei are concerned, it is particularly appropriate to divert our course to the controversial problem which is known as the spin anomaly [1]. To pose this problem, let us decompose the magnetic moment operator (9.145) into isoscalar and isovector terms:

$$\begin{aligned}\frac{1}{\mu_N}\boldsymbol{\mu} &= \sum_{i=1}^{A}\{\tfrac{1}{2}[1-(\tau_i)_z]\boldsymbol{l}_i + 5.586\tfrac{1}{2}[1-(\tau_i)_z]\boldsymbol{s}_i - 3.826\tfrac{1}{2}[1+(\tau_i)_z]\boldsymbol{s}_i\} \\ &= \tfrac{1}{2}\boldsymbol{L} + 0.88\boldsymbol{S} - \tfrac{1}{2}\sum_{i=1}^{A}(\tau_i)_z\boldsymbol{l}_i - 4.706\sum_{i=1}^{A}(\tau_i)_z\boldsymbol{s}_i \\ &= \tfrac{1}{2}\boldsymbol{J} + 0.38\boldsymbol{S} + \tfrac{1}{2}(\boldsymbol{L}_{\rm p} - \boldsymbol{L}_{\rm n}) + 4.706(\boldsymbol{S}_{\rm p} - \boldsymbol{S}_{\rm n}),\end{aligned} \tag{13.4}$$

where the g-factors (9.146) and the nucleonic projectors P_i^j (9.122) (j=p, n) have been inserted, and the proton and neutron (dimensionless) angular momenta are defined as

$$\boldsymbol{L}_j = \sum_{i=1}^{A} P_i^j \boldsymbol{l}_i, \quad \boldsymbol{S}_j = \sum_{i=1}^{A} P_i^j \boldsymbol{s}_i \quad (j = \rm p, n). \tag{13.5}$$

The first two terms in Eq. (13.4) are isoscalar, whereas the last two are isovector.

Let us assume that the two mirror nuclei show the charge symmetry, i.e., their g.s., Ψ_1 and Ψ_2, have the same orbital and spin structure.[4] Then the average of their magnetic moments has no contribution from the isovector part because of full cancellation, and the value of this average is simply

$$(2\mu_N)^{-1}[\mu(1) + \mu(2)] = 0.5J + 0.38\langle S_z\rangle, \tag{13.6}$$

[4]This can be expressed quantitatively by $\Psi_1 \sim (T_-)^n\Psi_2$, where Ψ_1 and Ψ_2 are the respective g.s., T_- is the isospin lowering operator and n is the number of p–n pairs in which the two nuclei differ.

with[5]

$$\langle S_z \rangle = \frac{J(J+1) - L(L+1) + S(S+1)}{2(J+1)}. \tag{13.7}$$

When the values of $\mu(1)$, $\mu(2)$, L and S are known, Eq. (13.6) can be used to test the charge symmetry. If, however, one is confident that the charge symmetry is satisfied, then one can extract information on the angular momentum structure of the mirror pair.

Let us apply this relation to the ^{9}Li–^{9}C pair to probe their spin structure. The empirical magnetic moments (cf. Table 13.10) result[6] in the average value of $\frac{1}{2}[\mu(^9\text{Li})+\mu(^9\text{C})]=1.02\,\mu_N$. A simple SM predicts a configuration with $L=1$ and $S=\frac{1}{2}$ to be dominant, and the correlated Gaussian cluster model confirms this conjecture. The single component with $L=1$, $S=\frac{1}{2}$ gives, for the isoscalar magnetic moment, $\frac{1}{2}[\mu(^9\text{Li})+\mu(^9\text{C})]=0.94\,\mu_N$, which deviates substantially from the experimental value. In fact, Eq. (13.6) requires an expectation value of $\langle S_z\rangle=0.71$ for $L=1$, which grossly disagrees with $\langle S_z\rangle=\frac{1}{2}$. This kind of discrepancy in the spin expectation value has caused some controversy dubbed the spin anomaly [1].

To look for a way out, let us confine our attention to what the model gives. The theoretical value of $\frac{1}{2}[\mu(^9\text{Li})+\mu(^9\text{C})]=0.97\,\mu_N$ (cf. Table 13.10) is rather close to the value of the pure configuration ($0.94\,\mu_N$). That they do not agree completely is due to the fact that the isovector parts do not cancel each other completely. This is so because the effect of the Coulomb interaction is appreciable for a mirror pair with so large a difference in $(N-Z)/A$ (it is $\frac{1}{3}$ for ^{9}Li and $-\frac{1}{3}$ for ^{9}C).

[5] The magnetic moment is defined as the expectation value of the z-component of the vector operator (9.145) in a state with fully aligned angular momentum ($|JM\rangle = |JJ\rangle$). The expression $\langle S_z\rangle = \langle (LS)JJ|S_z|(LS)JJ\rangle$ may be obtained by using a formula

$$\sum_{m_1 m_2} \langle j_1 m_1 j_2 m_2 | jm\rangle^2 m_2 = \frac{j(j+1) - j_1(j_1+1) + j_2(j_2+1)}{2j(j+1)} m.$$

The same result can be derived in a more elementary way. Combining the equalities $\boldsymbol{J}\cdot\boldsymbol{S}=\boldsymbol{L}\cdot\boldsymbol{S}+\boldsymbol{S}^2$ and $\boldsymbol{J}^2=\boldsymbol{L}^2+\boldsymbol{S}^2+2\boldsymbol{L}\cdot\boldsymbol{S}$ leads to a relation $\boldsymbol{J}\cdot\boldsymbol{S} = \frac{1}{2}(\boldsymbol{J}^2-\boldsymbol{L}^2+\boldsymbol{S}^2)$. Thus one has

$$\langle (LS)JJ|\boldsymbol{J}\cdot\boldsymbol{S}|(LS)JJ\rangle = \tfrac{1}{2}[J(J+1) - L(L+1) + S(S+1)].$$

The left hand side can be rewritten alternatively with the use of the relation $\boldsymbol{J}\cdot\boldsymbol{S}= \frac{1}{2}(J_-S_+ + J_+S_-)+J_zS_z = \frac{1}{2}([J_+,S_-]+J_-S_+ + S_-J_+)+J_zS_z = \frac{1}{2}(J_-S_+ + S_-J_+)+(J_z+1)S_z$, where the notations $J_\pm = J_x \pm iJ_y$ etc. and the commutation relation $[J_+,S_-]=2S_z$ have been used. Since $J_+|(LS)JJ\rangle = S_+|(LS)JJ\rangle=0$, the left hand side reduces to

$$\langle (LS)JJ|\boldsymbol{J}\cdot\boldsymbol{S}|(LS)JJ\rangle = (J+1)\langle (LS)JJ|S_z|(LS)JJ\rangle = (J+1)\langle S_z\rangle,$$

which proves Eq. (13.7).

[6] The sign of $\mu(^9\text{C})$ is assumed to be negative.

If the contribution of the charge symmetry violation to the magnetic moment is $0.03\,\mu_N$ as predicted by the model, the isoscalar magnetic moment is $0.99\,\mu_N$. This value can never be explained by excluding the $S=\frac{3}{2}$ components in the ^{9}Li–^{9}C pair. In fact, with $S=\frac{1}{2}$, the only other possible L value is $L=2$, which gives $\frac{1}{2}[\mu(^9\mathrm{Li})+\mu(^9\mathrm{C})]=0.636\,\mu_N$, so that a combination of these two configurations cannot produce $0.99\,\mu_N$. According to Eq. (13.7), $\langle S_z\rangle$ depends on L as well as on S. For $S=\frac{3}{2}$, there are three[7] values of L that yield $J=\frac{3}{2}$: $L=0$, 1, 2, which yield $\frac{1}{2}[\mu(^9\mathrm{Li})+\mu(^9\mathrm{C})]=1.32$, 1.168 and $0.864\,\mu_N$, respectively.

The two like nucleons in α+t+n+n and in α+h+p+p must couple to $S=1$ to yield total spin $S=\frac{3}{2}$. Table 13.4 shows that the percentage of the $S=1$ pair in ^{6}He is about 14%. It is not unreasonable to expect similar $S=\frac{3}{2}$ admixtures in the ^{9}Li–^{9}C pair. With accompanying $L=0$ and/or 1 configurations, it would not be difficult to explain the value of the isoscalar magnetic moment as the $\frac{1}{2}[\mu(^9\mathrm{Li})+\mu(^9\mathrm{C})]$ value is large enough for these configurations.

One may also pose the question to what extent the isovector contamination ($0.03\,\mu_\mathrm{N}$) in $\frac{1}{2}[\mu(^9\mathrm{Li})+\mu(^9\mathrm{C})]$ comes from the term $\boldsymbol{L}_\mathrm{p}-\boldsymbol{L}_\mathrm{n}$ and from $\boldsymbol{S}_\mathrm{p}-\boldsymbol{S}_\mathrm{n}$ in Eq. (13.4). Since in any realistic models the spins of the even numbers of like nucleons are largely paired off, it is probable that the spin expectation value of both ^{9}Li and ^{9}C comes mostly from the odd nucleon. This implies that $\boldsymbol{S}_\mathrm{p}-\boldsymbol{S}_\mathrm{n}$ cancels out in $\frac{1}{2}[\mu(^9\mathrm{Li})+\mu(^9\mathrm{C})]$ between them. Then the charge symmetry violation must be largely produced by an imperfect cancellation of the term $\boldsymbol{L}_\mathrm{p}-\boldsymbol{L}_\mathrm{n}$. This lack of cancellation is, of course, caused by the Coulomb force. A similar effect also appears in the densities: the proton density of ^{9}Li is not the same as the neutron density of ^{9}C (Fig. 13.13).

Thus there is a possibility that the spin anomaly is not an anomaly at all. It looks to be just a natural consequence of the mixing of different L, S components and the violation of the charge symmetry. These unevenly affect the magnetic moments of mirror nuclei.[8]

The charge symmetry violation does not necessarily imply isospin mixing. In the calculations for ^{9}Li and ^{9}C the isospin was taken to be pure $T=\frac{3}{2}$, and yet the charge symmetry has been manifestly broken.

[7]In the lowest SM configuration $L=3$ is ruled out. For ^{9}Li, this configuration is $(0s_{1/2})^4(0p_{3/2})^5$, in which the four p-shell neutrons having $J=0$ cannot have $S=2$ because $S=2$ would imply a totally symmetric spin configuration, which, in turn, implies a totally antisymmetric orbital part, and that is impossible for four neutrons in the 0p orbits. Thus the four p-shell neutrons have either $S(=L)=0$ or 1. Coupled with the proton angular momentum, the total L value can thus go up to 2.

[8]Since there remains a discrepancy in $\mu(^9\mathrm{C})$ between experiment and theory, it *may be* premature to declare a final conclusion. That may still depend on the way the discrepancy is resolved. But to advocate a spin 'crisis' before considering these effects *is certainly* premature.

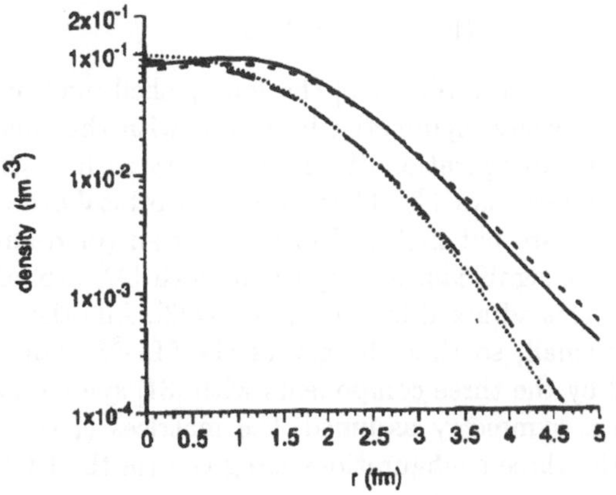

Figure 13.13: Proton density distributions of ^{9}C (short-dashed line) and of ^{9}Li (dotted line) and neutron density distributions of ^{9}C (long-dashed line) and of ^{9}Li (solid line).

In Sect. 13.3.3 we saw that the theoretical values of the magnetic moments of the ^{8}Li–^{8}B pair disagree with experiment. The isoscalar magnetic moment is $\frac{1}{2}[\mu(^8\text{Li})+\mu(^8\text{B})]$=1.34 μ_N experimentally, while it is 1.30 μ_N theoretically. Thus the isoscalar part is reasonably well accounted for by the theory. The discrepancy between theory and experiment comes from the isovector magnetic moment, $\frac{1}{2}[\mu(^8\text{Li})-\mu(^8\text{B})]$; the experimental value of the isovector part is 0.31 μ_N whereas the theoretical value is $-0.13\,\mu_N$. Even the sign of the isovector part is wrong, which indicates that the isovector term in the main component of the wave function, which is characterized by $L=1$, $S=1$ (cf. Table 13.12), also has the wrong sign. This problem has not been resolved in a quantum Monte Carlo calculation [374] either, which has been performed with a bare nucleon–nucleon interaction. This results in $\mu(^8\text{Li}) = 0.91\,\mu_N$, $\mu(^8\text{B})=1.71\,\mu_N$, i.e., for the isoscalar term, $\frac{1}{2}[\mu(^8\text{Li})+\mu(^8\text{B})]=1.31\,\mu_N$ and, for the isovector term, $\frac{1}{2}[\mu(^8\text{Li}) - \mu(^8\text{B})]=-0.40\,\mu_N$.

Let us now examine this problem further by decomposing the magnetic moment into contributions coming from the four operators corresponding to the second line of Eq. (13.4):

$$(\mu_N)^{-1}\mu = 0.5\langle l_z\rangle + 0.88\langle s_z\rangle - 0.5\langle \tau_z l_z\rangle - 4.706\langle \tau_z s_z\rangle, \qquad (13.8)$$

where summations over all nucleons in the expectation values $\langle\cdots\rangle$ are implicit. The most important LS components for the ^{8}Li–^{8}B pair are

$$\text{(i)} \qquad L=1,\ S=1,$$
$$\text{(ii)} \qquad L=2,\ S=0,$$

$$\text{(iii)} \qquad L=2,\ S=1$$

(cf. Table 13.12). In the SM [423] the four p-shell nucleons have predominantly [31] orbital symmetry, in accord with the cluster model.[9] There is another configuration of [31] symmetry with $L=3$, $S=1$, but its admixture is less than 1%. There are other orbital symmetries, [22] and [211], which are not included in the α+t+n (or α+h+p) cluster model. The most significant among them, with [22] orbital symmetry and $L=2$, $S=1$, is admixed by only 2–3% [423]. All other components are negligibly small, so that the g.s. of the ^{8}Li–^{8}B pair is very well approximated by the three components with [31] symmetry.

With charge symmetry assumed, the matrices $(\langle\cdots\rangle)$ of the four operators in the three configurations are given (in the HOSM) by

$$(\langle l_z\rangle) = \begin{pmatrix} 1 & 0 & 0 \\ 0 & 2 & 0 \\ 0 & 0 & \frac{5}{3} \end{pmatrix}, \qquad (\langle s_z\rangle) = \begin{pmatrix} 1 & 0 & 0 \\ 0 & 0 & 0 \\ 0 & 0 & \frac{1}{3} \end{pmatrix},$$

$$(\langle \tau_z l_z\rangle) = \begin{pmatrix} \frac{3}{4} & 0 & -\frac{1}{\sqrt{40}} \\ 0 & \frac{1}{3} & 0 \\ -\frac{1}{\sqrt{40}} & 0 & \frac{5}{18} \end{pmatrix}, \qquad (\langle \tau_z s_z\rangle) = \begin{pmatrix} 0 & 0 & 0 \\ 0 & 0 & -\sqrt{\frac{2}{3}} \\ 0 & -\sqrt{\frac{2}{3}} & 0 \end{pmatrix}. \tag{13.9}$$

As seen from Eq. (13.8), the isovector spin g-factor is much larger than the others, so that its contribution is usually dominant. The present case is, however, special in that the diagonal matrix elements of this term all vanish, which is a consequence of the [31] orbital symmetry.

The pure states predict the following magnetic moments:

$$\begin{aligned} &\text{(i)} && \mu(^8\text{Li}) = 1.005\,\mu_N, && \mu(^8\text{B}) = 1.755\,\mu_N, \\ &\text{(ii)} && \mu(^8\text{Li}) = 0.833\,\mu_N, && \mu(^8\text{B}) = 1.167\,\mu_N, \\ &\text{(iii)} && \mu(^8\text{Li}) = 0.988\,\mu_N, && \mu(^8\text{B}) = 1.266\,\mu_N. \end{aligned} \tag{13.10}$$

The non-central terms of the nucleon–nucleon interaction (such as the spin-orbit and tensor forces) mix the $L=2$ components into the dominant $L=1$ configuration. Let us denote the amplitudes of configurations (ii) and (iii) by ϵ and ϵ'. Since they are rather small, the amplitude of the main configuration can be approximated by $\sqrt{1-\epsilon^2-\epsilon'^2} \approx 1-\frac{1}{2}(\epsilon^2+\epsilon'^2)$. In this approximation the magnetic moments of the ^{8}Li–^{8}B pair are given (in units of μ_N) by

$$\mu(^8\text{Li}) \approx 1.005(1-\epsilon^2-\epsilon'^2) + 0.833\epsilon^2 + 0.988\epsilon'^2$$

[9]The symbol $[f_1 f_2 \cdots]$, with $f_1 \geq f_2 \geq \cdots$ denotes the belonging of a wave function to a well-defined representation of the group $S_{f_1+f_2+\cdots}$ of permutations; see App. I.

$$
\begin{aligned}
&+0.5\times 2\sqrt{\tfrac{1}{40}}(1-\tfrac{1}{2}\epsilon^2-\tfrac{1}{2}\epsilon'^2)\epsilon' + 4.706\times 2\sqrt{\tfrac{2}{3}}\epsilon\epsilon', \\
\mu(^8\mathrm{B}) &\approx 1.755(1-\epsilon^2-\epsilon'^2) + 1.167\epsilon^2 + 1.266\epsilon'^2 \\
&-0.5\times 2\sqrt{\tfrac{1}{40}}(1-\tfrac{1}{2}\epsilon^2-\tfrac{1}{2}\epsilon'^2)\epsilon' - 4.706\times 2\sqrt{\tfrac{2}{3}}\epsilon\epsilon'. \qquad (13.11)
\end{aligned}
$$

The isoscalar and isovector terms will then look like

$$
\begin{aligned}
\tfrac{1}{2}[\mu(^8\mathrm{Li}) + \mu(^8\mathrm{B})] &\approx 1.38 - 0.38\epsilon^2 - 0.253\epsilon'^2 \\
\tfrac{1}{2}[\mu(^8\mathrm{Li}) - \mu(^8\mathrm{B})] &\approx -0.375 + 0.208\epsilon^2 + 0.236\epsilon'^2 \\
&+\sqrt{\tfrac{1}{40}}\epsilon' + 9.412\sqrt{\tfrac{2}{3}}\epsilon\epsilon', \qquad (13.12)
\end{aligned}
$$

where the terms up to second order in ϵ and ϵ' have been kept. In order to obtain values close to the measured moments, the phases of ϵ and ϵ' must be the same and large enough. Since ϵ and ϵ' are expected to be about $-0.2\sim-0.3$, the last term is the really important one from among the isovector terms. For example, the values of $\epsilon = -0.19$ and $\epsilon' = -0.29$ predict $\mu(^8\mathrm{Li}) = 1.37\,\mu_N$ and $\mu(^8\mathrm{B}) = 1.32\,\mu_N$, which is a typical result of the p-shell SM [423, 424]. With this, the isovector magnetic moment ($0.03\,\mu_N$) is significantly improved but is still far from experiment ($0.31\,\mu_N$). In the cluster model the spin-orbit force may not be sufficiently well-defined, and the tensor force is missing, and those may be responsible for the missing admixture.

It is known that meson-exchange currents contribute with extra terms to the two-body magnetic moment operators. Meson-exchange currents arise due to the nucleons' interacting via the exchange of mesons, such as π and ρ, and via the nucleons' getting excited into Δ resonances, which can also exchange mesons with other nucleons. The former type of exchange currents is related, through the continuity equation, to the nucleon–nucleon interaction, whereas the latter type is not constrained by the continuity equation. The meson-exchange currents contribute to the isovector magnetic moment. A SM calculation was performed to investigate the effect of the meson-exchange currents on the magnetic moments in light nuclei [424]. Table 13.14 compares the magnetic moments calculated with and without meson-exchange currents. The extension of the model space to (0+2) $\hbar\omega$ gives a mild effect on the magnetic moment, and hardly alters the isoscalar magnetic moment. The inclusion of the meson-exchange currents is found to improve the isovector magnetic moment considerably.

Table 13.14 includes similar calculations [424] for ^{9}Li as well, but, unfortunately, there are no results for ^{9}C. Yet it can be seen that, unlike for ^{8}Li and ^{8}B, meson-exchange currents and the enlargement of the state space have opposite effects. Thus the discussion of ^{9}Li–^{9}C is probably not invalidated by meson-exchange effects.

Thus, while the problem of the magnetic moments of ^{8}Li–^{8}B seems to be straightened out, there is more uncertainty about ^{9}Li–^{9}C. To look for the possible causes of the remaining discrepancy, one should mention that the SM calculations do not take into account the charge-symmetry breaking effects of the Coulomb force (let alone, of the nuclear force) and are totally insensitive to the effects of the long tails of the wave functions, typical in light exotic nuclei.

13.3.6 Summary

Looking back to the results for the chain of nuclei ^{7,8,9}Li and their mirror pairs, one can fairly claim that the cluster model with correlated Gaussians does give a good overall description of them. The separation energies are reasonably reproduced. The calculated radii are mostly close to the model-dependent empirical values; for the nuclei ^{8}B and ^{9}C, however, the calculated values systematically exceed the empirical ones, while giving thicker proton skins as well. Other theoretical models [414, 417, 418] also tend to have similar predictions. The rms radii of ^{9}Li and ^{9}C are smaller than those of ^{8}Li and ^{8}B. This is likely to be a consequence of the closing of the $p_{3/2}$ neutron (proton) subshell of ^{9}Li (^{9}C). In the cluster-model picture the electric quadrupole moments are mostly influenced by the intercluster motion, thus they tend to be correct. The calculated magnetic dipole moments, on the contrary, are sensitive to the spin structure frozen in the schematic cluster internal wave functions, and may not be so reliable.

The neutron or the proton skins of these nuclei result mostly from a complicated interplay between several degrees of freedom and many configurations. Thus the nucleus cannot be separated into a core and a halo, and the skin is not too thick. There is one exception: ^{8}B, which is on the verge of having a one-proton halo. The halo-like structure results solely from the small proton separation energy, which puts the

Table 13.14: Magnetic moments (in μ_N) of the mirror nuclei ^{8}Li and ^{8}B calculated with and without meson-exchange current (MEC) corrections in $0\hbar\omega$ and $(0+2)\hbar\omega$ model spaces [424].

	$0\hbar\omega$	$(0+2)\hbar\omega$				Exp.
MEC	None	None	$+\pi$	$+\rho$	$+\Delta$	
^{8}Li	1.36	1.42	1.61	1.67	1.76	1.65
^{8}B	1.32	1.24	1.05	0.99	0.90	1.04
^{9}Li	3.20	3.01	3.23	3.39	3.48	3.44

last proton on an orbit with a pretty long tail, but is counteracted by the Coulomb barrier and is not amplified by the compactness of a closed-shell or closed-subshell core (like $N = 2$ in ^{6}He=^{4}He$_2$+n+n, and $N = 6$ in ^{11}Li=^{9}Li$_6$+n+n or in ^{11}Be=^{10}Be$_6$+n). But whether or not these nuclei have halos, their g.s. are rather soft and distortable. The distortability of ^{9}Li should be taken into account in a satisfactory description of ^{11}Li (cf. Sect. 13.7).

13.4 The mirror nuclei ^{9}Be and ^{9}B

13.4.1 Exposition

The beryllium isotopes may be counted among the most exotic light nuclei. The lightest isotope whose description is challenging theoretically is ^{9}Be. It is a nuclear analogue of the hydrogen molecular ion and is an ideal type of a three-cluster nucleus. At the same time, together with its isospin-mirror, ^{9}B, it is an introductory exercise to $^{10,\ldots,14}$Be, of which the first few are feasible to tackle in the framework of this approach. ^{9}Be has just one bound state, while its mirror, ^{9}B has none, but both have a few known resonances [387]. Thus, unlike the previous examples, they offer an opportunity to test the present approach in reproducing nuclear spectra. Furthermore, they provide a testing ground for the treatment of three-cluster resonances. Much work has been done concerning ^{9}Be in macroscopic [425–427] as well as in microscopic [428–431] cluster models. The correlated Gaussian model combines the merits of the best macroscopic and microscopic approaches: it is accurate for the intercluster motion, and it allows refinements in the description of the cluster intrinsic states as well. It is thus destined to improve our understanding of these systems.

^{9}Be is one of the best known exotic nuclei, which gives an opportunity to test the theory against numerous experimental results. Calculated energy spectra, together with the level widths, g.s. radii, static moments, electron scattering form factors and the transition matrix elements of β-decays from the ^{9}Li g.s. to two states of ^{9}Be will be presented.

Of particular interest is that both ^{9}Be and ^{9}B show parity-inverted spectra, like ^{11}Be and ^{11}N. The energy of the lowest-lying $\frac{1}{2}^+$ state is lower that that of the lowest $\frac{1}{2}^-$ state, which is difficult to understand in the SM. As will be shown, the microscopic multicluster model can reproduce the parity inversion in the $A = 9$ nuclei naturally. Related to this, the electric dipole transition from the first excited state ($\frac{1}{2}^+$) to the g.s. ($\frac{3}{2}^-$) [432, 433] is extremely strong. This is similar to that observed in the nucleus ^{11}Be, which may be brought about by the same mechanism. Another interest is the β-transition from the g.s. of ^{9}Li to the g.s. of ^{9}Be [434]. It will be shown that this β-decay takes place owing to a very small component in the wave function of ^{9}Be. The levels of ^{9}B are all particle-unbound and only a few of them have spin assignments [387]. It is mysterious and needs investigating that the analogue of the $\frac{1}{2}^+$ state of ^{9}Be is apparently missing [435, 436].

13.4.2 State spaces and energies

The HO size parameter for α has been set to $\beta = 0.52$ fm^{-2}. The first set of the spin-orbit parameters of Table 12.2 has been used and the Coulomb force has been included. With this, the exchange parameter u of the Minnesota interaction requires $u = 0.94$ in order to put the g.s. separation energy of ^{9}Be at the correct value. This is close to $u = 0.95$, which is optimal for the $\alpha+\alpha$ scattering, but is a compromise with respect to α+N scattering, which would prefer $u = 0.97$.

There are two types of cluster arrangements, $(\alpha\alpha)$N and $(\alpha\text{N})\alpha$, which mirror the T- and Y-type arrangements for α+N+N; just α and N have to be exchanged. The nucleon in the $(\alpha\alpha)$N arrangement corresponds to moving in a 'molecular' orbit around the two molecular centres, while in $(\alpha\text{N})\alpha$ it is, as it were, orbiting in one of the 'atomic' orbits.

The total spin is unique, $S = \frac{1}{2}$, so that the total orbital angular momentum can take $L = J \pm \frac{1}{2}$. The set of configurations to be included depends on the quantum numbers of the states. The bases used are listed in Table 13.15. To judge the importance of certain subspaces, it is instructive to omit some components. When all the nine sets belonging to $J^\pi = \frac{3}{2}^-$ are used for the g.s., the resulting energy (from the $\alpha+\alpha$+N threshold) is -1.431 MeV. When the three sets with $l_1 = 0$ and 2 of arrangement $(\alpha\text{N})\alpha$ are dropped, there is hardly any change, and the overlap between the wave functions of the two models is 0.9995. This is in accord with what was found for ^{8}Li and ^{9}Li, which we called the 'unfavoured status' of even l's in the intercluster relative motion in Sect. 13.3.3. If, however, the arrangement $(\alpha\text{N})\alpha$ is entirely excluded, then the energy increases to -0.32 MeV. Thus the arrangement $(\alpha\alpha)$N alone is incapable of describing the g.s., even though both the s and d waves of the α–α motion are taken into account, and, therefore, the models of Refs. [427, 431], which only include ^{8}Be+N configurations, cannot be considered fully satisfactory. On the other hand, if the $(\alpha\alpha)$N arrangement is omitted, the energy loss is merely 34 keV and the overlap is 0.9991. It should thus be concluded that the ^{5}He+α-type configuration with $l_1 = 1$ constitutes a very good

Table 13.15: Arrangements and angular momenta included in the three-cluster model calculation for ^{9}Be (N=n) and ^{9}B (N=p).

J^π	Arrangement	Angular momenta $(l_1, l_2)L$					
$\frac{1}{2}^-$	$(\alpha\alpha)$N	(0,1)1	(2,1)1	(2,3)1			
	$(\alpha$N$)\alpha$	(1,0)1	(1,2)1				
$\frac{1}{2}^+$	$(\alpha\alpha)$N	(0,0)0	(2,2)0	(2,2)1			
	$(\alpha N)\alpha$	(1,1)0	(1,1)1				
$\frac{3}{2}^-$	$(\alpha\alpha)$N	(0,1)1	(2,1)1	(2,1)2			
	$(\alpha$N$)\alpha$	(0,1)1	(1,0)1	(2,1)1	(1,2)1	(2,1)2	(1,2)2
$\frac{3}{2}^+$	$(\alpha\alpha)$N	(2,2)1	(0,2)2	(2,0)2	(2,2)2	(2,4)2	(4,2)2
	$(\alpha$N$)\alpha$	(1,1)1	(1,1)2	(1,3)2			
$\frac{5}{2}^-$	$(\alpha\alpha)$N	(2,1)2	(2,3)2	(0,3)3	(2,1)3	(2,3)3	
	$(\alpha$N$)\alpha$	(1,2)2	(1,2)3				
$\frac{5}{2}^+$	$(\alpha\alpha)$N	(0,2)2	(2,0)2	(2,2)2	(2,2)3		
	$(\alpha$N$)\alpha$	(1,1)2	(1,3)2	(1,3)3			
$\frac{7}{2}^-$	$(\alpha\alpha)$N	(2,1)3	(0,3)3	(2,3)3	(4,1)3	(2,3)4	(4,1)4
	$(\alpha$N$)\alpha$	(1,2)3	(1,4)3	(1,4)4			
$\frac{9}{2}^+$	$(\alpha\alpha)$N	(2,2)4	(0,4)4	(4,0)4	(2,4)4	(4,2)4	(4,4)4
		(2,4)5	(4,2)5	(4,4)5			
	$(\alpha$N$)\alpha$	(1,3)4	(1,5)4	(1,5)5			

approximation to the g.s. wave function.[10] In the $(\alpha$N$)\alpha$ arrangement l_1 is restricted to $l_1 = 1$ for all other states (cf. Table 13.15).

All resonance states have been treated with the CSM (Sect. 10.4.1) and some have been reconsidered with the ACCC method (Sect. 10.5) as well. For resonance states, the inclusion of high partial waves becomes important to yield stable resonance parameters in the CSM. To increase the stability of the method, the size parameters of the basis have been chosen as $\beta_k^\mu = \beta_0 p^{k-1}$ $(k = 1, \ldots, K)$, rather than randomly. The values of β_0, p and K have been varied for each resonance to get stable values for its energy and width, resulting in $K \simeq 10$. The basis dimension obtained in this way is K^2 times the number of the sets listed in Table 13.15, but for the ACCC calculations even larger bases have been used. (That is why the g.s. of ^{9}Be, which is a bound state, is also deeper with the ACCC basis than with the CSM basis, see Table 13.16.)

[10]It should be noted that, contrary to appearance, in the $(\alpha$N$)\alpha$ arrangement the two α's are still treated symmetrically; this symmetry would only be broken in a model with no parity projection.

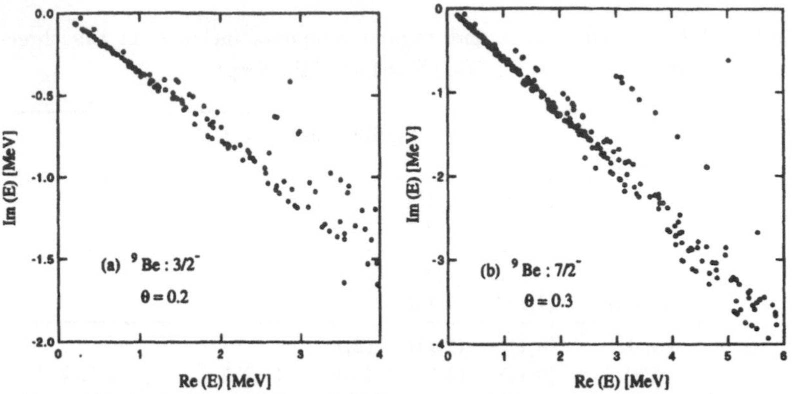

Figure 13.14: Complex eigenvalues for $J^\pi = \frac{3}{2}^-$ (a) and $\frac{7}{2}^-$ (b) of ^{9}Be. The rotation angle θ is in radians. The open circles correspond to resonances.

Figure 13.14 displays two examples for the complex scaled spectra of ^{9}Be. The non-resonant dots form shafts of straight lines parallel to those corresponding to the three-body continuum. These lines belong to continua starting from the discrete states of the two-body subsystems (most of which result just from the discretization of the two-body continua), as if they were genuine thresholds.

The calculated spectra of ^{9}Be and ^{9}B are compared with experiment in Table 13.16. The theoretical level sequence in ^{9}Be corresponds with the observed spectrum fairly well, and the widths are reproduced within a factor of 2, better than in Ref. [431]. There is a second $\frac{3}{2}^-$ resonance predicted, in agreement with other models [429–431]. Although no such state is cited in Ref. [387], the predicted resonance may correspond to the state at an excitation energy of 5.59 MeV mentioned in Ref. [437]. The two broad overlapping resonances with $\frac{7}{2}^-$ and with $\frac{9}{2}^+$ are borne out by recent experiments [437,439]. No $\frac{1}{2}^-$ resonance was found around 6.4 MeV, in accord, again, with Refs. [437,439], although such a state is quoted in Ref. [387] parenthetically. Instead, a $\frac{5}{2}^-$ resonance is obtained at 6.5 MeV, which agrees with other theoretical results [429,430].

The spectrum calculated for ^{9}B is similar to that of ^{9}Be. Based on a comparison with ^{9}Be, one can predict the energies and widths of a few resonances. For example, the calculation predicts a missing $\frac{1}{2}^-$ state at 2.73 MeV. This is in agreement with a recent ^{9}Be(p,n) experiment [438], which located the $\frac{1}{2}^-$ state at 3.11 MeV. The empirical spin assignment of a state at 3.065 MeV [387] is uncertain, but the present model is only consistent with one value, $\frac{5}{2}^+$.

Table 13.16: ^{9}Be and ^{9}B energies and widths. The energy is counted from the three-body threshold. The 3.065 MeV state of ^{9}B is assumed to be of $J^\pi = \frac{5}{2}^+$. ('M'd' stands for 'Method'.)

Nucl.	Experiment [387,437,438]		Theory		
J^π	E (MeV±keV)	Γ (MeV±keV)	M'd	E (MeV)	Γ (MeV)
^{9}Be					
$\frac{3}{2}^-$	−1.5735	0	CSM	−1.431	0
			ACCC	−1.501	0
$\frac{1}{2}^+$	0.111±7	0.217±10	CSM	—	—
			ACCC	1.02	2.62
$\frac{5}{2}^-$	0.8559±1.3	0.00077±0.15	CSM	0.84	0.001
			ACCC	0.838	0.001
$\frac{1}{2}^-$	1.21±120	1.080±110	CSM	1.20	0.46
			ACCC	1.17	0.59
$\frac{5}{2}^+$	1.476±9	0.282±11	CSM	1.98	0.6
			ACCC	1.98	0.6
$\frac{3}{2}^+$	3.131±25	0.743±55	CSM	3.3	1.6
$\frac{3}{2}^-$	4.02±100	1.33±360	CSM	2.9	0.8
$\frac{7}{2}^-$	4.81±60	1.21±230	CSM	5.03	1.2
$\frac{9}{2}^+$	5.19±60	1.33±90	CSM	4.9	2.9
$(\frac{1}{2}^-)$	6.37±80	∼1.0	CSM	—	—
$\frac{5}{2}^-$	—	—	CSM	6.5	2.1
^{9}B					
$\frac{3}{2}^-$	0.277	0.00054±0.21	CSM	0.30	0.004
			ACCC	0.288	0.001
$\frac{1}{2}^+$	(1.9)	∼0.7	CSM	—	—
			ACCC	2.3	2.7
$\frac{5}{2}^-$	2.638±5	0.081±5	CSM	2.55	0.044
			ACCC	2.56	0.050
$\frac{1}{2}^-$	3.11	3.1	CSM	2.73	1.0
			ACCC	2.66	1.15
$\frac{5}{2}^+$	3.065±30	0.550±40	CSM	3.5	1.2
$\frac{3}{2}^+$	—	—	CSM	4.6	2.7
$\frac{3}{2}^-$	—	—	CSM	4.2	1.4
$\frac{7}{2}^-$	7.25±60	2.0±200	CSM	7.0	1.7
$\frac{9}{2}^+$	—	—	CSM	6.6	3.3
$\frac{5}{2}^-$	—	—	CSM	8.4	2.4

The results obtained with the ACCC deserve some special attention. Like in Sect. 13.1, the coupling constant λ is again proportional to u. The energy of the α-particle is unaffected by the value of u; but since both the $\alpha+\alpha$ and the α+N energies depend on u, the two-body thresholds do also depend on u. The physical value of u is to be denoted by $\bar{u}$. Both ^{9}Be and ^{9}B are Borromean, so in the neighborhood of $\bar{u}$ the lowest-lying threshold is the three-body threshold, but that is not the case for larger u values that make the three-cluster states of present interest bound. In the procedure of analytic continuation this requires subtle considerations.

The analytic continuation is to be based on bound states, and the quantity to be extrapolated is the square root of the energy with respect to the threshold reached first from below by decreasing u, which is the ^{8}Be+N threshold. Thus the function extrapolated in this case has been $k(u) \sim \sqrt{E(^9\mathrm{Be})-E(^8\mathrm{Be})}$. The energy $E(^8\mathrm{Be})$ is complex in the u-region in which $\alpha+\alpha$ is unbound, i.e., for $u < 0.958$.

The agreement between the ACCC and CSM results is very good. The theory reproduces the measured resonance energies remarkably well, and the widths are also reasonably well reproduced, except for the $\frac{1}{2}^+$ state, which cannot be located with the CSM.

The results of the analytic continuation for the $\frac{1}{2}^+$ states of ^{9}Be and ^{9}B are shown in Fig. 13.15. The ^{9}Be trajectory shows the typical behaviour of an s-wave two-body state with an attractive potential. It is certainly different from the pattern of Fig. 10.14 expected for a Borromean system. The virtual-state section is easily understood by recalling that, from the origin to the point marked with $u=0.96$ the

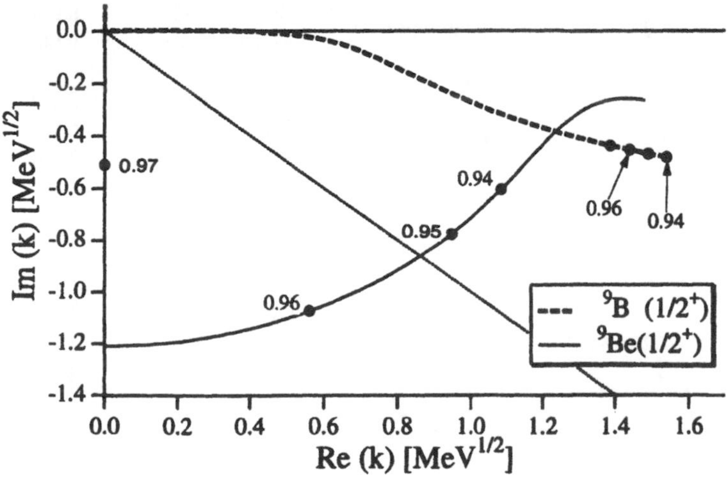

Figure 13.15: Trajectories of the $\frac{1}{2}^+$ states in ^{9}Be and ^{9}B as functions of the u parameter. Here k is identified with $E^{1/2}$. The dotted line marks Im k=$-$Re k.

model subsystem $\alpha+\alpha$ is indeed bound. But the behaviour of the curve is similar even beyond that point, which indicates that this state is indeed of ^{8}Be+n structure. The point corresponding to the physically correct potential ($\bar{u}=0.94$) is in the region of physical resonances, but is still close to the borderline of the unphysical region. The ^{9}B trajectory is, on the other hand, reminiscent of either a two-body resonance with a barrier, like the $B \neq 0$ case of Fig. 10.13, or of a true three-body resonance of a Borromean system, shown in Fig. 10.14. It is obvious that, owing to the higher Coulomb barrier and the sharp resonance in ^{8}Be, the α+α subsystem tends to be formed more dominantly than the α+p subsystem, but the trajectory alone cannot tell us whether the weight of the ^{8}Be+p configuration is large enough to prevail over the α+α+p three-body structure. In comparing the motion of the two poles, it is conspicuous that the rate of change of $k(u)$ along the trajectory is much larger for ^{9}Be than for ^{9}B, which makes the position of the ^{9}Be pole along the trajectory rather ill-defined. Thus it cannot be excluded that the $\frac{1}{2}^+$ state of ^{9}Be is actually a non-physical resonance or a virtual state. Since, however, there is no point on the trajectory that is consistent with the empirical data, theory and experiment cannot be fully reconciled. In particular, the calculated resonance cannot be made much narrower. Indeed, any improvement on the model is likely to make the resonances broader.

13.4.3 Radii and electromagnetic properties

The point rms radii, the electromagnetic moments and some reduced transition probabilities[11] are to be found in Table 13.17. In the correlated Gaussian calculations (SVM) bare operators are used. The charge radius of ^{9}Be fits the experimental value very accurately. The neutron rms radius is larger than the proton radius. Both the magnetic and the quadrupole moments of ^{9}Be are reproduced very well. The M1 and E2

[11]The reduced transition probabilities $B(\mathrm{E}L, J_1 \to J_2)$ and $B(\mathrm{M}L, J_1 \to J_2)$ for a $J_1 \to J_2$ transition are defined in Eq. (9.86). For electric multipoles the operator is given by the unit charge e times O_{LM} of Eq. (9.125), while for magnetic multipoles it is $\mu_N \sum_i [2(L+1)^{-1} g_{l_i} \boldsymbol{l}_i + g_{s_i} \boldsymbol{s}_i] \cdot \boldsymbol{\nabla}_i [r_i^L Y_{LM}(\hat{\boldsymbol{r}}_i)]$. The notations are consistent with those used in Sect. 9.5.1. The operator for $B(\mathrm{M1})$ can be expressed in terms of the magnetic dipole operator (9.145) in a standard way [141,173]. In Table 13.17 the $B(\mathrm{E}L)$ and $B(\mathrm{M}L)$ values are given in Weisskopf units. For an electric transition, the Weisskopf unit, $B_{\mathrm{W}}(\mathrm{E}L)$ [173], equals the value of $B(\mathrm{E}L)$ in a simplified s.p. model from state $j_1 = L+\frac{1}{2}$ to state $j_2 = \frac{1}{2}$ with constant radial wave functions up to radius $R=1.2A^{1/3}$ fm: $B_{\mathrm{W}}(\mathrm{E}L) = \frac{1}{4\pi} 1.2^{2L} \left(\frac{3}{L+3}\right)^2 A^{2L/3}\, e^2\, \mathrm{fm}^{2L}$. For a magnetic transition, the Weisskopf unit $B_{\mathrm{W}}(\mathrm{M}L)$ is defined similarly. When the $B_{\mathrm{W}}(\mathrm{E}L)$ and $B_{\mathrm{W}}(\mathrm{M}L)$ values are expressed in units of $e^2\,\mathrm{fm}^{2L}$ and $\mu_N^2\,\mathrm{fm}^{2L-2}$, respectively, the relationship between the two Weisskopf units is $B_{\mathrm{W}}(\mathrm{M}L) = 40 \left(\frac{L+2}{L+3}\right)^2 B_{\mathrm{W}}(\mathrm{E}\,L-1)$.

Table 13.17: Radii and electromagnetic properties of ^{9}Be. The reduced matrix elements belong to transitions to the $\frac{3}{2}^-$ g.s. The units are: E: MeV; r: fm; μ: μ_N; Q: e fm^2; σ_{reac}: mb; B: Weisskopf units. In the SVM calculations bare-nucleon charges and g-factors are used. Models: MO: α+α+n molecular orbital model; ^{8}Be+n: ^{8}Be+n-type cluster model. In the SM calculation [390, 439] effective charges have been used to calculate the quadrupole moment and the E2 strength.

		Theory					Experiment
		SVM	MO	^{8}Be+n	SM	SM	
		[440]	[430]	[431]	[439]	[390]	[387]
$\frac{3}{2}^-$	E	−1.431		−0.89			−1.5735
	r_m	2.50	2.62				
	r_p	2.39					2.37±0.01
	r_n	2.58					
	μ	−1.169	−1.23	−1.52	−1.27	−1.070	−1.1778±0.0009
	Q	5.13	5.76	4.77	4.35	4.66	5.3±0.3
	σ_{reac}	850					825±20 [441]
$\frac{5}{2}^-$	E	0.883		1.89			0.8559
	B(E2)	22.0	24.7	23.5	12.5	~ 7	24.4±1.8
	B(M1)	0.229		0.10	0.23		0.30±0.03
$\frac{1}{2}^+$	E	1.02[a]		0.05		0.75	0.111
	B(E1)	0.24[b]		0.68	0.03	0.03	0.22±0.09

[a]Calculated with the ACCC method; see Table 13.16.
[b]Estimated with the state $\frac{1}{2}^+$ bound by $u = 1$, which yields $E = -593$ keV.

transition probabilities from the $\frac{5}{2}^-$ state to the g.s. are also well reproduced, and the bound-state estimate for the E1 transition probability between the $\frac{1}{2}^+$ state and the g.s. also turns out to agree with the data.

Table 13.17 includes the results of other models as well. In one of the SM calculations [390] the effective interaction was fitted not only to the energies but also to the static electromagnetic moments of the p-shell nuclei. But a SM calculation in a (0+1) $\hbar\omega$ space cannot produce such a strong E2 or, let alone, E1 transition to the g.s., even with an effective charge of 0.35e [390]. Another SM calculation on a similar basis [439] reproduces the B(E2) value by using a large neutron effective charge, but again gives a very small B(E1) value. Cluster models [427,430,431] perform better, but none of them gives as good results as the SVM since they use more restricted state spaces.

The g.s. wave function of ^{9}Be is further tested by comparing its predictions with electron scattering data. The longitudinal electron-

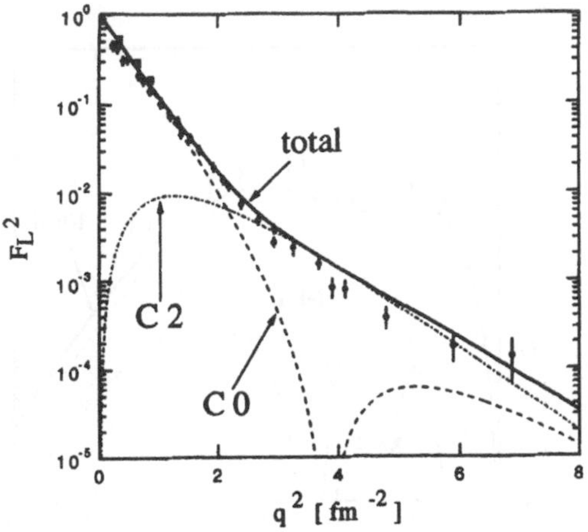

Figure 13.16: Elastic longitudinal form factor for ^{9}Be. The data are from Refs. [442–444].

scattering form factor is a sum of a charge monopole (C0) and quadrupole (C2) term, while the transverse form factor is composed of a magnetic dipole and octupole term. Figures 13.16 and 13.17 show the calculated longitudinal and transverse form factors compared with data from Refs. [442–444] and Refs. [445, 446], respectively. The correction for the finite proton size is taken into account by multiplying the form factor with the proton's form factor [403]. In accord with the results for the radius and the electromagnetic moments, the agreement between theory and experiment is good. It is clear that the quadrupole deformation of the charge density is important at high q^2 values, and so are the octupole components of the magnetization current. A SM calculation [439] needed a quenching factor of 0.7 for the transverse form factors, while no quenching is needed in the present model.

Interesting observations can be made for the structure of ^{9}Be by a closer look at the quadrupole moments and transition rates. The quadrupole moments of the proton and neutron distributions are 5.13 fm^2 and 3.86 fm^2, respectively. Thus, while the neutron radius is larger than the proton radius (see Table 13.17), the neutron deformation is smaller. The smaller neutron deformation suggests a picture of the single neutron moving equatorially, with the two α-clusters being at the poles. The larger neutron radius, on the other hand, suggests that the neutron is squeezed out from the centre to orbit at a distance.

The 2.43 MeV, $\frac{5}{2}^-$ and 6.38 MeV, $\frac{7}{2}^-$ states of ^{9}Be, together with the g.s., approximately follow a $J(J+1)$ rule and hence might be

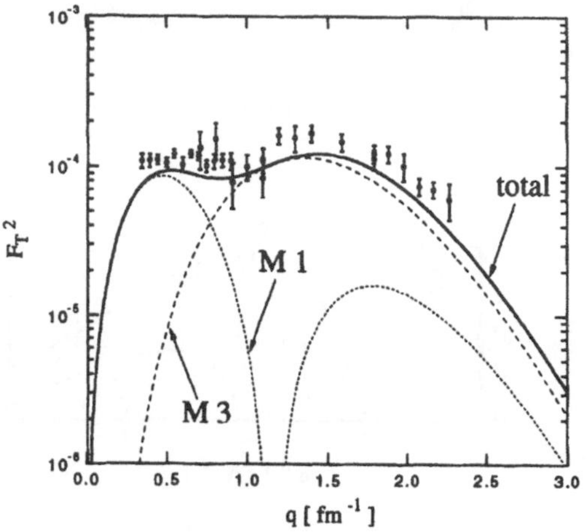

Figure 13.17: Elastic transverse form factor for ^{9}Be. The data are from Refs. [445, 446].

regarded as a rotational band with $K = \frac{3}{2}$ [437, 439]. This implies that the nucleus is permanently deformed. It is interesting to examine the self-consistency of the collective-model picture and its consistency with the cluster model.

The permanent deformation implies a well-defined intrinsic quadrupole moment of the band. From the experimental g.s. quadrupole moment the intrinsic quadrupole moment, Q_0, can be deduced by using the relation $Q = \frac{J(2J-1)}{(J+1)(2J+3)} Q_0 = \frac{1}{5} Q_0$ [174], which gives $Q_0 = 26.5$ fm^2. In the collective model the E2 transition probability within the band is related to Q_0 by

$$B(\text{E2}; KJ_1 \to KJ_2) = \frac{5}{16\pi} \mathrm{e}^2 Q_0^2 \langle J_1 K 20 | J_2 K \rangle^2. \tag{13.13}$$

This implies 23.9 e^2fm^4 and 9.98 e^2fm^4 for the B(E2) values of the $\frac{5}{2}^- \to \frac{3}{2}^-$ and $\frac{7}{2}^- \to \frac{3}{2}^-$ transitions, respectively. The corresponding experimental values, 27.1$\pm$2.0 e^2fm^4 and 7.0$\pm$3.0 e^2fm^4 [387], show the consistency of the model. It is thus meaningful to extract the intrinsic deformation parameter β_0 by using the relation $\beta_0 = \sqrt{\frac{\pi}{5}} Q_0 / (Z \langle r^2 \rangle)$. This gives $\beta_0 = 0.89$, and this value is in good agreement with the deformation of the cluster-model mass distribution.

One can conclude that the correlated Gaussian cluster model reproduces all electromagnetic properties of the g.s. of ^{9}Be.

13.4.4 Beta-decay of ^{9}Li to ^{9}Be

The g.s. of ^{9}Be is fed by β-decay from the g.s. of ^{9}Li. By the two states being thus connected, a stringent test is provided for the consistency of their description. The process is a Gamow–Teller transition, which is described by the reduced matrix element

$$M_{\rm GT}(i \to f) = \left\langle \Psi_{J_f}(^9\text{Be}) \left\| \sum_{k=1}^{9} t_-(k)\sigma(k) \right\| \Psi_{J_i}(^9\text{Li}) \right\rangle, \qquad (13.14)$$

where t_- is an isospin lowering operator. The experimental log ft value is as small as 5.31 [387,434] despite the transition being allowed. This is to be ascribed to the fact that the spatial symmetry of the main component of ^{9}Be is different from that of ^{9}Li [447]. Indeed, in the α+α+n cluster model of ^{9}Be the matrix element (13.14) to any state of ^{9}Be vanishes, regardless of the wave function of ^{9}Li. This can be seen by letting the Hermitean conjugate of the operator in Eq. (13.14) act on the ^{9}Be wave function. This operator has to transform a proton of ^{9}Be into a neutron, but if protons only exist within α-clusters, that is impossible since in the α-clusters all spin–isospin states are filled.

This indicates that the model space of ^{9}Be has to be augmented if this β-decay is to be accounted for. One can just hope that this will be small enough not to destroy the agreement found so far between experiment and theory. In the correlated Gaussian cluster model it is most straightforward to include the distortion of the α-cluster by admitting t+p and h+n components. Since the breaking up of an α-cluster is

Table 13.18: A set of arrangements and angular momenta included in the four-cluster model calculation for the ^{9}Be g.s. for the angular momenta l_1, l_2 and l_3. The spin of the nucleon clusters is coupled to s_{23}. The total spin S is restricted to $\frac{1}{2}$.

Arrangement	Orbital angular momenta $[(l_1,l_2)l_{12},l_3]L$			s_{23}
[(tp)α]n	[(0,0)0,1]1	[(0,2)2,1]1	[(0,2)2,1]2	1
(tp)(αn)	[(0,0)0,1]1	[(0,2)2,1]1	[(0,2)2,1]2	1
[(tp)n]α	[(0,1)1,0]1	[(0,1)1,2]1	[(0,1)1,2]2	1
[(tn)α]p	[(0,0)0,1]1	[(0,2)2,1]1	[(0,2)2,1]2	1
(tn)(αp)	[(0,0)0,1]1	[(0,2)2,1]1	[(0,2)2,1]2	1
[(tn)p]α	[(0,1)1,0]1	[(0,1)1,2]1	[(0,1)1,2]2	1
[(hn)α]n	[(0,0)0,1]1	[(0,2)2,1]1	[(0,2)2,1]2	0
(hn)(αn)	[(0,0)0,1]1	[(0,2)2,1]1	[(0,2)2,1]2	0
[(hn)n]α	[(0,1)1,0]1	[(0,1)1,2]1	[(0,1)1,2]2	0

not very probable, one can conjecture that a term, in which both are broken up will have a very small amplitude, and can be omitted. The state space that has been used is specified in Table 13.18. The excluded arrangements are those in which no nucleons are coordinated with t or h. To save computer time, compromise has been made on the angular momenta included. The intrinsic t and h states have the same size parameter as α. To get the correct four-body separation energy, the u parameter has been decreased to 0.88. The resulting overlap with the wave function of the α+α+n model is 97%.

For the calculation of the β-decay probability, the ^{9}Li g.s. wave function of Sect. 13.3.4 has been adopted. For $\log ft$, 5.60 has been obtained. The agreement with the experimental value, 5.31, may be said fair considering the delicacy of the effect. SM calculations [448] put the $\log ft$ value in the range of 4.86–5.64, depending on the interaction.

It may be mentioned that the mirror β-decay transitions ^{9}Li→^{9}Be and ^{9}C→^{9}B [387] also show the charge symmetry breaking discussed in Sect. 13.3.5 for the magnetic moment. The effect is most prominent for the transition between the $\frac{5}{2}^-$ states.

13.4.5 Summary

It has been demonstrated that the low-energy spectra and the main properties of ^{9}Be and ^{9}B can be well accounted for in an α+α+N three-cluster model. The correlated Gaussian model has now been applied to resonances as well, with success. The description of the three-body dynamics has obviously been adequate even for the highly complicated three-body resonances. Given the performance of the model for known states, its predictions for unknown states and unknown spins and parities are likely to be correct both in ^{9}Be and ^{9}B.

The most prominent successes of the theory are that it has very well reproduced the electromagnetic properties of the ^{9}Be g.s., such as the charge radius, the magnetic moment, the quadrupole moment and the elastic electron scattering form factors. The calculated g.s. density is also consistent with the total reaction cross section data in view of the fact that $\sigma_{\rm reac}$ has been calculated with the OLA (Sect. 3.3), which tends to overestimate the actual values [cf. Table 3.1].

There is one limitation of the α+α+N model encountered, that of its failure to allow β-transition from the g.s. of ^{9}Li to ^{9}Be. But that is not a limitation of the correlated Gaussian cluster model itself. By allowing one of the α-clusters to break up, a 3% admixture has been obtained, and that is found to account for the β-transition reasonably well.

The overall success of the α+α+n model for ^{9}Be gives an optimistic forecast for applications of an α+α+multi-neutron model to heavier Be isotopes.

13.5 The states of ^{10}Be

13.5.1 Exposition

The next member of the Be isotopic chain, ^{10}Be, has some specific properties, which make it a very interesting object for study [449]. Its structure must be dominated by α+α+n+n clustering, as ^{9}Be is described by the α+α+n model very well. If ^{9}Be is a nuclear analogue of the H_2^+ molecule, ^{10}Be is akin to H_2 itself since it consists of two identical heavy clusters and two identical lighter particles. Thus it must be a paradigmatic example for a nuclear molecule. Indeed, there are a few successful applications of a molecular-orbital model to Be isotopes, which assume that the two α-particles are stuck together by neutronic bonds [450–452]. It is to be noted, however, that the intrinsic shape may not be as rigid in such light nuclei as assumed in a molecular model, and a fully dynamical treatment is required to support or supersede the molecular-orbital picture.

Both three-cluster subsystems, ^{9}Be and ^{6}He, are successfully described as three-cluster nuclei, so it is a challenge to describe ^{10}Be as a four-cluster system. In terms of disintegration thresholds, ^{10}Be is the most stable of the Be isotopes. Moreover, it has a number of bound excited states. This is the first nucleus, discussed in this book, that can really be subjected to nuclear spectroscopy.

A delicate feature of the α+α+n+n system is that it consists of two bosons and two fermions. In the microscopic approach, however, there are nucleons only, and, for them, this symmetry derives from complete antisymmetry.

In the SM the second 0^+ state of ^{10}Be is considered an intruder state, which involves a substantial admixture of $2\hbar\omega$ excitations [390, 453]. This is corroborated by a $^{9}\mathrm{Be}(n,\gamma)^{10}\mathrm{Be}$ experiment [454], which shows that the second 0^+ state has a very small ^{9}Be+n spectroscopic factor. This state is of special interest since the α+α+n+n cluster model should be able to describe it on the same footing as the first 0^+ state.

The discussion of the structure of ^{10}Be is organized as follows. The α+α+n+n model will be presented with little detail in Sect. 13.5.2. The results for the spectroscopic properties will be summarized and compared with experiment in Sect. 13.5.3. The properties of the g.s. and the 0^+ excited state will be discussed through s.p. densities and intercluster motions in Sect. 13.5.4. Some general conclusions will be drawn in Sect. 13.5.5.

13.5.2 Model

The α+α+n+n clusters can be arranged in six different ways as shown by Fig. 13.18. Since all arrangements are included in the basis set, it suffices to have just a few angular momenta. The restrictions $l_i \leq 2$ and $l_1+l_2+l_3 \leq 4$ have been applied, and the angular momentum configurations have been randomly sampled along with the continuous parameters of the basis. This procedure has revealed that not all configurations are indispensable; e.g., for the $J^\pi=0^+$ states the SVM has singled out 16 configurations out of 48.

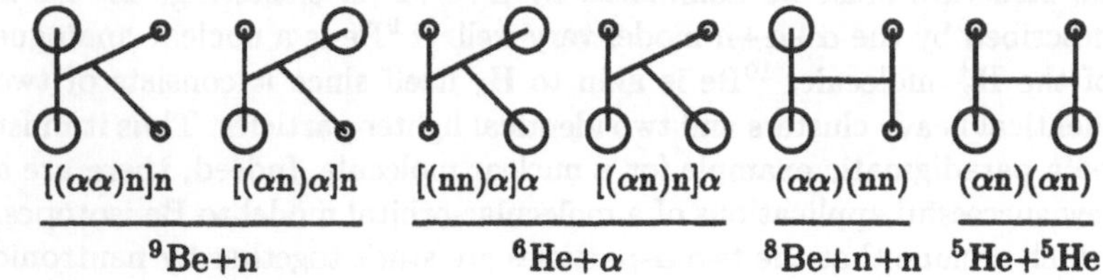

Figure 13.18: Cluster arrangements for ^{10}Be=α+α+n+n. The small circles represent neutrons and the big circles α-particles.

The problem of ^{10}Be is more challenging than any one discussed so far since the numbers of nucleons, clusters and states are all rather high. Since for some spin–parity–isospin values there are two states coming into play, the SVM has been modified to produce optimum bases for more than one lowest-lying states simultaneously. To obtain acceptable bases of dimensions of about $\mathcal{K}=400$–450, refinement cycles have also been used. Moreover, the special basis optimization procedures invented for excited states (cf. Sect. 10.2.4) have also had to be employed.

The intrinsic state Φ_α of the α-particle has been chosen to be the g.s. HO configuration of size parameter $\beta=0.52$ fm^{-2}. The Minnesota interaction has been used with a Coulomb and spin-orbit term. The lowest-lying thresholds for the ^{10}Be=α+α+n+n system are ^{9}Be+n, ^{6}He+α and ^{8}Be+n+n, and all other thresholds lie much higher. The description of ^{10}Be thus requires that the cluster separation energies of the subsystems ^{9}Be, ^{6}He and ^{8}Be be reproduced as well as possible simultaneously. As was seen in the previous sections, each of the subsystems ^{9}Be=α+α+n, ^{6}He=α+n+n and ^{8}Be=α+α is well described in the multicluster model, but with different u parameters, which makes it necessary to compromise in the case of ^{10}Be. The u value and the spin-orbit potential have been chosen such that (i) both ^{9}Be and ^{6}He may be bound, (ii) neither α+n nor α+α be bound. The values chosen are $u=0.91$ and spin-orbit force with $V_4=-629.4$ MeV, $\mu_4=2$ fm^{-2} (set c, Table 12.2). This spin-orbit potential is set III of Ref. [370], with its

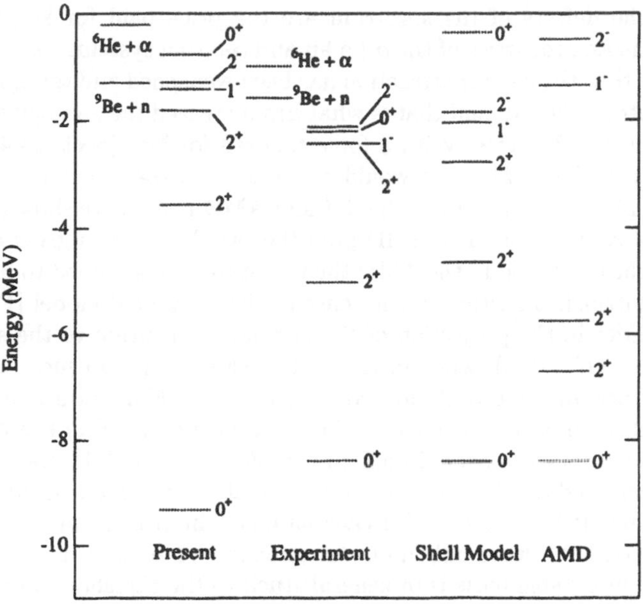

Figure 13.19: Energy spectra of ^{10}Be counted from the α+α+n+n threshold. Experiment: Ref. [387]; SM results: Refs. [390,453]; AMD results: Ref. [373]. The energy scales of the latter two are shifted so as to make the g.s. energies to agree with experiment.

strength V_4 increased by 40%. Fortunately, the potentials determined in this way lead to rather mild modifications in the wave functions of ^{9}Be and ^{6}He obtained in Sects. 13.1 and 13.4. The resulting ^{9}Be+n and ^{6}He+α threshold energies, counted from the α+α+n+n threshold, are -1.28 and -0.20 MeV, respectively, while the experimental values are -1.5735 and -0.9751 MeV [387].

13.5.3 Spectroscopy of states

The calculated energy spectra are compared with experiment and other theoretical models in Fig. 13.19. There are six bound states: two 0^+, two 2^+, one 1^- and one 2^-. In the four-cluster model all are bound except for the second 0^+ state. It will be evident later that the second 0^+ state is clustered predominantly as ^{6}He+α. It lies too high because the position obtained for the ^{6}He+α threshold is also too high.

The g.s. converges well and the resulting energy is deeper than the experimental value by about 0.9 MeV. The excitation energies are also put rather close to, but somewhat higher than, their experimental values. The defects of the spectrum are the price paid for keeping the Borromean properties of the α+n+n and α+α+n systems. It is for this reason that the central attraction has been weakened and the spin-orbit force strengthened, and that is what has increased the gap between the g.s. and the 2^+ state. With a force tailored for ^{9}Be (Sect. 13.4), both the g.s. and the 2^+ state would be closer to experiment (-9.03 and -4.32 MeV, respectively). The SM and AMD results are shown for the sake of comparison. The AMD puts the two 2^+ states a great deal off the right positions. In the AMD the two states are assumed to have the same intrinsic structure in the sense of the rotational model [174] and they differ in the projection of the angular momentum to the symmetry axis: ($K=0$, 2), while in the multicluster model no such 'intrinsic states' are introduced. As for the 0^+ states, the K quantum number is limited to zero, so that no second 0^+ state is predicted in the AMD.[12] As is mentioned in Sect. 13.5.1, in the SM the second 0^+ state cannot be considered anything but an intruder state and hence it is difficult to reproduce it by standard SM calculations. The present model renders the second 0^+ state unbound. An unbound state with an optimized square-integrable basis is in general stuck to the threshold and fills up the space defined by the span of the basis (cf. Sect. 10.4.1). The simultaneous optimization of the basis for the two 0^+ states, however, prevents the unbound state from sticking to the threshold and allows its properties to be qualitatively meaningful.

To expose the contrast between the structures of these states, in Table 13.19 we show a comparative analysis in terms of the occupancies (12.9) of the SM subspaces belonging to definite numbers of HO excitation quanta. The two states appear to be very different. While in the g.s. the lowest orbits are mostly occupied (in the AMD even more [373]), the distribution of oscillator quanta in the second 0^+ state is spread broadly over highly excited configurations. Since the protons are frozen into the α-clusters, the high probability of proton excitation can only be understood by assuming excitation in the relative motion of the clusters. This analysis indicates that in the second 0^+ state the

[12]This only applies to the simple basic version of the AMD, which works with angular-momentum projection after variation [373]. A variational solution without angular-momentum projection produces 'intrinsic states', which, if deformed, get multiplied into rotational bands upon subsequent angular-momentum projection. (Here the word 'intrinsic' has a specific meaning: 'in a body-fixed frame' (cf. p. 183.) This is an attractive property since it establishes or reveals the relationship between several states. It is in this way that the states of ^{10}Be can be organized into three rotational bands (with $K^\pi=0^+$, 1^- and 2^+), to which another 0^+ band should be added on account of the entirely different second 0^+ state. In the advanced version of AMD [455] angular-momentum projection before variation is applied, in which the simplicity is lost, but, with some effort, higher excited states can be produced for each quantum number, and the bound-state spectrum of ^{10}Be is improved.

Table 13.19: Occupation probability $\mathfrak{P}_Q$, per cent, of the $Q\,\hbar\omega$ subspace of the nucleon (N), proton (p) and neutron (n) motion in the g.s. (0_1^+) and first excited 0^+ (0_2^+) state of ^{10}Be. The value of $\hbar\omega$, 14.4 MeV, roughly corresponds to the size of ^{10}Be. Probabilities of less than 1% are indicated by asterisks. The average Q is given in the last column.

Q		0	1	2	3	4	5	6	7	8	9	10	11	12	13	14	Av.
	N	61	0	21	0	11	0	4	0	2	0	*	0	*	0	*	1.4
0_1^+	p	80	3	14	*	2	*	*	*	*	*	*	*	*	*	*	0.5
	n	71	2	16	1	7	*	2	*	*	*	*	*	*	*	*	0.9
	N	2	0	23	0	16	0	14	0	10	0	8	0	6	0	5	8.9
0_2^+	p	36	7	21	4	11	3	6	2	3	1	2	*	*	*	*	2.8
	n	4	1	32	3	16	4	11	3	6	2	4	2	3	1	2	6.1

Table 13.20: Point rms matter, proton and neutron radii of ^{10}Be and ^{9}Be. (Experiment: Refs. [10, 18].)

	J^π	r_m (fm)	r_p (fm)	r_n (fm)
^{10}Be	0_1^+	2.28 (2.30±0.02)	2.17 (2.24±0.02)	2.35 (2.34±0.02)
	0_2^+	3.31	3.06	3.47
	2_1^+	2.41	2.25	2.51
	2_2^+	2.58	2.36	2.72
	1_1^-	3.07	2.50	3.38
	2_1^-	3.17	2.50	3.50
^{9}Be	$\frac{3}{2}^-$	2.50	2.41	2.54

clusters are well separated, which is a severe challenge for the SM description.

The proton, neutron and matter rms radii are given in Table 13.20. The only information on the size of ^{10}Be comes from interaction cross section measurements. The g.s. radii are in reasonable agreement with the experimental values. The g.s. radii of ^{10}Be are slightly smaller than those of the ^{9}Be, showing that the additional neutron makes the nucleus more tightly bound. The second 0^+ state is seen to be dilated, which shows its resemblance to the well-clustered ^{8}Be. The proton radius, 3.06 fm, indicates that the two α-particles are well separated.[13]

[13]This shows that the dilated nature of this state is not a consequence of the model state's being unbound since a large spread in the positive-energy ^{9}Be+n channel would increase the neutron radius much more than the proton radius.

13.5.4 Density distributions

The properties of ^{10}Be can be best understood by comparing its various densities with those of ^{8}Be and ^{9}Be. The s.p. density distributions are defined in Eq. (9.127). For the $J=0$ states these functions are spherical, while for $J \neq 0$ states their monopole components are depicted.

Figure 13.20 shows the proton and neutron densities and neutron-excess density in the g.s. of ^{10}Be compared with those in the g.s. of ^{9}Be ($J^{\pi}=\frac{3}{2}^{-}$) and of ^{8}Be (0^{+}). The wave function of ^{9}Be is essentially the same as is discussed in Sect. 13.4. The ^{8}Be wave function is an $\alpha+\alpha$ wave function in a consistent bound-state approximation. In ^{8}Be the proton and neutron densities are, of course, the same. When a neutron is added, the proton density is contracted (Fig. 13.20a). This is so because the distribution of the additional neutron peaks around $r \sim 2.3$ fm (Fig. 13.20c) and pulls the α-particles, which are separated by more than 3 fm in ^{8}Be, inwards. When another neutron is added, the excess neutron peak doubles and the shrinking becomes more pronounced.

It is especially instructive to compare the densities of the two 0^{+} states. Both the proton and the neutron distributions reveal substantial differences between the two states (Fig. 13.21). The proton density of the second 0^{+} state is much flatter and resembles that of ^{8}Be. The g.s. neutron density has a peak around 1 fm reflecting the predominantly p-wave character of that state (Table 13.19). This dip also tallies with the molecular picture, in which the two 'valence' neutrons form a π-bond, which is a bond perpendicular to the line connecting the two 'atoms' [451, 452]. In the excited state there is no peak of this kind but a flatter falloff. In the SM the largest neutron component is of $2\hbar\omega$ character (Table 13.19), which can be reached by lifting two neutrons to the sd-shell. The peak of the density at the origin indicates that the 1s orbit is occupied mostly. In the molecular picture this peak implies that the 'interatomic' gap is filled up, which is characteristic of a σ-bond [451, 452].

The σ-bond results in an elongated shape, which is consistent with the flat density. Figure 13.21c shows, however, that the density of the excess neutrons is not concentrated in the centre of the nucleus; the second peak reveals that the neutrons actually prefer to stay in the tail region forming, as it were, a neutron skin. These features are perfectly borne out by the AMD as well [455], which is similar to the present approach in that no assumption on the molecular arrangement is introduced.[14]

[14]The AMD is even more unbiassed than the $\alpha+\alpha+$n+n model in one respect: no cluster ansatz is introduced. It is much more restrictive, however, in another respect: the intrinsic wave function in this version of the AMD is a combination of a few Slater determinants of Gaussian wave packets.

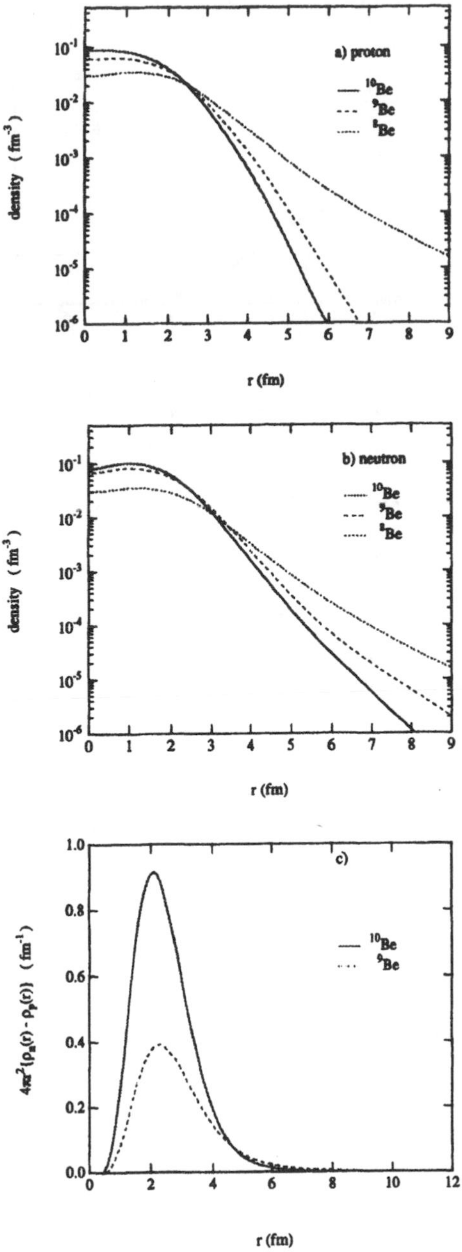

Figure 13.20: Ground-state proton (a) and neutron (b) density distributions of ^{8}Be, ^{9}Be and ^{10}Be and the neutron-excess densities of ^{9}Be and ^{10}Be (c).

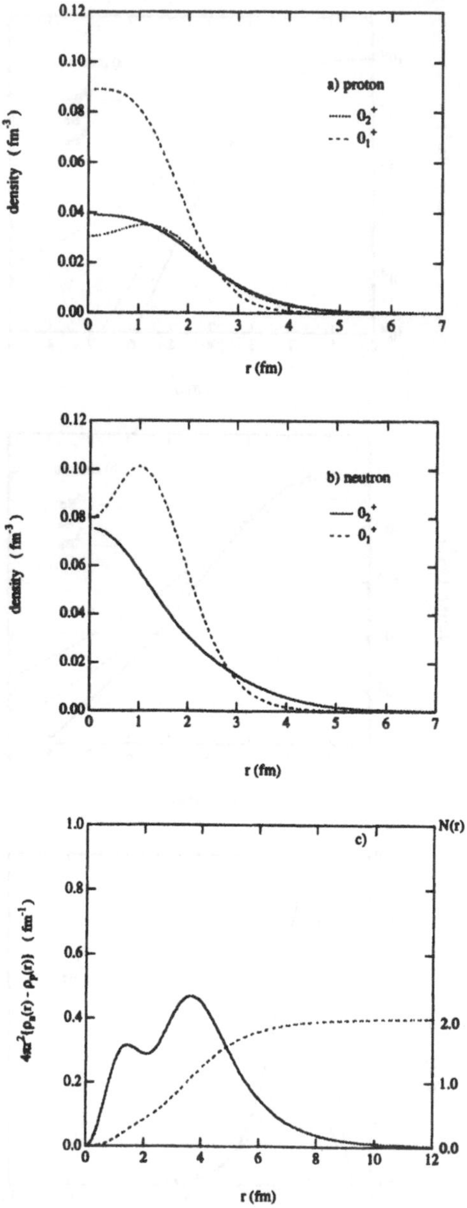

Figure 13.21: Proton (a) and neutron (b) densities of the g.s. (0_1^+) and the second 0^+ state (0_2^+) of ^{10}Be and the neutron-excess density of 0_2^+ (c). The dotted curve in (a) is the proton distribution of the ^{8}Be g.s. The dashed curve in (c) is $N(r) = 4\pi \int_0^r dr' \, r'^2 [\rho_n(r') - \rho_p(r')]$.

The structure studies cannot be complete without probing into some of the important relative motions. The two low-lying thresholds are ^{9}Be+n and ^{6}He+α and the corresponding relative motions will be examined through the spectroscopic amplitudes (11.11). The ^{9}Be and ^{6}He wave functions are, again, those calculated in the $\alpha+\alpha+$n and $\alpha+$n+n three-cluster models along the lines of Sects. 13.4 and 13.1, but with exactly the same potential as used for ^{10}Be.

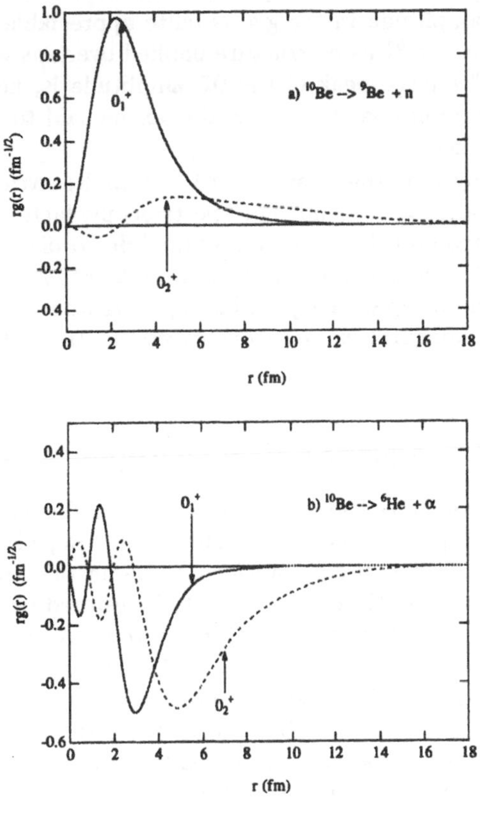

Figure 13.22: ^{9}Be($\frac{3}{2}^-$)+n (a) and ^{6}He+α (b) spectroscopic amplitudes of the g.s. and of the second 0^+ state of ^{10}Be.

The ^{9}Be+n spectroscopic amplitudes of the two 0^+ states for the disintegration into the g.s. ($J^\pi = \frac{3}{2}^-$) of ^{9}Be and a $p_{3/2}$-wave relative motion are displayed in Fig. 13.22a. With its large spectroscopic factor, 2.24, the g.s. behaves like a $0p_{3/2}$ neutron over the g.s. of ^{9}Be. In the second 0^+ state, on the contrary, the ^{9}Be+n relative motion is of little weight; its spectroscopic factor is 0.1, and the node of the amplitude shows a predominant $1p_{3/2}$ contribution.

The ^{6}He+α spectroscopic amplitudes in Fig. 13.22b reveal the back side of the same picture. These amplitudes belong to the $l=0$ relative partial wave. The g.s. and 0_2^+ spectroscopic factors are 0.42 and 0.72, respectively. The node numbers 2 and 3 carry $4\,\hbar\omega$ and $6\,\hbar\omega$, respectively (cf. discussion on p. 311). The energy $4\hbar\omega$ for the relative motion is the minimum required by the Pauli principle [cf. Eq. (11.102)], hence $6\hbar\omega$ in the 0_2^+ state is obviously consistent with $2\hbar\omega$ excitation over the g.s. The amplitude in the g.s. is quite appreciable but, owing to its short range, the ^{6}He+α structure implied overlaps with other configurations. The large peak of the 0_2^+ amplitude is, however, on the surface, which indicates that this state can be said to show clear-cut ^{6}He+α clustering.

From among the other states, the 1^- state deserves attention as a band-head of a $K^\pi=0^-$ band. The spectroscopic factor of the removal of an $s_{1/2}$ neutron is 0.79, while that of the ^{6}He+α disintegration (with $l=1$) is 0.01. The node number of the latter is 2, which is consistent with $1\hbar\omega$ excitation on top of the g.s. This supports its interpretation [455] as a SM-like configuration, with one excess neutron sitting in the 0p and one in the 1s subshell.

13.5.5 The sequence of Be isotopes

The structure of Be isotopes is influenced by α clustering and by the closure of the $0p_{3/2}$-subshell. The former causes ^{8}Be, which lies right on the line of deepest binding per nucleon, to be unstable against disintegration into two α-particles. The stability is restored by adding further neutrons. The highest stability of all Be isotopes reached at ^{10}Be may be interpreted as an effect of the closure of the $0p_{3/2}$-subshell, but clustering has a strong effect even on the closed-shell g.s. When one or two neutrons are lifted to higher orbits, the structure of the nucleus is loosened up, which can be interpreted to result from the increase of the distance between the centres of two α-clusters. The two extra neutrons act not only as a glue between the α-clusters but also as a cover around them. Owing to cluster correlation, the second 0^+ state actually involves much more than a simple $2\hbar\omega$ excitation. The correlated Gaussian approach does not assume any molecular-like arrangement and does not even introduce an 'intrinsic' frame, like AMD, in which 'intrinsic' shapes could be generated, but very much underpins the assumptions of the molecular-orbital models [450–452].

The competition between the filling of the p-shell and the opening of the next shells, which allows more pronounced clustering, is won by the former in ^{10}Be. The g.s. of the nucleus ^{11}Be, however, has spin–parity

$\frac{1}{2}^+$ instead of $\frac{1}{2}^-$, which would be the case if the $p_{1/2}$ shell were to be filled up further. This indicates that the odd neutron prefers to stay in the positive-parity orbits of the sd-shell, which can be energetically favourable only if aided by a more clustered configuration. Thus the famous parity-inversion problem of ^{11}Be is just one step further in the line of problems encountered now, and any model that explains the behaviour of ^{11}Be is only credible if it reproduces the behaviour of ^{10}Be on the same footing.

13.6 The parity inversion in the mirror nuclei ^{11}Be and ^{11}N

The nucleus ^{11}Be is in the heart of the most exotic region of light nuclei. It is next to the two-neutron-halo nucleus ^{11}Li among the $A=11$ isobars, and it is a member of the $Z=4$ isotopic chain, which is dominated by the two-centre structure imposed by the two α-clusters. No wonder, it is exotic in its own right: it is a paradigm for one-neutron-halo nuclei. Its neutron–core model has often been used in Part I to illustrate the reaction theories. Evidence for the halo structure comes from the interaction cross section, which is significantly larger than the geometrical cross section trend (cf. Figs. 1.1 and 3.6 and Ref. [93]) and by the narrow momentum distribution discussed in Sect. 6.3.1 (cf. Fig. 6.3). But the halo structure of ^{11}Be is by no means trivial. It is made puzzling by the parity inversion mentioned at the end of Sect. 13.5.5.

It sounds reasonable to assume that the core of ^{11}Be is the 0^+ g.s. of ^{10}Be, although this assumption is not completely supported by high-energy cross section data. In fact, if this assumption were correct, Eq. (4.16) should hold, but the data show a significant deviation from this rule (see Table 4.1). If ^{11}Be were simply ^{10}Be(g.s.)+n, the simplest SM would imply that the g.s. should be $\frac{1}{2}^-$ since the lowest available orbit for the extra neutron is a $p_{1/2}$ orbit. In fact, however, the g.s. has $J^\pi=\frac{1}{2}^+$, which implies an s-orbit for the halo neutron. The $\frac{1}{2}^-$ state does exist, but at a slightly higher energy. The anomalous sequence of the $\frac{1}{2}^+$ and $\frac{1}{2}^-$ states is usually referred to as a parity inversion. It is easy to produce a neutron halo of the right radius by putting, artificially, an $s_{1/2}$ orbit to the right position, but the structure of ^{11}Be is not understood without explaining the parity inversion.

The first phenomenological step in explaining the parity inversion was to recognize that the order of filling of the neutron shells may depend on the proton configuration [456]. The isotonic sequence of neutron number $N=7$ includes ^{12}B and ^{13}C. The g.s. and first excited state at $E_x=3.09$ MeV of ^{13}C are $\frac{1}{2}^-$ and $\frac{1}{2}^+$, respectively, in keeping with the normal shell filling. We shall extrapolate the trend between ^{13}C and ^{12}B to ^{11}Be and see that it does predict the parity inversion.

To follow up this trend, let us assume that these states of ^{13}C are $0p_{1/2}$ and $1s_{1/2}$ neutron s.p. states, respectively, and denote their energies by $E_{0p_{1/2}}(^{13}\mathrm{C})$, $E_{1s_{1/2}}(^{13}\mathrm{C})$. By removing a proton from the $0p_{3/2}$ orbit in these two states, one obtains $J^\pi = 1^+, 2^+$ or $J^\pi = 1^-, 2^-$ levels of ^{12}B, respectively. The ^{12}B energies $E_j(^{12}\mathrm{B})$ corresponding to $E_j(^{13}\mathrm{C})$ $(j = 0p_{1/2}, 1s_{1/2})$ are the centroids of the two pairs of levels relative to the corresponding s.p. level of ^{13}C. It can be shown that

$$E_j(^{12}\mathrm{B}) = E_j(^{13}\mathrm{C}) - \varepsilon(0p_{3/2}) - \bar{V}(j) + j\text{-independent potential energy terms,} \quad (13.15)$$

where $\varepsilon(0p_{3/2})$ is the s.p. energy of the $0p_{3/2}$ proton and

$$\bar{V}(j) = \frac{\sum_{J=1,2}(2J+1)\langle 0p_{3/2}\, j; JM|V|0p_{3/2}\, j; JM\rangle}{\sum_{J=1,2}(2J+1)}, \quad (13.16)$$

with V the full interaction.[15] The difference between the centroids is

$$E_{1s_{1/2}}(^{12}\mathrm{B}) - E_{0p_{1/2}}(^{12}\mathrm{B}) = E_{1s_{1/2}}(^{13}\mathrm{C}) - E_{0p_{1/2}}(^{13}\mathrm{C}) - [\bar{V}(1s_{1/2}) - \bar{V}(0p_{1/2})]. \quad (13.17)$$

The left-hand side may be taken from the lowest levels of ^{12}B with the assignments $|1^+\rangle = |(0p_{3/2})^{-1}0p_{1/2}; 1M\rangle$, $|2^+\rangle = |(0p_{3/2})^{-1}0p_{1/2}; 2M\rangle$, $|2^-\rangle = |(0p_{3/2})^{-1}1s_{1/2}; 2M\rangle$, $|1^-\rangle = |(0p_{3/2})^{-1}1s_{1/2}; 1M\rangle$. For the energy difference, this yields $E_{1s_{1/2}}(^{12}\mathrm{B}) - E_{0p_{1/2}}(^{12}\mathrm{B}) = 1.43$ MeV. Compared with $E_{1s_{1/2}}(^{13}\mathrm{C}) - E_{0p_{1/2}}(^{13}\mathrm{C})$=3.09 MeV, this means that the $1s_{1/2}$–$0p_{1/2}$ neutron shell spacing is effectively reduced by 1.66 MeV as a result of removing a $0p_{3/2}$ proton. Equation (13.17) indicates that the reduction is due to many-body dynamics.

By linear extrapolation, one obtains that removal of one more $0p_{3/2}$ proton shifts the $1s_{1/2}$ neutron orbit 0.23 MeV below the $0p_{1/2}$ neutron orbit. This is in accord with the empirical facts: the $\frac{1}{2}^+$ state of ^{11}Be is below the $\frac{1}{2}^-$ state by 0.316 MeV. This relationship paves the way for an explanation. It shows that the level inversion is correlated with the occupancy of the proton orbits.

[15]The ^{13}C states of our concern may be written as $\Psi_{jm} = a^\dagger_{jm}|^{12}\mathrm{C}(0^+)\rangle$, while the ^{12}B states are $\psi_{JM} = [a_{0p_{3/2}}a^\dagger_j]_{JM}|^{12}\mathrm{C}(0^+)\rangle$, where a, $a^\dagger$ are fermion annihilation and creation operators. If $|^{12}\mathrm{C}(0^+)\rangle$ is a normalized Slater determinant, these states are also normalized. Equation (13.15) can be obtained by the standard algebra of a, $a^\dagger$. The potential matrix element in Eq. (13.16) is an average of *antisymmetrized* two-particle matrix elements.

A similar relationship exists for the mirror nuclei of the $N=7$ isotones, which are the $Z=7$ isotopes, ^{11}N, ^{12}N and ^{13}N. The value of $\varepsilon(1s_{1/2})-\varepsilon(0p_{1/2})$ for ^{13}N is 2.36 MeV, and it is reduced to $\varepsilon'(1s_{1/2})-\varepsilon'(0p_{1/2})=0.82$ MeV for ^{12}N. Extrapolating this change to ^{11}N linearly suggests that the $1s_{1/2}$ proton is lower by 0.72 MeV than the $0p_{1/2}$ proton. The g.s. of ^{11}N is thus again a $\frac{1}{2}^+$ state[16] and not a $\frac{1}{2}^-$ state. A recent p+^{10}C elastic scattering experiment [457,458] with the radioactive ^{10}C beam has confirmed that the sequence of the resonance levels of the unbound ^{11}N nucleus is $J^\pi=\frac{1}{2}^+, \frac{1}{2}^-$ and $\frac{5}{2}^+$. They have been found at 1.27 MeV, 2.01 MeV, and 3.75 MeV above the p+^{10}C threshold, respectively. Not only the order but also the energy difference is in good agreement with the extrapolation.

It should be clear from these simple arguments that the problem of the s.p. energies underlying the parity inversion problem is an effect that must involve all nucleons. In fact, it is plausible to explain the dependence on the proton occupancies qualitatively as a collective clustering effect. While there are just four protons, the behaviour of the system is dominated by α-clustering, which is a deformed configuration. A deformed core usually causes level crossings, which may lead to the inverted level sequence. The more protons are added, the more the proton distribution must approximate the spherical shape imposed by the vicinity of the closed shell with $Z=8$, which then makes the level inversion disappear. The same considerations apply to the mirror nucleus, with reference to neutrons instead of protons.

Based on this reasoning, one should expect to find parity inversion in the spectrum of the only other neutron-rich odd Be isotope with partially filled neutron p shell, ^{9}Be; and, indeed, there is one (cf. Table 13.16): $E(\frac{1}{2}^+)=0.111$ MeV, $E(\frac{1}{2}^-)=1.21$ MeV. (These energies are measured from the α+α+n threshold.) This finding strongly corroborates the above arguments. What is more, this level sequence is reproduced by the correlated Gaussian model (cf. Table 13.16). Thus one can duly expect that this model will succeed in explaining the case of ^{11}Be as well.

These considerations make it natural to ask whether mean-field the-

---

[16]The parity inversion can also be observed in the spectra of ^{11}B and ^{11}C for the isobaric analogues of the $\frac{1}{2}^+, \frac{1}{2}^-$ parity doublet of ^{11}Be and ^{11}N, respectively. In ^{11}B the $J^\pi=\frac{1}{2}^+$ state (with a possible alternative assignment $\frac{3}{2}^+$) with $T=\frac{3}{2}$ is at 12.56 MeV, while the $J^\pi=\frac{1}{2}^-, T=\frac{3}{2}$ state is at 12.92 MeV. They are close to the ^{10}Be+p threshold (11.23 MeV) and to the ^{10}B+n threshold (11.45 MeV). In ^{11}C the corresponding energies are 12.16 MeV ($T=\frac{3}{2}$ but no spin assignment) and 12.51 MeV ($J^\pi=\frac{1}{2}^-, T=\frac{3}{2}$). Here the nucleon emission thresholds are displaced with respect to the ^{11}B case (the ^{10}B+p threshold at 8.69 MeV and the ^{10}C+n threshold at 13.12 MeV).

ories offer any solution. Spherical Hartree–Fock theories are, of course, unable to give the correct level order, and deformed Hartree–Fock mean fields with Skyrme interactions also fail [202]. In this model the g.s. turns out to be the $\frac{1}{2}^-$ state, and a configuration in which the last neutron occupies the $K^\pi = \frac{1}{2}^+$ orbit lies 3.7 MeV higher and is unstable with respect to neutron emission.

The AMD in its simplest version is basically a single-configuration theory with parity projection. No wonder that the only existing AMD calculation for ^{11}Be [373], in which this simplest version is used, fails to reproduce the parity inversion. The g.s. has $J^\pi = \frac{1}{2}^-$, and the $\frac{1}{2}^+$ state is found to be 7 MeV above the g.s. With a three-body force included, the excitation energy is reduced to about 5 MeV. The magnetic moment of the lowest $\frac{1}{2}^+$ state is $-1.9\,\mu_N$, which is to be compared with the experimental value of $-1.6816\,\mu_N$ [459].

In a pure core+s-orbit configuration the magnetic moment of ^{11}Be should be equal to that of the neutron (the 'Schmidt value'), i.e., $-1.91\,\mu_N$. The experimental value deviates from this, which is a direct indication that either the halo-neutron motion contains higher partial waves or the core is not a pure ^{10}Be(g.s.) configuration or both. In a decomposition of the ^{11}Be g.s. based on the ^{10}Be+n picture the most important configuration beyond ^{10}Be(g.s.)$\times s_{1/2}$ is ^{10}Be(2_1^+, 3.37 MeV)$\times d_{5/2}$. The weight of the higher admixtures can be determined both experimentally and theoretically.

The mechanism of the parity inversion has been studied in the SM [460, 461]. In the variational SM [461] the wave function is a superposition of Slater determinants constructed like in a usual SM, but with flexible s.p. orbits. The $\frac{1}{2}^-$ state comprises four nucleons in the 0s shell and seven in the 0p shell, whereas in the $\frac{1}{2}^+$ state one nucleon is lifted from the p shell to sd shell. With the Skyrme SIII interaction the variational SM reproduces the parity inversion. The major components of the g.s. are ^{10}Be(g.s.)$\times s_{1/2}$ (55%) and ^{10}Be(2^+)$\times d_{5/2}$ (40%). The admixing of the 2^+ core plays a vital role in producing the binding of the last neutron. This admixture may, however, be overdosed: the resulting magnetic moment is $-1.5\,\mu_N$, quenched too much with respect to the Schmidt value. Although the s.p. orbits are supposed to be flexible in the variational SM, the $\frac{1}{2}^- \to \frac{1}{2}^+$ E1 transition matrix element turns out to be too small to reproduce the experimental value [432], $B(\mathrm{E1}; \frac{1}{2}^- \to \frac{1}{2}^+) = 0.116\, e^2\,\mathrm{fm}^2$ (0.36 W.u.).

According to Eq. (7.2), the dipole operator for ^{11}Be can be expressed as a sum of two terms, one for the ^{10}Be core and one for the last neutron, which acquires an effective charge of $-\frac{2}{11}e$. When only the g.s. and the 2^+ state of ^{10}Be are included, the contribution of the core term is negligible, and the E1 matrix element will be determined by the single-neutron term. The fact that the variational SM gives an E1 matrix element which is too small clearly indicates that the spatial spread of the last neutron in the $\frac{1}{2}^+$ and/or $\frac{1}{2}^-$ state is not sufficiently large.

A macroscopic ^{10}Be+n cluster model has also been applied to ^{11}Be [462]. The core–neutron interaction has been assumed to be a Woods–Saxon potential with quadrupole deformation, which then leads to coupled equations for the 'channels' with core states ^{10}Be(0^+) and ^{10}Be(2^+). This model has been found to reproduce the dipole strength function reasonably well with fitted potential parameters. The resulting g.s. of ^{11}Be contains an $s_{1/2}$ neutron coupled to the ^{10}Be(0^+) core with a weight of 78% and a $d_{5/2}$ neutron coupled to the ^{10}Be(2^+) core with 20%. This result seems to be consistent with the reaction ^{9}Be(^{11}Be, ^{10}Beγ)X, in which only ^{10}Be and γ are observed, but neutron knockout from ^{11}Be is dominant [70].

A similar ^{10}Be+n coupled-channel cluster model has been worked out in a microscopic framework [463]. In this model ^{10}Be is described in a full multiconfiguration p-shell HOSM. The two lowest excited core states in this model have $J^\pi = 2^+$ in agreement with experiment, and their excitation energies, 3.4 MeV and 8.3 MeV, compare well with the empirical values, 3.37 MeV and 5.96 MeV. The ^{10}Be+n relative motion is represented by the GCM, virtually exactly. The model cannot reproduce the parity inversion if the same force is used for both states. This shows that there is some unknown mechanism, which would give more binding to the $\frac{1}{2}^+$ state. The calculation shows that both the $\frac{1}{2}^+$ and the $\frac{1}{2}^-$ state have halo structures, yielding a very large E1 transition rate, 40% larger than experiment. The $\frac{1}{2}^+$ state is found to be almost pure (91%) ^{10}Be(g.s.)$\times s_{1/2}$, while the $\frac{1}{2}^-$ state is, surprisingly, dominated by the ^{10}Be(2^+) component (71%). It is not easy to understand why the ^{10}Be+n microscopic cluster model and the variational SM give such inconsistent results despite their apparent similarity. In principle, the macroscopic cluster model is also very similar, but its phenomenological elements allow more freedom, so its better performance is not surprising.

To sum up, the estimates for the component admixed to the configuration ^{10}Be(g.s.)$\times s_{1/2}$ in the g.s. of ^{11}Be range from a few per cent to 40%. The p(^{11}Be, ^{10}Be)d reaction [71] supports a core excitation admixture of 16%. This result is, however, sensitive to the coupling of the 0d orbit to the deformed ^{10}Be core. The neutron knockout reaction [70] has shown that the partial cross section to the g.s. of ^{10}Be is 78% of the total cross section. This experiment indicates that the value of the ^{10}Be(g.s.)$\times s_{1/2}$ spectroscopic factor is as large as 0.8. In this experiment γ-rays from the negative-parity states of ^{10}Be have been observed. The spectroscopic factors to these states are fairly large [70], and, in fact, the partial cross sections to these states are comparable with that to the first 2^+ state.

The point just mentioned gives a hint that configurations that have not been explored so far may be important to explain the parity inversion. These are cluster configurations ^{10}Be(1^-) and ^{10}Be(2^-) coupled to appropriate neutron–core relative-motion states. With these, the g.s. would look like

$$\left|{}^{11}\mathrm{Be}\left(\tfrac{1}{2}^+\right)\right\rangle = c_1|{}^{10}\mathrm{Be}(0^+)\times \mathrm{s}_{1/2}\rangle + c_2|{}^{10}\mathrm{Be}(2^+)\times \mathrm{d}_{5/2}\rangle + c_3|{}^{10}\mathrm{Be}(1^-)\times \mathrm{p}_{1/2}\rangle + c_4|{}^{10}\mathrm{Be}(2^-)\times \mathrm{p}_{3/2}\rangle. \quad (13.18)$$

In the SM the main configuration of ^{10}Be(1^-) is $(0\mathrm{p}_{3/2})^{-1}1\mathrm{s}_{1/2}$, so the state $|{}^{10}\mathrm{Be}(1^-)\times\mathrm{p}_{1/2}\rangle$, has a configuration of $(0\mathrm{p}_{3/2})^{-1}0\mathrm{p}_{1/2}1\mathrm{s}_{1/2}$, i.e., $(0\mathrm{s})^4(0\mathrm{p}_{3/2})^5\,0\mathrm{p}_{1/2}1\mathrm{s}_{1/2}$. This configuration is independent of the configuration $^{10}\mathrm{Be}(0^+)\times\mathrm{s}_{1/2}$ even in the SM. With $|{}^{10}\mathrm{Be}(2^-)\times\mathrm{p}_{3/2}\rangle$ the case is somewhat different; although in the simple SM it coincides with $|{}^{10}\mathrm{Be}(0^+)\times\mathrm{s}_{1/2}\rangle$ since $|{}^{10}\mathrm{Be}(2^-)\rangle = |(0\mathrm{p}_{3/2})^{-1}1\mathrm{s}_{1/2}\rangle$, in the cluster model it does not. So the negative-parity core excitations in ^{10}Be do have a chance to provide the missing additional effect.

The microscopic multicluster model calculations for ^{10}Be (Sect. 13.5 and Ref. [449]) show that the main cluster arrangement of these negative-parity states is $^{9}\mathrm{Be}(\mathrm{g.s.}, \tfrac{3}{2}^-)\times\mathrm{s}_{1/2}$ and their matter rms radii are as large as 3 fm (see Table 13.20). Though the excitation energies of these negative-parity states are by 2.6–2.9 MeV higher than that of the $^{10}\mathrm{Be}(2_1^+)$ state, they have the advantage that they can accommodate a p-wave neutron, whose s.p. energy is expected to be considerably lower than that of a d-wave neutron. Therefore, the configurations involving the ^{10}Be core with negative parity seems to compete well with the admixtures involving the ^{10}Be(2^+) core. Another advantage of the negative-parity core states is their larger radii [the radius of ^{10}Be(2^+) is 2.41 fm], which may make them attract the halo neutron more (cf. Sect. 13.1.3 for a similar mechanism in ^{6}He).

The $\tfrac{1}{2}^-$ state of ^{11}Be should look like

$$\left|{}^{11}\mathrm{Be}\left(\tfrac{1}{2}^-\right)\right\rangle = c_1'|{}^{10}\mathrm{Be}(0^+)\times \mathrm{p}_{1/2}\rangle + c_2'|{}^{10}\mathrm{Be}(1^-)\times \mathrm{s}_{1/2}\rangle, \quad (13.19)$$

in which, however, the component $^{10}\mathrm{Be}(1^-)\times\mathrm{s}_{1/2}$ has probably no significant effect because it involves the excitation of not only the core but also of the last neutron with respect to the dominant component, $^{10}\mathrm{Be}(0^+)\times\mathrm{p}_{1/2}$.

The nucleus ^{11}Be as a five-cluster system α+α+n+n+n poses a formidable challenge to the correlated Gaussian cluster model. It is not at all hopeless though to cope with such a calculation. This is one of the most important issues that should be tackled in this model in the future. With the above arguments, a natural approximate scheme has been elaborated: that which is restricted to configurations ^{10}Be+n, with just a few ^{10}Be states involved.

The parity inversion and the one-proton-halo structure in ^{11}N can be discussed similarly.

To round off the discussion on ^{11}Be, it is appropriate to pay some attention to its neighbour, ^{12}Be, as well. There emerge a few questions: Is the g.s. of ^{12}Be formed by just putting another neutron into an orbit around ^{11}Be? Does the $s_{1/2}$ neutron orbit lie low again? How much is then the ^{11}Be core modified in ^{12}Be? If the coupling between ^{11}Be and the last neutron is weak, there should be two low-lying 0^+ states, formed by ^{11}Be($\frac{1}{2}^+$)$\times s_{1/2}$ and ^{11}Be($\frac{1}{2}^-$)$\times p_{1/2}$, respectively. Another possibility is that the (pairing) interaction between two neutrons is strong enough to tear off the odd neutron from the core; then ^{12}Be behaves as a ^{10}Be core plus two valence neutrons, and the level scheme will be reshuffled. Another interesting question is whether $N = 8$ is a magic number in ^{12}Be. Light $N = 8$ isotones include ^{10}He, ^{11}Li, ^{12}Be, ^{13}B and ^{14}C. The magic number seems to break down in ^{11}Li as halo structure is not consistent with closed-shell structure and $s_{1/2}$ orbits in the next shell seem to be significantly populated. The magic nature of $N = 8$ is undisputed at the other end of this sequence (for ^{14}C), and the question is what is in between. It had been predicted much before the exotic structures became accessible experimentally that in the $T = 2$ states of the $A = 12$ nuclei the sd shell is populated substantially [464]. The ^{12}Be g.s. is a member of such a multiplet.

Experiment has so far revealed three bound states in ^{12}Be, with 0^+(g.s.), 2^+($E_x = 2.10$ MeV) and 1^-($E_x = 2.70$ MeV). The spin and parity of the last state have been assigned at the time of the preparation of this book [465]. The B(E1; $1^- \to 0^+$) value is estimated to be 0.017 ± 0.004 e^2 fm^2, i.e., 0.050 ± 0.013 W.u. The apparent lack of a low-lying 0^+ excited state seems to indicate that the ^{12}Be g.s. is not a weakly coupled system of ^{11}Be and a neutron. It should rather be viewed as a ^{10}Be core with two valence neutrons.

In the simplest SM picture the 0^+ g.s. is likely to be a mixture of the closed-shell core and a $(0p_{1/2})^{-2}(1s_{1/2})^2$ structure. A projectile knockout experiment with the ^{12}Be beam [466] shows that the s- and p-wave spectroscopic factors for neutron removal from the g.s. are approximately equal. This indicates that the two components of the g.s. have comparable weights. As to the 2^+ state, it may predominantly involve two $0p_{1/2}$ holes and two sd particles, while the dominant component of the 1^- state may be $(0p_{1/2})^{-1}(1s_{1/2})$. Thus ^{12}Be seems to behave as a ^{10}Be+n+n system, with a partially broken-up p-shell.

It has been concluded in Sect. 13.5 that the second 0^+ state of ^{10}Be at 6.18 MeV shows ^{6}He+α clustering. It is therefore natural to conceive that the states which have ^{6}He+^{6}He or ^{8}He+α configurations lie at rather high excitation energies in ^{12}Be. (The ^{6}He+^{6}He threshold is at 11 MeV.) Experimental search for these molecular states has in fact been performed by measuring the fragments of ^{12}Be breakup [467].

13.7 The nuclei ^{10,11}Li

Table 12.1 suggests that the most intensively studied and yet least understood light exotic nucleus, ^{11}Li, should be described as a six-cluster system of α+t+n+n+n+n structure. Calculations as precise as those made for nuclei with $A \leq 9$ are as yet beyond the present technical possibilities, but perhaps not too far beyond. It is thus worthwhile to assess the possibilities and impediments and survey the preliminary results on the way towards such a description.

The understanding of ^{11}Li has to be preceded by understanding its subsystem ^{10}Li. A description of ^{11}Li may be satisfactory if it consistently describes ^{10}Li as well. For that, the spectrum of ^{10}Li should be known with confidence and should be well accounted for. Since ^{10}Li decays with neutron emission, a reliable model of ^{10}Li has to incorporate a ^{9}Li+n configuration. Thus the treatment of the n+^{9}Li system has to be realistic, which requires the understanding of ^{9}Li. We may terminate the chain of prerequisites here by saying that ^{9}Li is more compact than ^{8}Li, and is thus a better starting point. The density distribution of ^{9}Li (see Fig. 13.13) is, however, very different, particularly in the tail, from that produced by a naive HOSM [214], which points to the necessity of using a more sophisticated model. In Sect. 13.3 the g.s. of ^{9}Li was reasonably described, but the question of its distortability was also raised. Allowance for the distortion of ^{9}Li within ^{11}Li may require a description of its first few excited states.

In Sect. 13.7.1 some facts and speculations about the nuclei ^{9,10,11}Li will be surveyed, and in Sect. 13.7.2 some theoretical results will be shown.

13.7.1 Facts and speculations

^{9}Li has two bound states: the g.s with $J^\pi = \frac{3}{2}^-$, which is 4.063 MeV below the ^{8}Li+n threshold, and an excited state at $E_x = 2.691$ MeV, which has probably $J^\pi = \frac{1}{2}^-$ [387]. Some low-lying resonances are known, without spin–parity assignments. The dominant SM configurations of the g.s. and of the bound excited state are probably $(0s_{1/2})^4(0p_{3/2})^5$

and $(0s_{1/2})^4(0p_{3/2})^4 0p_{1/2}$, where the odd particle is, of course, a proton. The $0p_{3/2}$ neutron subshell is closed. The next state is expected to be a two-particle–two-hole neutron excitation, viz., $(p_{3/2})^{-2}(p_{1/2})^2$.

In spite of considerable effort, the structure of the unbound ^{10}Li is even less known. The spin, parity and location of the g.s. would be essential for the understanding of the halo structure of ^{11}Li [181]. They would give information on the strength of the ^{9}Li–n interaction. The simplest SM predicts the g.s. to be $(0s_{1/2})^4(0p_{3/2})^5 0p_{1/2}$, thus a $0p_{1/2}$ neutron coupled to a $0p_{3/2}$ proton would give rise either to a 1^+ or to a 2^+ state. If the SM prediction holds, the low-lying states of the Li isotopes may show similarity to those of B isotopes; then the counterpart of ^{10}Li is ^{12}B.

But one may as well expect a parity inversion similar to the case of ^{11}Be. It is difficult to say ***a priori*** whether ^{11}Be is exceptional. In ^{11}Be the $\alpha+\alpha$ 'frame' plays an important role, but perhaps the α+t frame works similarly. The parity inversion implies that the configuration $(0s_{1/2})^4(0p_{3/2})^5 1s_{1/2}$ lies lowest, thereby the g.s. spin–parity would be either 1^- or 2^-, coming from the coupling of a $1s_{1/2}$ neutron to the $0p_{3/2}$ proton. Since the s-wave neutron feels no barrier in ^{10}Li, these states are probably not resonances but virtual states (cf. Sect. 10.4.1). Virtual states may not be easily identifiable experimentally. That the g.s. of ^{10}Li may be a nonnormal-parity state ($J^\pi = 2^-$) had been conjectured long before the research of exotic nuclei were started [468].

The relative position of the s and p orbits in ^{10}Li may determine the angular-momentum composition of the halo of ^{11}Li. Since the momentum distributions of s- and p-wave halos give rise to different effects in the distribution of the momentum of the core fragment (Sect. 6.3), even the interpretation of the momentum distribution measurements could crucially depend on the composition of the halo.

The results for the low-lying resonance states of ^{10}Li are summarized in Table 13.21. Most measurements to date have been hindered by poor statistics, poor resolution and by the inherent difficulty in observing near-threshold unbound states. The observation of an s-wave neutron state, whether it is a resonance or a virtual state, is particularly difficult. At about 500 keV above the ^{9}Li+n threshold a state is identified in several different reactions. As far as the states at about 200 keV and at even lower energies are concerned, the results are contradictory. Nevertheless, there is a narrow peak at zero in the ^{9}Li–n relative velocity spectrum [471, 476], whose behaviour indicates an s-wave [473, 478].

The theoretical speculations on ^{10}Li starting from ^{11}Li seem to favour a parity inversion for ^{10}Li, but most calculations in the ^{9}Li+n cluster model apparently show no sign of it. A macroscopic ^{9}Li+n model

Table 13.21: Experimental data for the low-lying resonances of ^{10}Li. The energies ε_n (with respect to the n-emission threshold) and widths Γ are given in MeV.

ε_n	Γ	Reaction	Ref.
0.80±0.25	1.2 ±0.3	^{9}Be(^{9}Be, ^{8}B)^{10}Li	[469]
0.15±0.15	1.0	^{11}Be(π^-, p)^{10}Li	[470]
< 0.15 or ~2.5		^{18}O fragmentation	[471]
0.54 ± 0.06	0.36 ± 0.02	^{11}B(^{7}Li, ^{8}B)^{10}Li	[472]
< 0.10	< 0.23		
0.21 ± 0.05	0.12(+0.10, −0.05)	^{11}Li breakup	[473]
0.62 ± 0.10	0.6 ± 0.1		
0.24 ± 0.06		^{10}Be(^{12}C,^{12}N)^{10}Li	[474]
0.53 ± 0.06	0.30 ± 0.08	^{9}Be(^{13}C,^{12}N)^{10}Li	
0.4		p(^{11}Li, pn)^{10}Li	[475]
<0.05		^{18}O fragmentation	[476]
0.50±0.06	0.40±0.06	^{9}Be(^{9}Be, ^{8}B)^{10}Li	[477]

predicts no s-wave resonance either [479], but does not exclude a virtual state [480]. In a GCM model [481] ^{9}Li is described in the p-shell SM, and some of the lowest ^{9}Li states are included as coupled configurations. The solution of the problem produces the relative motion between ^{9}Li and n. The results favour the p-wave, so that the g.s. is found to be a 1^+ state, which happens to be bound, and the 2^+ state is a resonance. A similar model in the RGM framework [482] yields similar results. It may be as difficult to produce parity inversion in a microscopic model for ^{10}Li as for ^{11}Be.

As to the behaviour of an s-wave virtual state, the reader has experience for the case of ^{5}He (Sect. 13.1.2). The s-wave state has been found strongly model-dependent, and the case of ^{10}Li is similar. In the actual works this has been aggravated by the difficulties of finding a virtual state, and thus no definite conclusion has been drawn.

A best way to pin down the s-wave ^{9}Li+n interaction could be via the experimental finding of its isobaric analogue in the mirror nucleus ^{10}N. This state is presumably a resonance, owing to the Coulomb barrier. Then the potential that reproduces this resonance could be used (without the Coulomb term) in describing the ^{9}Li+n system.

The basic facts concerning ^{11}Li are the following.

Its g.s. energy with respect to the ^{9}Li+n+n disintegration threshold was found to be $\varepsilon = -0.340 \pm 0.05$ MeV [483] and -0.295 ± 0.035 MeV [484]. Due to the odd proton, it obviously has spin–parity $\frac{3}{2}^-$. The first few excited states that seem to be discernible are: (1) $E_x = 1.2$–1.3 MeV,

$\Gamma \sim 0.5$ MeV [485,486]; (2) $E_x = 2.5$–2.9 MeV, $\Gamma \sim 1.2$ MeV [486,487]; (3) $E_x = 4.8$ MeV, $\Gamma \leq 0.1$ MeV [486,487]; (4) $E_x = 6.2$–6.4 MeV, $\Gamma \leq 0.1$ MeV [487]. State (1) may be the soft dipole state. A most trivial assignment for state (2) could be $\frac{1}{2}^-$, the other member of the g.s. spin-orbit doublet. States (3) and (4) are just over the ^{8}Li+3n and ^{7}Li+4n thresholds and are pretty narrow. Such narrow structures are likely to arise from core excitations. There are excited states in ^{9}Li nearby: at 4.31 MeV and 6.43 MeV, respectively. There are theoretical arguments [488] for the existence of a low-lying narrow resonance to explain the low-energy ^{9}Li+n+n breakup, but there seems to be no candidate for such a state.

From interaction cross section measurements, the g.s. of ^{11}Li is deduced to have an rms matter radius of $r_m = 3.12$ fm [10,17]. At the same time, the proton and neutron rms radii are estimated to be $r_p = 2.88$ fm and $r_n = 3.21$ fm [10], respectively. Independent information on the ^{11}Li size comes from the elastic scattering of ^{11}Li on ^{28}Si at an incident energy of 29 A MeV [420]. An analysis of the angular distribution of this process with the double-folding model (cf. Sect. 5.1) supports the existence of a neutron halo: $r_m = 3.011$, $r_p = 2.235$ and $r_n = 3.255$ fm. Though both of the data give a very large rms matter radius exceeding 3 fm, the skin thicknesses defined as the difference between the neutron and proton rms radii significantly differ from each other: about 0.3 [10] and 1.0 fm [420]. In respect of the former analysis, one may question whether a single interaction cross section measurement provides sufficient information for extracting both the proton and neutron radii reliably. As for the latter analysis, it does not take into account the important effect of ^{11}Li breakup, which is certainly beyond the folding model (cf. Chap. 5). These values of the rms radii should therefore be considered as qualitative estimates.

The distribution of the ^{11}Li matter density has also been extracted form experiment, based on phenomenological model assumptions [17]. The momentum distribution of the neutrons in ^{11}Li, deduced from the transverse momentum distribution of the fragments [21], has a broad and a narrow component, in accord with the halo picture. The magnetic moment of ^{11}Li is 3.6673(25) μ_N [489] and its quadrupole moment is $|Q| = 3.12 \pm 0.45\, e\,\text{fm}^2$ [490]. These moments differ from those of ^{9}Li (see Table 13.10), which suggests that a naive picture in which two halo-neutrons have a spherically symmetric distribution around the ^{9}Li core does not hold.

Further empirical data are provided by the complex pattern of the β-decay transitions [491,492]. Various β-delayed particle (2n, d, 3n, t) emissions are possible (cf. Sec. 13.1.4 for ^{6}He). The threshold of ^{9}Li+d in ^{11}Be is just 2.7 MeV below the g.s. of ^{11}Li. Theoretical predictions for the β-delayed deuteron emission [493,494] gave a branching ratio of $\sim 10^{-4}$ and showed the sensitivity to the tail part of the halo structure of ^{11}Li. Subsequent experiment [495] put this ratio higher.

The $\log ft$ value for the transition $^{11}\text{Li}\to{}^{11}\text{Be}(\frac{1}{2}^-$, 0.32 MeV) is sensitive to the extension of the halo-neutron orbit and to the $(1s_{1/2})^2$ component in ^{11}Li [496]. In fact, in the simplest SM picture[17] the β-decay goes with the conversion of one of the halo neutrons into a proton in the p shell since the Gamow–Teller matrix element gets only contribution from the $(0p_{1/2})^2$ component. Therefore, the β-decay rate is expected to decrease by increasing the radius of the halo-neutron orbit and by increasing the $(1s_{1/2})^2$ component. The $\log ft$ data (5.67±0.04 [497] and 5.73±0.03 [498]) indicate that the Gamow–Teller matrix element is fairly strongly suppressed. It can be inferred that the weight of the $(0p_{1/2})^2$ component in the halo is about 50%. A similar conclusion has been drawn concerning the admixture of the $(1s_{1/2})^2$ component from the measured momentum distribution of the ^{10}Li fragment emitted in the $^{11}\text{Li}\to{}^{10}\text{Li}+\text{n}$ breakup of ^{11}Li and from the distribution of the neutron emitted by the fragment ^{10}Li [179].

The $(0p)^2$–$(1s)^2$ ratio has also been probed via the excitation of the isobaric analogue state of ^{11}Li by the p(^{11}Li, ^{11}Be)n reaction [499]. This state has been identified by calculating the Lorentz-invariant mass of the excited ^{11}Be from the momenta of the three decay particles, ^{9}Li, n and p. The isobaric analogue state has been found at $E_x = 21.16\pm0.02$ MeV, with a FWHM of 0.49±0.07 MeV. The excitation energy is close to a theoretical prediction [380], but its measured width is much larger than the theoretical estimate calculated by assuming a pure $(0p_{1/2})^2$ configuration for the halo neutrons. The $(1s_{1/2})^2$ admixture can, however, account for the discrepancy. The decay properties of this state are sensitive to the halo part of the ^{11}Li wave function [494].

13.7.2 Theoretical approaches

The theoretical approaches to the description of the halo structure of ^{11}Li are mostly macroscopic or semimicroscopic three-body calculations based on the $^9\text{Li}+\text{n}+\text{n}$ picture [188,189,195,196]. The occupancy of the core is generally taken into account by an orthogonality condition, so these models can be considered to be three-cluster OCM. It is investigated

[17]In the simplest SM including the s-component for the halo neutron, the g.s. of ^{11}Li is a mixture of the $(0s_{1/2})^4(0p_{3/2})^5(0p_{1/2})^2$ and $(0s_{1/2})^4(0p_{3/2})^5(1s_{1/2})^2$ configurations, while the $\frac{1}{2}^-$ excited state of ^{11}Be looks like $(0s_{1/2})^4(0p_{3/2})^6 0p_{1/2}$. Since the β-decay is governed by a one-body operator, it is obvious that the $(0s_{1/2})^4(0p_{3/2})^5(1s_{1/2})^2$ configuration cannot contribute to the decay; just a $0p_{1/2}$ neutron can be converted to a $0p_{3/2}$ proton.

(i) whether the binding of ^{11}Li can be reproduced by two-body forces appropriate for the description of the subsystems, ^{10}Li+n and n+n;
(ii) whether the known properties of ^{11}Li can be reproduced by a phenomenological adjustment of the interactions.

The binding-energy problem (i) is made more difficult by the insufficient knowledge of ^{10}Li, and the theoretical calculations giving credit to one or the other set of data can only have qualitative validity. A 1^+ and a 2^+ state of ^{10}Li can be produced in the range expected based on the experiments [500], but a disagreement in the widths points to the role of core excitations. With the same ^{9}Li–n interaction, ^{11}Li was found to be underbound by about 1 MeV [195]. This tallies with the case of ^{6}He (Sect. 13.1.3), and, similarly, must be due to the excitation of the core. It was pointed out that the effect of the excitation of the odd $p_{3/2}$ proton to the $p_{1/2}$ orbit, which corresponds to the first excited state of ^{9}Li, is not appreciable. The effect of the $(p_{3/2})^2 \rightarrow (p_{1/2})^2$ neutron excitation is, however, significant since the $(p_{1/2})^2$ neutrons block part of the $p_{1/2}$ state space for the extra neutron, and that enhances the energy of the $p_{1/2}$ orbit in ^{10}Li. For the sd-excitation, there is no blocking effect, thus the energy of the $p_{1/2}$ orbit is increased relative to the others. To restore the agreement with experiment, the ^{9}Li–n potential has to be deepened, which brings down the $1s_{1/2}$ state to the threshold [501]. It remains to be seen how much the binding of ^{11}Li is enhanced in a coupled-channel model in which the core excitations are taken into account and the ^{9}Li–n interaction is deepened accordingly.

Independently of the binding-energy problem, the answer to question (ii) is that the gross properties of ^{11}Li can be reproduced by a phenomenological three-body model whose parameters are adjusted to yield the correct ε_{2n}. A three-body calculation [195], which reproduces the energy as well as the phenomenological s.p. density of ^{11}Li, predicts the core and halo density shown in Fig. 13.23. These curves are thus likely to be realistic.

Problem (ii) has also been solved in a fully microscopic model with simple central interactions [481]. All bulk properties of ^{11}Li have been reproduced (apart from the total energy) in a ^{9}Li+n+n GCM model, in which ^{9}Li has been described by a full p-shell SM. Thus all core distortions that go with p-shell excitations are taken into account, but that was possible at the expense of the generator-coordinate configurations to be grossly simplified. The corresponding ^{9}Li+n model overbinds ^{10}Li, which shows that the binding-energy problem is not solved. Only positive-parity ^{10}Li states can be constructed in this way, so no information can be obtained on the s-wave state.

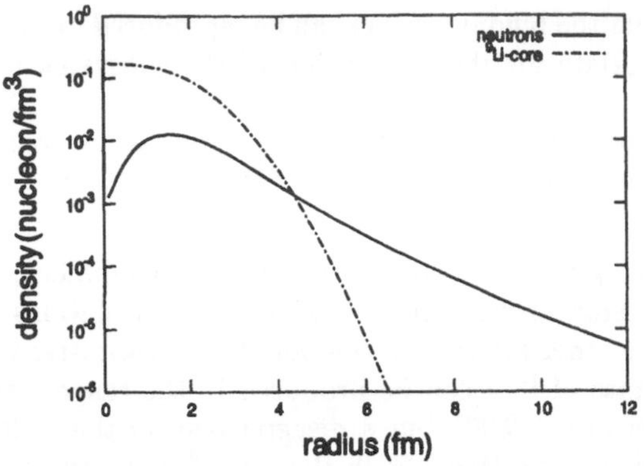

Figure 13.23: The core density (dash-dotted line) and the halo-neutron density (full line) of ^{11}Li in a phenomenological model that reproduces the bulk properties of ^{11}Li. Taken from Ref. [195].

Some preliminary results in the correlated Gaussian model look like a promising step towards the solution of the binding-energy problem as well.

The starting point is the description of ^{9}Li (Sect. 13.3.4). The wave function of ^{9}Li can be written as $\Psi^{JM}_{^9\mathrm{Li}} = \sum_i C_i \psi^{JM}_i(^9\mathrm{Li})$, with

$$\psi^{JM}_i(^9\mathrm{Li}) = \mathcal{A}\Big\{\Big[\Phi^i_S(^9\mathrm{Li})\mathrm{e}^{-\frac{1}{2}\tilde{\rho}A_i\rho}\theta_{l_1^i l_2^i l_3^i L}\Big]_{JM}\Big\},$$
$$\Phi^i_{SM_S}(^9\mathrm{Li}) = \Phi_\alpha(\xi_{1234})\Big[\Phi_{t\frac{1}{2}}(\xi_{567})\big[\chi_{\frac{1}{2}}(8)\chi_{\frac{1}{2}}(9)\big]_{s^i_{12}}\Big]_{SM_S}, \quad (13.20)$$

where Φ_j are the cluster intrinsic states, $\chi_{\frac{1}{2}m_s}$ are spin states (the isospin states are suppressed) and $\mathrm{e}^{-\frac{1}{2}\tilde{\rho}A_i\rho}\theta_{l_1^i l_2^i l_3^i LM_L}$ is a correlated Gaussian function [cf. Eq. (9.41)] of the three Jacobi coordinates describing the four-cluster relative motion. Rather than a cumbersome intercluster antisymmetrizer $\mathcal{A}_{(1234)(567)89}$, the full antisymmetrizer $\mathcal{A}$ is now written since the (partially) repeated antisymmetrization changes the wave function just by a constant factor. The summation i includes A_i, $\{s^i_{12}\}$ and $\{l^i_1 l^i_2 l^i_3\}$ corresponding to all arrangements and angular momenta included in Table 13.13. The quantum numbers of the model g.s. of ^{9}Li were fixed to be $(SL)J = (\frac{1}{2}1)\frac{3}{2}$. This configuration has been found to be dominant in the full model of ^{9}Li (Sect. 13.3.4).

The Minnesota force with $u = 1.00$ was used again, with the same spin-orbit term. A smaller and yet satisfactory basis was, however, attainable by giving more freedom to the elements of the matrix A_i (originally they were restricted to correspond to Jacobi arrangements; cf. Sect. 9.2.3) and by optimizing them more extensively. A compact basis helps to keep the basis size of the ^{11}Li calculation at a manageable level. The drawback of a smaller basis is that the asymptotic part of the wave function is not so accurate. The energy with respect to the α+t+n+n threshold was obtained to be $\varepsilon = -7.45$ MeV. This is comparable with the -8.05 MeV quoted in Sect. 13.3.4.

The nucleus ^{10}Li was then described as a ^{9}Li-like 50-configuration structure, coupled with a single neutron in a relative s or p state: $\Psi^{JM}_{^{10}\mathrm{Li}} = \sum_{ij} C'_{ij}\psi^{JM}_{ij}(^{10}\mathrm{Li})$, with

$$\psi^{JM}_{ij}(^{10}\mathrm{Li}) = \mathcal{A}\Big\{\Big[[\psi^{J_9}_i(^{9}\mathrm{Li})\chi_{\frac{1}{2}}(10)]_{S_i}\psi^{(\beta_j)}_{0l}(\rho_4)\Big]_{JM}\Big\}. \qquad (13.21)$$

The n–^{9}Li relative motion is described by a linear combination of nodeless HO states $\psi^{(\beta_j)}_{0lm}$. This is, of course, just a correlated Gaussian for a single n–core relative motion, apart from a constant factor. This motion was described by 8 Gaussians, the dimension totalling 400. The value $u = 1.00$, which was found to suit $^{7-9}$Li, gives rise to a bound ^{10}Li, so it has had to be changed to 1.03. The unbound-state energy was determined on a random Gaussian basis, which is, of course, square-integrable. The range of the parameter values was chosen to 'enclose' the system in a box by prescribing $(\beta_j)^{-1/2} \leq 6$ fm. It is very probable that the energy of the lowest-lying resonance state (if a resonance exists) in each quantum-number set agrees roughly with what could be obtained by imposing the proper outgoing boundary condition (cf. Sect. 10.4.1).[18] That is so because the position of a state representing a resonance in a box (cf. p. 265) depends weakly on the position of the wall of the box. It has been checked that the energies only slightly change in a larger box $[(\beta_j)^{-1/2} \leq 10$ fm$]$.

This model is not a frozen-cluster model. What is frozen is just those parameters of each basis state $\psi^{JM}_{ij}(^{10}\mathrm{Li})$ which belong to the core part, i.e., $\psi^{J_9M_9}_i(^{9}\mathrm{Li})$; the expansion coefficients C'_{ij} describing ^{10}Li are determined by full diagonalization of the Hamiltonian. This is thus a sophisticated cluster-distortion model, in which a large subspace of the ^{9}Li intrinsic motion is taken into account: that which carries the same angular-momentum quantum numbers as the g.s.

The level sequence obtained is as follows (the energy ε_n with respect to the ^{9}Li+n threshold in parenthesis): 1^+ (0.39 MeV), 2^+ (0.55 MeV), 2^- (1.56 MeV), 1^- (1.74 MeV), 0^+ (2.12 MeV). By increasing the box size, the two lowest levels slip slightly closer to the threshold (the 1^+

[18]Note that over the disintegration threshold no minimum principle holds for the levels. In the limit of large basis extension, the discrete levels obtained by a diagonalization over a square-integrable basis accumulate at the threshold. When the basis is of finite spread, the physically relevant levels distinguish themselves by their relative stability against changes of the basis (cf. Fig. 10.9). The box size is chosen so as to force the lowest-lying level to be a physically relevant level.

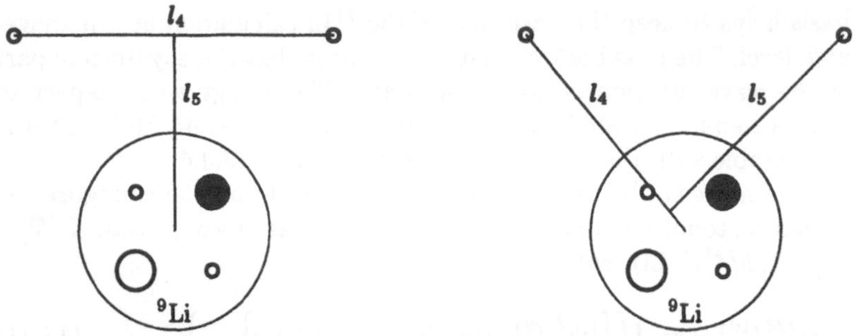

Figure 13.24: T- and Y-type arrangements for ^{9}Li+n+n, with the relative angular momenta belonging to the last two intrinsic Jacobi vectors shown. Notation: α: ●; t: ○; neutron: ∘. The description of ^{9}Li as α+t+n+n includes all arrangements and relative angular momenta l_1, l_2, l_3, given in Table 13.13 (cf. Fig. 13.12).

state is at 0.31 MeV) but their energy difference remains the same. The positive-parity states come from the p-orbits and should correspond to the experimentally observed levels. The appearance of a low-lying 0^+ state is surprising since in the SM it would imply a high-lying $1p_{3/2}$ orbit as the $0p_{3/2}$ orbits are filled. In the cluster model, however, the $0p_{3/2}$ orbits are allowed to be partially vacant, and the artefact of the square-integrable basis may also help to lower this level. The 2^-, 1^- states involve s-orbits, and neither the bound-state estimates nor the experimental values (Table 13.21) of such states are too reliable. They are most probably virtual states.

With square-integrable bases one can produce resonances or discretized continuum states. Virtual states cannot be produced as their 'wave functions' grow exponentially in the asymptotic region. If the position of a state is not stable against changes of the box, it cannot be regarded as a discrete physical state. The g.s. (1^+) and the first excited state (2^+) are resonance states, but one cannot draw definite conclusion for the other states from the present calculation. It should be remembered that states built on the excited states of ^{9}Li are not included in the basis, thus certain states of ^{10}Li are also excluded.

The wave function of ^{11}Li is constructed by adding two neutrons to ^{9}Li. The relationship of the ^{9}Li+n+n picture to the underlying α+t+n+n+n+n picture (see Fig. 13.24) can be formulated so as to conform to Eqs. (13.20) and (13.21). The wave function is expanded as $\Psi^{JM}_{^{11}\mathrm{Li}} = \sum_{ij} C''_{ij} \psi^{JM}_{ij}(^{11}\mathrm{Li})$, with

$$\psi^{JM}_{ij}(^{11}\mathrm{Li}) = \mathcal{A}\Big\{\Big[\psi^{J_9}_i(^9\mathrm{Li})\Big[\mathrm{e}^{-\frac{1}{2}\tilde{\rho}' A'_j \rho'}\theta'_{l^j_4 l^j_5 l^j_{45}}\chi_{s^j}\Big]_{J^j}\Big]_{JM}\Big\}, \quad (13.22)$$

where $\rho' = \{\rho_4 \rho_5\}$ are Jacobi coordinates of the ^{9}Li+n+n system; both Y- and T-type arrangements are included through the 2×2 matrices A'_j. The χ_s is a combined spin function of the two extra neutrons: $\chi_s = [\chi_{\frac{1}{2}}(10)\chi_{\frac{1}{2}}(11)]_s$. Each intermediate angular momentum is restricted to take only values of 0 or 1. The two-neutron states are then coupled to the $\frac{3}{2}^-$ state of ^{9}Li to form the $\frac{3}{2}^-$ g.s. of ^{11}Li with $J^\pi = \frac{3}{2}^-$.

The same u value, 1.03, was used for ^{11}Li as for ^{10}Li. The result for the energy of ^{11}Li below the ^{9}Li+n+n threshold is $\varepsilon = -0.54$ MeV, which is slightly deeper than the experimental value, $\varepsilon = -0.25 \sim -0.40$ MeV. At the low-precision level of the experimental ^{10}Li energy, the binding-energy problem seems to be solved. With this value of u, the nuclei $^{7-9}$Li become slightly underbound, but this value still looks a reasonable compromise. One can thus say that all Li isotopes are described by approximately the same force.

Table 13.22 shows the quantum numbers of the configurations included in the model and the ^{11}Li energies produced by the individual configurations without including the rest. Each relative-motion function was expanded in terms of three Gaussians, chosen according to a geometrical sequence along each relative coordinate. The 50 functions for the core, 8 configurations and 3×3 relative-motion Gaussians give 3600 basis elements altogether. The model space, as far as the three-cluster relative motion is concerned, is sufficiently large. The accuracy of the solution must be reasonably good. The various individual configurations, whether they involve T or Y arrangements, s- or p-orbits, produce very much the same energy, so it is fair to say that s- and p-orbits equally contribute to the binding.

Table 13.22: The energy of ^{11}Li (in MeV), with respect to the ^{9}Li+n+n threshold, as produced individually by each configuration used in the correlated Gaussian model.

Arrangement	l_4	l_5	l_{45}	s	ε
T	0	0	0	0	−0.32
T	1	1	0	1	−0.05
T	1	1	1	1	0.10
Y	0	0	0	0	0.08
Y	0	0	0	1	−0.38
Y	1	1	0	1	−0.27
Y	1	1	1	1	−0.25
Y	1	1	1	0	−0.22
all					−0.54

The proton, neutron and matter rms radii are found to be 2.36, 2.98 and 2.94 fm, respectively. These values are in qualitative agreement with the data of Ref. [420] (see p. 449). The rms matter radius is slightly smaller than the experimental estimate, but that is natural in view of the slight overbinding. The proton radius of ^{11}Li is somewhat larger than that of ^{9}Li (cf. Table 13.10), which can be accounted for by a c.m. oscillation within ^{11}Li. The neutron radius has increased by about 0.5 fm showing the pronounced neutron halo.

13.8 Overview of exotic structure

We have followed through the phenomenological arguments pointing to clustering in light exotic nuclei, and adopted a cluster model for their unified quantitative description. Nevertheless, throughout the detailed studies, many qualitative arguments have been borrowed from the SM. Recourse to the SM may not just be an *ad hoc* help illuminating particular cases; the SM provides a general vantage point to overview the structure of all exotic nuclei. While cluster structure is specific to each nucleus, from the point of view of the SM, there is no such qualitative difference between the p-shell nuclei of our primary concern. That is why the SM can provide a unified viewpoint. The results of the realistic model can be translated into the SM language, which illuminates another facet of nuclear structure. The SM formalism provides a measure for any nuclear states: the occupation probabilities of the shells. In this section some of the nuclei and models reviewed in the book will be analysed in terms of shell occupation probabilities.

The occupation probability $\mathfrak{P}_Q$ of a shell with Q excitation quanta $\hbar\omega$ is defined in Eq. (12.9); its calculation is discussed in App. E.4. To give a feeling for the meaning of this quantity, let us first see the occupancies of the α g.s. as produced by the p+p+n+n four-body model (cf. Fig. 13.3) with the Minnesota force. To facilitate a comparison with the SM [502], $\hbar\omega$ is chosen to be 14 MeV. The result is shown in Table 13.23. The average excitation is 1.1 $\hbar\omega$, which apparently shows a

Table 13.23: Nucleonic occupation probability $\mathfrak{P}_Q$ of the $Q\,\hbar\omega$ subspace with $\hbar\omega = 14$ MeV in the g.s. of the α-particle described in a p+p+n+n four-body model.

Q	0	2	4	6	8	$\langle Q \rangle$
$\mathfrak{P}_Q$ (%)	68	19	8	3	0.9	1.1

significant departure from the treatment of α in the cluster model. It should be mentioned, however, that the optimum $\hbar\omega$ for the α-particle used in the cluster model is about 22.7 MeV, substantially larger than the 14 MeV used in this analysis. The occupation probabilities agree well with those obtained in a no-core SM [502].

As a second example, let us analyse the occupancies in a simple α+α two-centre SM of ^{8}Be [231]. The two-centre SM is a cluster model and has been introduced in Sect. 11.7.3. A state in this model is a Slater determinant of Gaussian wave packets. The nucleons populate two HO potential wells separated by a fixed distance S. The nucleons tied to the two centres form two clusters, whose mean separation is S. In Sect. 12.3.1 it has been pointed out that such states are uncorrelated apart from the symmetry-restoring projections. The states to be considered are translation-invariant and have definite angular momenta (and parities). The width parameter of the wells is $\beta = 0.50$ fm^{-2}, which is realistic for an α-cluster. The two-centre SM will, of course, be analysed in terms of a usual single-centre SM. The $\hbar\omega$-value is chosen to be 12.4 MeV. This implies an oscillator constant $\gamma = 0.30$ fm^{-2} (cf. p. 520).

The occupation probabilities $\mathfrak{P}_Q$ for nucleons in the $L = 0$–6 states of ^{8}Be as functions of the mean intercluster separation S are shown in Fig. 13.25. If γ were chosen equal to β, $Q < 4$ would be strictly forbidden by the Pauli principle [cf. Eq. (11.102)]. Despite $\gamma \neq \beta$, the $Q < 4$ occupancies are very small and therefore are not shown in the figure. For each L value, the lowest-Q curve has a maximum at a finite distance, which shows that it is favourable energetically for the two clusters to stay apart. The value $\mathfrak{P}_4$ is not too high in the region around $S = 0$ since the Pauli principle disfavours a strong overlap of two identical clusters. (The value of $\mathfrak{P}_4(0) \neq 0$ because $S = 0$ allows the cluster c.m. to perform zero-point oscillation with respect to each other.) It is also remarkable that, for $Q = 4$, the larger the value of L, the closer the maximum is to the origin. This can also be understood as a Pauli effect: the higher the L-value, the weaker the exclusion effect is near the origin. This 'antistretching' effect of the relative angular momentum is a common property of cluster bands [503]. For small values of S, the small-Q curves run higher, but the tails of the high-Q curves stretch out farther. Therefore, for loosely bound systems, one expects that high-lying shells are occupied with appreciable probabilities.

The occupation probabilities for most nuclei and models reviewed in the present chapter are summarized in Table 13.24. The table contains a few data contained in Table 13.19 before. To show the correlation between the size and the shell occupancies, the rms radii are also included. In all models all cluster intrinsic motions are described by a common value of $\beta = 0.52$ fm^{-2}. The HO constant $\gamma = 0.34$ fm^{-2} ($\hbar\omega = 14.4$

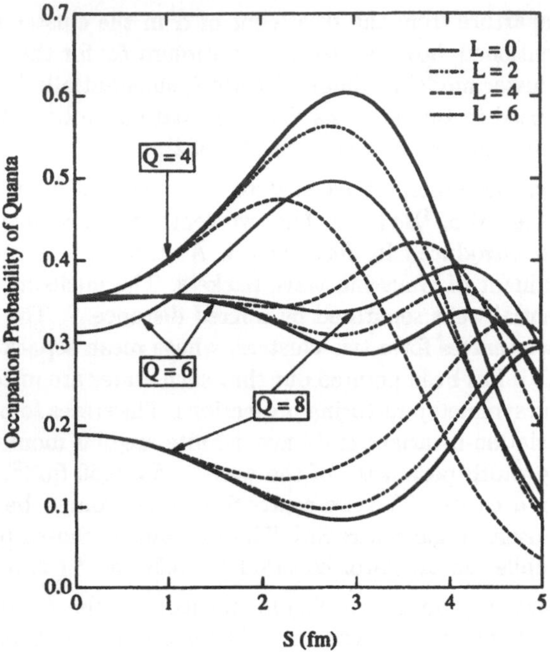

Figure 13.25: Nucleon occupation probability of shells characterized by the numbers Q of HO quanta $\hbar\omega$ for the $L = 0$–6 states of ^{8}Be described by an $\alpha+\alpha$ two-centre SM [231] as a function of the separation S of the SM potential centres.

MeV) is used. The occupation probabilities are now given as functions of $Q_{\mathrm{exc}} = Q - Q_{\mathrm{min}}$, where Q_{min} is the minimum number of HO quanta for the lowest Pauli-allowed configuration according to Eq. (11.102).

The 0 $\hbar\omega$ component is around 50–60% for most cases, and the 0, 2, and 4 $\hbar\omega$ components add up to about 90%. The admixtures of components higher than $Q_{\mathrm{exc}} = 4$ are also significant for weakly bound or unbound states. The unbound state ^{8}Be(g.s.) is treated as bound, but the results can yet be taken qualitatively correct. The occupation of high-lying shells is not a privilege of unbound states: it occurs for halo states (e.g., ^{6}He) and for any near-threshold states (e.g., ^{6}Li) as well. These results for $\langle Q_{\mathrm{exc}} \rangle$ are systematically higher than the corresponding results with the AMD [373], which is hence closer to simple SM configurations. The relationship of the proton and neutron occupation numbers to the nucleon occupation numbers is complicated. For example, the occupation of the 2 $\hbar\omega$ shell by nucleons is fed by 0+2, 1+1 as

Table 13.24: Percentage occupation probability $\mathfrak{P}_{Q_{exc}}$ of the $Q_{exc}\,\hbar\omega$ subspace of the nucleon (N), proton (p) and neutron (n) motion in the g.s. of some light nuclei as described by the correlated Gaussian model. The value of $\hbar\omega$ is 14.4 MeV. Where the neutron occupation probability is not shown, it is equal to the proton occupation probability. Probabilities of less than 1% are indicated by asterisks. In the third column the rms radii r (in fm), and in the last column the average excitations $\langle Q_{exc}\rangle$ are given.

Q_{exc}		r	0	1	2	3	4	5	6	7	8	9	10	11	12	13	14	$\langle Q_{exc}\rangle$
^{6}He	N	2.51	60	0	14	0	12	0	5	0	3	0	2	0	1	0	*	2.2
(0^+)	p	1.87	74	10	11	2	1	*	*	*	*	*	*	*	*	*	*	0.5
	n	2.78	67	3	8	5	7	2	2	1	1	*	*	*	*	*	*	1.7
^{6}Li	N	2.44	62	0	16	0	10	0	5	0	3	0	2	0	*	0	*	1.9
(1^+)	p	2.44	69	8	10	4	4	2	1	*	*	*	*	*	*	*	*	1.0
^{7}Li	N	2.34	63	0	20	0	9	0	4	0	2	0	*	0	*	0	*	1.4
$(\frac{3}{2}^-)$	p	2.28	77	2	16	*	4	*	*	*	*	*	*	*	*	*	*	0.6
	n	2.38	73	1	17	*	5	*	1	*	*	*	*	*	*	*	*	0.8
^{8}Li	N	2.45	61	0	18	0	11	0	4	0	2	0	1	0	*	0	*	1.7
(2^+)	p	2.19	79	6	11	1	2	*	*	*	*	*	*	*	*	*	*	0.4
	n	2.60	67	3	14	2	7	1	2	*	1	*	*	*	*	*	*	1.3
^{8}Be	N	3.27	36	0	18	0	12	0	7	0	5	0	4	0	3	0	2	7.6
(0^+)	p	3.27	47	0	21	0	11	0	6	0	4	0	3	0	2	0	1	3.8
^{9}Li	N	2.40	66	0	17	0	11	0	4	0	2	0	*	0	*	0	*	1.3
$(\frac{3}{2}^-)$	p	2.10	82	6	9	1	1	*	*	*	*	*	*	*	*	*	*	0.4
	n	2.54	71	3	12	2	6	1	2	*	*	*	*	*	*	*	*	1.0
^{9}C	N	2.52	60	0	17	0	12	0	5	0	3	0	1	0	*	0	*	1.8
$(\frac{3}{2}^-)$	p	2.68	65	4	12	3	7	1	2	*	1	*	*	*	*	*	*	1.4
	n	2.16	79	8	9	2	1	*	*	*	*	*	*	*	*	*	*	0.4
^{9}Be	N	2.50	54	0	21	0	12	0	5	0	3	0	2	0	*	0	*	2.1
$(\frac{3}{2}^-)$	p	2.39	71	3	17	1	5	*	1	*	*	*	*	*	*	*	*	0.8
	n	2.58	65	2	18	1	8	*	3	*	1	*	*	*	*	*	*	1.3
^{10}Be	N	2.28	61	0	21	0	11	0	4	0	2	0	*	0	*	0	*	1.4
(0^+)	p	2.24	80	3	14	*	2	*	*	*	*	*	*	*	*	*	*	0.5
	n	2.34	71	2	16	1	7	*	2	*	*	*	*	*	*	*	*	0.9

well as 2+0 p+n occupations, respectively. It is conspicuous that the components with odd Q_{exc} values for protons or neutrons are generally smaller than those with even Q_{exc}.

Comparing the results for $^{7-9}$Li, we see that the neutron occupan-

Table 13.25: Nucleonic occupation probability $\mathfrak{P}_{Q_{exc}}$ (in percentage) of the $Q_{exc}\hbar\omega$ subspace in the g.s. of ^{7}Li described in comparable α+t and α+p+n+n models.

Q_{exc}	0	2	4	6	8	$\langle Q_{exc}\rangle$
α+t	63	20	9	4	2	1.4
α+p+n+n	66	17	10	4	2	1.4

cies follow the neutron radius, which is consistent with the change of the neutron separation energy. The nucleus ^{8}Li has the smallest neutron separation energy as well as the broadest distribution and highest average value of neutron occupation numbers.

Just as the amounts of clustering (cf. Sect. 13.1.3), the shell occupation probabilities also lend themselves to testing the approach. As an example, the utility of the triton cluster can be tested through analysing the g.s. wave functions of ^{7}Li in the standard α+t model and an α+p+n+n four-cluster model. The state space in this latter model is similar to the state space of a large-space SM, with a ^{4}He core and three valence nucleons, which is free of spurious c.m. motion. With all parameters of this extended model chosen to be the same as used in the α+t model, the g.s. of ^{7}Li becomes slightly overbound, but its radius is hardly altered. The compositions of the wave functions are compared in Table 13.25. The two distributions are very similar. This indicates that the triton cluster can indeed be regarded as a useful substructure.

The shell occupation probabilities thus reveal that in light exotic nuclei there are substantial admixtures coming from high-lying shells even to g.s. This is an *a posteriori* justification of favouring the cluster models. The average excitation for the cases considered is about $2\,\hbar\omega$, which just contradicts the basic assumption of the SM. The pattern of occupation numbers comes about as an interplay of the occupations of many SM orbits, and the mixing of the configurations involved arise from complicated dynamics, which is best treated by the kind of sophisticated cluster model presented in this book.

13.9 Structure calculations with realistic nuclear forces

In all calculations presented so far effective nucleon–nucleon forces have been employed. In Sect. 12.2.3 their application was put forward as a necessity. It was argued that a nucleon–nucleon potential consistent with a truncated model space can only be an effective force.

The SVM, however, opens the way to numerically exact few-body calculations. In Sects. 12.3.2 and 13.1.3 we have seen calculations that can be called as such. True enough, the six- and eight-nucleon calculations presented in Sect. 12.3.2 are exact only in a restricted model space, but the 'exact' results for t, α and ^{6}He included in Table 13.7 can indeed be considered practically exact. The exact results with the Minnesota effective interaction impressively show how good the cluster models are.

In principle, an effective force should be tailored for the particular function space in which it is to be used, so the use of the full state space together with an effective force may be inconsistent. But, in practice, that does not lead to conceptual problems. In fact, the critical part of the function space is that which accommodates short-range correlations. Since the effective force has a much weaker repulsive core, the exact solution is weakly correlated at short internucleon distances, and thus does not go too much beyond the confines of the cluster-model space consistent with the effective interaction.

The aim of this section is to show to what extent and in what respects the *ab initio* calculations using realistic interactions can compete with the cluster models using effective interactions.

13.9.1 Realistic forces

An interaction may be called realistic if it is constructed to describe the bound states and scattering of free nucleons and few-nucleon systems in an exact framework. Realistic forces act between 'bare nucleons' (cf. Sect. 3.1).

The nucleons are composite particles, and their interaction should be derivable from that of their constituents. Such a derivation would require calculations in non-perturbative quantum chromodynamics, and that is not feasible. Realistic forces are thus constructed phenomenologically. The known qualitative properties (e.g., symmetries) of the underlying strong interaction restrict the form of the interaction, and the parameters can be determined by fitting to the two- and three-nucleon observables.

The potential between nucleons i and j may depend on their relative positions $\boldsymbol{r}=\boldsymbol{r}_i-\boldsymbol{r}_j$, on their spins and isospins, represented by the respective Pauli matrices $\boldsymbol{\sigma}_i$, $\boldsymbol{\sigma}_j$, $\boldsymbol{\tau}_i$ and $\boldsymbol{\tau}_j$, and on their relative momentum $\boldsymbol{p}=\frac{1}{2}(\boldsymbol{p}_i-\boldsymbol{p}_j)$. A most general form[19] of local two-body interactions that satisfy the translation, rotation, reflection, time-reversal and isospin invariance and have p- and l_{ij}-independent form factors are

[19] In fact, the symmetry principles would allow $(\boldsymbol{\sigma}_i\cdot\boldsymbol{p})(\boldsymbol{\sigma}_j\cdot\boldsymbol{p})$ terms as well [504].

of the following form:

$$V_{ij} = V_0(r)+V_\sigma(r)\boldsymbol{\sigma}_i\cdot\boldsymbol{\sigma}_j+V_\tau(r)\boldsymbol{\tau}_i\cdot\boldsymbol{\tau}_j+V_{\sigma\tau}(r)(\boldsymbol{\sigma}_i\cdot\boldsymbol{\sigma}_j)(\boldsymbol{\tau}_i\cdot\boldsymbol{\tau}_j) \quad (13.23)$$
$$+V_{LS}(r)\boldsymbol{l}_{ij}\cdot\frac{1}{2}(\boldsymbol{\sigma}_i+\boldsymbol{\sigma}_j)+V_{LS\tau}(r)\left[\boldsymbol{l}_{ij}\cdot\frac{1}{2}(\boldsymbol{\sigma}_i+\boldsymbol{\sigma}_j)\right](\boldsymbol{\tau}_i\cdot\boldsymbol{\tau}_j) \quad (13.24)$$
$$+V_{\mathrm{T}}(r)S_{ij}+V_{\mathrm{T}\tau}(r)S_{ij}\boldsymbol{\tau}_i\cdot\boldsymbol{\tau}_j \quad (13.25)$$
$$+V_{\mathrm{Q}}(r)Q_{ij}+V_{\mathrm{Q}\tau}(r)Q_{ij}\boldsymbol{\tau}_i\cdot\boldsymbol{\tau}_j$$
$$+V_{L^2}(r)\boldsymbol{l}_{ij}^2+V_{L^2\tau}(r)\boldsymbol{l}_{ij}^2\boldsymbol{\tau}_i\cdot\boldsymbol{\tau}_j$$
$$+V_{L^2\sigma}(r)\boldsymbol{l}_{ij}^2\boldsymbol{\sigma}_i\cdot\boldsymbol{\sigma}_j+V_{L^2\sigma\tau}(r)\boldsymbol{l}_{ij}^2(\boldsymbol{\sigma}_i\cdot\boldsymbol{\sigma}_j)(\boldsymbol{\tau}_i\cdot\boldsymbol{\tau}_j). \quad (13.26)$$

Line (13.23) contains the central interaction, which can be cast into the form of the first line of Eq. (12.10). Line (13.24) contains the spin-orbit interaction, which is a generalization of the second line of Eq. (12.10). Line (13.25) comprises the tensor potential, with

$$S_{ij} = \frac{3(\boldsymbol{r}\cdot\boldsymbol{\sigma}_i)(\boldsymbol{r}\cdot\boldsymbol{\sigma}_j)}{r^2} - \boldsymbol{\sigma}_i\cdot\boldsymbol{\sigma}_j. \quad (13.27)$$

The next line is the quadratic spin-orbit term, with

$$Q_{ij} = \frac{1}{2}[(\boldsymbol{\sigma}_i\cdot\boldsymbol{l}_{ij})(\boldsymbol{\sigma}_j\cdot\boldsymbol{l}_{ij}) + (\boldsymbol{\sigma}_j\cdot\boldsymbol{l}_{ij})(\boldsymbol{\sigma}_i\cdot\boldsymbol{l}_{ij})], \quad (13.28)$$

the next one is the $\boldsymbol{l}^2$ potential, and the last one contains a combination of $\boldsymbol{l}^2$ and spin–spin operators. The central interaction (13.23) contains four terms, and it is often referred to as the 'V4 potential'. Similarly, the sum of the central, tensor and spin-orbit terms, i.e., up to term (13.25), is called the 'V8 potential'.

The nuclear interaction should slightly depend on the charge, apart from the Coulomb force; otherwise the $T=1$ n–p and the p–p data as well as the properties of ^{3}H and ^{3}He could not be fitted equally well. Therefore, charge-dependent terms should be added to the interaction (13.26).

The radial factors $V_i(r)$ include the known one-pion exchange form [83] and fully phenomenological terms. Several realistic two-nucleon interactions have been constructed (e.g., the 'Argonne potential' [505]). They differ in detail, but their basic structures are similar, and they give comparably good fits to the scattering data. The important contributions to the binding energy come from the central and the tensor terms. The other terms have significant effects only on some phase shifts.

Precise few-body calculations on three- and four-nucleon systems have revealed an additional complication. The two-nucleon forces alone seem to be inadequate to explain the bound-state energies as well as some nucleon–deuteron and nucleon–triton scattering data. This indicates that the composite nature of the nucleons imply non-negligible multinucleon, at least three-nucleon, forces [207].

13.9.2 Stochastic variational solution

The SVM applied to the few-nucleon problems has to rival some precise and powerful methods: the solution of the Faddeev equations [506] for three-body problems and their generalizations to more than three bodies, the so-called Faddeev–Yakubovsky (FY) equations [507], the hyperspherical-harmonics method (HH) [508] (cf. Sect. 8.2) and the GMFC method.

In the few-nucleon SVM the full form of the correlated Gaussian formalism expounded in Sect. 9.2.3 has been used. The terms ψ_i of the trial function $\Psi=\sum_{i=1}^{\mathcal{K}} c_i\psi_i$ [cf. Eq. (9.3)] are defined in Eq. (9.41): $\psi_i = \mathcal{A}\left\{e^{-\frac{1}{2}\tilde{x}A_i x}\left[\theta_{l_iL_i}(\boldsymbol{x})\chi_{S_i}\right]_{JM}\eta_{TM_T}\right\}$. The successively coupled angle-dependent factor $\theta_{lLM_L}(\boldsymbol{x})$ as defined in Eq. (9.23) has been chosen. In keeping with that, the correlated Gaussians are constructed from uncorrelated terms (9.19), each belonging to a particular arrangement to be denoted by j. Although the formalism with the fully correlated Gaussians and the global-vector form (9.32) is simpler and more unbiassed, the terms belonging to well-defined arrangements and successively coupled functions (9.23) are more suitable for incorporating prior knowledge on the behaviour of the solution. In the form chosen there are a countable number of angular factors $\theta_{lLM_L}(\boldsymbol{x})$ to sample from, whereas in the global-vector representation one has to construct the angular part by optimizing additional continuous parameters.

Each basis state belongs to a specific arrangement j and a specific set of relative orbital angular momentum and intermediate coupling quantum numbers $l=\{l_1,\ldots,l_{A-1},L_{12},L_{123},\ldots,L\}$ and a similar set of intermediate spin and isospin coupling quantum numbers $\{st\}$. All possible arrangements (cf. Fig. 9.2) and intermediate angular momenta are included, with the orbital angular momenta $\{l_1,\ldots,l_{A-1}\}$ truncated such that $l_1+\cdots+l_{A-1}\le 4$. With this truncation, the number of configurations (arrangements and quantum-number combinations) included becomes large (several hundreds even for the α-particle) but finite.

The basis construction begins with a competitive selection procedure (cf. Sect. 10.2.2) followed by refinement steps (cf. Sect. 10.2). In each step of the competitive selection several different jl configurations are singled out randomly from the finite pool of quantum-number combinations included. For each jl, the nonlinear parameters are optimized by trial-and-error search for each possible st configurations.

The adopted basis state is that which lowers the total energy to the greatest extent; this implies a single *jlst* combination and a single set of nonlinear parameters. Once the basis constructed in this way tends to show saturation, the nonlinear parameters are further optimized by the refinement procedure.

13.9.3 The triton and the alpha-particle

As a first test, the Argonne V18 (AV18) interaction [509] is applied to the triton and the α-particle. This interaction contains the fourteen terms of the form of Eq. (13.26) and four charge-dependent terms.

Any computation with the full two-body interaction (13.26) is very time-consuming. As the major contributions come from the central and tensor terms, it suffices to optimize the basis with only these terms included. The full Hamiltonian can then be rediagonalized, which yields the g.s. energy with the full interaction.

The results with an eight-term version (AV8) of the AV18 force are summarized in Table 13.26. All results are in excellent agreement with the FY results (as well as with the GMFC results, not shown). The

Table 13.26: Description of the triton and the α-particle with the AV8 version of the AV18 interaction, with no Coulomb term. Given are the contributions of the kinetic-energy operator T, of the central (c), tensor (t), spin-orbit (so) potentials and of the full potential V to the energy, the total energy E itself (in MeV), the rms radii (in fm) and the $L = 0$, 1 and 2 probabilities (in %) and comparison is made with Faddeev–Yakubovsky (FY) results.

	t		α	
	FY [507]	SVM	FY [507]	SVM
$\langle T \rangle$	47.615	47.603	102.39	101.995
$\langle V_c \rangle$	−22.512	−22.505	−55.26	−55.05
$\langle V_t \rangle$	−30.867	−30.864	−68.35	−68.05
$\langle V_{so} \rangle$	−2.003	−2.003	−4.72	−4.75
$\langle V \rangle$	−55.381	−55.369	−128.34	-127.881
E	−7.767	−7.766	−25.94	−25.886
r_n	1.68[a]	1.68	1.49[a]	1.49
r_p	1.80[a]	1.81	1.49[a]	1.49
P_S	91.354	91.355	85.71	85.721
P_P	0.067	0.067	0.382	0.368
P_D	8.579	8.578	13.91	13.910

[a]Taken from a GMFC calculation [510].

Table 13.27: Energies of the triton and of the α-particle (in MeV) calculated with the full AV18 interaction with and without the UIX three-body term and with different techniques. The Coulomb potential is included. The SVM basis is optimized for the AV8 terms.

Force Ref.	Case	GFMC [510]	FY [507]	HH [508]	SVM	Exp.
AV18	t	−7.61	−7.62	−7.62	−7.61	−8.48
	α	−24.07	−24.28	−24.18	−24.16	−28.30
AV18 & UIX	t	−8.46	−8.48	−8.48	−8.43	−8.48
	α	−28.33	−28.50	−28.1	−28.21	−28.30

accuracy for the triton is better than 1 keV and for the α-particle it is about 20–30 keV. One can appreciate the performance of the schematic effective forces more if one sees how large a tensor-force contribution they have to mock up. The basis dimension used in the SVM is 150 for the triton and 300 for the α-particle. Very small bases already give quite acceptable results. For example, the α-particle energy is within 1 MeV from the exact value with 50 well-optimized basis states.

The result of calculations with the full AV18 interaction, with and without the Urbana IX (UIX) three-body term [207] are shown in Table 13.27. The agreement with other methods is still satisfactory, although the SVM basis is the same as that optimized for the AV8 two-body interaction. The inclusion of the three-nucleon interaction brings the theoretical and experimental values to perfect agreement.

This example demonstrates that the SVM with the correlated Gaussian basis can cope with the full nuclear interaction.

13.9.4 The description of ^{6}Li

The real challenge is, of course, to go beyond the α-particle. We will now show an attempt to describe ^{6}Li as a problem of six nucleons interacting with a realistic force.

The convergence of the energy for such a system becomes very slow. It would probably require more than 1000 basis states to attain an accuracy comparable with that shown for the α-particle. The basis construction is very time-consuming because of the great number of subspaces to be sampled and because of the complicated form of the interaction. The large basis is needed because of the strong repulsive core of the interaction. It is extremely difficult to obtain a bound ^{6}Li with respect to the α+d threshold. Although the GFMC method [510] produces $\varepsilon_{\alpha d}=-1.5$ MeV with the AV8 interaction, all other methods,

Table 13.28: Energies, produced by the SVM, for the A=2–6 nuclei with the V8 version of the SSC interaction, without the Coulomb force (in MeV).

Nucleus	d	t	α	^{6}Li
E	−1.896	−7.052	−24.203	−26.39

including the SVM for the time being, have failed to produce binding. It would be important to reproduce this result with variational methods since the GMFC does not produce a wave function[20] together with the energy.

It is more promising to make an attempt with another interaction, whose core is softer. Therefore, the V8 part of the so-called supersoft-core (SSC) interaction [511] has been adopted for the study of ^{6}Li. The SSC interaction is still realistic; it gives reasonably good energies and reproduces the properties of the A=2–4 nuclei. The results for the energies are given in Table 13.28. The binding energies, the D-state components of the wave functions as well as the contributions of the central, tensor and spin-orbit terms to the total energies are quite similar to those produced by other V8-type interactions.

The SVM calculation for ^{6}Li included in Table 13.28 comprises about 500 basis states and gives about $\varepsilon_{\alpha d}$=−0.3 MeV. (The experimental value is −1.44 MeV.) This energy has certainly not converged and the result is not accurate, but, with the binding attained, it is at least meaningful. Due to this underbinding, it is not surprising that the calculated radius (r_p=2.41 fm) is somewhat larger than the experimental value (2.32 fm [213]).

To give a feeling for the α+d clustering, in Fig. 13.26 we show the radial factor of the $l=0$ α+d spectroscopic amplitude. This function is defined as $u_l(R) = R\int d\hat{\boldsymbol{R}}\, Y^*_{lm}(\hat{\boldsymbol{R}})g(\boldsymbol{R})$, with $g(\boldsymbol{R})$ given by Eq. (11.11). (The $l=2$ component is very small and not shown.) For comparison, an α+d spectroscopic amplitude provided by an α+p+n cluster model is also shown [365]. The microscopic α+d spectroscopic amplitudes are all very much alike [366] provided $\varepsilon_{\alpha d}$ is correct, but depend strongly on the value of $\varepsilon_{\alpha d}$. The *ab initio* result agrees in the region of cluster overlap, but tends to zero much more rapidly. This is quite contrary to the normal behaviour of an underbound α+d system: this amplitude should have a much longer tail than the cluster-model amplitude function, whose asymptotics is correct. This shows that the basis is not complete enough to produce the correct asymptotics in the tail region

[20]The GFMC produces the modulus squares of the wave functions in the sampled points of the configuration space; that is not enough to calculate matrix elements.

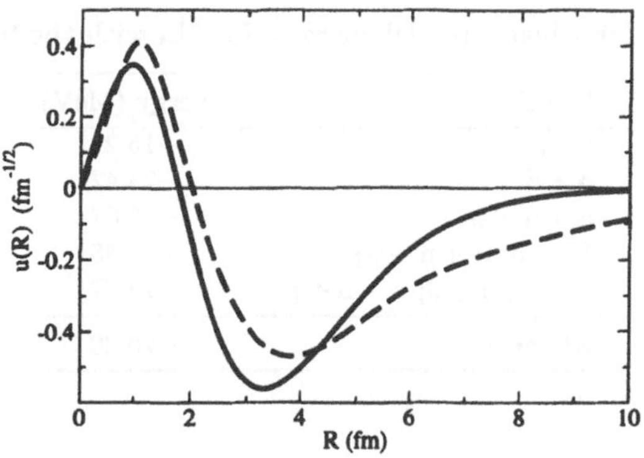

Figure 13.26: Two α+d spectroscopic amplitudes: *ab initio* calculation (solid line) and α+p+n cluster model with an effective force (dashed line) [365].

of the cluster-disintegration channel. The tail part of the wave function is known to give rather small corrections to the total energy, and, therefore, the tail behaviour is a stringent measure for the accuracy of the variational solution. The spectroscopic factors (11.26), however, which reflect the behaviour of the whole wave function, do agree well with each other (realistic force: 0.82, cluster model with an effective force: 0.88 [365]) and with experiment (0.85 [512]).

The relation of the *ab initio* calculation just presented to cluster models has been further studied by constructing cluster models with the SSC force. Five such models have been set up: a t+h, an α+d, an α+n+p model, a combination of the latter two: {α+d; α+n+p}, and a full cluster model, whose basis is a combination of the three pure models: {t+h; α+d; α+n+p}. The wave functions of these models will be denoted by $\Psi_{\rm th}$, $\Psi_{\alpha{\rm d}}$, $\Psi_{\alpha{\rm np}}$, $\Psi_{\alpha{\rm d}+\alpha{\rm np}}$ and $\Psi_{{\rm th}+\alpha{\rm d}+\alpha{\rm np}}$, respectively.

The α+d cluster-model basis has been constructed by taking the bases of α and d from their exact description as independent nuclei and assigning three $l=0$ Gaussians with judiciously chosen parameters for the relative motion. Each such basis function has of course been antisymmetrized. The Hamiltonian has been diagonalized, with some insignificant basis states discarded. The t+h model has been set up similarly. For the α+p+n model there is just one cluster, but two relative motions. A V-shaped basis has been chosen for that (see Fig. 8.1a), with angular momenta $S[(l_{\alpha p} l_{\alpha n})\, L]=$ 0[(00) 0], 1[(00) 0], 0[(11) 0], 1[(11) 0], 0[(11) 1], 1[(11) 1].

The model energies are listed in Table 13.29. None of the cluster models actually yields a bound state; the α+d threshold is open for all of them, though the α+p+n threshold is closed for the full {t+h;

Table 13.29: Cluster-model energies for ^{6}Li with the SSC force.

Model	Energy (MeV)
$\mathrm{t}+\mathrm{h}$	−15.21
$\alpha+\mathrm{d}$	−23.42
$\alpha+\mathrm{n}+\mathrm{p}$	−23.67
$\{\alpha+\mathrm{d};\, \alpha+\mathrm{n}+\mathrm{p}\}$	−24.48
$\{\mathrm{t}+\mathrm{h};\, \alpha+\mathrm{d};\, \alpha+\mathrm{n}+\mathrm{p}\}$	−24.77
Ab initio	−26.20

α+d; α+n+p} cluster model. This failure is again mostly due to the repulsive core of the force. The model allows for the short-range correlations within the clusters quite well, but the crude representation of the relative motion is not flexible enough to do the same between the nucleons belonging to two different clusters. Another possible source of this problem is that only the $l=0$ component of the relative motion is included.

It is seen that, apart from the t+h model, the cluster models give energies close to that of the *ab initio* calculation. Since, however, all cluster-model states are unbound, it is a trivial consequence of the variational principle that their energies must be close to the α-particle energy. The difficulty of the *ab initio* calculation, however, has shown that going beyond this truism is not so easy.

In conclusion, one can say that the first step has been made to close up the gap between *ab initio* calculations and cluster models. The reproduction of a clustering property by a realistic force provides an experience that the two worlds are not essentially different. It looks that the wave function produced by the realistic interaction behaves in the nuclear interior in much the same way as the cluster-model wave functions. But the wave function of the *ab initio* calculation in the asymptotic regions of the clustering channels is still far from being realistic. That is an indication that, for the description of loosely bound exotic nuclei, the *ab initio* calculations cannot yet compete.

13.10 Reaction calculations with correlated Gaussians

13.10.1 Exposition

In the present chapter the results of the structure theory have been compared with a great number of experimental data. Most of data come from collision experiments. To provide nuclear structure data, reaction experiments have to be analysed via the reaction theories as discussed in Part I. The reaction theory treats nuclear structure rather sketchily, which is not at all consistent with the rigour pursued in Part II. It is therefore important to make some sample analyses in which the elaborate structure model is used with some rigour. The goal of this section is to present some analyses of this kind.

All reaction calculations to be presented are concerned with ^{6}He, and the ^{6}He wave function of model 3 discussed in Sect. 13.1 is used, in a somewhat truncated form. The guideline in the comparisons will be to look for the imprint, on the cross sections, of exotic structure carried by the wave function and to see whether experiment confirms it. The reaction model will be used in its full-fledged form, but comparison with more approximate forms will also be made.

In Sect. 13.10.2 p+^{6}He elastic differential cross sections measured at a high ($\sim$ 700 MeV) energy will be compared with Glauber model calculations. In Sect. 13.10.3 a prediction will be made for the ^{6}He+^{12}C elastic differential cross section at 800 MeV per nucleon in the Glauber model and it will be compared with the α+^{12}C results at the same energy. In Sect. 13.10.4 α+^{6}He-type nuclear molecular states will be studied, which may appear in α+^{6}He scattering near the top of the Coulomb barrier.

13.10.2 High-energy p+^{6}He scattering

In Glauber's multiple scattering theory the scattering amplitude of the nucleus–nucleus collision has been derived in Sects. 3.2.4 and 3.2.5 of Part I. The approximations applied in the derivation are expected to be valid at high bombarding energies and can be interpreted as implying straight-line trajectories. This scattering amplitude is a two-dimensional Fourier transform of $e^{i\chi_{el}(\boldsymbol{b})}$ [cf. Eq. (3.57)], where $\chi_{el}(\boldsymbol{b})$ is the phase-shift function. The phase-shift function is related to the profile function $\Gamma(\boldsymbol{b})$ of the underlying nucleon–nucleon collision as follows:

$$e^{i\chi_{el}(\boldsymbol{b})} = \langle \Psi_0^{\mathrm{P}} \Psi_0^{\mathrm{T}} | \prod_{i\in \mathrm{P}} \prod_{j\in \mathrm{T}} [1 - \Gamma(\boldsymbol{b} + \bar{\boldsymbol{s}}_i^{\mathrm{P}} - \bar{\boldsymbol{s}}_j^{\mathrm{T}})] \, | \Psi_0^{\mathrm{P}} \Psi_0^{\mathrm{T}} \rangle, \qquad (13.29)$$

where Ψ_0^{P} and Ψ_0^{T} are the projectile and target intrinsic wave function, respectively, and $\bar{\boldsymbol{s}}_i$ is the projection of $\bar{\boldsymbol{r}}_i$ of Eq. (9.124) on the xy-plane. Equation (13.29) is the same as Eq. (3.78), but here the intrinsic coordinates are clearly distinguished from the extrinsic coordinates. The antisymmetry requirement between the projectile and target nucleons is neglected, which is a good approximation for high-energy collisions.

The phase-shift function can be calculated without approximation for a light system like α+α but is very complicated in general. Therefore, it is usually calculated in the OLA, introduced in Sect. 3.3, through the projectile and target densities:

$$\chi_{\mathrm{el}}^{\mathrm{OLA}}(\boldsymbol{b}) = i\int d\boldsymbol{r}^{\mathrm{P}}\int d\boldsymbol{r}^{\mathrm{T}}\,\rho_{\mathrm{P}}(\boldsymbol{r}^{\mathrm{P}})\rho_{\mathrm{T}}(\boldsymbol{r}^{\mathrm{T}})\Gamma(\boldsymbol{b}+\boldsymbol{s}^{\mathrm{P}}-\boldsymbol{s}^{\mathrm{T}}), \qquad (13.30)$$

where $\boldsymbol{s}^{\mathrm{P}}$, e.g., denotes the projection of $\boldsymbol{r}^{\mathrm{P}}$ on the xy-plane. When the target is a proton (a point particle), $\rho_{\mathrm{T}}(\boldsymbol{r})$ may be replaced by $\delta(\boldsymbol{r})$.

In the simplest case of a nucleon–nucleus collision the phase-shift function involves the evaluation of matrix elements of the many-body operator $\prod_{i=1}^{A}[1-\Gamma(\bar{\boldsymbol{s}}_i+\boldsymbol{b})]$ with A denoting the number of nucleons of the nucleus. Matrix elements of this type can be calculated with the method outlined in App. E.4.

When both the wave function and the profile function [cf. Eq. (3.58)] are built up from Gaussians, another method [58] can as well be employed. The phase-shift function (13.29), written for a proton target, can be expanded as

$$\begin{aligned} e^{i\chi_{\mathrm{el}}(\boldsymbol{b})} &= 1+\sum_{n=1}^{A}(-1)^n \sum_{i_1,\ldots,i_n} \langle\Psi_0|\Gamma(\bar{\boldsymbol{s}}_{i_1}+\boldsymbol{b})\cdots\Gamma(\bar{\boldsymbol{s}}_{i_n}+\boldsymbol{b})|\Psi_0\rangle \\ &= 1+\sum_{n=1}^{A}(-1)^n \frac{A!}{n!(A-n)!}\langle\Psi_0|\Gamma(\bar{\boldsymbol{s}}_1+\boldsymbol{b})\cdots\Gamma(\bar{\boldsymbol{s}}_n+\boldsymbol{b})|\Psi_0\rangle, \qquad (13.31) \end{aligned}$$

where the sum over $(i_1, ..., i_n)$ indicates any n target nucleons with different labels and, in the second equality, the antisymmetry of the wave function Ψ_0 is used. The terms that contain more than one Γ represent the multiple scattering of the nucleons. When Γ is of Gaussian form, any product of functions Γ is expressible in terms of the 2-dimensional generating function $g(u\boldsymbol{b}; \boldsymbol{s}, B)$ [cf. Eq. (9.59)], where $\tilde{\boldsymbol{s}}=(\boldsymbol{s}_1, ..., \boldsymbol{s}_A)$, u is a one-column matrix of A real numbers and B is an $A\times A$ symmetric matrix. With the use of this generating function, Eq. (13.31) can be calculated analytically even for correlated wave functions [58, 63].

The p+^{6}He elastic differential cross sections at forward angles were measured at $E_{\mathrm{p}}=717$ MeV [111]. A phenomenological fit to the experimental data was attempted by using the OLA, Eq. (13.30), in which the only unknown quantity to be fitted to experiment is the ^{6}He density. The fit produced 2.33±0.04 fm [68] for the matter rms radius. This is substantially smaller than the theoretical value, 2.46–2.51 fm (cf. Sect. 13.1.3). But the same reaction cross section data yield another value, 2.48±0.03 fm [10], depending on the actual details of the analysis. To resolve this problem, it is highly desirable to have a complete calculation for the phase-shift function with a realistic wave function.

Figure 13.27 shows the elastic differential cross section of the p+^{6}He scattering (a) and of the p+^{4}He scattering (b). To this end, the microscopic α+n+n wave function [213] discussed in Sect. 13.1 has been employed for ^{6}He, which gives 2.51 fm for the matter rms radius. A calculation of the p+^{4}He cross section is also shown for reference. The oscillator parameter β for the internal motion of the α-particle has been chosen to be 0.52 fm^{-2}, which is consistent with the well-known charge rms radius and is used in the α+n+n microscopic model as well. In Fig. 13.27 it is seen that in the full Glauber model these wave functions reproduce the elastic differential cross section very well, but the OLA gives 10% smaller differential cross sections and hence a clear deviation from the experimental data for both ^{6}He and ^{4}He.

To assess the sensitivity of the forward-angle cross section to the nuclear size, the nucleus ^{6}He has been dilated and contracted artificially by varying the u parameter of the Minnesota potential. A wave function tuned to give 2.33 fm for the ^{6}He rms radius produces larger elastic differential cross section than that belonging to 2.51 fm, whereas the wave function belonging to an rms radius of 2.63 fm predicts a smaller cross section. From these calculations it is well-founded to conclude that the matter rms radius of ^{6}He is not 2.33 fm but is, rather, around 2.51 fm, in agreement with other results [514].

As was mentioned above, the OLA tends to underestimate the elastic differential cross sections at forward angles. The tighter the binding, the larger the cross section is. Therefore, it is likely that the OLA with a smaller rms radius can give a good fit to experiment. Seeing just the OLA fit, one is tempted to give credit to the value 2.33 fm [111], but that is just an artefact.

To get a feeling for the sensitivity of the cross section to the type of the wave function, we also show Glauber-model calculations with a SM wave function, Ψ_{shell} and with an α+dineutron-like cluster-model wave function, $\Psi_{\text{dineutron}}$:

$$\Psi_{\text{shell}} = \mathcal{A}_{\alpha\text{nn}}\{\Phi_\alpha(\boldsymbol{x}_1,\boldsymbol{x}_2,\boldsymbol{x}_3)\Gamma^{(\beta_1\beta_2)}_{(11)00}(\boldsymbol{\rho}_1,\boldsymbol{\rho}_2)\chi_{00}\eta_{11}\}, \tag{13.32}$$

$$\Psi_{\text{dineutron}} = \mathcal{A}_{\alpha\text{nn}}\{\Phi_\alpha(\boldsymbol{x}_1,\boldsymbol{x}_2,\boldsymbol{x}_3)\Gamma^{(\beta_{1'}\beta_{2'})}_{(00)00}(\boldsymbol{\rho}_{1'},\boldsymbol{\rho}_{2'})\chi_{00}\eta_{11}\}, \tag{13.33}$$

where the ingredients are defined in App. H [cf. Eqs. (H.16), (H.17) and (H.18)]. The relative coordinates, $\boldsymbol{\rho}_1$ and $\boldsymbol{\rho}_2$, in Eq. (13.32) stand for the cluster Jacobi coordinates for the (αn)n (Y-type) arrangement. This gives the main (0s)^4(0p)^2 SM configuration if the values of $(\beta_1\beta_2)$ are chosen to be $(\beta_1,\beta_2)=(\frac{4}{5}\beta,\frac{5}{6}\beta)$, where β is the size parameter of the α-particle wave function Φ_α. Equation (13.32) exemplifies the relationship between the SM and the cluster model discussed in Sect. 11.6.

The coordinates, $\rho_{1'}$ and $\rho_{2'}$, in Eq. (13.33) emphasize another type of correlation corresponding to the (nn)α (T-type) arrangement. The $\beta_{1'}$ parameter determines the localization of the halo neutrons and its value is fixed to be 0.16 fm^{-2}, which corresponds to a rms distance of 3.1 fm. The oscillator parameter of the α-particle in Eq. (13.33) is set to 0.52 fm^{-2}. The parameter β of Ψ_{shell} and $\beta_{2'}$ of $\Psi_{dineutron}$ are determined so as to give the same rms radius (2.51 fm) as the α+n+n model.

As seen from Fig. 13.28a, neither the SM nor the dineutron-cluster model reproduces the experimental data. Both give steeper slopes although not to the same extent; the dineutron-cluster model gives a slightly better fit, particularly at very small angles. Figure 13.28b shows the difference between the three models in a wider range of angles. The three models predict quite different angular distributions, which shows the importance of using realistic wave functions in the analysis of the scattering. (Unfortunately, it is probably very difficult to measure the small cross sections at larger angles.)

The performance of the complete Glauber model calculation in reproducing elastic differential cross-section data has confirmed the validity of the α+n+n three-cluster model. In particular, the theoretical value of the rms radius of ^{6}He in the 2.46–2.51 fm range has been found to be consistent with all observations on ^{6}He.

13.10.3 High-energy ^{6}He+^{12}C scattering

An exact calculation of the phase-shift function (13.29) would require enormous computer time when the mass number A of either the projectile or the target approaches 10. A fast and accurate approximation method is thus badly needed. Such an approach is to base the Glauber theory on nucleon–target (NT) scattering considered an elementary process [63,65], as is explained in Sect. 3.5.2. With this, the phase-shift function (13.29) is simplified as [cf. Eq. (3.97)]

$$e^{i\tilde{\chi}_{el}(\boldsymbol{b})} = \langle \Psi_0^{P} | \prod_{i=1}^{A} [1 - \Gamma_{NT}(\boldsymbol{s}_i + \boldsymbol{b})] | \Psi_0^{P} \rangle, \qquad (13.34)$$

where Γ_{NT} is the nucleon–target profile function determined by fitting the form (3.96) to nucleon–target scattering data.

In Table 13.30 the OLA [χ_{el}^{OLA} of Eq. (13.30)] and a similar approximation introduced at the nucleon–nucleus level ['effective', Eq. (3.101)] are tested against $\tilde{\chi}_{el}$ of Eq. (13.34) ('exact') and against experimental reaction cross section data for the ^{4}He+^{12}C and ^{6}He+^{12}C collisions. The OLA cross sections are larger than the measured cross sections. The difference is modest for the stable projectile ^{4}He, but becomes

Table 13.30: Comparison of ^{4}He+^{12}C and ^{6}He+^{12}C theoretical reaction cross sections (in mb) with the corresponding measured interaction cross sections at 800 A MeV. The phase-shift function is calculated by Eq. (13.30) (OLA), Eq. (3.101) (effective) and Eq. (13.34) (exact).

Projectile	OLA	Effective	Exact	Exp. [8]
^{4}He	520	490	514	503±5
^{6}He	782	707	736	722±6

fairly large (about 10%) for the halo nucleus ^{6}He. The OLA cross section can thus be fitted with smaller nuclear radii [74], which results in empirical radii smaller than is realistic, especially for halo nuclei. The exact calculation with the microscopic projectile wave function leads to an excellent agreement with experiment, which indicates that one should adopt the rms radius of the microscopic model as the empirical value of the radius.

As to the two approximate profile functions, it is remarkable that the cross section predicted by Eq. (3.101) is much closer to experiment than that by the OLA. Notwithstanding that their definitions are similar and Eq. (3.101) is even simpler, it is markedly superior to the OLA since it is based on the correct empirical nucleon–nucleus profile function, while the OLA accumulates errors in both folding steps. Owing to its simplicity, the data coming from quite a few experiments have been analysed with it, and it has turned out to reproduce the reaction cross section at 800 A MeV much better (within a few per cent) than the conventional OLA formula (cf. Table 3.1).

As another example for the performance of the three approaches, Fig. 13.29 displays the ^{4}He+^{12}C scattering cross section at the intermediate energy of 342 MeV per nucleon. No p+^{12}C elastic scattering data are available here, and $\Gamma_{\rm NT}$ of Eq. (3.101) has been set by extrapolation from $E_{\rm p}=398$ MeV. The calculated ^{4}He+^{12}C cross section is nevertheless in fair agreement with experiment, actually better than the phenomenological fit [515]. The two approximate calculations are certainly inferior, which reflects the performance of the OLA in reproducing the underlying p+^{12}C elastic scattering.

Finally, in Fig. 13.30 exact [Eq. (13.34)] Glauber angular distributions are shown, viz. those of ^{6}He+^{12}C and ^{4}He+^{12}C elastic scattering at 800 A MeV. At small angles both cross sections significantly exceed the Rutherford cross sections. The difference in their diffraction pattern is due mainly to the structural differences between the two projectiles. In particular, the larger breakup probability of ^{6}He is bound to re-

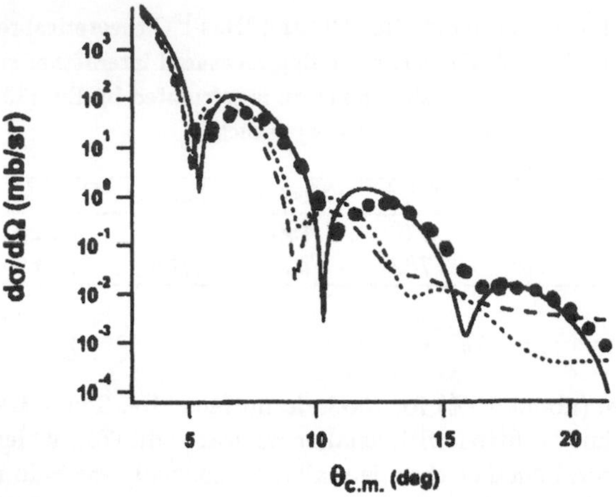

Figure 13.29: Elastic differential cross sections for ^{4}He+^{12}C scattering at E_α = 342 A MeV. The phase-shift functions have been calculated by Eq. (13.34) ($\tilde{\chi}_{\rm el}$, solid curve), by Eq. (3.101) ($\chi_{\rm eff}$, dashed curve) and by Eq. (13.30) ($\chi_{\rm el}^{\rm OLA}$, dotted curve). The data are from Ref. [64].

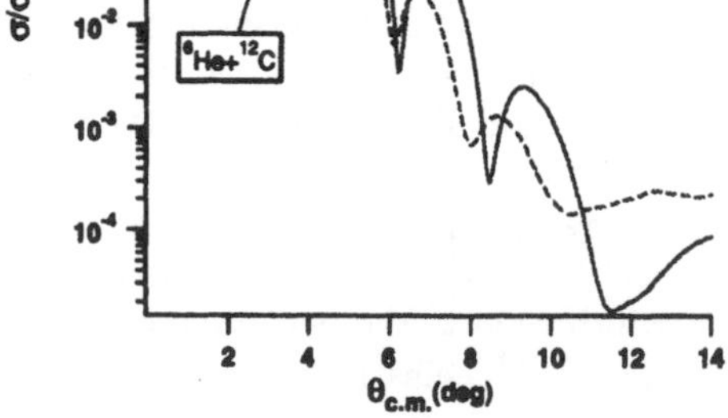

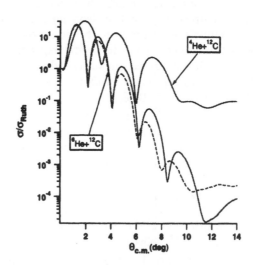

Figure 13.30: Elastic differential cross sections divided by the corresponding Rutherford cross section for ^{6}He+^{12}C and ^{4}He+^{12}C scattering at 800 A MeV. The solid curves result from the complete Glauber amplitudes [Eq. (13.34)], while the dashed curve is the approximation (3.101) for ^{6}He+^{12}C.

duce the ^{6}He+^{12}C elastic cross section, and that is what appears in the steeper slope of the angular distribution. The ^{6}He+^{12}C cross section calculated with the approximate phase-shift function (3.101) again follows the full calculation very well, especially at small angles.

13.10.4 Low-energy α+^{6}He scattering

Low-energy radioactive nuclear beams pose a new type of challenge to theory and experiment. Low-energy ^{6}He beams are of particular interest since ^{6}He is the simplest nucleus exhibiting an anomalously large matter radius indicative of a neutron halo. For example, the ^{6}He+p collision near the Coulomb barrier must show resonances of exotic structure. This collision must be dominated by the p wave because the lowest s-orbit is fully occupied in ^{6}He. In addition to $T=\frac{1}{2}$ states, the incoming proton, together with the two halo-neutrons, may form a loosely bound $T=\frac{3}{2}$ negative-parity configuration.

In fact, the ^{7}Li state at excitation energy $E_x = 11.24$ MeV, with width $\Gamma_{\rm c.m.} \simeq 260$ keV and $J^\pi = \frac{3}{2}^-$, located at 1.3 MeV above the ^{6}He+p threshold, is the lowest-lying $T=\frac{3}{2}$ state and is a candidate for incorporating such an extraordinary configuration. Being an analogue state of the g.s. of ^{7}He, this ^{7}Li state is expected to have a 2n–1p halo outside the α-particle. This interpretation seems to be supported by ^{6}Li(n,p)^{6}He and ^{6}Li(n,n')^{6}Li* experiments [516], which show that the reduced width (see footnote on p. 291) of the $T=\frac{3}{2}$ state to the decay to the g.s. of ^{6}Li is very small ($\theta_{\rm n}^2=2\times10^{-4}$), but is pretty large (0.16) for the decay to the ^{6}Li state with $E_x=3.56$ MeV, which is the analogue of the ^{6}He g.s. (cf. Sect. 13.1).

We show some microscopic calculations for low-energy α+^{6}He scattering. The cross sections have been measured at c.m. energies of 60.4, 11.6 and 15.9 MeV [517,518] with the expectation that several interesting physical effects, e.g., the elastic transfer of the halo neutrons, may be observed. Close to the Coulomb barrier two colliding nuclei of not very different masses often exhibit resonances, which can be interpreted as molecular states. These systems overlap weakly and, short-lived as they are, exhibit rotational-like level sequences. In the α+^{6}He case the large radius of ^{6}He may be expected to give rise to unusually extended nuclear molecules [519]. All these effects are related to the halo of ^{6}He and may provide new information on its properties.

A full quantum-mechanical treatment with antisymmetrization over all nucleons participating in the collision is particularly important in low-energy scattering. Such a treatment can be carried through in the framework of the RGM approach (cf. Sect. 11.5) adapted to scattering problems. In the RGM not only the antisymmetrization but also the angular-momentum and parity projections are treated exactly. In the RGM for the α+^{6}He scattering there are few model assumptions and parameters, and these are independent of the experimental data to be described.

The α+^{6}He wave function with angular momentum lm can be written as

$$\Psi_{lm} = \sum_i \mathcal{A}_{\alpha^6\text{He}} \left\{ \Phi_\alpha \Phi^{(i)}_{^6\text{He}} g^{(i)}_l(q) Y_{lm}(\hat{q}) \right\}, \tag{13.35}$$

where q is the relative coordinate between α and ^{6}He, and $\Phi^{(i)}_{^6\text{He}}$ is the intrinsic wave function of ^{6}He in state i. The ^{6}He cluster is described as in model 3 of Sect. 13.1, with $\text{T}_{0,(0,0)0}, \text{Y}_{0,(1,1)0}, \text{T}_{1,(1,1)1}, \text{Y}_{1,(1,1)1}$ configurations (cf. Table 13.2) included. The state with $i=0$ corresponds to the g.s., whereas the states with $i \geq 1$ are 'pseudostates', which are discrete positive-energy ^{6}He states produced by the diagonalization over a square-integrable basis.

The intrinsic wave functions of the α and ^{6}He clusters are kept fixed, but the ^{6}He cluster is allowed to be excited into its pseudostates. The pseudostates are expected to simulate the breakup effects of ^{6}He into the 0^+ continuum. The relative function $g^{(i)}_l(q)$ has to satisfy scattering boundary conditions, which can be derived from the partial-wave expansion of Eq. (3.10) or a generalization of Eq. (10.8). Accordingly, in the asymptotic region the function $g^{(i)}_l(q)$ is expressible in terms of the regular and outgoing Coulomb functions $F_l(k_i q)$ and $H^{(+)}_l(k_i q)$ or, alternatively, in terms of the incoming and outgoing Coulomb functions $H^{(-)}_l(k_i q)$ and $H^{(+)}_l(k_i q)$:

$$g^{(i)}_l(q) \sim F_l(k_i q)\delta_{i0} - f^{(l)}_{0i} H^{(+)}_l(k_i q) = \delta_{i0} H^{(-)}_l(k_i q) - S^{(l)}_{0i} H^{(+)}_l(k_i q), \quad q \to \infty, \tag{13.36}$$

where k_i are the respective wave numbers, $f^{(l)}_{0i}$ is a component of the scattering amplitude and $S^{(l)}_{0i}$ is an element of the S-matrix. In the internal region $g^{(i)}_l(q)$ is constructed as

$$g^{(i)}_l(q) Y_{lm}(\hat{q}) = \sum_n f^{(l)}_{in} \psi^{(\beta_n)}_{0lm}(q), \tag{13.37}$$

where $\psi^{(\beta)}_{0lm}(q)$ is given by Eq. (H.18) and $f^{(l)}_{in}$ are variable parameters. The parameters β_n determine the fall-off of the Gaussians. They have to be chosen appropriately so as to approximate the wave function well. Since ^{6}He is described in the α+n+n three-cluster model, the α+^{6}He wave function of Eq. (13.35) is an α+α+n+n four-cluster wave function, and the Gaussian expansion (13.37) ensures that it takes the same form as the multicluster basis for ^{10}Be discussed in Sect. 13.5. Thus the matrix elements can be obtained as in the bound-state problems discussed in the previous chapters. The scattering problem is, however, specific in one respect. The orbital angular momentum l must have a sharp value, which excludes the global-vector representation, Eq. (9.33). One must use, instead, the successive coupling, Eq. (9.23), which makes computations time-consuming, especially when l is large. This can be compensated for by keeping the basis dimension for ^{6}He as low as possible (15 elements now) and truncating the partial-wave expansion (to $l_{\rm max}=3$ now).

The scattering problem can be solved in various ways. It is beyond the scope of this book to discuss these ramifications in detail. It suffices to make a few general remarks here and refer the reader to the literature [301, 316, 317]. The simplest possible trick is to include the additional terms, $+\delta_{i0}\tilde{H}_l^{(-)}(k_iq)-S_{0i}^{(l)}\tilde{H}_l^{(+)}(k_iq)$, for each i and l, in Eq. (13.37), with variational parameters $S_{0i}^{(l)}$. The functions $\tilde{H}_l^{(-)}$ and $\tilde{H}_l^{(+)}$ are zero at the origin and differ from $H_l^{(-)}$ and $H_l^{(+)}$ only in the region, where some of $\psi_{0lm}^{(\beta_n)}(q)$ differ significantly from zero. [For closed channels, $\tilde{H}_l^{(+)}(k_iq)$ will reduce to the Whittaker function corresponding to the proper bound-state asymptotics.] By varying the linear parameters $f_{in}^{(l)}$ and $S_{0i}^{(l)}$, one obtains a set of linear inhomogeneous equations for the coefficients, which is solvable for all positive real energies.[21] The values of $S_{0i}^{(l)}$ are approximate S-matrix elements, which can be improved by adding to them simple correction terms, with which they become stationary [301, 316].

To eliminate some inaccuracies in the border region between the interaction region and the asymptotic region, one may divide the configuration space by a sharp surface, within and without which Eq. (13.37) and (13.36) holds, respectively. Then the above variational procedure still works in a modified form [301]. Alternatively, the problem can be rewritten into a dynamical problem for the inner region, in which the boundary condition on the surface is fulfilled by adding a so-called Bloch operator [520] to the Hamiltonian. This latter formalism is actually a reformulation of the R-matrix theory [520] and is called the microscopic R-matrix method. Both of these formulations involve matrix elements with integration over the inner region only. These are easily calculated by taking the integrals over the whole space and subtracting the contributions coming from the asymptotic region [521]. In the results to be presented these two prescriptions were used [521], with concording results.

---

[21]The non-trivial solutions of the corresponding homogeneous equations are the complex resonance energies [266].

The u parameter of the Minnesota force has been set to 0.95, which is suited to describing the $\alpha+\alpha$ scattering. The width parameter of the α internal motion has been chosen to be $\beta=0.52$ fm^{-2}, which is the standard value, and the spin-orbit force used in the description of ^{10}Be (Sect. 13.5) has been adopted. Owing to the large value of u (cf. $u=0.91$ in Sect. 13.5) and to the limited basis, the spectrum of ^{10}Be has been shifted up with respect to the bound-state calculations for ^{10}Be, but the energy difference between the g.s. (0^+) and the first excited state (2^+) is correct.

Figure 13.31 shows the elastic scattering phase shifts obtained in the multichannel calculation, with all ^{6}He pseudostates included. The negative-parity phase shifts show resonant behaviour near 0.47 MeV ($\Gamma_\alpha = 0.01$ MeV) and 3.7 MeV ($\Gamma_\alpha = 0.7$ MeV) for $l=1$ and $l=3$, respectively. The dimensionless reduced α-widths (footnote on p. 291) at a channel radius of 6 fm are $\theta_\alpha^2 = 0.09$ for $l=1$ and $\theta_\alpha^2 = 0.27$ for $l=3$. These large values, particularly the one for the 3^- state, are characteristic of molecular states, well-known in other systems, such as $\alpha+\alpha$, $\alpha+^{14}$C or $\alpha+^{16}$O [286, 522]. Additional resonances appear in the $l=0$ and $l=2$ partial waves ($E_{\text{c.m.}}=4.7$, 6.9 and 10.5 MeV). These states involve excited ^{6}He states, which are discrete in the model only, thus the $\alpha+^6$He resonances built upon them are most probably spurious.

The calculations suggest that $\alpha+^6$He molecular states exist near the Coulomb barrier. The 1^- and 3^- states apparently belong to a $K^\pi=0^-$ band in the language of the rotational model [174], and the 3^- state is a particularly good candidate for quasimolecular structure. The 3 MeV energy difference between the two yields a mean distance of 5.4 fm between α and ^{6}He. This value is much larger than the sum of the cluster

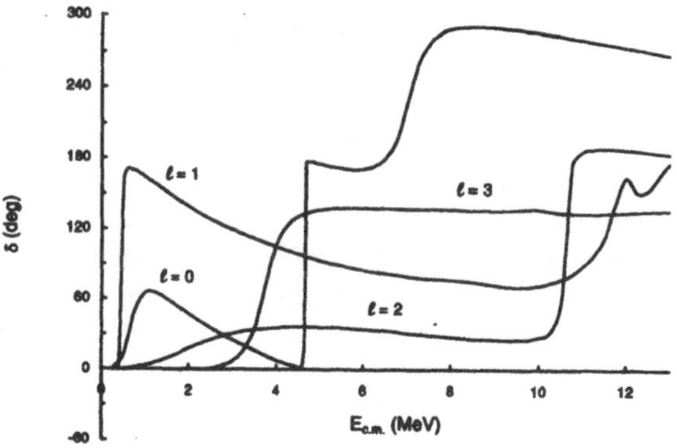

Figure 13.31: Real parts of multichannel $\alpha+^6$He phase shifts for $l=0$ to 3.

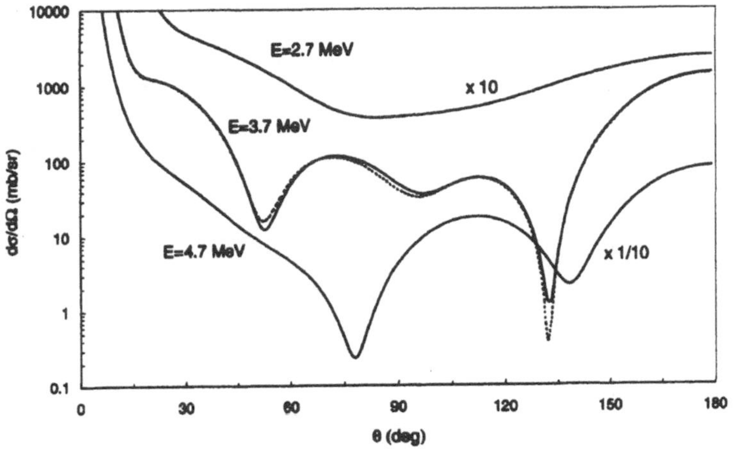

Figure 13.32: Elastic differential cross sections of the $\alpha+{}^6$He collision on the resonance (3.7 MeV) and off the resonance (2.7 and 4.7 MeV) with $l_{\max}=3$. The dotted line includes partial waves up to $l=8$.

radii and suggests a molecule-like shape. The rapid variation of the elastic differential cross section around $E=3.7$ MeV (Fig. 13.32) bears the imprint of the 3^- molecular state, so it should be possible to find it experimentally with facilities that provide low-energy ^{6}He beams. Of course, the prediction of the model may not be too accurate, but the indication for the existence of this state is certainly well-founded. The existence of molecular resonances is well-established in many other systems, and they lie in general close to the Coulomb barrier.

Appendix A

Overview of reaction theories

The outcome of a nucleus–nucleus collision event depends very much on the way the colliding partners overlap during the process. In fully quantum-mechanical treatments this can be parametrized by the relative angular momentum, which corresponds well to the semiclassical impact parameter (see p. 30).

When the impact parameter is close to the sum of the radii of the two colliding nuclei, only a few degrees of freedom participate in the reaction dynamics. Such a reaction mechanism is called a direct reaction. The smaller the impact parameters, the more degrees of freedom come into play. In the other extreme, the incident energy is distributed among all degrees of freedom statistically, and that is known as a compound nuclear reaction. In the course of the process more and more degrees of freedom get involved, and hence the information on the projectile nucleus is gradually lost. Therefore, structure information on the projectile can only be extracted from direct reactions, and the high-energy reaction theory discussed in this book is to be related to direct reaction theories.

The simplest direct reaction is elastic scattering, in which both projectile and target remain in their g.s. The elastic scattering is usually described by a one-body Schrödinger equation [cf. Eq. (2.1)] with an optical-model potential (cf. Sect. 5.1.1),

$$[T + U(\boldsymbol{r})]\psi_{\boldsymbol{k}}^{(+)}(\boldsymbol{r}) = E\psi_{\boldsymbol{k}}^{(+)}(\boldsymbol{r}), \tag{A.1}$$

where $T = -(\hbar^2/2\mu)\boldsymbol{\nabla}^2$, with μ the reduced mass, and $\psi_{\boldsymbol{k}}^{(+)}(\boldsymbol{r})$ is a wave function whose asymptotic form is a sum of the incident plane wave of momentum (actually wave number vector) $\boldsymbol{k}$ and of an outgoing wave [cf. Eq. (2.2)]. The optical potential $U(r)$ is complex. The real part is attractive and represents the refraction and reflection of the wave function. The imaginary part is also negative and its role is to take into account the loss of flux from the elastic channel. Flux is lost in the collision due to excitations (including those into the continuum) and due to particle transfer, which is significant only at low energies. The imaginary potential is so strong that the flux is fully absorbed when the relative distance is smaller than the sum of the radii of the two colliding nuclei.

One can obtain a perturbative estimate for the scattering amplitude (2.11) by replacing the wave function with a plane wave $e^{i\boldsymbol{k}'\cdot\boldsymbol{r}}$ propagating in the direction $\boldsymbol{k}'$:

$$f(\boldsymbol{q}) = -\frac{2\mu}{\hbar^2}\frac{1}{4\pi}\int d\boldsymbol{r}\, e^{-i\boldsymbol{q}\cdot\boldsymbol{r}}U(\boldsymbol{r}), \tag{A.2}$$

where $\boldsymbol{q}=\boldsymbol{k}'-\boldsymbol{k}$ [Eq. (2.34)]. This is called the Born approximation [cf. Eq. (2.47)]. It is, however, not a good approximation in nuclear physics mainly because the absorption is too strong. The eikonal approximation is, on the contrary, quite good so long as the incident energy is not very low. But, for a spherical optical potential, the solution of Eq. (A.1) is, in practice, easy without any approximation.

The simplest non-elastic process is inelastic scattering, in which the target and/or the projectile are excited, without the composition of the colliding nuclei changed. In rearrangement processes some nucleons are transferred between the two nuclei. At higher incident energies the cross sections of the rearrangement processes decrease because of momentum mismatch (see p. 48). At the same time, the projectile breakup reactions become significant.

In describing a transition to a particular final state one can often resort to a perturbative approximation. Consider a reaction A(a,b)B leading from incoming channel $\alpha=(\text{aA})$ to two-body outgoing channel $\beta=(\text{bB})$, where each symbol implies a well-defined internal state Φ_i, $(i=\text{A},\text{B},\text{a},\text{b})$. The perturbative approximation holds when the transition is much weaker than any of the elastic scattering processes, so that the latter can be described by optical potentials U_α, U_β. The transition amplitude can then be written as

$$\begin{aligned} T_{\text{DWBA}} &= \langle \psi^{(-)}_{\boldsymbol{k}_\beta}\Phi_\text{b}\Phi_\text{B}|V_\alpha - U_\alpha|\psi^{(+)}_{\boldsymbol{k}_\alpha}\Phi_\text{a}\Phi_\text{A}\rangle \\ &= \langle \psi^{(-)}_{\boldsymbol{k}_\beta}\Phi_\text{b}\Phi_\text{B}|V_\beta - U_\beta|\psi^{(+)}_{\boldsymbol{k}_\alpha}\Phi_\text{a}\Phi_\text{A}\rangle. \end{aligned} \tag{A.3}$$

This is called the distorted-wave Born approximation (DWBA). The relative wave functions $\psi^{(\pm)}_{\boldsymbol{k}}(\boldsymbol{r})$ are called the distorted waves and are solutions of Eq. (A.1) with appropriate boundary conditions [cf. Eq. (2.2)]:

$$\psi_{\boldsymbol{k}}^{(\pm)}(\boldsymbol{r}) = \mathrm{e}^{i\boldsymbol{k}\cdot\boldsymbol{r}} + f^{(\pm)}(\hat{\boldsymbol{r}}, \boldsymbol{k})\frac{\mathrm{e}^{\pm ikr}}{r} \qquad (r \to \infty). \tag{A.4}$$

The potentials V_α, V_β are sums of the two-body interactions between the nucleons in the nuclei, e.g., $V_\alpha = \sum_{i\in a}\sum_{j\in A} V_{ij}$. Since inside the nuclei the distorted waves are almost zero, this approximation does produce a transition that takes place at the nuclear surface. The DWBA is an analogue of the ordinary (plane-wave) Born approximation. It can be derived by writing the interaction, in each channel, like $V = U+(V-U)$, treating U exactly and $V-U$ perturbatively.

When a coupling to some specific channel or the effect of the transition back on elastic scattering is strong, one has to resort to a nonperturbative improved description called the coupled-channel method. Multistep transitions are also often described by the coupled-channel method. The coupled-channel method is derived by expanding the full wave function in terms of final states. Restricting the discussion to inelastic scattering, we may write the model wave function as

$$\Psi^{(+)}(\boldsymbol{r}, \xi) = \sum_\alpha u_\alpha(\boldsymbol{r})\Phi_{\mathrm{a}m}(\xi_{\mathrm{a}})\Phi_{\mathrm{A}n}(\xi_{\mathrm{A}}), \tag{A.5}$$

where ξ_i and $\boldsymbol{r}$ are intrinsic and relative coordinates, respectively, and $\alpha = \{m, n\}$ labels the channels, specified by the states $\{m=0,1,\ldots,n=0,1,\ldots\}$ of nuclei. By substitution of ansatz (A.5) into the total Schrödinger equation of the system,

$$\left(\sum_i T_i + \sum_{i<j} V_{ij}\right)\Psi^{(+)}(\boldsymbol{r}, \xi) = E\Psi^{(+)}(\boldsymbol{r}, \xi), \tag{A.6}$$

and by projection onto the intrinsic states $\Phi_{\mathrm{a}m}\Phi_{\mathrm{A}n}$, a set of coupled equations is obtained for $u_\alpha(\boldsymbol{r})$:

$$[E - E_\alpha - T_\alpha]\, u_\alpha(\boldsymbol{r}) = \sum_{\alpha'} U_{\alpha\alpha'}(\boldsymbol{r}) u_{\alpha'}(\boldsymbol{r}), \tag{A.7}$$

where E_α is the sum of the intrinsic energies, $E_\alpha = E_{\mathrm{a}m} + E_{\mathrm{A}n}$, and T_α is the kinetic-energy operator of the relative motion in channel α. The potential matrix $U_{\alpha\alpha'}(\boldsymbol{r})$ is defined as

$$U_{\alpha\alpha'}(\boldsymbol{r}) = \langle \Phi_{\mathrm{a}m}\Phi_{\mathrm{A}n}|V_\alpha|\Phi_{\mathrm{a}m'}\Phi_{\mathrm{A}n'}\rangle. \tag{A.8}$$

Asymptotically, $u_\alpha(\boldsymbol{r})$ has an incoming wave only in the entrance channel and has outgoing waves in all channels, which may be written schematically as

$$u_\alpha(\boldsymbol{r}) = \delta_{\alpha,00} e^{ik_\alpha z} + f_\alpha(\theta,\phi)\frac{e^{ik_\alpha r}}{r}. \tag{A.9}$$

The diagonal elements of the potential $U_{\alpha\alpha'}(\boldsymbol{r})$ can be considered optical potentials of some kind, whose imaginary parts allow for excitations that are not explicitly treated in the model space. These potentials do not coincide, however, with the ordinary optical potential since the strongly coupled excited states are taken into account explicitly. The off-diagonal elements are coupling terms, which are also complex in general. Their form is usually derived from phenomenological models of the nuclear states.

The coupled-channel treatment of rearrangement processes is somewhat similar but more complex owing to the nonorthogonality of the terms of Eq. (A.5) in that case.[1]

When the projectile is weakly bound, its (real or virtual[2]) breakup should be taken into account. In the coupled-channel treatment this can be done by including continuum states of the projectile in Eq. (A.5). The continuum is to be discretized into a finite number of momentum bins, and the method is called the continuum-discretized coupled-channel (CDCC) method [523,524].

The coupled-channel problem is significantly simplified if the collision can be assumed to excite a mode of intrinsic motion that is much slower than the projectile–target relative motion. Let us assume that the projectile has such a slow mode in coordinates $\xi^{\rm sl}$, and denote its other degrees of freedom by $\xi_{\rm a}^{\rm int}$, so that $\xi_{\rm a} = \{\xi^{\rm sl}\xi_{\rm a}^{\rm int}\}$. For rotational excitations, e.g., $\xi^{\rm sl}$ may stand for the orientation angles of the deformed projectile. For a halo nucleus, $\xi^{\rm sl}$ may be chosen to be the halo-nucleon–core relative coordinate(s), while $\xi_{\rm a}^{\rm int}$ denote the intrinsic coordinates of the core. The excitations of the slow motion take place through a component of the projectile intrinsic wave function that has a product form, $\Phi_{{\rm a}m}(\xi_{\rm a}) = \phi_m^{\rm sl}(\xi^{\rm sl})\Phi_{{\rm a}0}^{\rm int}(\xi_{\rm a}^{\rm int})$, and, for simplicity, the other components are omitted. Here $\Phi_{{\rm a}0}^{\rm int}$ stands for the g.s. in the other degrees

[1]In a two-channel problem one may choose to neglect the coupling in the first equation, which belongs to the elastic channel. Then the first equation can be solved independently, and the resulting function u_{00}, substituted into the second, will give rise to an inhomogeneous ('source') term there. That equation is also solvable as a single equation, and this approximation method is exactly equivalent to the DWBA. In a multichannel problem one may choose to treat some strongly coupled channels (e.g., the scattering channels) as explicitly coupled, and the weak transition between them (e.g., the rearrangements) perturbatively. Such a scheme is called the coupled-channel Born approximation. Alternatively, one may use a coupled-channel framework in which the backward couplings belonging to weak transitions are neglected. The two schemes are again strictly equivalent.

[2]The breakup is real if there is loss of flux into the breakup channel and virtual if the initial system is recombined.

of freedom. The wave functions $\{\phi_m^{\rm sl}\}$ are assumed to form a complete orthonormal set for the slow degree of freedom $\xi^{\rm sl}$.

In this simplified model the slow degree of freedom $\xi^{\rm sl}$ is treated as frozen during the collision. That is known as the adiabatic or sudden approximation. One can get to this approximation by writing the full wave function as

$$\Psi^{(+)}(\boldsymbol{r},\xi) = \chi(\boldsymbol{r},\xi^{\rm sl})\Phi_{\rm a0}^{\rm int}(\xi_{\rm a}^{\rm int})\Phi_{\rm A0}(\xi_{\rm A}) \tag{A.10}$$

and treating the dynamical variable $\xi^{\rm sl}$ as a parameter. The state $\chi(\boldsymbol{r},\xi^{\rm sl})$ describes the projectile–target relative motion with the projectile frozen into a particular configuration. For the rotational model this configuration is a particular orientation of the deformed body, while for the neutron-halo projectile it is a particular relative position of the halo-neutron–core system. From Eq. (A.10) it then follows that the function $\chi(\boldsymbol{r},\xi^{\rm sl})$ satisfies the equation

$$\left[E - E_0 - T_\alpha - V_\alpha(\boldsymbol{r},\xi^{\rm sl})\right]\chi(\boldsymbol{r},\xi^{\rm sl}) = 0, \tag{A.11}$$

where the intrinsic energy is put equal to the sum of the g.s. energies of the two colliding nuclei, E_0. This equation is derived by putting Eq. (A.10) into Eq. (A.6), and projecting it onto the intrinsic state $\Phi_{\rm a0}^{\rm int}\Phi_{\rm A0}$. The potential $V_\alpha(\boldsymbol{r},\xi^{\rm sl})$ is thus defined as

$$V_\alpha(\boldsymbol{r},\xi^{\rm sl}) = \langle \Phi_{\rm a0}^{\rm int}\Phi_{\rm A0}|V_\alpha|\Phi_{\rm a0}^{\rm int}\Phi_{\rm A0}\rangle. \tag{A.12}$$

In Eq. (A.11) the intrinsic coordinate $\xi^{\rm sl}$ appears only as a parameter, each value of which defines an intrinsic configuration. The scattering equation is to be solved with appropriate boundary conditions for all configurations of the intrinsic motion.[3] Note that the wave function (A.10) does contain the g.s. as well as all excited states involving one of

[3]Equation (A.11) is related to the set of coupled equations (A.7) as follows. Introduce channel wave function $u_m^{\rm adiab}(\boldsymbol{r})$ such that

$$\chi(\boldsymbol{r},\xi^{\rm sl}) = \sum_{m'} u_{m'}^{\rm adiab}(\boldsymbol{r})\phi_{m'}^{\rm sl}(\xi^{\rm sl}).$$

Putting this expression into Eq. (A.11) and projecting it onto the intrinsic state labelled m, we obtain the coupled equations

$$(E - E_0 - T_\alpha)u_m^{\rm adiab}(\boldsymbol{r}) = \sum_{m'} U_{mm'}(\boldsymbol{r})u_{m'}^{\rm adiab}(\boldsymbol{r}),$$

where $U_{mm'}(\boldsymbol{r})$ is defined by

$$U_{mm'}(\boldsymbol{r}) = \langle \phi_m^{\rm sl}|V_\alpha(\boldsymbol{r},\xi^{\rm sl})|\phi_{m'}^{\rm sl}\rangle.$$

This is analogous to Eq. (A.7). At a phenomenological level they may be identified if the excited-state energies, E_α in Eq. (A.7), are replaced with the g.s. energies, E_0.

$\{\phi_m^{\text{sl}}\}$. This is true to the extent to which the configurations $\{\chi(\boldsymbol{r}, \xi^{\text{sl}})\}$ span the state space $\{\phi_m^{\text{sl}}\}$.

Since no coupled equations are to be solved, the original problem is greatly simplified. The adiabatic approximation is useful when the excitation energies of the most important configurations are low enough, such as in the low-lying collective excitations and in the breakup of weakly-bound projectiles. This adiabatic approximation is akin to that introduced in Sect. 3.2.2 and used extensively in combination with the eikonal approximation in this book.

The direct reaction theories can in principle be applied to reactions of low as well as of high incident energies. In practice, however, the theory is less promising for high incident energies. This is partly because at high incident energies very many partial waves need to be included. To avoid this, one may apply WKB [26] or eikonal approximations for the distorted waves $\psi^{(\pm)}(\boldsymbol{r})$. Another impeding factor at high incident energies is the vast increase of the relevant channels.

What makes the high-energy collisions yet tractable is that the nucleon–nucleon collisions involved in them are not altered very much with respect to the free nucleon–nucleon collisions. This suggests that the transition amplitude could be calculated from the scattering amplitude t of the free nucleon–nucleon collisions. This t is related to the nuclear force v by $t = v + vG_0t$, with G_0 being the Green's function of the free motion. This approximation is called the impulse approximation. Replacing the nucleon–nucleon potential $V = \sum_{ij} V_{ij}$ with the free nucleon–nucleon t-matrix $\sum_{ij} t_{ij}$ in Eq. (A.3) is essentially what is known as the distorted-wave impulse approximation.

The high-energy approximation called the Glauber theory also relies on nucleon–nucleon collisions as the basic ingredients of the nucleus–nucleus collision. However, it departs from direct reaction theories first in that it does not attempt to describe transitions to final states one by one and to reconstruct the whole process as a sum of these transitions. The collision is viewed as a series of underlying nucleon–nucleon collisions, and the resulting wave function contains a copious number of excited states, which represent all possible exit channels. In addition, the high-energy nature of the process makes it possible to introduce semiclassical elements.

The basic ingredient is a semiclassical substitute for the scattering matrix, $e^{i\chi(\boldsymbol{b})}$. This is a function of the impact parameter, which plays the same role in semiclassical approaches as the orbital angular momentum does in quantum scattering theory. The nucleon–nucleon scattering involved in the nucleus–nucleus scattering is described in terms of the phase-shift function, $\chi(\boldsymbol{b})$. As a result of the full collision, the g.s. intrinsic wave functions of the colliding nuclei get multiplied by the product of the 'scattering matrices' for all pairs of nucleons as is seen on the left-hand side:

$$\prod_{i\in \mathrm{a}}\prod_{j\in \mathrm{A}} \mathrm{e}^{i\chi_{ij}(\boldsymbol{b}+\boldsymbol{s}_i-\boldsymbol{s}_j)}\Phi_{\mathrm{a}0}\Phi_{\mathrm{A}0} = \sum_{mn} C_{mn}(\boldsymbol{b})\Phi_{\mathrm{a}m}(\xi_{\mathrm{a}})\Phi_{\mathrm{A}n}(\xi_{\mathrm{A}}). \quad \text{(A.13)}$$

The vector $\boldsymbol{s}_i$ is the transverse component of a nucleon coordinate in the projectile or target measured from the respective nuclear centre. Then $\boldsymbol{b}+\boldsymbol{s}_i-\boldsymbol{s}_j$ is the impact parameter vector for the collision of two nucleons, i and j. On the right-hand side the wave function is expanded over the complete set of the projectile and target internal states. The reaction probability for the transition to any particular (elastic, inelastic or breakup) channel can be extracted by applying Eqs. (3.34)–(3.39) to the wave-function component belonging to the corresponding internal states. Thus the Glauber theory does produce all information expected from a reaction theory.

In the Glauber theory the reaction is viewed as a succession of free nucleon–nucleon collisions. In the course of the process the c.m. of the colliding nuclei are assumed to proceed on a straight-line trajectory, which is a semiclassical element, and the internal motion of the nucleus is frozen during the collision. No further assumption is necessary on the reaction mechanism. In particular, no perturbative approximation is invoked, so that multiple collisions are fully included.

Historically, direct reaction theories and high-energy reaction theories have been developed more or less independently. The direct reaction theories lend themselves best to the treatment of reactions leading to well-defined two-body final states, which can be measured mainly at low to medium incident energies. In direct reaction theories dynamical equations are to be solved, which require a number of model parameters. In the high-energy collisions, however, inclusive observables are usually measured. The Glauber theory does not require the solution of any dynamical equations. It is applicable primarily to inclusive processes, and the input parameters are nucleon–nucleon data, which are not specific to the system investigated. The Glauber theory can be applied to the exclusive processes as well. Although the direct reaction framework can in principle be applied to the reactions at high energies, the high-energy theories provide a less model-dependent and parameter-dependent description.

Appendix B

Conventional cluster Jacobi coordinates

The aim of this appendix is to summarize the formalism of the conventional Jacobi coordinates. There is a one-to-one correspondence between these Jacobi vectors and the type introduced in Sect. 9.1.3. Each vector differs from the corresponding one in the other formalism just in a factor. In few-body problems and cluster models the use of the Jacobi sets to be introduced now is more common since their meaning is more direct. Those introduced in Sect. 9.1.3 are, however, more convenient when HO (or Gaussian) bases are used, as in Part II of this book.

To make the formulation concise, let us assume that the A nucleons at positions $\boldsymbol{r}_i$ $(i = 1, \ldots, A)$ are arranged into n clusters, and each cluster contains a_i nucleons $(\sum_{i=1}^{n} a_i = A)$. When $n = A$, the formulae boil down to the no-cluster case. To label the new coordinates, it is convenient to introduce $\{j\}$ to denote the number of nucleons contained in the first $j-1$ clusters:

$$\{j\} = \sum_{m=1}^{j-1} a_m \quad (j = 2, \ldots, n+1), \quad \{1\} = 0. \tag{B.1}$$

Intrinsic Jacobi coordinates $\boldsymbol{t}_{\{j\}+i}$ and cluster c.m. coordinates $\bar{\boldsymbol{t}}_j$ may then be defined as

$$\boldsymbol{t}_{\{j\}+i} = \frac{1}{i} \sum_{m=1}^{i} \boldsymbol{r}_{\{j\}+m} - \boldsymbol{r}_{\{j\}+i+1} \qquad (i = 1, \ldots, a_j - 1,\ j = 1, \ldots, n), \tag{B.2}$$

$$\bar{\boldsymbol{t}}_j \equiv \boldsymbol{t}_{\{j\}+a_j} = \frac{1}{a_j} \sum_{i=1}^{a_j} \boldsymbol{r}_{\{j\}+i} \qquad (j = 1, \ldots, n). \tag{B.3}$$

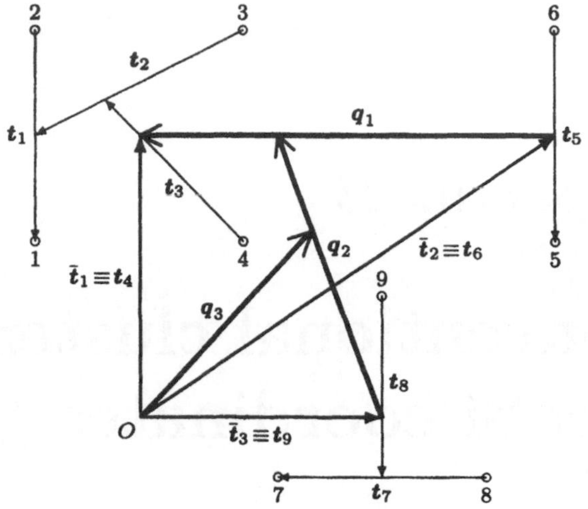

Figure B.1: The pattern of cluster Jacobi coordinates illustrated by a three-cluster case with $\{a_2, a_2, a_3\} = \{4, 2, 3\}$.

To illustrate the cluster Jacobi coordinates (B.3), a schematic sketch is shown in Fig. B.1. When $a_j = 1$, there is no need to define $\boldsymbol{t}_{\{j\}+i}$, and $\bar{\boldsymbol{t}}_j$ reduces to $\boldsymbol{r}_{\{j\}+1}$. The transformation from set $(\boldsymbol{r}_{\{j\}+1}, \ldots, \boldsymbol{r}_{\{j\}+a_j})$ to set $(\boldsymbol{t}_{\{j\}+1}, \ldots, \boldsymbol{t}_{\{j\}+a_j-1}, \bar{\boldsymbol{t}}_j)$ can also be defined by an $a_j \times a_j$ matrix:

$$T_j = \begin{pmatrix} 1 & -1 & 0 & \cdots & 0 \\ \frac{1}{2} & \frac{1}{2} & -1 & \cdots & 0 \\ \vdots & & & & \vdots \\ \frac{1}{a_j - 1} & \frac{1}{a_j - 1} & \frac{1}{a_j - 1} & \cdots & -1 \\ \frac{1}{a_j} & \frac{1}{a_j} & \frac{1}{a_j} & \cdots & \frac{1}{a_j} \end{pmatrix}, \tag{B.4}$$

whose inverse is

$$T_j^{-1} = \begin{pmatrix} \frac{1}{2} & \frac{1}{3} & \frac{1}{4} & \cdots & \frac{1}{a_j} & 1 \\ -\frac{1}{2} & \frac{1}{3} & \frac{1}{4} & \cdots & \frac{1}{a_j} & 1 \\ 0 & -\frac{2}{3} & \frac{1}{4} & \cdots & \frac{1}{a_j} & 1 \\ \vdots & & & & \frac{1}{a_j} & \vdots \\ 0 & 0 & 0 & \cdots & \frac{1}{a_j} & 1 \\ 0 & 0 & 0 & \cdots & -\frac{a_j - 1}{a_j} & 1 \end{pmatrix}. \tag{B.5}$$

From the T_j matrices one can build up a block-diagonal $A\times A$ matrix:

$$T = \begin{pmatrix} T_1 & 0 & \cdots & 0 \\ 0 & T_2 & \cdots & 0 \\ \vdots & & & \vdots \\ 0 & 0 & \cdots & T_n \end{pmatrix}. \tag{B.6}$$

With the help of these definitions the coordinate transformation can be written as

$$\boldsymbol{t} = T\boldsymbol{r}, \tag{B.7}$$

where $\tilde{\boldsymbol{t}}=(\boldsymbol{t}_1,\ldots,\boldsymbol{t}_A)$ and $\tilde{\boldsymbol{r}}=(\boldsymbol{r}_1,\ldots,\boldsymbol{r}_A)$, and the tilde denotes transposition. Note that the matrix T_j satisfies the following equalities

$$\sum_{i=1}^{a_j} (T_j)_{ki}\,(T_j)_{li} = \frac{1}{\mu_{\{j\}+k}}\delta_{kl}, \qquad \sum_{i=1}^{a_j} \left(T_j^{-1}\right)_{ik}\left(T_j^{-1}\right)_{il} = \mu_{\{j\}+k}\delta_{kl}, \tag{B.8}$$

where $\mu_{\{j\}+k}$ is the (dimensionless) reduced mass corresponding to $\boldsymbol{t}_{\{j\}+k}$ $(k=1,\ldots,a_j)$:

$$\mu_{\{j\}+i} = \begin{cases} \frac{i}{i+1} & \text{for } i = 1,\ldots,a_j-1, \\ a_j & \text{for } i = a_j. \end{cases} \tag{B.9}$$

To describe the motion of the clusters, intercluster Jacobi coordinates are defined in terms of $(\bar{\boldsymbol{t}}_1,\ldots,\bar{\boldsymbol{t}}_n)$:

$$\boldsymbol{q}_j = \frac{1}{\sum_{m=1}^{j} a_m}\sum_{m=1}^{j} a_m\bar{\boldsymbol{t}}_m - \bar{\boldsymbol{t}}_{j+1} \quad (j=1,\ldots,n-1). \tag{B.10}$$

The c.m. coordinate of the whole system can then be expressed either through the c.m. coordinates of the clusters or in terms of the s.p. coordinates:

$$\boldsymbol{q}_n = \frac{1}{\sum_{m=1}^{n} a_m}\sum_{m=1}^{n} a_m\bar{\boldsymbol{t}}_m = \frac{1}{A}\sum_{i=1}^{A}\boldsymbol{r}_i. \tag{B.11}$$

The transformation (B.10) from $\tilde{\bar{\boldsymbol{t}}}=(\bar{\boldsymbol{t}}_1,\ldots,\bar{\boldsymbol{t}}_n)$ to $\tilde{\boldsymbol{q}}=(\boldsymbol{q}_1,\ldots,\boldsymbol{q}_n)$ can be written in matrix form

$$\boldsymbol{q} = W\bar{\boldsymbol{t}}, \tag{B.12}$$

where the $n\times n$ matrix W is

$$W = \begin{pmatrix} 1 & -1 & 0 & \cdots & 0 \\ \frac{a_1}{\{3\}} & \frac{a_2}{\{3\}} & -1 & \cdots & 0 \\ \vdots & & & & \vdots \\ \frac{a_1}{\{n\}} & \frac{a_2}{\{n\}} & \frac{a_3}{\{n\}} & \cdots & -1 \\ \frac{a_1}{A} & \frac{a_2}{A} & \frac{a_3}{A} & \cdots & \frac{a_n}{A} \end{pmatrix}. \tag{B.13}$$

Then one can construct an $A \times A$ matrix V which transforms the coordinates $\tilde{t} = (t_{\{1\}+1}, \ldots, t_{\{1\}+a_1-1}, \bar{t}_1, t_{\{2\}+1}, \ldots, t_{\{n\}+a_n-1}, \bar{t}_n)$ into coordinates $\tilde{t}^{(q)} = (t_{\{1\}+1}, \ldots, t_{\{1\}+a_1-1}, q_1, t_{\{2\}+1}, \ldots, t_{\{n\}+a_n-1}, q_n)$, which is to act on the cluster c.m. coordinates only:

$$t^{(q)} = Vt, \tag{B.14}$$

with

$$V_{ij} = \begin{cases} W_{kl} & \text{if } i = \{k+1\} \text{ and } j = \{l+1\}, \\ \delta_{ij} & \text{otherwise,} \end{cases} \tag{B.15}$$

where $i, j = 1, \ldots, A$ and $k, l = 1, \ldots, n$. Combining Eqs. (B.7) and (B.14) leads to

$$t^{(q)} = Ur, \quad \text{where} \quad U = VT. \tag{B.16}$$

The transformation matrix U relates the s.p. coordinates to the cluster intrinsic and cluster relative coordinate sets. It is straightforward to prove that the matrix W satisfies the following equations:

$$\sum_{j=1}^{n} \frac{1}{a_j} W_{kj} W_{lj} = \frac{1}{\mu_{\{k+1\}}} \delta_{kl}, \qquad \sum_{j=1}^{n} a_j \left(W^{-1}\right)_{jk} \left(W^{-1}\right)_{jl} = \mu_{\{k+1\}} \delta_{kl}, \tag{B.17}$$

where the reduced-mass factor $\mu_{\{k+1\}}$ corresponds to the cluster relative Jacobi coordinate q_k:

$$\mu_{\{k+1\}} = \begin{cases} \frac{\{k+1\} a_{k+1}}{\{k+2\}} & \text{for } k = 1, \ldots, n-1, \\ A & \text{for } k = n. \end{cases} \tag{B.18}$$

The sum of the squared s.p. coordinates can be expressed in terms of the cluster intrinsic and cluster relative coordinates. In fact, by using the second equations of (B.8) and (B.17), one gets

$$\begin{aligned} \sum_{i=1}^{A} r_i^2 &= \sum_{j=1}^{n} \sum_{m=1}^{a_j} \sum_{k,l=1}^{a_j} (T_j)^{-1}_{mk} (T_j)^{-1}_{ml} t_{\{j\}+k} t_{\{j\}+l} = \sum_{j=1}^{n} \sum_{k=1}^{a_j} \mu_{\{j\}+k} t^2_{\{j\}+k} \\ &= \sum_{j=1}^{n} \sum_{k=1}^{a_j-1} \mu_{\{j\}+k} t^2_{\{j\}+k} + \sum_{j=1}^{n} a_j \sum_{k,l=1}^{n} (W_j)^{-1}_{jk} (W_j)^{-1}_{jl} q_k q_l \\ &= \sum_{j=1}^{n} \sum_{k=1}^{a_j-1} \mu_{\{j\}+k} t^2_{\{j\}+k} + \sum_{k=1}^{n} \mu_{\{k+1\}} q_k^2. \end{aligned} \tag{B.19}$$

Likewise, the kinetic-energy operator can be expressed in terms of the momenta, $\pi_{\{j\}+i}$ and ϖ_j, which are canonically conjugate to the coordinates $t_{\{j\}+i}$ and q_j, respectively. Use of the first equations of (B.8) and (B.17) leads to

$$\sum_{i=1}^{A}\frac{p_i^2}{2m}=\sum_{j=1}^{n}\sum_{k=1}^{a_j}\frac{p_{\{j\}+k}^2}{2m}=\sum_{j=1}^{n}\sum_{k=1}^{a_j-1}\frac{\pi_{\{j\}+k}^2}{2\mu_{\{j\}+k}m}+\sum_{j=1}^{n}\frac{\varpi_j^2}{2\mu_{\{j+1\}}m}. \tag{B.20}$$

Here the last term is the kinetic energy of the c.m. motion, $T_{\text{c.m.}}=\frac{\varpi_n^2}{2\mu_{\{n+1\}}m}$.

Appendix C

Borromean and Efimov states

The distinctive property of light exotic nuclei is that they show unusual few-body correlations. The most apparent few-body effect is Borromean binding. This has been defined as binding between three bodies without bound states existing in any of the two-body subsystems. There are several examples for Borromean systems mentioned in the main text, and two of them, ^{6}He and ^{11}Li, are discussed at length. The aim of this appendix is to give a generic quantitative analysis of this phenomenon. This gives an opportunity to discuss the Efimov states as well, which can be viewed as the limiting cases of Borromean states.

For a three-body system to be Borromean, the attractive part of the interaction between its constituents must be short-ranged. A system of more than two particles can be bound even if none of its subsystems is bound since the intrinsic kinetic energy is roughly proportional to $n-1$, where n is the number of particles, while the potential energy increases with the number of pairs, $\frac{1}{2}n(n-1)$.

A Borromean system may produce an infinite number of three-body bound states, which are called Efimov states, when at least two of the binary subsystems have s-states sharp at zero energy [525, 526]. They are accumulated at zero energy and their spatial extension is extremely large. When the states of the binary subsystems are slightly off zero energy, the three-body system may still have a number of very extensive bound states, which may also be called Efimov states [527]. The Efimov states will, however, disappear into the continuum if the strength of the attractive two-particle interaction is changed so as to make the two-body states fully bound or fully unbound. The specific property of the Efimov states is that they pass into the positive-energy region even if the potential is made more attractive.

The appearance of the Efimov states can be explained by a two+one-particle folding model (cf. Sect. 5.1.2). When one particle gets far from the other two, it feels the folding potential

$$V^{\mathrm{F}}(\boldsymbol{r}) = \int d\boldsymbol{r}' |\phi(\boldsymbol{r}')|^2 \left[V\left(\boldsymbol{r} - \frac{\boldsymbol{r}'}{2}\right) + V\left(\boldsymbol{r} + \frac{\boldsymbol{r}'}{2}\right) \right], \qquad \text{(C.1)}$$

where V is the interaction between the particles and $\phi(\boldsymbol{r}')$ is the relative zero-energy s-state of the other two, which falls off like r'^{-1}. Since V is short-ranged, contributions to the integral come from $\boldsymbol{r}' \simeq 2\boldsymbol{r}$ and $\boldsymbol{r}' \simeq -2\boldsymbol{r}$. Therefore, $V^{\mathrm{F}}(\boldsymbol{r})$ behaves at large distances as $V^{\mathrm{F}}(\boldsymbol{r}) \sim -r^{-2}$, and an attractive potential with a strong enough r^{-2} tail [528] supports an infinite number of bound states around zero energy, which gives rise to the Efimov effect. When a two-body subsystem is made fully bound, the tail of the effective potential will fall off more strongly. Moreover, there appears a two-body threshold, and the former Efimov states will lie over the threshold as unbound states.

Nuclear halos are often discussed in conjunction with Efimov states. Efimov's original idea was in fact to identify some states in the three-nucleon and three-α-particle systems with such states. More recently, there are speculations that the loosely-bound ^{18}C (^{16}C+n+n) and ^{20}C (^{18}C+n+n) systems may have Efimov states [529] (cf. Table 8.1).

To exemplify the Efimov effect, we consider a system of three identical bosons interacting with a two-body interaction. One can use a short-ranged potential of any form to create Efimov states by tuning the strength of the interaction so as to set the two-body g.s. energy to zero. As an illustration, we use a Gaussian potential:

$$V(r) = -V_0\, \mathrm{e}^{-\mu^2 r^2}. \qquad \text{(C.2)}$$

To get rid of unessential constants, we rescale the problem by the substitution $r \to \mu^{-1} r$, which amounts to using μ^{-1} as length units. (The new $\boldsymbol{r}$ is now dimensionless.) The Hamiltonian of two particles of mass m is transformed into

$$H = \frac{\hbar^2 \mu^2}{m}\left(-\nabla^2 - g\,\mathrm{e}^{-r^2}\right), \qquad \text{(C.3)}$$

where $g = mV_0/\hbar^2\mu^2$ is a dimensionless 'coupling constant'. The convenient energy unit $\hbar^2\mu^2/m$ will be used. This system of units will be referred to as 'atomic units'.

The minimum value of g required to produce a bound state in an n-particle system will be called a critical coupling constant and denoted by g_n. At $g = g_2$, the two-body system has a zero-energy state, and then

Efimov states appear in the three-body system. For $g \in (g_3, g_2)$, there is at least one three-body bound state, which is Borromean by definition. The relationship $g_3 < g_2$ is always true for bosonic systems, but not necessarily for fermionic systems because of the Pauli exclusion.

Figure C.1 shows the bound-state energies of the three-boson system for different values of the coupling constant between $g = 2.1$ and 3.0. (The lower the curve the deeper the binding.)

At the beginning of the interval shown ($g=2.1$) the potential is too weak to support a bound state in the two- and three-body systems. Beyond the critical value of $g_3 = 2.12$, the three-body system has a bound state, which is Borromean, and at about $g=2.6$ a second bound state also appears. The two-body s-state binding sets in at $g_2=2.6842$, and, at the same g value, in the three-body system a third bound state shows up, which must be a Efimov state. The energies of the normal three-body states change monotonically with g, but the Efimov state turns back and reaches the threshold at $g = 2.72$. This upturn is a unique quality of the Efimov states; it is caused by the turning of the long-ranged two-particle–one-particle effective force into short-ranged. The energies and rms radii of the three states are shown in Table C.1.

It should be clear that, if there is one Efimov state, then there must be an infinite number of them, but, to localize more than one, we need a very large and extensive basis stretching out to many thousand times the range of the force [187], which was not the aim here.

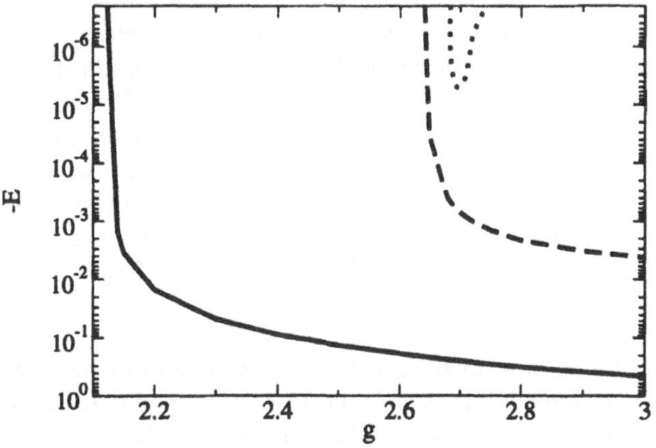

Figure C.1: Energies (in 'atomic units') of the g.s. (solid line), the first excited state (dashed line) and a Efimov state (dotted line) of the three-boson system as a function of the coupling constant. The critical values are $g_2=2.6842$ and $g_3=2.12$. The energies are measured from the lowest-lying threshold.

Table C.1: Three-boson bound-state energies E_k and rms radii r_k (in 'atomic units') produced by a Gaussian potential with $g = g_2$. The third state is a Efimov state.

E_1	r_1	E_2	r_2	E_3	r_3
−0.2395	1.21	-4.79×10^{-4}	22.06	-2.25×10^{-6}	380.92

With g increased further, more three-body bound states will appear. The next critical value is g_2^* at which the two-body system has a second zero-energy bound s-state. This state again creates an r^{-2}-like two-particle–one-particle effective potential. But this stronger force puts the two+one-particle threshold much deeper. So the three-body states produced are well out in the continuum, and they may not lead to dramatic effects.

On the other hand, Borromean systems with more than three particles may be produced by weakening the coupling constant. For $g < g_3$, there are no three-body bound states, but four-body bound states may exist for g exceeding a critical value g_4. Within the interval of $g \in (g_4, g_3)$, the four-body bound states are Borromean since none of their two- and three-particle subsystems is bound (see Table C.2). One can continue to create even larger systems with no bound subsystems. These exercises show that multiparticle binding may be produced even by very weak potentials. Although no example has been found for such systems, exotic nuclei may yet provide such surprises.

The above observations are quite general and do not depend on the details of the interaction. To show this, we used the Morse potential $V(r) = V_0\left[e^{-2\mu(r-r_0)} - 2e^{-\mu(r-r_0)}\right]$, which simulates the interaction between two composite particles (e.g., noble-gas atoms or nucleons). An additional parameter, r_0, controls the height and thickness of a strong repulsive core and a relatively long-ranged but weak attractive tail. This potential can also be rescaled. With the choice of $r_0 = 1$, the critical

Table C.2: Energies of three- and four-boson systems in the Gaussian and Morse ($r_0 = 1$) potentials in 'atomic units'. The energy $E_k(g_{k-1})$ is that of the k-boson system at the critical coupling constant for the $k-1$-boson system.

Potential	g_2	$E_3(g_2)$	g_3/g_2	$E_4(g_3)$	g_4/g_3
Gaussian	2.68415	−0.2395	0.790	−0.438	0.80
Morse	0.36875	−0.0325	0.797	−0.0636	0.79

strength is obtained as $g_2 = 0.369$. With $g = g_2$, an SVM calculation puts the three-body g.s. at -0.0325, while the energy of the Efimov state is at -0.00006. Table C.2 shows the four-body energy corresponding to $g = g_3$ as well.

The last example for a Borromean system to be considered is ^{12}C in a microscopic $\alpha+\alpha+\alpha$ cluster model. The Minnesota interaction (see Table 12.2) was used for its description, without the Coulomb force, with each α-cluster described as a single 0s HO state of parameter $\beta = 0.52\,\text{fm}^{-2}$. In this model the energy of the α-cluster is -24.814 MeV. By setting the u parameter of the Minnesota potential to $u = 0.7791$, the two α-clusters form a zero-energy state. (In reality, owing to the Coulomb barrier, the g.s. of ^{8}Be is a narrow resonance lying just slightly above the $\alpha+\alpha$ threshold.)

In the microscopic three-cluster calculation (cf. Chap. 12) two bound 0^+ states have been found. There may be other bound states but, due to the closeness of the $\alpha+\alpha+\alpha$ threshold, it is extremely difficult to find them numerically. The g.s. is at -1.2 MeV with respect to the $\alpha+\alpha+\alpha$ threshold, and the first excited state, a Efimov state, is at -0.004 MeV. In reality, the 0^+ excited state lies 0.38 MeV above the $\alpha+\alpha+\alpha$ threshold. This exercise shows that Borromean, and even Efimov, states may appear in the relative motions of three composite clusters as well, and their microscopic description is possible.

Appendix D

Antisymmetrization

The Pauli principle says that any quantum state has to be antisymmetric with respect to the particle labels of any two of identical fermions. For instance, if $\Phi(1,2)$ is to be a physically meaningful state of nucleons 1 and 2, it must have the property $\Phi(1,2) = -\Phi(2,1)$. Mathematically, this implies that a state should be invariant with respect to even permutations and should change sign with respect to odd permutations of the labels of identical fermions.

This is an absolutely strict principle for strictly identical elementary particles, but nucleons are elementary only for low-energy nuclear physics, and protons are identical to neutrons only approximately. Yet the charge independence of the nuclear forces is almost perfect, and even the Coulomb force causes just a minor symmetry breaking. Therefore, it is permissible to treat protons and neutrons as identical particles, and it is a great simplification indeed. The tool developed for this purpose is the isospin formalism, and this formalism is used throughout this book. Thus the labels to be exchanged in examining the permutation symmetry should include not only the spatial and spin coordinates, but also the isospin coordinates.

Antisymmetry is in general imposed by the antisymmetrization operator or antisymmetrizer $\mathcal{A}$ defined by Eq. (9.40):

$$\mathcal{A} = \frac{1}{\sqrt{A!}} \sum_{P}^{A!} \text{sign}(P)P, \tag{D.1}$$

where the sum runs over all $A!$ permutations of the A nucleonic labels and $\text{sign}(P)$ stands for the sign of the permutation P.

The antisymmetrizer is essentially a projector; more precisely, $\mathcal{A}' = A!^{-1/2}\mathcal{A} = A!^{-1}\sum_P \text{sign}(P)P$ is a projector. To show this it suffices to know that $\mathcal{A}'$ is Hermitean and that $\mathcal{A}'^2 = \mathcal{A}'$. The latter statement can also be made plausible by noting that $P\mathcal{A}' = \mathcal{A}'P = \text{sign}(P)\mathcal{A}'$ for any permutation P, so that

$$\mathcal{A}'^2\{\phi(1,\ldots,A)\} = \frac{1}{A!}\sum_P \text{sign}(P)P\mathcal{A}'\{\phi(1,\ldots,A)\} = \mathcal{A}'\{\phi(1,\ldots,A)\}. \tag{D.2}$$

It follows from this theorem that the state space of antisymmetric functions is part of the direct product of s.p. spaces. For the square of the operator $\mathcal{A}$, one has

$$\mathcal{A}^2 = \sqrt{A!}\mathcal{A}. \tag{D.3}$$

The effect of $\mathcal{A}$ can be illustrated by its acting on a three-nucleon function $\phi(1,2,3)$:

$$\mathcal{A}\phi(1,2,3) = \frac{1}{\sqrt{6}}\,[\phi(1,2,3) - \phi(1,3,2) + \phi(2,3,1) - \phi(2,1,3) + \phi(3,1,2) - \phi(3,2,1)]. \tag{D.4}$$

If the state ϕ is a product of s.p. states, $\phi = \psi_1(1)\psi_2(2)\psi_3(3)$, then $\mathcal{A}$ will turn it into a determinant:

$$\mathcal{A}\phi(1,2,3) = \frac{1}{\sqrt{6}}\det\{\psi_1(1)\psi_2(2)\psi_3(3)\} = \frac{1}{\sqrt{6}}\begin{vmatrix} \psi_1(1) & \psi_1(2) & \psi_1(3) \\ \psi_2(1) & \psi_2(2) & \psi_2(3) \\ \psi_3(1) & \psi_3(2) & \psi_3(3) \end{vmatrix}. \tag{D.5}$$

The determinants of s.p. states are called Slater determinants.

For a system of n clusters,[1] $a_1,\ldots,a_n$, one can introduce subsystem (or 'intracluster') antisymmetrizers $\mathcal{A}_{\{1\}},\ldots,\mathcal{A}_{\{n\}}$ in complete analogy with Eq. (D.1). A product of antisymmetrical cluster wave functions can be made fully antisymmetrical by means of an intercluster antisymmetrizer, $\mathcal{A}_{\{1\}\cdots\{n\}}$, introduced in Eq. (11.1). It antisymmetrizes between the fragments such that, acting upon the internally antisymmetrized state, it will make them fully antisymmetrical. It will contain $A!/a_1!\cdots a_n!$ terms, and each term will contain the permutation operators and constant factors of $\mathcal{A}$ that are not contained in the subsystem antisymmetrizers $\mathcal{A}_{\{1\}},\ldots,\mathcal{A}_{\{n\}}$, i.e.,

$$\mathcal{A} = \mathcal{A}_{\{1\}\cdots\{n\}}\mathcal{A}_{\{1\}}\cdots\mathcal{A}_{\{n\}}. \tag{D.6}$$

It can thus be written in the explicit form

$$\mathcal{A}_{\{1\}\cdots\{n\}} = \sqrt{\frac{a_1!\cdots a_n!}{A!}}\sum_{P_\leftrightarrow}^{N}\text{sign}(P_\leftrightarrow)P_\leftrightarrow, \text{ with } N = \frac{A!}{a_1!\cdots a_n!}, \tag{D.7}$$

[1]Since in this appendix both particle labels and cluster labels occur plentifully, we distinguish them by putting the cluster labels in { } except when they are applied directly to the symbols that denote the clusters and their mass numbers: $a_1,\ldots,a_n$.

where $P_{\leftrightarrow}$ runs over all *non-equivalent* permutations which involve at least one exchange between the subsets of labels $\{1,\ldots,a_1\},\ldots,\{A-a_n+1,\ldots,A\}$ and also includes the unit operator. We call two intercluster permutations, $P^{(1)}$ and $P^{(2)}$, equivalent if there are intracluster permutations $P_{\{1\}},\ldots,P_{\{n\}}$ and $P'_{\{1\}},\ldots,P'_{\{n\}}$ that turn them into each other in the sense that $P^{(1)}P_{\{1\}}\cdots P_{\{n\}}=P^{(2)}P'_{\{1\}}\cdots P'_{\{n\}}$. For example, for a four-particle system divided into two clusters (12) and (34), the intercluster antisymmetrizer will be

$$\begin{aligned}\mathcal{A}_{(12)(34)} &= \frac{1}{\sqrt{6}}(1-P_{13}-P_{14}-P_{23}-P_{24}+P_{13}P_{24}) \\ &= \frac{1}{\sqrt{6}}(1-P_{13}-P_{14}-P_{23}-P_{24}+P_{14}P_{23}), \qquad \text{(D.8)}\end{aligned}$$

where the last term in the second line is equivalent to the last term in the first line, so they should be included alternatively. Their equivalence is established by finding intracluster permutations that connect them. In fact, $P_{13}P_{24}P_{12}P_{34}=P_{13}P_{14}P_{24}P_{34}=P_{14}P_{34}P_{24}P_{34}=P_{14}P_{23}1$.

For the 4+2 and 3+3 clusterizations of six nucleons one of the equivalent forms of the two intercluster antisymmetrizers is this:

$$\begin{aligned}\mathcal{A}_{(1234)(56)} &= \frac{1}{\sqrt{15}}(1-P_{15}-P_{16}-P_{25}-P_{26}-P_{35}-P_{36}-P_{45}-P_{46} \\ &\quad +P_{15}P_{26}+P_{15}P_{36}+P_{15}P_{46}+P_{25}P_{36}+P_{25}P_{46} \\ &\quad +P_{35}P_{46}), \qquad \text{(D.9)} \\ \mathcal{A}_{(123)(456)} &= \frac{1}{\sqrt{20}}(1-P_{14}-P_{15}-P_{16}-P_{24}-P_{25}-P_{26}-P_{34}-P_{35}-P_{36} \\ &\quad +P_{14}P_{25}+P_{14}P_{26}+P_{14}P_{35}+P_{14}P_{36}+P_{15}P_{26} \\ &\quad +P_{16}P_{35}+P_{34}P_{25}+P_{25}P_{36}+P_{34}P_{26} \\ &\quad -P_{14}P_{25}P_{36}). \qquad \text{(D.10)}\end{aligned}$$

As a final example for the intercluster antisymmetrizer, let us consider the simplest non-trivial three-cluster case, that of four nucleons arranged into 2+1+1-nucleon clusters. Then, along with the familiar transpositions, three-particle permutations will also appear:

$$\begin{aligned}\mathcal{A}_{(12)34} &= \frac{1}{\sqrt{12}}[1-P_{13}-P_{14}-P_{23}-P_{24}-P_{34} \\ &\quad +P_{13}P_{24}+P_{23}P_{14} \\ &\quad +(134)+(143)+(234)+(243)], \qquad \text{(D.11)}\end{aligned}$$

where $(i_1 i_2\cdots i_k)$ stands for a cyclic permutation $\begin{pmatrix} i_1 & i_2 & \ldots & i_{k-1} & i_k \\ i_2 & i_3 & \ldots & i_k & i_1\end{pmatrix}$.

As an example for Eq. (D.6), let us consider the case of the two-cluster system of three particles 1+(23). The state $\mathcal{A}_{1(23)}\{\psi_1(1)\Phi(2,3)\}$ is constructed from $\Phi(2,3) = \mathcal{A}_{(23)}\{\psi_2(2)\psi_3(3)\}$ as follows:

$$\begin{aligned}
&\mathcal{A}_{1(23)}\{\psi_1(1)\Phi(2,3)\}\\
&\qquad = \frac{1}{\sqrt{3}}(1 - P_{12} - P_{13})\psi_1(1)\Phi(2,3)\\
&\qquad = \frac{1}{\sqrt{3}}[\psi_1(1)\Phi(2,3) - \psi_1(2)\Phi(1,3) - \psi_1(3)\Phi(2,1)]\\
&\qquad = \frac{1}{\sqrt{6}}\{[\psi_1(1)\psi_2(2)\psi_3(3) - \psi_1(1)\psi_2(3)\psi_3(2)]\\
&\qquad\quad -[\psi_1(2)\psi_2(1)\psi_3(3) - \psi_1(2)\psi_2(3)\psi_3(1)]\\
&\qquad\quad -[\psi_1(3)\psi_2(2)\psi_3(1) - \psi_1(3)\psi_2(1)\psi_3(2)]\},
\end{aligned} \tag{D.12}$$

which is indeed equal to the expression (D.5).

Note that $\mathcal{A}_{\{1\}\cdots\{n\}}$ does not commute with any of $\mathcal{A}_{\{1\}}, \ldots, \mathcal{A}_{\{n\}}$, nor is it proportional to a projector: $\mathcal{A}^2_{\{1\}\cdots\{n\}} \neq \text{const.} \times \mathcal{A}_{\{1\}\cdots\{n\}}$. Nevertheless, in the subspace of internally antisymmetrized states, in which it is mostly applied, it has this projection operator property. This restriction can be expressed by sandwiching it between two operators $\mathcal{A}_{\{1\}}\cdots\mathcal{A}_{\{n\}}$:

$$\begin{aligned}
\mathcal{A}_{\{1\}}\cdots\mathcal{A}_{\{n\}}\mathcal{A}^2_{\{1\}\cdots\{n\}}\mathcal{A}_{\{1\}}\cdots\mathcal{A}_{\{n\}} &= \mathcal{A}_{\{1\}}\cdots\mathcal{A}_{\{n\}}\mathcal{A}_{\{1\}\cdots\{n\}}\mathcal{A}\\
&= \sqrt{\frac{A!}{a_1!\cdots a_n!}}\mathcal{A}_{\{1\}}\cdots\mathcal{A}_{\{n\}}\mathcal{A}\\
&= \sqrt{\frac{A!}{a_1!\cdots a_n!}}\mathcal{A}_{\{1\}}\cdots\mathcal{A}_{\{n\}}\mathcal{A}_{\{1\}\cdots\{n\}}\mathcal{A}_{\{1\}}\cdots\mathcal{A}_{\{n\}}.
\end{aligned} \tag{D.13}$$

Here the first and third equalities follow from Eq. (D.6), while the second one is a consequence of (D.7) and $P\mathcal{A} = \text{sign}(P)\mathcal{A}$. One can tackle the problem of intercluster antisymmetrization in a different way as well. One can introduce an alternative intercluster antisymmetrizer $\mathcal{A}'_{\{1'\}\cdots\{n'\},\{1\}\cdots\{n\}}$ with the definition

$$\mathcal{A} = \mathcal{A}_{\{1'\}}\cdots\mathcal{A}_{\{n'\}}\mathcal{A}'_{\{1'\}\cdots\{n'\},\{1\}\cdots\{n\}}\mathcal{A}_{\{1\}}\cdots\mathcal{A}_{\{n\}}, \tag{D.14}$$

where the partitioning $a_{1'}, \ldots, a_{n'}$ may or may not be the same as $a_1, \ldots, a_n$. The significance of this is that the bra and ket may describe different clusterizations, and by finding $\mathcal{A}'_{\{1'\}\cdots\{n'\},\{1\}\cdots\{n\}}$, one formulates the antisymmetrization in the bra and the ket simultaneously. Since $\mathcal{A}'_{\{1'\}\cdots\{n'\},\{1\}\cdots\{n\}}$ appears in the definition (D.14) in between two projectors, (i.e., only its projections are defined), this definition is not unique. For example, for the special case when the partitioning is the same on both sides (i.e., for $\mathcal{A}'_{\{1\}\cdots\{n\},\{1\}\cdots\{n\}}$), the operator $(a_1!\cdots a_n!)^{-1/2}\mathcal{A}_{\{1\}\cdots\{n\}}$ certainly satisfies the definition (D.14), but so do others, e.g., $(a_1!\cdots a_n!)^{-1}\mathcal{A}_{\{1\}}\cdots\mathcal{A}_{\{n\}}\mathcal{A}_{\{1\}\cdots\{n\}}$.

The advantage of this operator is that its least complicated form can be constructed with rigorous mathematics [530], and this form does usually contain substantially fewer than $A!/a_1!\cdots a_n!$ terms. With this, the non-trivial identification of non-equivalent permutations can be avoided. The price one has to pay is to learn a very intricate mathematical theorem.

The complicated procedure of antisymmetrization may reduce to simpler procedures if the spatial orbits have other symmetries. The antisymmetric total wave function Φ can be constructed as $\sum_i \Phi_i(\text{space})\mathcal{X}_i$, where $\mathcal{X}_i$ are spin–isospin state vectors and neither $\Phi_i(\text{space})$ nor $\mathcal{X}_i$ is necessarily antisymmetric with respect to the permutation of their labels, but they have specific permutation symmetries which are in a kind of 'adjoint' relationship with each other. The substates of suitable permutation symmetries can be constructed with group theoretical methods. In the description of light exotic nuclei we use this adjoint relationship explicitly for the trivial case of cluster states only, when this relationship reduces to one term: $\Phi = \Phi(\text{space})\mathcal{X}$. In such a factored state either $\Phi(\text{space})$ has to be fully symmetric and $\mathcal{X}$ antisymmetric or *vice versa*. In particular, in the simple g.s. SM configuration of the 0s shell nuclei, $\Phi(\text{space})$ is symmetric and $\mathcal{X} \equiv \mathcal{X}_{SM_STM_T}$ is antisymmetric, and the latter may carry the correct spin and isospin values without explicit spin–isospin coupling:

$$
\begin{aligned}
&\mathcal{X}_{1100}(1,2) = \mathcal{A}\{\chi_{\frac{1}{2}\frac{1}{2}}(1)\eta_{\frac{1}{2}\frac{1}{2}}(1)\chi_{\frac{1}{2}\frac{1}{2}}(2)\eta_{\frac{1}{2}-\frac{1}{2}}(2)\},\\
&\mathcal{X}_{0011}(1,2) = \mathcal{A}\{\chi_{\frac{1}{2}\frac{1}{2}}(1)\eta_{\frac{1}{2}\frac{1}{2}}(1)\chi_{\frac{1}{2}-\frac{1}{2}}(2)\eta_{\frac{1}{2}\frac{1}{2}}(2)\},\\
&\mathcal{X}_{\frac{1}{2}\frac{1}{2}\frac{1}{2}\frac{1}{2}}(1,2,3) = \mathcal{A}\{\chi_{\frac{1}{2}\frac{1}{2}}(1)\eta_{\frac{1}{2}\frac{1}{2}}(1)\chi_{\frac{1}{2}-\frac{1}{2}}(2)\eta_{\frac{1}{2}\frac{1}{2}}(2)\chi_{\frac{1}{2}\frac{1}{2}}(3)\eta_{\frac{1}{2}-\frac{1}{2}}(3)\}\\
&\mathcal{X}_{0000}(1,2,3,4)\\
&= \mathcal{A}\{\chi_{\frac{1}{2}\frac{1}{2}}(1)\eta_{\frac{1}{2}\frac{1}{2}}(1)\chi_{\frac{1}{2}-\frac{1}{2}}(2)\eta_{\frac{1}{2}\frac{1}{2}}(2)\chi_{\frac{1}{2}\frac{1}{2}}(3)\eta_{\frac{1}{2}-\frac{1}{2}}(3)\chi_{\frac{1}{2}-\frac{1}{2}}(4)\eta_{\frac{1}{2}-\frac{1}{2}}(4)\}.
\end{aligned}
\tag{D.15}
$$

The fact that these antisymmetric spin–isospin states automatically have definite spin and isospin values shows an example for the connection between permutational and other (here the rotational) symmetries. (But for some other states $\mathcal{X}_{SM_STM_T}$ of the same systems, e.g., $\mathcal{X}_{1000}(1,2)$, have to be coupled in spin and/or isospin explicitly!)

Appendix E

Matrix elements between Slater determinants

In this appendix the formulae (9.108)–(9.111) will be derived for the matrix elements of the most important operators between the Slater determinants (9.67) and some other matrix elements as well. The Slater determinants consist of s.p. Gaussian wave packets $\hat{\varphi}_{\boldsymbol{s}_i m_s^i m_t^i}$, of the same size, whose centres are pinned down by the generator coordinate vectors $\boldsymbol{s}_i$. The resulting matrix elements will depend on the $2A$ generator coordinate vectors compressed in the one-column matrix $\boldsymbol{t}$ defined in Eq. (9.112).

E.1 Unit operator

From $\mathcal{A}^2 = \sqrt{A!}\mathcal{A}$ it follows that the overlap of two Slater determinants is equal to the determinant of the matrix of the s.p. overlaps:

$$\langle \phi_\kappa(\boldsymbol{s}_1, \ldots, \boldsymbol{s}_A) | \phi_{\kappa'}(\boldsymbol{s}'_1, \ldots, \boldsymbol{s}'_A) \rangle = \det\{B\}, \tag{E.1}$$

where

$$B_{ij} = \langle \hat{\varphi}_{\boldsymbol{s}_i m_s^i m_t^i} | \hat{\varphi}_{\boldsymbol{s}'_j m_s^{j\prime} m_t^{j\prime}} \rangle \qquad (i, j = 1, \ldots, A). \tag{E.2}$$

By using the definition of a determinant, this can be rewritten as

$$\begin{aligned} &\langle \phi_\kappa(\boldsymbol{s}_1, \ldots, \boldsymbol{s}_A) | \phi_{\kappa'}(\boldsymbol{s}'_1, \ldots, \boldsymbol{s}'_A) \rangle \\ &= \sum_P \operatorname{sign}(P) \langle \hat{\varphi}_{\boldsymbol{s}_1 m_s^1 m_t^1} | \hat{\varphi}_{\boldsymbol{s}'_{p_1} m_s^{p_1\prime} m_t^{p_1\prime}} \rangle \cdots \langle \hat{\varphi}_{\boldsymbol{s}_A m_s^A m_t^A} | \hat{\varphi}_{\boldsymbol{s}'_{p_A} m_s^{p_A\prime} m_t^{p_A\prime}} \rangle, \end{aligned} \tag{E.3}$$

where $(p_1, \ldots, p_A)$ is the permutation P of the set $(1, \ldots, A)$. Substitution of the overlap, Eq. (9.95), of the s.p. wave packets into Eq. (E.3) yields an explicit formula for the overlap of the Slater determinants. Each $\langle \hat{\varphi}_{s_i m_s^i m_t^i} | \hat{\varphi}_{s'_{p_i} m_s^{p_i\prime} m_t^{p_i\prime}} \rangle$ is a Gaussian, and the product of such factors gives an exponential function with a quadratic form of the vector set $\boldsymbol{t}$ in the exponent:

$$-\frac{\beta}{4}\sum_{i=1}^{A}(\boldsymbol{s}_i \cdot \boldsymbol{s}_i + \boldsymbol{s}'_{p_i} \cdot \boldsymbol{s}'_{p_i} - 2\boldsymbol{s}_i \cdot \boldsymbol{s}'_{p_i}) = -\frac{1}{2}\sum_{i=1}^{2A}\sum_{j=1}^{2A}(A_P^{(\mathrm{o})})_{ij}\boldsymbol{t}_i \cdot \boldsymbol{t}_j, \tag{E.4}$$

where

$$\begin{aligned}
(A_P^{(\mathrm{o})})_{ij} &= (A_P^{(\mathrm{o})})_{A+i,A+j} = \frac{\beta}{2}\delta_{ij}, \\
(A_P^{(\mathrm{o})})_{i,A+j} &= (A_P^{(\mathrm{o})})_{A+j,i} = -\frac{\beta}{2}\delta_{j,p_i} \qquad (1 \le i, j \le A).
\end{aligned} \tag{E.5}$$

For future reference, it is worth noting that, by reordering the terms, the quadratic form $\tilde{\boldsymbol{t}}A_P^{(\mathrm{o})}\boldsymbol{t}$ can be written as

$$\tilde{\boldsymbol{t}}A_P^{(\mathrm{o})}\boldsymbol{t} = \frac{\beta}{2}\sum_{j=1}^{A}(\boldsymbol{s}_j \cdot \boldsymbol{s}_j + \boldsymbol{s}'_j \cdot \boldsymbol{s}'_j - 2\boldsymbol{s}_j \cdot \boldsymbol{s}'_{p_j}). \tag{E.6}$$

With the form (E.4), however, the overlap can be cast into the form

$$\langle \phi_\kappa(\boldsymbol{s}_1, \ldots, \boldsymbol{s}_A) | \phi_{\kappa'}(\boldsymbol{s}'_1, \ldots, \boldsymbol{s}'_A) \rangle = \sum_P C_P^{(\mathrm{o})} \mathrm{e}^{-\frac{1}{2}\tilde{\boldsymbol{t}}A_P^{(\mathrm{o})}\boldsymbol{t}}, \tag{E.7}$$

where

$$C_P^{(\mathrm{o})} = \mathrm{sign}(P)\delta_{m_s^1 m_s^{p_1\prime}}\delta_{m_t^1 m_t^{p_1\prime}} \cdots \delta_{m_s^A m_s^{p_A\prime}}\delta_{m_t^A m_t^{p_A\prime}}. \tag{E.8}$$

The orthogonality of the spin–isospin functions indicates that the matrix B can be made block-diagonal in the four spin–isospin states, which greatly reduces the number of terms in the summation over P.

E.2 One-body operators

The most important matrix elements of this type are the kinetic energy and the sum of the squared position vectors.

The matrix element of the kinetic-energy operator can be made more explicit in the following way. Since $[\mathcal{A}, \sum_i T_i] = 0$, the antisymmetrizer in the bra can be shifted to the ket, where $\mathcal{A}^2 = \sqrt{A!}\mathcal{A}$ can be used, whereby the bra and the ket are, respectively, reduced to a product and to a determinant of the s.p. functions. The bra-ket is thus itself a determinant, similar to Eq. (E.1), but with one column replaced by s.p. matrix elements of the kinetic energy. Expanding this determinant, we reduce the matrix element as

$$\sum_{i=1}^{A}\langle\phi_{\kappa}(s_1,\ldots,s_A)|T_i|\phi_{\kappa'}(s'_1,\ldots,s'_A)\rangle$$
$$=\sum_{i=1}^{A}\sum_{j=1}^{A}\langle\hat{\varphi}_{s_i m_s^i m_t^i}|T|\hat{\varphi}_{s'_j m_s^{j'} m_t^{j'}}\rangle(-1)^{i+j}\det\{B^{ij}\}, \tag{E.9}$$

where B^{ij} is obtained by omitting the ith row and the jth column of the matrix B defined in Eq. (E.2). Substitution of the s.p. overlaps (9.95) and the s.p. matrix elements (9.96) of the kinetic-energy operator can lead us to the desired formula. Alternatively, one can derive the many-particle kinetic-energy matrix element similarly to the s.p. matrix element, using a many-particle analogue of Eq. (9.100). Abbreviating $\phi_{\kappa}(s_1,\ldots,s_A)$ and $\hat{\varphi}_{s_i m_s^i m_t^i}$ as ϕ_{κ} and $\hat{\varphi}_{s_i}$, respectively, we can cast the matrix element into the following form:

$$\sum_{i=1}^{A}\langle\phi_{\kappa}|T_i|\phi_{\kappa'}\rangle=\sum_{i=1}^{A}\langle T_i\prod_{j=1}^{A}\hat{\varphi}_{s_j}(\boldsymbol{r}_j)|\det\{\hat{\varphi}_{s'_k}(\boldsymbol{r}_k)\}\rangle$$
$$=\sum_{i}\begin{vmatrix}\langle\hat{\varphi}_{s_1}|\hat{\varphi}_{s'_1}\rangle & \cdots & \langle\hat{\varphi}_{s_i}|T|\hat{\varphi}_{s'_1}\rangle & \cdots & \langle\hat{\varphi}_{s_A}|\hat{\varphi}_{s'_1}\rangle\\ \vdots & & \vdots & & \vdots\\ \langle\hat{\varphi}_{s_1}|\hat{\varphi}_{s'_A}\rangle & \cdots & \langle\hat{\varphi}_{s_i}|T|\hat{\varphi}_{s'_A}\rangle & \cdots & \langle\hat{\varphi}_{s_A}|\hat{\varphi}_{s'_A}\rangle\end{vmatrix}. \tag{E.10}$$

One should find ways to reduce the differentiations with respect to the physical coordinates, as are in the kinetic-energy operator, to differentiations with respect to parameters: the Gaussian centres or the width. It is instructive to show both ways.

In the first procedure one should first use the symmetry of $\hat{\varphi}_{s'_k}(\boldsymbol{r}_k)$ with respect to $\boldsymbol{r}_k$ and $\boldsymbol{s}'_k$ in Eq. (E.7) and then change the sequence of the differentiation and the integration. One can thus write

$$\sum_{i=1}^{A}\langle\phi_{\kappa}|T_i|\phi_{\kappa'}\rangle=-\frac{\hbar^2}{2m}\sum_{i=1}^{A}\Delta_{s'_i}\langle\phi_{\kappa}|\phi_{\kappa'}\rangle$$
$$=-\frac{\hbar^2}{2m}\sum_{P}C_P^{(\mathrm{o})}\sum_{i=1}^{A}\Delta_{s'_i}\mathrm{e}^{-\frac{1}{2}\tilde{t}A_P^{(\mathrm{o})}t} \tag{E.11}$$

In calculating the effect of the Laplacian on the exponential function, one needs

$$\nabla_{s'_i} e^{-\frac{1}{2}\tilde{t}A_P^{(o)}t} = -\frac{1}{2} e^{-\frac{1}{2}\tilde{t}A_P^{(o)}t} \nabla_{s'_i} \tilde{t}A_P^{(o)}t = -\frac{\beta}{2}(s'_i - s_{p_i^{-1}}) e^{-\frac{1}{2}\tilde{t}A_P^{(o)}t}, \tag{E.12}$$

where Eq. (E.6) has been used. The label p_i^{-1} is an element of the permuted set $(p_1^{-1}, \ldots, p_A^{-1})$, which is defined by $P(p_1^{-1}, \ldots, p_A^{-1}) = (1, \ldots, A)$. The evaluation of $\sum_{i=1}^{A} \nabla_{s'_i}^2 e^{-\frac{1}{2}\tilde{t}A_P^{(o)}t}$ then proceeds as follows:

$$\begin{aligned}
\sum_{i=1}^{A} \nabla_{s'_i}^2 e^{-\frac{1}{2}\tilde{t}A_P^{(o)}t} &= \sum_{i=1}^{A} \nabla_{s'_i} \cdot \left[-\frac{\beta}{2}(s'_i - s_{p_i^{-1}}) e^{-\frac{1}{2}\tilde{t}A_P^{(o)}t} \right] \\
&= -\frac{\beta}{2} \left[3A - \frac{\beta}{2} \sum_{i=1}^{A} (s'_i - s_{p_i^{-1}}) \cdot (s'_i - s_{p_i^{-1}}) \right] e^{-\frac{1}{2}\tilde{t}A_P^{(o)}t} \\
&= -\frac{\beta}{2} \left[3A - \frac{\beta}{2} \sum_{j=1}^{A} (s_j \cdot s_j + s'_j \cdot s'_j - 2 s_j \cdot s'_{p_j}) \right] e^{-\frac{1}{2}\tilde{t}A_P^{(o)}t} \\
&= -\frac{\beta}{2} \left(3A - \tilde{t}A_P^{(o)}t \right) e^{-\frac{1}{2}\tilde{t}A_P^{(o)}t},
\end{aligned} \tag{E.13}$$

where, in the third step, two of the three summations have been reordered, and, in the fourth step, Eq. (E.6) has been used. Inserting Eq. (E.13) into Eq. (E.11), one obtains

$$\begin{aligned}
&\sum_{i=1}^{A} \langle \phi_\kappa(s_1, \ldots, s_A) | T_i | \phi_{\kappa'}(s'_1, \ldots, s'_A) \rangle \\
&\qquad = \frac{\hbar^2}{2m} \frac{\beta}{2} \sum_P C_P^{(o)} \left(3A - \tilde{t}A_P^{(o)}t \right) e^{-\frac{1}{2}\tilde{t}A_P^{(o)}t},
\end{aligned} \tag{E.14}$$

which agrees with Eq. (9.109). It has become clear now that the coefficients $C_P^{(o)}$ and the matrices $A_P^{(o)}$ in this formula are indeed the same as those entering into the formula of the overlap.

In the alternative derivation of Eq. (9.109) one should relate $T\hat{\varphi}_{s'_k}$ to $\partial\hat{\varphi}_{s'_k}/\partial\beta$:

$$T_i \hat{\varphi}_{s'_k} = -\frac{\hbar^2}{2m}[-3\beta + \beta^2(r_i - s'_k)^2]\hat{\varphi}_{s'_k} = \frac{\hbar^2\beta}{m}\left(\frac{3}{4} + \beta\frac{\partial}{\partial\beta}\right)\hat{\varphi}_{s'_k}. \tag{E.15}$$

Since $\partial\langle\hat{\varphi}_s|\hat{\varphi}_{s'}\rangle/\partial\beta = \langle\partial\hat{\varphi}_s/\partial\beta|\hat{\varphi}_{s'}\rangle + \langle\hat{\varphi}_s|\partial\hat{\varphi}_{s'}/\partial\beta\rangle$, the s.p. matrix element is expressible as

$$\langle\hat{\varphi}_{s_i}|T|\hat{\varphi}_{s'_k}\rangle = \frac{\hbar^2\beta}{m}\left(\frac{3}{4} + \frac{\beta}{2}\frac{\partial}{\partial\beta}\right)\langle\hat{\varphi}_s|\hat{\varphi}_{s'}\rangle. \tag{E.16}$$

The differentiation of the s.p. matrix element can be conveniently inserted in the determinant:

$$\sum_{i=1}^{A}\langle\phi_\kappa|T_i|\phi_{\kappa'}\rangle$$

$$=\frac{\hbar^2\beta}{m}\sum_{i=1}^{A}\begin{vmatrix}\langle\hat{\varphi}_{s_1}|\hat{\varphi}_{s'_1}\rangle & \cdots & \left(\frac{3}{4}+\frac{\beta}{2}\frac{\partial}{\partial\beta}\right)\langle\hat{\varphi}_{s_i}|\hat{\varphi}_{s'_1}\rangle & \cdots & \langle\hat{\varphi}_{s_A}|\hat{\varphi}_{s'_1}\rangle\\ \vdots & & \vdots & & \vdots\\ \langle\hat{\varphi}_{s_1}|\hat{\varphi}_{s'_A}\rangle & \cdots & \left(\frac{3}{4}+\frac{\beta}{2}\frac{\partial}{\partial\beta}\right)\langle\hat{\varphi}_{s_i}|\hat{\varphi}_{s'_A}\rangle & \cdots & \langle\hat{\varphi}_{s_A}|\hat{\varphi}_{s'_A}\rangle\end{vmatrix}$$

$$=\frac{\hbar^2\beta}{m}\left(\frac{3}{4}A+\frac{\beta}{2}\frac{\partial}{\partial\beta}\right)\langle\phi_\kappa|\phi_{\kappa'}\rangle. \tag{E.17}$$

The matrix $A_P^{(\mathrm{o})}$ in Eq. (E.7) is proportional to β for each P, thus the operation $\partial/\partial\beta$ results in $\partial(\mathrm{e}^{-\frac{1}{2}\tilde{t}A_P^{(\mathrm{o})}t})/\partial\beta=-\frac{1}{2\beta}\tilde{t}A_P^{(\mathrm{o})}t\mathrm{e}^{-\frac{1}{2}\tilde{t}A_P^{(\mathrm{o})}t}$. Inserted into Eq. (E.17), this results, again, in Eq. (E.14).

Equation (9.110) can be derived, e.g., by using the relationship between the s.p. matrix elements of r^2 and T,

$$\langle\varphi_{s_1}|r_1^2|\varphi_{s_2}\rangle=\frac{2m}{\hbar^2\beta^2}\langle\varphi_{s_1}|T_1|\varphi_{s_2}\rangle+\frac{1}{2}(s_1^2+s_2^2)\langle\varphi_{s_1}|\varphi_{s_2}\rangle, \tag{E.18}$$

which follows from Eqs. (9.96) and (9.97). Substitution of Eq. (E.14) results in

$$\sum_{i=1}^{A}\langle\phi_\kappa(s_1,\ldots,s_A)|r_i^2|\phi_{\kappa'}(s'_1,\ldots,s'_A)\rangle$$

$$=\frac{2m}{\hbar^2\beta^2}\sum_{i=1}^{A}\langle\phi_\kappa(s_1,\ldots,s_A)|T_i|\phi_{\kappa'}(s'_1,\ldots,s'_A)\rangle$$

$$+\frac{1}{2}\sum_{i=1}^{A}(s_i^2+s_i'^2)\det\{\langle\hat{\varphi}_{s_i m_s^i m_t^i}|\hat{\varphi}_{s'_j m_s^{j\prime} m_t^{j\prime}}\rangle\}$$

$$=\frac{1}{2\beta}\sum_P C_P^{(\mathrm{o})}\left(3A-\tilde{t}A_P^{(\mathrm{o})}t+\beta\tilde{t}t\right)\mathrm{e}^{-\frac{1}{2}\tilde{t}A_P^{(\mathrm{o})}t}. \tag{E.19}$$

E.3 Two-body operators

The most important two-body operator is the interaction potential, and we now turn to the problem of the potential matrix element. With the use of $\delta(r_1-r_2-r)$ in Eq. (9.98), one can define the correlation operator $\sum_{i<j}^{A}\delta(r_i-r_j-r)O_{ij}^{\mathrm{p}}$, where O_{ij}^{p} is specific to the type of the potential

considered. The matrix element of $\sum_{i<j}^{A} \delta(\boldsymbol{r}_i - \boldsymbol{r}_j - \boldsymbol{r}) O^{\mathrm{p}}_{ij}$ between Slater determinants may be called the correlation function of type p:

$$\sum_{i<j}^{A} \langle \phi_{\kappa}(s_1, \ldots, s_A) | \delta(\boldsymbol{r}_i - \boldsymbol{r}_j - \boldsymbol{r}) O^{\mathrm{p}}_{ij} | \phi_{\kappa'}(s'_1, \ldots, s'_A) \rangle. \tag{E.20}$$

The matrix elements of the two-body interactions can be expressed with the correlation function. The correlation function, in turn, can be expressed in terms of the two-body matrix elements, (9.98), of $\delta(\boldsymbol{r}_1 - \boldsymbol{r}_2 - \boldsymbol{r})$ and of the s.p. overlaps, (9.95). To show this, one should shift $\mathcal{A}$ in the bra in Eq. (E.20) through the operator $\sum_{i<j}^{A} \delta(\boldsymbol{r}_i - \boldsymbol{r}_j - \boldsymbol{r}) O^{\mathrm{p}}_{ij}$, which is permutation-invariant, use $\mathcal{A}^2 = \sqrt{A!}\mathcal{A}$ and the Laplace theorem[1] for the expansion of the determinant in the ket. The result is

$$\begin{aligned} &\sum_{i<j}^{A} \langle \phi_{\kappa}(s_1, \ldots, s_A) | \delta(\boldsymbol{r}_i - \boldsymbol{r}_j - \boldsymbol{r}) O^{\mathrm{p}}_{ij} | \phi_{\kappa'}(s'_1, \ldots, s'_A) \rangle \\ &= \sum_{i<j}^{A} \sum_{k,l=1}^{A} \langle \hat{\varphi}_{s_i m_s^i m_t^i} \hat{\varphi}_{s_j m_s^j m_t^j} | \delta(\boldsymbol{r}_i - \boldsymbol{r}_j - \boldsymbol{r}) O^{\mathrm{p}}_{ij} | \hat{\varphi}_{s'_k m_s^{k\prime} m_t^{k\prime}} \hat{\varphi}_{s'_l m_s^{l\prime} m_t^{l\prime}} \rangle \\ &\times (-1)^{i+j+k+l} \det\{B^{ijkl}\}, \end{aligned} \tag{E.21}$$

where B^{ijkl} is obtained by omitting the ith and jth rows and the kth and lth columns of the matrix B.

For a general central potential term, O^{p}_{ij} is

$$O^{\mathrm{p}}_{ij} = w + bP^{\sigma}_{ij} - hP^{\tau}_{ij} - mP^{\sigma}_{ij}P^{\tau}_{ij}, \tag{E.22}$$

where P^{σ}_{ij} and P^{τ}_{ij} are the spin- and isospin-exchange operators, respectively. Looking at the ingredients of Eq. (E.21) for this case, viz. Eqs. (9.95) and (9.98), one sees that the matrix element will again be a combination of exponential functions of quadratic forms. Furthermore, since O^{p}_{ij} only acts on the spin–isospin factor, it will only have a trivial effect on the coefficient of each term. Since Eq. (9.98) contains the s.p. overlaps $\langle \varphi_{s_1} | \varphi_{s_3} \rangle \langle \varphi_{s_2} | \varphi_{s_4} \rangle$, the quadratic form that replaces $\tilde{\boldsymbol{t}} A^{(\mathrm{o})}_{P} \boldsymbol{t}$ in Eq. (E.7) will contain all terms of $\tilde{\boldsymbol{t}} A^{(\mathrm{o})}_{P} \boldsymbol{t}$, together with additional

[1]The Laplace theorem gives an expansion of an $(m+n)\times(m+n)$ determinant in terms of products of $m\times m$ and $n\times n$ determinants. The $m\times m$ determinant factor in each term is obtained by excluding n rows and n columns, and the complementary $n\times n$ factor is obtained by excluding the rows and columns included in the first factor. The m columns are fixed, and the summation goes over all rank-m combinations of the $m+n$ rows; the roles of the rows and columns can be exchanged. The sign of the term in which the $m \times m$ determinant includes rows $i_1, \ldots, i_m$ and columns $j_1, \ldots, j_m$ will be $(-1)^{i_1+\cdots+i_m+j_1+\cdots+j_m}$.

terms. Thus the result will have the form

$$\sum_{i<j}^{A} \langle \phi_\kappa(\boldsymbol{s}_1, \ldots, \boldsymbol{s}_A) | \delta(\boldsymbol{r}_i - \boldsymbol{r}_j - \boldsymbol{r}) O_{ij}^{\mathrm{p}} | \phi_{\kappa'}(\boldsymbol{s}_1', \ldots, \boldsymbol{s}_A') \rangle$$

$$= \left(\frac{\beta}{2\pi}\right)^{3/2} \mathrm{e}^{-\frac{1}{2}\beta r^2} \sum_P \sum_{i<j}^{A} C_P^{(\mathrm{c})(ij)} \mathrm{e}^{-\frac{1}{2}\tilde{\boldsymbol{t}}(A_P^{(\mathrm{o})} + B_P^{(ij)})\boldsymbol{t} + \boldsymbol{d}_P^{(ij)} \cdot \boldsymbol{r}}. \tag{E.23}$$

The factor $\mathrm{e}^{-\frac{1}{2}\beta r^2}$ and the additional terms $-\frac{1}{2}\tilde{\boldsymbol{t}} B_P^{(ij)} \boldsymbol{t} + \boldsymbol{d}_P^{(ij)} \cdot \boldsymbol{r}$ in the exponent come from the expression $-\frac{1}{2}\beta[\boldsymbol{r} - \frac{1}{2}(\boldsymbol{s}_i - \boldsymbol{s}_j + \boldsymbol{s}_{p_i}' - \boldsymbol{s}_{p_j}')]^2$ involved in Eq. (9.98) when written for the present parameters. Noting that, according to Eq. (9.112), $\boldsymbol{s}_k' = \boldsymbol{t}_{A+k}$, one has

$$\beta\left[\boldsymbol{r} - \frac{1}{2}(\boldsymbol{s}_i - \boldsymbol{s}_j + \boldsymbol{s}_{p_i}' - \boldsymbol{s}_{p_j}')\right]^2 = \beta r^2 + \frac{\beta}{4}\Big[(\boldsymbol{t}_i^2 + \boldsymbol{t}_j^2 + \boldsymbol{t}_{A+p_i}^2 + \boldsymbol{t}_{A+p_j}^2)$$
$$+ 2(\boldsymbol{t}_i \cdot \boldsymbol{t}_{A+p_i} + \boldsymbol{t}_j \cdot \boldsymbol{t}_{A+p_j} - \boldsymbol{t}_i \cdot \boldsymbol{t}_j - \boldsymbol{t}_i \cdot \boldsymbol{t}_{A+p_j}$$
$$- \boldsymbol{t}_j \cdot \boldsymbol{t}_{A+p_i} - \boldsymbol{t}_{A+p_i} \cdot \boldsymbol{t}_{A+p_j})\Big] - \beta(\boldsymbol{t}_i - \boldsymbol{t}_j + \boldsymbol{t}_{A+p_i} - \boldsymbol{t}_{A+p_j}) \cdot \boldsymbol{r}. \tag{E.24}$$

Thus the matrix $B_P^{(ij)}$ and the vector $\boldsymbol{d}_P^{(ij)}$ are given by

$$(B_P^{(ij)})_{kl} = (B_P^{(ij)})_{lk} = \begin{cases} \frac{\beta}{4} & \text{for } (kl) = (ii), (jj), (A+p_i, A+p_i), \\ & \quad (A+p_j, A+p_j), (i, A+p_i), \\ & \quad (j, A+p_j); \\ -\frac{\beta}{4} & \text{for } (kl) = (ij), (i, A+p_j), (j, A+p_i), \\ & \quad (A+p_i, A+p_j); \\ 0 & \text{otherwise} \end{cases} \tag{E.25}$$

and

$$\boldsymbol{d}_P^{(ij)} = \frac{\beta}{2}(\boldsymbol{t}_i - \boldsymbol{t}_j + \boldsymbol{t}_{A+p_i} - \boldsymbol{t}_{A+p_j}). \tag{E.26}$$

The coefficients $C_P^{(\mathrm{c})(ij)}$ differ from $C_P^{(\mathrm{o})}$ given in Eq. (E.8) in that the product $\delta_{m_s^i m_s^{p_i\prime}} \delta_{m_t^i m_t^{p_i\prime}} \delta_{m_s^j m_s^{p_j\prime}} \delta_{m_t^j m_t^{p_j\prime}}$ is replaced with

$$\langle \chi_{\frac{1}{2}m_s^i} \eta_{\frac{1}{2}m_t^i} \chi_{\frac{1}{2}m_s^j} \eta_{\frac{1}{2}m_t^j} | O_{12}^{\mathrm{p}} | \chi_{\frac{1}{2}m_s^{p_i\prime}} \eta_{\frac{1}{2}m_t^{p_i\prime}} \chi_{\frac{1}{2}m_s^{p_j\prime}} \eta_{\frac{1}{2}m_t^{p_j\prime}} \rangle$$
$$= w\delta_{m_s^i m_s^{p_i\prime}} \delta_{m_t^i m_t^{p_i\prime}} \delta_{m_s^j m_s^{p_j\prime}} \delta_{m_t^j m_t^{p_j\prime}}$$
$$+ b\delta_{m_s^i m_s^{p_j\prime}} \delta_{m_t^i m_t^{p_i\prime}} \delta_{m_s^j m_s^{p_i\prime}} \delta_{m_t^j m_t^{p_j\prime}}$$
$$- h\delta_{m_s^i m_s^{p_i\prime}} \delta_{m_t^i m_t^{p_j\prime}} \delta_{m_s^j m_s^{p_j\prime}} \delta_{m_t^j m_t^{p_i\prime}}$$
$$- m\delta_{m_s^i m_s^{p_j\prime}} \delta_{m_t^i m_t^{p_j\prime}} \delta_{m_s^j m_s^{p_i\prime}} \delta_{m_t^j m_t^{p_i\prime}}. \tag{E.27}$$

For the spin-orbit potential, the operator $O_{ij}^{\mathrm{s.o.}}$ entering into the correlation operator is

$$O_{ij}^{\mathrm{s.o.}} = \boldsymbol{l}_{ij} \cdot \frac{1}{2}(\boldsymbol{\sigma}_i + \boldsymbol{\sigma}_j), \tag{E.28}$$

where $\boldsymbol{\sigma}_i$ is a Pauli vector and the relative orbital angular momentum $\boldsymbol{l}_{ij}$ is defined in Eq. (9.94). A matrix element of the spin-orbit correlation operator is calculated in close analogy with Eq. (E.23):

$$\begin{aligned}
&\sum_{i<j}^{A} \langle \phi_{\kappa}(s_1,\ldots,s_A) | \delta(\boldsymbol{r}_i - \boldsymbol{r}_j - \boldsymbol{r}) O_{ij}^{\mathrm{s.o.}} | \phi_{\kappa'}(s'_1,\ldots,s'_A) \rangle \\
&\quad = -\frac{i}{2}\beta \left(\frac{\beta}{2\pi}\right)^{3/2} \mathrm{e}^{-\frac{1}{2}\beta r^2} \sum_{P} \sum_{i<j}^{A} \left[\boldsymbol{r} \times (\boldsymbol{t}_{A+p_i} - \boldsymbol{t}_{A+p_j})\right] \cdot \boldsymbol{C}_P^{(ls)(ij)} \\
&\quad\quad \times \mathrm{e}^{-\frac{1}{2}\tilde{\boldsymbol{t}}(A_P^{(\mathrm{o})} + B_P^{(ij)})\boldsymbol{t} + \boldsymbol{d}_P^{(ij)} \cdot \boldsymbol{r}}.
\end{aligned} \tag{E.29}$$

The factor $-\frac{i}{2}\beta \boldsymbol{r} \times (\boldsymbol{t}_{A+p_i} - \boldsymbol{t}_{A+p_j})$ comes from the two-particle matrix element, Eq. (9.99). The coefficients $\boldsymbol{C}_P^{(ls)(ij)}$ are Cartesian vectors since $O_{ij}^{\mathrm{s.o.}}$ of Eq. (E.28) is a vector in the spin space. They are obtained by replacing $\delta_{m_s^i m_s^{p_i\prime}} \delta_{m_t^i m_t^{p_i\prime}} \delta_{m_s^j m_s^{p_j\prime}} \delta_{m_t^j m_t^{p_j\prime}}$ in Eq. (E.8) with the two-particle matrix element of $\frac{1}{2}(\boldsymbol{\sigma}_i + \boldsymbol{\sigma}_j)$ in the spin–isospin space. For the μ tensor component $(\boldsymbol{C}_P^{(ls)(ij)})_\mu$ of $\boldsymbol{C}_P^{(ls)(ij)}$ this replacement factor is

$$\begin{aligned}
&\delta_{m_t^i m_t^{p_i\prime}} \delta_{m_t^j m_t^{p_j\prime}} \langle \chi_{\frac{1}{2} m_s^i} \chi_{\frac{1}{2} m_s^j} | \frac{1}{2}(\boldsymbol{\sigma}_i + \boldsymbol{\sigma}_j)_\mu | \chi_{\frac{1}{2} m_s^{p_i\prime}} \chi_{\frac{1}{2} m_s^{p_j\prime}} \rangle \\
&\quad = \frac{1}{2} \delta_{m_t^i m_t^{p_i\prime}} \delta_{m_t^j m_t^{p_j\prime}} \\
&\quad\quad \times \left[\delta_{m_s^j m_s^{p_j\prime}} \langle \chi_{\frac{1}{2} m_s^i} | (\boldsymbol{\sigma}_i)_\mu | \chi_{\frac{1}{2} m_s^{p_i\prime}} \rangle + \delta_{m_s^i m_s^{p_i\prime}} \langle \chi_{\frac{1}{2} m_s^j} | (\boldsymbol{\sigma}_j)_\mu | \chi_{\frac{1}{2} m_s^{p_j\prime}} \rangle\right] \\
&\quad = \frac{\sqrt{3}}{2} \delta_{m_t^i m_t^{p_i\prime}} \delta_{m_t^j m_t^{p_j\prime}} \\
&\quad\quad \times \left[\delta_{m_s^j m_s^{p_j\prime}} \langle \tfrac{1}{2} m_s^{p_i\prime} 1\mu | \tfrac{1}{2} m_s^i \rangle + \delta_{m_s^i m_s^{p_i\prime}} \langle \tfrac{1}{2} m_s^{p_j\prime} 1\mu | \tfrac{1}{2} m_s^j \rangle\right],
\end{aligned} \tag{E.30}$$

where the last line can be obtained by the use of the Wigner–Eckart theorem (9.85).

E.4 Many-body operators

There are some many-body operators of physical interest: those of the product form, with each factor belonging to one particle. Since we wish to work in an intrinsic frame, we use particle positions with respect to the c.m., $\bar{\boldsymbol{r}}_i = \boldsymbol{r}_i - \boldsymbol{R}_{\mathrm{c.m.}}$, like in Eq. (9.138):

$$O(\boldsymbol{r}) = \prod_{i=1}^{A} f_i, \quad \text{with} \quad f_i = f(\boldsymbol{r}_i - \boldsymbol{R}_{\mathrm{c.m.}} - \boldsymbol{r}) P_i, \tag{E.31}$$

where P_i is one of the operators P introduced in Eq. (9.122). It acts in the spin–isospin space. As a first step, a relationship is set up between a matrix element of the operator (E.31),

$$o(\boldsymbol{r}) = \langle \Psi | O(\boldsymbol{r}) | \Psi' \rangle, \tag{E.32}$$

and a matrix element that can be directly evaluated in terms of the s.p. coordinates $\boldsymbol{r}_i$.

Let us associate c.m. wave functions Θ and Θ' with the intrinsic wave functions Ψ and Ψ'. Then the total wave functions are

$$\Phi = \Theta(\boldsymbol{R}_{\rm c.m.})\Psi \quad \text{and} \quad \Phi' = \Theta'(\boldsymbol{R}_{\rm c.m.})\Psi'. \tag{E.33}$$

Let us define $I(\boldsymbol{r})$ as

$$I(\boldsymbol{r}) = \langle \Phi | \delta(\boldsymbol{R}_{\rm c.m.}) O(\boldsymbol{r}) | \Phi' \rangle. \tag{E.34}$$

Both the two wave functions and the operator involved in $I(\boldsymbol{r})$ can be expressed in terms of s.p. coordinates. Substitution of Eq. (E.33) into Eq. (E.34) reveals that the two matrix elements are related as

$$o(\boldsymbol{r}) = \frac{I(\boldsymbol{r})}{\Theta(0)\Theta'(0)}. \tag{E.35}$$

Thus the calculation of $o(\boldsymbol{r})$ reduces to that of $I(\boldsymbol{r})$, which, as will be immediately seen, can indeed be evaluated by integration over the s.p. coordinates.

Because of the δ-function in Eq. (E.34), the variable $\boldsymbol{R}_{\rm c.m.}$ in f_i can be set equal to zero, so that, in the evaluation of $I(\boldsymbol{r})$, one can replace f_i by a translation-noninvariant operator $f_i' = f(\boldsymbol{r}_i - \boldsymbol{r})P_i$. One thus obtains

$$\begin{aligned} I(\boldsymbol{r}) = \langle \Phi | \delta(\boldsymbol{R}_{\rm c.m.}) \prod_{i=1}^{A} f_i' | \Phi' \rangle &= \frac{1}{(2\pi)^3} \int d\boldsymbol{k}\, \langle \Phi | \mathrm{e}^{i\boldsymbol{k}\cdot\boldsymbol{R}_{\rm c.m.}} \prod_{i=1}^{A} f_i' | \Phi' \rangle \\ &= \frac{1}{(2\pi)^3} \int d\boldsymbol{k}\, \langle \Phi | \prod_{j=1}^{A} \mathrm{e}^{i\frac{1}{A}\boldsymbol{k}\cdot\boldsymbol{r}_j} f_j' | \Phi' \rangle. \end{aligned} \tag{E.36}$$

In this way the calculation of the desired matrix element is reduced to that of the product of the s.p. operators $\exp(i\frac{1}{A}\boldsymbol{k}\cdot\boldsymbol{r}_j)f_j'$ and the integration involved can be performed over the s.p. coordinates.

When Ψ is a translation-invariant Gaussian wave packet, it is appropriate to choose $\Theta(\boldsymbol{R}_{\rm c.m.})$ as $\Theta(\boldsymbol{R}_{\rm c.m.}) = \varphi_{S_A}(\boldsymbol{x}_A)$ [cf. Eq. (9.72)], and Φ will be a Slater determinant. A matrix element of the operator $\prod_{j=1}^{A} \exp(i\frac{1}{A}\boldsymbol{k}\cdot\boldsymbol{r}_j)f_j'$ between two Slater determinants (9.67) is

$$\langle \phi_{\kappa}(\boldsymbol{s}_1, \ldots, \boldsymbol{s}_A) | \prod_{j=1}^{A} \mathrm{e}^{i\frac{1}{A}\boldsymbol{k}\cdot\boldsymbol{r}_j} f_j' | \phi_{\kappa'}(\boldsymbol{s}_1', \ldots, \boldsymbol{s}_A') \rangle = \det\{B'\}, \tag{E.37}$$

where

$$B'_{jl} = \langle \hat{\varphi}_{s_j m_s^j m_l^j} | e^{i\frac{1}{A}\boldsymbol{k}\cdot\boldsymbol{r}_1} f'_1 | \hat{\varphi}_{s'_l m_s^{l'} m_l^{l'}} \rangle \qquad (j, l = 1, \ldots, A). \qquad \text{(E.38)}$$

The matrix element (E.37) is thus a determinant of s.p. quantities, just as in the case of the overlap (E.1). The difference is that now the elements of the determinant are s.p. matrix elements rather than simple overlaps. But that is not an essential difference as one can redefine the s.p. states in either ϕ_κ or in $\phi_{\kappa'}$ as $\exp(i\frac{1}{A}\boldsymbol{k}\cdot\boldsymbol{r}_j)f'_j\hat{\varphi}_{s_j m_s^j m_l^j}(\boldsymbol{r}_j)$.

Very similar arguments apply to the matrix element of a one-body operator of the form $\sum_{i=1}^{A} f_i$:

$$\langle \Psi | \sum_{i=1}^{A} f_i | \Psi' \rangle = [\Theta(0)\Theta'(0)]^{-1} \frac{1}{(2\pi)^3} \int d\boldsymbol{k} \, \langle \Phi | e^{i\boldsymbol{k}\cdot\boldsymbol{R}_{\text{c.m.}}} \sum_{i=1}^{A} f'_i | \Phi' \rangle. \qquad \text{(E.39)}$$

Under similar conditions each term of this matrix element can be expressed as an overlap of two Slater determinants. In one of them the jth s.p. orbit is multiplied by $\exp(i\frac{1}{A}\boldsymbol{k}\cdot\boldsymbol{r}_j)$. This procedure for deriving the s.p. matrix element is an alternative to that given in Sect. 9.5.1.

A most important application of the present formulae is the matrix elements of the Glauber multiple-scattering operator for the proton–nucleus scattering. When the incident momentum is chosen to be parallel with the z-axis, the operator f_i takes the form (see Sect. 3.2.4 and Sect. 13.10.2).

$$f_i = 1 - \Gamma[\boldsymbol{b} - (\boldsymbol{r}_i - \boldsymbol{R}_{\text{c.m.}})_\perp] \quad \text{with} \quad \Gamma(\boldsymbol{x}) = c e^{-\omega \boldsymbol{x}^2}, \qquad \text{(E.40)}$$

where $\boldsymbol{b}$ is the impact parameter. This is the vector that connects the line of the incident proton momentum with the c.m. of the target nucleus perpendicularly to the momentum. The subscript $\perp$ indicates a vector component perpendicular to the incident momentum. [In Part I $(\boldsymbol{r}_i)_\perp$ and $(\boldsymbol{R}_{\text{c.m.}})_\perp$ are denoted by $\boldsymbol{s}_i$ and $\boldsymbol{S}_{\text{c.m.}}$, respectively. That notation is abandoned here to avoid confusion with the generator coordinates introduced in Part II.] The profile function Γ contains information on the collisions of the proton with the individual target nucleons, and it is a key vehicle in the Glauber theory. It is parametrized with a complex number c and a real number ω. The s.p. matrix element in Eq. (E.38) can be calculated easily with the f'_i belonging to the f_i of Eq. (E.40), and then it is straightforward to evaluate Eq. (E.37).

To avoid treating many-body matrix elements, it is customary to introduce approximations in the calculation of the Glauber amplitude. In particular, the optical-limit approximation (OLA) introduced in Sect. 3.3 of Part I, reduces the many-body matrix element to a one-body matrix element. In accord with the framework applied in Part I, the derivation there has been performed in the fixed c.m. approximation. Here it will be shown that the same formula can be derived by treating the nucleon coordinates properly.

The phase-shift function for nucleon–nucleus scattering, defined in Eq. (3.61), can be rewritten with respect to the c.m. frame as

$$e^{i\chi_{el}(\boldsymbol{b})} = \langle\Psi|\prod_{i=1}^{A}[1-\Gamma[\boldsymbol{b}-(\boldsymbol{r}_i-\boldsymbol{R}_{\rm c.m.})_\perp]\,|\Psi\rangle. \tag{E.41}$$

Following the logic of Sect. 3.3, the OLA is introduced as a first-order approximation to the Taylor series of a parametric function $\ln G(\boldsymbol{b},\lambda)$ around $\lambda=0$, where [with notation $G(\boldsymbol{b},\lambda)\equiv G(\lambda)$]

$$\begin{aligned} G(\lambda) &= \langle\Psi|\prod_{i=1}^{A}\{1-\lambda\Gamma[\boldsymbol{b}-(\boldsymbol{r}_i-\boldsymbol{R}_{\rm c.m.})_\perp]\}|\Psi\rangle \\ &= \frac{1}{\Theta^2(0)}\frac{1}{(2\pi)^3}\int d\boldsymbol{k}\,\langle\Phi|\prod_{j=1}^{A}e^{i\frac{1}{A}\boldsymbol{k}\cdot\boldsymbol{r}_j}\{1-\lambda\Gamma[\boldsymbol{b}-(\boldsymbol{r}_i)_\perp]\}|\Phi\rangle. \end{aligned} \tag{E.42}$$

It follows from $e^{i\chi_{el}(\boldsymbol{b})}=G(1)$ that $\lambda=1$ is to be substituted into the approximant. Expanding $\ln G(\lambda)$ and omitting all terms beyond first order in λ, one obtains

$$\ln G(1) \approx \left[\ln G(0) + \left.\frac{G'(\lambda)}{G(\lambda)}\right|_{\lambda=0}\lambda+\cdots\right]\Bigg|_{\lambda=1} = G'(\lambda)|_{\lambda=0}\equiv G'(0). \tag{E.43}$$

Thus the OLA approximant to $e^{i\chi_{el}(\boldsymbol{b})}$ is $e^{i\chi_{el}^{\rm OLA}(\boldsymbol{b})}=e^{G'(0)}$, with

$$\begin{aligned} G'(0) &= -\frac{1}{\Theta^2(0)}\frac{1}{(2\pi)^3}\int d\boldsymbol{k}\,\langle\Phi|e^{i\boldsymbol{k}\cdot\boldsymbol{R}_{\rm c.m.}}\sum_i\Gamma[\boldsymbol{b}-(\boldsymbol{r}_i)_\perp]|\Phi\rangle \\ &= -\frac{1}{\Theta^2(0)}\langle\Phi|\delta(\boldsymbol{R}_{\rm c.m.})\sum_i\Gamma[\boldsymbol{b}-(\boldsymbol{r}_i)_\perp]|\Phi\rangle \\ &= -\sum_i\langle\Psi|\Gamma[\boldsymbol{b}-(\boldsymbol{r}_i-\boldsymbol{R}_{\rm c.m.})_\perp]|\Psi\rangle, \end{aligned} \tag{E.44}$$

so that

$$\begin{aligned} e^{i\chi_{el}(\boldsymbol{b})} \longrightarrow e^{i\chi_{el}^{\rm OLA}(\boldsymbol{b})} &= \exp\left\{-\langle\Psi|\sum_i\Gamma[\boldsymbol{b}-(\boldsymbol{r}_i-\boldsymbol{R}_{\rm c.m.})_\perp]|\Psi\rangle\right\} \\ &= \exp\left[-\int d\boldsymbol{r}\rho(\boldsymbol{r})\Gamma(\boldsymbol{b}-\boldsymbol{r}_\perp)\right]. \end{aligned} \tag{E.45}$$

This formula is very similar to Eq. (3.67). The main difference is that the density $\rho(\boldsymbol{r})$ there was defined by Eq. (3.63) without regard to the c.m. (cf. footnote on p. 46), whereas here it is defined precisely, as in Eq. (9.127). However, if one reinterprets the coordinates in Eqs. (3.63) and (3.67) as though they had been defined with respect to the c.m. and the A-fold integration in them as integrations in terms of A–1 independent variables, then the treatment of the c.m. in Eq. (3.67) becomes entirely correct. This is the way the formulae used in the reaction theory should be reinterpreted (cf. footnote in Sect. 3.2.1). Finally, it should be noted that Eq. (3.67) is derived in a non-antisymmetrical framework, whereas now we have arrived at Eq. (E.45) without compromising on the antisymmetrization. Sometimes the non-antisymmetrized formulae can also be made correct by a simple reinterpretation, but we warn the reader against unwarranted generalizations of this observation.

The Glauber matrix element $\langle\Psi|\prod_i\{1-ce^{-\omega[\boldsymbol{b}-(\boldsymbol{r}_i-\boldsymbol{R}_{\text{c.m.}})_\perp]^2}\}|\Psi'\rangle$ can as well be calculated in a more direct alternative way by expressing the operator in terms of the generating function (9.59) of the correlated Gaussians [58]. The functions Ψ and Ψ' are also generated by similar functions, so that the matrix element is expressible in terms of an integral of the product of three generating functions (9.59). This integral is a simple Gaussian multiple integral, which can be evaluated in one step via Eq. (9.61). (Cf. Sect. 13.10.2.)

Another example for the many-body operators presented in this text is the projector $\mathcal{P}_Q$ onto the subspace of $Q\,\hbar\omega$ SM excitation quanta, Eq. (12.7). This is useful to analyse a (non-SM) state in terms of its SM components. Equation (12.7) can be rewritten as

$$\mathcal{P}_Q = \frac{1}{2\pi}\int_0^{2\pi} d\theta\; \mathrm{e}^{-iQ\theta}\prod_{j=1}^{A}\exp\left\{i\theta P_j[(\hbar\omega)^{-1}H_{\text{HO}}(j)-\tfrac{3}{2}]\right\}. \quad \text{(E.46)}$$

Here $H_{\text{HO}}(j)$ is the three-dimensional HO Hamiltonian belonging to energy quantum $\hbar\omega = (\hbar^2\gamma/m)$ (which defines γ), and P_j projects onto protons, neutrons or both. The probability calculated with Eq. (E.46) contains no contribution from c.m. excitations if the wave function is free from c.m. excitations. In all approaches discussed in Part II that is the case. The operator (E.46) contains the c.m. coordinate, therefore, the intrinsic wave function should be multiplied by the HO g.s. $\varphi_0^{(\gamma)}(\boldsymbol{x}_A)$ [cf. Eq. (9.66)]. Since $\varphi_0^{(\gamma)}(\boldsymbol{x}_A)$ carries no excitation, the probability calculated will be a purely intrinsic quantity.

The matrix element of the operator $\mathcal{P}_Q$ between Slater determinants is given by

$$\langle\phi_\kappa(\boldsymbol{s}_1,\ldots,\boldsymbol{s}_A)|\mathcal{P}_Q|\phi_{\kappa'}(\boldsymbol{s}'_1,\ldots,\boldsymbol{s}'_A)\rangle = \frac{1}{2\pi}\int_0^{2\pi} d\theta\; \mathrm{e}^{-iQ\theta}\det\{B''\}, \quad \text{(E.47)}$$

where the elements of the matrix B'' are defined by

$$B''_{ij} = \langle \hat{\varphi}^{(\beta)}_{s_i m_s^i m_t^i} | \exp\{i\theta P[(\hbar\omega)^{-1} H_{\rm HO} - \tfrac{3}{2}]\} | \hat{\varphi}^{(\beta)}_{s'_j m_s^{j'} m_t^{j'}} \rangle \qquad (i,\, j = 1, \ldots, A). \qquad \text{(E.48)}$$

Since the constants β and γ are in general different, in calculating B''_{ij} one has to be transformed into the other. The transformation is given by

$$\varphi_s^{(\beta)}(\boldsymbol{r}) = \left[\frac{\beta\gamma^3}{4\pi^2(\beta-\gamma)^2}\right]^{3/4} \int d\boldsymbol{t} \exp\left[-\frac{\beta\gamma}{2(\gamma-\beta)}(\boldsymbol{t}-\boldsymbol{s})^2\right] \varphi_t^{(\gamma)}(\boldsymbol{r}), \qquad \text{(E.49)}$$

which can be easily proven by using Eq. (9.50). Equation (E.49) expresses the well-known fact that the convolution of two Gaussians is also a Gaussian. Once all Gaussians have the same parameter γ, one can act with $\exp\{i\theta[(\hbar\omega)^{-1}H_{\rm HO} - \frac{3}{2}]\}$ on $\varphi_t^{(\gamma)}(\boldsymbol{r})$. With Eq. (9.48), one has $\varphi_t^{(\gamma)}(\boldsymbol{r}) = {\rm e}^{-\frac{1}{4}\gamma t^2} g^{(\gamma)}(\boldsymbol{t}, \boldsymbol{r})$ and, using Eq. (9.49), one arrives at the required relationship:

$$\begin{aligned} &\exp\{i\theta[(\hbar\omega)^{-1}H_{\rm HO} - \tfrac{3}{2}]\}\, \varphi_t^{(\gamma)}(\boldsymbol{r}) \\ &= {\rm e}^{-\frac{1}{4}\gamma t^2} \sum_{nlm} (-1)^n \sqrt{\frac{B_{nl}}{2^{2n+l}(2n+l)!}} (\sqrt{\gamma}t)^{2n+l} Y_{lm}^*(\hat{\boldsymbol{t}}) {\rm e}^{i\theta(2n+l)} \psi_{nlm}^{(\gamma)}(\boldsymbol{r}) \\ &= {\rm e}^{-\frac{1}{4}\gamma t^2} {\rm e}^{\frac{1}{4}\gamma (zt)^2} \varphi_{zt}^{(\gamma)}(\boldsymbol{r}) = {\rm e}^{-\frac{1}{4}\gamma(1-z^2)t^2} \varphi_{zt}^{(\gamma)}(\boldsymbol{r}), \end{aligned} \qquad \text{(E.50)}$$

where $z = {\rm e}^{i\theta}$. Now, using Eqs. (E.49), (E.50) and (9.50), one can evaluate the s.p. matrix element involved in B''_{ij}:

$$\begin{aligned} &\langle \hat{\varphi}_{s m_s m_t} | \exp\{i\theta P[(\hbar\omega)^{-1} H_{\rm HO} - \tfrac{3}{2}]\} | \hat{\varphi}_{s' m'_s m'_t} \rangle \\ &= \left[\frac{4\beta\gamma}{(\beta+\gamma)^2 - (\beta-\gamma)^2\bar{z}^2}\right]^{3/2} \\ &\quad \times \exp\left[-\frac{\beta\gamma}{2} \frac{\beta+\gamma+(\beta-\gamma)\bar{z}^2}{(\beta+\gamma)^2-(\beta-\gamma)^2\bar{z}^2}(\boldsymbol{s}^2 + \boldsymbol{s}'^2) \right. \\ &\quad \left. + \frac{2\beta^2\gamma\bar{z}}{(\beta+\gamma)^2-(\beta-\gamma)^2\bar{z}^2} \boldsymbol{s}\cdot\boldsymbol{s}'\right] \delta_{m_s m'_s} \delta_{m_t m'_t}, \end{aligned} \qquad \text{(E.51)}$$

where $\bar{z} = z$ or 1 depending on whether $\langle m_t | P | m'_t \rangle = 1$ or 0. The value of β is usually chosen to give the cluster an appropriate size, while γ is most often chosen to correspond to the size of the whole nucleus. Hence the value of β is usually larger than that of γ. Then the exponential function in Eq. (E.49) diverges, but the integral still exists and Eq. (E.51) can also be used safely.

Appendix F

Matrix elements between correlated Gaussians

This appendix concludes the derivation of the overlap and Hamiltonian matrix elements between basis functions defined in Eq. (9.53) or (9.84). The generic forms of the matrix elements between Slater determinants have been derived in App. E. The translation-invariant forms of these matrix elements have been obtained in Sect. 9.4.4 [see Eqs. (9.116)–(9.119)]. It remains to carry through the operations prescribed in Eq. (9.91) explicitly. Although not difficult, this is perhaps the most laborious part of the correlated Gaussian formalism. Nevertheless, we shall illustrate the application of Eq. (9.91) by a thoroughgoing derivation: by the derivation of the overlap. The evaluation of the Hamiltonian matrix element involves just more formula crunching but no extra conceptual difficulty. Therefore, that derivation will be sketchier than any others before.

Upon setting $O=1$, one has to substitute Eq. (9.116) into Eq. (9.91):

$$\langle \mathcal{A}\{f_{KLM}(u,\boldsymbol{x},A)\chi_{SM_S}\eta_{TM_T}\}|\mathcal{A}\{f_{K'L'M'}(u',\boldsymbol{x},A')\chi_{S'M'_S}\eta_{T'M'_T}\}\rangle$$
$$= \frac{1}{B_{KL}B_{K'L'}}\left[(4\pi\beta)^{A-1}\det C\right]^{-3/2}\int d\hat{e}\int d\hat{e}'\, Y^*_{LM}(\hat{e})Y_{L'M'}(\hat{e}')$$
$$\times\left[\frac{\partial^{\kappa+\kappa'}}{\partial\alpha^{\kappa}\partial\alpha'^{\kappa'}}\left(\mathrm{e}^{-\frac{1}{2}\tilde{X}CX}\right.\right.$$
$$\left.\left.\left.\times\int d\boldsymbol{T}\,\mathrm{e}^{-\frac{1}{2}\tilde{T}QT+\tilde{X}T}\sum_i C_i^{(\mathrm{o})}\mathrm{e}^{-\frac{1}{2}\tilde{T}B_i^{(\mathrm{o})}T}\right)\right]\right|_{\alpha=\alpha'=0}, \tag{F.1}$$

with

$$\kappa = 2K+L, \qquad \kappa' = 2K'+L'. \tag{F.2}$$

Remember [cf. Eqs. (9.75) and (9.93)] that the matrices $\mathcal{C}$ and $\mathcal{Q}$ are related to the matrices A and A' of the correlated Gaussians as follows:

$$\begin{aligned} \mathcal{C} &= \begin{pmatrix} C & 0 \\ 0 & C' \end{pmatrix} = \begin{pmatrix} \beta^{-1}(I-\beta^{-1}A) & 0 \\ 0 & \beta^{-1}(I-\beta^{-1}A') \end{pmatrix}, \\ \mathcal{Q} &= \begin{pmatrix} C^{-1}-\beta I & 0 \\ 0 & C'^{-1}-\beta I \end{pmatrix}, \end{aligned} \tag{F.3}$$

and $\boldsymbol{X}$ is a one-column matrix of $2A-2$ vectors, i.e.,

$$\tilde{\boldsymbol{X}} = (\alpha\tilde{\omega}\boldsymbol{e}, \alpha'\tilde{\omega}'\boldsymbol{e}'), \tag{F.4}$$

where $|\boldsymbol{e}|=|\boldsymbol{e}'|=1$ and

$$\begin{aligned} \omega_k &= \beta^{-1}\sum_{j=1}^{A-1}(C^{-1})_{kj}u_j, \\ \omega'_k \equiv \omega_{k+A-1} &= \beta^{-1}\sum_{j=1}^{A-1}(C'^{-1})_{kj}u'_j \quad (k=1,\ldots,A-1). \end{aligned} \tag{F.5}$$

By integrating over the generator coordinates $\boldsymbol{T}$ using Eq. (9.61), one easily obtains

$$\begin{aligned} &\langle\mathcal{A}\{f_{KLM}(u,\boldsymbol{x},A)\chi_{SM_S}\eta_{TM_T}\}|\mathcal{A}\{f_{K'L'M'}(u',\boldsymbol{x},A')\chi_{S'M'_S}\eta_{T'M'_T}\}\rangle \\ &\quad = \frac{1}{B_{KL}B_{K'L'}}\left[(4\pi\beta)^{A-1}\det\mathcal{C}\right]^{-3/2}\int d\hat{\boldsymbol{e}}\int d\hat{\boldsymbol{e}}'\, Y^*_{LM}(\hat{\boldsymbol{e}})Y_{L'M'}(\hat{\boldsymbol{e}}') \\ &\qquad \times\sum_i C_i^{(o)}\left[\frac{(2\pi)^{2A-2}}{\det(B_i^{(o)}+\mathcal{Q})}\right]^{3/2} \\ &\qquad \times\left(\frac{\partial^{\kappa+\kappa'}}{\partial\alpha^{\kappa}\partial\alpha'^{\kappa'}}e^{p_i\alpha^2+p'_i\alpha'^2+q_i\alpha\alpha'\boldsymbol{e}\cdot\boldsymbol{e}'}\right)\Bigg|_{\alpha=\alpha'=0}, \end{aligned} \tag{F.6}$$

where, to display the α- and α'-dependence of the resulting expression, we introduced the abbreviations

$$\begin{aligned} p_i &= \frac{1}{2}\sum_{j,k=1}^{A-1}\omega_j\left[(B_i^{(o)}+\mathcal{Q})^{-1}-\mathcal{C}\right]_{jk}\omega_k, \\ p'_i &= \frac{1}{2}\sum_{j,k=A}^{2A-2}\omega_j\left[(B_i^{(o)}+\mathcal{Q})^{-1}-\mathcal{C}\right]_{jk}\omega_k, \\ q_i &= \sum_{j=1}^{A-1}\sum_{k=A}^{2A-2}\omega_j\left[(B_i^{(o)}+\mathcal{Q})^{-1}-\mathcal{C}\right]_{jk}\omega_k. \end{aligned} \tag{F.7}$$

The differentiations with respect to α and α' can be performed by expanding the exponential function into a MacLaurin series. To simplify the formulae, let us omit the label i and introduce $r=qe\cdot e'$ temporarily. The κth derivative of the Nth term with respect to α may then be written as

$$\frac{1}{N!}\frac{\partial^\kappa}{\partial\alpha^\kappa}\left[(p\alpha^2+p'\alpha'^2+r\alpha\alpha')^N\right]=\sum_{\lambda=\max\{0,\kappa-N\}}^{[\kappa/2]}\frac{(2\lambda-1)!!}{(N-\kappa+\lambda)!}\binom{\kappa}{2\lambda}$$
$$\times(2p)^\lambda(p\alpha^2+p'\alpha'^2+r\alpha\alpha')^{N-\kappa+\lambda}(2p\alpha+r\alpha')^{\kappa-2\lambda}, \qquad \text{(F.8)}$$

where the convention $(-1)!!=1$ has been used, and the symbol $[\cdots]$ in the summation limit stands for the integer part. The limits of the summations come from the fact that $1/(-m)!=0$ if m is a positive integer. This formula can be easily verified by induction with respect to κ. Now the mixed derivative can be calculated by using

$$\frac{\partial^{\kappa'}}{\partial\alpha'^{\kappa'}}\left[(p\alpha^2+p'\alpha'^2+r\alpha\alpha')^{N-\kappa+\lambda}(2p\alpha+r\alpha')^{\kappa-2\lambda}\right]=\sum_{\mu=0}^{\kappa'}\binom{\kappa'}{\mu}$$
$$\times\frac{\partial^{\kappa'-\mu}}{\partial\alpha'^{\kappa'-\mu}}\left[(p\alpha^2+p'\alpha'^2+r\alpha\alpha')^{N-\kappa+\lambda}\right]\frac{\partial^\mu}{\partial\alpha'^\mu}\left[(2p\alpha+r\alpha')^{\kappa-2\lambda}\right]. \qquad \text{(F.9)}$$

For the differentiations in the first factor, Eq. (F.8) can be used again, while the differentiations in the second are trivial. All these put together will give, for the mixed derivative,

$$\frac{1}{N!}\frac{\partial^{\kappa+\kappa'}}{\partial\alpha^\kappa\partial\alpha'^{\kappa'}}\left[(p\alpha^2+p'\alpha'^2+r\alpha\alpha')^N\right]$$
$$=\sum_{\lambda=\max\{0,\kappa-N\}}^{[\kappa/2]}\sum_{\mu=0}^{\kappa'}\sum_{\nu=0}^{[(\kappa'-\mu)/2]}(p\alpha^2+p'\alpha'^2+r\alpha\alpha')^{N-\kappa-\kappa'+\lambda+\mu+\nu}$$
$$\times(2p\alpha+r\alpha')^{\kappa-\mu-2\lambda}(2p'\alpha'+r\alpha)^{\kappa'-\mu-2\nu}$$
$$\times\frac{\kappa!\kappa'!p^\lambda p'^\nu r^\mu}{(N-\kappa-\kappa'+\lambda+\mu+\nu)!(\kappa-\mu-2\lambda)!(\kappa'-\mu-2\nu)!\lambda!\nu!\mu!}. \qquad \text{(F.10)}$$

The next step is to take the L-projection denominated in Eq. (F.1) and to set $\alpha=\alpha'=0$. To perform the L-projecting, one has to substitute $r=qe\cdot e'$ into Eq. (F.10) and use Eq. (9.58) for $(e\cdot e')^\mu$. Equation (9.58) sets $\mu=2n+l$, and the orthonormality of the spherical harmonics implies that the non-vanishing l-terms are those with $l=L=L'$. When $\alpha=\alpha'=0$ is taken, all terms vanish except those in which the power of each α, α'-dependent factor is zero, i.e., $\kappa-\mu=2\lambda$, $\kappa'-\mu=2\nu$, $N=\kappa+\kappa'-\lambda-\mu-\nu=\frac{1}{2}(\kappa+\kappa')$. Recalling that $\kappa=2K+L$, $\kappa'=2K'+L'$, we see that, for the non-vanishing terms, $N = K + K' + L$. One can thus write

$$\int d\hat{e} \int d\hat{e}' \, Y^*_{LM}(\hat{e}) Y_{L'M'}(\hat{e}')$$

$$\times \left[\frac{\partial^{\kappa+\kappa'}}{\partial\alpha^{\kappa}\partial\alpha'^{\kappa'}} \sum_{N=0}^{\infty} \frac{(p\alpha^2 + p'\alpha'^2 + qe\cdot e'\alpha\alpha')^N}{N!} \right] \Bigg|_{\alpha=\alpha'=0}$$

$$= \delta_{LL'}\delta_{MM'} \sum_{N=0}^{\infty} \sum_{\lambda=\max\{0,\kappa-N\}}^{[\kappa/2]} \sum_{\mu=0}^{\kappa'} \sum_{\nu=0}^{[(\kappa'-\mu)/2]} \sum_{k,l} B_{kl}$$

$$\times \delta_{2N,\kappa+\kappa'} \delta_{2\lambda,\kappa-\mu} \delta_{2\nu,\kappa'-\mu}$$

$$\times \frac{\kappa!\kappa'! p^{\lambda} p'^{\nu} q^{\mu} \delta_{Ll} \delta_{2k+l,\mu}}{(N-\kappa-\kappa'+\lambda+\mu+\nu)!(\kappa-\mu-2\lambda)!(\kappa'-\mu-2\nu)!\lambda!\nu!\mu!} \quad \text{(F.11)}$$

$$= \delta_{LL'}\delta_{MM'} \kappa!\kappa'! \sum_{k=0}^{\infty} \sum_{\mu=0}^{\kappa'} B_{kL} \frac{p^{(\kappa-\mu)/2} p'^{(\kappa'-\mu)/2} q^{\mu} \delta_{2k+l,\mu}}{[\frac{1}{2}(\kappa-\mu)]! \, [\frac{1}{2}(\kappa'-\mu)]! \mu!} \quad \text{(F.12)}$$

$$= \delta_{LL'}\delta_{MM'} (2K+L)!(2K'+L)! \sum_{k=0}^{\min\{K,K'\}} B_{kL} \frac{p^{K-k} p'^{K'-k} q^{2k+L}}{(K-k)!(K'-k)!(2k+L)!}. \quad \text{(F.13)}$$

In Eq. (F.11) one can see that in the non-vanishing terms $\kappa-\mu$ and $\kappa'-\mu$ are always even, so that the symbol $[\cdots]$ in the upper limits become immaterial. One can also easily verify that the value $\lambda = \frac{1}{2}(\kappa-\mu)$ is within the range of the λ-summation. The sum in Eq. (F.13) is limited by the vanishing of the quantity to be summed owing either to $[(K-k)!]^{-1} = 0$ or to $[(K'-k)!]^{-1} = 0$.

Finally, Eq. (F.13) should be substituted into Eq. (F.6), to obtain

$$\langle \mathcal{A}\{f_{KLM}(u, x, A)\chi_{SM_S}\eta_{TM_T}\} | \mathcal{A}\{f_{K'L'M'}(u', x, A')\chi_{S'M'_S}\eta_{T'M'_T}\}\rangle$$

$$= \delta_{LL'}\delta_{MM'} \frac{(2K+L)!(2K'+L)!}{B_{KL}B_{K'L}} \left[(4\pi\beta)^{A-1} \det C\right]^{-3/2}$$

$$\times \sum_i C_i^{(o)} \left[\frac{(2\pi)^{2A-2}}{\det(B_i^{(o)} + Q)} \right]^{3/2} \sum_{k=0}^{\min(K,K')} B_{kL} \frac{{p_i}^{K-k} {p'_i}^{K'-k} {q_i}^{2k+L}}{(K-k)!(K'-k)!(2k+L)!}. \quad \text{(F.14)}$$

This formula is remarkably simple. The coefficients $C_i^{(o)}$ are to be determined for each physical problem. Usually the number of terms is just a fraction of the combinatorial limit, $A!$, owing to the orthogonality of the spin–isospin s.p. states belonging to many permutations.

The matrix elements of the kinetic-energy operator can be derived in exactly the same manner as the overlap matrix element. Substitution of Eq. (9.117) into Eq. (9.91) yields

$$
\begin{aligned}
&\langle \mathcal{A}\{f_{KLM}(u,x,A)\chi_{SM_S}\eta_{TM_T}\}| \\
&\qquad\qquad \sum_{i=1}^{A-1} \frac{\pi_i^2}{2m} |\mathcal{A}\{f_{K'L'M'}(u',x,A')\chi_{S'M'_S}\eta_{T'M'_T}\}\rangle \\
&= \frac{1}{B_{KL}B_{K'L'}} \left[(4\pi\beta)^{A-1}\det C\right]^{-3/2} \int d\hat{e} \int d\hat{e}'\, Y^*_{LM}(\hat{e}) Y_{L'M'}(\hat{e}') \\
&\quad \times \frac{\hbar^2\beta}{4m} \left(\frac{\partial^{\kappa+\kappa'}}{\partial\alpha^\kappa \partial\alpha'^{\kappa'}} \left\{ e^{-\frac{1}{2}\tilde{X}CX} \int dT\, e^{-\frac{1}{2}\tilde{T}QT + \tilde{X}T} \right.\right. \\
&\quad \left.\left.\left. \times \sum_i C_i^{(o)} \left[3(A-1) - \tilde{T}B_i^{(o)}T\right] e^{-\frac{1}{2}\tilde{T}B_i^{(o)}T} \right\} \right) \right|_{\alpha=\alpha'=0} . \qquad \text{(F.15)}
\end{aligned}
$$

In performing the T-integration, Eqs. (9.61) and (9.63) are to be used:

$$
\begin{aligned}
&\int dT \left[3(A-1) - \tilde{T}B_i^{(o)}T\right] e^{-\frac{1}{2}\tilde{T}(B_i^{(o)}+Q)T + \tilde{X}T} \\
&= \left[\frac{(2\pi)^{2A-2}}{\det(B_i^{(o)} + Q)}\right]^{3/2} e^{\frac{1}{2}\tilde{X}(B_i^{(o)}+Q)^{-1}X} \\
&\quad \times \left\{ 3(A-1) - 3\sum_{j,k} \left[(B_i^{(o)} + Q)^{-1}\right]_{kj} (B_i^{(o)})_{jk} \right. \\
&\qquad \left. + \tilde{X}(B_i^{(o)} + Q)^{-1} B_i^{(o)} (B_i^{(o)} + Q)^{-1} X \right\}. \qquad \text{(F.16)}
\end{aligned}
$$

Substituted in Eq. (F.15), this gives

$$
\begin{aligned}
&\langle \mathcal{A}\{f_{KLM}(u,x,A)\chi_{SM_S}\eta_{TM_T}\}| \\
&\qquad\qquad \sum_{i=1}^{A-1} \frac{\pi_i^2}{2m} |\mathcal{A}\{f_{K'L'M'}(u',x,A')\chi_{S'M'_S}\eta_{T'M'_T}\}\rangle \\
&= \frac{1}{B_{KL}B_{K'L'}} \left[(4\pi\beta)^{A-1}\det C\right]^{-3/2} \frac{\hbar^2\beta}{2m} \int d\hat{e} \int d\hat{e}'\, Y^*_{LM}(\hat{e}) Y_{L'M'}(\hat{e}') \\
&\quad \times \sum_i C_i^{(o)} \left[\frac{(2\pi)^{2A-2}}{\det(B_i^{(o)} + Q)}\right]^{3/2} \left\{ \frac{\partial^{\kappa+\kappa'}}{\partial\alpha^\kappa \partial\alpha'^{\kappa'}} (R_i + P_i\alpha^2 + P'_i\alpha'^2 + Q_i\alpha\alpha' e\cdot e') \right. \\
&\qquad \left.\left. \times\, e^{p_i\alpha^2 + p'_i\alpha'^2 + q_i\alpha\alpha' e\cdot e'} \right\} \right|_{\alpha=\alpha'=0} , \qquad \text{(F.17)}
\end{aligned}
$$

where

$$
R_i = \frac{3}{2}(A-1) - \frac{3}{2}\mathrm{Tr}\left[(B_i^{(o)} + Q)^{-1} B_i^{(o)}\right],
$$

$$P_i = -\frac{1}{2}\sum_{j,k=1}^{A-1}\omega_j\left[(B_i^{(o)}+Q)^{-1}B_i^{(o)}(B_i^{(o)}+Q)^{-1}\right]_{jk}\omega_k,$$

$$P_i' = -\frac{1}{2}\sum_{j,k=A}^{2A-2}\omega_j\left[(B_i^{(o)}+Q)^{-1}B_i^{(o)}(B_i^{(o)}+Q)^{-1}\right]_{jk}\omega_k,$$

$$Q_i = -\sum_{j=1}^{A-1}\sum_{k=A}^{2A-2}\omega_j\left[(B_i^{(o)}+Q)^{-1}B_i^{(o)}(B_i^{(o)}+Q)^{-1}\right]_{jk}\omega_k. \quad \text{(F.18)}$$

The differentiation with respect to α and α', the L-projection and the substitution $\alpha=\alpha'=0$ repeat the steps of Eqs. (F.8)–(F.13), with some extra complications. The result is

$$\begin{aligned}&\langle\mathcal{A}\{f_{KLM}(u,x,A)\chi_{SM_S}\eta_{TM_T}\}|\\ &\qquad\sum_{i=1}^{A-1}\frac{\pi_i^2}{2m}|\mathcal{A}\{f_{K'L'M'}(u',x,A')\chi_{S'M_S'}\eta_{T'M_T'}\}\rangle\\ &=\delta_{LL'}\delta_{MM'}\frac{(2K+L)!(2K'+L)!}{B_{KL}B_{K'L}}\left[(4\pi\beta)^{A-1}\det C\right]^{-3/2}\frac{\hbar^2\beta}{2m}\\ &\times\sum_i C_i^{(o)}\left[\frac{(2\pi)^{2A-2}}{\det(B_i^{(o)}+Q)}\right]^{3/2}\sum_{k=0}^{\min(K,K')}B_{kL}\frac{p_i^{K-k}p_i'^{K'-k}q_i^{2k+L}}{(K-k)!(K'-k)!(2k+L)!}\\ &\times\left[R_ip_ip_i'q_i+(K-k)P_ip_i'q_i+(K'-k)p_iP_i'q_i+(2k+L)p_ip_i'Q_i\right]. \quad \text{(F.19)}\end{aligned}$$

The matrix element of the kinetic energy can also be derived by using the analogy between its starting formula, Eq. (F.15), and that of the overlap, Eq. (F.1). One can see that the terms in Eq. (F.15) are equal to those in Eq. (F.1) times $[3(A-1)-\tilde{\boldsymbol{T}}B_i^{(o)}\boldsymbol{T}]$. The expression $-\tilde{\boldsymbol{T}}B_i^{(o)}\boldsymbol{T}$ also appears in the exponent of the exponential function in the same term, and it can be 'pulled down' by differentiation with respect to a suitable variable. Therefore, the result (F.19) can be obtained by a differentiation of Eq. (F.14). Let us denote the (right-hand side of) Eq. (F.14) by $F(B_1^{(o)},\ldots,B_{n_o}^{(o)})$, where n_o is the number of the i-terms in any of these formulae. The kinetic-energy matrix element will be

$$\begin{aligned}&\langle\mathcal{A}\{f_{KLM}(u,x,A)\chi_{SM_S}\eta_{TM_T}\}|\\ &\qquad\sum_{i=1}^{A-1}\frac{\pi_i^2}{2m}|\mathcal{A}\{f_{K'L'M'}(u',x,A')\chi_{S'M_S'}\eta_{T'M_T'}\}\rangle\\ &=\frac{\hbar^2\beta}{4m}\left\{\left[3(A-1)+2\frac{\partial}{\partial z}\right]F(zB_1^{(o)},\ldots,zB_{n_o}^{(o)})\right\}\Big|_{z=1}. \quad \text{(F.20)}\end{aligned}$$

A matrix element of a central two-body potential is obtained by putting $P_i(\boldsymbol{T},\boldsymbol{r})=1$ and $\boldsymbol{D}_i=\tilde{\widehat{D}}_{(i)}\boldsymbol{T}$ into Eq. (9.119) and substituting

Eq. (9.119) into Eq. (9.91):

$$\langle \mathcal{A}\{f_{KLM}(u,x,A)\chi_{SM_S}\eta_{TM_T}\}| \sum_{i<j}^{A} V_{ij} |\mathcal{A}\{f_{K'L'M'}(u',x,A')\chi_{S'M'_S}\eta_{T'M'_T}\}\rangle$$
$$= \frac{1}{B_{KL}B_{K'L'}}\left[(4\pi\beta)^{A-1}\det C\right]^{-3/2}\left(\frac{\beta}{2\pi}\right)^{3/2}\int d\boldsymbol{r}\, V(r)\mathrm{e}^{-\frac{1}{2}\beta r^2}$$
$$\times\int d\hat{e}\int d\hat{e}'\, Y^*_{LM}(\hat{e})Y_{L'M'}(\hat{e}')\sum_i C_i^{(\mathrm{p})}$$
$$\times\left[\frac{\partial^{\kappa+\kappa'}}{\partial\alpha^{\kappa}\partial\alpha'^{\kappa'}}\left(\mathrm{e}^{-\frac{1}{2}\tilde{X}CX}\int d\boldsymbol{T}\mathrm{e}^{-\frac{1}{2}\tilde{T}QT+\tilde{X}T}\mathrm{e}^{-\frac{1}{2}\tilde{T}B_i^{(\mathrm{p})}T+D_i\cdot\boldsymbol{r}}\right)\right]\Bigg|_{\alpha=\alpha'=0}. \tag{F.21}$$

With the use of Eq. (9.61), the $\boldsymbol{T}$-integration yields

$$\int d\boldsymbol{T}\,\mathrm{e}^{-\frac{1}{2}\tilde{T}(B_i^{(\mathrm{p})}+Q)T+(\tilde{X}+\tilde{\hat{D}}_{(i)}\boldsymbol{r})\cdot\boldsymbol{T}}$$
$$= \left[\frac{(2\pi)^{2A-2}}{\det(B_i^{(\mathrm{p})}+Q)}\right]^{3/2}\mathrm{e}^{\frac{1}{2}(\tilde{X}+\tilde{\hat{D}}_{(i)}\boldsymbol{r})(B_i^{(\mathrm{p})}+Q)^{-1}(X+\hat{D}_{(i)}\boldsymbol{r})}. \tag{F.22}$$

With this inserted into Eq. (F.21), one has

$$\langle \mathcal{A}\{f_{KLM}(u,x,A)\chi_{SM_S}\eta_{TM_T}\}| \sum_{i<j}^{A} V_{ij} |\mathcal{A}\{f_{K'L'M'}(u',x,A')\chi_{S'M'_S}\eta_{T'M'_T}\}\rangle$$
$$= \frac{1}{B_{KL}B_{K'L'}}\left[(4\pi\beta)^{A-1}\det C\right]^{-3/2}\left(\frac{\beta}{2\pi}\right)^{3/2}\int d\boldsymbol{r}\, V(r)$$
$$\times\int d\hat{e}\int d\hat{e}'\, Y^*_{LM}(\hat{e})Y_{L'M'}(\hat{e}')\sum_i C_i^{(\mathrm{p})}\left[\frac{(2\pi)^{2A-2}}{\det(B_i^{(\mathrm{p})}+Q)}\right]^{3/2}\mathrm{e}^{-\frac{1}{2}c_i r^2}$$
$$\times\left[\frac{\partial^{\kappa+\kappa'}}{\partial\alpha^{\kappa}\partial\alpha'^{\kappa'}}\left(\mathrm{e}^{\bar{p}_i\alpha^2+\bar{p}'_i\alpha'^2+\bar{q}_i\alpha\alpha'\boldsymbol{e}\cdot\boldsymbol{e}'+\rho_i\alpha\boldsymbol{e}\cdot\boldsymbol{r}+\rho'_i\alpha'\boldsymbol{e}'\cdot\boldsymbol{r}}\right)\right]\Bigg|_{\alpha=\alpha'=0}, \tag{F.23}$$

where

$$c_i = \beta - \tilde{\hat{D}}_{(i)}(B_i^{(p)}+Q)^{-1}\hat{D}_{(i)},$$
$$\bar{p}_i = \frac{1}{2}\sum_{j,k=1}^{A-1}\omega_j\left[(B_i^{(p)}+Q)^{-1}-C\right]_{jk}\omega_k,$$

$$
\begin{aligned}
\bar{p}'_i &= \frac{1}{2}\sum_{j,k=A}^{2A-2}\omega_j\left[(B_i^{(p)}+\mathcal{Q})^{-1}-\mathcal{C}\right]_{jk}\omega_k,\\
\bar{q}_i &= \sum_{j=1}^{A-1}\sum_{k=A}^{2A-2}\omega_j\left[(B_i^{(p)}+\mathcal{Q})^{-1}-\mathcal{C}\right]_{jk}\omega_k,\\
\rho_i &= \sum_{j=1}^{A-1}\omega_j\left[(B_i^{(p)}+\mathcal{Q})^{-1}\widehat{\mathcal{D}}_{(i)}\right]_j,\\
\rho'_i &= \sum_{j=A}^{2A-2}\omega_j\left[(B_i^{(p)}+\mathcal{Q})^{-1}\widehat{\mathcal{D}}_{(i)}\right]_j.
\end{aligned}
\tag{F.24}
$$

One can prove that c_i is always positive.

This is again followed by the performing of the differentiations with respect to α and α', the evaluation of the $\alpha=\alpha'=0$ limit and the integration over $\hat{\mathbf{e}}$ and $\hat{\mathbf{e}}'$. The last step is more involved now, since the powers of all scalar products, $\boldsymbol{e}\cdot\boldsymbol{e}'$, $\boldsymbol{e}\cdot\boldsymbol{r}$ and $\boldsymbol{e}'\cdot\boldsymbol{r}$, are to be expanded according to Eq. (9.58), but is essentially similar to the previous cases. The details can be found in the monograph [187]. The final result is

$$
\begin{aligned}
&\langle\mathcal{A}\{f_{KLM}(u,\boldsymbol{x},A)\chi_{SM_S}\eta_{TM_T}\}|\\
&\qquad\sum_{i<j}^{A}V_{ij}|\mathcal{A}\{f_{K'L'M'}(u',\boldsymbol{x},A')\chi_{S'M'_S}\eta_{T'M'_T}\}\rangle\\
&=\delta_{LL'}\delta_{MM'}\frac{(2K+L)!(2K'+L)!}{B_{KL}B_{K'L}}\left[(4\pi\beta)^{A-1}\det\mathcal{C}\right]^{-3/2}\\
&\times\left(\frac{\beta}{2\pi}\right)^{3/2}\sum_i C_i^{(p)}\left[\frac{(2\pi)^{2A-2}}{\det(B_i^{(p)}+\mathcal{Q})}\right]^{3/2}\\
&\times\sum_{\substack{k_1,k_2,k_3,l_1,l_2\\2k_1+l_1+2k_2+l_2\le 2K+L\\2k_1+l_1+2k_3+l_2\le 2K'+L}}B_{k_1l_1}B_{k_2l_2}B_{k_3l_2}\frac{(2l_1+1)(2l_2+1)}{4\pi(2L+1)}\langle l_1 0 l_2 0|L0\rangle^2\\
&\times\frac{\bar{p}_i^{\,K-k_1-k_2-\frac{1}{2}(l_1+l_2-L)}\,\bar{p}_i'^{\,K'-k_1-k_3-\frac{1}{2}(l_1+l_2-L)}}{(K-k_1-k_2-\frac{1}{2}(l_1+l_2-L))!\,(K'-k_1-k_3-\frac{1}{2}(l_1+l_2-L))!}\\
&\times\frac{\bar{q}_i^{\,2k_1+l_1}\,\rho_i^{\,2k_2+l_2}\,\rho_i'^{\,2k_3+l_2}}{(2k_1+l_1)!\,(2k_2+l_2)!\,(2k_3+l_2)!}I(2k_2+l_2+2k_3+l_2+2,c_i).
\end{aligned}
\tag{F.25}
$$

The last factor in Eq. (F.25) is an integral of the form

$$
I(k,c)=\int_0^\infty dr\,V(r)r^k\mathrm{e}^{-\frac{1}{2}cr^2},
\tag{F.26}
$$

which can be calculated analytically for a certain class of potentials. In particular, when the radial part of $V(r)$ is given in the form

$$V(r) = r^n e^{-ar^2-br}, \tag{F.27}$$

for $n \geq -k$ a closed expression exists for $I(k,c)$:

$$\int_0^\infty dr\, r^n e^{-ar^2-br} = \sqrt{\pi}(-1)^n \left(\frac{1}{2\sqrt{a}}\right)^{n+1} \sum_{k=0}^{n} \binom{n}{k} f_k g_{n-k}, \tag{F.28}$$

with

$$\begin{aligned} f_k &= \sum_{i=0}^{[\frac{k}{2}]} \frac{k!}{i!(k-2i)!} \left(\frac{b}{\sqrt{a}}\right)^{k-2i}, \\ g_0 &= \exp\left(\frac{b^2}{4a}\right) \text{erfc}\left(\frac{b}{2\sqrt{a}}\right), \\ g_k &= (-1)^k \frac{2}{\sqrt{\pi}} H_{k-1}\left(\frac{b}{2\sqrt{a}}\right) \quad (k \geq 1). \end{aligned} \tag{F.29}$$

Here $H_n(x)$ is an Hermite polynomial and $\text{erfc}(x) = 1 - \text{erf}(x)$. When the potential form factor does not allow analytic evaluation, one has to rely on numerical integration in Eq. (F.26), but such integrations are not time-consuming.

Although the above derivations are fairly involved, the formulae themselves are very simple. The simplicity of the calculation of the matrix elements becomes especially evident if one compares the work involved in the above procedures with the work involved when the function $\theta_{lLM_L}(\boldsymbol{x})$ is composed of partial waves of the relative coordinates as in Eq. (9.23). In that case one has to treat the spherical harmonics belonging to each relative vector explicitly, and one has to cope with the complicated angular momentum algebra implied (cf. App. G.1). It should be noted, however, that the calculation of the matrix elements of this latter type gets greatly simplified if the function $\theta_{lLM_L}(\boldsymbol{x})$ of Eq. (9.23) is expressed as a linear combination of the terms of Eq. (9.32) with appropriate u-vectors.

All matrix elements can be cast into similar closed analytic forms and their numerical computation as functions of the nonlinear parameters is therefore straightforward. The values of K and K' can usually be chosen small in practical cases and the sums over k (and k_i) in Eqs. (F.14), (F.19) and (F.25) are limited to a few terms.

Appendix G

Other matrix elements

G.1 Successive coupling

In the main text we have worked our way through the correlated Gaussian formalism in the 'global vector representation', in which the angle dependent factor of the trial function is the function $\theta_{lLM_L}(\boldsymbol{x})$ defined in Eq. (9.32). Here the formalism of successive coupling is outlined rather sketchily since there is a close analogy between the two cases. We just indicate, without proof, how the formalism changes when the angular function is chosen to be the successively coupled function (9.23),

$$\theta_{lLM_L}(\boldsymbol{x})=\Big[\dots[[\mathcal{Y}_{l_1}(\boldsymbol{x}_1)\mathcal{Y}_{l_2}(\boldsymbol{x}_2)]_{L_{12}}\,\mathcal{Y}_{l_3}(\boldsymbol{x}_3)]_{L_{123}}\cdots\mathcal{Y}_{l_{A-1}}(\boldsymbol{x}_{A-1})\Big]_{LM_L},$$
$$\text{with } l=\{l_1,\dots,l_{A-1},L_{12},L_{123},\dots\}. \tag{G.1}$$

Further details can be found in App. A.2 of the book [187].

With this the basis function (9.53) will be

$$\psi=\mathcal{A}\{f_{lLM_L}(\boldsymbol{x},A)\chi_{SM_S}\eta_{TM_T}\}, \tag{G.2}$$

where

$$f_{lLM_L}(\boldsymbol{x},A)=\theta_{lLM_L}(\boldsymbol{x})\mathrm{e}^{-\frac{1}{2}\tilde{\boldsymbol{x}}A\boldsymbol{x}}, \tag{G.3}$$

with $\theta_{lLM_L}(\boldsymbol{x})$ of Eq. (G.1). The function $f_{lLM_L}(\boldsymbol{x},A)$ is a linear combination of terms:

$$f_{lLM_L}(\boldsymbol{x},A)=\exp\left(-\frac{1}{2}\tilde{\boldsymbol{x}}A\boldsymbol{x}\right)\sum_{\kappa=\{m_1,m_2,\dots,m_{A-1}\}}c_\kappa\prod_{i=1}^{A-1}\mathcal{Y}_{l_im_i}(\boldsymbol{x}_i), \tag{G.4}$$

where

$$c_\kappa=\langle l_1m_1l_2m_2|L_{12}\,m_1+m_2\rangle\langle L_{12}\,m_1+m_2\;l_3m_3|L_{123}\,m_1+m_2+m_3\dots\rangle$$
$$\times\langle L_{12\dots A-2}\,m_1+m_2+\dots+m_{A-2}\,l_{A-1}m_{A-1}|LM_L\rangle \tag{G.5}$$

is a combination of Clebsch-Gordan coefficients, which couple the orbital angular momenta to the specified quantum numbers.

A term of Eq. (G.4) can be expressed in terms of the generating function $g[(\alpha|\boldsymbol{t});\boldsymbol{x},A]$ as

$$\exp\left(-\frac{1}{2}\tilde{\boldsymbol{x}}A\boldsymbol{x}\right)\prod_{i=1}^{A-1}\mathcal{Y}_{l_i m_i}(\boldsymbol{x}_i)$$
$$=\left\{\left(\prod_{i=1}^{A-1}\frac{1}{C_{l_i m_i}}\frac{\partial^{l_i}}{\partial\alpha_i{}^{l_i}}\frac{\partial^{l_i-m_i}}{\partial\tau_i{}^{l_i-m_i}}\right)g[(\alpha|\boldsymbol{t});\boldsymbol{x},A]\right\}\Bigg|_{\substack{\alpha_1=0,\ldots,\alpha_{A-1}=0\\ \tau_1=0,\ldots,\tau_{A-1}=0}}, \tag{G.6}$$

where $g(\varsigma;\boldsymbol{x},A)$ has the same functional form as before [cf. Eq. (9.59)]

$$g(\varsigma;\boldsymbol{x},A)=\exp\left(-\frac{1}{2}\tilde{\boldsymbol{x}}A\boldsymbol{x}+\tilde{\varsigma}\boldsymbol{x}\right), \tag{G.7}$$

but its parameter is $\varsigma=(\alpha|\boldsymbol{t})$. The set $\boldsymbol{t}$ of Cartesian vectors $\boldsymbol{t}_j$ is defined by $\boldsymbol{t}_j=(1-\tau_j{}^2, i(1+\tau_j{}^2), -2\tau_j)$, and the symbol $(\alpha|\boldsymbol{t})$ is a column vector whose ith element is $\alpha_i\boldsymbol{t}_i$. The coefficient C_{lm} is

$$C_{lm}=(-2)^l\, l!\sqrt{\frac{4\pi(l-m)!}{(2l+1)(l+m)!}}. \tag{G.8}$$

The derivation of the generating function relationship (G.6) requires the generating function relationship for the solid spherical harmonic $\mathcal{Y}_{lm}(\boldsymbol{r})$:

$$\left(\frac{\partial^l}{\partial\alpha_i^l}\frac{\partial^{l-m}}{\partial\tau_i^{l-m}}e^{\alpha_i\boldsymbol{t}_i\cdot\boldsymbol{r}}\right)\Bigg|_{\substack{\alpha_i=0\\ \tau_i=0}}=C_{lm}\mathcal{Y}_{lm}(\boldsymbol{r}). \tag{G.9}$$

Matrix elements in the basis of successive coupling may be obtained in much the same way as in the global basis:

1. calculate the matrix element in the generator coordinate basis;
2. transform the generator coordinate basis into the generating function basis by the integral transformation (9.74);
3. differentiate the matrix element in the generating function basis according to Eq. (G.6) to get the matrix elements between the terms in Eq. (G.4);
4. combine them with coefficients c_κ as prescribed in Eq. (G.4).

G.2 Unnatural parity states

The aim of this section is to show how to calculate matrix elements involving unnatural parity states. Two simplifications are adopted with respect to the rest of the book. First, we restrict ourselves to basis states of fully stretched l-coupling, i.e., to those with $K=0$ [cf. Eq. (9.32) and the remark after Eq. (9.35)]. This is a physically relevant restriction, which we adopt to keep the balance between the amount of formalism and the importance of the problem in the context of this book. Second, we will calculate matrix elements between spatial functions $f_{KLM_L}(\boldsymbol{u},\boldsymbol{x},A)$ (with $K=0$) of Eq. (9.55), rather than between full basis elements $[\psi_{L;S;TM_T}(\boldsymbol{x})]_{JM}$ of Eq. (9.53). This implies that we will not treat the intrinsic degrees of freedom of the nucleons and disregard antisymmetry as well. This may be viewed as stopping short of completing the derivation. But the aim is just to give the rudimentary outline of this formalism. Further details can be found elsewhere [531].

One should start with a generating function g which has the same form as that of Eq. (9.59), but its parameter ς has been replaced by $\alpha_1\boldsymbol{e}_1u_1+\alpha_2\boldsymbol{e}_2u_2$:

$$g(\alpha_1\boldsymbol{e}_1u_1+\alpha_2\boldsymbol{e}_2u_2;\boldsymbol{x},A)=\exp\left[-\frac{1}{2}\tilde{\boldsymbol{x}}A\boldsymbol{x}+(\alpha_1\boldsymbol{e}_1\widetilde{u_1}+\alpha_2\boldsymbol{e}_2\widetilde{u_2})\boldsymbol{x}\right]. \tag{G.10}$$

Here α_i are parameters, $\boldsymbol{e}_i$ are unit vectors ($|\boldsymbol{e}_i|=1$) and u_i are two column vectors of $A-1$ elements ($i=1,2$). The vectors u_i define the global vectors $\boldsymbol{v}_i$ as $\boldsymbol{v}_i=\widetilde{u}_i\boldsymbol{x}=\sum_{j=1}^{A-1}(u_i)_j\boldsymbol{x}_j$ ($j=1,2$). As a generalization of Eq. (9.60), the basis function of total orbital angular momentum L can be defined by

$$\begin{aligned}
&f_{0(L_1L_2)LM}(u_1,u_2,\boldsymbol{x},A)\\
&\quad\equiv\exp\left(-\frac{1}{2}\tilde{\boldsymbol{x}}A\boldsymbol{x}\right)[\mathcal{Y}_{L_1}(\boldsymbol{v}_1)\mathcal{Y}_{L_2}(\boldsymbol{v}_2)]_{LM}\\
&\quad=\frac{1}{B_{0L_1}B_{0L_2}}\int d\hat{\mathbf{e}}_1\int d\hat{\mathbf{e}}_2\,[Y_{L_1}(\hat{\mathbf{e}}_1)Y_{L_2}(\hat{\mathbf{e}}_2)]_{LM}\\
&\quad\times\left[\frac{d^{L_1+L_2}}{d\alpha_1{}^{L_1}d\alpha_2{}^{L_2}}g(\alpha_1\boldsymbol{e}_1u_1+\alpha_2\boldsymbol{e}_2u_2;\boldsymbol{x},A)\right]\Bigg|_{\alpha_1=0,\alpha_2=0},
\end{aligned} \tag{G.11}$$

where B_{kL} is defined in Eq. (9.35).

To make the formula more explicit, one has to expand the exponential function $\exp[(\alpha_1\boldsymbol{e}_1\widetilde{u_1}+\alpha_2\boldsymbol{e}_2\widetilde{u_2})\boldsymbol{x}]$ into its power series. The derivatives in Eq. (G.11) cause that only the terms with $(\alpha_1\boldsymbol{e}_1)^{L_1}$ and $(\alpha_2\boldsymbol{e}_2)^{L_2}$ can differ from 0. These terms contain $Y_{l_1}(\hat{\mathbf{e}}_1)$ and $Y_{l_2}(\hat{\mathbf{e}}_2)$ with l_1,l_2 values up to L_1 and L_2, respectively, since each vector $\boldsymbol{e}_i$ carries angular momentum 1, and the L_i-power is a combination of all values up to power L_i. The integrations eliminate all terms but those with $l_1=L_1$ and $l_2=L_2$, respectively, i.e., just the term with fully stretched coupling gives a non-vanishing contribution. This special property makes the calculation of the matrix element fairly simple.

We evaluate matrix elements of the unit operator, the kinetic-energy and the potential-energy operators. They are rotationally invariant, so their matrix elements are diagonal in LM.

The use of Eq. (G.11) results in

$$\begin{aligned}\langle f'|O|f\rangle &\equiv \langle f_{0(L_3L_4)LM}(u_3,u_4,x,A')|O|f_{0(L_1L_2)LM}(u_1,u_2,x,A)\rangle \\ &= \left(\prod_{i=1}^{4}\frac{1}{B_{0L_i}}\int d\hat{e}_i\right)[Y_{L_3}(\hat{e}_3)Y_{L_4}(\hat{e}_4)]^*_{LM}[Y_{L_1}(\hat{e}_1)Y_{L_2}(\hat{e}_2)]_{LM} \\ &\times\left(\prod_{i=1}^{4}\frac{d^{L_i}}{d\alpha_i{}^{L_i}}\langle g(\varsigma';x,A')|O|g(\varsigma;x,A)\rangle\Big|_{\alpha_i=0}\right),\end{aligned} \tag{G.12}$$

where $\varsigma=\alpha_3e_3u_3+\alpha_4e_4u_4$ and $\varsigma'=\alpha_1e_1u_1+\alpha_2e_2u_2$ have been introduced. The matrix element between the generating functions can be obtained by using Eqs. (9.61) and (9.63).

The overlap between the generating functions is obtained as

$$\langle g(\varsigma';x,A')|g(\varsigma;x,A)\rangle = \left[\frac{(2\pi)^{A-1}}{\det B}\right]^{3/2} e^{\frac{1}{2}\tilde{z}B^{-1}z}, \tag{G.13}$$

where

$$B=A+A', \qquad z=\sum_{i=1}^{4}\alpha_ie_iu_i. \tag{G.14}$$

The exponent of Eq. (G.13) can be expressed more explicitly as

$$\frac{1}{2}\tilde{z}B^{-1}z=\frac{1}{2}\sum_i\rho_{ii}\alpha_i^2+\sum_{i<j}\alpha_i\alpha_j\rho_{ij}(e_i\cdot e_j) \tag{G.15}$$

with $\rho_{ij}=\tilde{u}_iB^{-1}u_j$. The first term of Eq. (G.15) can be dropped since it has no effect on Eq. (G.12). In fact, the terms that come in Eq. (G.12) from $\frac{1}{2}\sum_i\rho_{ii}\alpha_i^2$ will vanish after the differentiations either when $\alpha_i=0$ is taken or upon the subsequent angular-momentum projection. Then the expansion of the exponential function gives rise to terms like

$$\sum_{p_{ij}}\prod_{i<j}\frac{\rho_{ij}^{p_{ij}}}{p_{ij}!}[\alpha_i\alpha_j(e_i\cdot e_j)]^{p_{ij}}. \tag{G.16}$$

After the differentiations, the only terms that contribute to Eq. (G.12) are those with all p_{ij} non-negative and all satisfying the conditions

$$\begin{aligned}&p_{12}+p_{13}+p_{14}=L_1, \qquad p_{12}+p_{23}+p_{24}=L_2,\\ &p_{13}+p_{23}+p_{34}=L_3, \qquad p_{14}+p_{24}+p_{34}=L_4.\end{aligned} \tag{G.17}$$

These conditions imply that only two of p_{ij}, e.g., p_{12} and p_{13}, are independent and their values are limited by the inequalities

$$p_{12} \geq \max\left[0, \frac{1}{2}(L_1 + L_2 - L_3 - L_4)\right],$$
$$p_{13} \geq \max\left[0, \frac{1}{2}(L_1 - L_2 + L_3 - L_4)\right],$$
$$p_{12} + p_{13} \leq \min\left[L_1, \frac{1}{2}(L_1 + L_2 + L_3 - L_4)\right]. \tag{G.18}$$

As was noted above, $(e_i \cdot e_j)^{p_{ij}}$ gives a contribution only through its term that carries the maximum angular momentum, p_{ij}. Expanding $(e_i \cdot e_j)^{p_{ij}}$, and writing explicitly only this term, we have

$$(e_i \cdot e_j)^{p_{ij}} = (-1)^{p_{ij}}\sqrt{2p_{ij}+1}B_{0p_{ij}}[Y_{p_{ij}}(\hat{e}_i)Y_{p_{ij}}(\hat{e}_j)]_{00} + \cdots \tag{G.19}$$

Finally, to locate the terms that do not vanish when projected onto $[Y_{L_1}(\hat{e}_1)Y_{L_2}(\hat{e}_2)]^*_{LM}$ and $[Y_{L_3}(\hat{e}_3)Y_{L_4}(\hat{e}_4)]_{LM}$ in Eq. (G.12), we put together and recouple the functions $Y_{p_{ij}}$ with the same arguments. This is done in two steps. First $[Y_{p_{13}}(\hat{e}_1)Y_{p_{13}}(\hat{e}_3)]_{00}$, $[Y_{p_{14}}(\hat{e}_1)Y_{p_{14}}(\hat{e}_4)]_{00}$, $[Y_{p_{23}}(\hat{e}_2)Y_{p_{23}}(\hat{e}_3)]_{00}$ and $[Y_{p_{24}}(\hat{e}_2)Y_{p_{24}}(\hat{e}_4)]_{00}$ are recoupled. It can be shown that the coefficient $X_L(p_{ij})$ of the only non-vanishing term, $[[Y_{p_{13}+p_{14}}(\hat{e}_1)Y_{p_{23}+p_{24}}(\hat{e}_2)]_L[Y_{p_{13}+p_{23}}(\hat{e}_3)Y_{p_{14}+p_{24}}(\hat{e}_4)]_L]_{00}$, is

$$\begin{aligned} X_L(p_{ij}) = {} & \sqrt{\frac{2L+1}{(2p_{13}+1)(2p_{14}+1)(2p_{23}+1)(2p_{24}+1)}} \\ & \times C(p_{13}p_{14}; L_1 - p_{12})C(p_{23}p_{24}; L_2 - p_{12}) \\ & \times C(p_{13}p_{23}; L_3 - p_{34})C(p_{14}p_{24}; L_4 - p_{34}) \\ & \times \begin{bmatrix} p_{13} & p_{14} & L_1 - p_{12} \\ p_{23} & p_{24} & L_2 - p_{12} \\ L_3 - p_{34} & L_4 - p_{34} & L \end{bmatrix}, \end{aligned} \tag{G.20}$$

where

$$C(l_1 l_2; l_3) = \sqrt{\frac{(2l_1+1)(2l_2+1)}{4\pi(2l_3+1)}}\langle l_1 0 l_2 0 | l_3 0\rangle \tag{G.21}$$

come from the coupling of two spherical harmonics with the same argument and $[\cdots]$ is the unitary $9j$ coefficient defined by

$$\begin{bmatrix} j_1 & j_2 & J_{12} \\ j_3 & j_4 & J_{34} \\ J_{13} & J_{24} & J \end{bmatrix} \equiv \langle (j_1 j_2)J_{12}, (j_3 j_4)J_{34}; JM | (j_1 j_3)J_{13}, (j_2 j_4)J_{24}; JM\rangle$$

$$= \sqrt{(2J_{12}+1)(2J_{34}+1)(2J_{13}+1)(2J_{24}+1)} \begin{Bmatrix} j_1 & j_2 & J_{12} \\ j_3 & j_4 & J_{34} \\ J_{13} & J_{24} & J \end{Bmatrix}, \tag{G.22}$$

where $\{\cdots\}$ is the conventional $9j$ symbol [cf. Eq. (9.88)]. Then the coupling in $[[Y_{p_{13}+p_{14}}(\hat{e}_1)\,Y_{p_{23}+p_{24}}(\hat{e}_2)]_L[Y_{p_{13}+p_{23}}(\hat{e}_3)Y_{p_{14}+p_{24}}(\hat{e}_4)]_L]_{00}$ is to be rearranged and the functions of the same arguments are to be coupled with those in $[Y_{p_{12}}(\hat{e}_1)Y_{p_{12}}(\hat{e}_2)]_{00}$ and $[Y_{p_{34}}(\hat{e}_3)Y_{p_{34}}(\hat{e}_4)]_{00}$. We finally obtain

$$\begin{aligned}\langle f'|f\rangle = {} & \frac{(-1)^{L_1+L_2}}{\sqrt{2L+1}}\left(\prod_{i=1}^{4}\frac{L_i!}{B_{0L_i}}\right)\left(\frac{(2\pi)^{A-1}}{\det B}\right)^{3/2}\\ & \times\sum_{p_{ij}}\left(\prod_{i<j}(-1)^{p_{ij}}\sqrt{2p_{ij}+1}B_{0p_{ij}}\frac{\rho_{ij}^{p_{ij}}}{p_{ij}!}\right)\\ & \times C(p_{12}\,L_1-p_{12};L_1)C(p_{12}\,L_2-p_{12};L_2)\\ & \times C(p_{34}\,L_3-p_{34};L_3)C(p_{34}\,L_4-p_{34};L_4)X_L(p_{ij})\\ & \times\begin{bmatrix} p_{12} & p_{12} & 0\\ L_1-p_{12} & L_2-p_{12} & L\\ L_1 & L_2 & L\end{bmatrix}\begin{bmatrix} p_{34} & p_{34} & 0\\ L_3-p_{34} & L_4-p_{34} & L\\ L_3 & L_4 & L\end{bmatrix}.\end{aligned} \tag{G.23}$$

The intrinsic kinetic energy can be expressed as $T-T_{\rm cm}=\frac{1}{2}\tilde{\pi}\Lambda\pi$, where π_j is $-i\hbar\frac{\partial}{\partial x_j}$ and Λ is an $(A-1)\times(A-1)$ diagonal matrix [cf. Eq. (B.20)] or at least symmetric if x are more general coordinates. The matrix element of the kinetic energy between the generating functions is given by

$$\begin{aligned}&\langle g(\varsigma';x,A')|\frac{1}{2}\tilde{\pi}\Lambda\pi|g(\varsigma;x,A)\rangle\\ &=\frac{\hbar^2}{2}[3\mathrm{Tr}(B^{-1}A'\Lambda A)-\tilde{d}\Lambda d]\langle g(\varsigma';x,A')|g(\varsigma;x,A)\rangle,\end{aligned} \tag{G.24}$$

with

$$d=A'B^{-1}\varsigma'-AB^{-1}\varsigma=\sum_{i=1}^{4}A_iB^{-1}\alpha_i e_i u_i. \tag{G.25}$$

Here $A_1=A_2=A'$ and $A_3=A_4=-A$. In $\tilde{d}\Lambda d$, the only $\alpha_i\alpha_j$ terms that contribute are those with $i\neq j$. Noting this, we can get the kinetic-energy matrix elements in exactly the same way as the overlap, and the result is

$$\langle f'|T-T_{\rm cm}|f\rangle=\frac{\hbar^2}{2}\left[3\mathrm{Tr}(B^{-1}A'\Lambda A)-\sum_{i<j}Q_{ij}\frac{\partial}{\partial\rho_{ij}}\right]\langle f'|f\rangle, \tag{G.26}$$

where

$$Q_{ij}=2\tilde{u}_iB^{-1}A_i\Lambda A_jB^{-1}u_j. \tag{G.27}$$

Since the ρ_{ij}-dependence of $\langle f'|f\rangle$ is simple, the ρ_{ij}-derivative is easily calculated.

At last, we consider the matrix element of a central potential of the Gaussian form. It is useful to express the relative distance vector between particles k and l in terms of $\boldsymbol{x}$:

$$\boldsymbol{r}_k - \boldsymbol{r}_l = \sum_{i=1}^{A-1} w_i \boldsymbol{x}_i = \tilde{w}\boldsymbol{x}. \tag{G.28}$$

Then the matrix element of the Gaussian potential between the generating functions becomes

$$\begin{aligned}&\langle g(\boldsymbol{\varsigma}'; \boldsymbol{x}, A')|\mathrm{e}^{-\frac{1}{2}\mu(\boldsymbol{r}_k-\boldsymbol{r}_l)^2}|g(\boldsymbol{\varsigma}; \boldsymbol{x}, A)\rangle \\ &= \left(\frac{c}{c+\mu}\right)^{3/2} \exp\left[-\frac{1}{2}\frac{c\mu}{c+\mu}(\tilde{w}B^{-1}\boldsymbol{z})^2\right] \langle g(\boldsymbol{\varsigma}'; \boldsymbol{x}, A')|g(\boldsymbol{\varsigma}; \boldsymbol{x}, A)\rangle,\end{aligned} \tag{G.29}$$

where the constant c, which depends on the particle labels k and l through w, is defined by $c^{-1} = \tilde{w}B^{-1}w$. We note that in

$$\frac{1}{2}(\tilde{w}B^{-1}\boldsymbol{z})^2 = \frac{1}{2}\sum_i (\tilde{w}B^{-1}u_i)^2 \alpha_i^2 + \sum_{i<j} (\tilde{w}B^{-1}u_i)(\tilde{w}B^{-1}u_j)\alpha_i\alpha_j(\boldsymbol{e}_i \cdot \boldsymbol{e}_j), \tag{G.30}$$

again, the first term of $\frac{1}{2}(\tilde{w}B^{-1}\boldsymbol{z})^2$ does not contribute to the matrix element. Then, by comparing Eq. (G.13) with Eq. (G.29) and Eq. (G.15) with Eq. (G.30), one can recognize that the matrix elements of the Gaussian potential are given by the same formula as gives the overlap matrix element (G.23), with the following modification:

1. multiply Eq. (G.23) by $\left(\frac{c}{c+\mu}\right)^{3/2}$;
2. replace ρ_{ij} with $\rho_{ij} - \frac{c\mu}{c+\mu}(\tilde{w}B^{-1}u_i)(\tilde{w}B^{-1}u_j)$.

G.3 Calculation of the amplitudes related to clustering

The spectroscopic amplitudes (11.11), clustering amplitudes (11.28), their Fourier transforms (11.14), (11.29) and norm squares (11.26), (11.32) contain overlaps of the nuclear wave function with the test state $\Psi_{\boldsymbol{R}_1,\ldots,\boldsymbol{R}_{n-1}}$. A test state, which is defined in Eq. (11.9) or (11.31), contains one or more δ-functions of the coordinates describing a particular clustering (or fragmentation) pattern. The clustering amplitudes and the amounts of clustering contain, in addition, the norm operator $\mathcal{N}$

defined in (11.17). The calculation of these quantities requires specific considerations, but is quite straightforward.

A general recipe for the treatment of the Dirac-δ's in the overlap is to substitute Eq. (11.34) or (11.126). This will cause Gaussian wave packets $\varphi_{\boldsymbol{S}_1}, \ldots, \varphi_{\boldsymbol{S}_{n-1}}$ of some intercluster relative coordinates $\boldsymbol{q}_1, \ldots, \boldsymbol{q}_{n-1}$ to appear. This can be identically rewritten such that $\boldsymbol{q}_n$ may also appear by multiplication with the c.m. wave packet $\varphi_{\boldsymbol{S}_n}$ and integration over $\boldsymbol{S}_n$ [cf. Eq. (11.85)]. The exponent in the product of these Gaussians is a diagonal quadratic form of $\{\boldsymbol{q}_1, \ldots, \boldsymbol{q}_n\}$, which can be transformed into a quadratic form of the cluster c.m. vectors $\{\bar{\boldsymbol{t}}_1, \ldots, \bar{\boldsymbol{t}}_n\}$ via the inverse, $\bar{\boldsymbol{t}} = W^{-1}\boldsymbol{q}$, of Eq. (B.12). The resulting exponential function is a correlated Gaussian for the cluster c.m. motions. The product of this with the cluster internal wave functions in $\Psi_{\boldsymbol{R}_1,\ldots,\boldsymbol{R}_{n-1}}$ is a correlated Gaussian with a quadratic form of $\{\boldsymbol{t}_1, \ldots, \boldsymbol{t}_A\}$ in the exponent. This set of coordinates is defined in Eqs. (B.2) and (B.3), and it consists of the intrinsic cluster Jacobi coordinates and the cluster c.m. coordinates. This set transforms into the Cartesian set $\boldsymbol{r}$ as $\boldsymbol{r} = T^{-1}\boldsymbol{t}$, which is just Eq. (B.7) inverted. With that transformation performed, $\Psi_{\boldsymbol{R}_1,\ldots,\boldsymbol{R}_{n-1}}$ is expressed in terms of Slater determinants, thus the evaluation of an overlap with $\Psi_{\boldsymbol{R}_1,\ldots,\boldsymbol{R}_{n-1}}$ fits into the general scheme of the calculation of matrix elements. The additional integrals implied by the Dirac-δ's (11.34) and by the elimination of the $\boldsymbol{S}_n$-dependence via Eq. (11.85) are to be performed as usual, with the aid of Eq. (9.61).

The amount of clustering $\mathfrak{P}$ can be directly calculated as $\mathfrak{P} = \langle \phi | g \rangle$ [cf. Eq.(11.33)] provided not only the covariant component g (i.e., the spectroscopic amplitude) has been determined but also the contravariant component ϕ. From the covariant amplitudes g the contravariant amplitudes can be calculated in various ways. One possibility is to cast Eq. (11.17) into the form of an inhomogeneous integral equation by substituting $\mathcal{N} = 1 - \mathcal{K}$ [cf. Eq. (11.64)]. For a binary fragmentation, this integral equation will look like

$$\phi(\boldsymbol{R}) = g(\boldsymbol{R}) + \int d\boldsymbol{R}' K(\boldsymbol{R}, \boldsymbol{R}') \phi(\boldsymbol{R}'), \tag{G.31}$$

which can be solved iteratively, starting with $\phi(\boldsymbol{R}) \approx g(\boldsymbol{R})$ [392]. Another method is using the inverted equation (11.19) directly, which is applicable if the spectral decomposition (11.51) of $\mathcal{N}$ is known.

The methods discussed so far are quite general; they are not specific to the correlated Gaussian cluster model. Relying more on the correlated Gaussian formalism, one can calculate the clustering properties with almost no effort beyond what is anyway needed for the solution of the dynamical problem. It would be too complicated to write down generally applicable explicit formulae of this kind. Rather, for the spectroscopic amplitude, we show explicit formulae, complete with angular-momentum algebra, for a simplest non-trivial particular case [365]: an $a_1+a_2+a_3$ three-cluster system with three possible arrangements. For the amount of clustering, the formalism to be presented will be more general but less explicit.

Let us first see how to calculate a spectroscopic amplitude. Consider the fragmentation of a three-cluster system $a_1+a_2+a_3$ into fragments $(a_1a_2)+a_3$. Let the model space be defined by the expansion

$$\Psi_{JM} = \sum_i C_i \psi^{(\nu_1^i\nu_2^i),(a^i)}_{\{[(S_1^iS_2^i)S_0^iS_3^i]S^i,(l_1^il_2^i)L^i\}JM}(\xi_1^i,\xi_2^i,\xi_3^i,\boldsymbol{q}_1^i,\boldsymbol{q}_2^i), \tag{G.32}$$

where the basis state is defined, with the label i dropped, as follows:

$$\psi^{(\nu_1\nu_2),(a)}_{\{[(S_1S_2)S_0S_3]S,(l_1l_2)L\}JM}(\xi_1,\xi_2,\xi_3,\boldsymbol{q}_1,\boldsymbol{q}_2)$$
$$= \mathcal{A}_{123}\left\{\left[\left[[\Phi^{(a)}_{S_1}(\xi_1)\Phi^{(a)}_{S_2}(\xi_2)]_{S_0}\Phi^{(a)}_{S_3}(\xi_3)\right]_S\left[\psi^{(\nu_1)}_{0l_1}(\boldsymbol{q}_1)\psi^{(\nu_2)}_{0l_2}(\boldsymbol{q}_2)\right]_L\right]_{JM}\right\}, \tag{G.33}$$

where $\mathcal{A}_{123}$ is the intercluster antisymmetrizer between clusters a_1, a_2 and a_3 [cf. Eq. (D.7)], the superscript $a \equiv a^i$ specifies a permutation of $\{a_1a_2a_3\}$, the functions $\Phi^{(a)}_{S_1M_{S1}}(\xi_1)$, $\Phi^{(a)}_{S_2M_{S2}}(\xi_2)$, $\Phi^{(a)}_{S_3M_{S3}}(\xi_3)$ are the cluster intrinsic states, S_1, S_2, S_3 are the cluster spins, $\psi^{(\nu_1)}_{0l_1m_1}(\boldsymbol{q}_1)$, $\psi^{(\nu_2)}_{0l_2m_2}(\boldsymbol{q}_2)$ are nodeless s.p. HO wave functions [cf. Eq. (9.51)], of parameters ν_1, ν_2, considered to be functions of the conventional intercluster Jacobi coordinates $\boldsymbol{q}_1$, $\boldsymbol{q}_2$ [cf. Eq.(B.10)]. This function is a correlated Gaussian basis function in the successive coupling [cf. App. G.1].

Let the intrinsic wave function of fragment (a_1a_2) be also given by the correlated Gaussian cluster model: $\Psi'_{I'M'} = \sum_j C'_j\phi_j$, with a basis state $\phi_j \equiv \phi^{(\nu_1^j)}_{[(S_1S_2)S_0^j,l_1^j]I'M'}$ defined (with the label j temporarily dropped) by

$$\phi^{(\nu_1)}_{[(S_1S_2)S_0,l_1]I'M'}(\xi_1,\xi_2,\boldsymbol{q}_1) = \mathcal{A}_{12}\left\{\left[[\Phi_{S_1}(\xi_1)\Phi_{S_2}(\xi_2)]_{S_0}\psi^{(\nu_1)}_{0l_1}(\boldsymbol{q}_1)\right]_{I'M'}\right\}. \tag{G.34}$$

Here the operator $\mathcal{A}_{12}$ antisymmetrizes between a_1 and a_2. Note that only S_0^j, l_1^j, ν_1^j are labelled by j since S_1, S_2, ξ_1, ξ_2, $\boldsymbol{q}_1$ are the same for all terms. The test function with appropriate angular-momentum coupling is

$$\Psi_{\boldsymbol{R}_2} = \mathcal{A}_{(12)3}\left\{[[\Psi'_{I'}\Phi_{S_3}(\xi_3)]_I\delta_{l_2}(\boldsymbol{R}_2-\boldsymbol{q}_2)]_{JM}\right\}, \tag{G.35}$$

with $\delta_{lm}(\boldsymbol{R}-\boldsymbol{q}) = (Rq)^{-1}\delta(R-q)Y_{lm}(\hat{q})$, and the amplitude is defined as

$$\begin{aligned} g_{l_2}(R_2) &= {}_{R_2}\langle\Psi_{\boldsymbol{R}_2}|\Psi\rangle \\ &= {}_{R_2}\langle\mathcal{A}_{(12)3}\{[[\Psi'_{I'}\Phi_{S_3}]_I\delta_{l_2}(\boldsymbol{R}_2-\boldsymbol{q}_2)]_{JM}\}|\Psi_{JM}\rangle. \end{aligned} \tag{G.36}$$

The operator $\mathcal{A}_{(12)3}$ antisymmetrizes between fragments (a_1a_2) and a_3.

The coupling sequences in the bra and ket are not the same. In the wave function Ψ_{JM} it is $\{[(S_1S_2)S_0S_3](l_1l_2)L\}JM$, while in $\Psi_{\boldsymbol{R}_2}$ it is $\{[[(S_1S_2)S_0l_1]I'S_3]Il_2\}JM$. The coupling $\{[(S_0l_1)I'S_3]Il_2\}JM$ in $\Psi_{\boldsymbol{R}_2}$ has to be changed to $[(S_0S_3)S(l_1l_2)L]JM$. This can be done in two steps: $[(S_0l_1)I'S_3]I \to [(S_0S_3)Sl_1]I$, then $[(Sl_1)Il_2]J \to [S(l_1l_2)L]J$ (cf. Fig. G.1):

$$\begin{aligned} &\mathcal{A}_{123}\left\{\left[\left[\left[[\Phi_{S_1}\Phi_{S_2}]_{S_0}\psi_{0l_1}^{(\nu_1)}\right]_{I'}\Phi_{S_3}\right]_I\delta_{l_2}(\boldsymbol{R}_2-\boldsymbol{q}_2)\right]_{JM}\right\} \\ &\quad = \sum_{SL}(-1)^{S_0+I-I'-S}U(l_1S_0IS_3;I'S)U(Sl_1Jl_2;IL) \\ &\quad\quad \times \mathcal{A}_{123}\left\{\left[[[\Phi_{S_1}\Phi_{S_2}]_{S_0}\Phi_{S_3}]_S\left[\psi_{0l_1}^{(\nu_1)}\delta_{l_2}(\boldsymbol{R}_2-\boldsymbol{q}_2)\right]_L\right]_{JM}\right\}, \end{aligned} \tag{G.37}$$

where U are unitary Racah coefficients, which may be expressed in terms of the $6j$ symbols as

$$\begin{aligned} U(j_1j_2Jj_3;J_{12}J_{23}) &= \langle(j_1j_2)J_{12},j_3;JM|j_1,(j_2j_3)J_{23};JM\rangle \\ &= (-1)^{j_1+j_2+J+j_3}\sqrt{(2J_{12}+1)(2J_{23}+1)}\begin{Bmatrix} j_1 & j_2 & J_{12} \\ j_3 & J & J_{23} \end{Bmatrix}. \end{aligned} \tag{G.38}$$

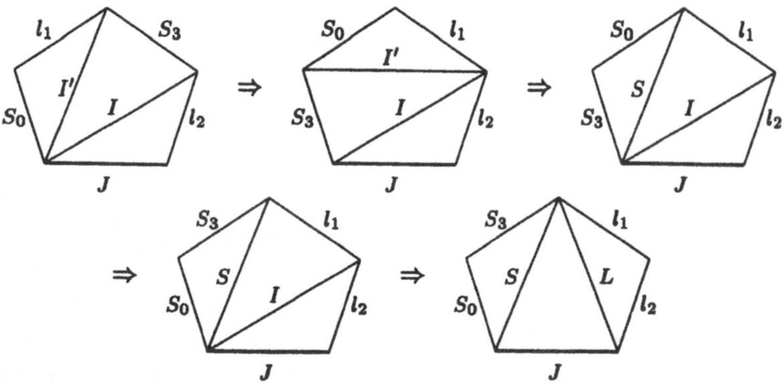

Figure G.1: Graphical representation of the double recoupling and change of sequence of angular momenta.

In Eq. (G.37) the factor $(-1)^{S_0+I-I'-S}U(l_1S_0IS_3;I'S)$ comes from the recoupling in the first step and $U(Sl_1Jl_2;IL)$ from the second step. Substituting Eq. (G.37) into Eq. (G.36) yields (with the labels i and j recovered)

$$\frac{g_{l_2}(R_2)}{R_2} = \sum_{ji} C_j^{l*} C_i \sum_{S^iL^iSL} (-1)^{S_0+I-I'-S}\delta_{SS^i}\delta_{LL^i} \times U(l_1^jS_0^jIS_3;I'S)U(Sl_1^jJl_2;IL)\left(\mathcal{N}_{ji}\psi_{0l_2^i}^{(\nu_2^i)}\right)(\boldsymbol{R}_2), \quad \text{(G.39)}$$

where the kernel of integral operator $\mathcal{N}_{ji}$ is

$$\begin{aligned} N_{ji}(\boldsymbol{R}_2,\boldsymbol{R}_2') = \Big\langle \mathcal{A}_{123}\Big\{\Big[\Big[[\Phi_{S_1}(\xi_1)\Phi_{S_2}(\xi_2)]_{S_0^j}\Phi_{S_3}(\xi_3)\Big]_S \\ \times\Big[\psi_{0l_1^j}^{(\nu_1^j)}(\boldsymbol{q}_1)\delta_{l_2}(\boldsymbol{R}_2-\boldsymbol{q}_2)\Big]_L\Big]_{JM}\Big\}\Big| \\ \times\mathcal{A}_{123}\Big\{\Big[\Big[[\Phi_{S_1^i}^{(a^i)}(\xi_1^i)\Phi_{S_2^i}^{(a^i)}(\xi_2^i)]_{S_0^i}\Phi_{S_3^i}^{(a^i)}(\xi_3^i)\Big]_{S^i} \\ \times\Big[\psi_{0l_1^i}^{(\nu_1^i)}(\boldsymbol{q}_1^i)\delta_{l_2^i}(\boldsymbol{R}_2'-\boldsymbol{q}_2^i)\Big]_{L^i}\Big]_{JM}\Big\}\Big\rangle. \end{aligned} \quad \text{(G.40)}$$

Equation (G.39) has a remarkably transparent form: it shows that the basic ingredient of g is an asymmetric matrix element of a norm operator. One can express this matrix element in terms of the overlap matrix of the basis functions, $\langle\psi_{0l_2'}^{(\nu_2')}|\mathcal{N}_{ji}|\psi_{0l_2}^{(\nu_2)}\rangle$. To this end, one has to define a unit operator expressed in terms of $\{\psi_{0l_2}^{(\nu_2)}\}$. This expression of the unit operator can be obtained by diagonalizing the matrix $\langle\psi_{0l_2'}^{(\nu_2')}|\psi_{0l_2}^{(\nu_2)}\rangle$, i.e., solving its eigenvalue problem,

$$\sum_n \langle\psi_{0l_2^m}^{(\nu_2^m)}|\psi_{0l_2^n}^{(\nu_2^n)}\rangle c_n^{(k)} = a_k c_m^{(k)}. \quad \text{(G.41)}$$

Due to the orthonormality of the eigenvectors, $\sum_m c_m^{(k)*}c_m^{(k')} = \delta_{kk'}$, the state vectors

$$|k\rangle = a_k^{-1/2}\sum_m c_m^{(k)}|\psi_{0l_2^m}^{(\nu_2^m)}\rangle \quad \text{(G.42)}$$

also form an orthonormal set. We can get an approximant to the unit operator by writing

$$1 \approx \sum_k |k\rangle\langle k| = \sum_k a_k^{-1}\sum_{mn} c_m^{(k)}|\psi_{0l_2^m}^{(\nu_2^m)}\rangle\langle\psi_{0l_2^n}^{(\nu_2^n)}|c_n^{(k)*}. \quad \text{(G.43)}$$

Employing this yields

$$\begin{aligned}\left(\mathcal{N}_{ji}\psi_{0l_2^i}^{(\nu_2^i)}\right)(\boldsymbol{R}_2) &= 1\left(\mathcal{N}_{ji}\psi_{0l_2^i}^{(\nu_2^i)}\right)(\boldsymbol{R}_2)\\ &\approx \sum_k a_k^{-1}\sum_{mn} c_m^{(k)}c_n^{(k)*}\psi_{0l_2^m}^{(\nu_2^m)}(\boldsymbol{R}_2)\langle\psi_{0l_2^n}^{(\nu_2^n)}|\mathcal{N}_{ji}|\psi_{0l_2^i}^{(\nu_2^i)}\rangle\\ &= \sum_k a_k^{-1}\sum_{mn} c_m^{(k)}c_n^{(k)*}N_{jn;ii}\,\psi_{0l_2^m}^{(\nu_2^m)}(\boldsymbol{R}_2),\end{aligned} \tag{G.44}$$

where $N_{jn;ii}=\langle\psi_{0l_2^n}^{(\nu_2^n)}|\mathcal{N}_{ji}|\psi_{0l_2^i}^{(\nu_2^i)}\rangle$ is an overlap between the basis states, a submatrix of the overlap matrix between the elements of the basis:

$$\begin{aligned}N_{jn;ii} = \Bigg\langle \mathcal{A}_{123}\bigg\{\Big[\big[[\Phi_{S_1}\Phi_{S_2}]_{S_0^j}\Phi_{S_3}\big]_S\big[\psi_{0l_1^j}^{(\nu_1^j)}\psi_{0l_2^n}^{(\nu_2^n)}\big]_L\Big]_{JM}\bigg\}\Bigg|\\ \times\mathcal{A}_{123}\bigg\{\Big[\big[[\Phi_{S_1^i}^{(a^i)}\Phi_{S_2^i}^{(a^i)}]_{S_0^i}\Phi_{S_3^i}^{(a^i)}\big]_{S^i}\big[\psi_{0l_1^i}^{(\nu_1^i)}\psi_{0l_2^i}^{(\nu_2^i)}\big]_{L^i}\Big]_{JM}\bigg\}\Bigg\rangle.\end{aligned} \tag{G.45}$$

Thus the calculation of the spectroscopic amplitude demands hardly more than calculating the overlap of the basis functions. The only extra work is the diagonalization of the matrix $\langle\psi_{0l_2^m}^{(\nu_2^m)}|\psi_{0l_2^n}^{(\nu_2^n)}\rangle$.

The building blocks of the spectroscopic factor $S_{l_2}=\langle g_{l_2}|g_{l_2}\rangle$ turn out to be $\int d\boldsymbol{R}_2\,(\mathcal{N}_{ji}\psi_{0l_2^i}^{(\nu_2^i)})^*(\boldsymbol{R}_2)(\mathcal{N}_{ji'}\psi_{0l_2^{i'}}^{(\nu_2^{i'})})(\boldsymbol{R}_2)$. This expression can be recognized to be the $\boldsymbol{R}_2$-representation of $\langle\psi_{0l_2^i}^{(\nu_2^i)}|\mathcal{N}_{ij}^\dagger\mathcal{N}_{ji'}|\psi_{0l_2^{i'}}^{(\nu_2^{i'})}\rangle$. Such an expression can again be evaluated by insertion of Eq. (G.43):

$$\begin{aligned}\int d\boldsymbol{R}_2\left(\mathcal{N}_{ji}\psi_{0l_2^i}^{(\nu_2^i)}\right)^*(\boldsymbol{R}_2)\left(\mathcal{N}_{ji'}\psi_{0l_2^{i'}}^{(\nu_2^{i'})}\right)(\boldsymbol{R}_2) &\equiv \langle\psi_{0l_2^i}^{(\nu_2^i)}|\mathcal{N}_{ij}^\dagger\mathcal{N}_{ji'}|\psi_{0l_2^{i'}}^{(\nu_2^{i'})}\rangle\\ &= \sum_k\sum_{mn} N_{ii;jm}^\dagger c_m^{(k)}a_k^{-1}c_n^{(k)*}N_{jn;i'i'}.\end{aligned} \tag{G.46}$$

Because of the truncation of the unit operator, Eq. (G.43), it appears that there is some approximation in this scheme. That is, however, not necessarily the case. The truncation is actually a projection, and a similar projection is already included in the model. Since the repetition of a projection causes no approximation, it is possible that no approximation is actually implied. It has been assumed that the set of states $\{\psi_{0l_2^m}^{(\nu_2^m)}\}$ in Eq. (G.43) is exactly the same as that representing the $(a_1a_2)+a_3$ relative motion in the wave function. The question is then whether the $(a_1a_2)+a_3$ relative motion is adequately covered by the set $\{\psi_{0l_2^m}^{(\nu_2^m)}\}$. If so, then it is likely that it faithfully represents the contribution of other (nonorthogonal) arrangements as well, and, in that case, there is practically no extra approximation implied by Eq. (G.43).

To calculate the amount of some clustering μ, one should define the particular type of clustering through a basis in the μ subspace: $\{\psi_i^\mu(\boldsymbol{r}_1, \ldots, \boldsymbol{r}_{A-1})\}$. Label μ implies some particular arrangement(s) and, possibly, some particular angular momentum coupling(s). For example, the ^{8}Be+n clustering in ^{9}Be implies a T-type arrangement (with the neutron at the bottom of T) with any angular momenta, but a ^{8}Be(g.s.)+n clustering restricts the angular momenta as well. In the correlated Gaussian model the basis for μ may be a subset of the full basis if the sequential coupling is used for the relative motion.

Then the nuclear wave function Ψ and the projection $\Psi^{\text{proj.}} \equiv \Psi^\mu = \mathcal{P}^\mu \Psi$ of state Ψ onto subspace μ can be written as

$$\Psi = \sum_{j\nu} F_j^\nu \psi_j^\nu, \qquad \Psi^\mu = \sum_i f_i^\mu \psi_i^\mu, \tag{G.47}$$

where f_i^μ are coefficients analogous to the contravariant component ϕ in Eq. (11.21) and the functions ψ_j^ν with $\nu \neq \mu$ may be any other basis states. They may or may not belong to other well-defined clusterizations and are, in general, not orthogonal to the basis states ψ_i^μ:

$$N_{ij}^{\mu\nu} \equiv \langle \psi_i^\mu | \psi_j^\nu \rangle \not\propto \delta_{\mu\nu}. \tag{G.48}$$

According to Eqs. (11.33) and (11.22), the amount of μ-clustering can be written as

$$\mathfrak{P}_\mu = \langle \Psi | \mathcal{P}^\mu | \Psi \rangle = \langle \Psi^\mu | \Psi \rangle = \sum_\varrho \sum_{kl} f_k^{\mu *} N_{kl}^{\mu\varrho} F_l^\varrho. \tag{G.49}$$

The coefficients f_i^μ can be determined by minimizing the norm of the component orthogonal to Ψ^μ (cf. p. 293):

$$\delta \langle \Psi - \Psi^\mu | \Psi - \Psi^\mu \rangle = 0, \quad \text{with} \quad \langle \Psi | \Psi \rangle = \langle \Psi^\mu | \Psi^\mu \rangle = 1. \tag{G.50}$$

If Eq. (G.50) is satisfied, then it indeed gives a minimum [284]. Performing the variation, one obtains $\langle \psi_i^\mu | \Psi - \Psi^\mu \rangle = 0$, i.e., $\langle \psi_i^\mu | \Psi^\mu \rangle = \langle \psi_i^\mu | \Psi \rangle$ (for all i), which can be written explicitly by using Eqs. (G.47) and (G.48):

$$\sum_j N_{ij}^{\mu\mu} f_j^\mu = \sum_\nu \sum_j N_{ij}^{\mu\nu} F_j^\nu. \tag{G.51}$$

This is a set of linear equations for f_j^μ. It can be solved formally, resulting in $f_k^\mu = \sum_\nu \sum_{ij} (N^{\mu\mu})_{kj}^{-1} N_{ji}^{\mu\nu} F_i^\nu$, which, substituted into Eq. (G.49), gives

$$\mathfrak{P}_\mu = \sum_{\nu\varrho} \sum_{ijkl} F_i^{\nu *} N_{ij}^{\nu\mu} (N^{\mu\mu})_{jk}^{-1} N_{kl}^{\mu\varrho} F_l^\varrho. \tag{G.52}$$

Thus the amount of clustering can be calculated from the overlaps between the various basis states and the combination coefficients obtained by the dynamical solution of the problem. The representing of Ψ^{μ} with the same states as are included in the basis for Ψ, again, may not be an additional approximation.

It should be observed that Eq. (G.52) has exactly the same structure as Eq. (11.32). Indeed, since $\sum_{i\nu} N^{\mu\nu}_{ji} F^{(\nu)}_i = \langle \psi^{\mu}_j | \Psi \rangle$ and $\sum_{l\varrho} N^{\mu\varrho}_{kl} F^{(\varrho)}_l = \langle \psi^{\mu}_k | \Psi \rangle$, Eq. (G.52) can be cast into the form

$$\mathfrak{P}_{\mu} = \sum_{jk} \langle \Psi | \psi^{\mu}_j \rangle (N^{\mu\mu})^{-1}_{jk} \langle \psi^{\mu}_k | \Psi \rangle, \tag{G.53}$$

where $N^{\mu\mu}_{mn} = \langle \psi^{\mu}_m | \psi^{\mu}_n \rangle$. This is obviously the same as Eq. (11.32); the difference is that in (the binary fragmentation version of) Eq. (11.32) the relative motion is represented by $\Psi_{\boldsymbol{R}}$ ($\boldsymbol{R} \in$ full space), while here by the discrete set $\{\psi^{(\nu^i_2)}_{0l^i_2 m^i_2}(\boldsymbol{R})\}$.

Appendix H

An α+n+n three-cluster model for ^{6}He

The aim of this appendix is to illustrate the implementation of the multicluster correlated Gaussian formalism as is described mainly in Sects. 9.3, 9.4 and 12.1. We do this through the example of the α+n+n three-cluster model for ^{6}He. For the sake of simplicity, we assume that the α-particle is described by the $(0s)^4$ HO function and that the two halo-neutrons are in a spin-singlet state. The discussion of the α+n+n model will give us an opportunity to point out the relation of the correlated Gaussian approach to the GCM presented in Sect. 11.7.1.

The constructing of the basis function starts from the three-cluster Gaussian wave packet, which is a special case of Eqs. (12.4) and (9.68):

$$\begin{aligned}\widehat{\Phi}(\boldsymbol{s}_1,\boldsymbol{s}_2,\boldsymbol{s}_3) &= \Phi(\boldsymbol{s}_1,\boldsymbol{s}_1,\boldsymbol{s}_1,\boldsymbol{s}_1,\boldsymbol{s}_2,\boldsymbol{s}_3)\\ &= \mathcal{A}\{\varphi_{\boldsymbol{s}_1}(\boldsymbol{r}_1)\varphi_{\boldsymbol{s}_1}(\boldsymbol{r}_2)\varphi_{\boldsymbol{s}_1}(\boldsymbol{r}_3)\varphi_{\boldsymbol{s}_1}(\boldsymbol{r}_4)\varphi_{\boldsymbol{s}_2}(\boldsymbol{r}_5)\varphi_{\boldsymbol{s}_3}(\boldsymbol{r}_6)\mathcal{X}_{0011}(1,\dots,6)\},\end{aligned} \tag{H.1}$$

where $\varphi_{\boldsymbol{s}_i}$ are the s.p. orbits given in Eq. (9.66) and $\mathcal{X}_{SM_STM_T}(1,\dots,6)$ is a spin–isospin function coupled to total spin SM_S and isospin TM_T, composed of the s.p. spin–isospin states $\mathcal{X}_{m_s^i m_t^i}(i)=\chi_{\frac{1}{2}m_s^i}(i)\eta_{\frac{1}{2}m_t^i}(i)$. As the product $\varphi_{\boldsymbol{s}_1}(\boldsymbol{r}_1)\cdots\varphi_{\boldsymbol{s}_1}(\boldsymbol{r}_4)$ is symmetric with respect to any permutations over the four spatial coordinates, the spin–isospin state of the first four nucleons must be fully antisymmetrical, which implies spin 0 and isospin 0 (cf. Eq. D.15). It thus follows that $\mathcal{X}_{0011}(1,\dots,6)$ must be proportional to $\mathcal{X}_{0000}(1,\dots,4)\frac{1}{\sqrt{2}}[\mathcal{X}_{\frac{1}{2}\frac{1}{2}}(5)\mathcal{X}_{-\frac{1}{2}\frac{1}{2}}(6)-\mathcal{X}_{-\frac{1}{2}\frac{1}{2}}(5)\mathcal{X}_{\frac{1}{2}\frac{1}{2}}(6)]$, where $\mathcal{X}_{0000}(1,\dots,4) = \mathcal{A}_{(1234)}\{\mathcal{X}_{-\frac{1}{2}\frac{1}{2}}(1)\mathcal{X}_{-\frac{1}{2}-\frac{1}{2}}(2)\mathcal{X}_{\frac{1}{2}\frac{1}{2}}(3)\mathcal{X}_{\frac{1}{2}-\frac{1}{2}}(4)\}$. Since, however, $\mathcal{A}\mathcal{A}_{(1234)}\sim\mathcal{A}$, the operator $\mathcal{A}_{(1234)}$ can be omitted behind the full antisymmetrizer $\mathcal{A}$, so that the fully coupled function $\mathcal{X}_{0011}$

can be replaced by the partially coupled function $\mathcal{X}$ [cf. Eqs. (D.15)]:

$$\widehat{\Phi}(\boldsymbol{s}_1, \boldsymbol{s}_2, \boldsymbol{s}_3) = \mathcal{A}\{\varphi_{\boldsymbol{s}_1}(\boldsymbol{r}_1)\varphi_{\boldsymbol{s}_1}(\boldsymbol{r}_2)\varphi_{\boldsymbol{s}_1}(\boldsymbol{r}_3)\varphi_{\boldsymbol{s}_1}(\boldsymbol{r}_4)\varphi_{\boldsymbol{s}_2}(\boldsymbol{r}_5)\varphi_{\boldsymbol{s}_3}(\boldsymbol{r}_6)\mathcal{X}\}, \tag{H.2}$$

where

$$\mathcal{X} = \mathcal{X}_{-\frac{1}{2}\frac{1}{2}}(1)\mathcal{X}_{-\frac{1}{2}-\frac{1}{2}}(2)\mathcal{X}_{\frac{1}{2}\frac{1}{2}}(3)\mathcal{X}_{\frac{1}{2}-\frac{1}{2}}(4) \times\frac{1}{\sqrt{2}}\left[\mathcal{X}_{\frac{1}{2}\frac{1}{2}}(5)\mathcal{X}_{-\frac{1}{2}\frac{1}{2}}(6) - \mathcal{X}_{-\frac{1}{2}\frac{1}{2}}(5)\mathcal{X}_{\frac{1}{2}\frac{1}{2}}(6)\right]. \tag{H.3}$$

To construct the wave function as in Sect. 12.1, let us introduce the c.m. coordinate of the α-particle, $\bar{\boldsymbol{\xi}}_4 = \frac{1}{\sqrt{4}}(\boldsymbol{r}_1 + \boldsymbol{r}_2 + \boldsymbol{r}_3 + \boldsymbol{r}_4)$ [cf. Eqs. (9.14), (11.2)] and the Jacobi coordinates (9.8), (9.9):

$$\boldsymbol{x}_1 = \frac{1}{\sqrt{2}}(\boldsymbol{r}_1 - \boldsymbol{r}_2), \qquad \boldsymbol{x}_2 = \frac{1}{\sqrt{6}}(\boldsymbol{r}_1 + \boldsymbol{r}_2 - 2\boldsymbol{r}_3),$$
$$\boldsymbol{x}_3 = \frac{1}{\sqrt{12}}(\boldsymbol{r}_1 + \boldsymbol{r}_2 + \boldsymbol{r}_3 - 3\boldsymbol{r}_4), \qquad \boldsymbol{x}_4 = \frac{1}{\sqrt{20}}\left(\sqrt{4}\bar{\boldsymbol{\xi}}_4 - 4\boldsymbol{r}_5\right),$$
$$\boldsymbol{x}_5 = \frac{1}{\sqrt{30}}(\sqrt{4}\bar{\boldsymbol{\xi}}_4 + \boldsymbol{r}_5 - 5\boldsymbol{r}_6), \qquad \boldsymbol{x}_6 = \frac{1}{\sqrt{6}}(\sqrt{4}\bar{\boldsymbol{\xi}}_4 + \boldsymbol{r}_5 + \boldsymbol{r}_6). \tag{H.4}$$

The set of cluster intrinsic coordinates for the α-cluster, ξ_α, and the intercluster relative coordinates, $\boldsymbol{\rho}_i$ [cf. Eq. (11.5)], may be defined as

$$\xi_\alpha \equiv \{\boldsymbol{x}_1, \boldsymbol{x}_2, \boldsymbol{x}_3\}, \qquad \boldsymbol{\rho}_1 \equiv \boldsymbol{x}_4, \qquad \boldsymbol{\rho}_2 \equiv \boldsymbol{x}_5. \tag{H.5}$$

This new set of coordinates makes it possible to express the product of the Gaussian wave packets in Eq. (H.1) as follows:

$$\varphi_{\boldsymbol{s}_1}(\boldsymbol{r}_1)\varphi_{\boldsymbol{s}_1}(\boldsymbol{r}_2)\varphi_{\boldsymbol{s}_1}(\boldsymbol{r}_3)\varphi_{\boldsymbol{s}_1}(\boldsymbol{r}_4)\varphi_{\boldsymbol{s}_2}(\boldsymbol{r}_5)\varphi_{\boldsymbol{s}_3}(\boldsymbol{r}_6) = \Phi^x_\alpha(\boldsymbol{x}_1, \boldsymbol{x}_2, \boldsymbol{x}_3)\varphi_{\boldsymbol{S}_1}(\boldsymbol{\rho}_1)\varphi_{\boldsymbol{S}_2}(\boldsymbol{\rho}_2)\varphi_{\boldsymbol{S}_3}(\boldsymbol{x}_6), \tag{H.6}$$

where $\{\boldsymbol{S}_1, \boldsymbol{S}_2, \boldsymbol{S}_3\}$ are defined in exactly the same way as $\{\boldsymbol{\rho}_1, \boldsymbol{\rho}_2, \boldsymbol{x}_6\}$ with $\{\frac{1}{\sqrt{4}}\bar{\boldsymbol{\xi}}_4, \boldsymbol{r}_5, \boldsymbol{r}_6\}$ replaced by $\{\boldsymbol{s}_1, \boldsymbol{s}_2, \boldsymbol{s}_3\}$. A clumsy scaling by $\sqrt{4}$ appears henceforth since the c.m. coordinate of α is scaled to comply with the definition of the *cluster* Jacobi coordinates, while the corresponding generator coordinate has been defined by letting the four nucleonic generator coordinates coincide without rescaling. (Cf. the transition from the s.p. generator coordinates $\boldsymbol{s}_i$ to conventional cluster generator coordinates $\underline{\boldsymbol{s}}_{[i]}$ and to normalized cluster generator coordinates $\boldsymbol{s}_{[i]}$ on p. 353.) The new generator coordinates are thus $\boldsymbol{S}_1 = \sqrt{\frac{4}{5}}(\boldsymbol{s}_1 - \boldsymbol{s}_2)$, $\boldsymbol{S}_2 = \frac{1}{\sqrt{30}}(4\boldsymbol{s}_1 + \boldsymbol{s}_2 - 5\boldsymbol{s}_3)$, $\boldsymbol{S}_3 = \frac{1}{\sqrt{6}}(4\boldsymbol{s}_1 + \boldsymbol{s}_2 + \boldsymbol{s}_3)$. In Eq. (H.6) Φ^x_α stands for the orbital part of the α-particle wave function,

$$\Phi^x_\alpha(\boldsymbol{x}_1, \boldsymbol{x}_2, \boldsymbol{x}_3) = \left(\frac{\beta}{\pi}\right)^{9/4} \mathrm{e}^{-\frac{1}{2}\beta(\boldsymbol{x}_1^2 + \boldsymbol{x}_2^2 + \boldsymbol{x}_3^2)}, \tag{H.7}$$

and use is made of the relationships

$$\sum_{i=1}^{4}(\boldsymbol{r}_i - \boldsymbol{s}_1)^2 = \boldsymbol{x}_1^2 + \boldsymbol{x}_2^2 + \boldsymbol{x}_3^2 + (\bar{\boldsymbol{\xi}}_4 - \sqrt{4}\boldsymbol{s}_1)^2,$$
$$(\bar{\boldsymbol{\xi}}_4 - \sqrt{4}\boldsymbol{s}_1)^2 + (\boldsymbol{r}_5 - \boldsymbol{s}_2)^2 + (\boldsymbol{r}_6 - \boldsymbol{s}_3)^2$$
$$= (\boldsymbol{\rho}_1 - \boldsymbol{S}_1)^2 + (\boldsymbol{\rho}_2 - \boldsymbol{S}_2)^2 + (\boldsymbol{x}_6 - \boldsymbol{S}_3)^2. \qquad \text{(H.8)}$$

The size of the α-cluster is controlled by the parameter β. Since the function $\varphi_{\boldsymbol{S}_3}(\boldsymbol{x}_6)$ remains unchanged under antisymmetrization, the Gaussian wave packet Eq. (H.1) can be recast into the form

$$\widehat{\Phi}(\boldsymbol{s}_1, \boldsymbol{s}_2, \boldsymbol{s}_3) = \varphi_{\boldsymbol{S}_3}(\boldsymbol{x}_6)\Psi(\boldsymbol{S}_1, \boldsymbol{S}_2), \qquad \text{(H.9)}$$

where the function $\Psi(\boldsymbol{S}_1, \boldsymbol{S}_2)$ is similar to the basis states employed in the GCM (cf. Sect. 11.7.1):

$$\Psi(\boldsymbol{S}_1, \boldsymbol{S}_2) = \mathcal{A}\{\Phi_\alpha^x(\boldsymbol{x}_1, \boldsymbol{x}_2, \boldsymbol{x}_3)\varphi_{\boldsymbol{S}_1}(\boldsymbol{\rho}_1)\varphi_{\boldsymbol{S}_2}(\boldsymbol{\rho}_2)\mathcal{X}\}. \qquad \text{(H.10)}$$

Through Eq. (H.9) the matrix element of a translation-invariant operator O between intrinsic functions of the type of Eq. (H.10) can be expressed in terms of Gaussian wave packets [cf. Eq. (9.103)]:

$$\langle\Psi(\boldsymbol{S}_1, \boldsymbol{S}_2)|O|\Psi(\boldsymbol{S}_1', \boldsymbol{S}_2')\rangle = \frac{\langle\widehat{\Phi}(\boldsymbol{s}_1, \boldsymbol{s}_2, \boldsymbol{s}_3)|O|\widehat{\Phi}(\boldsymbol{s}_1', \boldsymbol{s}_2', \boldsymbol{s}_3')\rangle}{\langle\varphi_{\boldsymbol{S}_3}(\boldsymbol{x}_6)|\varphi_{\boldsymbol{S}_3'}(\boldsymbol{x}_6)\rangle}$$
$$= \mathrm{e}^{\frac{1}{4}\beta(\boldsymbol{S}_3 - \boldsymbol{S}_3')^2}\langle\widehat{\Phi}(\boldsymbol{s}_1, \boldsymbol{s}_2, \boldsymbol{s}_3)|O|\widehat{\Phi}(\boldsymbol{s}_1', \boldsymbol{s}_2', \boldsymbol{s}_3')\rangle, \qquad \text{(H.11)}$$

where, in the second step, Eq. (9.95) has been used. The vectors $\boldsymbol{S}_1', \boldsymbol{S}_2'$ and $\boldsymbol{S}_3'$ are defined in terms of the s.p. vectors $\boldsymbol{s}_1', \boldsymbol{s}_2'$ and $\boldsymbol{s}_3'$ just as $\boldsymbol{S}_1, \boldsymbol{S}_2, \boldsymbol{S}_3$ are defined in terms of $\boldsymbol{s}_1, \boldsymbol{s}_2, \boldsymbol{s}_3$.

A microscopic description of ^{6}He involves matrix elements of the type of the left-hand side of Eq. (H.11). Its direct calculation is in general very tedious because the antisymmetrization mixes up the relative coordinates over which the integrations are to be performed. Equation (H.11), however, makes it expressible in terms of a matrix element that involves integration over s.p. coordinates. The matrix element on the right-hand side is a function of the six vectors $\boldsymbol{s}_1, \ldots, \boldsymbol{s}_3'$ whose indirect dependence on $\boldsymbol{S}_3$ and $\boldsymbol{S}_3'$ consists in a factor $\mathrm{e}^{-\frac{1}{4}\beta(\boldsymbol{S}_3 - \boldsymbol{S}_3')^2}$, so that the left-hand side will not depend on $\boldsymbol{S}_3$ and $\boldsymbol{S}_3'$.

The relative coordinates $\boldsymbol{\rho}_1, \boldsymbol{\rho}_2$ defined in Eq. (H.5) are Jacobi coordinates for the clusters α, n and n. Another sequence of the clusters implies another set of such coordinates, $\{\boldsymbol{\rho}_{1'}, \boldsymbol{\rho}_{2'}\}$ [cf. Eqs. (9.16) and (11.5)]:

$$\boldsymbol{\rho}_{1'} = \frac{1}{\sqrt{2}}(\boldsymbol{r}_5 - \boldsymbol{r}_6), \qquad \boldsymbol{\rho}_{2'} = \frac{1}{\sqrt{3}}(\boldsymbol{r}_5 + \boldsymbol{r}_6 - \bar{\boldsymbol{\xi}}_4). \qquad \text{(H.12)}$$

The sets $\{\boldsymbol{\rho}_1, \boldsymbol{\rho}_2\}$ and $\{\boldsymbol{\rho}_{1'}, \boldsymbol{\rho}_{2'}\}$ emphasize two different types of correlation; e.g., since $\boldsymbol{\rho}_{1'}$ is the distance vector between the two halo-neutrons, the latter emphasizes the correlation between the two neutrons. The corresponding generator-coordinate vectors $\{\boldsymbol{S}_{1'}, \boldsymbol{S}_{2'}\}$ are defined through $\{\boldsymbol{s}_1, \boldsymbol{s}_2, \boldsymbol{s}_3\}$ in exactly the same way as $\{\boldsymbol{\rho}_{1'}, \boldsymbol{\rho}_{2'}\}$ are defined in Eq. (H.12) via $\{\frac{1}{\sqrt{4}}\bar{\boldsymbol{\xi}}_4, \boldsymbol{r}_5, \boldsymbol{r}_6\}$, i.e., $\boldsymbol{S}_{1'} = \frac{1}{\sqrt{2}}(\boldsymbol{s}_2 - \boldsymbol{s}_3)$, $\boldsymbol{S}_{2'} = \frac{1}{\sqrt{3}}(\boldsymbol{s}_2 + \boldsymbol{s}_3 - \sqrt{4}\boldsymbol{s}_1)$. They are related to $\{\boldsymbol{S}_1, \boldsymbol{S}_2\}$ as follows:

$$\begin{pmatrix} \boldsymbol{S}_1 \\ \boldsymbol{S}_2 \end{pmatrix} = \begin{pmatrix} -\sqrt{\frac{2}{5}} & -\sqrt{\frac{3}{5}} \\ \sqrt{\frac{3}{5}} & -\sqrt{\frac{2}{5}} \end{pmatrix} \begin{pmatrix} \boldsymbol{S}_{1'} \\ \boldsymbol{S}_{2'} \end{pmatrix}. \tag{H.13}$$

The primed analogue of Eq. (H.10) is

$$\Psi(\boldsymbol{S}_{1'}, \boldsymbol{S}_{2'}) = \mathcal{A}\{\Phi_\alpha^x(\boldsymbol{x}_1, \boldsymbol{x}_2, \boldsymbol{x}_3)\varphi_{\boldsymbol{S}_{1'}}(\boldsymbol{\rho}_{1'})\varphi_{\boldsymbol{S}_{2'}}(\boldsymbol{\rho}_{2'})\mathcal{X}\}. \tag{H.14}$$

A matrix element $\langle\Psi(\boldsymbol{S}_{1'}, \boldsymbol{S}_{2'})|O|\Psi(\boldsymbol{S}'_{1'}, \boldsymbol{S}'_{2'})\rangle$ can be calculated from $\langle\Psi(\boldsymbol{S}_1, \boldsymbol{S}_2)|O|\Psi(\boldsymbol{S}'_1, \boldsymbol{S}'_2)\rangle$ via the transformation (H.13). With that, the calculation of $\langle\Psi(\boldsymbol{S}_{1'}, \boldsymbol{S}_{2'})|O|\Psi(\boldsymbol{S}'_{1'}, \boldsymbol{S}'_{2'})\rangle$ is also traced back to that of $\langle\widehat{\Phi}(\boldsymbol{s}_1, \boldsymbol{s}_2, \boldsymbol{s}_3)|O|\widehat{\Phi}(\boldsymbol{s}'_1, \boldsymbol{s}'_2, \boldsymbol{s}'_3)\rangle$. The function $\widehat{\Phi}(\boldsymbol{s}_1, \boldsymbol{s}_2, \boldsymbol{s}_3)$ is a combination of Slater determinants, so $\langle\widehat{\Phi}(\boldsymbol{s}_1, \boldsymbol{s}_2, \boldsymbol{s}_3)|O|\widehat{\Phi}(\boldsymbol{s}'_1, \boldsymbol{s}'_2, \boldsymbol{s}'_3)\rangle$ can be calculated by the standard analytic technique discussed in App. E.

The functions $\Psi(\boldsymbol{S}_1, \boldsymbol{S}_2)$ and $\Psi(\boldsymbol{S}_{1'}, \boldsymbol{S}_{2'})$ of Eqs. (H.10) and (H.14) may serve as model wave functions themselves: they are intrinsic two-centre SM states (cf. Sect. 11.7.3). In these states the intercluster relative motion is represented by a product of the Gaussian wave packets like $\varphi_{\boldsymbol{S}_1}(\boldsymbol{\rho}_1)\varphi_{\boldsymbol{S}_2}(\boldsymbol{\rho}_2)$. The width parameters of these Gaussians are rigidly linked with β, the width parameter of the cluster intrinsic motion, which is a disadvantage. To go beyond this simple description, one may take a linear combination of functions $\Psi(\boldsymbol{S}_1, \boldsymbol{S}_2)$ with suitable sets of the position parameters $\boldsymbol{S}_1$, $\boldsymbol{S}_2$. With this improvement, one obtains the GCM of Sect. 11.7.1. It is to be noted that in each GCM basis element the two intercluster motions are uncorrelated, i.e., its dependence on $\boldsymbol{\rho}_1$ and $\boldsymbol{\rho}_2$ is factored (behind the antisymmetrizer). It should also be noted that the GCM functions mix up different values of conserved quantum numbers, such as parity and angular momentum, which has to be remedied by projection techniques.

The correlated Gaussian approach is more powerful. For this approach $\Psi(\boldsymbol{S}_1, \boldsymbol{S}_2)$ is not a basis state but an ingredient of the generating function of the basis states, which are correlated and carry definite quantum numbers. The spatial part of the correlated Gaussian basis element, the function $f_{KLM_L}(u, \rho, A)$, is defined in Eq. (9.55). For our three-cluster problem, the spherical factor of $f_{KLM_L}(u, \rho, A)$ is $\exp(-\frac{1}{2}A_{11}\boldsymbol{\rho}_1^2 - \frac{1}{2}A_{22}\boldsymbol{\rho}_2^2 - A_{12}\boldsymbol{\rho}_1 \cdot \boldsymbol{\rho}_2)$, in which the $\boldsymbol{\rho}_1$- and $\boldsymbol{\rho}_2$-dependences are explicitly correlated; furthermore, the width-like variational parameters A_{ij} are not tied to β, and the angle-dependent factor carries good angular momentum and parity. The basis function of the correlated Gaussian approach is obtained by replacing $\varphi_{S_1}(\rho_1)\varphi_{S_2}(\rho_2)$ in Eq. (H.10) with this function:

$$\psi_{KLM_L}(u, A) = \mathcal{A}\{\Phi_\alpha^x(x_1, x_2, x_3) f_{KLM_L}(u, \rho, A)\mathcal{X}\}. \tag{H.15}$$

To calculate matrix elements between such states, one should make use of the fact that $f_{KLM_L}(u, \rho, A)$ is generated from $\varphi_{S_1}(\rho_1)\varphi_{S_2}(\rho_2)$ by an integral transformation as is given in Eq. (9.83). The matrix element is obtained from $\langle\Psi(S_1, S_2)|O|\Psi(S_1', S_2')\rangle$ via Eqs. (9.91) and (H.11). To be feasible and accurate, the integrations and differentiations involved have to be done analytically. This is an elementary, yet rather tedious operation.

As was shown in Sect. 9.2, the stochastic variational method can also be formulated in terms of relative coordinates, with expansions over all relative angular momenta [cf. Eqs. (9.19) and (9.20)]. In that case it is desirable to include basis states pertinent to more than one set of relative coordinates. The coordinates $\{\rho_1, \rho_2\}$ are suitable for describing single-particle motions of the neutrons around the α-particle, while the set $\{\rho_{1'}, \rho_{2'}\}$ is adapted to dineutron-like correlated motion. The basis states of this type are defined as follows:

$$\begin{aligned}
\psi_{(l_1 l_2)LM_L}^{(\nu_1\nu_2),(\alpha n)n} &= \mathcal{A}\{\Phi_\alpha(x_1, x_2, x_3)\Gamma_{(l_1 l_2)LM_L}^{(\nu_1\nu_2)}(\rho_1, \rho_2)\mathcal{X}\}, \\
\psi_{(l_{1'} l_{2'})LM_L}^{\prime(\nu_{1'}\nu_{2'}),\alpha(nn)} &= \mathcal{A}\{\Phi_\alpha(x_1, x_2, x_3)\Gamma_{(l_{1'} l_{2'})LM_L}^{(\nu_{1'}\nu_{2'})}(\rho_{1'}, \rho_{2'})\mathcal{X}\},
\end{aligned} \tag{H.16}$$

where the intercluster motion is represented by

$$\Gamma_{(l_1 l_2)LM_L}^{(\nu_1\nu_2)}(\rho_1, \rho_2) = [\psi_{0l_1}^{(\nu_1)}(\rho_1)\psi_{0l_2}^{(\nu_2)}(\rho_2)]_{LM_L}, \tag{H.17}$$

with

$$\psi_{0lm}^{(\nu)}(\rho) = \left[\frac{2^{l+2}\nu^{l+\frac{3}{2}}}{\sqrt{\pi}(2l+1)!!}\right]^{1/2} \rho^l e^{-\frac{1}{2}\nu\rho^2} Y_{lm}(\hat{\rho}). \tag{H.18}$$

The function $\psi_{0lm}^{(\nu)}(\rho)$ is just the normalized zero-node HO eigenfunction with angular momentum lm [cf. Eq. (9.51)]. The two basis states in Eqs. (H.16) are manifestly of the type of Eqs. (9.19) and (9.20), as they belong to two different arrangements, and there is no term $\rho_1 \cdot \rho_2$ and $\rho_{1'} \cdot \rho_{2'}$, respectively, in the Gaussian exponents. One should have an educated guess for the potentially important partial waves l_1, l_2 to be included. The width parameters ν_1, ν_2 are again variational parameters. Though in Eq. (H.17) $\rho_1 \cdot \rho_2$ does not appear, the function $\Gamma_{(l_{1'} l_{2'})LM_L}^{(\nu_{1'}\nu_{2'})}(\rho_{1'}, \rho_{2'})$ does in fact contain such a term in the exponent, which can be verified by expressing $\{\rho_{1'}, \rho_{2'}\}$ in terms of $\{\rho_1, \rho_2\}$.

The matrix elements can be derived from $\langle\Psi(S_1,S_2)|O|\Psi(S_1',S_2')\rangle$ by an integral transformation even when the basis (H.16) is used. To see this, note that the function $\Gamma^{(\nu_1\nu_2)}_{(l_1l_2)LM_L}(\rho_1,\rho_2)$ is obtained from $\varphi_{S_1}(\rho_1)\varphi_{S_2}(\rho_2)$ [301] as

$$\Gamma^{(\nu_1\nu_2)}_{(l_1l_2)LM_L}(\rho_1,\rho_2)=\int dS_1\int dS_2\, F^{(\nu_1\nu_2)}_{(l_1l_2)LM_L}(S_1,S_2)\varphi_{S_1}(\rho_1)\varphi_{S_2}(\rho_2), \tag{H.19}$$

where

$$F^{(\nu_1\nu_2)}_{(l_1l_2)LM_L}(S_1,S_2)=[F^{(\nu_1)}_{l_1}(S_1)F^{(\nu_2)}_{l_2}(S_2)]_{LM_L}, \tag{H.20}$$

with

$$F^{(\nu)}_{lm}(S)=\sqrt{\frac{2^{l-1}}{(2l+1)!!}\frac{\beta^{l+\frac{9}{4}}}{\pi}} \times\left(\frac{\sqrt{\nu}}{\beta-\nu}\right)^{l+\frac{3}{2}}S^lY_{lm}(\widehat{S})\exp\left(-\frac{1}{2}\frac{\beta\nu}{\beta-\nu}S^2\right)\quad(\beta\neq\nu). \tag{H.21}$$

Note that Eq. (H.21) yields a well-behaved function only if $\beta>\nu$, but this is not a cause for worry. The Gaussian wave packets in Eq. (H.19) make $\Gamma^{(\nu_1\nu_2)}_{(l_1l_2)LM_L}(\rho_1,\rho_2)$ well-behaved even if $\beta<\nu_1,\nu_2$. The calculation of matrix elements with functions (H.16) is discussed in detail in App. A.2 of the book [187].

Appendix I

The nuclear SU(3) symmetry

The three-dimensional harmonic-oscillator (HO) Hamiltonian plays a vital role in nuclear physics, particularly in the nuclear shell model (SM). The underlying reason for this is that the nuclear mean field has some resemblance to the HO potential except for the surface region. In an extreme model in which the nucleons move independently in an HO potential, the model wave function is given by an antisymmetrized product of the HO wave functions supplemented by the spin and isospin functions.

The HO Hamiltonian is invariant with respect to the transformations that constitute the SU(3) group, and the complete HO state space can be sliced into subspaces, each spanning a particular irreducible representation of the SU(3) group. Moreover, there is an close connection between the representations of the SU(3) and of the symmetric group S_n, the latter being the group of all permutations of n (particle) labels. This makes it convenient to choose the HO bases of n nucleons so that its elements belong to SU(3) irreducible representations. The decomposition of a reducible representation into irreducible representations by finding a suitable linear transformation is called reduction. The SU(3) group was introduced in nuclear physics as a tool to explain rotational features in light nuclei on the basis of the HOSM [314].

The aim of this appendix is to summarize some basic elements of the nuclear SU(3) model relevant to this book and clarify their use. Within its limitations it cannot be complete in any restrictive sense. The definitions, the theorems as well as their explanations will be given in broad, approximate terms, without proof. Details and exhaustive explanations can be found in Refs. [532,533] for SU(3) and in Refs. [534,

535] for group theory.

The three-dimensional HO Hamiltonian is invariant under the unitary transformations of the dynamical variables $\{\boldsymbol{r}, \boldsymbol{p}\}$, which rotate the HO quanta in the three-dimensional space. These transformations span the U(3) (three-dimensional unitary) group. The unitary transformations are those which conserve the norm when acting on the nuclear states. The determinants of the matrices of such transformations may be ± 1. The transformations that rotate the excitation quanta mix the particle coordinates and momenta, thus invariance with respect to them is called a *dynamical* symmetry. The transformations in the U(3) group can be constructed by the consecutive applications of infinitesimal transformations

$$U = 1 + i \sum_{j,k=1}^{3} \delta\varrho_{jk} C_{jk}, \tag{I.1}$$

where $\delta\varrho_{jk}$ are infinitesimal quantities and the 'generators' C_{jk} are

$$C_{jk} = \frac{1}{2}(b_j^\dagger b_k + b_k b_j^\dagger), \tag{I.2}$$

where $b_j^\dagger$ and b_j, respectively, creates and annihilates an HO quantum in direction $j=1$ (i.e., x), $j=2$ (i.e., y) and $j=3$ (i.e., z):

$$b_j^\dagger = \left(\frac{m\omega}{2\hbar}\right)^{1/2} \left(r_j - \frac{i}{m\omega} p_j\right), \qquad b_j = (b_j^\dagger)^\dagger, \tag{I.3}$$

with $\dagger$ denoting Hermitean conjugation. The action of the generators C_{jk} is plausible: they move a quantum from direction k to direction j. From the well-known commutation relations $[b_j, b_k^\dagger] = \delta_{jk}$, $[b_j, b_k] = 0$, $[b_j^\dagger, b_k^\dagger] = 0$ it follows that the generators C_{jk} obey the commutation rules

$$[C_{jk}, C_{lm}] = \delta_{kl} C_{jm} - \delta_{jm} C_{lk}. \tag{I.4}$$

These are the commutation relations of the Lie algebra associated with the U(3) group. The group can as well be defined by any nine independent linear combinations of the C_{jk}. One can combine spherical tensors out of them: one scalar, $\hbar\omega \sum_{j=1}^{3} C_{jj}$ (which, apart from the term $\frac{3}{2}\hbar\omega$, is the HO Hamiltonian itself and contains $\sum_{j=1}^{3} C_{jj}$, which is the operator of the number of oscillator quanta), the three spherical components of the orbital angular momentum vector and five components of a symmetric second-rank tensor [a 'quadrupole' operator, which contains $r^2 Y_{2\mu}(\hat{\boldsymbol{r}})$ and $p^2 Y_{2\mu}(\hat{\boldsymbol{p}})$ symmetrically]. By omitting the trivial combination $\sum_{j=1}^{3} C_{jj}$, the remaining eight combinations of the

U(3) generators span the three-dimensional special unitary group [the SU(3) group, which is also called unimodular since the determinant of any transformation belonging to the SU(3) group is unity].

The U(3) group in an abstract form is defined by the commutation relations (I.4) of its associated algebra. Thus a generalization to n independent oscillators is straightforward by defining new generators as

$$C_{jk} = \sum_{i=1}^{n} C_{jk}(i) = \sum_{i=1}^{n} \frac{1}{2}[b_j^\dagger(i)b_k(i) + b_k(i)b_j^\dagger(i)], \tag{I.5}$$

where i is the particle index. It is easy to verify that these operators satisfy the commutation rules (I.4) and thus form the U(3) algebra. A set of generators for the HO intrinsic motion can be defined by replacing the oscillator creation and annihilation operators in the above equation with

$$b_j^\dagger(i) \longrightarrow b_j^\dagger(i) - \frac{1}{n}\sum_{k=1}^{n} b_j^\dagger(k), \qquad b_j(i) \longrightarrow b_j(i) - \frac{1}{n}\sum_{k=1}^{n} b_j(k). \tag{I.6}$$

As a next step, the s.p. coordinates may be transformed into an appropriate system of relative coordinates, such as the intrinsic Jacobi coordinates or the cluster intrinsic and cluster relative coordinates. The HO creation and annihilation operators can be defined in terms of such coordinates, and then the index i may be interpreted as referring to them. For a two-cluster system made up of fragments of mass number f and $A-f$, the generators of Eq. (I.5) may be decomposed into three terms:

$$C_{jk} = C_{jk}^{(\mathrm{f1})} + C_{jk}^{(\mathrm{f2})} + C_{jk}^{(\mathrm{rel})} = \sum_{i=1}^{f-1} C_{jk}(i) + \sum_{i=f}^{A-2} C_{jk}(i) + C_{jk}(A-1), \tag{I.7}$$

where each term forms the U(3) algebra.[1]

Let us consider the problem of putting n nucleons into d s.p. orbits φ_ν. Since φ_ν contains no spin–isospin, the nucleons may be treated as distinguishable so far as no more than 4 are put into the same orbit φ_ν. There are altogether d^n n-particle product wave functions Ψ. If all the s.p. functions are subjected to a unitary transformation

[1] The precise way of doing this is via returning to the s.p. operators $b_j^\dagger(i), b_j(i)$. They are to be grouped into two sets containing f and $A-f$ elements, respectively. One can then remove the c.m. for $n=f$ and $n=A-f$ separately as in Eq. (I.6) and, as a second step, remove the total c.m. through a similar substitution. This is analogous to the procedure of introducing cluster c.m., relative and intrinsic coordinates; cf., e.g., App. B.

$\varphi'_p = U\varphi_p = \sum_q U_{pq}\varphi_q$, where $U^*_{pq} = (U^{-1})_{qp}$ by the definition of unitarity, the set of n-particle functions transforms into itself, and hence spans a representation of the d-dimensional unitary group U(d). The problem is to decompose the d^n functions into irreducible representations of U(d) by finding suitable linear transformations. An irreducible representation is a set of functions that cannot be decomposed any further by linear transformations into smaller sets of functions each of which transforms into itself.

Since the unitary transformations are totally symmetric in the particle labels, they preserve the permutational symmetry of the functions they act on. Therefore, the transformations belonging to U(d) transform any state into a state of the same permutational symmetry. It is a non-trivial consequence of this fact that the labels of an irreducible representation of S_n are suited for labelling a representation of U(d) as well. In fact, there is a way of choosing these U(d) representations irreducible.

It is convenient to denote the irreducible representation of S_n by a symbol which expresses some properties of the representation. Such a symbol is a 'partition' $[f] = [f_1 f_2 \cdots f_d]$, in which f_i are integer numbers such that $f_1 \geq f_2 \geq \cdots \geq f_d \geq 0$, $f_1 + f_2 + \cdots + f_d = n$ and $f_1 \leq 4$. (If $d > n$, $f_{n+1} = \cdots = f_d = 0$.) A partition can be represented alternatively by a so-called Young tableau consisting of n squares. The squares are ordered in rows of length $f_1, f_2, \ldots$ and the rows are flushed left. For example, the partition [431] is represented by the tableau 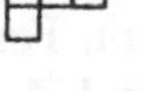. A basis spanning the representation $[f_1 f_2 \cdots]$ may be a set of states which have been symmetrized within subsets of $f_1, f_2, \ldots$ (i.e., within each row of the corresponding Young tableau) and then antisymmetrized column by column of the Young tableau (whereby the symmetry attained in the first step is in general lost). The limitation $f_1 \leq 4$ comes from the Pauli principle: since there are four orthogonal spin–isospin states, the orbital wave function of at most four nucleons can be symmetrical to permutation. The unitary transformation U(d) preserves the shape of the tableaux.

Suppose that the set of d s.p. states consists of all HO wave functions belonging to a major shell with $\hbar\omega(Q + \frac{3}{2})$ $[d = \frac{1}{2}(Q+1)(Q+2)]$. As a unitary transformation acting on these s.p. orbits, consider that generated by the infinitesimal transformation (I.1), where $\delta\varrho_{jk} = \delta\varrho^*_{kj}$ is required by the unitarity of U (note that $C^\dagger_{jk} = C_{kj}$). This transformation shifts the direction of the HO quanta. The generators of the transformation are indeed symmetric with respect to the permutations of particle labels, so the set of functions belonging to a particular tableau $[f]$ denoting an irreducible representation of group U(d) spans a representation of SU(3) as well, which is in general reducible. [The group U(3) is a subgroup of U(d), i.e., U(3)$\subseteq$U(d).] The decomposition of the irreducible representations $[f]$ of U(d) into those of U(3) can be attained via a recursive procedure [535].

The bases of the irreducible representations of U(3) can be set up in a Cartesian frame, with each basis function $|g_x g_y g_z\rangle$ carrying a definite number of quanta g_i in each direction. Since the generators C_{jk} cannot change the total number of quanta, $Q=g_x+g_y+g_z$, each basis function of a particular representation must carry the same Q. The quanta may be distributed unevenly among the directions, and the distribution of the most anisotropic of the functions belonging to a particular representation is characteristic of the representation. The most anisotropic of these states, the so-called highest-weight state, can be defined[2] by $C_{jk}|g_x^0 g_y^0 g_z^0\rangle = C_{ji}|g_x^0 g_y^0 g_z^0\rangle = C_{ik}|g_x^0 g_y^0 g_z^0\rangle = 0$, where $\{jki\}$ is one particular permutation of $\{xyz\}$ and the definition is unique, apart from a permutation of the directions. To remove this freedom of choice, the set of g_x^0, g_y^0, g_z^0 is reordered and denoted by $[g_1 g_2 g_3]$ such that $g_1 \geq g_2 \geq g_3$. This set may be regarded as a partition in the 'quantum symmetry' of U(3) and a Young tableau can be associated with it. The partition $[g_1 g_2 g_3]$ may be used to label the irreducible representations of U(3). Of the irreducible representations of SU(3) only the differences of g_i are characteristic; in nuclear physics Elliott's convention, the double label $(\lambda\mu)$, is used, in which $\lambda = g_1 - g_2$ and $\mu = g_2 - g_3$.

If a group G is a symmetry group of a Hamiltonian, then so are its subgroups. The group of three-dimensional spatial rotations, SO(3) or, with alternative notation, R(3), is a subgroup of SU(3): SO(3) $\subset$ SU(3). Its generators are the three components of the (dimensionless) angular momentum, which, of course, can be expressed via the SU(3) generators:

$$L_x = \frac{1}{i}(C_{yz} - C_{zy}), \quad L_y = \frac{1}{i}(C_{zx} - C_{xz}), \quad L_z = \frac{1}{i}(C_{xy} - C_{yx}). \quad \text{(I.8)}$$

The significance of this relationship is that the basis of any irreducible representation of SU(3) can be subdivided into those of the SO(3). In this U(d) $\supset$ SU(3) $\supset$ SO(3) reduction the basis elements are 'classified' by $[f](\lambda\mu)\kappa L M_L$, where κ is an additional quantum number, whose choice is non-unique [536]; its role is to distinguish multiple occurrences of a given L for a particular $(\lambda\mu)$.

[2]An operator C_{jk} may have a null effect either because there are no more k quanta or because the Pauli principle forbids putting more quanta in the j direction. The usual definition of the highest-weight state involves a definite orientation, $i=x$, $k=y$, $j=z$.

The group SU(3) has another subgroup chain, that which involves SU(2)×U(1); i.e., SU(3) ⊃ SU(2)×U(1). The generator of U(1) is

$$Q_0 = 2C_{zz} - C_{xx} - C_{yy}, \tag{I.9}$$

while the generators of the SU(2) subgroup are three operators which satisfy the commutation relations of the components $L_0 = L_z$, $L_\pm = L_x \pm iL_y$ of the angular momentum

$$\Lambda_0 = \frac{1}{2}(C_{xx} - C_{yy}), \quad \Lambda_+ = C_{xy}, \quad \Lambda_- = C_{yx}. \tag{I.10}$$

In fact, there is a very close relationship between the groups SU(2) and SO(3). In the irreducible representation belonging to this group chain the basis functions are labelled by $[f]$ and by the eigenvalues $\epsilon, \Lambda(\Lambda+1), M_\Lambda$ of $Q_0, \Lambda^2, \Lambda_0$, respectively, i.e., by $[f](\lambda\mu)\epsilon\Lambda M_\Lambda$. This latter group chain is used in the classification of hadrons in the SU(3)-flavour quark model by 'decomposing' the representations of SU(3) into the SU(2) isospin and the U(1) hypercharge [535]. Two other possible SU(3) ⊃ SU(2)×U(1) reductions may be obtained by a cyclic permutation of x, y, z. In what follows we use α to denote any of the convenient sets of SU(3) subgroup labels.

A trivial but important example for the above considerations is the SU(3) classification of s.p. HO states in the Q'th major shell. There are $\frac{1}{2}(Q+1)(Q+2)$ states and they are represented by $\{|n_x n_y n_z\rangle\}$ ($Q = n_x + n_y + n_z$) in the Cartesian coordinates, or $\{|nlm\rangle\}$ ($Q = 2n + l$) in the polar coordinates. These s.p. states span an irreducible representation of the SU(3) group. This is irreducible because all of the elements in this set can be obtained from a single function $|00Q\rangle$ through

$$|n_x n_y n_z\rangle = \frac{1}{(Q-n_z)!}\sqrt{\frac{n_x!\,n_z!}{Q!\,n_y!}}(C_{yx})^{n_y}(C_{xz})^{Q-n_z}|00Q\rangle, \tag{I.11}$$

and it is a representation since the SU(3) generators conserve Q. Equation (I.11) is easily proved by using the well-known laddering relationships $(b^\dagger)^p|n\rangle = \sqrt{(n+p)!/n!}\,|n+p\rangle$, $b^p|n\rangle = \sqrt{n!/(n-p)!}\,|n-p\rangle$. What SU(3) representation labels characterize this set? These states have Q oscillator quanta which are distributed among the three different directions. This distribution determines which irreducible representation a particular state belongs to. According to the rules of the U(3) classification given above, the state $|00Q\rangle$ belongs to the partition $[Q00]$, and thus must belong to a $(\lambda\mu) = (Q0)$ irreducible representation of SU(3). Then all states (I.11) belong to the representation $(\lambda\mu) = (Q0)$. The degree of degeneracy of the HO s.p. states, $\frac{1}{2}(Q+1)(Q+2)$, is a special case of the formula for the dimension of SU(3) representation $(\lambda\mu)$:

$$\dim(\lambda\mu) = \frac{1}{2}(\lambda+1)(\mu+1)(\lambda+\mu+2). \tag{I.12}$$

The HO s.p. wave function (I.11) is an example for a basis state belonging to an SU(3) $\supset$ SU(2)$\times$U(1) reduction; the quantum numbers can be read from the definitions (I.9) and (I.10): $\epsilon = 2n_z - n_x - n_y$, $M_\Lambda = \frac{1}{2}(n_x - n_y)$. The value of Λ can be obtained by acting with $\Lambda^2 = \Lambda_+\Lambda_- + \Lambda_0^2 - \Lambda_0$ on the state (I.11), which results in $\Lambda = \frac{1}{2}(n_x + n_y)$. Thus the state $|n_x n_y n_z\rangle$ can be labelled by $|n_x n_y n_z\rangle = |(\lambda\mu)\epsilon\Lambda M_\Lambda\rangle = |(Q0)2n_z - n_x - n_y, \frac{1}{2}(n_x + n_y), \frac{1}{2}(n_x - n_y)\rangle$. The polar-coordinate form of the HO basis is an example for the SU(3) $\supset$ SO(3) reduction, $|nlm\rangle = |(Q0)lm\rangle$, where κ is redundant and is suppressed.

The irreducible representations of SU(3) for a composite system can be built up from those of its constituents. The resulting representations will form a product of the two subsystems' representations. The rule for reducing the product of two irreducible representations of the SU(3) group is rather complicated for arbitrary $(\lambda_1\mu_1)$ and $(\lambda_2\mu_2)$. When, however, both λ and μ are small or one of them is zero, it is not so difficult. Two examples are given here:

$$\begin{aligned} (11) \times (11) &= (22) + (30) + (11) + (11) + (03) + (00), \\ (\lambda\mu) \times (Q0) &= \sum_{\substack{0 \le k, 0 \le l \le \lambda \\ Q-\mu \le k+l \le Q}} (\lambda + k - l, \mu + k + 2l - Q). \end{aligned} \tag{I.13}$$

The first line shows that the (11) symmetry appears twice in the coupling of (11)$\times$(11). The product can be visualized graphically in terms of Young tableaux. For example, the second line looks like

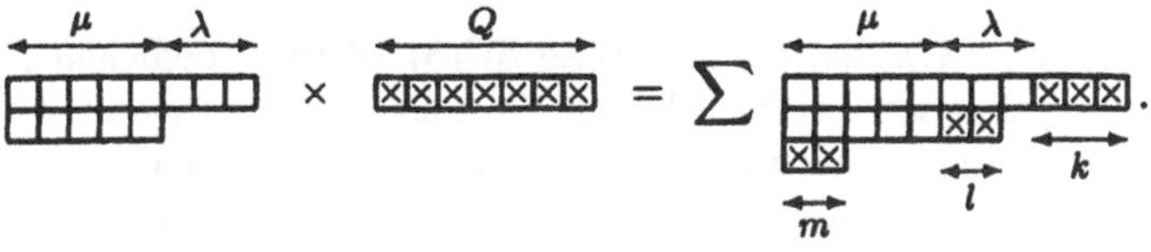

(I.14)

The multiplication by the second factor amounts to adding, to the first factor, k boxes in the first row, l boxes in the second row and m boxes in the third row $(k+l+m = Q)$. The restrictions in the summation mean $0 \le l \le \lambda$ and $0 \le m \le \mu$, so that the boxes coming from the second factor are never put on top of each other. The resulting U(3) representation is $[\lambda+\mu+k, \mu+l, m]$, which implies, for SU(3), $(\lambda+k-l, \mu+k+2l-Q)$. One can check the formulae (I.13) by counting the dimensions on both sides of the equations. The reduction of the coupling $(\mu\lambda) \times (0Q)$ can be obtained by exchanging the two SU(3) labels in each symbol in the second line of Eq. (I.13).

A set of linearly independent operators $T_\alpha^{(\lambda\mu)}$, whose number is equal to the dimension of the irreducible representation $(\lambda\mu)$, is called an SU(3) tensor operator if it satisfies the following relation:

$$[C_{jk}, T_\alpha^{(\lambda\mu)}] = \sum_{\alpha'} \langle(\lambda\mu)\alpha'|C_{jk}|(\lambda\mu)\alpha\rangle T_{\alpha'}^{(\lambda\mu)}. \tag{I.15}$$

This is a generalization of the notion of a spherical tensor operator well-known for the SO(3) algebra [26,54] (cf. footnote on p. 221). These operators are called irreducible if they cannot be decomposed into sums of smaller sets of SU(3) tensor operators. The HO creation operators have the following SU(3)⊃SU(2)×U(1) tensor character

$$b_x^\dagger = T^{(10)}_{-1\frac{1}{2}\frac{1}{2}}, \quad b_y^\dagger = T^{(10)}_{-1\frac{1}{2}-\frac{1}{2}}, \quad b_z^\dagger = T^{(10)}_{200}, \tag{I.16}$$

which is in accord with the labelling of the HO wave function, i.e., $b_x^\dagger|000\rangle \sim |100\rangle = |(10){-1}\tfrac{1}{2}\tfrac{1}{2}\rangle$ etc. The annihilation operators, the Hermitean conjugates of the creation operators, have (01) tensor character. An operator which commutes with the eight SU(3) generators is called SU(3)-scalar; it has $(\lambda\mu)=(00)$ tensor character. An important example for SU(3)-scalar operators is the antisymmetrization operator $\mathcal{A}$.

New tensors can be constructed by combining products of two SU(3) tensor operators just as in the case of SO(3) (i.e., spherical) tensors. The appropriate combination is a tensor product, which is defined for two SU(3) tensor operators, acting in different spaces 1 and 2, as

$$\begin{aligned}&[T^{(\lambda_1\mu_1)}(1) \times T^{(\lambda_2\mu_2)}(2)]_\alpha^{(\lambda\mu)\rho} \\ &\quad = \sum_{\alpha_1\alpha_2} \langle(\lambda_1\mu_1)\alpha_1;(\lambda_2\mu_2)\alpha_2|(\lambda\mu)\alpha\rangle_\rho\, T_{\alpha_1}^{(\lambda_1\mu_1)}(1) T_{\alpha_2}^{(\lambda_2\mu_2)}(2).\end{aligned} \tag{I.17}$$

Here $\langle(\lambda_1\mu_1)\alpha_1;(\lambda_2\mu_2)\alpha_2|(\lambda\mu)\alpha\rangle_\rho$ is an SU(3) Wigner coefficient [532, 536], a generalization of the Clebsch–Gordan coefficient, and ρ is a label which distinguishes multiple occurrences of a particular $(\lambda\mu)$ in the coupling $(\lambda_1\mu_1)\times(\lambda_2\mu_2)$. [See, e.g., Eq. (I.13).] Its definition is non-unique, but conventions exist.

One can also examine the tensorial character of the (uncoupled) product of two tensors. With the use of the orthogonality of the Wigner coefficients, from Eq. (I.17) one can get

$$\begin{aligned}T_{\alpha_1}^{(\lambda_1\mu_1)}(1) T_{\alpha_2}^{(\lambda_2\mu_2)}(2) = &\sum_{(\lambda\mu)\rho\alpha} \langle(\lambda_1\mu_1)\alpha_1;(\lambda_2\mu_2)\alpha_2|(\lambda\mu)\alpha\rangle_\rho \\ &\times[T^{(\lambda_1\mu_1)}(1) \times T^{(\lambda_2\mu_2)}(2)]_\alpha^{(\lambda\mu)\rho}.\end{aligned} \tag{I.18}$$

The Wigner–Eckart theorem for SU(3) gives the dependence of the matrix elements of SU(3) tensor operators between SU(3) basis states on the SU(3) labels other than the representation labels $(\lambda\mu)$. It states that such a matrix element can be expressed as a sum over the multiplicity label ρ of the product of a ρ-dependent reduced matrix element multiplied by the corresponding SU(3) Wigner coefficient:

$$\langle(\lambda_2\mu_2)\alpha_2|O_\alpha^{(\lambda\mu)}|(\lambda_1\mu_1)\alpha_1\rangle$$
$$= \sum_\rho \langle(\lambda_1\mu_1)\alpha_1;(\lambda\mu)\alpha|(\lambda_2\mu_2)\alpha_2\rangle_\rho \, \langle(\lambda_2\mu_2)\|O^{(\lambda\mu)}\|(\lambda_1\mu_1)\rangle_\rho. \quad \text{(I.19)}$$

For an SU(3)-scalar operator, the Wigner coefficient is 1 for $(\lambda_1\mu_1)\alpha_1 = (\lambda_2\mu_2)\alpha_2$ and 0 otherwise, so the diagonal matrix element of a scalar operator is obviously independent of the subgroup labels. This property, combined with SU(3) recoupling techniques, has been extensively used in calculating the norm and overlap matrix elements of complex cluster systems in HO cluster models [224].

The HO two-cluster wave function can be written in an SU(3)-coupled basis following the decomposition (I.7) as

$$\Psi = \mathcal{A}_{12}\left\{\left[[\Phi_1^{(\lambda_1\mu_1)}(\xi_1)\times\Phi_2^{(\lambda_2\mu_2)}(\xi_2)]^{(\lambda_c\mu_c)\rho_c}\times\psi^{(Q0)}(\boldsymbol{q})\right]_\alpha^{(\lambda\mu)}\right\}, \quad \text{(I.20)}$$

where ξ_1 and ξ_2 denote the cluster intrinsic coordinates of the respective clusters and $\boldsymbol{q}$ is the relative distance vector between them.

The simplest examples of cluster wave functions are closed-shell configurations; these are, e.g., the α-particle and ^{16}O. Whether the c.m. motion is eliminated or not, they form the one-dimensional representation $(\lambda\mu) = (00)$ because they are invariant to the effect of the U(3) generators. When the c.m.-free closed-shell α and ^{16}O wave functions are used with the same HO parameter in the HO cluster-model of ^{20}Ne, the resultant wave function belongs to the $(Q0)$ representation of SU(3), where Q equals the $2N+L$ combination of the quantum numbers of the relative motion. Another example is ^{12}C. If the orbital symmetry $[f]$ of ^{12}C is assumed to be [444] in an $(0s)^4(0p)^8$ configuration, its SU(3) labels are (04) (Ref. [314]). Thus the wave functions of the α+^{12}C cluster model for ^{16}O will belong to the $(04)\times(Q0) = (Q-k, 4-k)$ $[k=0,\ldots,\min(Q,4)]$ representation. If two ^{12}C nuclei are put together as in Eq. (I.20) to form cluster-model states of ^{24}Mg, they may couple to SU(3) labels $(\lambda_c\mu_c) = (04)\times(04)$, which include (08), (24), and (40) (ρ_c is redundant in this case). [This should yet be coupled to the relative-motion state of $(Q0)$.] The other SU(3) labels, like (16) and (32), are not allowed because the two ^{12}C fragments are identical. [The forbiddenness of (16) and (32) is similar to that of the relative motion of two identical bosons with odd partial waves.]

The generators C_{jk} of Eq. (I.5) are bilinear combinations of HO creation and annihilation operators and conserve the number of HO quanta. One can construct other similar bilinear combinations that do not conserve the HO quanta:

$$B^\dagger_{jk} = \sum_{i=1}^{A} \frac{1}{2}[b^\dagger_j(i)b^\dagger_k(i) + b^\dagger_k(i)b^\dagger_j(i)],$$

$$B_{jk} = \sum_{i=1}^{A} \frac{1}{2}[b_j(i)b_k(i) + b_k(i)b_j(i)]. \qquad \text{(I.21)}$$

The operators $B^\dagger_{jk}$ and B_{jk} ladder the oscillator quanta by 2 and −2. They are SU(3) irreducible tensor operators with $(\lambda\mu)=(20)$ and (02), respectively. There are altogether 21 bilinear combinations of creation and annihilation operators, which constitute an algebra with the commutation rules

$$[B_{jk}, C_{lm}] = \delta_{kl}B_{jm} + \delta_{jl}B_{km},$$

$$[B_{jk}, B^\dagger_{lm}] = \delta_{km}C_{lj} + \delta_{jm}C_{lk} + \delta_{kl}C_{mj} + \delta_{jl}C_{mk}. \qquad \text{(I.22)}$$

These generators are known to be infinitesimal operators of the non-compact[3] real symplectic group, Sp(6, R) [sometimes also denoted by Sp(3, R)]. It is evident from its construction that SU(3) ⊂ Sp(6, R). This group is the basic ingredient of a theory of collective quadrupole motion. (See, e.g., Refs. [537–540].) To eliminate the spurious c.m. excitations, one has to make a replacement similar to Eq. (I.6) in Eq. (I.21).

The c.m.-free HOSM space provides a representation space for the Sp(6, R) group. This space can be decomposed into infinite-dimensional irreducible representation subspaces [538, 539]. Each set of physical states produced in this way is called a symplectic band. While the Sp(6, R) group provides a microscopic theory of the nuclear collective motion with quadrupole collectivity, the cluster model is a microscopic theory which describes arbitrary excitations in the relative motion between clusters. While there is obviously overlap between the two types of motions for low excitations, the relationship of the two models is essentially complementarity [541].

---

[3] Roughly speaking, a Lie group, whose elements depend on continuous parameters, is non-compact if the range of the parameters is unbounded. The irreducible representations of non-compact groups are generally infinite-dimensional. See Ref. [534].

Bibliography

[1] I. Tanihata, *J. Phys. G* **22**, 157 (1996).

[2] I. Tanihata, *Prog. Part. Nucl. Phys.* **35**, 505 (1995).

[3] P. G. Hansen, A. S. Jensen and B. Jonson, *Ann. Rev. Nucl. Part. Sci.* **45**, 591 (1995).

[4] K. Riisager, *Rev. Mod. Phys.* **66**, 1105 (1994).

[5] C. A. Bertulani, L. F. Canto and M. S. Hussein, *Phys. Rep.* **226**, 281 (1993).

[6] B. Jonson and K. Riisager, *Phil. Trans. Roy. Soc. (London) A* **358**, 2063 (1998).

[7] I. Tanihata, in *Treatise on Heavy-Ion Science*, edited by D. A. Bromley (Plenum, New York, 1989), Vol. 8, p. 443.

[8] I. Tanihata, H. Hamagaki, O. Hashimoto, Y. Shida, N. Yoshikawa, K. Sugimoto, O. Yamakawa, T. Kobayashi and N. Takahashi, *Phys. Rev. Lett.* **55**, 2676 (1985).

[9] I. Tanihata *et al.*, *Phys. Lett. B* **160**, 380 (1985).

[10] I. Tanihata, T. Kobayashi, O. Yamakawa, S. Shimoura, K. Ekuni, K. Sugimoto, N. Takahashi, T. Shimoda and H. Sato, *Phys. Lett. B* **206**, 592 (1988); A. Ozawa, T. Suzuki and I. Tanihata, *Nucl. Phys. A* **693**, 32 (2001).

[11] P. G. Hansen and B. Jonson, *Europhys. Lett.* **4**, 409 (1987).

[12] T. Suzuki *et al.*, *Phys. Rev. Lett.* **75**, 3241 (1995); *Nucl. Phys. A* **630**, 661 (1998).

[13] D. Bazin *et al.*, *Phys. Rev. Lett.* **74**, 3569 (1995).

[14] T. Baumann *et al.*, *Phys. Lett. B* **439**, 256 (1998).

[15] T. Nakamura *et al.*, *Phys. Rev. Lett.* **83**, 1112 (1999).

[16] A. Ozawa *et al.*, *Nucl. Phys. A* **691**, 599 (2001).

[17] I. Tanihata *et al.*, *Phys. Lett. B* **287**, 307 (1992).

[18] A. Ozawa, I. Tanihata, T. Kobayashi, Y. Sugahara, O. Yamakawa, K. Omata, K. Sugimoto, D. Olson, W. Christie and H. Wieman, *Nucl. Phys. A* **608**, 63 (1996).

[19] M. M. Obuti, T. Koyabashi, D. Hirata, Y. Ogawa, A. Ozawa, K. Sugimoto, I. Tanihata, D. Olson, W. Christie and H. Wieman, *Nucl. Phys. A* **609**, 74 (1996).

[20] T. Suzuki *et al.*, *Nucl. Phys. A* **658**, 313 (1999).

[21] T. Kobayashi, O. Yamakawa, K. Omata, K. Sugimoto, T. Shimoda, N. Takahashi and I. Tanihata, *Phys. Rev. Lett.* **60**, 2599 (1988).

[22] B. Blank *et al.*, *Z. Phys. A* **343**, 375 (1992).

[23] J. Hüfner and M. C. Nemes, *Phys. Rev. C* **23**, 2538 (1981).

[24] A. S. Goldhaber and H. H. Heckman, *Ann. Rev. Nucl. Part. Sci.* **28**, 161 (1978).

[25] T. Kobayashi *et al.*, *Phys. Lett. B* **232**, 51 (1989).

[26] A. Messiah, *Quantum Mechanics* (North-Holland, Amsterdam, 1961).

[27] J. J. Kolata *et al.*, *Phys. Rev. Lett.* **69**, 2631 (1992).

[28] C.-B. Moon *et al.*, *Phys. Lett. B* **297**, 39 (1992).

[29] M. Lewitowicz *et al.*, *Nucl. Phys. A* **562**, 301 (1993).

[30] A. A. Korsheninnikov *et al.*, *Phys. Lett. B* **316**, 38 (1993).

[31] A. A. Korsheninnikov *et al.*, *Phys. Lett. B* **343**, 53 (1995).

[32] T. Teranishi *et al.*, *Phys. Lett. B* **407**, 110 (1997).

[33] A. Yoshida *et al.*, *Phys. Lett. B* **389**, 457 (1996).

[34] J. J. Kolata *et al.*, *Phys. Rev. Lett.* **81**, 4580 (1998).

[35] N. Takigawa and H. Sagawa, *Phys. Lett. B* **265**, 23 (1991).

[36] M. S. Hussein, *Nucl. Phys. A* **531**, 192 (1991).

[37] C. H. Dasso and R. Donangelo, *Phys. Lett. B* **276**, 1 (1991).

[38] R. J. Glauber, *in Lectures in Theoretical Physics* (Interscience, New York, 1959), Vol. 1, p. 315.

[39] N. Austern, *Direct Nuclear Reaction Theories* (Wiley, New York, 1970).

[40] G. R. Satchler, *Direct Nuclear Reactions* (Clarendon, Oxford, 1983).

[41] L. I. Schiff, *Quantum Mechanics* (McGraw-Hill, New York, 1968).

[42] J. R. Taylor, *Scattering Theory* (Wiley, New York, 1972).

[43] *Handbook of Mathematical Functions*, edited by M. Abramowitz and I. A. Stegun (National Bureau of Standards, Washington, DC, 1964).

[44] S. J. Wallace, *Phys. Rev. Lett.* **27**, 622 (1971); *Phys. Rev. D* **8**, 1846 (1973).

[45] D. Waxman, C. Wilkin, J.-F. Germond and R. J. Lombard, *Phys. Rev. C* **24**, 578 (1981).

[46] A. Vitturi and F. Zardi, *Phys. Rev. C* **36**, 1404 (1987).

[47] S. M. Lenzi, A. Vitturi and F. Zardi, *Phys. Rev. C* **40**, 2114 (1989).

[48] L. Ray, *Phys. Rev. C* **20**, 1857 (1979).

[49] M. S. Hussein, R. A. Rego and C. A. Bertulani, *Phys. Rep.* **201**, 279 (1991).

[50] CNS DAC Services (SAID Program), http://gwdac.phys.gwu.edu.

[51] P. J. Karol, *Phys. Rev. C* **11**, 1203 (1975).

[52] R. M. de Vries and J. C. Peng, *Phys. Rev. C* **22**, 1055 (1980).

[53] S. Kox *et al.*, *Phys. Rev. C* **35**, 1678 (1987).

[54] D. M. Brink and G. R. Satchler, *J. Phys. G* **7**, 43 (1981).

[55] N. J. DiGiacomo, R. M. DeVries and J. C. Peng, *Phys. Rev. Lett.* **45**, 527 (1980); *Phys. Lett. B* **101**, 383 (1981).

[56] J. Matero, *Z. Phys. A* **351**, 29 (1995).

[57] J. Berger *et al.*, *Nucl. Phys. A* **338**, 421 (1980).

[58] B. Abu-Ibrahim, K. Fujimura and Y. Suzuki, *Nucl. Phys. A* **657**, 391 (1999).

[59] B. Abu-Ibrahim, PhD thesis, Cairo University (2000).

[60] G. S. Blanpied *et al.*, *Phys. Rev. Lett.* **39**, 1447 (1977).

[61] K. W. Jones *et al.*, *Phys. Rev. C* **33**, 17 (1986).

[62] H. O. Meyer, P. Schwandt, G. L. Moake and P. P. Singh, *Phys. Rev. C* **23**, 616 (1981).

[63] B. Abu-Ibrahim and Y. Suzuki, *Phys. Rev. C* **62**, 034608 (2000).

[64] A. Chaumeaux *et al.*, *Nucl. Phys. A* **267**, 413 (1976).

[65] B. Abu-Ibrahim and Y. Suzuki, *Phys. Rev. C* **61**, 051601 (2000).

[66] J. Jaros *et al.*, *Phys. Rev. C* **18**, 2273 (1978).

[67] Y. Ogawa, K. Yabana and Y. Suzuki, *Nucl. Phys. A* **543**, 722 (1992).

[68] I. Tanihata, D. Hirata, K. Kobayashi, S. Shimoura, K. Sugimoto and H. Toki, *Phys. Lett. B* **289**, 261 (1992).

[69] Y. Suzuki, T. Kido, Y. Ogawa, K. Yabana and D. Baye, *Nucl. Phys. A* **567**, 957 (1994).

[70] T. Aumann *et al.*, *Phys. Rev. Lett.* **84**, 35 (2000).

[71] S. Fortier *et al.*, *Phys. Lett. B* **461**, 22 (1999).

[72] J. A. Tostevin, *J. Phys. G* **25**, 735 (1999).

[73] G. F. Bertsch, H. Esbensen and A. Sustich, *Phys. Rev. C* **42**, 758 (1990).

[74] J. S. Al-Khalili and J. A. Tostevin, *Phys. Rev. Lett.* **76**, 3903 (1996); J. S. Al-Khalili, J. A. Tostevin and I. J. Thompson, *Phys. Rev. C* **54**, 1843 (1996).

[75] T. Kobayashi, in *Proc. 1st Int. Conf. on Radioactive Nuclear Beams, Berkeley*, edited by W. D. Myers, J. M. Nitschke and E. B. Norman (World Scientific, Singapore, 1990), p. 325.

[76] G. F. Bertsch, B. A. Brown and H. Sagawa, *Phys. Rev. C* **39**, 1154 (1989).

[77] K. Soutome, S. Yamaji and M. Sano, *Nucl. Phys. A* **538**, 383c (1992).

[78] C. A. Bertulani, G. Baur and M. S. Hussein, *Nucl. Phys. A* **526**, 751 (1991).

[79] K. Yabana, Y. Ogawa and Y. Suzuki, *Nucl. Phys. A* **539**, 295 (1992).

[80] K. Yabana, Y. Ogawa and Y. Suzuki, *Phys. Rev. C* **45**, 2909 (1992).

[81] H. Esbensen and G. F. Bertsch, *Phys. Rev. C* **46**, 1552 (1992).

[82] G. F. Bertsch, K. Hencken and H. Esbensen, *Phys. Rev. C* **57**, 1366 (1998).

[83] A. deShalit and H. Feshbach, *Theoretical Nuclear Physics* (Wiley, New York, 1974), Vol. I: Nuclear Structure.

[84] J. P. Jeukenne, A. Lejeune and C. Mahaux, *Phys. Rep.* **25**, 83 (1976).

[85] F. Brieva and R. J. Rook, *Nucl. Phys. A* **291**, 299; 317 (1977).

[86] M. E. Brandan and G. R. Satchler, *Phys. Rep.* **285**, 143 (1997).

[87] G. R. Satchler and W. G. Love, *Phys. Rep.* **55**, 184 (1979).

[88] Y. Sakuragi, M. Yahiro and M. Kamimura, *Prog. Theor. Phys. Suppl.* No. **89**, 136 (1986).

[89] R. C. Johnson and P. J. R. Soper, *Phys. Rev. C* **1**, 976 (1970).

[90] M. Yahiro, Y. Iseri, H. Kameyama, M. Kamimura and M. Kawai, *Prog. Theor. Phys. Suppl.* No. **89**, 32 (1986).

[91] F. Negoita *et al.*, *Phys. Rev. C* **59**, 2082 (1999).

[92] F. Negoita *et al.*, *Phys. Rev. C* **54**, 1787 (1996).

[93] M. Fukuda *et al.*, *Phys. Lett. B* **268**, 339 (1991).

[94] M. D. Cortina-Gil, PhD thesis, Université de Caen (1996).

[95] R. C. Johnson, J. S. Al-Khalili and J. A. Tostevin, *Phys. Rev. Lett.* **79**, 2771 (1997).

[96] M. Zahar *et al.*, *Phys. Rev. C* **49**, 1540 (1994).

[97] M. D. Cortina-Gil *et al.*, *Phys. Lett. B* **401**, 9 (1997).

[98] M. C. Mermaz, *Phys. Rev. C* **47**, 2213 (1993).

[99] S. G. Cooper and R. S. Mackintosh, *Nucl. Phys. A* **582**, 283 (1995).

[100] S. Hirenzaki, H. Toki and I. Tanihata, *Nucl. Phys. A* **552**, 57 (1993).

[101] R. Kanungo and C. Samanta, *Nucl. Phys. A* **617**, 265 (1997).

[102] Y. Suzuki, K. Yabana and Y. Ogawa, *Phys. Rev. C* **47**, 1317 (1993).

[103] I. J. Thompson, J. S. Al-Khalili, J. A. Tostevin and J. M. Bang, *Phys. Rev. C* **47**, R1364 (1993).

[104] J. S. Al-Khalili, *Nucl. Phys. A* **581**, 315 (1995).

[105] J. S. Al-Khalili, I. J. Thompson and J. A. Tostevin, *Nucl. Phys. A* **581**, 331 (1995).

[106] J. S. Al-Khalili, J. A. Tostevin and J. M. Brooke, *Phys. Rev. C* **55**, R1018 (1997).

[107] J. A. Christley, J. S. Al-Khalili, J. A. Tostevin and R. C. Johnson, *Nucl. Phys. A* **624**, 275 (1997).

[108] R. C. Johnson, *J. Phys. G* **24**, 1583 (1998).

[109] H. Esbensen and G. F. Bertsch, *Phys. Rev. C* **59**, 3240 (1999).

[110] R. Crespo, J. A. Tostevin and I. J. Thompson, *Phys. Rev. C* **54**, 1867 (1996); J. A. Thompson, F. M. Nunes and I. J. Thompson, *ibid.* **63**, 024617 (2001).

[111] G. D. Alkhazov *et al.*, *Phys. Rev. Lett.* **78**, 2313 (1997).

[112] A. A. Korsheninnikov *et al.*, *Phys. Rev. Lett.* **78**, 2317 (1997).

[113] S. Karataglidis, P. G. Hansen, B. A. Brown, K. Amos and P. J. Dortmans, *Phys. Rev. Lett.* **79**, 1447 (1997).

[114] D. Aleksandrov *et al.*, *Nucl. Phys. A* **669**, 51 (2000).

[115] R. Serber, *Phys. Rev.* **72**, 1008 (1947).

[116] R. J. Glauber, *Phys. Rev.* **99**, 1515 (1955).

[117] K. Van Bibber *et al.*, *Phys. Rev. Lett.* **43**, 840 (1979).

[118] D. E. Greiner, P. J. Lindstrom, H. H. Heckman, B. Cork and F. S. Bieser, *Phys. Rev. Lett.* **35**, 152 (1975).

[119] A. S. Goldhaber, *Phys. Lett. B* **53**, 306 (1974).

[120] W. A. Friedman, *Phys. Rev. C* **27**, 569 (1983).

[121] E. J. Moniz, I. Sick, R. R. Whitney, J. R. Ficenec, R. D. Kephart and W. P. Trower, *Phys. Rev. Lett.* **26**, 445 (1971).

[122] T. Fujita and J. Hüfner, *Nucl. Phys. A* **343**, 493 (1980).

[123] N. A. Orr, *Nucl. Phys. A* **616**, 155c (1997).

[124] E. Garrido, D. V. Fedorov and A. S. Jensen, *Phys. Rev. C* **53**, 3159 (1996); *ibid.* **55**, 1327 (1997).

[125] S. N. Ershov, B. V. Danilin, T. Rogde and J. S. Vaagen, *Phys. Rev. Lett.* **82**, 908 (1999); S. N. Ershov, B. V. Danilin and J. S. Vaagen, *Phys. Rev. C* **62**, R041001 (2000).

[126] C. A. Bertulani and K. W. McVoy, *Phys. Rev. C* **46**, 2638 (1992).

[127] F. Barranco, E. Vigezzi and R. A. Broglia, *Phys. Lett. B* **319**, 387 (1993).

[128] H. Esbensen, *Phys. Rev. C* **53**, 2007 (1996).

[129] K. Hencken, G. F. Bertsch and H. Esbensen, *Phys. Rev. C* **54**, 3043 (1996).

[130] M. S. Hussein and K. W. McVoy, *Nucl. Phys. A* **445**, 124 (1985).

[131] R. Anne *et al.*, *Nucl. Phys. A* **575**, 125 (1994).

[132] S. Grevy *et al.*, *Nucl. Phys. A* **650**, 47 (1999).

[133] A. Bonaccorso and D. M. Brink, *Phys. Rev. C* **57**, R22 (1998); *ibid.* **58**, 2864 (1998).

[134] T. Minamisono *et al.*, *Phys. Rev. Lett.* **69**, 2058 (1992).

[135] J. H. Kelley *et al.*, *Phys. Rev. Lett.* **77**, 5020 (1996).

[136] P. G. Hansen, *Phys. Rev. Lett.* **77**, 1016 (1996).

[137] Y. Ogawa and I. Tanihata, *Nucl. Phys. A* **616**, 239c (1997).

[138] M. H. Smedberg *et al.*, *Phys. Lett. B* **452**, 1 (1999).

[139] W. Schwab *et al.*, *Z. Phys. A* **350**, 283 (1995).

[140] D. Cortina-Gil *et al.*, private communication (2001).

[141] P. Ring and P. Schuck, *The Nuclear Many-Body Problem* (Springer, Berlin, 1980).

[142] K. Ikeda, NS report, JHP-7INS (1988).

[143] Y. Suzuki, K. Ikeda and H. Sato, in *Proc. 1st Int. Conf. on Radioactive Nuclear Beams, Berkeley*, edited by W. D. Myers, J. M. Nitschke and E. B. Norman (World Scientific, Singapore, 1990), p. 279; *Prog. Theor. Phys.* **83**, 180 (1990).

[144] K. Ikeda, *Nucl. Phys. A* **538**, 355c (1992).

[145] Y. Suzuki and Y. Tosaka, *Nucl. Phys. A* **517**, 599 (1990).

[146] G. Bertsch and J. Foxwell, *Phys. Rev. C* **41**, 1300 (1990).

[147] B. Gyarmati, A. M. Lane and J. Zimányi, *Phys. Lett. B* **50**, 316 (1974).

[148] T. Nakamura *et al.*, *Phys. Lett. B* **331**, 296 (1994).

[149] D. Sackett *et al.*, *Phys. Rev. C* **48**, 118 (1993).

[150] S. Shimoura, T. Nakamura, M. Ishihara, N. Inabe, T. Kobayashi, T. Kubo, R. H. Siemssen, I. Tanihata and Y. Watanabe, *Phys. Lett. B* **348**, 29 (1995).

[151] T. Aumann *et al.*, *Phys. Rev. C* **59**, 1252 (1999).

[152] A. Csótó, *Phys. Rev. C* **49**, 3035 (1994).

[153] B. V. Danilin, I. J. Thompson, M. V. Zhukov, J. S. Vaagen and J. M. Bang, *Phys. Lett. B* **333**, 299 (1994).

[154] H. Kurasawa and T. Suzuki, *Prog. Theor. Phys.* **94**, 931 (1995).

[155] Y. Alhassid, M. Gai and G. F. Bertsch, *Phys. Rev. Lett.* **49**, 1482 (1982).

[156] H. Sagawa and M. Honma, *Phys. Lett. B* **251**, 17 (1990).

[157] Y. Suzuki, *Nucl. Phys. A* **528**, 395 (1991).

[158] G. F. Bertsch and H. Esbensen, *Ann. Phys. (N. Y)* **209**, 327 (1991).

[159] C. A. Bertulani and G. Baur, *Phys. Rep.* **163**, 299 (1988).

[160] A. Winther and K. Alder, *Nucl. Phys. A* **319**, 518 (1979).

[161] T. Motobayashi *et al.*, *Phys. Rev. Lett.* **73**, 2680 (1994).

[162] N. Iwasa *et al.*, *J. Phys. Soc. Japan* **65**, 1256 (1996).

[163] H. Esbensen and G. F. Bertsch, *Phys. Lett. B* **359**, 13 (1995); *Nucl. Phys. A* **600**, 37 (1996).

[164] S. Typel, H. H. Wolter and G. Baur, *Nucl. Phys. A* **613**, 147 (1997).

[165] K. Ieki *et al.*, *Phys. Rev. Lett.* **70**, 730 (1993).

[166] J. E. Bush *et al.*, *Phys. Rev. Lett.* **81**, 61 (1998).

[167] T. Kido, K. Yabana and Y. Suzuki, *Phys. Rev. C* **50**, 1276 (1994).

[168] T. Kido, K. Yabana and Y. Suzuki, *Phys. Rev. C* **53**, 2296 (1996).

[169] H. Esbensen, G. F. Bertsch and C. A. Bertulani, *Nucl. Phys. A* **581**, 107 (1995).

[170] L. F. Canto, R. Donangelo, A. Romanelli, M. S. Hussein and A. F. R. de Toledo Piza, *Phys. Rev. C* **55**, 570 (1997).

[171] V. S. Melezhik and D. Baye, *Phys. Rev. C* **59**, 3232 (1999).

[172] S. Typel and G. Baur, *Nucl. Phys. A* **573**, 486 (1994); *Phys. Rev. C* **54**, 2104 (1994).

[173] A. Bohr and B. R. Mottelson, *Nuclear Structure* (Benjamin, New York, 1969), Vol. I.

[174] A. Bohr and B. R. Mottelson, *Nuclear Structure* (Benjamin, New York, 1975), Vol. II.

[175] G. Audi, O. Bersillon, J. Blachot and A. H. Wapstra, *Nucl. Phys. A* **624**, 1 (1997).

[176] N. Fukunishi, T. Otsuka and I. Tanihata, *Phys. Rev. C* **48**, 1648 (1993).

[177] V. Maddalene *et al.*, *Phys. Rev. C* **63**, 024613 (2001).

[178] A. Ozawa, T. Kobayashi, T. Suzuki, K. Yoshida and I. Tanihata, *Phys. Rev. Lett.* **84**, 5493 (2000).

[179] H. Simon *et al.*, *Phys. Rev. Lett.* **83**, 496 (1999).

[180] Y. Ogawa, Y. Suzuki and K. Yabana, *Nucl. Phys. A* **571**, 784 (1994).

[181] I. J. Thompson and M. V. Zhukov, *Phys. Rev. C* **49**, 1904 (1994).

[182] M. Labiche *et al.*, *Phys. Rev. Lett.* **86**, 600 (2001).

[183] B. Buck, A. C. Merchant and S. M. Perez, *Phys. Rev. C* **58**, 2049 (1998).

[184] P. J. Brussaard and P. W. M. Glaudemans, *Shell-model Applications in Nuclear Spectroscopy* (North-Holland, Amsterdam, 1977).

[185] Y. Suzuki and K. Ikeda, *Phys. Rev. C* **38**, 410 (1988).

[186] Y. Suzuki and J. J. Wang, *Phys. Rev. C* **41**, 736 (1990).

[187] Y. Suzuki and K. Varga, *Stochastic Variational Approach to Quantum-Mechanical Few-Body Problems* (Springer, Berlin, 1998).

[188] Y. Tosaka and Y. Suzuki, *Nucl. Phys. A* **512**, 46 (1990).

[189] Y. Tosaka, Y. Suzuki and K. Ikeda, *Prog. Theor. Phys.* **83**, 1140 (1990).

[190] S. Hara, K. Ogawa and Y. Suzuki, *Prog. Theor. Phys.* **88**, 329 (1992).

[191] B. H. Wildenthal, in *Progress in Particle and Nuclear Physics*, edited by D. Wilkinson (Pergamon, Oxford, 1984), Vol. 11, p. 5.

[192] B. A. Brown, W. A. Richter, R. E. Julies and B. H. Wildenthal, *Ann. Phys. (N. Y.)* **182**, 191 (1988).

[193] F. Ajzenberg-Selove, *Nucl. Phys. A* **460**, 1 (1986); *ibid.* **475**, 1 (1987).

[194] P. M. Endt, *Nucl. Phys. A* **521**, 1 (1990); *ibid.* **633**, 1 (1998).

[195] S. Mukai, S. Aoyama, K. Katō and K. Ikeda, *Prog. Theor. Phys.* **99**, 381 (1999).

[196] M. V. Zhukov, B. V. Danilin, D. V. Fedorov, J. M. Bang, I. J. Thompson and J. S. Vaagen, *Phys. Rep.* **231**, 150 (1993).

[197] V. I. Kukulin, V. M. Krasnopol'sky, V. T. Voronchev and P. B. Sazonov, *Nucl. Phys. A* **417**, 128 (1984); *ibid.* **453**, 365 (1986); V. I. Kukulin, V. N. Pomerantsev, Kh. D. Rasikov, V. T. Voronchev and G. G. Ryzhikh, *ibid.* **586**, 151 (1995).

[198] H. Horiuchi, K. Ikeda and Y. Suzuki, *Prog. Theor. Phys. Suppl.* No. **52**, 89 (1972).

[199] K. Ikeda *et al.*, Comprehensive Study of Structure of Light Nuclei (*Prog. Theor. Phys. Suppl.* No. **68**, Kyoto, 1980).

[200] A. T. Kruppa, P.-H. Heenen, H. Flocard and R. J. Liotta, *Phys. Rev. Lett.* **79**, 2217 (1997).

[201] S. Takami, K. Yabana and K. Ikeda, *Prog. Theor. Phys.* **94**, 1011 (1995).

[202] X. Li and P.-H. Heenen, *Phys. Rev. C* **54**, 1617 (1996).

[203] H. Feldmeier, *Nucl. Phys. A* **515**, 147 (1990).

[204] H. Horiuchi, *Nucl. Phys. A* **522**, 257c (1991).

[205] S. F. Boys, *Proc. Roy. Soc. A* **258**, 402 (1960); K. Singer, *ibid.* **258**, 412 (1960).

[206] K. Varga and Y. Suzuki, *Phys. Rev. C* **52**, 2885 (1995); *Phys. Rev. A* **53**, 1907 (1996).

[207] B. S. Pudliner, V. R. Pandharipande, J. Carlson and R. B. Wiringa, *Phys. Rev. Lett.* **74**, 4396 (1995).

[208] N. Metropolis, A. Rosenbluth, M. Rosenbluth, A. Teller and E. Teller, *J. Chem. Phys.* **21**, 1087 (1953).

[209] B. Jeziorski, R. Bukowski and K. Szalewicz, *Int. J. Quantum Chem.* **61**, 769 (1997).

[210] M. Moshinsky and Yu. F. Smirnov, *The Harmonic Oscillator in Modern Physics* (Harwood Academic, Amsterdam, 1996).

[211] M. Kamimura and H. Kameyama, *Nucl. Phys. A* **508**, 17 (1990); H. Kameyama, M. Kamimura and Y. Fukushima, *Phys. Rev. C* **40**, 974 (1989).

[212] K. Varga, Y. Suzuki and Y. Ohbayasi, *Phys. Rev. C* **50**, 189 (1994).

[213] K. Arai, Y. Suzuki and K. Varga, *Phys. Rev. C* **51**, 2488 (1995).

[214] K. Varga, Y. Suzuki and I. Tanihata, *Phys. Rev. C* **52**, 3013 (1995).

[215] Y. Akaishi, in *International Review of Nuclear Physics* (World Scientific, Singapore, 1986), Vol. 4, p. 259.

[216] A. Kievsky, M. Viviani and S. Rosati, *Nucl. Phys. A* **501**, 503 (1989); *ibid.* **551**, 241 (1993); *ibid.* **577**, 511 (1994).

[217] J. Carlson and V. R. Pandharipande, *Nucl. Phys. A* **371**, 301 (1981).

[218] R. B. Wiringa, *Nucl. Phys. A* **543**, 199c (1992).

[219] S. Fantoni, L. Panattoni and S. Rosati, *Nouvo Cimento A* **69**, 81 (1970).

[220] S. A. Alexander, H. J. Monkhorst and K. Szalewicz, *J. Chem. Phys.* **85**, 5821 (1986), *ibid.* **87**, 3976 (1987); *ibid.* **89**, 355 (1988).

[221] D. A. Varshalovich, A. N. Moskalev and V. K. Khersonskii, *Quantum Theory of Angular Momentum* (World Scientific, Singapore, 1988).

[222] K. T. Hecht, *The Vector Coherent State Method and Its Application to Problems of Higher Symmetries* (Springer, Berlin, 1987).

[223] V. Bargmann, *Commun. Pure and Appl. Math.* **14**, 187 (1961).

[224] K. T. Hecht, E. J. Reske, T. H. Seligman and W. Zahn, *Nucl. Phys. A* **356**, 146 (1981).

[225] Y. Suzuki, *Nucl. Phys. A* **405**, 40 (1983).

[226] A. deShalit and I. Talmi, *Nuclear Shell Theory* (Academic, New York, 1963).

[227] R. D. Lawson, *Theory of the Nuclear Shell Model* (Clarendon, Oxford, 1980).

[228] D. Hill and J. A. Wheeler, *Phys. Rev.* **89**, 1102 (1953).

[229] J. Yoccoz, in *Proc. Int. School of Physics "Enrico Fermi", Course XXXVI*, edited by C. Bloch (Academic, New York, 1966), p. 474.

[230] P. O. Löwdin, *Phys. Rev.* **97**, 1490 (1955).

[231] D. M. Brink, in *Proc. Int. School of Physics "Enrico Fermi", Course XXXVI*, edited by C. Bloch (Academic, New York, 1966), p. 247.

[232] H. Horiuchi, *Prog. Theor. Phys. Suppl.* No. **62**, 90 (1977).

[233] A. N. Antonov, P. E. Hodgson and I. Zh. Petkov, *Nucleon Momentum and Density Distributions in Nuclei* (Clarendon, Oxford, 1988).

[234] T. de Forest, Jr and J. D. Walecka, *Adv. Phys.* **15**, 1 (1966).

[235] V. I. Kukulin and V. M. Krasnopol'sky, *J. Phys. G* **3**, 795 (1977).

[236] K. Varga, Y. Suzuki and R. G. Lovas, *Nucl. Phys. A* **571**, 447 (1994).

[237] K. Varga, Y. Ohbayasi and Y. Suzuki, *Phys. Lett. B* **396**, 1 (1997).

[238] Y. Suzuki, J. Usukura and K. Varga, *J. Phys. B* **31**, 31 (1998).

[239] W. H. Press, S. A. Teukolsky, W. T. Vetterling and B. P. Flanney, *Numerical Recipes in FORTRAN*, 2nd ed. (Cambridge University Press, New York, 1992).

[240] D. Cvijovic and J. Klinowski, *Science* **267**, 664 (1995).

[241] P. Serra, F. Stanton and S. Kais, *Phys. Rev. E* **55**, 1162 (1997).

[242] S. Kirkpatrick, C. D. Gelett, Jr and M. P. Vecchi, *Science* **220**, 671 (1983).

[243] D. E. Goldberg, *Genetic Algorithms in Search, Optimization and Machine Learning* (Addison Wesley, Reading, Massachusetts, 1989).

[244] M. W. Schmidt and K. Ruedenberg, *J. Chem. Phys.* **71**, 3951 (1979).

[245] M. Kamimura, *Phys. Rev. A* **38**, 621 (1988).

[246] A. J. Thakker and V. H. Smith, Jr, *Phys. Rev. A* **15**, 1 (1977), *ibid.* **15**, 16 (1977).

[247] S. A. Alexander and H. J. Monkhorst, *Phys. Rev. A* **38**, 26 (1988).

[248] R. A. Malfliet and J. A. Tjon, *Nucl. Phys. A* **127**, 161 (1969).

[249] K. Varga, J. Usukura and Y. Suzuki, *Phys. Rev. Lett.* **80**, 1876 (1998).

[250] J. Usukura, K. Varga and Y. Suzuki, *Phys. Rev. A* **58**, 1918 (1998).

[251] G. G. Ryzhikh, J. Mitroy and K. Varga, *J. Phys. B* **31**, 265 (1998).

[252] L. Ya. Glozman, Z. Papp, W. Plessas, K. Varga and R. F. Wagenbrunn, *Nucl. Phys. A* **623**, 90 (1997); *Phys. Rev. C* **57**, 3406 (1998); L. Ya. Glozman, W. Plessas, K. Varga, and R. F. Wagenbrunn, *Phys. Rev. D* **58**, 094030 (1998).

[253] J. Usukura, Y. Suzuki and K. Varga, *Phys. Rev. B* **59**, 5652 (1999).

[254] K. Varga, P. Navratil, J. Usukura and Y. Suzuki, *Phys. Rev. B* **63**, 205308 (2001).

[255] S. Nakaichi-Maeda and Y. Akaishi, *Prog. Theor. Phys.* **84**, 1025 (1990).

[256] H. Nemura, Y. Suzuki, Y. Fujiwara and C. Nakamoto, *Prog. Theor. Phys.* **103**, 929 (2000).

[257] E. Hiyama, M. Kamimura, T. Motoba, T. Yamada and Y. Yamamoto, *Prog. Theor. Phys.* **97**, 881 (1997).

[258] M. M. Nagels, T. A. Rijken and J. J. de Swart, *Phys. Rev. D* **15**, 2547 (1977).

[259] H. De Raedt and M. Frick, *Phys. Rep.* **231**, 109 (1993).

[260] J. Horáček V. I. Kukulin, V. M. Krasnopol'sky, *Theory of Resonances: Principles and Applications* (Kluwer, Dordrecht, 1989).

[261] E. Hernandez, A. Jauregui and A. Mondragon, *J. Phys. A* **33**, 4507 (2000).

[262] R. G. Newton, *Scattering Theory of Waves and Particles*, 2nd ed. (Springer, Berlin, 1982).

[263] A. G. Sitenko, *Scattering Theory* (Springer, Berlin, 1991), p. 100.

[264] A. Csótó, R. G. Lovas and A. T. Kruppa, *Phys. Rev. Lett.* **70**, 1389 (1993).

[265] T. Vertse, K. F. Pál and Z. Balogh, *Comput. Phys. Commun.* **27**, 309 (1982).

[266] B. G. Giraud, M. V. Mihailović, R. G. Lovas and M. A. Nagarajan, *Ann. Phys. (N. Y.)* **140**, 29 (1982).

[267] A. U. Hazi and H. S. Taylor, *Phys. Rev. A* **1**, 1109 (1970).

[268] R. G. Lovas and M. A. Nagarajan, *J. Phys. A* **15**, 2383 (1982).

[269] E. Holøien and J. Midtdal, *J. Chem. Phys.* **45**, 2209 (1966).

[270] A. T. Kruppa and K. Arai, *Phys. Rev. A* **59**, 3556 (1999).

[271] P. Descouvemont and M. Vincke, *Phys. Rev. A* **42**, 3835 (1990).

[272] J. Usukura and Y. Suzuki, Proc. Int. Workshop on Resonances in Few-Body Systems, Sárospatak, *Few-Body Systems Suppl.* **13**, 56 (2001).

[273] J. Aguilar and J. M. Combes, *Commun. Math. Phys.* **22**, 269; E. Balslev and J. M. Combes, *ibid.* 280 (1971).

[274] Y. K. Ho, *Phys. Rep.* **99**, 1 (1983).

[275] A. Csótó, *Phys. Rev. C* **49**, 2244 (1994).

[276] V. I. Kukulin and V. M. Krasnopol'sky, *J. Phys. A* **10**, 33 (1977); V. I. Kukulin, V. M. Krasnopol'sky and M. Miselkhi, *Sov. J. Nucl. Phys.* **29**, 421 (1979).

[277] H. M. Nussenzweig, *Nucl. Phys.* **11**, 499 (1959); H. M. Nussenzweig, *Causality and Dispertion Relations* (Academic, New York, 1972).

[278] N. Tanaka, Y. Suzuki, K. Varga and R. G. Lovas, *Phys. Rev. C* **59**, 1391 (1999).

[279] N. Tanaka, Y. Suzuki and K. Varga, *Phys. Rev. C* **56**, 562 (1997).

[280] R. J. Eden and J. R. Taylor, *Phys. Rev. B* **133**, 1575 (1964).

[281] A. Csótó, B. Gyarmati, A. T. Kruppa, K. F. Pál and N. Moiseyev, *Phys. Rev. A* **41**, 3469 (1990).

[282] Y. Fujiwara, H. Horiuchi, K. Ikeda, M. Kamimura, K. Katō, Y. Suzuki and E. Uegaki, *Prog. Theor. Phys. Suppl.* No. **68**, 29 (1980).

[283] K. Wildermuth and Y. C. Tang, *A Unified Theory of the Nucleus* (Vieweg, Braunschweig, 1977).

[284] R. Beck, F. Dickmann and R. G. Lovas, *Ann. Phys. (N. Y.)* **173**, 1 (1987).

[285] H. Feshbach, *Ann. Phys. (N. Y.)* **19**, 287 (1962).

[286] H. Horiuchi and Y. Suzuki, *Prog. Theor. Phys.* **49**, 1974 (1973).

[287] Y. Suzuki, *Prog. Theor. Phys.* **55**, 1751 (1976); *ibid.* **56**, 111 (1976); Y. Suzuki, T. Ando and B. Imanishi, *Nucl. Phys. A* **295**, 365 (1978); Y. Suzuki and B. Imanishi, *Phys. Rev. C* **23**, 2414 (1981).

[288] A. Săndulescu, E. W. Schmid and G. Spitz, *Few-Body Systems* **5**, 107 (1988).

[289] R. G. Lovas, *Z. Phys. A* **322**, 589 (1985).

[290] R. G. Lovas, in *Proc. 7th Int. Conf. on Clustering Aspects of Nuclear Structure and Dynamics*, edited by Z. Basrak, R. Čaplar and M. Korolija (World Scientific, Singapore, 2000), p. 45.

[291] K. F. Pál and R. G. Lovas, in *Proc. Int. Symp. on In-Beam Nuclear Spectroscopy*, edited by T. Fényes (Akadémiai, Budapest, 1984), p. 507.

[292] A. M. Lane and R. G. Thomas, *Rev. Mod. Phys.* **30**, 257 (1958).

[293] R. G. Lovas, R. J. Liotta, A. Insolia, K. Varga and D. S. Delion, *Phys. Rep.* **294**, 265 (1998).

[294] K. Varga and R. G. Lovas, *Phys. Rev. C* **37**, 2906 (1988).

[295] E. W. Schmid and H. Fiedeldey, *Phys. Rev. C* **39**, 2170 (1989).

[296] K. Ikeda, N. Takigawa and H. Horiuchi, *Prog. Theor. Phys. Suppl.* **Extra Number**, 464 (1968).

[297] K. Varga, R. G. Lovas and R. J. Liotta, *Nucl. Phys. A* **550**, 421 (1992).

[298] Y. Abe, Y. Kondō and T. Matsuse, *Prog. Theor. Phys. Suppl.* No. **68**, 303 (1980).

[299] F. Iachello, *Phys. Rev. C* **23**, 2778 (1981).

[300] J. Cseh and G. Lévai, *Ann. Phys. (N. Y.)* **230**, 165 (1994).

[301] M. Kamimura, *Prog. Theor. Phys. Suppl.* No. **62**, 236 (1977).

[302] S. Saito, *Prog. Theor. Phys.* **40**, 893 (1968); *ibid.* **41**, 705 (1969).

[303] E. W. Schmid, A. Faessler, H. Ito and G. Spitz, *Few-Body Systems* **5**, 45 (1988).

[304] S. Saito, *Prog. Theor. Phys. Suppl.* No. **62**, 11 (1977).

[305] M. Kruglanski and D. Baye, *Nucl. Phys. A* **548**, 39 (1992).

[306] W. Walliser, T. Fliessbach and Y. C. Tang, *Nucl. Phys. A* **437**, 367 (1985).

[307] M. Kamimura and T. Matsuse, *Prog. Theor. Phys.* **51**, 438 (1974).

[308] N. Wiener, *The Fourier Integral and Certain of its Applications* (Cambridge Univ. Press, Cambridge, 1933), p. 100.

[309] G. F. Filippov, V. S. Vasilevskii and L. L. Chopovskii, *Sov. J. Part. Nucl.* **15**, 600 (1984).

[310] H. A. Bethe and M. E. Rose, *Phys. Rev.* **51**, 283 (1937).

[311] J. P. Elliott and T. H. R. Skyrme, *Proc. Roy. Soc. A* **232**, 561 (1955).

[312] V. G. Neudatchin and Yu. F. Smirnov, *Nuklonnie assotsiatsii v legkikh yadrakh* (Nauka, Moscow, 1969).

[313] O. F. Nemets, V. G. Neudatchin, A. T. Rudchik, Yu. F. Smirnov and Yu. M. Chuvil'ski, *Nuklonnie assotsiatsii v atomnikh yadrakh i yadernie reaktsii mnogonuklonnikh peredatch* (Naukova Dumka, Kiev, 1988).

[314] J. P. Elliott, *Proc. Roy. Soc. A* **245**, 128; 562 (1958).

[315] H. Horiuchi and K. Ikeda, *Prog. Theor. Phys.* **40**, 277 (1968).

[316] M. V. Mihailović, L. J. B. Goldfarb and M. A. Nagarajan, *Nucl. Phys. A* **273**, 207 (1976).

[317] D. Baye, P.-H. Heenen and M. Libert-Heinemann, *Nucl. Phys. A* **291**, 230 (1977); D. Baye and P. Descouvemont, *Nucl. Phys. A* **407**, 77 (1983); Proc. 5th Int. Conf. on Clustering Aspects in Nuclear and Subnuclear Systems, *J. Phys. Soc. Jpn. Suppl.* **58**, 103 (1989).

[318] R. Beck, R. Krivec and M. V. Mihailović, *Nucl. Phys. A* **363**, 365 (1981).

[319] B. Jancovici and D. H. Schiff, *Nucl. Phys.* **58**, 678 (1964).

[320] H. Ui and L. C. Biedenharn, *Phys. Lett. B* **27**, 608 (1968).

[321] D. Brink and A. Weiguny, *Nucl. Phys. A* **120**, 59 (1968).

[322] W. Glöckle, *Nucl. Phys. A* **211**, 372 (1973).

[323] C. W. Wong, *Phys. Rep.* **15**, 283 (1975).

[324] Y. Suzuki, *Prog. Theor. Phys.* **50**, 341 (1973).

[325] Y. Suzuki, E. J. Reske and K. T. Hecht, *Nucl. Phys. A* **381**, 77 (1982).

[326] R. Beck, F. Dickmann and A. T. Kruppa, *Phys. Rev. C* **30**, 1044 (1984).

[327] A. B. Volkov, *Nucl. Phys.* **74**, 33 (1965).

[328] R. G. Lovas and A. T. Kruppa, in *Developments of Nuclear Cluster Dynamics*, edited by Y. Akaishi *et al.* (World Scientific, Singapore, 1989), p. 39.

[329] J. P. Vary and C. B. Dover, *Phys. Rev. Lett.* **31**, 1510 (1973); B. Buck, C. B. Dover and J. P. Vary, *Phys. Rev. C* **11**, 1803 (1975); B. Buck, A. C. Merchant and S. M. Perez, *Phys. Rev. Lett.* **76**, 380 (1996).

[330] Y. Suzuki and H. Nemura, *Prog. Theor. Phys.* **102**, 203 (1999).

[331] A. Tohsaki-Suzuki, M. Kamimura and K. Ikeda, *Prog. Theor. Phys. Suppl.* No. **68**, 359 (1980).

[332] H. Friedrich and K. Langanke, in *Advances in Nuclear Physics*, edited by J. W. Negele and E. Vogt (Plenum, New York, 1986), Vol. 17, p. 223.

[333] A. T. Kruppa and K. Katō, *Prog. Theor. Phys.* **84**, 1145 (1990).

[334] W. Timm, H. R. Fiebig and H. Friedrich, *Phys. Rev. C* **25**, 79 (1982).

[335] H. Horiuchi, in *Trends in Theoretical Physics*, edited by P. J. Ellis and Y. C. Tang (Addison-Wesley, New York, 1991), Vol. 2, p. 277.

[336] K. Katō, Proc. 5th Int. Conf. on Clustering Aspects in Nuclear and Subnuclear Systems, *J. Phys. Soc. Jpn. Suppl.* **58**, 49 (1989).

[337] H. Friedrich, *Phys. Rep.* **74**, 209 (1981).

[338] B. Buck, H. Friedrich and C. Wheatley, *Nucl. Phys. A* **275**, 246 (1977).

[339] F. Perey and B. Buck, *Nucl. Phys.* **32**, 353 (1962).

[340] B. Buck and R. Lipperheide, *Nucl. Phys. A* **368**, 141 (1981).

[341] H. Horiuchi, *Prog. Theor. Phys.* **64**, 184 (1980).

[342] H. Horiuchi, *Prog. Theor. Phys.* **69**, 516 (1983).

[343] Y. Suzuki and K. T. Hecht, *Phys. Rev. C* **27**, 299 (1983).

[344] R. G. Lovas and K. F. Pál, *Nucl. Phys. A* **424**, 143 (1984).

[345] S. Ali and A. R. Bodmer, *Nucl. Phys.* **80**, 99 (1966).

[346] D. Baye, *Phys. Rev. Lett.* **58**, 2738 (1987).

[347] D. R. Lehman, *Phys. Rev. C* **25**, 3146 (1982).

[348] D. Baye, C. Sauwens, P. Descouvemont and S. Keller, *Nucl. Phys. A* **529**, 467 (1991).

[349] H. Horiuchi, *Prog. Theor. Phys.* **69**, 886 (1983).

[350] K. F. Pál and R. G. Lovas, *Phys. Lett. B* **96**, 19 (1980).

[351] S. Ohkubo *et al.*, Alpha-clustering and Molecular Structure of Medium-Weight and Heavy Nuclei (*Prog. Theor. Phys. Suppl.* No. **132**, Kyoto, 1998).

[352] B. Buck, A. C. Merchant and S. M. Perez, *Phys. Rev. Lett.* **72**, 1326 (1994).

[353] S. Ohkubo, *Phys. Rev. Lett.* **74**, 2176 (1995).

[354] B. Buck, A. C. Merchant and S. M. Perez, *J. Phys. G* **17**, 1223 (1991).

[355] B. Buck, A. C. Merchant and S. M. Perez, *J. Phys. G* **20**, 351 (1994).

[356] J. Lomnitz-Adler, University of Illinois (Urbana-Champaign), ILL–(TH)–78–58 (1978).

[357] E. Schmid, *Z. Phys. A* **297**, 105 (1980).

[358] E. Schmid, *Z. Phys. A* **302**, 311 (1981).

[359] E. W. Schmid, M. Orlowski and Bao Cheng-guang, *Z. Phys. A* **308**, 237 (1982).

[360] S. Nakaichi-Maeda and E. W. Schmid, *Z. Phys. A* **318**, 171 (1984).

[361] K. Varga and R. G. Lovas, *Phys. Rev. C* **43**, 1201 (1991).

[362] Y. Suzuki, K. Arai, Y. Ogawa and K. Varga, *Phys. Rev. C* **54**, 2073 (1996).

[363] K. Suzuki and S. Y. Lee, *Prog. Theor. Phys.* **64**, 2091 (1980).

[364] T. Kajino, T. Matsuse and A. Arima, *Nucl. Phys. A* **413**, 323 (1984) *ibid.* **414** 185, (1984).

[365] A. Csótó and R. G. Lovas, *Phys. Rev. C* **46**, 576 (1992).

[366] R. G. Lovas, A. T. Kruppa, R. Beck and F. Dickmann, *Nucl. Phys. A* **474**, 451 (1987).

[367] D. M. Brink and E. Boeker, *Nucl. Phys. A* **91**, 1 (1967).

[368] A. Hasegawa and S. Nagata, *Prog. Theor. Phys.* **45**, 1786 (1971); F. Tanabe, A. Tohsaki and R. Tamagaki, *ibid.* **53**, 677 (1975).

[369] D. R. Thompson, M. LeMere and Y. C. Tang, *Nucl. Phys. A* **286**, 53 (1977).

[370] I. Reichstein and Y. C. Tang, *Nucl. Phys. A* **158**, 529 (1970).

[371] O. Dumbrajs, R. Koch, H. Pilkuhn, G. C. Oades H. Behrens, J. J. de Swart and P. Kroll, *Nucl. Phys. B* **216**, 277 (1983).

[372] P. M. Kozlowski and L. Adamowicz, *Phys. Rev. A* **48**, 1903 (1993).

[373] Y. Kanada-En'yo, H. Horiuchi and A. Ono, *Phys. Rev. C* **52**, 628 (1995); Y. Kanada-En'yo and H. Horiuchi, *ibid.* **52**, 647 (1995).

[374] R. B. Wiringa, S. Pieper, J. Carlson and V. R. Pandharipande, *Phys. Rev. C* **62**, 014001 (2000).

[375] R. Krivec and M. V. Mihailović, *J. Phys. G* **8**, 821 (1982).

[376] A. T. Kruppa, R. G. Lovas, R. Beck and F. Dickmann, *Phys. Lett. B* **179**, 317 (1987).

[377] A. T. Kruppa, R. Beck and F. Dickmann, *Phys. Rev. C* **36**, 327 (1987).

[378] R. G. Lovas, A. T. Kruppa and J. B. J. M. Lanen, *Nucl. Phys. A* **516**, 325 (1990).

[379] H. Kanada, T. Kaneko, S. Saito and Y. C. Tang, *Nucl. Phys. A* **444**, 209 (1985).

[380] Y. Suzuki and K. Yabana, *Phys. Lett. B* **272**, 173 (1991).

[381] D. Baye, Y. Suzuki and P. Descouvemont, *Prog. Theor. Phys.* **91**, 271 (1994).

[382] A. Csótó and D. Baye, *Phys. Rev. C* **49**, 818 (1994).

[383] M. Unkelbach and H. M. Hofmann, *Few-Body Systems* **11**, 143 (1991).

[384] A. Csótó, *Phys. Rev. C* **48**, 165 (1993).

[385] R. G. Lovas, K. Arai, Y. Suzuki and K. Varga, *Nuovo Cimento A* **110**, 907 (1998).

[386] K. Arai, Y. Suzuki and R. G. Lovas, *Phys. Rev. C* **59**, 1432 (1999).

[387] F. Ajzenberg-Selove, *Nucl. Phys. A* **490**, 1 (1988).

[388] G. M. Hale, R. E. Brown and N. Jarmie, *Phys. Rev. Lett.* **59**, 763 (1987).

[389] A. Csótó and G. M. Hale, *Phys. Rev. C* **55**, 536 (1997).

[390] A. G. M. van Hees and P. W. M. Glaudemans, *Z. Phys. A* **315**, 223 (1984); A. G. M. van Hees, A. A. Wolters and P. W. M. Glaudemans, *Nucl. Phys. A* **476**, 61 (1988).

[391] D. C. Zheng, J. P. Vary and B. R. Barrett, *Phys. Rev. C* **50**, 2841 (1994).

[392] R. Beck, F. Dickmann and R. G. Lovas, *Nucl. Phys. A* **446**, 701 (1985).

[393] B. V. Danilin, M. V. Zhukov, A. A. Korshenninikov and L. V. Chulkov, *Sov. J. Nucl. Phys.* **53**, 45 (1991); B. V. Danilin, M. V. Zhukov, S. N. Ershov, F. A. Gareev, R. S. Kurmanov, J. S. Vaagen and J. Bang, *Phys. Rev. C* **43**, 2835 (1991).

[394] D. R. Lehman and W. C. Parke, *Few-Body Systems* **1**, 193 (1986).

[395] J. Bang and C. Gignoux, *Nucl. Phys. A* **313**, 119 (1979).

[396] H. de Vries, C. W. de Jager and C. de Vries, *Atomic Data and Nuclear Data Tables* **36**, 495 (1987).

[397] D. R. Tilley, H. R. Weller and G. M. Hale, *Nucl. Phys. A* **541**, 1 (1992).

[398] F. Petrovich, S. K. Yoon, M. J. Threapleton, R. J. Philpott and J. A. Carr, *Nucl. Phys. A* **563**, 387 (1993).

[399] R. R. Kiziah, M.D. Brown, C. J. Harvey, D. S. Oakley, W. B. Cottingame, R. W. Garnett, S. J. Greene and D. B. Holtkamp, *Phys. Rev. C* **30**, 1643 (1984).

[400] M. J. G. Borge, L. Johannsen, B. Jonson, W. Kurcewicz, T. Nilsson, G. Nyman, K. Riisager, O. Tengblad and K. Wilhelmsen, *Nucl. Phys. A* **560**, 664 (1993).

[401] Particle Data Group, *Eur. Phys. J. C* **15**, 1 (2000).

[402] G. C. Li, I. Sick, R. R. Whitney and M. R. Yearian, *Nucl. Phys. A* **162**, 583 (1971).

[403] T. Janssens, R. Hofstadter, E. B. Hughes and M. R. Yearian, *Phys. Rev.* **142**, 922 (1966).

[404] J. C. Bergstrom, S. B. Kowalski and R. Neuhausen, *Phys. Rev. C* **25**, 1156 (1982).

[405] J. C. Bergstrom, I. P. Auer and R.S. Hicks, *Nucl. Phys. A* **251**, 401 (1975); J. C. Bergstrom, U. Deutschman and R. Neuhausen, *Nucl. Phys. A* **327**, 439 (1979).

[406] F. A. Bumiller, F. R. Buskirk, J. N. Dyer and W. A. Monson, *Phys. Rev. C* **5**, 391 (1972).

[407] S. Weber, M. Kachelriess, M. Unkelbach and H. M. Hofmann, *Phys. Rev. C* **50**, 1492 (1994).

[408] R. B. Wiringa and R. Schiavilla, *Phys. Rev. Lett.* **81**, 4317 (1998).

[409] G. G. Ryzhikh, R. A. Eramzhyan, V. I. Kukulin and Y. M. Tchuvil'sky, *Nucl. Phys. A* **563**, 247 (1993).

[410] Y. Fujiwara and Y. C. Tang, *Phys. Rev. C* **28**, 1869 (1983).

[411] K. F. Pál, R. G. Lovas, M. A. Nagarajan, B. Gyarmati and T. Vertse, *Nucl. Phys. A* **403**, 114 (1983).

[412] N. Sharma and M. A. Nagarajan, *J. Phys. G* **10**, 1703 (1984).

[413] T. Mertelmeier and H. M. Hofmann, *Nucl. Phys. A* **459**, 387 (1986).

[414] H. Kitagawa and H. Sagawa, *Phys. Lett. B* **299**, 1 (1993).

[415] H. Nakada and T. Otsuka, *Phys. Rev. C* **49**, 886 (1994).

[416] Y. Fujiwara and Y. C. Tang, *Phys. Rev. C* **41**, 28 (1990).

[417] D. Baye, P. Descouvemont and N. K. Timofeyuk, *Nucl. Phys. A* **577**, 624 (1994); P. Descouvemont and D. Baye, *Phys. Lett. B* **292**, 235 (1992).

[418] A. Csótó, *Phys. Lett. B* **315**, 24 (1993).

[419] K. Matsuta *et al.*, *Nucl. Phys. A* **588**, 153c (1995).

[420] Yu. E. Penionzhkevich, *Nucl. Phys. A* **616**, 247c (1997).

[421] H.-G. Voelk and D. Fick, *Nucl. Phys. A* **530**, 475 (1991).

[422] E. Arnold, J. Bonn, W. Neu, R. Neugart, E.-W. Otten and the ISOLDE Collaboration, *Z. Phys. A* **331**, 295 (1988).

[423] D. J. Millener, private communication (2001).

[424] J. G. Booten, A. G. M. van Hees, P. W. M. Glaudemans and R. Wervelman, *Phys. Rev. C* **43**, 335 (1991).

[425] A. C. Fonseca, J. Revai and A. Matveenko, *Nucl. Phys. A* **326**, 182 (1979); J. Révai and A. Matveenko, *ibid.* **339**, 448 (1980).

[426] M. C. Orlowski, Bao Cheng-guang and Liu-yuen, *Z. Phys. A* **305**, 249 (1982).

[427] V. T. Voronchev, V. I. Kukulin, V. N. Pomerantsev, Kh. D. Razikov and G. G. Ryzhikh, *Yad. Fiz.* **57**, 1964 (1994).

[428] W. Zahn, *Nucl. Phys. A* **269**, 138 (1976).

[429] S. Okabe, Y. Abe and H. Tanaka, *Prog. Theor. Phys.* **57**, 866 (1977); S. Okabe and Y. Abe, *ibid.* **59**, 315 (1978); *ibid.* **61**, 1049 (1979).

[430] H. Furutani, H. Kanada, T. Kaneko, S. Nagata, H. Nishioka, S. Okabe, S. Saito, T. Sakuda and M. Seya, *Prog. Theor. Phys. Suppl.* No. **68**, 193 (1980).

[431] P. Descouvemont, *Phys. Rev. C* **39**, 1557 (1989).

[432] D. J. Millener, J. W. Olness, E. K. Warburton and S. S. Hanna, *Phys. Rev. C* **28**, 497 (1983).

[433] F. C. Barker, *Aust. J. Phys.* **37**, 267 (1984).

[434] G. Nyman *et al.*, *Nucl. Phys. A* **510**, 189 (1990).

[435] R. Sherr and G. Bertsch, *Phys. Rev. C* **32**, 1809 (1985).

[436] M. A. Tiede *et al.*, *Phys. Rev. C* **52**, 1315 (1995).

[437] S. Dixit *et al.*, *Phys. Rev. C* **43**, 1758 (1991).

[438] B. Pugh, *quoted by M. A. Tiede et al.*, *Phys. Rev. C* **52**, 1315 (1995).

[439] J. P. Glickman *et al.*, *Phys. Rev. C* **43**, 1740 (1991).

[440] K. Arai, Y. Ogawa, Y. Suzuki and K. Varga, *Phys. Rev. C* **54**, 132 (1996).

[441] M. G. Saint-Laurent *et al.*, *Z. Phys. A* **332**, 457 (1989).

[442] U. Meyer-Berkhout, K. W. Ford and A. E. S. Green, *Ann. Phys. (N. Y.)* **8**, 119 (1959).

[443] M. Bernheim, T. Stovall and D. Vinciguerra, *Nucl. Phys. A* **97**, 488 (1967); M. Bernheim, R. Riskalla, T. Stovall and D. Vinciguerra, *Phys. Lett. B* **30**, 412 (1969).

[444] A. G. Slight, T. E. Drake and G. R. Bishop, *Nucl. Phys. A* **208**, 157 (1973).

[445] R. E. Rand, R. Frosch and M. R. Yearian, *Phys. Rev.* **144**, 859 (1966).

[446] L. Lapikás, G. Box and H. de Vries, *Nucl. Phys. A* **253**, 324 (1975).

[447] S. Cohen and D. Kurath, *Nucl. Phys.* **73**, 1 (1965).

[448] D. Mikolas *et al.*, *Phys. Rev. C* **37**, 766 (1988).

[449] Y. Ogawa, K. Arai, Y. Suzuki and K. Varga, *Nucl. Phys. A* **673**, 122 (2000).

[450] M. Seya, M. Kohno and S. Nagata, *Prog. Theor. Phys.* **65**, 204 (1981).

[451] W. von Oertzen, *Z. Phys. A* **354**, 37 (1996).

[452] N. Itagaki and S. Okabe, *Phys. Rev C* **61**, 044306 (2000).

[453] A. A. Wolters, A. G. M. van Hees and P. W. M. Glaudemans, *Phys. Rev. C* **42**, 2062 (1990).

[454] S. F. Mughabghab, *Phys. Rev. Lett.* **54**, 986 (1985).

[455] Y. Kanada-En'yo, H. Horiuchi and A. Doté, *Phys. Rev. C* **60**, 064304 (1999).

[456] I. Talmi and I. Unna, *Phys. Rev. Lett.* **4**, 469 (1960).

[457] L. Axelsson *et al.*, *Phys. Rev. C* **54**, 1511 (1996).

[458] K. Markenroth *et al.*, *Phys. Rev. C* **62**, 034308 (2000).

[459] W. Geithner *et al.*, *Phys. Rev. Lett.* **83**, 3792 (1999).

[460] H. Sagawa, B. A. Brown and H. Esbensen, *Phys. Lett. B* **309**, 1 (1993).

[461] T. Otsuka, N. Fukunishi and H. Sagawa, *Phys. Rev. Lett.* **70**, 1385 (1993).

[462] F. M. Nunes, I. J. Thompson and R. C. Johnson, *Nucl. Phys. A* **596**, 171 (1996).

[463] P. Descouvemont, *Nucl. Phys. A* **615**, 261 (1997).

[464] F. C. Barker, *J. Phys. G* **2**, 45 (1976).

[465] H. Iwasaki *et al.*, *Phys. Lett. B* **491**, 8 (2000).

[466] A. Navin *et al.*, *Phys. Rev. Lett.* **85**, 266 (2000).

[467] M. Freer *et al.*, *Phys. Rev. Lett.* **82**, 1383 (1999); *Phys. Rev. C* **63**, 034301 (2001).

[468] F. C. Barker and G. T. Hickey, *J. Phys. G* **3**, L23 (1977).

[469] K. H. Wilcox, R. B. Weisenmiller, G. J. Wozniak, N. A. Jelley, D. Ashery and J. Cerny, *Phys. Lett. B* **59**, 142 (1975).

[470] A. I. Amelin *et al.*, *Sov. J. Nucl. Phys.* **52**, 782 (1990).

[471] R. A. Kryger *et al.*, *Phys. Rev. C* **47**, R2439 (1993).

[472] B. M. Young *et al.*, *Phys. Rev. C* **49**, 279 (1994).

[473] M. Zinser *et al.*, *Phys. Rev. Lett.* **75**, 1719 (1995); *Nucl. Phys. A* **619**, 151 (1997).

[474] H. G. Bohlen, W. von Oertzen, Th. Stolla, R. Kalpakchieva, B. Gebauer, M. Wilpert, Th. Wilpert, A. N. Ostrowski, S. M. Grimes and T. N. Massey, *Nucl. Phys. A* **616**, 254c (1997).

[475] T. Kobayashi, K. Yoshida, A. Ozawa, I. Tanihata, A. Korsheninnikov, E. Nikolski and T. Nakamura, *Nucl. Phys. A* **616**, 223c (1997).

[476] M. Thoennessen *et al.*, *Phys. Rev. C* **59**, 111 (1999).

[477] J. A. Caggiano, D. Bazin, W. Benenson, B. Davids, B. M. Sherill, M. Steiner, J. Yurkon, A. F. Zeller and B. Blank, *Phys. Rev. C* **60**, 064322 (1999).

[478] M. Chartier *et al.*, *Phys. Lett. B* **510**, 24 (2001).

[479] S. Aoyama, K. Katō and K. Ikeda, *Phys. Lett. B* **414**, 13 (1997).

[480] H. Masui, S. Aoyama, T. Myo, K. Katō and K. Ikeda, *Nucl. Phys. A* **673**, 207 (2000).

[481] P. Descouvemont, *Nucl. Phys. A* **626**, 647 (1997).

[482] J. Wurzer and H. M. Hofmann, *Z. Phys. A* **354**, 135 (1996).

[483] T. Kobayashi, *Nucl. Phys. A* **538**, 343c (1992).

[484] B. M. Young *et al.*, *Phys. Rev. Lett.* **71**, 4124 (1993).

[485] T. Kobayashi, *Nucl. Phys. A* **553**, 465c (1993).

[486] A. A. Korsheninnikov *et al.*, *Phys. Rev. C* **53**, R537 (1996).

[487] H. G. Bohlen *et al.*, *Z. Phys. A* **351**, 7 (1995).

[488] M. V. Zhukov, L. V. Chulkov, D. V. Fedorov, B. V. Danilin, J. M. Bang, J. S. Vaagen and I. J. Thompson, *J. Phys. G* **20**, 201 (1994).

[489] E. Arnold, J. Bonn, R. Gegenwart, W. Neu, R. Neugart, E.-W. Otten, G. Ulm, K. Wendt and ISOLDE Collaboration, *Phys. Lett. B* **197**, 311 (1987).

[490] E. Arnold, J. Bonn, A. Klein, P. Lievens, R. Neugart, M. Neuroth, E.-W. Otten, H. Reich and W. Widdra, *Z. Phys. A* **349**, 337 (1994).

[491] M. J. G. Borge *et al.*, *Nucl. Phys. A* **613**, 199 (1997).

[492] D. J. Morrissey *et al.*, *Nucl. Phys. A* **627**, 222 (1997).

[493] Y. Ohbayasi and Y. Suzuki, *Phys. Lett. B* **346**, 223 (1995).

[494] M. V. Zhukov, B. V. Danilin, L. V. Grigorenko and J. S. Vaagen, *Phys. Rev. C* **52**, 2461 (1995).

[495] I. Mukha *et al.*, *Phys. Lett. B* **367**, 65 (1996).

[496] T. Suzuki and T. Otsuka, *Phys. Rev. C* **50**, R555 (1994); *ibid.* **56**, 847 (1997).

[497] N. Aoi *et al.*, *Nucl. Phys. A* **616**, 181c (1997).

[498] M. J. G. Borge *et al.*, *Phys. Rev. C* **55**, R8 (1997).

[499] S. Shimoura *et al.*, *Nucl. Phys. A* **630**, 387c (1998).

[500] K. Katō and K. Ikeda, *Prog. Theor. Phys.* **89**, 623 (1993).

[501] K. Katō, T. Yamada and K. Ikeda, *Prog. Theor. Phys.* **101**, 119 (1999).

[502] D. C. Zheng, B. R. Barrett, J. P. Vary, W. C. Haxton and C. L. Song, *Phys. Rev. C* **52**, 2488 (1995).

[503] A. Arima, H. Horiuchi, K. Kubodera and N. Takigawa, in *Advances in Nuclear Physics*, edited by M. Baranger and E. Vogt (Plenum, New York, 1972), Vol. 5, p. 345.

[504] L. Fonda and G. C. Ghirardi, *Symmetry Principles in Quantum Mechanics* (Dekker, New York, 1970).

[505] R. B. Wiringa, V. G. J. Stoks and R. Schiavilla, *Phys. Rev. C* **51**, 38 (1995).

[506] C. R. Chen, G. L. Payne, J. L. Friar and B. F. Gibson, *Phys. Rev. C* **31**, 2266 (1985).

[507] W. Glöckle, H. Witała, D. Hüber, H. Kamada and J. Golak, *Phys. Rep.* **274**, 107 (1996).

[508] A. Kievsky, S. Rosati, W. Tornow and M. Viviani, *Nucl. Phys. A* **607**, 402 (1996).

[509] R. B. Wiringa, S. C. Pieper, J. Carlson and V. R. Pandharipande, *Phys. Rev. C* **62**, 014001 (2000).

[510] B. S. Pudliner, V. R. Pandharipande, J. Carlson, S.C. Pieper and R. B. Wiringa, *Phys. Rev. C* **56**, 1720 (1997).

[511] R. De Tourreil, B. Rouben and D. W. L. Sprung, *Nucl. Phys. A* **242**, 465 (1975).

[512] R. G. H. Robertson, P. Dyer, R. A. Warner, R. C. Melin, T. J. Bowles, A. B. McDonald, G. C. Ball, W. G. Davies and E. D. Earle, *Phys. Rev. Lett.* **47**, 1867 (1981).

[513] O. G. Grebenjuk, A. V. Khanadeev, G. A. Korolev, S. I. Manayenkov, J. Saudinos, G. N. Velichko and A. A. Vorobyov, *Nucl. Phys. A* **500**, 637 (1989).

[514] J. S. Al-Khalili and J. A. Tostevin, *Phys. Rev. C* **57**, 1846 (1998).

[515] I. Ahmad and M. A. Alvi, *Phys. Rev. C* **28**, 2543 (1983).

[516] G. Presser, R. Bass and K. Krüger, *Nucl. Phys. A* **131**, 679 (1969).

[517] G. M. Ter-Akopian *et al.*, *Phys. Lett. B* **426**, 251 (1998).

[518] R. Raabe *et al.*, *Phys. Lett. B* **458**, 1 (1999).

[519] N. Soić *et al.*, *Europhys. Lett.* **34**, 7 (1996).

[520] A. M. Lane and D. Robson, *Phys. Rev.* **178**, 1715 (1969).

[521] K. Fujimura, D. Baye, P. Descouvemont, Y. Suzuki and K. Varga, *Phys. Rev. C* **59**, 817 (1999).

[522] P. Descouvemont and D. Baye, *Phys. Rev. C* **31**, 2274 (1985).

[523] M. Kamimura *et al.*, *Prog. Theor. Phys. Suppl.* No. **89**, 1 (1986).

[524] N. Austern *et al.*, *Phys. Rep.* **154**, 125 (1987).

[525] V. Efimov, *Phys. Lett. B* **33**, 563 (1970); *Nucl. Phys. A* **210**, 157 (1973).

[526] D. V. Fedorov and A. S. Jensen, *Phys. Rev. Lett.* **71**, 4103 (1993).

[527] Yu. N. Ovchinnikov and I. M. Sigal, *Ann. Phys. (N. Y.)* **123**, 274 (1979).

[528] L. D. Landau and E. M. Lifshitz, *Quantum Mechanics. Non-Relativistic Theory*, Vol. 3 of *Course of Theoretical Physics* (Pergamon Press, London, 1958), p. 120.

[529] D. V. Fedorov, A. S. Jensen and K. Riisager, *Phys. Rev. Lett.* **73**, 2817 (1994).

[530] P. Kramer, G. John and D. Schenzle, *Group Theory and the Interaction of Composite Nucleon Systems* (Vieweg, Braunschweig, 1981).

[531] Y. Suzuki and J. Usukura, *Nucl. Inst. Meth. in Phys. Res. B* **171**, 67 (2000).

[532] K. T. Hecht, *Nucl. Phys.* **62**, 1 (1965).

[533] M. Harvey, in *Advances in Nuclear Physics*, edited by M. Baranger and E. Vogt (Plenum, New York, 1968), Vol. 1, p. 67.

[534] B. G. Wybourne, *Classical Groups for Physicists* (Wiley, New York, 1974).

[535] J. P. Elliott and P. G. Dawber, *Symmetry in Physics* (Macmillan, Hong Kong, 1979).

[536] J. P. Draayer and Y. Akiyama, *J. Math. Phys.* **14**, 1904 (1973).

[537] F. Arickx, *Nucl. Phys. A* **268**, 347 (1976); *ibid.* **284**, 264 (1977).

[538] G. Rosensteel and D. J. Rowe, *Phys. Rev. Lett.* **38**, 10 (1977); *Ann. Phys. (N. Y.)* **126**, 343 (1980).

[539] D. R. Peterson and K. T. Hecht, *Nucl. Phys. A* **344**, 361 (1980).

[540] Y. Suzuki and K. T. Hecht, *Nucl. Phys. A* **455**, 315 (1986).

[541] K. T. Hecht, in Int. Conf. on Nuclear Structure, Tokyo, 1977 (*J. Phys. Soc. Jpn. Suppl.* **44**, Tokyo, 1978), p. 232; K. T. Hecht and D. Braunschweig, *Nucl. Phys. A* **295**, 34 (1978); Y. Suzuki, *Nucl. Phys. A* **448**, 395 (1986).

Glossary of symbols

Nucleon numbers

Symbol	Range	Explanation in words	First occurrence
A		Mass No. of a nucleus	
n	1 to A	No. of clusters	p. 281
a_i	1 to A	Mass No. of a cluster	p. 281
$\{j\}$	0 to A	$\sum_{m=1}^{j-1} a_m$	(B.1)

Coordinates

Physical coordinates

$\boldsymbol{r}_j$, $j=1,\ldots,A$: Cartesian s.p. coordinates

$\boldsymbol{x}_j$, $j=1,\ldots,A-1$: normalized intrinsic Jacobi coordinates of particles, Eq. (9.8)

$$\boldsymbol{x}_j = \sqrt{\frac{j}{j+1}}\left(\frac{1}{j}\sum_{i=1}^{j}\boldsymbol{r}_i - \boldsymbol{r}_{j+1}\right)$$

$\boldsymbol{x}'_j$, $j=1,\ldots,A-1$: normalized (intrinsic) relative coordinates of particles, Eq. (9.25)

$\boldsymbol{x}_A$: normalized c.m. Jacobi coordinate of particles, Eq. (9.9)

$$\boldsymbol{x}_A = \frac{1}{\sqrt{A}}\sum_{i=1}^{A}\boldsymbol{r}_i$$

$\boldsymbol{y}_j$, $j=1,\ldots,A-1$: unnormalized intrinsic Jacobi coordinates of particles, Eq. (9.10)

$$\boldsymbol{y}_j = \frac{1}{j}\sum_{i=1}^{j}\boldsymbol{r}_i - \boldsymbol{r}_{j+1}$$

$\boldsymbol{R}_{\text{c.m.}}$: unnormalized c.m. Jacobi coordinate of particles, Eq. (9.11)

$$\boldsymbol{R}_{\text{c.m.}} \equiv \boldsymbol{y}_A = \frac{1}{A}\sum_{i=1}^{A} \boldsymbol{r}_i$$

ξ: all intrinsic coordinates of a nucleus, including spin and isospin

ξ_j, $j=1, a_i$: all intrinsic coordinates of cluster (or nucleon) i, including spin and isospin

$\bar{\boldsymbol{\xi}}_j$, $j=1,\ldots,n$: normalized c.m. coordinate of cluster j, Eq. (9.14)

$$\bar{\boldsymbol{\xi}}_j = \frac{1}{\sqrt{a_j}}\sum_{i=1}^{a_j} \boldsymbol{r}_{\{j\}+i}$$

$\boldsymbol{t}_{\{j\}+i}$, $i=1,\ldots,a_j-1$, $j=1,\ldots,n$: unnormalized intrinsic Jacobi coordinates of cluster j, Eq. (B.2)

$$\boldsymbol{t}_{\{j\}+i} = \frac{1}{i}\sum_{m=1}^{i} \boldsymbol{r}_{\{j\}+m} - \boldsymbol{r}_{\{j\}+i+1}$$

$\bar{\boldsymbol{t}}_j$, $j=1,\ldots,n$: unnormalized c.m. coordinates of cluster j, Eq. (B.3)

$$\bar{\boldsymbol{t}}_j = \boldsymbol{t}_{\{j\}+a_j} = \frac{1}{a_j}\sum_{i=1}^{a_j} \boldsymbol{r}_{\{j\}+i}$$

$\boldsymbol{\rho}$: normalized relative Jacobi coordinate between two clusters, Eq. (9.16)

$$\boldsymbol{\rho} = \sqrt{\frac{a_1 a_2}{a_1 + a_2}}\left(\frac{1}{\sqrt{a_1}}\bar{\boldsymbol{\xi}}_1 - \frac{1}{\sqrt{a_2}}\bar{\boldsymbol{\xi}}_2\right)$$

$\boldsymbol{\rho}_j$, $j=1,\ldots,n-1$: normalized relative Jacobi coordinates between n clusters, Eq. (11.5)

$$\boldsymbol{\rho}_j = \sqrt{\frac{a_{j+1}\sum_{i=1}^{j} a_i}{\sum_{i=1}^{j+1} a_i}} \times \left(\frac{1}{\sum_{i=1}^{j} a_i}\sum_{i=1}^{j}\sqrt{a_i}\bar{\boldsymbol{\xi}}_i - \frac{1}{\sqrt{a_{j+1}}}\bar{\boldsymbol{\xi}}_{j+1}\right)$$

$\boldsymbol{\rho}_n$: normalized c.m. Jacobi coordinate of n clusters, Eq. (11.5)

$$\boldsymbol{\rho}_n = \frac{1}{\sqrt{\sum_{i=1}^{n} a_i}}\sum_{i=1}^{n}\sqrt{a_i}\bar{\boldsymbol{\xi}}_i = \boldsymbol{x}_A = \frac{1}{\sqrt{A}}\sum_{i=1}^{A}\boldsymbol{r}_i$$

$\boldsymbol{q}$: unnormalized relative Jacobi coordinate between two clusters, Eq. (9.17)

$$\boldsymbol{q} = \bar{\boldsymbol{t}}_1 - \bar{\boldsymbol{t}}_2$$

$\boldsymbol{q}_j$, $j=1,\ldots,n-1$: unnormalized relative Jacobi coordinates between n clusters, Eq. (11.3)

$$\boldsymbol{q}_j = \frac{1}{\sum_{i=1}^{j} a_i} \sum_{i=1}^{j} a_i \bar{\boldsymbol{t}}_i - \bar{\boldsymbol{t}}_{j+1}$$

$\boldsymbol{q}_n$: unnormalized c.m. Jacobi coordinate of n clusters, Eq. (11.4)

$$\boldsymbol{q}_n = \frac{1}{\sum_{i=1}^{n} a_i} \sum_{i=1}^{n} a_i \bar{\boldsymbol{t}}_i = \frac{1}{A} \sum_{i=1}^{A} \boldsymbol{r}_i \equiv \boldsymbol{R}_{\text{c.m.}}$$

Generator coordinates

$\boldsymbol{s}_j$, $j=1,\ldots,A$: Cartesian s.p. generator coordinates, Eq. (9.67)

$\boldsymbol{S}_j$, $j=1,\ldots,A-1$: normalized intrinsic Jacobi generator coordinates of particles, Eq. (9.69)

$$\boldsymbol{S}_j = \sqrt{\frac{j}{j+1}} \left(\frac{1}{j} \sum_{i=1}^{j} \boldsymbol{s}_i - \boldsymbol{s}_{j+1} \right)$$

$\boldsymbol{S}_A$: normalized c.m. Jacobi generator coordinate of particles, Eq. (9.69)

$$\boldsymbol{S}_A = \frac{1}{\sqrt{A}} \sum_{i=1}^{A} \boldsymbol{s}_i$$

$\boldsymbol{S}$: normalized relative Jacobi generator coordinate between two clusters, Eq. (11.78)

$$\boldsymbol{S} = \sqrt{\frac{a_1 a_2}{a_1 + a_2}} \left(\frac{1}{\sqrt{a_1}} \bar{\boldsymbol{s}}_1 - \frac{1}{\sqrt{a_2}} \bar{\boldsymbol{s}}_2 \right)$$

$\boldsymbol{S}_A$: normalized c.m. Jacobi generator coordinate of two clusters, Eq. (11.78)

$$\boldsymbol{S}_A = \frac{1}{\sqrt{\sum_{i=1}^{n} a_i}} \sum_{i=1}^{2} \sqrt{a_i} \boldsymbol{s}_i$$

$\underline{s}_{[j]}$, $j=1,\ldots,n$: Cartesian generator coordinates of n clusters, Eq. (12.3)

$$\underline{s}_{[j]} = \frac{1}{a_j}\sum_{i=1}^{a_j} s_{\{j\}+i} \quad (s_{\{j\}+1} = \cdots = s_{\{j\}+a_j})$$

$s_{[j]}$, $j=1,\ldots,n$: normalized generator coordinates of n clusters, Eq. (12.3)

$$s_{[j]} = \frac{1}{\sqrt{a_j}}\sum_{i=1}^{a_j} s_{\{j\}+i} \quad (s_{\{j\}+1} = \cdots = s_{\{j\}+a_j})$$

$S_{[j]}$, $j=1,\ldots,n-1$: normalized relative Jacobi generator coordinates between n clusters, Eq. (12.5)

$$S_{[j]} = \sqrt{\frac{a_{j+1}\sum_{i=1}^{j} a_i}{\sum_{i=1}^{j+1} a_i}} \times \left(\frac{1}{\sum_{i=1}^{j} a_i}\sum_{i=1}^{j}\sqrt{a_i}s_{[i]} - \frac{1}{\sqrt{a_{j+1}}}s_{[j+1]}\right)$$

$S_{[n]}$: normalized c.m. Jacobi generator coordinate of n clusters, Eq. (12.5)

$$S_{[n]} = \frac{1}{\sqrt{\sum_{i=1}^{n} a_i}}\sum_{i=1}^{n}\sqrt{a_i}s_{[i]} = S_A$$

Miscellaneous notations

Symbol	Explanation	First occurrence
A_k	matrix of nonlinear parameters characterizing a correlated Gaussian	(9.28)
$\mathcal{A}$	antisymmetrization operator	(9.40)
$\boldsymbol{b}$	impact parameter	p. 29
$B(\mathrm{E}L)$	reduced transition probability of electric 2^L-pole transitions	(7.10)
$B(\kappa, J' \to J)$	reduced transition probability from J' to J due to a rank-κ operator	(9.86)
$dB(\mathrm{E1})/dE$	electric-dipole strength function per unit energy	(7.9)
E_R	resonance energy	p. 263
$f(\theta,\phi)$	scattering amplitude	(2.2)
$f_{KLM_L}(u, \boldsymbol{x}, A)$	spatial part of a correlated Gaussian	(9.55)
$[f_1 f_2 \cdots]$	representation of the symmetric group $\mathrm{S}_{f_1+f_2+\cdots}$	p. 414

$g(\boldsymbol{R})$	spectroscopic amplitude	(11.11)
$g(s,x)$	generating function for one-dimensional HO wave functions	(9.46)
$g(\boldsymbol{s},\boldsymbol{x})$	generating function for three-dimensional HO wave functions	(9.48)
$g(\varsigma;\boldsymbol{x},A)$	generating function for the spatial part of a correlated Gaussian	(9.59)
$G(\boldsymbol{r}-\boldsymbol{r}')$	Green's function	(2.5)
$H(\boldsymbol{R},\boldsymbol{R}')$	kernel of Hamiltonian operator $\mathcal{H}$	(11.43)
$\boldsymbol{j}$	probability current density	(2.12)
$\boldsymbol{j}(\boldsymbol{r})$	current density of electrodynamics	(9.152)
$N(\boldsymbol{R},\boldsymbol{R}')$	kernel of norm operator $\mathcal{N}$	(11.15)
$O^{(W)}(\boldsymbol{R},\boldsymbol{p})$	Wigner transform of a nonlocal operator O	(11.167)
$P(\boldsymbol{r}\gamma,\boldsymbol{r}_0\gamma_0)$	pair correlation function for nucleon intrinsic states γ and γ_0	(12.16)
P_i^{p}, P_i^{n}, P_i^{N}	projection operators of particle i onto proton, neutron and nucleon	(9.122)
$P_{\mathrm{x}}(\boldsymbol{b})$	absorption, elastic scattering and reaction probabilities for x = abs, el and reac	p. 32, (3.34)
$P_{\parallel}$, $\boldsymbol{P}_{\perp}$	longitudinal and transverse momentum (wave number)	p. 132
$\mathcal{P}$	projection operator onto a cluster subspace	(11.22)
$\mathcal{P}_{\boldsymbol{p}}$	projection operator onto momentum $\boldsymbol{p}$	(9.106)
$\mathcal{P}_Q$	projection operator onto $Q\hbar\omega$ HOSM space	(12.7)
$\boldsymbol{r}_i^{\mathrm{P}}$, $\boldsymbol{r}_i^{\mathrm{T}}$	coordinate of particle i in projectile and target	p. 57
r_{p}, r_{n}, r_{m}	point rms proton, neutron and matter radii	p. 400
$\hat{\boldsymbol{r}}$	unit vector pointing to $\boldsymbol{r}$	p. 21
S	spectroscopic factor	(11.26)
S_1	energy-weighted electric dipole sum	(7.11)
S_n	symmetric group (group of permutations) of n elements	p. 414
$\boldsymbol{s}_i^{\mathrm{P}}$, $\boldsymbol{s}_i^{\mathrm{T}}$	projection of $\boldsymbol{r}_i^{\mathrm{P}},\boldsymbol{r}_i^{\mathrm{T}}$ onto the xy-plane	p. 58
$\mathcal{Y}_{lm}(\boldsymbol{r})$	solid spherical harmonic	(9.2)
$\hat{\boldsymbol{z}}$	unit vector pointing to z-direction	p. 23

$\chi(\boldsymbol{b})$	phase-shift function	(2.38)
$\chi_C(\boldsymbol{b})$	Coulomb phase-shift function	(2.87)
$\chi_{\frac{1}{2}m_s}$	spin function of a nucleon	p. 34
χ_{SM_S}	spin function of a nucleus	(9.39)
Γ	width of a resonance	p. 263
$\Gamma(\boldsymbol{b})$	profile function	(3.56)
η	Sommerfeld parameter	p. 41
$\eta_{\frac{1}{2}m_t}$	isospin function of a nucleon	p. 210
η_{TM_T}	isospin function of a nucleus	(9.39)
$(\lambda\mu)$	Elliott's label of SU(3) representation	p. 312
$\boldsymbol{\mu}$	magnetic dipole operator	(9.145)
$\phi_\kappa(\boldsymbol{s}_1, \ldots, \boldsymbol{s}_A)$	Slater determinant of Gaussian wave packets	(9.67)
$\varphi_{\boldsymbol{s}}(\boldsymbol{r})$	Gaussian wave packet	(9.66)
$\hat{\varphi}_{\boldsymbol{s}m_sm_t}(\boldsymbol{r})$	s.p. Gaussian wave packet (or shifted Gaussian)	(9.65)
$\Phi_{SM_STM_T}$	spin–isospin-coupled antisymmetrized Gaussian wave packets	(9.68)
ψ_{nlm}	three-dimensional HO wave function	(9.49)
$\Psi_{SM_STM_T}$	intrinsic function with c.m. motion removed from $\Phi_{SM_STM_T}$	(9.72)
$\Psi_{\boldsymbol{R}}$	test state for the cluster model	(11.9)
$\varrho(\boldsymbol{k})$	momentum density	(9.131)
$\rho(\boldsymbol{r})$	s.p. density	(3.63) or (9.127)
$\rho(\boldsymbol{r}, \boldsymbol{r}')$	two-particle density or pair correlation function	(3.75) or (9.160)
σ_{x}	absorption, elastic scattering, interaction, reaction and total cross sections for x = abs, el, int, reac and tot	p. 20, p. 25, (3.33), (3.54)
$\sigma_{-x\mathrm{n}}$	x-neutron removal cross section	(4.11)
$\sigma_{-\mathrm{n}}^{\mathrm{el,\,inel}}$	elastic and inelastic neutron-removal cross section	(4.34)
$\theta_{lLM_L}(\boldsymbol{x})$	angular factor of a correlated Gaussian	(9.20) or (9.32)
$\mathfrak{P}$	probability of clustering	(11.24)
$\mathfrak{P}_Q$	occupation probability of $Q\hbar\omega$ HOSM space	(12.9)
$\mid\ \rangle\!\rangle$	'ket' defined for intercluster motion in parameter-coordinate space	(11.25)

Index

Abbreviations

ACCC	analytic continuation in coupling constant
α	α-particle (^{4}He), α-cluster
AMD	antisymmetrized molecular dynamics
c.m.	centre of mass
CDCC	continuum-discretized coupled-channel (method)
CGCM	complex generator-coordinate method
CSM	complex scaling method
DWBA	distorted-wave Born approximation
EL	electric 2^L-pole
FWHM	full width at half maximum
GCM	generator-coordinate method (model)
GFMC	Green's function Monte Carlo (method)
g.s.	ground state
h	helion (^{3}He)
HO	harmonic oscillator
HOSM	harmonic-oscillator shell model
ML	magnetic 2^L-pole
N	nucleon
NT	nucleon–(target-)nucleus
OCM	orthogonality-condition model
OLA	optical-limit approximation
n	neutron
p	proton
RGM	resonating-group method (model)
rms	root mean square
SM	shell model
SO(3)	group of three-dimensional special orthogonal transformations
s.p.	single-particle
SU(3)	group of three-dimensional special unitary transformations
SVM	stochastic variational method (with correlated Gaussian bases)
t	triton (^{3}H)
U(d)	group of d-dimensional unitary transformations
WKB	Wentzel–Kramers–Brillouin
W.u.	Weisskopf unit